MONOGRAPHS AND RESEARCH NOTES IN MATHEMATICS

Cremona Groups and the Icosahedron

Ivan Cheltsov
University of Edinburgh
Scotland, UK

Constantin Shramov
Steklov Mathematical Institute and Higher School of Economics
Moscow, Russia

CRC Press
Taylor & Francis Group
Boca Raton London New York

CRC Press is an imprint of the
Taylor & Francis Group, an **informa** business

A CHAPMAN & HALL BOOK

MONOGRAPHS AND RESEARCH NOTES IN MATHEMATICS

Series Editors

John A. Burns
Thomas J. Tucker
Miklos Bona
Michael Ruzhansky

Published Titles

Application of Fuzzy Logic to Social Choice Theory, John N. Mordeson, Davender S. Malik and Terry D. Clark

Blow-up Patterns for Higher-Order: Nonlinear Parabolic, Hyperbolic Dispersion and Schrödinger Equations, Victor A. Galaktionov, Enzo L. Mitidieri, and Stanislav Pohozaev

Cremona Groups and Icosahedron, Ivan Cheltsov and Constantin Shramov

Difference Equations: Theory, Applications and Advanced Topics, Third Edition, Ronald E. Mickens

Dictionary of Inequalities, Second Edition, Peter Bullen

Iterative Optimization in Inverse Problems, Charles L. Byrne

Modeling and Inverse Problems in the Presence of Uncertainty, H. T. Banks, Shuhua Hu, and W. Clayton Thompson

Monomial Algebras, Second Edition, Rafael H. Villarreal

Partial Differential Equations with Variable Exponents: Variational Methods and Qualitative Analysis, Vicenţiu D. Rădulescu and Dušan D. Repovš

A Practical Guide to Geometric Regulation for Distributed Parameter Systems, Eugenio Aulisa and David Gilliam

Signal Processing: A Mathematical Approach, Second Edition, Charles L. Byrne

Sinusoids: Theory and Technological Applications, Prem K. Kythe

Special Integrals of Gradshetyn and Ryzhik: the Proofs – Volume I, Victor H. Moll

Forthcoming Titles

Actions and Invariants of Algebraic Groups, Second Edition, Walter Ferrer Santos and Alvaro Rittatore

Analytical Methods for Kolmogorov Equations, Second Edition, Luca Lorenzi

Complex Analysis: Conformal Inequalities and the Bierbach Conjecture, Prem K. Kythe

Computational Aspects of Polynomial Identities: Volume I, Kemer's Theorems, 2nd Edition Belov Alexey, Yaakov Karasik, Louis Halle Rowen

Geometric Modeling and Mesh Generation from Scanned Images, Yongjie Zhang

Groups, Designs, and Linear Algebra, Donald L. Kreher

Handbook of the Tutte Polynomial, Joanna Anthony Ellis-Monaghan and Iain Moffat

Forthcoming Titles (continued)

Lineability: The Search for Linearity in Mathematics, Juan B. Seoane Sepulveda, Richard W. Aron, Luis Bernal-Gonzalez, and Daniel M. Pellegrinao

Line Integral Methods and Their Applications, Luigi Brugnano and Felice Iaverno

Microlocal Analysis on R^n and on NonCompact Manifolds, Sandro Coriasco

Nonlinear Functional Analysis in Banach Spaces and Banach Algebras: Fixed Point Theory Under Weak Topology for Nonlinear Operators and Block Operators with Applications Aref Jeribi and Bilel Krichen

Practical Guide to Geometric Regulation for Distributed Parameter Systems, Eugenio Aulisa and David S. Gilliam

Reconstructions from the Data of Integrals, Victor Palamodov

Special Integrals of Gradshetyn and Ryzhik: the Proofs – Volume II, Victor H. Moll

Stochastic Cauchy Problems in Infinite Dimensions: Generalized and Regularized Solutions, Irina V. Melnikova and Alexei Filinkov

Symmetry and Quantum Mechanics, Scott Corry

CRC Press
Taylor & Francis Group
6000 Broken Sound Parkway NW, Suite 300
Boca Raton, FL 33487-2742

© 2016 by Taylor & Francis Group, LLC
CRC Press is an imprint of Taylor & Francis Group, an Informa business

No claim to original U.S. Government works

Printed on acid-free paper
Version Date: 20150617

International Standard Book Number-13: 978-1-4822-5159-3 (Hardback)

This book contains information obtained from authentic and highly regarded sources. Reasonable efforts have been made to publish reliable data and information, but the author and publisher cannot assume responsibility for the validity of all materials or the consequences of their use. The authors and publishers have attempted to trace the copyright holders of all material reproduced in this publication and apologize to copyright holders if permission to publish in this form has not been obtained. If any copyright material has not been acknowledged please write and let us know so we may rectify in any future reprint.

Except as permitted under U.S. Copyright Law, no part of this book may be reprinted, reproduced, transmitted, or utilized in any form by any electronic, mechanical, or other means, now known or hereafter invented, including photocopying, microfilming, and recording, or in any information storage or retrieval system, without written permission from the publishers.

For permission to photocopy or use material electronically from this work, please access www.copyright.com (http://www.copyright.com/) or contact the Copyright Clearance Center, Inc. (CCC), 222 Rosewood Drive, Danvers, MA 01923, 978-750-8400. CCC is a not-for-profit organization that provides licenses and registration for a variety of users. For organizations that have been granted a photocopy license by the CCC, a separate system of payment has been arranged.

Trademark Notice: Product or corporate names may be trademarks or registered trademarks, and are used only for identification and explanation without intent to infringe.

Visit the Taylor & Francis Web site at
http://www.taylorandfrancis.com

and the CRC Press Web site at
http://www.crcpress.com

Contents

Preface xi

Acknowledgments xv

Notation and conventions xvii

1 Introduction 1
 1.1 Conjugacy in Cremona groups 1
 1.2 Three-dimensional projective space 5
 1.3 Other rational Fano threefolds 8
 1.4 Statement of the main result 11
 1.5 Outline of the book 12

I Preliminaries 21

2 Singularities of pairs 23
 2.1 Canonical and log canonical singularities 23
 2.2 Log pairs with mobile boundaries 26
 2.3 Multiplier ideal sheaves 28
 2.4 Centers of log canonical singularities 30
 2.5 Corti's inequality 33

3 Noether–Fano inequalities 39
 3.1 Birational rigidity 39
 3.2 Fano varieties and elliptic fibrations 40
 3.3 Applications to birational rigidity 46
 3.4 Halphen pencils 49

4 Auxiliary results · 53
    4.1 Zero-dimensional subschemes 53
    4.2 Atiyah flops 54
    4.3 One-dimensional linear systems 59
    4.4 Miscellanea 61

II Icosahedral group   65

5 Basic properties   67
    5.1 Action on points and curves 67
    5.2 Representation theory 71
    5.3 Invariant theory 73
    5.4 Curves of low genera 77
    5.5 $SL_2(\mathbb{C})$ and $PSL_2(\mathbb{C})$ 79
    5.6 Binary icosahedral group 87
    5.7 Symmetric group 89
    5.8 Dihedral group 91

6 Surfaces with icosahedral symmetry   95
    6.1 Projective plane 95
    6.2 Quintic del Pezzo surface 105
    6.3 Clebsch cubic surface 117
    6.4 Two-dimensional quadric 123
    6.5 Hirzebruch surfaces 136
    6.6 Icosahedral subgroups of $Cr_2(\mathbb{C})$ 140
    6.7 $K3$ surfaces 143

III Quintic del Pezzo threefold   151

7 Quintic del Pezzo threefold   153
    7.1 Construction and basic properties 153
    7.2 $PSL_2(\mathbb{C})$-invariant anticanonical surface 159
    7.3 Small orbits 166
    7.4 Lines 170
    7.5 Orbit of length five 174
    7.6 Five hyperplane sections 178
    7.7 Projection from a line 183
    7.8 Conics 185

CONTENTS

8 Anticanonical linear system — 189
- 8.1 Invariant anticanonical surfaces — 189
- 8.2 Singularities of invariant anticanonical surfaces — 198
- 8.3 Curves in invariant anticanonical surfaces — 203

9 Combinatorics of lines and conics — 207
- 9.1 Lines — 207
- 9.2 Conics — 213

10 Special invariant curves — 221
- 10.1 Irreducible curves — 221
- 10.2 Preliminary classification of low degree curves — 230

11 Two Sarkisov links — 239
- 11.1 Anticanonical divisors through the curve $\mathcal{L}_6$ — 239
- 11.2 Rational map to $\mathbb{P}^4$ — 244
- 11.3 A remarkable sextic curve — 255
- 11.4 Two Sarkisov links — 260
- 11.5 Action on the Picard group — 269

IV Invariant subvarieties — 279

12 Invariant cubic hypersurface — 281
- 12.1 Linear system of cubics — 281
- 12.2 Curves in the invariant cubic — 282
- 12.3 Bring's curve in the invariant cubic — 288
- 12.4 Intersecting invariant quadrics and cubic — 293
- 12.5 A remarkable rational surface — 306

13 Curves of low degree — 313
- 13.1 Curves of degree 16 — 313
- 13.2 Six twisted cubics — 320
- 13.3 Irreducible curves of degree 18 — 326
- 13.4 A singular curve of degree 18 — 332
- 13.5 Bring's curve — 337
- 13.6 Classification — 339

14 Orbits of small length   347
     14.1 Orbits of length 20 347
     14.2 Ten conics 360
     14.3 Orbits of length 30 363
     14.4 Fifteen twisted cubics 370

15 Further properties of the invariant cubic   375
     15.1 Intersections with low degree curves 375
     15.2 Singularities of the invariant cubic 377
     15.3 Projection to Clebsch cubic surface 388
     15.4 Picard group 397

16 Summary of orbits, curves, and surfaces   405
     16.1 Orbits vs. curves 405
     16.2 Orbits vs. surfaces 406
     16.3 Curves vs. surfaces 406
     16.4 Curves vs. curves 410

V   Singularities of linear systems   413

17 Base loci of invariant linear systems   415
     17.1 Orbits of length 10 415
     17.2 Linear system $\mathcal{Q}_3$ 416
     17.3 Isolation of orbits in $\mathscr{S}$ 423
     17.4 Isolation of arbitrary orbits 428
     17.5 Isolation of the curve $\mathcal{L}_{15}$ 434

18 Proof of the main result   443
     18.1 Singularities of linear systems 443
     18.2 Restricting divisors to invariant quadrics 444
     18.3 Exclusion of points and curves different from $\mathcal{L}_{15}$ 451
     18.4 Exclusion of the curve $\mathcal{L}_{15}$ 453
     18.5 Alternative approach to exclusion of points 456
     18.6 Alternative approach to the exclusion of $\mathcal{L}_{15}$ 461

19 Halphen pencils and elliptic fibrations   465
     19.1 Statement of results 465
     19.2 Exclusion of points 469
     19.3 Exclusion of curves 471

19.4 Description of non-terminal pairs 475
19.5 Completing the proof 478

Bibliography   **485**

Index   **495**

Preface

A complex projective space $\mathbb{P}^n$ is one of the most fundamental objects of algebraic geometry. It provides a motivation for the study of an exceptionally complicated object, the group $\mathrm{Cr}_n(\mathbb{C})$ of its birational automorphisms, called a *Cremona group* of rank n. This book deals with the Cremona group $\mathrm{Cr}_3(\mathbb{C})$ of rank 3, describing the beautiful appearances of the icosahedral group $\mathfrak{A}_5$ in it.

Most questions about the group $\mathrm{Cr}_1(\mathbb{C})$ are easy to answer, because it coincides with the group of biregular selfmaps of the projective line, which is isomorphic to $\mathrm{PGL}_2(\mathbb{C})$. The group $\mathrm{Cr}_2(\mathbb{C})$ is more complicated, since it does contain non-biregular transformations. The first example of such transformation, a circle inversion, was used by Apollonius of Perga to find a circle that is tangent to three given circles. This group has been intensively studied over the last two centuries, and many facts about it were established. These range from classical results about generators of $\mathrm{Cr}_2(\mathbb{C})$ due to Noether and Castelnuovo, and about relations between these generators due to Gizatullin, up to the action of $\mathrm{Cr}_2(\mathbb{C})$ on an infinite-dimensional hyperbolic space due to Cantat and Lamy, and complete classification of finite subgroups due to Dolgachev and Iskovskikh.

The structure of the group $\mathrm{Cr}_3(\mathbb{C})$ becomes way more complicated. It is known that it does not admit any "reasonable" set of generators. This group still resists any attempts to study its "global" structure, but one can access it on the level of finite subgroups, which became possible thanks to recent achievements in three-dimensional birational geometry. This accessibility is based on a general observation that a birational action of a finite group G on the projective space can be regularized, that is, replaced by a *regular* action of this group on some more complicated rational variety. This transfers the discussion into a rich world of varieties with large groups of symmetries.

At the moment it seems hardly possible to obtain a complete classification of finite subgroups in the Cremona group of rank 3. Nevertheless, Prokhorov managed to find all finite *simple* subgroups of $\mathrm{Cr}_3(\mathbb{C})$. He proved that the six groups $\mathfrak{A}_5$, $\mathrm{PSL}_2(\mathbf{F}_7)$, $\mathfrak{A}_6$, $\mathrm{SL}_2(\mathbf{F}_8)$, $\mathfrak{A}_7$, and $\mathrm{PSp}_4(\mathbf{F}_3)$ are the only non-abelian finite simple subgroups of $\mathrm{Cr}_3(\mathbb{C})$. The former three of these six groups actually admit embeddings to $\mathrm{Cr}_2(\mathbb{C})$, and the group $\mathfrak{A}_5$ is also realized as a subgroup of $\mathrm{PGL}_2(\mathbb{C})$, while the latter three groups are new three-dimensional artefacts.

Given a subgroup G of $\mathrm{Cr}_3(\mathbb{C})$ (or of any other group) it is natural to ask how many non-conjugate subgroups isomorphic to G are contained in the group $\mathrm{Cr}_3(\mathbb{C})$. It appears that methods of birational geometry fit very well to answer such questions. They allow classifying all embeddings of $\mathrm{SL}_2(\mathbf{F}_8)$, $\mathfrak{A}_7$, and $\mathrm{PSp}_4(\mathbf{F}_3)$ into $\mathrm{Cr}_3(\mathbb{C})$ up to conjugation (and actually there are very few of them). As a next step one can construct many non-conjugate embeddings of $\mathrm{PSL}_2(\mathbf{F}_7)$ and $\mathfrak{A}_6$ into $\mathrm{Cr}_3(\mathbb{C})$, although a complete answer is not known. The last remaining case that has not been studied yet is the smallest non-abelian simple group, the icosahedral group $\mathfrak{A}_5$, which is remarkable on its own. This book grew out of an attempt to fill this gap.

Being a group-theoretic counterpart of the icosahedron, the most symmetric of Platonic solids, the group $\mathfrak{A}_5$ may boast one of the longest histories of appearances in many areas of mathematics. A recognition of its significance is a famous book by Klein, centered mostly around this group. In connection with the discussed problem it behaves in an interesting way as well. On the one hand, there are many rational threefolds admitting icosahedral symmetry, including the projective space $\mathbb{P}^3$ itself, the three-dimensional quadric, and also classically studied Segre cubic, Igusa and Burkhardt quartics, and the double cover of $\mathbb{P}^3$ branched along Barth sextic surface. On the other hand, it is currently unknown whether the corresponding icosahedral subgroups of $\mathrm{Cr}_3(\mathbb{C})$ are conjugate or not. Moreover, even the most powerful method to study questions of this kind, the theory of birational rigidity, is usually not applicable here. One of the goals of our book is to expand the frontiers of its applicability, and in particular to present an example of an $\mathfrak{A}_5$-birationally rigid rational threefold.

At this point the second main character of the book enters the scene. Among the rational threefolds with an action of the icosahedral group there is one remarkable smooth variety, a quintic del Pezzo threefold V_5. Apart from having a rich group of symmetries, it has been studied from many points of view such as explicit birational transformations, Kähler–Einstein metrics, exceptional collections in derived categories and instanton bundles. However, its $\mathfrak{A}_5$-equivariant geometry has never been explored. We focus on this problem. We manage to describe explicitly a huge number of interesting $\mathfrak{A}_5$-invariant subvarieties of V_5, including all $\mathfrak{A}_5$-orbits, a long list of low degree curves, a pencil of invariant anticanonical $K3$ surfaces, and a mildly singular surface of general type that is a degree-five cover of the diagonal Clebsch cubic surface. Furthermore, we discover two birational selfmaps of V_5 that commute with $\mathfrak{A}_5$-action and use them to describe the whole group of $\mathfrak{A}_5$-birational automorphisms. Finally, we prove our main result: the variety V_5 is $\mathfrak{A}_5$-birationally rigid, which means in particular that it cannot be $\mathfrak{A}_5$-equivariantly birationally transformed to a Fano threefold with mild singularities or a threefold fibered by rational curves or surfaces. As an application, we return to the starting point of our journey and produce three non-conjugate icosahedral subgroups in the Cremona group $\mathrm{Cr}_3(\mathbb{C})$, one of them arising from the threefold V_5.

One thing that we find really impressive is that all our classification results go hand in hand with each other, so that the purely birational objects like explicit birational selfmaps help to classify invariant curves, and they in turn help to deal with birational transformations.

The book has a clear motivational problem that is finally solved. While working on it, we discovered many relevant facts that are interesting on their own. Although some of them are not used in the proof of $\mathfrak{A}_5$-birational rigidity of V_5, we decided to include many of them, because we find them at least equally interesting as the initial birational question. We believe that these results can provide the same kind of aesthetic feeling as one that possibly stood behind the classical works of Klein, Maschke, Blichfeldt, and many others regarding symmetry groups. We hope that the reader will enjoy and appreciate them as well.

Acknowledgments

This book originated from the authors' conversations with Joseph Cutrone, Nicholas Marshburn, and Vyacheslav Shokurov at Johns Hopkins University in April 2011. It was carried out during the authors' visits to the Max Planck Institut für Mathematik in Bonn, the Center for Geometry and Physics in Pohang, Centro Internazionale per la Ricerca Matematica in Trento, the National Center for Theoretical Sciences in Taipei, the International Centre for Mathematical Sciences in Edinburgh, the Institute for the Physics and Mathematics of the Universe in Kashiwa, the Johann Radon Institute for Computational and Applied Mathematics in Linz, East China Normal University in Shanghai, the University Centre in Svalbard, Mathematisches Forschungsinstitut in Oberwolfach, Fazenda Siriúba in Ilhabela, Bethlemi Hut near Stepantsminda, Nugget Inn in Nome, and the Courant Institute of Mathematical Sciences in New York in the period 2011–2015. Both authors appreciate their excellent environments and hospitality.

We are indebted to many people who explained to us various mathematical and historical aspects related to the material in the book. We are especially grateful to Hamid Ahmadinezhad, Harry Braden, Thomas Breuer, Joseph Cutrone, Tim Dokshitzer, Igor Dolgachev, Sergey Gorchinskiy, Kenji Hashimoto, Igor Krylov, Alexander Kuznetsov, Nicholas Marshburn, Viacheslav Nikulin, Jihun Park, Vladimir Popov, Yuri Prokhorov, Jürgen Richter-Gebert, Francesco Russo, Leonid Rybnikov, Dmitrijs Sakovics, Giangiacomo Sanna, Evgeny Smirnov, Vyacheslav Shokurov, Nadezhda Timofeeva, Andrey Trepalin, Vadim Vologodsky, and Michael Wemyss.

Notation and conventions

Unless explicitly stated otherwise, all varieties are assumed to be projective and normal (but are allowed to be reducible). Everything is defined over the field $\mathbb{C}$ of complex numbers. Of course, the field $\mathbb{C}$ can be replaced by the reader's favorite algebraically closed field of zero characteristic.

By curves and surfaces we mean algebraic varieties of pure dimension 1 and 2, respectively. In particular, they can be reducible but are always reduced. By a conic we usually mean an irreducible plane curve of degree 2; if we need to allow a reducible curve, we try to indicate this explicitly.

An irreducible divisor is a divisor whose support is an irreducible variety; in particular, such a divisor is not necessarily prime. In many cases we do not make a distinction between divisors and classes of divisors. For example, given a Cartier divisor D on a variety X we may speak about D as an element of the Picard group of X. If X is a variety with a fixed embedding into $\mathbb{P}^n$, and D is a divisor on X that is cut out by a hypersurface in $\mathbb{P}^n$ of degree d, then we often refer to D simply as a hypersurface of degree d (or as quadric and cubic hypersurface for $d = 2$ and 3, respectively). By K_X we always denote the canonical class of X. Throughout the book we use the standard language of the singularities of pairs (see [30], [73], [74]).

Given two cycles Z_1 and Z_2 on a variety X, we denote their intersection cycle by $Z_1 \cdot Z_2$. If the latter is a zero-cycle, we will often use the same notation for its degree.

A complete linear system on a variety X is a projective space parameterizing all divisors that are linearly equivalent to a given (Weil) divisor D on X; we consider this notion only on normal varieties for simplicity. A linear system is a projective subspace of some complete linear system. The dimension of a linear system is the dimension of the corresponding projective space; in particular, pencils are linear systems of dimension 1. A mobile linear system is a non-empty linear system that does not have fixed components.

A projectivization $\mathbb{P}(V)$ of a vector space V is the projective space of all lines in V. A projectivization of a vector bundle is defined in a similar way. The Hirzebruch surface $\mathbb{F}_n$, $n \geqslant 0$, is defined as the projectivization of the vector bundle $\mathcal{O}_{\mathbb{P}^1} \oplus \mathcal{O}_{\mathbb{P}^1}(n)$ on $\mathbb{P}^1$. In particular, one has $\mathbb{F}_0 \cong \mathbb{P}^1 \times \mathbb{P}^1$, and $\mathbb{F}_1$ is the blow-up of $\mathbb{P}^2$ in one point.

Given a variety X and a sheaf $\mathcal{F}$ on X, we write $H^i(\mathcal{F})$ instead of $H^i(X, \mathcal{F})$ for brevity. The same applies to the notation $\chi(\mathcal{F})$ for the Euler characteristic of $\mathcal{F}$. We write $\chi_{top}(X)$ for the topological Euler characteristic of a variety X defined over $\mathbb{C}$. By $p_a(C)$ we denote the arithmetic genus $1 - \chi(\mathcal{O}_C)$ of a curve C.

A plane curve singularity given locally by an equation $x^2 = y^{n+1}$ for some $n \geqslant 1$ is called a *singularity of type* A_n. A *node* is a singularity of type A_1, i.e., one given locally by an equation $x^2 = y^2$. An *ordinary cusp* is a singularity of type A_2, i.e., one given locally by an equation $x^2 = y^3$. Also, we say that a variety X has a singularity of type A_n (or an ordinary cusp, respectively) along a smooth subvariety D of codimension 1, if the singularity locally in analytic topology looks like a product of D by a singularity of type A_n (or an ordinary cusp, respectively).

Let G be an algebraic group, and X be a variety with an action of G. We say that a subvariety of X is G-irreducible, if it is G-invariant and is not a union of two G-invariant subvarieties. If Y is a subvariety of X, we allow a small abuse of terminology and speak about a G-orbit of Y meaning the minimal G-invariant subvariety that contains Y. By a general point of Y we mean a point in a *dense* open subset in Y. This does not make much sense as an abstract notion, but is useful in some specific situations. In particular, a statement that a stabilizer of a general point of Y in G is trivial does make sense (it means the triviality of a stabilizer of a general point of every irreducible component of Y), although there is no concept of a stabilizer of a general point of Y. Similarly, we may say that a stabilizer of a general point of Y is isomorphic to some (fixed) group F, although there may be no way to identify stabilizers of points on different irreducible components of Y (which applies even to the case when Y is G-irreducible). Also, one can speak about a general point of Y being smooth. Suppose that Y is G-irreducible, its general point is smooth, and Z is a G-invariant subvariety, or a G-invariant divisor, or a G-invariant linear system on X. Then one can define the multiplicity of Z at a general point of Y as a local intersection index of Z with an appropriate number of general divisors from some very ample linear system passing through the corresponding point; we will denote the latter by $\mathrm{mult}_Y(Z)$. When we speak about a general curve on a variety without an action of a group, we mean a general curve of sufficiently large degree with respect to some ample divisor.

If $\mathcal{V}$ is a vector bundle on X, we say that $\mathcal{V}$ is G-invariant if for any element $g \in G$ the pull-back $g^*\mathcal{V}$ is isomorphic to $\mathcal{V}$; we say that $\mathcal{V}$ is a G-equivariant vector bundle when we have chosen a lifting of the action of G on X to $\mathcal{V}$.

If U is a representation of a finite group G, we denote by $U^\vee$ the dual representation of G. If X is a variety in $\mathbb{P}(U)$, we denote by $X^\vee$ the projectively dual variety in $\mathbb{P}(U^\vee)$.

By $\boldsymbol{\mu}_m$ we denote a cyclic group of order m. A dihedral group of order m is denoted by $\mathfrak{D}_m$. Note that $\mathfrak{D}_6 \cong \mathfrak{S}_3$ and $\mathfrak{D}_4 \cong \boldsymbol{\mu}_2 \times \boldsymbol{\mu}_2$. A symmetric group of rank n is denoted by $\mathfrak{S}_n$, and an alternating group of rank n is denoted by $\mathfrak{A}_n$. To describe particular elements of the groups $\mathfrak{S}_n$ and $\mathfrak{A}_n$, we assume that these groups permute the numbers $1, \ldots, n$. Then we denote by $(i_1\ i_2 \ldots i_k)$ the cycle that sends i_1 to i_2 and so on, up to i_k that is sent to i_1. By $(i_{1,1} \ldots i_{k_1,1}) \ldots (i_{1,r} \ldots i_{k_r,r})$ we denote a composition of the corresponding cycles.

If Γ is some group, $G_1, \ldots, G_r$ are subgroups of Γ, and $g_1, \ldots, g_r$ are elements of Γ, then we denote by $\langle G_1, \ldots, G_s, g_1, \ldots, g_r \rangle$ the subgroup in Γ generated by $G_1, \ldots, G_s$ and $g_1, \ldots, g_r$.

Nearly every concept of algebraic geometry has its counterpart in the situation when all objects involved are acted on by some finite group G. If X is a variety with an action of G, we will sometimes say that X is a G-variety. A rational map $\phi\colon X \dashrightarrow X'$ between G-varieties X and X' is G-equivariant, if for each element g of the group G the diagram

$$\begin{array}{ccc} X & \dashrightarrow^{\phi} & X' \\ {\scriptstyle g}\downarrow & & \downarrow{\scriptstyle g} \\ X & \dashrightarrow^{\phi} & X' \end{array}$$

commutes. We say that ϕ is a G-rational map (or G-map), if there is an automorphism u of the group G such that for each element g of the group G the diagram

$$\begin{array}{ccc} X & \dashrightarrow^{\phi} & X' \\ {\scriptstyle g}\downarrow & & \downarrow{\scriptstyle u(g)} \\ X & \dashrightarrow^{\phi} & X' \end{array}$$

commutes. If a G-rational map ϕ is a birational map or a morphism, we sometimes say that ϕ is a G-birational map or a G-morphism, respectively. Similarly, if there exists a biregular G-morphism between two G-varieties, we will say that they are G-biregular. The reader should be aware that this terminology is not universally accepted, and in many sources a G-map means just a G-equivariant map. The maps that we call G-rational are referred to as rational maps of G-varieties in [35, §3].

Note that any G-equivariant rational map is a G-map. Contrary to this, if one takes two copies of a variety X with the same action of G, then a non-central element $g \in G$ defines a G-morphism $g\colon X \to X$ that is not G-equivariant. On the other hand, any element $g \in G$ defines a G-equivariant morphism $g\colon X \to X$, where the action on the second copy of X is constructed as the initial action twisted by an inner automorphism of G given by conjugation by g.

In general, if X' is a variety with an action of G, and u is an automorphism of G, one can produce another action of G on X' as the initial action twisted by u. Denote the variety X' with this twisted action by X'_u. Let X be some other variety with an action of G. We see that each G-rational map

$$\phi\colon X \dashrightarrow X'$$

gives rise to a G-equivariant rational map $\phi_u\colon X \dashrightarrow X'_u$ for a suitably chosen automorphism u of the group G. Since the actions of G on X' and X'_u have many properties in common (say, G has fixed points on X' if and only if it has fixed points on X'_u), it will sometimes be more convenient for us to work with G-equivariant maps, although in most cases our motivation will come from studying G-rational maps.

We denote by $\mathrm{Aut}(X)$ the group of (biregular) automorphisms of X, and by $\mathrm{Aut}^G(X)$ the group of G-automorphisms of X. If X is irreducible, we denote by $\mathrm{Bir}(X)$ the group of birational automorphisms of X, and by $\mathrm{Bir}^G(X)$ the group of G-birational automorphisms of X. We write $\mathrm{Cr}_n(\mathbb{C})$ for $\mathrm{Bir}(\mathbb{P}^n)$.

Since the largest part of the book is devoted to geometry of one particular variety, and we have to keep track of some objects for a long time, we naturally try to keep the same notation for such objects throughout the book. For example, we denote the quintic del Pezzo threefold by V_5 everywhere, denote the surface in V_5 that is the closure of the unique two-dimensional $\mathrm{PSL}_2(\mathbb{C})$-orbit in V_5 by $\mathscr{S}$, and denote the unique one-dimensional $\mathrm{PSL}_2(\mathbb{C})$-orbit in V_5 by $\mathscr{C}$ (which is one of the two $\mathfrak{A}_5$-invariant rational curves of degree 6 in V_5) starting from Chapter 7 up to the very end. On the other hand, in some special situations we may use one and the same symbol for two different objects in non-overlapping parts of the book, even when in one of the instances it is used for something basic for a long time; this happens when we feel that there is a similarity between the geometry of two varieties, which justifies similar notation for different objects that behave

NOTATION AND CONVENTIONS

in a similar manner in these two situations. For example, we also use the symbol $\mathscr{C}$ to denote one of the two $\mathfrak{A}_5$-invariant rational curves of degree 6 in the Clebsch cubic surface. These similarities become most visible in Theorems 6.3.18 and 13.6.1 classifying low degree $\mathfrak{A}_5$-invariant curves in the Clebsch cubic surface and the quintic del Pezzo threefold, respectively. We tried to do our best not to cause any confusion for the reader with this convention.

Chapter 1

Introduction

In this chapter we briefly explain the motivation, the main results, and the structure of the book.

1.1 Conjugacy in Cremona groups

To prove or disprove that two given isomorphic subgroups in the Cremona group $\mathrm{Cr}_n(\mathbb{C}) = \mathrm{Bir}(\mathbb{P}^n)$ are conjugate is a problem that can be restated purely in terms of birational geometry. Indeed, let G be a finite subgroup in $\mathrm{Cr}_n(\mathbb{C})$. Then its embedding into $\mathrm{Cr}_n(\mathbb{C})$ arises from a biregular action of the group G on some rational projective variety X of dimension n. The normalizer of the subgroup G in $\mathrm{Cr}_n(\mathbb{C})$ coincides with the group of G-birational selfmaps of X, which is usually denoted by $\mathrm{Bir}^G(X)$, while the centralizer of G in $\mathrm{Cr}_n(\mathbb{C})$ coincides with the group of G-equivariant birational selfmaps of X (see the section "Notation and conventions" for the relevant terminology). Similarly, another embedding of G into the group $\mathrm{Cr}_n(\mathbb{C})$ is given by a biregular action of the group G on another rational projective variety X' of dimension n, where a priori X' can be isomorphic to X. Then the two corresponding subgroups are conjugate in $\mathrm{Cr}_n(\mathbb{C})$ if and only if there exists a G-birational map between X and X'. We refer the reader to [35] and [96] for details.

Currently, we do not know many obstructions to the existence of such a birational map. Let us describe three of them.

The first obstruction is very naive and, nevertheless, very powerful. It is well known and implied by the results of János Kollár and Endre Szabó on fixed points of abelian groups (see [103, Proposition A.4]). For the reader's convenience we reproduce the details in a particular case that is most suitable in our situation.

Theorem 1.1.1. Let X and X' be smooth varieties acted on by a finite abelian group A. Suppose that there exists an A-equivariant birational map
$$\psi\colon X \dashrightarrow X'.$$
Then X contains a point fixed by A if and only if X' contains a point fixed by A.

Proof. It is enough to prove the "only if" part of the assertion, and then apply the same argument to the birational map ψ^{-1}. Thus, we assume that there is an A-fixed point $P \in X$. If ψ is regular in a neighborhood of the point P, then $\psi(P)$ is an A-fixed point in X'. Thus, we may assume that ψ is undefined at P. Regularizing the rational map ψ, we obtain a commutative diagram

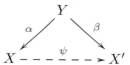

where β is a morphism, and α is a sequence of A-equivariant blow-ups at smooth subvarieties. By induction, we may assume that α is a single blow-up of an A-invariant irreducible smooth subvariety $Z \subset X$ that contains the point P.

Let $F \subset Y$ be the preimage of the point P with respect to α. Then F is identified with the projectivization of the quotient of Zariski tangent spaces
$$V = T_P(X)/T_P(Z).$$
The vector space $T_P(X)$ carries an action of A. Moreover, $T_P(Z) \subset T_P(X)$ is an A-invariant subspace, so that the group A acts on the vector space V as well. Since A is abelian, V splits as a sum of A-invariant one-dimensional subspaces, which means that F has an A-fixed point. □

Remark 1.1.2. If X is a (projective) smooth rational variety, then every finite cyclic group acting on X has a fixed point by the holomorphic Lefschetz fixed point formula (see, for example, [49, §3.4]). In particular, Theorem 1.1.1 does not give any obstruction for existence of an A-equivariant birational map between two rational varieties in case of a cyclic group A.

Remark 1.1.3. Another popular observation is that the whole set of points fixed by the elements of a group G acting on a variety X may provide G-birational invariants. For example, if X is a surface, then any non-rational curve whose points are G-fixed gives such invariant (cf. [8] and [98]).

The second obstruction has been classically known in the context of rationality questions over non algebraically closed fields (see, e. g., [79, § IV.7]), but was explicitly established in the context of equivariant birational maps only recently by Fedor Bogomolov and Yuri Prokhorov in [11].

Theorem 1.1.4 ([11, Corollary 2.3]). Let X and X' be smooth varieties acted on by a finite group G. Suppose that there exists a G-birational map

$$X \times \mathbb{P}^n \dashrightarrow X' \times \mathbb{P}^m,$$

where G acts trivially on the second factors of $X \times \mathbb{P}^n$ and $X' \times \mathbb{P}^m$. Then

$$H^1\bigl(G, \mathrm{Pic}(X)\bigr) \cong H^1\bigl(G, \mathrm{Pic}(X')\bigr).$$

In particular, the latter condition holds if there exists a G-equivariant birational map $X \dashrightarrow X'$.

Note that [11, Corollary 2.3] considers G-equivariant birational maps, but the proof works literally in the same way for G-birational maps.

The third obstruction comes from G-equivariant Minimal Model Program.

Definition 1.1.5. A variety Y endowed with an action of a finite group G and a G-equivariant surjective morphism $\phi \colon Y \to S$ with connected fibers to some variety S with an action of G is a *G-Mori fiber space* if it has terminal singularities, all G-invariant Weil divisors on Y are Cartier divisors, the dimension of S is strictly smaller than the dimension of Y, the anticanonical class $-K_Y$ is ϕ-ample, and

$$\mathrm{rk}\,\mathrm{Pic}\,(Y)^G = \mathrm{rk}\,\mathrm{Pic}\,(S)^G + 1.$$

If S is a point, then Y is called a *G-Fano variety*.

Suppose that Y is a G-Fano variety. Then Y is said to be *G-birationally rigid* if Y is not G-equivariantly birational to any other G-Mori fiber space. We refer the reader to Definition 3.1.1 for a precise definition of G-birational rigidity, and to Chapter 3 for a more detailed survey of basic notions and methods of the related theory. Note that the definition of birational rigidity can be formulated in a wider context of G-Mori fiber spaces (i. e., not only for G-Fano varieties, cf. [18, Definition A.3]). However, in this case the theory is much less applicable to our initial problem about conjugacy in Cremona groups.

Now let us return to our rational varieties X and X' with an action of a finite group G. Taking G-equivariant resolutions of singularities of the varieties X and X' and applying G-equivariant Minimal Model Program, we may assume that both X and X' are G-Mori fiber spaces. If one is lucky enough to get a situation where, say, X is a G-Fano variety, and to show that X is G-birationally rigid and X' is not G-biregular to X, then it becomes possible to conclude that there is no G-birational map between X and X', and thus the corresponding embeddings of the group G into the Cremona groups are not conjugate.

Let us consider several examples to illustrate the three approaches described above.

Example 1.1.6. Suppose that X is a general smooth del Pezzo surface of degree 4. Then $\mathrm{Aut}(X) \cong \boldsymbol{\mu}_2^4$ (cf. [34, §8.6.4]). Let A be a subgroup in $\mathrm{Aut}(X)$ such that $A \cong \boldsymbol{\mu}_2 \times \boldsymbol{\mu}_2$. Then $\mathrm{Pic}(X)^A \cong \mathbb{Z}$ (see [35, Theorem 6.9]). Moreover, one can show that A has a fixed point on X. This implies that X can be A-equivariantly birationally transformed into an A-equivariant conic bundle. In particular, X is not A-birationally rigid. Suppose that $X' \cong \mathbb{P}^2$. Then one has an embedding $A \subset \mathrm{Aut}(X')$. The group A has a fixed point on X'. A blow-up of this point provides an A-equivariant $\mathbb{P}^1$-bundle over $\mathbb{P}^1$, which is a non-trivial A-Mori fiber space A-birational to X'. In particular, the surface X' is not A-birationally rigid. Clearly, one has $H^1(A, \mathrm{Pic}(X')) = 0$. On the other hand, $H^1(A, \mathrm{Pic}(X)) \neq 0$ by [100, Theorem 1.2].

Example 1.1.7. Suppose that X is a smooth del Pezzo surface of degree 2, and $A \cong \boldsymbol{\mu}_2$ acts on X by an involution of the anticanonical double cover. Suppose that $X' \cong \mathbb{P}^2$. Then X is A-birationally rigid. This is well known (see, for example, [35, §8] or Example 3.3.4). In particular, there exists no A-birational map $X \dashrightarrow X'$. This also follows from Theorem 1.1.4, since

$$\boldsymbol{\mu}_2^6 \cong H^1(A, \mathrm{Pic}(X)) \neq H^1(A, \mathrm{Pic}(\mathbb{P}^2)) = 0$$

by [11, Theorem 1.1]. On the other hand, we cannot apply Theorem 1.1.1 here, since there are A-fixed points both on X and X'.

Example 1.1.8. Suppose that X is a smooth del Pezzo surface of degree 5, and $X' \cong \mathbb{P}^2$. Recall that X and X' are both acted on by the icosahedral group $G \cong \mathfrak{A}_5$. The surface X is G-birationally rigid (see Lemma 6.6.4). Moreover, X' is G-birationally rigid as well (see Lemma 6.6.5). Thus, there exists no G-birational map $X \dashrightarrow X'$. On the other hand, for any subgroup $H \subset G$ one has

$$H^1(H, \mathrm{Pic}(X)) = H^1(H, \mathrm{Pic}(X')) = 0$$

by [100, Theorem 1.2], which means that Theorem 1.1.4 is not applicable here. Moreover, any cyclic subgroup of G has a fixed point both on X and X' by Remark 1.1.2. The only non-cyclic abelian subgroup of G is $A \cong \boldsymbol{\mu}_2 \times \boldsymbol{\mu}_2$. It is straightforward to check that A has fixed points both on X and X' (see Lemma 5.3.1(i) and Theorem 6.2.12(iv)).

Example 1.1.9. Suppose that $X \cong \mathbb{P}^1 \times \mathbb{P}^1$, and $X' \cong \mathbb{P}^2$. Recall that a central extension of the group $A \cong \boldsymbol{\mu}_2 \times \boldsymbol{\mu}_2$ by a group $\boldsymbol{\mu}_2$ has a faithful two-dimensional representation, which gives a faithful action of the group A on $\mathbb{P}^1$. Consider a diagonal action of the group A on X, i.e., a simultaneous faithful action of A on two copies of $\mathbb{P}^1$ (actually, we need only that this action preserves both projections of X to $\mathbb{P}^1$ and is faithful on each of the factors). Also one has an embedding $A \subset \mathrm{Aut}(X')$. The group A has a fixed point on X'. In particular, the surface X' is not A-birationally rigid (cf. Example 1.1.6). Furthermore, one can check that X is not A-birationally rigid either. Moreover, for any subgroup $H \subset A$ one has

$$H^1\big(H, \mathrm{Pic}(X)\big) = H^1\big(H, \mathrm{Pic}(X')\big) = 0$$

by [100, Theorem 1.2]. On the other hand, X does not have A-fixed point, and X' has an A-fixed point. Thus, there exists no A-birational map $X \dashrightarrow X'$ by Theorem 1.1.1.

1.2 Three-dimensional projective space

Starting from this section our main character will be the icosahedral group $\mathfrak{A}_5$. Recall that $\mathfrak{A}_5$ has a unique irreducible five-dimensional representation W_5, a unique irreducible four-dimensional representation W_4, and two irreducible three-dimensional representations, one of which we denote by W_3. By I we denote the trivial representation of $\mathfrak{A}_5$. Note that $I \oplus W_4$ is the standard five-dimensional permutation representation of $\mathfrak{A}_5$. The group $\mathfrak{A}_5$ has a non-trivial central extension $2.\mathfrak{A}_5$ by a group of order 2. By U_2 and U_2' we denote two different two-dimensional irreducible representations of the group $2.\mathfrak{A}_5$, and by U_4 we denote the faithful four-dimensional irreducible representation of this group. We refer the reader to Chapter 5 for more details on the basic properties of the icosahedral group, including representations of $\mathfrak{A}_5$ and $2.\mathfrak{A}_5$.

Up to conjugation, the group $\mathrm{Aut}(\mathbb{P}^3) \cong \mathrm{PGL}_4(\mathbb{C})$ contains five subgroups isomorphic to $\mathfrak{A}_5$ (cf. [10, Chapter VII]). Namely, the projective space $\mathbb{P}^3$ equipped with the corresponding actions of the group $\mathfrak{A}_5$ can be identified with one of the following:

- a projectivization $\mathbb{P}(W_4)$ of the irreducible four-dimensional representation of $\mathfrak{A}_5$;

- a projectivization $\mathbb{P}(I \oplus W_3)$ of a reducible four-dimensional representation of $\mathfrak{A}_5$;

- a projectivization $\mathbb{P}(U_2 \oplus U_2)$ of a reducible faithful four-dimensional representation of $2.\mathfrak{A}_5$;

- a projectivization $\mathbb{P}(U_2 \oplus U_2')$ of a reducible faithful four-dimensional representation of $2.\mathfrak{A}_5$;

- a projectivization $\mathbb{P}(U_4)$ of the faithful four-dimensional irreducible representations of $2.\mathfrak{A}_5$.

Remark 1.2.1. Choose a subgroup $G \cong \boldsymbol{\mu}_2 \times \boldsymbol{\mu}_2$ of the group $\mathfrak{A}_5$. Then G acts on the projective space $\mathbb{P}^3 \cong \mathbb{P}(I \oplus W_3)$ with a fixed point, while its action on the projective space $\mathbb{P}^3 \cong \mathbb{P}(U_2 \oplus U_2')$ does not have fixed points. Therefore, Theorem 1.1.1 tells us that there is no $\mathfrak{A}_5$-birational map between these two copies of $\mathbb{P}^3$, because there is no $\boldsymbol{\mu}_2 \times \boldsymbol{\mu}_2$-equivariant birational map between them. Thus, the corresponding two embeddings of the group $\mathfrak{A}_5$ into the Cremona group $\mathrm{Cr}_3(\mathbb{C})$ are not conjugate. The same holds if we replace $\mathbb{P}(I \oplus W_3)$ by $\mathbb{P}(W_4)$, or replace $\mathbb{P}(U_2 \oplus U_2')$ by $\mathbb{P}(U_2 \oplus U_2)$.

The following examples show that $\mathbb{P}^3$ is never $\mathfrak{A}_5$-birationally rigid. Namely, for each of the possible actions of $\mathfrak{A}_5$ on $\mathbb{P}^3$ we provide an $\mathfrak{A}_5$-birational map from $\mathbb{P}^3$ to another $\mathfrak{A}_5$-Mori fiber space.

Example 1.2.2. Put $\mathbb{P}^3 = \mathbb{P}(I \oplus W_3)$. Then $\mathbb{P}^3$ has an $\mathfrak{A}_5$-fixed point. Blowing up this point, we obtain an $\mathfrak{A}_5$-equivariant $\mathbb{P}^1$-bundle over $\mathbb{P}^2$ that is a projectivization of the vector bundle $\mathcal{O}_{\mathbb{P}^2} \oplus \mathcal{O}_{\mathbb{P}^2}(1)$. The latter $\mathbb{P}^1$-bundle is an $\mathfrak{A}_5$-Mori fiber space that is $\mathfrak{A}_5$-birational to $\mathbb{P}^3$.

Example 1.2.3. Put $\mathbb{P}^3 = \mathbb{P}(I \oplus W_3)$. Then there is an $\mathfrak{A}_5$-fixed point P and an $\mathfrak{A}_5$-invariant plane $\Pi \cong \mathbb{P}(W_3)$ contained in $\mathbb{P}^3$. There is a smooth $\mathfrak{A}_5$-invariant conic C contained in Π (see Lemma 5.3.1(iii)). Blowing up C and contracting the proper transform of Π to a point we obtain an $\mathfrak{A}_5$-equivariant birational transformation from $\mathbb{P}^3$ to a smooth quadric in $\mathbb{P}^4$, which is thus an $\mathfrak{A}_5$-Fano variety that is $\mathfrak{A}_5$-birational to $\mathbb{P}^3$. Note that the latter quadric has two $\mathfrak{A}_5$-invariant points: one of them is the image of the proper transform of Π, and the other is the image of the point P.

Example 1.2.4 ([97, Proposition 4.7]). Put $\mathbb{P}^3 = \mathbb{P}(W_4)$. Then $\mathbb{P}^3$ contains an $\mathfrak{A}_5$-orbit Ω of length 5 (see Lemma 5.3.3(i)). Twisted cubics passing through Ω define a commutative diagram

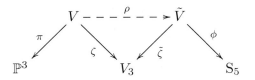

Here V_3 is the Segre cubic hypersurface in $\mathbb{P}^4$ (cf. Example 1.3.4), and S_5 is the del Pezzo surface of degree 5. The morphism π is the blow-up of Ω, the morphism ζ is the contraction of the proper transforms of the 10 lines in $\mathbb{P}^3$ passing through pairs of points in Ω (cf. Lemma 5.3.4(i)), the rational map ρ is a composition of Atiyah flops in these 10 curves, and ϕ is a conic bundle. Thus, the $\mathfrak{A}_5$-Mori fiber space $\phi \colon \tilde{V} \to S_5$ is $\mathfrak{A}_5$-birational to $\mathbb{P}^3$.

Example 1.2.5. Put $\mathbb{P}^3 = \mathbb{P}(U_2 \oplus U_2)$ or $\mathbb{P}^3 = \mathbb{P}(U_2 \oplus U_2')$. Then there is an $\mathfrak{A}_5$-invariant line in $\mathbb{P}^3$. Blowing up this line one obtains an $\mathfrak{A}_5$-equivariant $\mathbb{P}^2$-bundle over $\mathbb{P}^1$. The latter $\mathbb{P}^2$-bundle is an $\mathfrak{A}_5$-Mori fiber space that is $\mathfrak{A}_5$-birational to $\mathbb{P}^3$.

Example 1.2.6. Put $\mathbb{P}^3 = \mathbb{P}(U_2 \oplus U_2)$ or $\mathbb{P}^3 = \mathbb{P}(U_2 \oplus U_2')$. Then one can choose two skew $\mathfrak{A}_5$-invariant lines, say L_1 and L_2, in $\mathbb{P}^3$. Blowing up these lines one obtains an $\mathfrak{A}_5$-equivariant $\mathbb{P}^1$-bundle over $L_1 \times L_2 \cong \mathbb{P}^1 \times \mathbb{P}^1$. The latter $\mathbb{P}^1$-bundle is an $\mathfrak{A}_5$-Mori fiber space that is $\mathfrak{A}_5$-birational to $\mathbb{P}^3$.

Example 1.2.7. Put $\mathbb{P}^3 = \mathbb{P}(U_4)$. Then one has an $\mathfrak{A}_5$-equivariant identification
$$\mathbb{P}^3 \cong \mathbb{P}\big(\mathrm{Sym}^3(U_2)\big),$$
see Lemma 5.6.3(i). Thus $\mathbb{P}^3$ contains an $\mathfrak{A}_5$-invariant twisted cubic curve C. Blowing up this curve, we obtain an $\mathfrak{A}_5$-equivariant $\mathbb{P}^1$-bundle over
$$\mathrm{Sym}^2(C) \cong \mathbb{P}^2$$
that is a projectivization of a stable rank two vector bundle $\mathcal{E}$ on $\mathbb{P}^2$ with $c_1(\mathcal{E}) = 0$ and $c_2(\mathcal{E}) = 2$ defined by the exact sequence
$$0 \longrightarrow \mathcal{O}_{\mathbb{P}^2}(-1)^{\oplus 2} \longrightarrow \mathcal{O}_{\mathbb{P}^2}^{\oplus 4} \longrightarrow \mathcal{E} \otimes \mathcal{O}_{\mathbb{P}^2}(1) \longrightarrow 0$$
(see [117, Application 1]). The latter $\mathbb{P}^1$-bundle is an $\mathfrak{A}_5$-Mori fiber space that is $\mathfrak{A}_5$-birational to $\mathbb{P}^3$.

Example 1.2.8 ([118, 2.13.1],[58, §8.1],[97, §2]). Put $\mathbb{P}^3 = \mathbb{P}(W_4)$. There is a unique $\mathfrak{A}_5$-invariant smooth cubic surface $S_3 \subset \mathbb{P}^3$, known as the *Clebsch diagonal cubic surface* (see §6.3). It contains two smooth $\mathfrak{A}_5$-invariant rational sextic curves. Denote one of them by $\mathscr{C}$. The surface S_3 also contains an $\mathfrak{A}_5$-irreducible curve that is a disjoint union of six bi-tangent lines of the curve $\mathscr{C}$. These facts are well known (cf. [9, Proposition 4.2]) and can be derived from Lemma 6.3.17. We denote the latter sixtuple of lines by $\mathcal{L}_6'$. There exists a commutative diagram

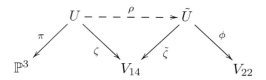

Here V_{14} is a Fano threefold of anticanonical degree $-K_{V_{14}}^3 = 14$ with six ordinary double points, and V_{22} is a smooth Fano threefold of anticanonical degree $-K_{V_{22}}^3 = 22$ which admits a faithful action of $\mathrm{PSL}_2(\mathbb{C})$ and is known as the Mukai–Umemura threefold (see [86], [95]). The morphism π is the blow-up of the curve $\mathscr{C}$, the morphism ζ is the contraction of the proper transforms of the irreducible components of the curve $\mathcal{L}_6'$, the rational map ρ is a composition of Atiyah flops in these 6 curves, and ϕ is a contraction of the proper transform of the surface S_3 to a point P. The Fano variety V_{22} is anticanonically embedded into $\mathbb{P}^{13}$, and the rational map $\pi \circ \rho^{-1} \circ \phi^{-1}$ is given by the linear system of hyperplane sections of V_{22} that have multiplicity at least 3 at the point P. Thus, the threefold V_{22} is an $\mathfrak{A}_5$-Fano variety that is $\mathfrak{A}_5$-birational to $\mathbb{P}^3$.

1.3 Other rational Fano threefolds

Of course, $\mathbb{P}^3$ is not the only rational Fano threefold that admits an icosahedral symmetry. Below we give some examples of Fano threefolds with an action of the group $\mathfrak{A}_5$, including several interesting singular varieties. Note that we do not know the rationality of some of them.

Example 1.3.1. Put $\mathbb{P}^4 = \mathbb{P}(I \oplus W_4)$. Then $\mathbb{P}^4$ contains a pencil of $\mathfrak{A}_5$-invariant quadrics. A general element of this pencil is a smooth quadric with an action of $\mathfrak{A}_5$.

Example 1.3.2. Put $\mathbb{P}^4 = \mathbb{P}(W_5)$. Then $\mathbb{P}^4$ contains a unique $\mathfrak{A}_5$-invariant quadric. This quadric is smooth, and is acted on by $\mathfrak{A}_5$.

Example 1.3.3. Put $\mathbb{P}^4 = \mathbb{P}(I \oplus I \oplus W_3)$. Then $\mathbb{P}^4$ contains a two-dimensional linear system of $\mathfrak{A}_5$-invariant quadrics. A general quadric Q in this linear system is smooth. Note that Q contains two $\mathfrak{A}_5$-invariant points. This action of $\mathfrak{A}_5$ on the quadric threefold already appeared in Example 1.2.3.

Example 1.3.4 (cf. Example 1.2.4)**.** Let V_3 be the complete intersection in $\mathbb{P}^5$ with homogeneous coordinates $x_0, \ldots, x_5$ given by

$$\sum_{i=0}^{5} x_i = \sum_{i=0}^{5} x_i^3 = 0.$$

The threefold V_3 is known as the Segre cubic (see [59, §3.2]). It has exactly 10 isolated ordinary double points (and hence V_3 is rational), and there exists a natural embedding $\mathfrak{S}_6 \hookrightarrow \mathrm{Aut}(V_3)$ which is actually an isomorphism. Furthermore, one has $\mathrm{Cl}(V_3) \cong \mathbb{Z}^6$ (see [44]). Up to conjugation, the group $\mathfrak{S}_6$ has two subgroups isomorphic to $\mathfrak{A}_5$, given by the standard and the non-standard embeddings of the latter group. The Segre cubic with an action of $\mathfrak{A}_5$ corresponding to the standard embedding is an $\mathfrak{A}_5$-Fano variety. We do not know whether it is $\mathfrak{A}_5$-birationally rigid (cf. Example 3.3.7). The Segre cubic with an action of $\mathfrak{A}_5$ corresponding to the non-standard embedding is $\mathfrak{A}_5$-birational to $\mathbb{P}^3$ (cf. Example 1.2.4).

Example 1.3.5. Put $\mathbb{P}^3 = \mathbb{P}(I \oplus W_3)$. Then there exists a unique $\mathfrak{A}_5$-invariant sextic surface S_6 in $\mathbb{P}^3$ with 65 isolated singular points; the surface S_6 is given in appropriate homogeneous coordinates x, y, z, t by equation

$$4(\lambda^2 x^2 - y^2)(\lambda^2 y^2 - z^2)(\lambda^2 z^2 - x^2) = t^2(1 + 2\lambda)(x^2 + y^2 + z^2 - t^2)^2,$$

where $\lambda = \frac{1+\sqrt{5}}{2}$, and it has only ordinary double singularities (see [5]). The surface S_6 is known as the Barth sextic. Let X_2 be a double cover of $\mathbb{P}^3$ branched over S_6. One can check that the action of $\mathfrak{A}_5$ lifts to an action on X_2. Also, one can show that $\mathrm{Cl}(X_2) \cong \mathbb{Z}^{14}$, and there exists a determinantal quartic threefold $Y_4 \subset \mathbb{P}^4$ with 42 ordinary double points such that the diagram

$$\begin{array}{ccc} Y_4 & \hookrightarrow & \mathbb{P}^4 \\ \rho \downarrow & & \downarrow \gamma \\ X_2 & \xrightarrow{\pi} & \mathbb{P}^3 \end{array}$$

commutes (see [43, Example 3.7], [92]). Here γ is the projection from a singular point of the quartic Y_4, and ρ is a birational map. In particular, the variety X_2 is rational, because determinantal quartics are rational. We expect that X_2 is $\mathfrak{A}_5$-birationally rigid (cf. [20, Theorem A]).

Example 1.3.6. Put $\mathbb{P}^4 = \mathbb{P}(I \oplus W_4)$. Let Q be the three-dimensional $\mathfrak{A}_5$-invariant quadric in $\mathbb{P}^4$ (see Example 1.3.1), and X_4 be an irreducible $\mathfrak{A}_5$-invariant quartic in $\mathbb{P}^4$. Let X_4' be a double cover of Q branched over $Q \cap X_4$. Then X_4' is a Fano threefold such that $-K_{X_4'}^3 = 4$. One can check that the action of $\mathfrak{A}_5$ lifts to X_4'. If X_4' is smooth, it is known to be non-rational by [63, Theorem 3.8]. It is easy to show that a general threefold X_4' is smooth. We do not know whether there are rational examples or not.

Example 1.3.7. Put $\mathbb{P}^4 = \mathbb{P}(W_5)$. Let $Q \subset \mathbb{P}^4$ be an $\mathfrak{A}_5$-invariant quadric (see Example 1.3.2), and $X_4 \subset \mathbb{P}^4$ be an irreducible $\mathfrak{A}_5$-invariant quartic. Similarly to Example 1.3.6, we can construct a double cover of Q branched over $Q \cap X_4$. One can show that it is also a Fano threefold that is acted on by $\mathfrak{A}_5$. This variety is singular, and we do not know whether it is rational or not.

Example 1.3.8. Put $\mathbb{P}^3 \cong \mathbb{P}(W_4)$. There is a pencil $\mathcal{Q}$ of $\mathfrak{A}_5$-invariant quartics in $\mathbb{P}^3$. Let $S \subset \mathbb{P}^3$ be a (reduced) quartic from the pencil $\mathcal{Q}$, and let X be a double cover of $\mathbb{P}^3$ branched over S. One can check that X is a Fano threefold with an action of $\mathfrak{A}_5$. If S is a general element of the pencil $\mathcal{Q}$, then S and X are smooth, so that X is non-rational by [120, Corollary 4.7(b)]. Except one non-reduced divisor, the pencil $\mathcal{Q}$ contains four singular members: one $K3$ surface with 5 ordinary double singularities, two $K3$ surfaces with 10 ordinary double singularities, and one $K3$ surface with 15 ordinary double singularities (see [54]). Thus we have four cases when X is singular. By [99, Theorem 8.1], the threefold X is rational in the case when S has 15 singular points. We do not know whether X is rational or not in the remaining three singular cases, but we expect it is not. We see [112, Conjecture 10.3] and [121, Theorem 1.1] as strong evidence that supports our expectation in the case when S has 5 singular points. The case of 10 singular points may be related to the famous Artin–Mumford construction of [1].

Example 1.3.9. The group $\mathfrak{A}_5$ acts on the threefolds $\mathbb{P}^1 \times \mathbb{P}^2$, $\mathbb{P}^1 \times \mathbb{P}^1 \times \mathbb{P}^1$ and $\mathbb{P}^1 \times \mathbb{S}_5$, where $\mathbb{S}_5$ is a del Pezzo surface of degree 5. Actually, there are several actions like this (cf. §6.4). One can use Theorem 1.1.1 to show that at least some of these varieties are not $\mathfrak{A}_5$-birational to each other. This is the case, say, for $\mathbb{P}^1 \times \mathbb{P}^1 \times \mathbb{P}^1$ with an action of $\mathfrak{A}_5$ that preserves all three projections from $\mathbb{P}^1 \times \mathbb{P}^1 \times \mathbb{P}^1$ and is faithful on each of the factors, and $\mathbb{P}^1 \times \mathbb{P}^2$ with a trivial action of $\mathfrak{A}_5$ on $\mathbb{P}^1$ and a non-trivial action on $\mathbb{P}^2$.

1.4 Statement of the main result

As we saw in §1.2 and §1.3, it is indeed hard to find a rational $\mathfrak{A}_5$-Fano threefold that has a chance to be $\mathfrak{A}_5$-birationally rigid. The motivation of our further work is to present such variety, and to work out the methods to check birational rigidity using this example.

The threefold we will be mostly working with in this book is called a *quintic del Pezzo threefold*, and is denoted by V_5. By definition, it is a smooth intersection of the Grassmannian $\mathrm{Gr}(2,5) \subset \mathbb{P}^9$ in its Plücker embedding with a linear subspace of codimension 3. Thus, V_5 is a smooth Fano threefold of degree 5 in $\mathbb{P}^6$. The variety V_5 is rational. The Picard rank of V_5 equals 1, so that V_5 is a G-Fano variety with respect to any finite group G acting on it. The automorphism group of V_5 is isomorphic to $\mathrm{PSL}_2(\mathbb{C})$, which provides an action of the group $\mathfrak{A}_5$ on V_5. We refer the reader to §7.1 for details.

The goal that we are going to pursue and that will shape all our work is the following assertion, that we regard as a single main result of the book.

Theorem 1.4.1. *The threefold V_5 is $\mathfrak{A}_5$-birationally rigid, and the group $\mathrm{Bir}^{\mathfrak{A}_5}(V_5)$ of $\mathfrak{A}_5$-birational selfmaps of V_5 is isomorphic to $\mathfrak{S}_5 \times \boldsymbol{\mu}_2$.*

We expect that V_5 is the only smooth rational $\mathfrak{A}_5$-Fano threefold that is $\mathfrak{A}_5$-birationally rigid.

As was explained in §1.1, an immediate consequence of Theorem 1.4.1 is the fact that the embedding of $\mathfrak{A}_5$ into the Cremona group $\mathrm{Cr}_3(\mathbb{C})$ given by the action of $\mathfrak{A}_5$ on V_5 is not conjugate to any embedding of $\mathfrak{A}_5$ to $\mathrm{Cr}_3(\mathbb{C})$ given by an action of $\mathfrak{A}_5$ on another $\mathfrak{A}_5$-Mori fiber space. Thus, Theorem 1.4.1 together with Remark 1.2.1 or Example 1.3.9 imply:

Corollary 1.4.2. *The group $\mathrm{Cr}_3(\mathbb{C})$ contains at least three non-conjugate subgroups isomorphic to $\mathfrak{A}_5$.*

We emphasize that if we wanted to obtain just two non-conjugate icosahedral subgroups in $\mathrm{Cr}_3(\mathbb{C})$, it would be enough to use Remark 1.2.1 or Example 1.3.9. Moreover, due to the fact that the order of the group $\mathfrak{A}_5$ is relatively small, and the anticanonical degree of V_5 is relatively large, Theorem 1.4.1 is indeed not easy to prove. Nevertheless, we find it well motivated, since it gives us a reason both to refine the methods of proving $\mathfrak{A}_5$-birational rigidity, and to study a beautiful world of icosahedral symmetries of the variety V_5. The largest part of the book is occupied by such studies.

Many of the results proved in this book find their application during the proof of Theorem 1.4.1 in Chapter 18, as well as in the closely related classification of $\mathfrak{A}_5$-invariant Halphen pencils on V_5 in Chapter 19. However, some of the facts about $\mathfrak{A}_5$-equivariant geometry of V_5 that we establish are not directly used in the proofs, and in general we regard such results as having their internal value and beauty. All in all, we feel that the only reasonable way to prove Theorem 1.4.1 and similar assertions is to get a deep understanding of the biregular geometry of the underlying variety, and then to proceed to its birational geometry. This is basically what we do in the rest of the book.

1.5 Outline of the book

Let us briefly describe the structure of our book.

In **Part I** we collect various technical notions and facts that allow one to study birational geometry of higher dimensional varieties.

In **Chapter 2** we recall the basic theory of singularities of pairs, mostly following [73], [74], and [29]. In §2.1 we start with the definitions of various classes of singularities in the simplest setting of pairs that consist of varieties and $\mathbb{Q}$-divisors. In §2.2 we formulate similar notions for pairs whose boundary may be a linear system. In §2.3 we recall the definitions and main properties of multiplier ideal sheaves, that provide an important tool to work with log canonical singularities. In particular, we discuss Nadel's vanishing theorem and related connectedness result. In §2.4 we discuss further properties of centers of log canonical singularities, mostly based on Kawamata's subadjunction theorem. In §2.5 we formulate a local inequality due to Alessio Corti concerning the multiplicities of linear systems on threefolds, and give its proof.

Chapter 3 deals with birational transformations between G-Fano varieties and G-Mori fiber spaces for an arbitrary finite group G, and some varieties with similar properties. In §3.1 we recall the notions of G-birational rigidity and superrigidity for G-Fano varieties. In §3.2 we formulate and prove a slightly generalized version of Noether–Fano inequalities for G-Fano varieties, that is the main tool to work with birational rigidity. In our setting we start with a weak G-Fano variety, and describe a property of linear systems on it that gives an obstruction for existence of G-birational maps to other weak G-Fano varieties and fibrations with relatively ample or relatively trivial canonical class. As with the classical Noether–Fano inequalities, this property is, roughly speaking, a requirement that mobile linear systems of small degree cannot have large singularities. In §3.3, we use results obtained in §3.2 to derive criteria for G-birational rigidity and superrigidity for G-Fano varieties, and consider some examples. In §3.4 we replace a weak G-Fano variety by a variety with semi-ample anticanonical divisor, and develop a similar approach to study G-invariant Halphen pencils, that are pencils of divisors of zero Kodaira dimension on our variety.

Chapter 4 is a collection of some auxiliary results that will be useful in the sequel. In §4.1 we present estimates for dimensions of cohomology groups on some zero-dimensional schemes, communicated to us by Nadezhda Timofeeva. In §4.2 we briefly discuss Atiyah flops. In §4.3 we collect some technical facts about one-dimensional linear systems. In §4.4 we gather several sporadic useful results that will be applied in the rest of the book.

The purpose of **Part II** is to discuss in detail various elementary properties of the icosahedral subgroup $\mathfrak{A}_5$, and to understand its actions on some curves and surfaces as a warming up before more serious three-dimensional problems.

Chapter 5 describes the most basic properties of the icosahedral group. We start with the group structure itself in §5.1, including obvious but useful information about subgroups and conjugacy classes, and also less obvious but still elementary restrictions on genera of curves that admit an action of $\mathfrak{A}_5$. In §5.2 we collect the information about representations of the group $\mathfrak{A}_5$. We include here restrictions to subgroups, and also results of some specific computations with representations that will later help us to describe $\mathfrak{A}_5$-equivariant geometry of surfaces and threefolds. In §5.3 we recall the well-known facts about actions of the group $\mathfrak{A}_5$ on low-dimensional projective spaces. In §5.4 we discuss in detail several curves of low genera with an action of $\mathfrak{A}_5$; these curves will appear along our way surprisingly often in later chapters. In §5.5 we recall the representation theory of the group $\mathrm{SL}_2(\mathbb{C})$, and collect several results about varieties with the action of $\mathrm{PSL}_2(\mathbb{C})$. In §5.6 we recall a couple of facts about representations of the central extension $2.\mathfrak{A}_5$ of the group $\mathfrak{A}_5$. In §5.7 we present two computational results concerning representations of $\mathfrak{A}_5$ that actually have a structure of representations of the symmetric group $\mathfrak{S}_5$; these computations will be used later to study the del Pezzo surface of degree 5. We note that although one can approach the group $\mathfrak{A}_5$ working out the properties of some of the larger groups, like $\mathrm{SL}_2(\mathbb{C})$, $2.\mathfrak{A}_5$ or $\mathfrak{S}_5$ first, we do not use such an option because all other groups will play only an auxiliary role in our further geometrical investigations. We conclude Chapter 5 with §5.8 where we make few easy observations about dihedral subgroups $\mathfrak{S}_3$ and $\mathfrak{D}_{10}$ in $\mathfrak{A}_5$.

Chapter 6 is devoted to interesting surfaces with an action of the group $\mathfrak{A}_5$. In §6.1 we consider the simplest instance of this, the projective plane viewed as a projectivization of a three-dimensional $\mathfrak{A}_5$-representation. From one point of view it is an auxiliary object that will be widely used later when we start working with threefolds. On the other hand, we think that this example, although very simple, is interesting on its own due to an impressive supply of charmingly symmetric configurations it provides. In §6.2 we work with the del Pezzo surface of degree 5. As before, it is going to be used later in our three-dimensional studies, but we also note that this surface is a very exact two-dimensional match for the threefold V_5 we are mostly aiming at. We describe $\mathfrak{A}_5$-orbits on this surface and recover classical results concerning the unique pencil of $\mathfrak{A}_5$-invariant bi-canonical curves, which were obtained by William Edge in [39]. In §6.3 we pass to another surface with the action of $\mathfrak{A}_5$, a famous Clebsch diagonal cubic surface introduced in [25]. We spend some time to describe interesting $\mathfrak{A}_5$-invariant subvarieties of the Clebsch surface, mainly $\mathfrak{A}_5$-orbits and invariant curves of low degree, and discuss some configurations of lines and conics. Since the Clebsch surface is a midpoint of one important $\mathfrak{A}_5$-birational selfmap of a projective plane, we also do some preparational work that will be used later to study $\mathfrak{A}_5$-birational rigidity of $\mathbb{P}^2$. In §6.4 we study the actions of the group $\mathfrak{A}_5$ on a smooth quadric surface. This case is definitely less beautiful than the previous one, but has many applications in a threefold setting; due to this §6.4 mostly deals with $\mathfrak{A}_5$-invariant curves of low bi-degree that are rarely involved in any interesting configurations, but that will make an impressive appearance later on the threefold V_5. In §6.5 we classify the actions of $\mathfrak{A}_5$ on other Hirzebruch surfaces, and study $\mathfrak{A}_5$-birational maps between them. Namely, we show that the group $\mathfrak{A}_5$ does not act on the surfaces $\mathbb{F}_n$ with odd n, and all $\mathfrak{A}_5$-actions on the surfaces $\mathbb{F}_n$ with even n are related by $\mathfrak{A}_5$-birational maps. In §6.6 we use the results of the previous sections to completely describe the actions of $\mathfrak{A}_5$ on rational surfaces up to $\mathfrak{A}_5$-birational equivalence, and thus recover the description of all embeddings of $\mathfrak{A}_5$ into the Cremona group $\mathrm{Cr}_2(\mathbb{C})$ up to conjugation obtained by Shinzo Bannai and Hiro-o Tokunaga in [4]. In §6.7 we recall some well-known facts about $K3$ surfaces with an action of the group $\mathfrak{A}_5$, and derive some consequences that will largely shape our understanding of the three-dimensional case.

Starting from **Part III** we concentrate on geometry of the quintic del Pezzo threefold V_5 embedded to $\mathbb{P}^6$. This variety is interesting from many points of view. Due to its remarkable properties reflected in several different areas of algebraic geometry, we feel that the rich world of $\mathfrak{A}_5$-invariant configurations we find in it is not just a coincidence.

In **Chapter 7** we deal with the most basic properties of the variety V_5

itself and of the action of the group $\mathfrak{A}_5$ on it. In §7.1 we recall the construction of V_5 and explain the action of the automorphism group of V_5, which is isomorphic to $\mathrm{PSL}_2(\mathbb{C})$. Everything here is widely known, but we still find it useful to reproduce the proofs of several relevant facts. In §7.2 we survey the geometry of the unique $\mathrm{PSL}_2(\mathbb{C})$-invariant surface $\mathscr{S}$ in V_5. In §7.3 we classify all $\mathfrak{A}_5$-orbits in V_5 of length less than 20; it appears that there is only a finite number of them, and they all can be clearly described. In §7.4 we start our investigation of low degree $\mathfrak{A}_5$-invariant curves in V_5 by describing all $\mathfrak{A}_5$-invariant curves that are unions of at most 20 lines. In §7.5 we deal with the unique $\mathfrak{A}_5$-orbit of length five in V_5. We also discuss two $\mathfrak{A}_5$-equivariant linear projections of V_5. One of them gives rise to an elliptic fibration over $\mathbb{P}^2$, and the other is a five-to-one cover of $\mathbb{P}^3$. In §7.6 we discuss the properties of a unique $\mathfrak{A}_5$-invariant configuration of five hyperplane sections of V_5. In §7.7 we recall a (non-equivariant) birational transformation between V_5 and a three-dimensional quadric defined by a projection from a line in V_5. In §7.8 we proceed with low degree curves by describing $\mathfrak{A}_5$-invariant unions of less than 10 conics.

In **Chapter 8** we study $\mathfrak{A}_5$-invariant surfaces in V_5. This provides one of our main tools to study the geometry of V_5. In §8.1 we write down the action of $\mathfrak{A}_5$ on the space of global sections of the anticanonical line bundle of V_5 (which is actually the line bundle $\mathcal{O}_{V_5}(2H)$, where H is a hyperplane section of V_5 in $\mathbb{P}^6$). We find that there is a pencil $\mathcal{Q}_2$ of $\mathfrak{A}_5$-invariant anticanonical surfaces and completely describe its base curve. In particular, the pencil $\mathcal{Q}_2$ contains the surface $\mathscr{S}$ studied in §7.2, which is not just $\mathfrak{A}_5$-invariant but also $\mathrm{PSL}_2(\mathbb{C})$-invariant. In §8.2 we describe all singular surfaces of this pencil; it appears that all of them except the $\mathrm{PSL}_2(\mathbb{C})$-invariant surface $\mathscr{S}$ are $K3$ surfaces, and apart from four surfaces with ordinary double points all other members of this pencil are smooth $K3$ surfaces. We find many similarities with a pencil of $\mathfrak{A}_5$-invarian quartic $K3$ surfaces in $\mathbb{P}^3$ that was extensively studied by Kenji Hashimoto in [54]. In §8.3 we use the theory of $K3$ surfaces, and in particular results from §6.7, to obtain restrictions on low degree $\mathfrak{A}_5$-invariant curves in V_5. In particular, it appears that many such curves are contained in the surfaces of the pencil $\mathcal{Q}_2$. This allows us to demonstrate that there is only a finite number of $\mathfrak{A}_5$-invariant curves of degree less than 30 and justifies our further attempts to obtain a classification of invariant curves of low degree.

In **Chapter 9** we describe combinatorial properties of $\mathfrak{A}_5$-invariant curves in V_5 that are unions of small numbers of lines or conics. In particular, we show that some of them are disjoint unions of lines or conics, and describe intersections in several other situations. Note that some of the results obtained in §9.1 are supported by the data collected in §6.1, which emphasizes the amplitude of connections between varieties with icosahedral symmetry.

In **Chapter 10** we produce a preliminary classification of $\mathfrak{A}_5$-invariant low degree curves in V_5, reaching degree 15 with a surprising (temporary) gap in degree 6. In §10.1 we do some preliminary work excluding several particularly tough possibilities, and in §10.2 present the classification itself.

Chapter 11 is our first essentially birational attempt to deal with the variety V_5. In §11.1 we study a linear system of anticanonical surfaces passing through the unique $\mathfrak{A}_5$-invariant curve $\mathcal{L}_6$ in V_5 that is a union of six lines. We find that it defines a rational map to $\mathbb{P}^4$. In §11.2 we study this map in detail, and describe its image that appears to be a quartic threefold with one ordinary double point and five singular points of type cD_4, for which we manage to find an explicit equation. In §11.3 we use these results to produce an extra $\mathfrak{A}_5$-invariant rational normal sextic curve in V_5, that removes an uncertainty in our list of invariant curves of low degree obtained in Chapter 10. It is worth mentioning that while our construction of the above rational map is based on a partial classification of invariant low degree curves, the properties of this rational map help to advance the classification itself. In §11.4 we present two important $\mathfrak{A}_5$-birational self-maps of V_5, one of them arising from the above rational map to $\mathbb{P}^4$, and another defined by the $\mathfrak{A}_5$-invariant rational normal sextic curve that was found in §11.3. Later we will see that they generate the whole group of $\mathfrak{A}_5$-birational selfmaps of V_5, which will enable us to describe its structure. Both of these selfmaps give rise to $\mathfrak{A}_5$-Sarkisov links, and one of them is actually a Sarkisov link in the usual sense. An amazing point here is that we do not know any other way to construct the latter link, although its existence was already predicted by Joseph Cutrone and Nicholas Marshburn in [31]. In §11.5 we compute the action of the two birational selfmaps on the Picard groups of their regularizations.

In **Part IV** we describe further properties of some interesting invariant subvarieties of V_5.

In **Chapter 12** we deal with invariant hypersurfaces of low degree in V_5 (or, to be more precise, divisors in V_5 that are cut out by such hypersurfaces in $\mathbb{P}^6$). In §12.1 we write down the action of $\mathfrak{A}_5$ on $H^0(\mathcal{O}_{V_5}(3H))$. It appears that there is a unique $\mathfrak{A}_5$-invariant cubic hypersurface S_{Cl} in V_5. In §12.2 we discuss the first properties of the surface S_{Cl}, including some easy facts about orbits and curves contained in it. It also turns out that S_{Cl} is mapped to the Clebsch diagonal cubic surface thoroughly described in §6.3 by the $\mathfrak{A}_5$-equivariant linear projection of V_5 to $\mathbb{P}^3$ constructed in §7.5. In §12.3 we use

CREMONA GROUPS AND THE ICOSAHEDRON 17

the properties of the surface S_{Cl} to construct a special $\mathfrak{A}_5$-invariant smooth curve of genus 4 and degree 18 in V_5, isomorphic to a classically studied Bring's curve. It appears as an irreducible component of an intersection of S_{Cl} with a certain $K3$ surface from the pencil $\mathcal{Q}_2$. In §12.4 we describe intersections of S_{Cl} with all other $K3$ surfaces from the pencil $\mathcal{Q}_2$ in order to use it later in §15.2 to describe the singularities of S_{Cl}. We conclude the chapter with §12.5 where we describe some properties of a special $\mathfrak{A}_5$-invariant quartic hypersurface in V_5 that appears in the description of a Sarkisov link from §11.4.

In **Chapter 13** we proceed with our classification of invariant low degree curves in V_5. We aim to cover all degrees less than 20, and find that the most difficult cases are degrees 16 and 18. In §13.1 we describe all $\mathfrak{A}_5$-invariant curves of degree 16. In §13.2 we produce an $\mathfrak{A}_5$-invariant curve that is a union of six twisted cubics, and show that it is the only curve of degree 18 that is reducible but $\mathfrak{A}_5$-irreducible. In §13.3 we prove some auxiliary assertion about irreducible invariant curves of degree 18. In §13.4 we find a singular irreducible invariant curve of degree 18 and show that it is a unique curve with these properties. In §13.5 we study the properties of the Bring's curve in V_5 that was found in §12.3. Finally, in §13.6 we write down a classification of $\mathfrak{A}_5$-invariant curves of degree less than 20 in V_5.

In **Chapter 14** we deal with $\mathfrak{A}_5$-orbits in V_5 whose points have non-trivial stabilizers. Since we have already described all orbits of length less than 20 in §7.3, it remains to describe orbits of length 20 and 30. In §14.1 we find that all orbits of length 20 except one are contained in a certain curve $\mathcal{G}_{10}$ of degree 20 that is a union of 10 conics. We also show that every $\mathfrak{A}_5$-irreducible curve in V_5 of degree less than 30 is contained in some surface of the pencil $\mathcal{Q}_2$, with the exception of $\mathcal{G}_{10}$ and the unique $\mathfrak{A}_5$-irreducible curve $\mathcal{L}_{15}$ that is a union of 15 lines. We use this information to sketch an approach to a classification of $\mathfrak{A}_5$-irreducible curves of degree 20 in V_5. In §14.2 we describe the intersections of the curve $\mathcal{G}_{10}$ with other invariant curves of low degree. In §14.3 we show that any $\mathfrak{A}_5$-orbit of length 30 is contained either in the curve $\mathcal{L}_{15}$, or in a certain curve $\mathcal{T}_{15}$ that is a union of 15 twisted cubics. Both in §14.1 and in §14.3 the main tool that enables us to describe the orbits is provided by the properties of invariant $K3$ surfaces of the pencil $\mathcal{Q}_2$ which we studied in Chapter 8. We conclude Chapter 14 by §14.4 where we describe the intersections of the curve $\mathcal{T}_{15}$ with other invariant curves of low degree.

In **Chapter 15** we return to the invariant cubic hypersurface S_{Cl}. In §15.1, we list all low degree $\mathfrak{A}_5$-invariant curves contained in the surface S_{Cl}, and describe its intersection with some other low degree curves.

In §15.2 we prove that its singular locus consists of 12 ordinary double points. As a consequence we show that S_{Cl} is a surface of general type. In §15.3 we describe the images in the Clebsch diagonal cubic surface of all $\mathfrak{A}_5$-irreducible curves of small degree contained in S_{Cl}. In particular, we show that S_{Cl} contains exactly 18 lines. In §15.4 we find several relations between curves contained in S_{Cl} and use them to estimate the rank of its Picard group.

Chapter 16 contains several tables that summarize our knowledge of incidence relations between orbits of small length, invariant curves of low degree and invariant surfaces of low degree. The curves that appear here include those classified in Chapter 13, and some other curves important for us, like the base curve of the pencil $\mathcal{Q}_2$, and the curves $\mathcal{G}_{10}$ and $\mathcal{T}_{15}$. The surfaces include the $\mathrm{PSL}_2(\mathbb{C})$-invariant anticanonical surface $\mathscr{S}$, all other singular surfaces in the pencil $\mathcal{Q}_2$, and the surface S_{Cl}.

In **Part V** we return to the problem that motivated us to start studying $\mathfrak{A}_5$-equivariant geometry of the threefold V_5. Namely, we use the results obtained in previous chapters to prove that V_5 is $\mathfrak{A}_5$-birationally rigid, and to describe the group of its $\mathfrak{A}_5$-birational selfmaps. Moreover, we describe all $\mathfrak{A}_5$-invariant Halphen pencils on V_5 and $\mathfrak{A}_5$-rational maps from V_5 with elliptic fibers. An approach to birational rigidity and to structures of the latter type discussed in Chapter 3 suggests that we need a tool to prove that an $\mathfrak{A}_5$-invariant mobile linear system of low degree on V_5 cannot have large singularities. Our strategy is to suppose that such a linear system of small degree and with large singularities does exist and consider the possible centers of these singularities (which usually leaves us with a large, sometimes infinite list of subvarieties). Then we take an $\mathfrak{A}_5$-orbit Z of some a priori possible center and try to *isolate* it, that is to find some auxiliary linear system that has controllable singularities along Z, and does not have base points in a neighborhood of Z except on Z itself. After this it is often possible to exclude most possibilities by considering some explicit intersections.

In **Chapter 17** we prove auxiliary facts about base loci of various $\mathfrak{A}_5$-invariant linear systems that will be used later in the process of isolation of centers of bad singularities. In §17.1 we show that the points of $\mathfrak{A}_5$-orbits of length 10 in V_5 are in some sense in sufficiently general position. In §17.2 we use this information to determine the base locus of the unique two-dimensional $\mathfrak{A}_5$-invariant subsystem of the anticanonical linear system on V_5, which helps to isolate several $\mathfrak{A}_5$-orbits of small length. In §17.3 we work out an isolation procedure for all $\mathfrak{A}_5$-orbits contained in the $\mathrm{PSL}_2(\mathbb{C})$-invariant anticanonical surface $\mathscr{S}$, and in §17.4 we complete the isolation for all $\mathfrak{A}_5$-orbits in V_5. In §17.5 we start isolating the curve $\mathcal{L}_{15}$, that is a unique $\mathfrak{A}_5$-invariant curve of degree 15 in V_5. This case is somewhat special, because $\mathcal{L}_{15}$ is one of the very few invariant curves of low degree in V_5 that is not contained in any quadric hypersurface. Thus we work with the linear system of all cubic hypersurfaces passing through $\mathcal{L}_{15}$; we do not literally carry out the plan described above in this case, but obtain a result that will be enough for applications in the last two chapters.

In **Chapter 18** we prove $\mathfrak{A}_5$-birational rigidity of V_5. In §18.1 we state a technical reformulation of our problem in terms of centers of singularities, and derive birational rigidity from it. In §18.2 we discuss multiplicities of mobile linear systems restricted to the surfaces of the pencil $\mathcal{Q}_2$ along curves contained in these surfaces. In §18.3 we exclude all possible centers of singularities except the irreducible components of the curve $\mathcal{L}_{15}$. In §18.4, we deal with the curve $\mathcal{L}_{15}$ keeping in the framework of the method of test class (cf. [65]). In §18.5 we replace one step of the proof concerning exclusion of "long" $\mathfrak{A}_5$-orbits by an alternative argument using centers of log canonical singularities and multiplier ideal sheaves. In §18.6 we provide another approach to exclusion of the curve $\mathcal{L}_{15}$.

In **Chapter 19** we apply the results of Chapter 3 to study $\mathfrak{A}_5$-invariant Halphen pencils on V_5 and $\mathfrak{A}_5$-rational maps from V_5 with elliptic fibers. Namely, we show that, up to $\mathfrak{A}_5$-birational selfmaps, the elliptic fibration described in §7.5 is the unique one that is $\mathfrak{A}_5$-birational to V_5, and the pencil $\mathcal{Q}_2$ described in Chapter 8 is the only $\mathfrak{A}_5$-invariant Halphen pencil on V_5. We also classify all Fano threefolds with canonical singularities that are $\mathfrak{A}_5$-birational to V_5 (it appears that there are four of them apart from V_5 itself). In §19.1 we formulate the answer to this problem, and reduce it to a technical assertion about singularities of linear systems appearing from Noether–Fano inequalities. In the remaining four sections, we prove this assertion. In §19.2 we exclude the possibility of zero-dimensional centers of singularities, with explicitly described exceptions. In §19.3 we exclude curves, again except several explicitly described exceptions. In §19.4 we work out the above exceptions in more detail. Finally, in §19.5 we put everything together to complete the proof of the technical assertion from §19.1.

Part I
Preliminaries

Chapter 2

Singularities of pairs

In this chapter we collect basic results about singularities of pairs that consist of a variety and a $\mathbb{Q}$-divisor on it. In some cases, the $\mathbb{Q}$-divisor is replaced by a formal linear combination of mobile linear systems. We also present some well-known results about multiplier ideal sheaves and centers of log canonical singularities. Our primary sources are [29], [30], [72], [73], [74] and [77].

2.1 Canonical and log canonical singularities

Let X be a normal variety such that K_X is a $\mathbb{Q}$-Cartier divisor, i.e., there exists an integer $n > 0$ such that the divisor nK_X is Cartier. Let

$$\pi\colon \bar{X} \to X$$

be its resolution of singularities. Denote the π-exceptional divisors by $E_1, \ldots, E_m$. Then

$$K_{\bar{X}} \sim_{\mathbb{Q}} \pi^*(K_X) + \sum_{i=1}^{m} e_i E_i \qquad (2.1.1)$$

for some rational numbers $e_1, \ldots, e_n$. These numbers are called discrepancies. If X is smooth, they are all positive. In general, the numbers $e_1, \ldots, e_m$ measure how singular the variety X is. However, if one of them is smaller than -1, then we can replace π by another resolution of singularities of X such that the new discrepancies can be as small (negative) as we wish. If each e_i is greater than or equal to -1, then we can give

Definition 2.1.2 ([73, Definition 3.5]). The variety X has terminal (resp., canonical, Kawamata log terminal, log canonical) singularities if all numbers $e_1, \ldots, e_m$ are positive (resp., non-negative, greater than -1, is at least -1).

This definition is consistent, i.e., one can show that it does not depend on the choice of the morphism π. Thus, it gives us four very natural classes of singularities. By Definition 2.1.2, smooth varieties have terminal singularities.

Remark 2.1.3. Kawamata log terminal singularities are rational by [73, Theorem 11.1]. Also, if X is a surface, then it follows from [73, Theorem 3.6] that X has terminal (resp., canonical, Kawamata log terminal) singularities if the surface X is smooth (resp., has du Val singularities, has quotient singularities).

Canonical and terminal singularities play a crucial role in the Minimal Model Program. They behave very well with respect to birational maps that appear there. This is why they are very important.

Example 2.1.4. Let X be a surface. Suppose it has quotient singularities, so that X has Kawamata log terminal singularities by Remark 2.1.3. Take an integer $n > 0$ such that nK_X is Cartier. If the linear system $|mnK_X|$ is not empty for some positive integer m, it gives a rational map

$$\phi_m \colon X \dashrightarrow \mathbb{P}^N,$$

where $N = h^0(\mathcal{O}_X(mnK_X)) - 1$. In this case, we can try to define the Kodaira dimension of X as

$$\kappa(X) = \sup_{m \in \mathbb{N}} \Big(\dim \big(\phi_m(X) \big) \Big).$$

If $|mnK_X|$ is empty for all positive integers m, we can put $\kappa(X) = -\infty$. Then $\kappa(\bar{X}) \leqslant \kappa(X)$. The equality $\kappa(X) = \kappa(\bar{X})$ does not hold in general. However, if X has canonical singularities, then we always have $\kappa(X) = \kappa(\bar{X})$ by Definition 2.1.2, so that the number $\kappa(X)$ is, actually, a birational invariant.

In this book, we mostly deal with smooth surfaces and smooth threefolds. So, why bother to study terminal, canonical, Kawamata log terminal, and log canonical singularities? The reason is simple: Definition 2.1.2 makes sense for *pairs* consisting of a variety and a $\mathbb{Q}$-divisor on it. Even if the underlying variety is smooth, the pair can have very bad singularities, because the $\mathbb{Q}$-divisor can be very singular.

Starting from now, we assume that X has Kawamata log terminal singularities. In particular, X has rational singularities. Let B_X be some $\mathbb{Q}$-divisor on X, i.e., B_X is a formal $\mathbb{Q}$-linear combination

$$B_X = \sum_{i=1}^{r} a_i B_i$$

such that each B_i is a prime Weil divisor on X, and each a_i is a rational number. Usually, the numbers $a_1, \ldots, a_r$ are assumed to be non-negative. If this is the case, B_X is said to be effective. However, in many applications, we must take negative coefficients into account. So, for a moment, we assume that the numbers $a_1, \ldots, a_r$ are just any rational numbers.

The pair (X, B_X) is sometimes called a *log pair*, and the formal sum $K_X + B_X$ is said to be its log canonical divisor.

Suppose that B_X is also a $\mathbb{Q}$-Cartier divisor. Let us generalize Definition 2.1.2 for the pair (X, B_X) as follows. First we define rational numbers $d_1, \ldots, d_m$ by

$$K_{\bar{X}} + \sum_{i=1}^{r} a_i \bar{B}_i + \sum_{i=1}^{m} d_i E_i \sim_{\mathbb{Q}} \pi^*\left(K_X + B_X\right). \tag{2.1.5}$$

Remark 2.1.6. The π-exceptional divisors in (2.1.5) appear on the different side compared to (2.1.1). So, if $B_X = 0$, then we have $e_i = -d_i$ for each of them. This is done deliberately, because it provides better shapes for many formulas.

Suppose now that B_X is also a $\mathbb{Q}$-Cartier divisor. Denote the proper transforms of the divisors $B_1, \ldots, B_r$ on $\bar{X}$ by $\bar{B}_1, \ldots, \bar{B}_r$, respectively. Replace, if necessary, the resolution of singularities π by a better one such that the divisor

$$\sum_{i=1}^{r} \bar{B}_i + \sum_{i=1}^{m} E_i$$

has simple normal crossing. Such resolution of singularities is usually called a *log resolution* of the pair (X, B_X). By Hironaka theorem log resolutions do exist. Now we are ready to give

Definition 2.1.7 ([73, Definition 3.5]). The log pair (X, B_X) has Kawamata log terminal (resp., log canonical) singularities if each a_i and each d_j is smaller than 1 (resp., does not exceed 1).

This definition is also consistent in the sense that it does not depend on the choice of the log resolution $\pi\colon \bar{X} \to X$. Note that we can also define terminal and canonical singularities for pairs in a similar way (see [73, Definition 3.5]). However, we decided not to do this here (cf. Definition 2.2.3).

As we already mentioned, divisor B_X is usually assumed to be effective. If it is not effective, many important vanishing type results fail to be true. However, there is one very simple reason why we should not do this right away: the log pair

$$\left(\bar{X}, \sum_{i=1}^{r} a_i \bar{B}_i + \sum_{i=1}^{m} d_i E_i\right) \qquad (2.1.8)$$

inherits many properties of the log pair (X, B_X). Indeed, it follows from Definition 2.1.2 and (2.1.5) that (X, B_X) has Kawamata log terminal (resp., log canonical) singularities if and only if (2.1.8) has Kawamata log terminal (resp., log canonical) singularities. In fact, this is still true if we replace $\pi\colon \bar{X} \to X$ by an *arbitrary* birational morphism. Because of this, the pair (2.1.8) is usually called *the log pull back* of the pair (X, B_X).

2.2 Log pairs with mobile boundaries

Let X be a variety such that K_X is $\mathbb{Q}$-Cartier, and X has at most Kawamata log terminal singularities (see Definition 2.1.2). In many problems arising from birational geometry, we have to consider pairs $(X, \lambda \mathcal{M})$, where λ is a positive rational number, and $\mathcal{M}$ is a mobile linear system, i.e., a linear system without fixed components. Such log pairs are usually called *mobile*. Similar to the log pairs considered in §2.1, we can generalize both Definitions 2.1.2 and 2.1.7 for mobile pairs. In this section, we consider log pairs of both types at once. There are many important reasons for doing so. One of them is that the log pull back of the mobile log pair is not mobile anymore, simply because exceptional divisors do not move. The drawback of this approach is that it is a bit bulky.

Let B_X be a formal $\mathbb{Q}$-linear combination

$$\sum_{i=1}^{r} a_i B_i + \sum_{j=1}^{s} c_j \mathcal{M}_j,$$

where each B_i is a prime Weil divisor, each $\mathcal{M}_j$ is a mobile linear system, and each a_i and c_j are rational numbers. We say that B_X is effective if all a_i and c_j are non-negative. Similarly, we say that B_X and (X, B_X) are mobile if every a_i is zero. Sometimes, we will refer to B_X as *boundary*.

Remark 2.2.1. We can work with the pair (X, B_X) absolutely in the same way as with usual log pairs. In fact, by replacing each linear system $\mathcal{M}_j$ with its very general member, we always can handle the boundary B_X as a $\mathbb{Q}$-divisor. Moreover, it follows from [73, Theorem 4.8] that Definition 2.1.7 can be generalized for (X, B_X) by replacing each linear system $\mathcal{M}_j$ by the $\mathbb{Q}$-divisor
$$\frac{M_j^1 + M_j^2 + \ldots + M_j^N}{N},$$
where each $M_j^1, \ldots, M_j^N$ are general members in $\mathcal{M}_j$ and N is a sufficiently big number. However, there is a much simpler way of dealing with this problem (see, for example, [73, Definition 4.6]).

Suppose that the boundary B_X is $\mathbb{Q}$-Cartier. Choose a resolution of singularities $\pi\colon \bar{X} \to X$. Denote by $E_1, \ldots, E_m$ its exceptional divisors. Denote by $\bar{B}_i$ the proper transforms of B_i on $\bar{X}$ via π. Similarly, denote by $\bar{\mathcal{M}}_j$ the proper transforms of $\mathcal{M}_j$ on $\bar{X}$. Replace, if necessary, the resolution π by a resolution such that the divisor
$$\sum_{i=1}^{r} \bar{B}_i + \sum_{i=1}^{m} E_i$$
is a divisor with simple normal crossing, and each linear system $\bar{\mathcal{M}}_j$ is free from base points. Define rational numbers $d_1, \ldots, d_m$ by
$$K_{\bar{X}} + \sum_{i=1}^{r} a_i \bar{B}_i + \sum_{j=1}^{s} c_j \bar{\mathcal{M}}_j + \sum_{i=1}^{m} d_i E_i \sim_{\mathbb{Q}} \pi^*\Big(K_X + B_X\Big).$$

Definition 2.2.2 ([73, Definition 3.5]). The pair (X, B_X) has Kawamata log terminal (resp., log canonical) singularities if all numbers $a_1, \ldots, a_r, d_1, \ldots, d_m$ are less than 1 (resp., do not exceed 1).

Definition 2.2.3 (cf. [73, Definition 3.5]). The pair (X, B_X) has terminal (resp., canonical) singularities if all numbers $a_1, \ldots, a_r, d_1, \ldots, d_m$ are smaller than 0 (resp., do not exceed 0).

Neither of these definitions depends on the choice of the resolution $\pi\colon \bar{X} \to X$ provided that it satisfies assumptions we imposed on it.

Remark 2.2.4. Note that Definition 2.2.3 implies the mobility of B_X in the case when B_X is effective and has terminal or canonical singularities. In particular, our Definition 2.2.3 differs slightly from [73, Definition 3.5].

Let P be a point in X.

Definition 2.2.5. The pair (X, B_X) has log canonical (resp., Kawamata log terminal, canonical, terminal) singularities at P if the following two conditions are satisfied:

- every a_i does not exceed 1 (resp., is less than 1, does not exceed 0, is less than 0) provided that $P \in B_i$;

- every d_i does not exceed 1 (resp., is less than 1, does not exceed 0, is less than 0) provided that $P \in \pi(E_i)$.

This definition is just a localized version of Definitions 2.2.2 and 2.2.3.

Lemma 2.2.6. Suppose that X is smooth at P and B_X is effective. Then the following assertions hold:

(i) if $\mathrm{mult}_P(B_X) \leqslant 1$, then (X, B_X) is log canonical at P;

(ii) if $\mathrm{mult}_P(B_X) < 1$, then (X, B_X) is Kawamata log terminal at P;

(iii) if $\mathrm{mult}_P(B_X) > \dim(X)$, then (X, B_X) is not log canonical at P;

(iv) if $\mathrm{mult}_P(B_X) \geqslant \dim(X)$, then (X, B_X) is not Kawamata log terminal at P.

Proof. The proof is easy and left to the reader (see, for example, [73, Lemma 8.10] or [30, Exercise 6.18]). □

Let us conclude this section with the following useful and simple looking:

Lemma 2.2.7 ([30, Proposition 5.14]). Suppose that $\dim(X) = 2$, the surface X is smooth at P, and the boundary B_X is effective and mobile. Then $\mathrm{mult}_P(B_X) \leqslant 1$ (resp., $\mathrm{mult}_P(B_X) < 1$) if and only if (X, B_X) is canonical at P (resp., terminal at P).

Proof. The proof is very easy and left to the reader. □

2.3 Multiplier ideal sheaves

Let us use the assumptions and notation of §2.2. In particular, we assume that X has Kawamata log terminal singularities. Suppose, in addition, that B_X is effective. Put

$$\mathrm{LCS}\Big(X, B_X\Big) = \Big(\bigcup_{a_i \geqslant 1} B_i\Big) \cup \Big(\bigcup_{d_i \geqslant 1} \pi(E_i)\Big) \subsetneq X.$$

Definition 2.3.1. The set $\mathrm{LCS}(X, B_X)$ is called the locus of log canonical singularities of the log pair (X, B_X).

Note that $\mathrm{LCS}(X, B_X) = \varnothing$ if and only if the log pair (X, B_X) is Kawamata log terminal. Moreover, the locus $\mathrm{LCS}(X, B_X)$ can be naturally equipped with a subscheme structure. Namely, put

$$\mathcal{I}(X, B_X) = \pi_*\left(\mathcal{O}_{\bar{X}}\left(-\sum_{i=1}^{m}\lfloor d_i\rfloor E_i - \sum_{i=1}^{r}\lfloor a_i\rfloor B_i\right)\right).$$

Definition 2.3.2 (cf. [77, §9.2]). The ideal sheaf $\mathcal{I}(X, B_X)$ is said to be *the multiplier ideal sheaf* of the log pair (X, B_X).

Note that $\mathcal{I}(X, B_X)$ is indeed an ideal sheaf, because we assumed that B_X is effective. Let $\mathcal{L}(X, B_X)$ be the subscheme of X defined by $\mathcal{I}(X, B_X)$.

Definition 2.3.3. The subscheme $\mathcal{L}(X, B_X)$ is called *the log canonical singularities subscheme* of the log pair (X, B_X).

Note that the support of the scheme $\mathcal{L}(X, B_X)$ is exactly the locus $\mathrm{LCS}(X, B_X)$ by Definition 2.3.1.

Remark 2.3.4. If the log pair (X, B_X) is log canonical, then the subscheme $\mathcal{L}(X, B_X)$ is reduced.

The following assertion, known as Nadel's vanishing theorem, plays a crucial role in the proofs of many results in this book.

Theorem 2.3.5 ([77, Theorem 9.4.8]). Let H be a nef and big $\mathbb{Q}$-divisor on X such that
$$K_X + B_X + H \sim_{\mathbb{Q}} D$$
for some Cartier divisor D on the variety X. Then
$$H^i\Big(\mathcal{I}(X, B_X) \otimes \mathcal{O}_X(D)\Big) = 0$$
for every $i \geqslant 1$.

Corollary 2.3.6 ([72, Theorem 17.4], [74, Corollary 5.49]). Suppose that the divisor $-(K_X + B_X)$ is big and nef. Then $\mathrm{LCS}(X, B_X)$ is connected.

Let us conclude this section by the following very powerful:

Theorem 2.3.7 ([74, Theorem 5.50]). Suppose that $r \geq 1$, $a_1 = 1$, B_1 is a Cartier divisor, and B_1 has at most Kawamata log terminal singularities. Then the following assertions are equivalent:

- the log pair (X, B_X) is log canonical in a neighborhood of the divisor B_1;

- the singularities of the log pair

$$\left(B_1, \sum_{i=2}^{r} a_i B_i\bigg|_{B_1} + \sum_{j=1}^{s} c_j \mathcal{M}_j|_{B_1}\right)$$

  are log canonical.

2.4 Centers of log canonical singularities

Let us use assumptions and notation of §2.2. In particular, we assume that X has Kawamata log terminal singularities. Let Z be an irreducible subvariety of X.

Definition 2.4.1 (cf. [68, Definition 1.3]). The subvariety Z is a center of log canonical (resp., non-log canonical, canonical, non-canonical) singularities of the log pair (X, B_X) if one of the following conditions is satisfied:

- $Z = B_i$ for some i and $a_i \geq 1$ (resp., $a_i > 1$, $a_i \geq 0$, $a_i > 0$);
- $Z = \pi(E_i)$ for some i and $d_i \geq 1$ (resp., $d_i > 1$, $d_i \geq 0$, $d_i > 0$);

for *some choice* of the resolution $\pi\colon \bar{X} \to X$.

Let us denote by $\mathbb{LCS}(X, B_X)$ (resp., by $\mathbb{NLCS}(X, B_X)$, by $\mathbb{CS}(X, B_X)$, by $\mathbb{NCS}(X, B_X)$) the sets of centers of log canonical (resp., non-log canonical, canonical, non-canonical) singularities of the log pair (X, B_X). Then

$$Z \in \mathbb{LCS}\Big(X, B_X\Big) \Longrightarrow Z \subseteq \mathrm{LCS}\Big(X, B_X\Big)$$

by Definition 2.3.1 (note that the inverse implication often fails). Moreover, it follows from Definition 2.4.1 that

$$\mathbb{NLCS}(X, B_X) \subseteq \mathbb{LCS}(X, B_X) \subseteq \mathbb{NCS}(X, B_X) \subseteq \mathbb{CS}(X, B_X).$$

Remark 2.4.2. Let $\mathcal{H}$ be a linear system on X that has no base points, and let H be a sufficiently general divisor in it. Put $B_H = B_X|_H$, and let

$$Z\Big|_H = \sum_{i=1}^m Z_i,$$

where each Z_i is an irreducible subvariety. Then Z is a center in $\mathbb{LCS}(X, B_X)$ (resp., a center in $\mathbb{NLCS}(X, B_X)$, a center in $\mathbb{CS}(X, B_X)$, a center in $\mathbb{NCS}(X, B_X)$) if and only if each Z_i is a center in $\mathbb{LCS}(H, B_H)$ (resp., a center in $\mathbb{NLCS}(H, B_H)$, a center in $\mathbb{CS}(H, B_H)$, a center in $\mathbb{NCS}(H, B_H)$).

The following two assertions are well known and very easy to prove.

Lemma 2.4.3. Suppose that X is smooth at general point of Z, the boundary B_X is effective, and the codimension of Z in X is at least two. Suppose that $Z \in \mathbb{LCS}(X, B_X)$. Then $\operatorname{mult}_Z(B_X) > 1$.

Proof. This follows from Definition 2.4.1. $\square$

Lemma 2.4.4 (cf. [30, Proposition 5.14]). Suppose that X is smooth at general point of Z, the boundary B_X is effective and mobile. Then the following assertions hold:

(i) if the codimension of Z in X is at least three and $Z \in \mathbb{CS}(X, B_X)$, then $\operatorname{mult}_Z(B_X) > 1$;

(ii) if the codimension of Z in X is two, then $Z \in \mathbb{NCS}(X, B_X)$ if and only if $\operatorname{mult}_Z(B_X) > 1$;

(iii) if the codimension of Z in X is two, then $Z \in \mathbb{CS}(X, B_X)$ if and only if $\operatorname{mult}_Z(B_X) \geqslant 1$.

Proof. Assertions (ii) and (iii) follow from Lemma 2.2.7. Assertion (i) is also very easy to prove. It follows almost immediately from Definition 2.4.1 and Lemma 2.2.7. Because of this, we leave its proof to the reader. $\square$

The following lemma is merely an observation. Surprisingly, it is very useful, because it links together non-canonical and non-log canonical singularities.

Lemma 2.4.5. Suppose that X is smooth at a general point of the subvariety Z, and B_X is effective. If $Z \in \mathbb{CS}(X, B_X)$, then $Z \subset \mathbb{LCS}(X, 2B_X)$. Similarly, if $Z \in \mathbb{NCS}(X, B_X)$, then $Z \subset \mathbb{NLCS}(X, 2B_X)$.

Proof. The required assertions follow from Lemma 2.4.4(i),(ii) and Definition 2.4.1. □

Recall that we assume that X has Kawamata log terminal singularities. From now on and until the end of this section, we assume additionally that B_X is effective and (X, B_X) has log canonical singularities in every point of Z. We need these two assumptions, because centers of log canonical singularities behave much better under them. It can be illustrated by:

Lemma 2.4.6 ([68, Proposition 1.5])**.** *Let Z' be an irreducible subvariety in X that is different from Z. Suppose that Z and Z' are both contained in $\mathbb{LCS}(X, B_X)$. Then every irreducible component of the intersection $Z \cap Z'$ is also contained in $\mathbb{LCS}(X, B_X)$.*

Thus, for every point $P \in Z$, the set $\mathbb{LCS}(X, B_X)$ has a unique *minimal* element that contains P. Here minimal is understood in the following sense:

Definition 2.4.7. Suppose that Z is a center of log canonical singularities of the log pair (X, B_X). It is said to be minimal if it does not contain a proper irreducible subvariety that is also a center of log canonical singularities of the log pair (X, B_X).

The following result is the celebrated Kawamata's subadjunction theorem.

Theorem 2.4.8 ([69, Theorem 1])**.** *Suppose that Z is a minimal center in $\mathbb{LCS}(X, B_X)$. Then Z is normal and has at most rational singularities. Moreover, for every ample $\mathbb{Q}$-Cartier $\mathbb{Q}$-divisor Δ on X, there exists an effective $\mathbb{Q}$-divisor B_Z on the variety Z such that*

$$\left(K_X + B_X + \Delta\right)\Big|_Z \sim_{\mathbb{Q}} K_Z + B_Z,$$

and the log pair (Z, B_Z) has Kawamata log terminal singularities.

Suppose, in addition, that X is faithfully acted on by a finite group G such that the boundary B_X is G-invariant. Note that this does not mean that each B_i is G-invariant or that each $\mathcal{M}_j$ is G-invariant, since the group G can permute the divisors $B_1, \ldots, B_r$ and it can permute the linear systems $\mathcal{M}_1, \ldots, \mathcal{M}_s$.

Remark 2.4.9. Let g be an element in G. Suppose that Z is a minimal center in $\mathbb{LCS}(X, B_X)$. Then $g(Z) \in \mathbb{LCS}(X, B_X)$. By Lemma 2.4.6, we have

$$Z \cap g(Z) \neq \varnothing \iff Z = g(Z).$$

The following useful result is known as the Kawamata–Shokurov trick.

Lemma 2.4.10. Suppose that B_X is ample. Let ϵ be any positive rational number. Then there is a G-invariant linear system $\mathcal{B}$ on the variety X that has no fixed components, and there are rational numbers $1 \gg \epsilon_1 \geqslant 0$ and $1 \gg \epsilon_2 \geqslant 0$ such that

$$(1 - \epsilon_1)B_X + \epsilon_2 \mathcal{B} \sim_{\mathbb{Q}} (1 + \epsilon)B_X,$$

the log pair $(X, (1 - \epsilon_1)B_X + \epsilon_2 \mathcal{B})$ is log canonical in every point of $g(Z)$ for every $g \in G$, one has

$$\mathrm{NLCS}\Big(X, (1 - \epsilon_1)B_X + \epsilon_2 \mathcal{B}\Big) \subseteq \mathrm{NLCS}(X, B_X),$$

and the set $\mathbb{LCS}(X, (1 - \epsilon_1)B_X + \epsilon_2 \mathcal{B})$ is a disjoint union

$$\left(\bigsqcup_{g \in G} \{g(Z)\}\right) \bigsqcup \mathrm{NLCS}\Big(X, (1 - \epsilon_1)B_X + \epsilon_2 \mathcal{B}\Big).$$

Proof. See the proofs of [68, Theorem 1.10] and [69, Theorem 1]. □

2.5 Corti's inequality

The goal of this section is to prove the following result of Alessio Corti.

Theorem 2.5.1 ([29, Theorem 3.1]). Let S be a surface, let O be its smooth point, and let $\mathcal{M}$ be a mobile linear system on S. Suppose that $(S, \epsilon \mathcal{M})$ is not log canonical at O for some positive rational number ϵ. Then

$$\mathrm{mult}_O\Big(M \cdot M'\Big) > \frac{4}{\epsilon^2},$$

where M and M' are general curves in $\mathcal{M}$.

This result and Theorem 2.3.7 imply the following:

Theorem 2.5.2 ([29, Corollary 3.4],[30, Theorem 5.20]). Let X be a threefold, let O be its smooth point, and let $\mathcal{M}$ be a mobile linear system on X. Suppose that $O \in \mathrm{NCS}(X, \epsilon \mathcal{M})$ for some positive rational number ϵ. Then

$$\mathrm{mult}_O\Big(M \cdot M'\Big) > \frac{4}{\epsilon^2},$$

where M and M' are general elements in $\mathcal{M}$.

Proof. Let S be a general hyperplane section of X that passes through O. Then
$$O \in \mathrm{NLC}\Big(X, S + \mathcal{M}|_S\Big)$$
by Definition 2.4.1. In particular, the log pair $(X, S + \mathcal{M}|_S)$ is not log canonical at O. Then $(S, \mathcal{M}|_S)$ is not log canonical at O by Theorem 2.3.7. Now Theorem 2.5.1 implies the required claim. $\square$

Theorem 2.5.1 follows from:

Theorem 2.5.3 ([29], [64, Lemma 3.3]). *Let S be a surface, let O be its smooth point, and let $\mathcal{M}$ be a mobile linear system on S. Let a_1 and a_2 be rational numbers, let Δ_1 and Δ_2 be irreducible curves on S such that $O \in \Delta_1 \cap \Delta_2$. Suppose that both Δ_1 and Δ_2 are smooth at O, and Δ_1 intersects the curve Δ_2 transversally at the point O. Finally, suppose that*
$$\Big(S, \epsilon\mathcal{M} + a_1\Delta_1 + a_2\Delta_2\Big)$$
is not Kawamata log terminal at the point O. Then
$$\mathrm{mult}_O\Big(M_1 \cdot M_2\Big) \geqslant \begin{cases} \dfrac{4(1-a_1)(1-a_2)}{\epsilon^2} & \text{if } a_1 \geqslant 0 \text{ or } a_2 \geqslant 0, \\ \dfrac{4(1-a_1-a_2)}{\epsilon^2} & \text{if } a_1 \leqslant 0 \text{ and } a_2 \leqslant 0. \end{cases} \quad (2.5.4)$$

Moreover, if (2.5.4) is an equality, then
$$\mathrm{mult}_O(\mathcal{M}) = \frac{2(a_1-1)}{\epsilon},$$
the log pair $(S, \epsilon\mathcal{M} + a_1\Delta_1 + a_2\Delta_2)$ is log canonical, and $a_1 = a_2 \geqslant 0$.

Proof. Let $\phi \colon \bar{S} \to S$ be a blow-up of the point O, let E be the exceptional curve of ϕ. Then
$$K_{\bar{S}} + \epsilon\bar{\mathcal{M}} + a_1\bar{\Delta}_1 + a_2\bar{\Delta}_2 + \Big(\epsilon\mathrm{mult}_O(\mathcal{M}) + a_1 + a_2 - 1\Big)E \sim_{\mathbb{Q}}$$
$$\sim_{\mathbb{Q}} \phi^*\Big(K_S + \epsilon\mathcal{M} + a_1\Delta_1 + a_2\Delta_2\Big),$$
where $\bar{\mathcal{M}}$, $\bar{\Delta}_1$, $\bar{\Delta}_2$ are proper transforms of $\mathcal{M}$, Δ_1, Δ_2 on the surface $\bar{S}$, respectively. Let M_1 and M_2 be general curves in the linear system $\mathcal{M}$, let $\bar{M}_1$ and $\bar{M}_2$ be the proper transforms of the curves $\bar{M}_1$ and $\bar{M}_2$ on the surface $\bar{S}$, respectively.

Suppose that $\operatorname{\epsilon mult}_O(\mathcal{M}) \geqslant 2 - a_1 - a_2$. Then

$$\operatorname{mult}_O\left(M_1 \cdot M_2\right) \geqslant \operatorname{mult}_O^2(\mathcal{M}) \geqslant \frac{(2-a_1-a_2)^2}{\epsilon^2} \geqslant$$

$$\geqslant \begin{cases} \dfrac{4(1-a_1)(1-a_2)}{\epsilon^2} & \text{if } a_1 \geqslant 0 \text{ or } a_2 \geqslant 0, \\ \dfrac{4(1-a_1-a_2)}{\epsilon^2} & \text{if } a_1 \leqslant 0 \text{ and } a_2 \leqslant 0, \end{cases}$$

which implies inequality (2.5.4) in the case when $\operatorname{\epsilon mult}_O(\mathcal{M}) \geqslant 2 - a_1 - a_2$.

To complete the proof we assume that $\operatorname{\epsilon mult}_O(\mathcal{M}) < 2 - a_1 - a_2$. Then the log pair

$$\left(\bar{S},\ \epsilon\bar{\mathcal{M}} + a_1\bar{\Delta}_1 + a_2\bar{\Delta}_2 + \left(\operatorname{\epsilon mult}_O(\mathcal{M}) + a_1 + a_2 - 1\right)E\right)$$

is not Kawamata log terminal at some point $Q \in E$. Thus

$$\operatorname{mult}_O\left(M_1 \cdot M_2\right) \geqslant \operatorname{mult}_O^2(\mathcal{M}) + \operatorname{mult}_Q\left(\bar{M}_1 \cdot \bar{M}_2\right). \qquad (2.5.5)$$

To complete the proof, we must consider the following possible cases:

- $Q \notin \bar{\Delta}_1 \cup \bar{\Delta}_2$;
- $Q = E \cap \bar{\Delta}_1$;
- $Q = E \cap \bar{\Delta}_2$.

Suppose that $Q \notin \bar{\Delta}_1 \cup \bar{\Delta}_2$. Then the log pair

$$\left(\bar{S}, \epsilon\bar{\mathcal{M}} + \left(\operatorname{\epsilon mult}_O(\mathcal{M}) + a_1 + a_2 - 1\right)E\right)$$

is not Kawamata log terminal at the point Q. Thus, we may assume that

$$\operatorname{mult}_Q\left(\bar{M}_1 \cdot \bar{M}_2\right) \geqslant \frac{4\left(2 - \operatorname{\epsilon mult}_O(\mathcal{M}) - a_1 - a_2\right)}{\epsilon^2}$$

by induction, replacing the curve Δ_1 by E, the number a_1 by the number $\operatorname{\epsilon mult}_O(\mathcal{M}) + a_1 + a_2 - 1$, and the number a_2 by 0. Thus, it follows from inequality (2.5.5) that

$$\operatorname{mult}_O\left(M_1 \cdot M_2\right) \geqslant \operatorname{mult}_O^2(\mathcal{M}) + \frac{4\left(2 - \operatorname{\epsilon mult}_O(\mathcal{M}) - a_1 - a_2\right)}{\epsilon^2} \geqslant$$

$$\geqslant \frac{4(1-a_1-a_2)}{\epsilon^2},$$

which implies inequality (2.5.4) in the case when $a_1 \leqslant 0$ and $a_2 \leqslant 0$. Hence, it follows from the equality

$$\frac{4(1-a_1-a_2)}{\epsilon^2} = \frac{4(1-a_1)(1-a_2)}{\epsilon^2} - \frac{a_1 a_2}{\epsilon^2}$$

that we may assume that $a_1 > 0$ and $a_2 > 0$. Then

$$\mathrm{mult}_O\left(M_1 \cdot M_2\right) \geqslant \mathrm{mult}_O^2(\mathcal{M}) + \frac{4(2-\epsilon\mathrm{mult}_O(\mathcal{M}) - a_1 - a_2)}{\epsilon^2} >$$

$$> \frac{4(1-a_1)(1-a_2)}{\epsilon^2} + (a_1 - a_2)^2,$$

because $\epsilon\mathrm{mult}_O(\mathcal{M}) < 2 - a_1 - a_2$. This completes the proof in the case when $Q \notin \bar{\Delta}_1 \cup \bar{\Delta}_2$.

Without loss of generality, we may assume that $Q = E \cap \bar{\Delta}_1$. Then the log pair

$$\left(\bar{S},\ \epsilon\bar{\mathcal{M}} + a_1\bar{\Delta}_1 + (\epsilon\mathrm{mult}_O(\mathcal{M}) + a_1 + a_2 - 1)E\right)$$

is not Kawamata log terminal at the point Q. By induction, we may assume that

$$\mathrm{mult}_Q\left(\bar{M}_1 \cdot \bar{M}_2\right) \geqslant \frac{4(1-a_1)(2-\epsilon\mathrm{mult}_O(\mathcal{M}) - a_1 - a_2)}{\epsilon^2}$$

if $a_1 \geqslant 0$ or $\epsilon\mathrm{mult}_O(\mathcal{M}) + a_1 + a_2 \geqslant 1$, and we may assume that

$$\mathrm{mult}_Q\left(\bar{M}_1 \cdot \bar{M}_2\right) \geqslant \frac{4(2-\epsilon\mathrm{mult}_O(\mathcal{M}) - 2a_1 - a_2)}{\epsilon^2}$$

if $a_1 \leqslant 0$ and $\epsilon\mathrm{mult}_O(\mathcal{M}) + a_1 + a_2 \leqslant 1$. To complete the proof, we must consider the following possible cases:

- $a_2 \geqslant 0$ and either $a_1 \geqslant 0$ or $\epsilon\mathrm{mult}_O(\mathcal{M}) + a_1 + a_2 \geqslant 1$;
- $a_2 \geqslant 0$ and $a_1 \leqslant 0$ and $\epsilon\mathrm{mult}_O(\mathcal{M}) + a_1 + a_2 \leqslant 1$;
- $a_2 \leqslant 0$ and either $a_1 \geqslant 0$ or $\epsilon\mathrm{mult}_O(\mathcal{M}) + a_1 + a_2 \geqslant 1$;
- $a_2 \leqslant 0$ and $a_1 \leqslant 0$ and $\epsilon\mathrm{mult}_O(\mathcal{M}) + a_1 + a_2 \leqslant 1$.

CREMONA GROUPS AND THE ICOSAHEDRON

Suppose that $a_2 \geqslant 0$ and either $a_1 \geqslant 0$ or $\epsilon\mathrm{mult}_O(\mathcal{M}) + a_1 + a_2 \geqslant 1$. Then

$$\mathrm{mult}_Q\left(\bar{M}_1 \cdot \bar{M}_2\right) \geqslant \frac{4(1-a_1)(2-\epsilon\mathrm{mult}_O(\mathcal{M}) - a_1 - a_2)}{\epsilon^2},$$

and it follows from inequality (2.5.5) that

$$\mathrm{mult}_O\left(M_1 \cdot M_2\right) \geqslant \mathrm{mult}_O^2(\mathcal{M}) + \frac{4(1-a_1)(2-\epsilon\mathrm{mult}_O(\mathcal{M}) - a_1 - a_2)}{\epsilon^2} \geqslant$$

$$\geqslant \frac{4(1-a_1)(1-a_2)}{\epsilon^2} + \frac{4a_1^2}{\epsilon^2},$$

which completes the proof in the case when $a_2 \geqslant 0$ and either $a_1 \geqslant 0$ or $\epsilon\mathrm{mult}_O(\mathcal{M}) + a_1 + a_2 \geqslant 1$.

Suppose that $a_2 \geqslant 0$ and $a_1 \leqslant 0$ and $\epsilon\mathrm{mult}_O(\mathcal{M}) + a_1 + a_2 \leqslant 1$. Then

$$\mathrm{mult}_Q\left(\bar{M}_1 \cdot \bar{M}_2\right) \geqslant \frac{4(2-\epsilon\mathrm{mult}_O(\mathcal{M}) - 2a_1 - a_2)}{\epsilon^2},$$

and it follows from inequality (2.5.5) that

$$\mathrm{mult}_O\left(M_1 \cdot M_2\right) \geqslant \mathrm{mult}_O^2(\mathcal{M}) + \frac{4(2-\epsilon\mathrm{mult}_O(\mathcal{M}) - 2a_1 - a_2)}{\epsilon^2} \geqslant$$

$$\geqslant \frac{4(1-a_1)(1-a_2)}{\epsilon^2} - \frac{4a_1(1+a_2)}{\epsilon^2},$$

which completes the proof in the case when $a_2 \geqslant 0$ and $a_1 \leqslant 0$ and $\epsilon\mathrm{mult}_O(\mathcal{M}) + a_1 + a_2 \leqslant 1$.

Suppose that $a_2 \leqslant 0$ and $a_1 \leqslant 0$ and $\epsilon\mathrm{mult}_O(\mathcal{M}) + a_1 + a_2 \leqslant 1$. Then

$$\mathrm{mult}_Q\left(\bar{M}_1 \cdot \bar{M}_2\right) \geqslant \frac{4(2-\epsilon\mathrm{mult}_O(\mathcal{M}) - 2a_1 - a_2)}{\epsilon^2},$$

and it follows from inequality (2.5.5) that

$$\mathrm{mult}_O\left(M_1 \cdot M_2\right) \geqslant \mathrm{mult}_O^2(\mathcal{M}) + \frac{4(2-\epsilon\mathrm{mult}_O(\mathcal{M}) - 2a_1 - a_2)}{\epsilon^2} \geqslant$$

$$\geqslant \frac{4(1-a_1-a_2)}{\epsilon^2} - \frac{4a_1}{\epsilon^2},$$

which completes the proof of the theorem in the case when $a_2 \leqslant 0$, $a_1 \leqslant 0$ and $\epsilon\mathrm{mult}_O(\mathcal{M}) + a_1 + a_2 \leqslant 1$.

Thus, to complete the proof, we may assume that $a_2 \leqslant 0$ and either $a_1 \geqslant 0$ or $\epsilon\mathrm{mult}_O(\mathcal{M}) + a_1 + a_2 \geqslant 1$. Then

$$\mathrm{mult}_Q\left(\bar{M}_1 \cdot \bar{M}_2\right) \geqslant \frac{4(1-a_1)(2 - \epsilon\mathrm{mult}_O(\mathcal{M}) - a_1 - a_2)}{\epsilon^2}.$$

and it follows from inequality (2.5.5) that

$$\mathrm{mult}_O\left(M_1 \cdot M_2\right) \geqslant \mathrm{mult}_O^2(\mathcal{M}) + \frac{4(1-a_1)(2 - \epsilon\mathrm{mult}_O(\mathcal{M}) - a_1 - a_2)}{\epsilon^2} \geqslant$$
$$\geqslant \frac{4 - 4(1-a_1)(a_1 + a_2)}{\epsilon^2}.$$

In the case when $a_1 \geqslant 0$, we have

$$\mathrm{mult}_O\left(M_1 \cdot M_2\right) \geqslant \frac{4 - 4(1-a_1)(a_1 + a_2)}{\epsilon^2} = \frac{4(1-a_1)(1-a_2)}{\epsilon^2} + \frac{4a_1^2}{\epsilon^2}.$$

In the case when $a_1 \leqslant 0$, we have

$$\mathrm{mult}_O\left(M_1 \cdot M_2\right) \geqslant \frac{4-4(1-a_1)(a_1+a_2)}{\epsilon^2} = \frac{4(1-a_1-a_2)}{\epsilon^2} + \frac{4a_1(a_1+a_2)}{\epsilon^2},$$

which completes the proof of Theorem 2.5.3. □

Arguing as in the proof of Theorem 2.5.2 and using Theorem 2.5.3 instead of Theorem 2.5.1, we obtain:

Theorem 2.5.6 ([16, Theorem 1.7.18]). Let X be a threefold, let O be its smooth point, and let $\mathcal{M}$ be a mobile linear system on X. Suppose that $O \in \mathbb{CS}(X, \epsilon\mathcal{M})$ for some positive rational number ϵ. Then

$$\mathrm{mult}_O\left(M \cdot M'\right) \geqslant \frac{4}{\epsilon^2}, \qquad (2.5.7)$$

where M and M' are general elements in $\mathcal{M}$. Moreover, if (2.5.7) is an equality, then $\mathrm{mult}_O(\mathcal{M}) = \frac{2}{\epsilon}$.

Chapter 3

Noether–Fano inequalities

In this chapter, we study obstructions to the existence of equivariant birational maps between varieties with semi-ample anticanonical divisors and varieties fibered into varieties of negative or zero Kodaira dimension. These obstructions are the most important global tools used in the proof of Theorem 1.4.1, which is the main result of this book. They will also be used in the proofs of Theorems 19.1.1, 19.1.3, and 19.1.4.

3.1 Birational rigidity

Let G be a finite group, and let V be a G-Fano variety (see Definition 1.1.5). In this book, V will always be either some rational surface or some rational threefold, and the group G will almost always be $\mathfrak{A}_5$. But the methods we are going to describe work in a wider context.

Definition 3.1.1. The G-Fano variety V is said to be G-birationally rigid if the following two conditions are satisfied

(A) there is no G-birational map $V \dashrightarrow V'$ such that V' is a G-Mori fiber space over a positive dimensional variety;

(B) if there is a G-birational map $\xi \colon V \dashrightarrow V'$ such that V' is a G-Fano variety, then

  (B1) the variety V' is isomorphic to V;

  (B2) there is a G-birational map $\rho \in \mathrm{Bir}^G(V)$ such that the composition $\xi \circ \rho \colon V \dashrightarrow V'$ is a biregular G-morphism.

The variety V is said to be G-birationally superrigid if both conditions (A) and (B) are satisfied and, in addition, the following condition holds:

(C) the group $\mathrm{Bir}^G(V)$ coincides with the group $\mathrm{Aut}^G(V)$.

Note that condition (B2) implies condition (B1). We decided to include both of them to emphasize the importance of the G-action in Definition 3.1.1. Note also that this definition implies:

Corollary 3.1.2. Let V' be a G-Fano variety. If there exists a G-birational map $V \dashrightarrow V'$ and V is not isomorphic to V', then neither V nor V' is G-birationally rigid.

In the next section, we will give a technical criterion for V to be G-birationally rigid. To prove it, we need the following well-known:

Lemma 3.1.3 ([72, Lemma 2.19]). Let X and Y be varieties, and let

$$f \colon Y \to X$$

be a birational map. Denote f-exceptional divisors by $F_1, \ldots, F_k$. Let L be a line bundle on X, and let M be a f-nef line bundle on Y. Let $\sum_{i=1}^{l} b_i G_i$ be an effective $\mathbb{Q}$-divisor on Y such that $G_1, \ldots, G_l$ are prime Weil divisors on Y, and none of them is f-exceptional. Suppose that

$$f^*(L) \sim_{\mathbb{Q}} M + \sum_{i=1}^{l} b_i G_i + \sum_{i=1}^{k} a_i F_i$$

for some rational numbers $a_1, \ldots, a_k$. Then $a_i \geqslant 0$ for every $i \in \{1, \ldots, k\}$.

3.2 Fano varieties and elliptic fibrations

In this section we prove two results that play a crucial role in the proof of Theorem 1.4.1. These results are Theorems 3.2.1 and 3.2.6 below. They are also important for the proofs of Theorems 19.1.3 and 19.1.4.

Let X be a variety with at most terminal singularities such that the divisor $-K_X$ is nef and big. Suppose that X is faithfully acted on by some finite group G. Then there is an integer $m > 0$ such that the linear system $|-mK_X|$ is free from base points and gives a G-birational morphism

$$\eta \colon X \to U$$

such that U is a Fano variety with at most canonical singularities. This follows from the Kawamata–Shokurov contraction theorem (see, for example, [74, Theorem 3.3]).

CREMONA GROUPS AND THE ICOSAHEDRON

Theorem 3.2.1 (cf. [30, Theorem 5.5]). Let V be a Fano variety with at most canonical singularities that is faithfully acted on by the group G. Suppose that there exists a G-birational map $\nu\colon X \dashrightarrow V$. Denote by $\mathcal{M}$ the proper transform via ν^{-1} of the linear system $|-nK_V|$ on the variety X, where n is a positive integer such that the divisor $-nK_V$ is very ample. Suppose, in addition, that there is a positive rational number λ such that

$$K_X + \lambda \mathcal{M} \sim_{\mathbb{Q}} 0,$$

and the log pair $(X, \lambda\mathcal{M})$ has canonical singularities. Then the following assertions hold:

(i) if the log pair $(X, \lambda\mathcal{M})$ has terminal singularities, then there exists a commutative diagram

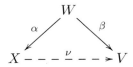

(3.2.2)

for some G-birational morphism $\phi\colon U \to V$;

(ii) if $-K_X$ is ample and V is G-Fano variety, then the map ν is biregular.

Proof. There exists a commutative diagram

$$\begin{array}{ccc} & W & \\ {}^{\alpha}\swarrow & & \searrow^{\beta} \\ X & \dashrightarrow^{\nu} & V \end{array}$$

where α and β are G-birational morphisms and W is a variety with at most canonical singularities that is faithfully acted on by the group G. Denote by $\mathcal{H}$ the proper transform of the linear system $\mathcal{M}$ on the variety W, denote by $F_1, \ldots, F_k$ the G-irreducible α-exceptional divisor, and denote by $E_1, \ldots, E_l$ the G-irreducible β-exceptional divisors. Then

$$\alpha^*\bigl(K_X + \lambda\mathcal{M}\bigr) + \sum_{j=1}^{k} a_j F_j \sim_{\mathbb{Q}} K_W + \lambda\mathcal{H} \sim_{\mathbb{Q}}$$

$$\sim_{\mathbb{Q}} \beta^*\bigl((1-\lambda n)K_V\bigr) + \sum_{i=1}^{l} b_i E_i \quad (3.2.3)$$

for some rational numbers $a_1, \ldots, a_k, b_1, \ldots, b_l$. Note that each a_i is non-negative, since $(X, \lambda \mathcal{M})$ has canonical singularities by assumption. Moreover, if the log pair $(X, \lambda \mathcal{M})$ has terminal singularities, then each a_j is positive by Definition 2.2.3. Also, each b_i is also non-negative, because $|-nK_V|$ is free from base points (recall that we assume that $-nK_V$ is very ample). In fact, the base point freeness of $|-nK_V|$ shows that the numbers $b_1, \ldots, b_l$ do not depend on λ, so that

$$K_W \sim_{\mathbb{Q}} \beta^*(K_V) + \sum_{i=1}^{l} b_i E_i.$$

On the other hand, the numbers $a_1, \ldots, a_k$ may depend on λ, because $\mathcal{M}$ may have base points. In fact, if the linear system $\mathcal{M}$ indeed has base points, then at least one of the numbers $a_1, \ldots, a_k$ depends on λ non-trivially.

Take a sufficiently general curve Z in the threefold X. Then we have $K_X \cdot Z < 0$, since $-K_X$ is assumed to be nef and big. Denote by $\hat{Z}$ the proper transform of Z on the variety W. We may assume that $\hat{Z}$ is not contained in any β-exceptional divisor, which implies that $\beta(\hat{Z})$ is a curve in V. Then $K_V \cdot \beta(\hat{Z}) < 0$, since $-K_V$ is ample. Thus (3.2.3) gives

$$0 = \left(K_X + \lambda \mathcal{M}\right) \cdot Z = \left(\alpha^*\left(K_X + \lambda \mathcal{M}\right) + \sum_{j=1}^{k} a_j F_j\right) \cdot \hat{Z} =$$

$$= \left(K_W + \lambda \mathcal{H}\right) \cdot \hat{Z} =$$

$$= \left(\beta^*\left((1 - \lambda n)K_V\right) + \sum_{i=1}^{l} b_i E_i\right) \cdot \hat{C} =$$

$$= (1 - \lambda n) K_V \cdot \beta(\hat{Z}) + \sum_{i=1}^{l} b_i E_i \cdot \hat{C} \geqslant (1 - \lambda n) K_V \cdot \beta(\hat{Z}),$$

which implies that $\lambda \leqslant \frac{1}{n}$.

Now we take a sufficiently general curve C in the threefold V. Then one has $K_V \cdot C < 0$, since $-K_V$ is ample. Denote by $\hat{C}$ the proper transform of C on the variety W. We may assume that $\hat{C}$ is not contained in any

α-exceptional divisor. Then (3.2.3) gives

$$0 \leqslant \sum_{j=1}^{k} a_j F_j \cdot \hat{C} = \left(\alpha^* \Big(K_X + \lambda \mathcal{M} \Big) + \sum_{j=1}^{k} a_j F_j \right) \cdot \hat{C} =$$

$$= \Big(K_W + \lambda \mathcal{H} \Big) \cdot \hat{C} =$$

$$= \left(\beta^* \Big((1 - \lambda n) K_V \Big) + \sum_{i=1}^{l} b_i E_i \right) \cdot \hat{C} = \Big((1 - \lambda n) K_V \Big) \cdot C.$$

Thus, one has $\lambda \geqslant \frac{1}{n}$, so that $\lambda = \frac{1}{n}$, because we already proved that $\lambda \leqslant \frac{1}{n}$.

We have

$$\sum_{j=1}^{k} a_j F_j \sim_{\mathbb{Q}} \sum_{i=1}^{l} b_i E_i$$

by (3.2.3). Since each a_j is non-negative and each b_i is non-negative, this gives

$$\sum_{j=1}^{k} a_j F_j = \sum_{i=1}^{l} b_i E_i \qquad (3.2.4)$$

by Lemma 3.1.3.

Now we take some positive rational number $\lambda' > \lambda$ such that λ' is slightly bigger than λ. Then

$$\left(1 - \frac{\lambda'}{\lambda} \right) \alpha^*(K_X) + \sum_{j=1}^{k} a'_j F_j \sim_{\mathbb{Q}}$$

$$\sim_{\mathbb{Q}} \alpha^* \Big(K_X + \lambda' \mathcal{M} \Big) + \sum_{j=1}^{k} a'_j F_j \sim_{\mathbb{Q}}$$

$$\sim_{\mathbb{Q}} K_W + \lambda' \mathcal{H} \sim_{\mathbb{Q}} \beta^* \Big((1 - \lambda' n) K_V \Big) + \sum_{i=1}^{l} b_i E_i \sim_{\mathbb{Q}}$$

$$\sim_{\mathbb{Q}} \left(1 - \frac{\lambda'}{\lambda} \right) \beta^*(K_V) + \sum_{i=1}^{l} b_i E_i \qquad (3.2.5)$$

for some rational numbers $a'_1, \ldots, a'_k$. If each a'_j is positive, then

$$\sum_{j=1}^{k} a'_j F_j = \sum_{i=1}^{l} b_i E_i = \sum_{j=1}^{k} a_j F_j$$

by (3.2.4) and Lemma 3.1.3, which implies, in particular, that $a'_j = a_j$. The latter can only happen if the linear system $\mathcal{M}$ is free from base points and, therefore, the rational map ν is a morphism!

Can we choose $\lambda' > \lambda$ such that each a'_j is positive? Yes, we can, provided that $(X, \lambda \mathcal{M})$ has terminal singularities. Indeed, if $(X, \lambda \mathcal{M})$ has terminal singularities, then we always can choose $\lambda' > \lambda$ such that the singularities of the log pair $(X, \lambda' \mathcal{M})$ are still terminal, which implies, in particular, that each a'_j is positive. This proves assertion (i).

Now we are going to prove assertion (ii). Thus, we assume that $-K_X$ is ample, and V is G-Fano variety. We have to show that the birational map $\nu \colon X \dashrightarrow V$ is, actually, biregular.

Note that we proved that ν is a morphism provided that $(X, \lambda \mathcal{M})$ has terminal singularities. Moreover, if ν is a morphism, then we can put α to be an identity map, so that $\beta = \nu$ and $k = 0$. On the other hand, if α is an identity map, then it follows from (3.2.5) that

$$\left(\frac{\lambda'}{\lambda} - 1\right)\eta^*(-K_U) \sim_{\mathbb{Q}} \left(\frac{\lambda'}{\lambda} - 1\right)(-K_X) \sim_{\mathbb{Q}}$$

$$\sim_{\mathbb{Q}} \left(\frac{\lambda'}{\lambda} - 1\right)\beta^*(-K_V) + \sum_{i=1}^{l} b_i E_i,$$

which implies that each b_i equals zero (or $l = 0$) by Lemma 3.1.3, so that the commutative diagram (3.2.2) exists, because both divisors $-K_U$ and $-K_V$ are ample.

Since V has terminal singularities, each b_i is positive. Thus, (3.2.4) implies that $k \geqslant l$. On the other hand, we have

$$\operatorname{rk} \operatorname{Cl}(X)^G + k = \operatorname{rk} \operatorname{Cl}(W)^G = \operatorname{rk} \operatorname{Pic}(V)^G + l = l + 1,$$

which implies that $\operatorname{rk} \operatorname{Cl}(X)^G = 1$ and $k = l$. In particular, (3.2.4) implies that each a_j is also positive. Then the singularities of the log pair $(X, \lambda \mathcal{M})$ are terminal (see Definition 2.2.3). But we already proved that this condition implies that ν is a morphism. Hence, we have

$$-K_X \sim \nu^*(-K_V),$$

where both divisors $-K_X$ and $-K_V$ are ample! This is only possible if ν is a biregular map. This gives assertion (ii) and completes the proof of Theorem 3.2.1. □

Theorem 3.2.6 (cf. [30, Theorem 5.5]). Let V be a variety with at most canonical singularities that is faithfully acted on by the group G. Suppose that there exists a surjective G-morphism $\psi\colon V \to Z$ such that either $-K_V|_{\mathcal{F}}$ is ample or $-K_V|_{\mathcal{F}} \sim_{\mathbb{Q}} 0$, where $\mathcal{F}$ is a sufficiently general fiber of the morphism ψ. Suppose that there exists a G-birational map $\nu\colon X \dashrightarrow V$. Let H be a very ample divisor on Z such that its class in $\mathrm{Pic}(Z)$ is G-invariant. Denote by $\mathcal{M}$ the proper transform via ν^{-1} of the linear system $|\psi^*(H)|$ on the variety X. Suppose, in addition, that there is a positive rational number λ such that

$$K_X + \lambda\mathcal{M} \sim_{\mathbb{Q}} 0.$$

Then the following assertions hold:

(i) the singularities of the log pair $(X, \lambda\mathcal{M})$ are not terminal;

(ii) if $-K_V|_{\mathcal{F}}$ is ample, then the singularities of the log pair $(X, \lambda\mathcal{M})$ are not canonical.

Proof. There exists a commutative diagram

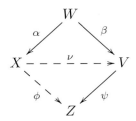

where α and β are G-birational morphisms, ϕ is the rational map given by the linear system $\mathcal{M}$, and W is a smooth variety.

Denote by $\mathcal{H}$ the proper transform of the linear system $\mathcal{M}$ on the variety W, denote by $F_1, \ldots, F_k$ the G-irreducible α-exceptional divisor, and denote by $E_1, \ldots, E_l$ the G-irreducible β-exceptional divisors. Then

$$\alpha^*\Big(K_X + \lambda\mathcal{M}\Big) + \sum_{j=1}^{k} a_j F_j \sim_{\mathbb{Q}} K_W + \lambda\mathcal{H} \sim_{\mathbb{Q}}$$

$$\sim_{\mathbb{Q}} \beta^*\Big(K_V + \lambda\psi^*(H)\Big) + \sum_{i=1}^{l} b_i E_i$$

for some rational numbers $a_1, \ldots, a_k, b_1, \ldots, b_l$. Since W is smooth and $\mathcal{H}$ is free from base points (by construction), each a_i is non-negative (resp., positive) if and only if the log pair $(X, \lambda \mathcal{M})$ has canonical (resp., terminal) singularities.

Let $\mathcal{F}$ be a sufficiently general fiber of the morphism ψ, and let C be a general curve in it. Then either $K_V \cdot C < 0$ or $K_V \cdot C = 0$, because either $-K_V|_{\mathcal{F}}$ is ample or $-K_V|_{\mathcal{F}} \sim_{\mathbb{Q}} 0$ by assumption. In both cases, we have $K_V \cdot C \leqslant 0$.

Denote by $\hat{C}$ the proper transform of C on W. We may assume that $\hat{C}$ is not contained in any α-exceptional divisor. Then

$$\sum_{j=1}^{k} a_j F_j \cdot \hat{C} = \left(\alpha^* \left(K_X + \lambda \mathcal{M} \right) + \sum_{j=1}^{k} a_j F_j \right) \cdot \hat{C} =$$

$$= \left(K_W + \lambda \mathcal{H} \right) \cdot \hat{C} = \left(\beta^* \left(K_V + \lambda \psi^*(H) \right) + \sum_{i=1}^{l} b_i E_i \right) \cdot \hat{C} =$$

$$= \left(K_V + \lambda \psi^*(D) \right) \cdot C = K_V \cdot C \leqslant 0, \quad (3.2.7)$$

which implies that at least one number among $a_1, \ldots, a_k$ is not positive. Thus, the singularities of the log pair $(X, \lambda \mathcal{M})$ are not terminal, which proves assertion (i). Similarly, if $K_V \cdot C < 0$, then (3.2.7) implies that at least one number among $a_1, \ldots, a_k$ is negative. Hence, if $-K_V|_{\mathcal{F}}$ is ample, then the singularities of the log pair $(X, \lambda \mathcal{M})$ are not canonical, which implies assertion (ii). □

3.3 Applications to birational rigidity

In this section, we give an application of Theorems 3.2.1 and 3.2.6 and consider few examples.

Let G be a finite group, and let X be a G-Fano variety.

Theorem 3.3.1 ([18, Corollary A.21]). *Let Γ be a subgroup in $\mathrm{Bir}^G(X)$ that contains $\mathrm{Aut}^G(X)$. Suppose that for every G-invariant linear system $\mathcal{M}$ on the variety X that has no fixed components, there exists $\xi \in \Gamma$ such that the log pair $(X, \lambda \xi(\mathcal{M}))$ has canonical singularities, where λ is a positive rational number such that*

$$K_X + \lambda \xi(\mathcal{M}) \sim_{\mathbb{Q}} 0.$$

Then X is G-birationally rigid and $\mathrm{Bir}^G(X) = \Gamma$.

Proof. Apply Theorems 3.2.1(ii) and 3.2.6(ii). □

CREMONA GROUPS AND THE ICOSAHEDRON

Corollary 3.3.2. Suppose that for every G-invariant linear system $\mathcal{M}$ on the variety X that has no fixed components, there exists $\xi \in \mathrm{Bir}^G(X)$ such that the log pair $(X, \lambda\xi(\mathcal{M}))$ has canonical singularities, where λ is a positive rational number such that

$$K_X + \lambda\xi(\mathcal{M}) \sim_{\mathbb{Q}} 0.$$

Then X is G-birationally rigid.

Corollary 3.3.3 (cf. [30, Corollary 5.10]). Suppose that for every G-invariant linear system $\mathcal{M}$ on the variety X that has no fixed components, the log pair $(X, \lambda\mathcal{M})$ has canonical singularities, where λ is a positive rational number such that

$$K_X + \lambda\mathcal{M} \sim_{\mathbb{Q}} 0.$$

Then X is G-birationally superrigid.

Let us present several examples that show how to apply Theorem 3.3.1 and its corollaries.

Example 3.3.4 ([35, §8]). Let S be a smooth del Pezzo surface such that $K_S^2 \leqslant 3$, and let G be any finite subgroup in $\mathrm{Aut}(S)$ such that $\mathrm{rk}\,\mathrm{Pic}\,(S)^G = 1$. For every G-invariant linear system $\mathcal{M}$ on S that has no fixed components, one can explicitly construct $\xi \in \mathrm{Bir}^G(S)$ such that the log pair $(S, \lambda\xi(\mathcal{M}))$ has canonical singularities, where λ is a positive rational number defined by

$$K_S + \lambda\xi(\mathcal{M}) \sim_{\mathbb{Q}} 0.$$

Thus, S is G-birationally rigid by Corollary 3.3.2. Moreover, if $K_S^2 = 1$, then ξ can be chosen to be an identity map, so that S is G-birationally superrigid Corollary 3.3.3.

Example 3.3.5 ([23, §4]). Recall that $\mathfrak{A}_6$ acts faithfully on $\mathbb{P}^3$. Moreover, there are two $\mathfrak{A}_6$-irreducible curves, say $\mathcal{L}_6$ and $\mathcal{L}_6'$, that are disjoint unions of 6 lines. Let $\alpha\colon U \to \mathbb{P}^3$ be a blow-up of the curve $\mathcal{L}_6$. Then there is a commutative diagram

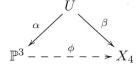

where β is a $\mathfrak{A}_6$-birational morphism that contracts finitely many curves, X_4 is a singular determinantal quartic threefold in $\mathbb{P}^4$ such that $\mathrm{Aut}(X_4) \cong \mathfrak{S}_6$, and ϕ is a birational map that is given by the linear system of all quartic surfaces on $\mathbb{P}^3$ containing $\mathcal{L}_6$. Take any odd permutation $\theta \in \mathrm{Aut}(X_4)$. Put

$$\iota = \phi^{-1} \circ \theta \circ \phi \in \mathrm{Bir}^{\mathfrak{A}_6}(\mathbb{P}^3).$$

Similarly, we can construct $\iota' \in \mathrm{Bir}^{\mathfrak{A}_6}(\mathbb{P}^3)$ using $\mathcal{L}_6'$. Let Γ be a subgroup in $\mathrm{Bir}^{\mathfrak{A}_6}(\mathbb{P}^3)$ that is generated by $\mathfrak{A}_6$, ι and ι'. By [23, Lemma 4.7], for every $\mathfrak{A}_6$-invariant linear system $\mathcal{M}$ on $\mathbb{P}^3$ that has no fixed components, there exists $\xi \in \Gamma$ such that $(\mathbb{P}^3, \lambda\xi(\mathcal{M}))$ has canonical singularities, where λ is a positive rational number such that

$$K_{\mathbb{P}^3} + \lambda\xi(\mathcal{M}) \sim_{\mathbb{Q}} 0.$$

Thus, $\mathbb{P}^3$ is $\mathfrak{A}_6$-birationally rigid and $\mathrm{Bir}^{\mathfrak{A}_6}(\mathbb{P}^3) = \Gamma$ by Theorem 3.3.1, because $\mathrm{Aut}^{\mathfrak{A}_6}(\mathbb{P}^3) = \mathfrak{A}_6$.

Example 3.3.6 ([24, §5]). Recall that $\mathrm{PSL}_2(\mathbf{F}_7)$ acts faithfully on $\mathbb{P}^3$. Then there exists a smooth $\mathrm{PSL}_2(\mathbf{F}_7)$-invariant irreducible curve C_6 of genus 3 and degree 6 in $\mathbb{P}^3$. Let $\alpha \colon U \to \mathbb{P}^3$ be a blow-up of the curve C_6, and let E be its exceptional divisor. Then there is a $\mathrm{PSL}_2(\mathbf{F}_7)$-biregular involution $\theta \in \mathrm{Aut}(U)$ such that $\theta(E) \neq E$. Put

$$\iota = \alpha \circ \theta \circ \alpha^{-1}.$$

By [24, Theorem 1.10], for every $\mathrm{PSL}_2(\mathbf{F}_7)$-invariant linear system $\mathcal{M}$ on $\mathbb{P}^3$ that has no fixed components, either $(\mathbb{P}^3, \lambda\mathcal{M})$ or $(\mathbb{P}^3, \lambda'\iota(\mathcal{M}))$ has canonical singularities, where λ and λ' are positive rational numbers such that

$$K_{\mathbb{P}^3} + \lambda\mathcal{M} \sim_{\mathbb{Q}} K_{\mathbb{P}^3} + \lambda'\iota(\mathcal{M}) \sim_{\mathbb{Q}} 0.$$

Thus, $\mathbb{P}^3$ is $\mathrm{PSL}_2(\mathbf{F}_7)$-birationally rigid by Theorem 3.3.1.

Example 3.3.7 ([23, §5]). Let V_3 be the Serge cubic hypersurface in $\mathbb{P}^4$ (see Example 1.3.4). Then $\mathrm{Aut}(V_3) \cong \mathfrak{S}_6$, so that V_3 is faithfully acted on by the group $\mathfrak{A}_6$. Furthermore, it follows from [23, §5], that for every $\mathfrak{A}_6$-invariant linear system $\mathcal{M}$ on V_3 that has no fixed components, the log pair $(V_3, \lambda\mathcal{M})$ has canonical singularities, where λ is a positive rational number such that

$$K_{V_3} + \lambda\mathcal{M} \sim_{\mathbb{Q}} 0.$$

Thus, V_3 is $\mathfrak{A}_6$-birationally rigid and $\mathrm{Bir}^{\mathfrak{A}_6}(\mathbb{P}^3) = \mathrm{Aut}(V_3)$ by Corollary 3.3.3.

One can show that the assertion of Corollary 3.3.2 is also a sufficient condition for X to be G-birationally rigid (see [21, Theorem 1.26]). Similarly, Corollary 3.3.3 is also a sufficient condition for X to be G-birationally superrigid (see [18, Corollary A.17]). Let us show now how to apply the latter corollary.

Lemma 3.3.8 ([35, Corollary 7.11],[18, Lemma A.18]). *Suppose that* $\dim(X) = 2$ *and* X *contains no* G-*orbits of length less than* K_X^2. *Then* X *is* G-*birationally superrigid.*

Proof. Suppose that X is not G-birationally superrigid. Then it follows from Corollary 3.3.3 that there is a G-invariant linear system $\mathcal{M}$ on the surface X such that $\mathcal{M}$ does not have fixed curves, and the log pair $(X, \lambda\mathcal{M})$ is not canonical at some point $O \in X$, where λ is a positive rational number such that
$$K_X + \lambda\mathcal{M} \sim_{\mathbb{Q}} 0.$$

Let Σ be the G-orbit of the point O. Then it follows from Lemma 2.2.7 that
$$\mathrm{mult}_P(\mathcal{M}) > \frac{1}{\lambda}$$
for every point $P \in \Sigma$. Let M_1 and M_2 be sufficiently general curves in $\mathcal{M}$. Then
$$\frac{K_X^2}{\lambda^2} = M_1 \cdot M_2 \geqslant \sum_{P \in \Sigma} \mathrm{mult}_P\left(M_1 \cdot M_2\right) \geqslant \sum_{P \in \Sigma} \mathrm{mult}_P^2(\mathcal{M}) > \frac{|\Sigma|}{\lambda^2} \geqslant \frac{K_X^2}{\lambda^2},$$
which is absurd. □

Example 3.3.9 ([18, Theorem A.19]). Recall that $\mathfrak{A}_6$ acts faithfully on $\mathbb{P}^2$. Since $\mathbb{P}^2$ does not contain $\mathfrak{A}_6$-orbits of length less than 12, it is $\mathfrak{A}_6$-birationally superrigid by Lemma 3.3.8.

Example 3.3.10 ([18, Theorem B.8]). Recall that $\mathrm{PSL}_2(\mathbf{F}_7)$ acts faithfully on $\mathbb{P}^2$. Since $\mathbb{P}^2$ does not contain $\mathrm{PSL}_2(\mathbf{F}_7)$-orbits of length less than 21, it is $\mathrm{PSL}_2(\mathbf{F}_7)$-birationally superrigid by Lemma 3.3.8.

3.4 Halphen pencils

In this section, we prove one generalization of Theorem 3.2.6 that will be used in the proof of Theorem 19.1.1. In fact, its proof is almost identical to the proof of Theorem 3.2.6. Nevertheless, we prefer to prove it separately to emphasize the following:

Definition 3.4.1. A mobile pencil on a variety is said to be a *Halphen pencil* if its general member is birational to a smooth variety of Kodaira dimension zero.

Halphen pencils on $\mathbb{P}^2$ have been studied by Georges Halphen in [51]. They were completely described by Igor Dolgachev in [32].

Let X be a variety that has at most terminal singularities such that the divisor $-K_X$ is nef. Suppose that X is faithfully acted on by a finite group G. Suppose, in addition, that there exists a positive integer m such that the linear system $|-mK_X|$ is free from base points and is not composed of a pencil. Then $|-mK_X|$ gives a surjective G-morphism

$$\eta\colon X \to S,$$

where S is a variety of dimension at least 2 acted on (not necessarily faithfully) by the group G. The following result is an obstruction to the existence of G-invariant Halphen pencils on the variety X.

Theorem 3.4.2. Let $\mathcal{M}$ be a G-invariant Halphen pencil on X. Suppose that there is a positive rational number λ such that

$$K_X + \lambda \mathcal{M} \sim_{\mathbb{Q}} 0.$$

Then the singularities of the log pair $(X, \lambda\mathcal{M})$ are not terminal.

Proof. Suppose that the singularities of the log pair $(X, \lambda\mathcal{M})$ are terminal. Then there exists a positive rational number $\lambda' > \lambda$ such that the singularities of the log pair $(X, \lambda'\mathcal{M})$ are still terminal. Let us show that the latter is impossible.

Let $\phi\colon X \dashrightarrow \mathbb{P}^1$ be a dominant rational map that is given by the pencil $\mathcal{M}$. Then there exists a commutative diagram

$$\begin{array}{ccc} & W & \\ {}^{\alpha}\swarrow & & \searrow^{\beta} \\ X \dashrightarrow\!\!\!\!\!\!\!{}^{\phi}\!\!\!\!\!\!\!\!\dashrightarrow & & \mathbb{P}^1, \end{array}$$

where α and β are G-birational morphisms and W is smooth. Denote the proper transform of the pencil $\mathcal{M}$ on the threefold W by $\mathcal{H}$, denote α-exceptional divisors by $F_1, \ldots, F_k$, and denote β-exceptional divisors by $E_1, \ldots, E_l$. Then

$$K_W + \lambda'\mathcal{H} \sim_{\mathbb{Q}} \alpha^*\Big(K_X + \lambda'\mathcal{M}\Big) + \sum_{j=1}^{k} a'_j F_j \sim_{\mathbb{Q}} \left(1 - \frac{\lambda'}{\lambda}\right)\alpha^*(K_X) + \sum_{j=1}^{k} a'_j F_j$$

for some positive rational numbers $a'_1, \ldots, a'_k$.

Let H be a general divisor in $\mathcal{H}$. Then H is a smooth irreducible variety of Kodaira dimension zero. Moreover, we have

$$K_H \sim K_W\Big|_H \sim_{\mathbb{Q}} -\lambda' H\Big|_H + \left(1 - \frac{\lambda'}{\lambda}\right)\alpha^*(K_X)\Big|_H + \sum_{j=1}^{k} a'_j F_j\Big|_H \sim_{\mathbb{Q}}$$

$$\sim_{\mathbb{Q}} \left(1 - \frac{\lambda'}{\lambda}\right)\alpha^*(K_X)\Big|_H + \sum_{j=1}^{k} a'_j F_j\Big|_H \quad (3.4.3)$$

by the adjunction formula. Note that the divisor

$$\sum_{j=1}^{k} a'_j F_j|_H$$

is effective by construction, because all numbers $a'_1, \ldots, a'_k$ are positive.

Let N be a sufficiently big and sufficiently divisible positive integer. We may assume that $m = N(\frac{\lambda'}{\lambda} - 1)$ and all numbers $Na'_1, \ldots, Na'_k$ are positive integers. By (3.4.3), we have

$$mK_H \sim \alpha^*(-mK_X)\Big|_H + \sum_{j=1}^{k} Na'_j F_j\Big|_H,$$

which implies that the linear system $|-mK_X|$ is composed of a pencil, because H is of Kodaira dimension zero. The latter contradicts our assumption. $\square$

The reader is encouraged to reprove Dolgachev's results from [32] using Theorem 3.4.2. One can also use Theorem 3.4.2 to find all finite groups faithfully acting on $\mathbb{P}^2$ that leave some Halphen pencil invariant.

Chapter 4

Auxiliary results

In this chapter we collect some sporadic auxiliary results that are used in various parts of the book.

4.1 Zero-dimensional subschemes

Let X be a smooth variety, $\mathcal{I}$ be an ideal sheaf on X, and H be an ample (Cartier) divisor on X. Let Z be a reduced subscheme of X, and let $\mathcal{I}_Z$ be the ideal sheaf of Z.

Lemma 4.1.1. Suppose that $\mathcal{I} \subset \mathcal{I}_Z$, the support of the sheaf $\mathcal{I}_Z/\mathcal{I}$ is zero-dimensional, and $h^1(\mathcal{I} \otimes \mathcal{O}_X(H)) = 0$. Then

$$h^0(\mathcal{O}_Z \otimes \mathcal{O}_X(H)) = h^0(\mathcal{O}_X(H)) - h^0(\mathcal{I}_Z \otimes \mathcal{O}_X(H)) =$$
$$= h^0(\mathcal{O}_X(H)) - h^0(\mathcal{I} \otimes \mathcal{O}_X(H)) - h^0(\mathcal{I}_Z/\mathcal{I}).$$

Proof. Consider an exact sequence of sheaves

$$0 \to \mathcal{I} \otimes \mathcal{O}_X(H) \to \mathcal{I}_Z \otimes \mathcal{O}_X(H) \to \mathcal{I}_Z/\mathcal{I} \to 0.$$

There is an exact sequence of cohomology groups

$$0 \to H^0\big(\mathcal{I}\otimes\mathcal{O}_X(H)\big) \to H^0\big(\mathcal{I}_Z\otimes\mathcal{O}_X(H)\big) \to H^0\big((\mathcal{I}_Z/\mathcal{I})\otimes\mathcal{O}_X(H)\big) \to$$
$$\to H^1\big(\mathcal{I}\otimes\mathcal{O}_X(H)\big) \to H^1\big(\mathcal{I}_Z\otimes\mathcal{O}_X(H)\big) \to H^1\big((\mathcal{I}_Z/\mathcal{I})\otimes\mathcal{O}_X(H)\big).$$

Recall that $h^1(\mathcal{I}\otimes\mathcal{O}_X(H)) = 0$ by assumption. On the other hand, we have

$$h^1\big((\mathcal{I}_Z/\mathcal{I}) \otimes \mathcal{O}_X(H)\big) = 0$$

and
$$h^0(\mathcal{I}_Z/\mathcal{I} \otimes \mathcal{O}_X(H)) = h^0(\mathcal{I}_Z/\mathcal{I})$$
because the support of the sheaf $\mathcal{I}_Z/\mathcal{I}$ is zero-dimensional. Thus from the above exact sequence we conclude that
$$h^0\big(\mathcal{I}_Z \otimes \mathcal{O}_X(H)\big) = h^0\big(\mathcal{I} \otimes \mathcal{O}_X(H)\big) + h^0\big((\mathcal{I}_Z/\mathcal{I}) \otimes \mathcal{O}_X(H)\big) \qquad (4.1.2)$$
and $h^1(\mathcal{I}_Z \otimes \mathcal{O}_X(H)) = 0$.

Consider an exact sequence of sheaves
$$0 \to \mathcal{I}_Z \otimes \mathcal{O}_X(H) \to \mathcal{O}_X(H) \to \mathcal{O}_Z \otimes \mathcal{O}_X(H) \to 0.$$
Since $h^1(\mathcal{I}_Z \otimes \mathcal{O}_X(H)) = 0$, there is an exact sequence of cohomology groups
$$0 \to H^0\big(\mathcal{I}_Z \otimes \mathcal{O}_X(H)\big) \to H^0\big(\mathcal{O}_X(H)\big) \to H^0\big(\mathcal{O}_Z \otimes \mathcal{O}_X(H)\big) \to 0.$$
Therefore, one has
$$h^0\big(\mathcal{I}_Z \otimes \mathcal{O}_X(H)\big) = h^0\big(\mathcal{O}_X(H)\big) - h^0\big(\mathcal{O}_Z \otimes \mathcal{O}_X(H)\big). \qquad (4.1.3)$$
Equations (4.1.2) and (4.1.3) imply what we need. $\square$

Corollary 4.1.4. In the assumptions of Lemma 4.1.1 one has
$$h^0(\mathcal{O}_Z \otimes \mathcal{O}_X(H)) \leqslant h^0(\mathcal{O}_X(H)).$$

Corollary 4.1.5. Let S be a subscheme of X defined by the ideal sheaf $\mathcal{I}$. Suppose that $Z = S_{red}$, the scheme S is reduced outside of finitely many points, and $h^1(\mathcal{I} \otimes \mathcal{O}_X(H)) = 0$. Then
$$h^0(\mathcal{O}_Z \otimes \mathcal{O}_X(H)) \leqslant h^0(\mathcal{O}_X(H)).$$

4.2 Atiyah flops

Let Y be a smooth projective variety, and let C be a smooth rational curve such that its normal bundle in Y is isomorphic to $\mathcal{O}_{\mathbb{P}^1}(-1) \oplus \mathcal{O}_{\mathbb{P}^1}(-1)$. Then there exists a birational morphism $f\colon Y \to X$ that contracts C to an isolated ordinary double point $P \in X$ such that f is an isomorphism away from C (see, for example, [105, Corollary 5.6]). Note that X is not necessarily projective.

CREMONA GROUPS AND THE ICOSAHEDRON

Remark 4.2.1. The threefold X is uniquely determined in the following sense. Let $f'\colon Y \to X'$ be a birational morphism that contracts C to an isolated normal singularity $P' \in X'$, and f' is an isomorphism away from C. Then there exists a commutative diagram

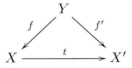

where t is an isomorphism.

The following lemma shows that Y is *almost* uniquely determined by X.

Lemma 4.2.2. Let Y be a smooth threefold. Let $f'\colon Y' \to X$ be a birational morphism such that X is normal, and f' contracts a union of finitely many curves $C_1, \ldots, C_r$ to an ordinary double point $P \in X$. Suppose that f' induces an isomorphism

$$Y' \setminus (C_1 \cup \ldots \cup C_r) \xrightarrow{\sim} X \setminus P.$$

Then $r = 1$ and C_1 is a smooth rational curve with normal bundle

$$\mathcal{N}_{C_1/Y'} \cong \mathcal{O}_{\mathbb{P}^1}(-1) \oplus \mathcal{O}_{\mathbb{P}^1}(-1).$$

Proof. Let H_X be a general hyperplane section of X that contains P. Denote by $H_{Y'}$ its proper transform on Y'. Then $H_{Y'}$ contains all curves $C_1, \ldots, C_r$. On the other hand, it follows from [105, Theorem 1.14(iii)] that $H_{Y'}$ is smooth along $C_1 \cup \ldots \cup C_r$ and the induced morphism $H_{Y'} \to H_X$ is a minimal resolution of the singular point $P \in H_X$. This shows that $r = 1$ and $C_1 \cong \mathbb{P}^1$, since H_X has an ordinary double point at P.

By [26, Corollary 16.3], the normal bundle of $C_1 \cong \mathbb{P}^1$ in Y' is either $\mathcal{O}_{\mathbb{P}^1}(-1) \oplus \mathcal{O}_{\mathbb{P}^1}(-1)$, or $\mathcal{O}_{\mathbb{P}^1} \oplus \mathcal{O}_{\mathbb{P}^1}(-2)$ or $\mathcal{O}_{\mathbb{P}^1}(1) \oplus \mathcal{O}_{\mathbb{P}^1}(-3)$. Since $H_{Y'}$ is smooth along C_1 and $C_1^2 = -2$ on $H_{Y'}$, we have

$$\mathcal{N}_{C_1/H_{Y'}} \cong \mathcal{O}_{\mathbb{P}^1}(-2).$$

On the other hand, there is an exact sequence of sheaves

$$0 \to \mathcal{N}_{C_1/H_{Y'}} \to \mathcal{N}_{C_1/Y'} \to \mathcal{N}_{H_{Y'}/Y'}|_{C_1} \to 0,$$

and one has

$$\mathcal{N}_{H_{Y'}/Y'}|_{C_1} \cong \mathcal{O}_{\mathbb{P}^1}.$$

Thus, the normal bundle of C_1 in Y' is either $\mathcal{O}_{\mathbb{P}^1}(-1) \oplus \mathcal{O}_{\mathbb{P}^1}(-1)$, or $\mathcal{O}_{\mathbb{P}^1} \oplus \mathcal{O}_{\mathbb{P}^1}(-2)$, i.e., C_1 is a (-2)-curve in the sense of [105, Definition 5.1]. If
$$\mathcal{N}_{C_1/Y'} \cong \mathcal{O}_{\mathbb{P}^1} \oplus \mathcal{O}_{\mathbb{P}^1}(-2),$$
then C_1 can be contracted to an isolated terminal cA_1 singularity that is not an ordinary double point (see the proof of [105, Corollary 5.6]). Similarly to Remark 4.2.1, the latter contradicts the assumption that P is an ordinary double point. Therefore, we have
$$\mathcal{N}_{C_1/H_{Y'}} \cong \mathcal{O}_{\mathbb{P}^1}(-1) \oplus \mathcal{O}_{\mathbb{P}^1}(-1).$$
□

There exists a commutative diagram

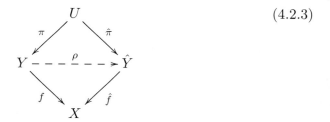

(4.2.3)

such that π is a blow-up of C, the morphism f is a contraction of C to an isolated ordinary double point, $\hat{\pi}$ is the contraction of the π-exceptional surface (isomorphic to $\mathbb{P}^1 \times \mathbb{P}^1$) to a smooth rational curve $\hat{C}$ such that ρ is not an isomorphism, and $\hat{f}$ is a contraction of C to the isolated ordinary double point $f(C)$.

Definition 4.2.4. The birational map $\rho \colon Y \dashrightarrow \hat{Y}$ in (4.2.3) is called an *Atiyah flop* in the curve C.

The construction of the Atiyah flop first appeared in [2]. Later it was explicitly introduced by Victor Kulikov in [75, §4.2] as *perestroika I*.

Remark 4.2.5. The commutative diagram (4.2.3) is unique in the following sense. Suppose that there exists a non-biregular birational map
$$\rho' \colon Y \dashrightarrow \hat{Y}'$$
such that $\hat{Y}$ is smooth and ρ' induces an isomorphism
$$Y \setminus C \xrightarrow{\sim} \hat{Y}' \setminus (\hat{C}_1 \cup \ldots \cup \hat{C}_r)$$

for some curves $\hat{C}_1, \ldots, \hat{C}_r$ in $\hat{Y}'$. Then Lemma 4.2.2 implies that $r = 1$, the curve $\hat{C}_1$ is smooth and rational, and there exists a birational morphism

$$\hat{f}' \colon \hat{Y}' \to X$$

that contracts $\hat{C}_1$ to the ordinary double point P and fits into a commutative diagram

$$\begin{array}{ccc} \hat{Y}' & \xrightarrow{s} & \hat{Y} \\ \hat{f}' \downarrow & & \downarrow \hat{f} \\ X & \xrightarrow{t} & X \end{array}$$

where s and t are isomorphisms.

The following result is a good illustration of Atiyah flops.

Lemma 4.2.6. Let V be a smooth threefold, let L be a smooth rational curve in V, let C be a smooth (possibly reducible) curve in V such that C intersects L transversally in two points. Let $\pi \colon W \to V$ be a blow-up of the curve C, and let $\tilde{\pi} \colon \tilde{V} \to V$ be the blow-up of the curve L. Denote by $\tilde{E}$ the $\tilde{\pi}$-exceptional divisor. Suppose that

$$\mathcal{N}_{L/V} \cong \mathcal{O}_{\mathbb{P}^1}(a) \oplus \mathcal{O}_{\mathbb{P}^1}(a)$$

for some integer a, so that $\tilde{E} \cong \mathbb{P}^1 \times \mathbb{P}^1$. Denote by $\tilde{C}$ the proper transform of the curve $\tilde{C}$ on the threefold $\tilde{V}$, denote by $\tilde{P}_1$ and $\tilde{P}_2$ the two points of the intersection $\tilde{C} \cap \tilde{E}$, and denote by $\bar{L}$ the proper transform of the curve L on the threefold W. Then either

$$\mathcal{N}_{\bar{L}/W} \cong \mathcal{O}_{\mathbb{P}^1}(a-1) \oplus \mathcal{O}_{\mathbb{P}^1}(a-1)$$

or

$$\mathcal{N}_{\bar{L}/W} \cong \mathcal{O}_{\mathbb{P}^1}(a) \oplus \mathcal{O}_{\mathbb{P}^1}(a-2).$$

Moreover, the latter case happens if and only if the points $\tilde{P}_1$ and $\tilde{P}_2$ are contained in one section of the natural projection $\tilde{E} \to L$ that has self-intersection 0.

Proof. We have

$$\mathcal{N}_{\bar{L}/W} \cong \mathcal{O}_{\mathbb{P}^1}(b) \oplus \mathcal{O}_{\mathbb{P}^1}(c)$$

for some integers b and c such that $b \leqslant c$. Then

$$b + c = -K_W \cdot \bar{L} + 2g(\bar{L}) - 2 = -K_W \cdot \bar{L} - 2 = -K_V \cdot L - 4 = 2a - 2.$$

Put $n = c-b$. Let $\hat{\pi}\colon \hat{V} \to W$ be the blow-up of the curve $\bar{L}$. Denote by $\hat{E}$ the $\hat{\pi}$-exceptional divisor. Then $\hat{E} \cong \mathbb{F}_n$. We have to show that either $n = 0$ or $n = 2$, and the equality $n = 2$ is possible if and only if $\tilde{P}_1$ and $\tilde{P}_2$ are contained in one section of the natural projection $\tilde{E} \to L$ that has self-intersection 0.

Denote by $\tilde{l}_1$ and $\tilde{l}_2$ the fibers of the natural projection $\tilde{E} \to L$ that pass through the points $\tilde{P}_1$ and $\tilde{P}_2$, respectively. Let $\check{\pi}\colon \check{V} \to \tilde{V}$ be the blow-up of the curve $\tilde{C}$. Denote by $\check{l}_1$ and $\check{l}_2$ the proper transforms of the curves $\tilde{l}_1$ and $\tilde{l}_2$ on the threefold $\check{V}$, respectively. Then there exists a commutative diagram

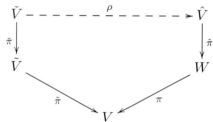

where ρ is a composition of Atiyah flops in the curves $\check{l}_1$ and $\check{l}_2$.

Denote by $\check{E}$ the proper transform of the surface $\tilde{E}$ on the threefold $\check{V}$. Then the induced morphism
$$\check{\pi}|_{\check{E}}\colon \check{E} \to \tilde{E}$$
is just a blow-up of the points $\tilde{P}_1$ and $\tilde{P}_2$. Moreover, the birational map ρ induces a birational morphism
$$\check{E} \to \hat{E} \cong \mathbb{F}_n$$
that contracts the curves $\check{l}_1$ and $\check{l}_2$. Since $\tilde{E} \cong \mathbb{P}^1 \times \mathbb{P}^1$, we see that either $n = 0$ or $n = 2$. Furthermore, one has $n = 2$ if and only if the points $\tilde{P}_1$ and $\tilde{P}_2$ are contained in one section of the natural projection $\tilde{E} \to L$ that has self-intersection 0. □

Remark 4.2.7. Let V be a smooth threefold, let T be a smooth curve on V, and let Q be a point in T. Let $\pi\colon W \to V$ be a blow-up of the curve T, and let $\sigma\colon \bar{V} \to V$ be a blow-up of the point Q. Denote by F the π-exceptional surface, denote by G the σ-exceptional surface, and denote by $\bar{T}$ the proper transform of the curve T on the threefold $\bar{V}$. Let L be a fiber of the restriction $\pi|_E\colon E \to T$ over the point $Q \in T$. Then there exists a commutative

diagram

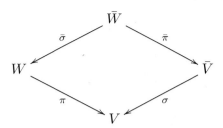

Here $\bar\pi$ is the blow-up of the curve $\bar T$, and $\bar\sigma$ is the blow-up of the curve L. Furthermore, $\bar\pi$ contracts the proper transform of the surface F to the curve $\bar T$, and $\bar\sigma$ contracts the proper transform of the surface G to the curve L. In particular, this implies that the $\bar\pi$-exceptional surface is isomorphic to the surface F.

4.3 One-dimensional linear systems

In this section we collect some auxiliary facts about pencils and linear systems composed of pencils.

Lemma 4.3.1. Let X be a threefold, and let $\mathcal{M}$ be a linear system on X that does not have fixed components. Suppose that there exists a proper Zariski closed subset $\Sigma \subsetneq X$ such that

$$S_1 \cap S_2 \subset \Sigma \subsetneq X$$

for every sufficiently general divisors S_1 and S_2 in the linear system $\mathcal{M}$. Then $\mathcal{M}$ is composed of a pencil.

Proof. Suppose that $\mathcal{M}$ is not composed of a pencil, and there is a proper Zariski closed subset $\Sigma \subset X$ such that the (set-theoretic) intersection of the sufficiently general divisors S_1 and S_2 of the linear system $\mathcal{M}$ is contained in the set Σ. Let $\rho\colon X \dashrightarrow \mathbb{P}^n$ be a rational map induced by the linear system $\mathcal{M}$, where n is the dimension of $\mathcal{M}$. Then there is a commutative diagram

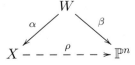

Here W is a smooth variety, α is a birational morphism, and β is some morphism. Let Y be the image of the morphism β. Then $\dim(Y) \geqslant 2$, because the linear system $\mathcal{M}$ is not composed of a pencil.

Let Λ be a Zariski closed subset of the variety W such that the morphism

$$\alpha|_{W\setminus\Lambda}\colon W \setminus \Lambda \longrightarrow X \setminus \alpha(\Lambda)$$

is an isomorphism, and let Δ be a union of the subset $\Lambda \subset W$ and the closure of the proper transform of the set $\Sigma \setminus \alpha(\Lambda)$ on the variety W. Then Δ is a Zariski closed proper subset of W.

Let B_1 and B_2 be general hyperplane sections of the variety Y. Let D_1 and D_2 be proper transforms of the divisors B_1 and B_2 on the variety W, respectively. Then $\alpha(D_1)$ and $\alpha(D_2)$ are general divisors of the linear system $\mathcal{M}$. Hence, we have

$$\varnothing \neq \beta^{-1}\Big(B_1 \cap B_2\Big) = D_1 \cap D_2 \subset \Delta \subsetneq W,$$

because $\dim(Y) \geqslant 2$. The latter is absurd. $\square$

Lemma 4.3.1 can be generalized as follows.

Lemma 4.3.2 ([19, Theorem 2.2]). *Let X be a threefold, let $\mathcal{M}$ be a pencil on X that does not have fixed components, and let $\mathcal{B}$ be a mobile linear system on the threefold X such that its general surface is irreducible. Suppose that there is a Zariski closed proper subset $\Sigma \subset X$ such that*

$$M \cap B \subseteq \Sigma,$$

where M and B are general divisors of the pencil $\mathcal{M}$ and the linear system $\mathcal{B}$, respectively. Then $\mathcal{B} = \mathcal{M}$.

Proof. Let $\rho\colon X \dashrightarrow \mathbb{P}^1$ be the rational map induced by the pencil $\mathcal{M}$, and let $\xi\colon X \dashrightarrow \mathbb{P}^r$ be the rational map given by the linear system $\mathcal{B}$. Let us consider a simultaneous resolution of both rational maps ρ and ξ as follows:

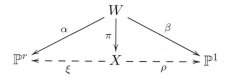

Here W is a smooth variety, π is a birational morphism, α and β are some morphisms.

Let Λ be a Zariski closed subset of the variety W such that the morphism

$$\pi\Big|_{W\setminus\Lambda}\colon W \setminus \Lambda \longrightarrow X \setminus \pi(\Lambda)$$

is an isomorphism and Δ is the union of the set Λ and the closure of the proper transform of the set $\Sigma \setminus \pi(\Lambda)$ on the variety W. Then the set Δ is a proper Zariski closed subset of the variety W.

Suppose that the pencil $\mathcal{M}$ is different from the linear system $\mathcal{B}$. Let B_W be the pull-back of a general hyperplane of $\mathbb{P}^r$ by the morphism α and let M_W be a general fiber of the morphism β. Then the intersection $M_W \cap B_W$ is not empty and is not contained in Δ. Hence, we have

$$\pi(M_W) \cap \pi(B_W) \not\subseteq \Sigma.$$

Since $\pi(M_W)$ and $\pi(B_W)$ are general divisors in the linear systems $\mathcal{M}$ and $\mathcal{B}$, respectively, this gives a contradiction. $\square$

4.4 Miscellanea

Let X be an irreducible variety that is faithfully acted on by a finite group G. Let P be a point in X that is G-invariant.

Lemma 4.4.1 (see, e. g., [45, Lemma 2.7(b)] or [94, Lemma 4]). *The natural linear action of G on the Zariski tangent space $T_P(X)$ is faithful.*

Lemma 4.4.2. *Let $C \subset X$ and $C' \subset X$ be two G-invariant curves passing through the point $P \in X$ such that both C and C' are smooth at P. Suppose that the action of G on C is not faithful, and the action of G on C' is faithful. Then the curves C and C' are transversal at P.*

Proof. Suppose that C and C' are tangent at P. Then there is a natural identification of the Zariski tangent spaces $T_P(C)$ and $T_P(C')$. By Lemma 4.4.1, the group G acts faithfully on $T_P(C')$, while G acts non-faithfully on $T_P(C)$, which is a contradiction. $\square$

Lemma 4.4.3. *Suppose that X is smooth at P, and the Zariski tangent space $T_P(X)$ contains a G-invariant hyperplane $\mathcal{H}$ such that G acts trivially on $\mathcal{H}$. Then there exists a G-invariant subvariety of codimension one $H \subset X$ such that H is smooth at P and G acts trivially on H.*

Proof. Let $U \subset X$ be an open G-invariant affine neighborhood of P. One can choose a G-equivariant embedding $U \hookrightarrow V$, where V is a vector space acted on by G; we will identify U and P with their images in V. Due to a G-equivariant identification $T_P(V) \cong V$ there is a G-equivariant embedding

$$T_P(U) \hookrightarrow V.$$

One can assume that P is the zero in the vector space V. Then there is a linear function F on V such that G acts on F by a character, and F cuts out the hyperplane $\mathcal{H}$ in $T_P(U)$. Now the subvariety of U given by equation $F = 0$ provides the desired hypersurface H. □

Let $\mathscr{W}$ be a $(n+1)$-dimensional G-representation. Suppose there is a G-equivariant embedding

$$X \hookrightarrow \mathbb{P}(\mathscr{W}) \cong \mathbb{P}^n.$$

Then one has a G-equivariant embedding

$$T_P(X) \hookrightarrow T_P(\mathbb{P}^n).$$

Remark 4.4.4 (cf. [23, Remark 5.1]). Let $\mathscr{W}_P$ be the one-dimensional G-subrepresentation of $\mathscr{W}$ that corresponds to the G-fixed point $P \in \mathbb{P}^n$. Then there exists a G-equivariant identification

$$T_P(\mathbb{P}^n) \cong \mathscr{W}/\mathscr{W}_P \otimes \mathscr{W}_P^\vee.$$

Indeed, let $x_0, \ldots, x_n$ be coordinates in $\mathscr{W}$ such that the subspace $\mathscr{W}_P$ is cut out by equations

$$x_1 = \ldots = x_n = 0.$$

Then $x_0, \ldots, x_n$ may be interpreted as homogeneous coordinates in $\mathbb{P}^n$. Let $U \subset \mathbb{P}^n$ be the affine neighborhood of the point P given by $x_0 \neq 0$. Then $U \cong \mathbb{A}^n$ is G-invariant, and $x_1/x_0, \ldots, x_n/x_0$ may be interpreted as coordinates in U. It remains to notice that there is a G-equivariant identification $T_P(\mathbb{P}^n) \cong U$.

Remark 4.4.5. Let Y be a variety with an action of a finite group G, and $\mathcal{L}$ be a G-equivariant line bundle on Y. Let Y_1 be a connected component of Y, and suppose that Y is a G-orbit of Y_1. Let $G_1 \subset G$ be a stabilizer of Y_1. Then there is an identification of G-representations

$$H^0(\mathcal{L}) \cong \operatorname{Ind}_{G_1}^G H^0(\mathcal{L}|_{Y_1}).$$

In the sequel we will denote by $p_a(C)$ the arithmetic genus $1 - \chi(\mathcal{O}_C)$ of a curve C.

Lemma 4.4.6 ([53, Exercise IV.1.8(a)], [107, §3.6]). Let C be an irreducible curve acted on by a finite group G. Let g be the genus of the normalization of C. Write

$$\operatorname{Sing}(C) = \Omega_1 \sqcup \ldots \sqcup \Omega_r,$$

CREMONA GROUPS AND THE ICOSAHEDRON

where Ω_i are G-orbits. Put $s_i = |\Omega_i|$. Then

$$p_a(C) = g + \sum_{i=1}^{r} m_i s_i$$

for some positive integers $m_1, \ldots, m_r$. Moreover, one has $m_i = 1$ for all i if and only if all singularities of C are either nodes or ordinary cusps.

The following remark is nearly trivial but rather useful.

Remark 4.4.7. Let $X \subset \mathbb{P}^n$ be a variety that is not contained in any hyperplane in $\mathbb{P}^n$, and $\Pi \subset \mathbb{P}^n$ be a linear subspace of dimension m. Let S be one of the irreducible components of $X \cap \Pi$. Suppose that all other irreducible components of $X \cap \Pi$ have dimensions that do not exceed $\dim(S)$. Then

$$\deg(S) + (n - \dim(X)) - (m - \dim(S)) \leqslant \deg(X).$$

Indeed, take a general linear subspace $\Pi' \subset \Pi$ of dimension $m - \dim(S)$. The intersection $X \cap \Pi'$ is finite, and the intersection $S \cap \Pi'$ consists of $\deg(S)$ points. Now take a general linear subspace Π'' of dimension $n - \dim(X)$ passing through Π' and general $n - \dim(X) - (m - \dim(S))$ points in $X \setminus \Pi$. We see that the intersection $X \cap \Pi''$ is finite and consists of at most $\deg(X)$ points.

The following result will be used in §13.2.

Lemma 4.4.8. Let Q be a smooth quadric threefold in $\mathbb{P}^4$, and let C be a rational twisted cubic curve in Q. Denote by Π the linear span of C. Suppose that the intersection $\Pi \cap Q$ is smooth. Then the normal bundle of $C \cong \mathbb{P}^1$ in Q is $\mathcal{O}_{\mathbb{P}^1}(3) \oplus \mathcal{O}_{\mathbb{P}^1}(4)$.

Proof. Put $S = \Pi \cap Q$. Since S is smooth by assumption, we see that

$$S \cong \mathbb{P}^1 \times \mathbb{P}^1.$$

There is an exact sequence

$$0 \to \mathcal{N}_{C/S} \to \mathcal{N}_{C/Q} \to \mathcal{N}_{S/Q}|_C \to 0.$$

The curve $C \cong \mathbb{P}^1$ is a divisor of bi-degree $(1, 2)$ on the surface S. Therefore, one has $\mathcal{N}_{C/S} \cong \mathcal{O}_{\mathbb{P}^1}(4)$ and $\mathcal{N}_{S/Q}|_C \cong \mathcal{O}_{\mathbb{P}^1}(3)$, which implies the assertion of the lemma. □

Part II
Icosahedral group

Chapter 5

Basic properties

In this chapter we collect some basic facts about the alternating group $\mathfrak{A}_5$ and some other relevant groups.

5.1 Action on points and curves

The following assertion can be routinely checked using nothing more than definitions.

Lemma 5.1.1. Any non-trivial proper subgroup of $\mathfrak{A}_5$ is isomorphic to one of the groups $\mathfrak{A}_4$, $\mathfrak{D}_{10}$, $\mathfrak{S}_3$, $\boldsymbol{\mu}_5$, $\boldsymbol{\mu}_2 \times \boldsymbol{\mu}_2$, $\boldsymbol{\mu}_3$ or $\boldsymbol{\mu}_2$. Furthermore, all subgroups of each of these types are conjugate in $\mathfrak{A}_5$.

For any non-trivial proper subgroup $H \subset \mathfrak{A}_5$ the order $|H|$ of H and the number $c(H)$ of subgroups of $\mathfrak{A}_5$ conjugate to H (which is the same as the number of subgroups isomorphic to H) is given in Table 5.1.

Table 5.1: Subgroups of the icosahedral group

H	$\mathfrak{A}_4$	$\mathfrak{D}_{10}$	$\mathfrak{S}_3$	$\boldsymbol{\mu}_5$	$\boldsymbol{\mu}_2 \times \boldsymbol{\mu}_2$	$\boldsymbol{\mu}_3$	$\boldsymbol{\mu}_2$
$|H|$	12	10	6	5	4	3	2
$c(H)$	5	6	10	6	5	10	15

An immediate consequence of Lemma 5.1.1 is the following:

Corollary 5.1.2. If $\mathfrak{A}_5$ acts transitively on a finite set Σ, then

$$|\Sigma| \in \{1, 5, 6, 10, 12, 15, 20, 30, 60\}.$$

We will need some information about the splitting of various sets with transitive action of $\mathfrak{A}_5$ into the orbits of subgroups of $\mathfrak{A}_5$.

Lemma 5.1.3. Let $\mathfrak{A}_5$ act transitively on a finite set Σ. Then the following assertions hold:

(i) if $|\Sigma| = 5$, then the action of $\mathfrak{A}_5$ on Σ is triply transitive;

(ii) if $|\Sigma| = 6$, then the action of $\mathfrak{A}_5$ on Σ is doubly transitive;

(iii) if $|\Sigma| = 5$, then the action of a subgroup $\mathfrak{D}_{10}$ of $\mathfrak{A}_5$ on Σ is transitive;

(iv) if $|\Sigma| = 10$, then Σ contains an orbit of length 2 of a subgroup $\boldsymbol{\mu}_2 \times \boldsymbol{\mu}_2$ of $\mathfrak{A}_5$;

(v) if $|\Sigma| = 12$, then Σ is a union of one orbit of length 2 and one orbit of length 10 of a subgroup $\mathfrak{D}_{10}$ of $\mathfrak{A}_5$;

(vi) if $|\Sigma| = 12$, then the action of a subgroup $\mathfrak{A}_4$ of $\mathfrak{A}_5$ on Σ is transitive;

(vii) if $|\Sigma| = 10$, then the set $\Sigma^{(2)}$ of 45 non-ordered pairs of different elements of Σ splits as a union of an $\mathfrak{A}_5$-orbit of length 15 and an $\mathfrak{A}_5$-orbit of length 30.

Proof. Straightforward. □

More detailed information about subgroups of the group $\mathfrak{A}_5$ is contained in Table 5.2. We use the following notation. The first row of Table 5.2 represents the conjugacy classes in the group $\mathfrak{A}_5$: the symbol id denotes the identity element, the symbols $[2,2]$, $[3]$, $[5]$, and $[5]'$ denote the classes of elements of the corresponding cycle type. The symbols $[5]$ and $[5]'$ denote two different conjugacy classes of elements of order 5; note that if $g \in [5]$, then $g^4 \in [5]$, while $g^2, g^3 \in [5]'$. Other rows list the number of elements in each conjugacy class for $\mathfrak{A}_5$ and its subgroups.

Our next goal is to study smooth irreducible curves with an action of the group $\mathfrak{A}_5$.

Lemma 5.1.4. Suppose that C is a smooth irreducible curve with a non-trivial action of the group $\mathfrak{A}_5$, and $\Sigma \subset C$ is an $\mathfrak{A}_5$-orbit. Then

$$|\Sigma| \in \{12, 20, 30, 60\}.$$

Table 5.2: Elements of subgroups of the icosahedral group

	id	[2, 2]	[3]	[5]	[5]'
$\mathfrak{A}_5$	1	15	20	12	12
$\mathfrak{A}_4$	1	3	8	0	0
$\mathfrak{D}_{10}$	1	5	0	2	2
$\mathfrak{S}_3$	1	3	2	0	0
μ_5	1	0	0	2	2
$\mu_2 \times \mu_2$	1	3	0	0	0
μ_3	1	0	2	0	0
μ_2	1	1	0	0	0

Proof. Let P be a point of Σ, and G_P be the stabilizer of P. Then G_P acts faithfully on the one-dimensional Zariski tangent space $T_P(C)$ by Lemma 4.4.1. Thus G_P is a cyclic group, and the assertion follows from Lemma 5.1.1. □

We will also need information about the action of $\mathfrak{A}_5$ on other curves of small genus.

Lemma 5.1.5. *Suppose that C is a smooth irreducible curve of genus $g \leqslant 34$ with a non-trivial action of the group $\mathfrak{A}_5$. Then*

$$g \in \{0, 4, 5, 6, 9, 10, 11, 13, 15, 16, 19,$$
$$20, 21, 24, 25, 26, 28, 29, 30, 31, 33, 34\},$$

and the possible numbers of $\mathfrak{A}_5$-orbits in C consisting of 12, 20 and 30 points are contained in Table 5.3.

Proof. It follows from the non-solvability of the group $\mathfrak{A}_5$ that $g \neq 1$.

Let $\Omega \subset C$ be an $\mathfrak{A}_5$-orbit. Then $|\Omega| \in \{12, 20, 30, 60\}$ by Lemma 5.1.4.

Put $\bar{C} = C/\mathfrak{A}_5$. Then $\bar{C}$ is a smooth curve. Let $\bar{g}$ be the genus of the curve $\bar{C}$. The Riemann–Hurwitz formula gives

$$2g - 2 = 60(2\bar{g} - 2) + 30a_{30} + 40a_{20} + 48a_{12},$$

Table 5.3: Possible small genera of curves with $\mathfrak{A}_5$-action

g	0	4	5	6	9	10	11	13	15	16
a_{12}	1	2	1	0	2	1	0	3	1	0
a_{20}	1	0	2	1	1	0	2	0	1	3
a_{30}	1	1	0	3	0	3	2	0	2	1

g	16	16	19	20	21	21	21	24	25	25
a_{12}	0	0	2	1	0	0	0	2	1	1
a_{20}	0	0	0	2	4	1	1	1	3	0
a_{30}	5	1	2	1	0	4	0	1	0	4

g	25	26	28	29	30	31	31	31	33	34
a_{12}	1	0	3	2	1	0	0	0	3	2
a_{20}	0	2	0	2	1	3	0	0	1	0
a_{30}	0	3	1	0	3	2	6	2	0	3

where a_k is the number of $\mathfrak{A}_5$-orbits in C of length k.

Since $a_k \geqslant 0$ and $g \leqslant 34$, one has $\bar{g} \leqslant 1$. If $\bar{g} = 1$, then

$$2g - 2 = 30a_{30} + 40a_{20} + 48a_{12}.$$

If $\bar{g} = 0$, then

$$2g - 2 = -120 + 30a_{30} + 40a_{20} + 48a_{12}.$$

Going through the values $0 \leqslant g \leqslant 34$, $g \neq 1$, and solving these two equations case by case we obtain the solutions listed in Table 5.3. $\square$

Remark 5.1.6. The method we use to prove Lemma 5.1.5 is rather primitive and does not allow one to find out if all cases listed in Table 5.3 are actually possible. However, using more advanced techniques one can show that all genera listed in Lemma 5.1.5 are indeed possible, and even to find *all* genera of curves that admit a non-trivial action of the group $\mathfrak{A}_5$. See [14] for the relevant background, and [87] for closely related computations for some groups including $\mathfrak{A}_5$ (note that in [87, Table 4] one has $\mu(\mathfrak{A}_5) = 4$). We will not use this approach since the information provided by Lemma 5.1.5 is enough for our purposes, but we will work out some examples that are important for us in §5.4.

5.2 Representation theory

The group $\mathfrak{A}_5$ has exactly five non-isomorphic irreducible representations (see, e.g., [27, p. 2] or [46, Exercise 3.5]), which we will denote by I, W_3, W_3', W_4, and W_5 until the end of the book. The values of the characters of these representations are listed in Table 5.4. We use the following notation. The first row of Table 5.4 represents the conjugacy classes in the group $\mathfrak{A}_5$ (see Table 5.2). Other rows list the values of the characters of irreducible representations. By b_5 we denote the number $1/2 - \sqrt{5}/2$, and by b_5' we denote the number $1/2 + \sqrt{5}/2$.

Table 5.4: Representations of the icosahedral group

	id	[2,2]	[3]	[5]	[5]'
I	1	1	1	1	1
W_3	3	-1	0	b_5	b_5'
W_3'	3	-1	0	b_5'	b_5
W_4	4	0	1	-1	-1
W_5	5	1	-1	0	0

Looking at Table 5.4, we can derive the following:

Corollary 5.2.1. *All representations of the group $\mathfrak{A}_5$ are self-dual. Furthermore, the following isomorphisms of the $\mathfrak{A}_5$-representations hold:*

$$\mathrm{Sym}^2(W_3) \cong \mathrm{Sym}^2(W_3') \cong I \oplus W_5,$$
$$\Lambda^2(W_5) \cong W_3' \oplus W_3 \oplus W_4,$$
$$\mathrm{Sym}^2(W_4) \cong I \oplus W_4 \oplus W_5,$$
$$\mathrm{Sym}^2(W_5) \cong I \oplus W_4 \oplus W_5^{\oplus 2},$$
$$\mathrm{Sym}^2(W_3 \oplus W_4) \cong I^{\oplus 2} \oplus W_3' \oplus W_4^{\oplus 2} \oplus W_5^{\oplus 3},$$
$$\mathrm{Sym}^6(W_3) \cong I^{\oplus 2} \oplus W_3 \oplus W_4^{\oplus 2} \oplus W_5^{\oplus 3}.$$

Comparing Tables 5.2 and 5.4, one easily obtains the following corollaries.

Corollary 5.2.2. *Let Γ be a subgroup of $\mathfrak{A}_5$. After restriction to Γ the $\mathfrak{A}_5$-representation W_3 (resp., W_3') splits as*

(o) a sum of one trivial and two non-trivial representations, if $\Gamma \cong \boldsymbol{\mu}_2$;

(i) a sum of three different one-dimensional representations, if $\Gamma \cong \boldsymbol{\mu}_3$;

(ii) a sum of three different non-trivial one-dimensional representations, if $\Gamma \cong \boldsymbol{\mu}_2 \times \boldsymbol{\mu}_2$;

(iii) a sum of a trivial and two different non-trivial one-dimensional representations, if $\Gamma \cong \boldsymbol{\mu}_5$;

(iv) a sum of an irreducible two-dimensional and a non-trivial one-dimensional representation, if $\Gamma \cong \mathfrak{S}_3$;

(v) a sum of an irreducible two-dimensional and a non-trivial one-dimensional representation, if $\Gamma \cong \mathfrak{D}_{10}$.

(vi) If $\Gamma \cong \mathfrak{A}_4$, the representation W_3 (resp., W_3') remains irreducible after restriction to Γ.

Corollary 5.2.3. Let Γ be a subgroup of $\mathfrak{A}_5$. After restriction to Γ the $\mathfrak{A}_5$-representation W_4 splits as

(i) a sum of two trivial and two different non-trivial one-dimensional representations, if $\Gamma \cong \boldsymbol{\mu}_3$;

(ii) a sum of four different one-dimensional representations, if $\Gamma \cong \boldsymbol{\mu}_2 \times \boldsymbol{\mu}_2$;

(iii) a sum of four different non-trivial one-dimensional representations, if $\Gamma \cong \boldsymbol{\mu}_5$;

(iv) a sum of an irreducible two-dimensional, one trivial and one non-trivial one-dimensional representation, if $\Gamma \cong \mathfrak{S}_3$;

(v) a sum of two different irreducible two-dimensional representations, if $\Gamma \cong \mathfrak{D}_{10}$;

(vi) a sum of an irreducible three-dimensional and a trivial representation, if $\Gamma \cong \mathfrak{A}_4$.

Corollary 5.2.4. Let Γ be a subgroup of $\mathfrak{A}_5$. After restriction to Γ the $\mathfrak{A}_5$-representation W_5 splits as

(i) a sum of a trivial one-dimensional and two different irreducible two-dimensional representations, if $\Gamma \cong \mathfrak{D}_{10}$;

(ii) a sum of two different non-trivial one-dimensional and one three-dimensional irreducible representation, if $\Gamma \cong \mathfrak{A}_4$.

5.3 Invariant theory

The following facts are well known.

Lemma 5.3.1. Put $\mathbb{P}^2 = \mathbb{P}(W_3)$ or $\mathbb{P}^2 = \mathbb{P}(W_3')$. Then

(i) if an $\mathfrak{A}_5$-orbit in $\mathbb{P}^2$ has length $r \leqslant 20$, then $r \in \{6, 10, 12, 15, 20\}$, and for any $r \in \{6, 10, 12, 15, 20\}$ there is a unique $\mathfrak{A}_5$-orbit Σ_r of length r in $\mathbb{P}^2$;

(ii) if $\mathcal{L}$ is an $\mathfrak{A}_5$-irreducible curve that is a union of $r \leqslant 20$ lines in $\mathbb{P}^2$, then $r \in \{6, 10, 12, 15, 20\}$, and for any $r \in \{6, 10, 12, 15, 20\}$ there is a unique $\mathfrak{A}_5$-irreducible curve $\mathcal{L}_r$ that is a union of r lines in $\mathbb{P}^2$;

(iii) there is a unique $\mathfrak{A}_5$-invariant effective divisor $\mathfrak{C}$ of degree 2 on $\mathbb{P}^2$, and $\mathfrak{C}$ is a smooth conic;

(iv) if C is an $\mathfrak{A}_5$-irreducible curve of degree $\deg(C) \leqslant 9$ in $\mathbb{P}^2$, then either $\deg(C) = 2$ and $A = \mathfrak{C}$, or $\deg(C) = 6$;

(v) the $\mathfrak{A}_5$-orbit Σ_{15} is not contained in any curve in $\mathbb{P}^2$ of degree at most 3;

(vi) all $\mathfrak{A}_5$-invariant effective divisors of degree 6 form a pencil.

Proof. Assertions (i) and (ii) follow from Corollary 5.2.2 (cf. [36, Proposition 2]). Note that while for any subgroup $G \cong \mu_3$ of $\mathfrak{A}_5$ there are exactly three points fixed by G by Corollary 5.2.2(i), one of these points is actually fixed by a larger subgroup $\mathfrak{S}_3$ containing G (and thus gives rise to an $\mathfrak{A}_5$-orbit of length 10), and two others are contained in the same $\mathfrak{A}_5$-orbit of length 20. Similarly, while for any subgroup $H \cong \mu_5$ of $\mathfrak{A}_5$ there are exactly three points fixed by H by Corollary 5.2.2(i), one of these points is actually fixed by a larger subgroup $\mathfrak{D}_{10}$ containing H (and thus gives rise to an $\mathfrak{A}_5$-orbit of length 6), and two others are contained in the same $\mathfrak{A}_5$-orbit of length 12.

The existence and uniqueness of the divisor $\mathfrak{C}$ mentioned in assertion (iii) follows from Corollary 5.2.1. If $\mathfrak{C}$ was non-reduced or singular, then there would exist an $\mathfrak{A}_5$-invariant union of at most two lines in $\mathbb{P}^2$, which is impossible by (ii). This completes the proof of assertion (iii).

Suppose that $C \subset \mathbb{P}^2$ is an $\mathfrak{A}_5$-irreducible curve of degree $d \leqslant 9$ such that C does not contain the conic $\mathfrak{C}$. Then the intersection $C \cap \mathfrak{C}$ is an $\mathfrak{A}_5$-invariant subset of a smooth curve $\mathfrak{C}$ that consists of at most $2d \leqslant 18$ points. This implies that $d = 6$ by Lemma 5.1.4, which gives assertion (iv).

To prove assertion (v), suppose that there exists a curve $R \subset \mathbb{P}^2$ of degree $d \leqslant 3$ that contains Σ_{15}. Let $\mathcal{M}$ be the linear subsystem in $|\mathcal{O}_{\mathbb{P}^2}(d)|$ consisting of curves that contain Σ_{15}. Then $\mathcal{M}$ must have fixed components, because $|\Sigma_{15}| > d^2$. Since the fixed part of $\mathcal{M}$ must be $\mathfrak{A}_5$-invariant, we see that it coincides with the unique $\mathfrak{A}_5$-invariant conic $\mathfrak{C}$ by assertion (iv). But the conic $\mathfrak{C}$ does not contain Σ_{15} by Lemma 5.1.4. Hence, Σ_{15} is contained in the base locus of the mobile part of $\mathcal{M}$, which immediately leads to a contradiction.

Existence and uniqueness of the pencil mentioned in assertion (vi) follows from Corollary 5.2.1 (also, one can construct this pencil explicitly — it is spanned by the divisor $3\mathfrak{C}$ and the curve $\mathcal{L}_6$). □

Remark 5.3.2. Since the conic $\mathfrak{C}$ described in Lemma 5.3.1(iii) is $\mathfrak{A}_5$-invariant, it follows from Lemma 5.3.1(i),(ii) that the points of the $\mathfrak{A}_5$-orbit Σ_6 (resp., $\Sigma_{10}, \Sigma_{12}, \Sigma_{15}, \Sigma_{20}$) are polar with respect to $\mathfrak{C}$ to the lines that are irreducible components of the curve $\mathcal{L}_6$ (resp., $\mathcal{L}_{10}, \mathcal{L}_{12}, \mathcal{L}_{15}, \mathcal{L}_{20}$).

Lemma 5.3.3. *Put* $\mathbb{P}^3 = \mathbb{P}(W_4)$. *Then*

(i) *if an $\mathfrak{A}_5$-orbit on $\mathbb{P}^3$ has length $r < 15$, then $r \in \{5, 10, 12\}$;*

(ii) *an $\mathfrak{A}_5$-orbit on $\mathbb{P}^3$ of length $r = 5$ is unique;*

(iii) *there are exactly two $\mathfrak{A}_5$-orbits on $\mathbb{P}^3$ of length 10;*

(iv) *there are exactly two $\mathfrak{A}_5$-orbits on $\mathbb{P}^3$ of length 12;*

(v) *there exists a unique $\mathfrak{A}_5$-invariant effective divisor B_2 on $\mathbb{P}^3$ of degree 2, and B_2 is a smooth surface;*

(vi) *the surface B_2 contains both $\mathfrak{A}_5$-orbits of length 12 contained in $\mathbb{P}^3$, and does not contain $\mathfrak{A}_5$-orbits of length 5 and 10;*

(vii) *there exists a unique $\mathfrak{A}_5$-invariant effective divisor B_3 on $\mathbb{P}^3$ of degree 3, and B_3 is a smooth surface (cf. §6.3);*

(viii) *the intersection $B_2 \cap B_3$ is a smooth curve of degree 6 and genus 4 (cf. Remark 5.4.2);*

(ix) *there are no $\mathfrak{A}_5$-invariant curves of degree at most 3 in $\mathbb{P}^3$;*

(x) *all $\mathfrak{A}_5$-invariant effective divisors in $\mathbb{P}^3$ of degree 4 form a pencil $\mathcal{Q}$ that contains the divisor $2B_2$, and does not contain any other non-reduced divisors;*

(xi) for any surface $B_4 \in \mathcal{Q}$ different from $2B_2$, the intersection $B_2 \cap B_4$ is a smooth curve of degree 8 and genus 9.

Proof. Assertions (i), (ii), (iii), and (iv) follow from Corollary 5.2.3. Assertions (v), (vii), (viii), (x), and (xi) are easy exercises in symmetric polynomials.

To prove assertion (vi), note that the action of $\mathfrak{A}_5$ on the quadric $B_2 \cong \mathbb{P}^1 \times \mathbb{P}^1$ is faithful. Since $\mathbb{P}^1$ does not contain $\mathfrak{A}_5$-orbits of length 5 and 10 by Lemma 5.1.4, neither does B_2. Also, it is easy to see that $\mathbb{P}^1 \times \mathbb{P}^1$ contains at least two $\mathfrak{A}_5$-orbits of length 12 (cf. Lemma 6.4.1 below). This implies that B_2 contains both $\mathfrak{A}_5$-orbits of length 12 contained in $\mathbb{P}^3$ by (iv).

Now let us prove assertion (ix). Suppose that there exists an $\mathfrak{A}_5$-invariant curve of degree at most 3 in $\mathbb{P}^3$. Denote it by C. By Corollary 5.1.2 one may assume that C is irreducible. Since W_4 is an irreducible $\mathfrak{A}_5$-representation, the only possibility is that C is a twisted cubic. Since $B_2 \cap B_3$ is a smooth curve of degree 6 by (viii), this intersection does not contain C. Intersecting C with one of these surfaces, we find an $\mathfrak{A}_5$-orbit on C of length at most 9, which is impossible by Lemma 5.1.4. $\square$

Lemma 5.3.4. Put $\mathbb{P}^3 = \mathbb{P}(W_4)$. Let $P_1, \ldots, P_5$ be the points of the unique $\mathfrak{A}_5$-orbit of length 5 in $\mathbb{P}^3$, let $G_1, \ldots, G_5$ be the stabilizers in $\mathfrak{A}_5$ of the points $P_1, \ldots, P_5$, respectively, and let H_i, $1 \leqslant i \leqslant 5$, be the unique plane in $\mathbb{P}^3$ that is G_i-invariant, see Corollary 5.2.3(vi). Then

(i) no three points among $P_1, \ldots, P_5$ are collinear;

(ii) no four points among $P_1, \ldots, P_5$ are coplanar;

(iii) the planes $H_1, \ldots, H_5$ are disjoint from the points $P_1, \ldots, P_5$.

Proof. Suppose that there are three points among $P_1, \ldots, P_5$ that are contained in a line L. Then all of them are contained in L by Lemma 5.1.3(i), which contradicts the irreducibility of the $\mathfrak{A}_5$-representation W_4. This proves assertion (i).

Suppose that there are four points among $P_1, \ldots, P_5$ that are contained in a plane Π. Then there are four other points among $P_1, \ldots, P_5$ that are contained in some (possibly different) plane Π'. The intersection $\Pi \cap \Pi'$ contains at least three points among $P_1, \ldots, P_5$. By assertion (i), we conclude that $\Pi = \Pi'$, which contradicts the irreducibility of the $\mathfrak{A}_5$-representation W_4. This proves assertion (ii).

Note that the plane H_1 does not contain the point P_1 by Corollary 5.2.3(vi). Suppose that H_1 contains a point among $P_2,\ldots,P_5$. Then it contains all four of them by Lemma 5.1.3(i), which contradicts assertion (ii). This proves assertion (iii). □

Note that W_4 may be considered as a representation of the symmetric group $\mathfrak{S}_5$. Thus the projective space $\mathbb{P}(W_4)$ is also acted on by the group $\mathfrak{S}_5$, cf. (5.7.1) below.

Lemma 5.3.5. Put $\mathbb{P}^3 = \mathbb{P}(W_4)$. Then

(i) there does not exist an $\mathfrak{A}_5$-invariant set of 6 planes in $\mathbb{P}^3$;

(ii) if $\{Q_1,\ldots,Q_r\}$ is an $\mathfrak{A}_5$-invariant r-tuple of quadrics in $\mathbb{P}^3$, and $r \leqslant 6$, then $r \in \{1,5,6\}$;

(iii) there is a unique $\mathfrak{A}_5$-invariant quadric Q_0 in $\mathbb{P}^3$;

(iv) there is a unique $\mathfrak{A}_5$-invariant sixtuple of quadrics $\{Q_1,\ldots,Q_6\}$ in $\mathbb{P}^3$;

(v) the quadric Q_0 and the sixtuple $\{Q_1,\ldots,Q_6\}$ are $\mathfrak{S}_5$-invariant.

Proof. Assertion (i) follows from Lemma 5.3.3(i). Recall that one has

$$\mathrm{Sym}^2(W_4) \cong I \oplus W_4 \oplus W_5$$

by Corollary 5.2.1. Therefore, assertions (ii), (iii), and (iv) are implied by Lemma 5.1.1 together with Corollaries 5.2.3 and 5.2.4 (note that Q_0 coincides with the quadric B_2 of Lemma 5.3.3). Finally, assertion (v) follows from uniqueness of the quadric Q_0 and the sixtuple $\{Q_1,\ldots,Q_6\}$. □

Lemma 5.3.6. Put $\mathbb{P}^4 = \mathbb{P}(I \oplus W_4)$. Let $\mathcal{H} \cong \mathbb{P}^3 \subset \mathbb{P}^4$ be the hyperplane that corresponds to the $\mathfrak{A}_5$-subrepresentation W_4, and $P \in \mathbb{P}^4$ be the point that corresponds to the $\mathfrak{A}_5$-subrepresentation I. Let $L \subset \mathbb{P}^4$ be a line, and let $\mathcal{L}$ be the $\mathfrak{A}_5$-orbit of L. Suppose that $\mathcal{L}$ is a union of 12 lines. Then either $P \in L$, or $L \subset \mathcal{H}$.

Proof. Let $G_L \subset \mathfrak{A}_5$ be the stabilizer of the line L. Then $G_L \cong \boldsymbol{\mu}_5$ by Lemma 5.1.1. The $\mathfrak{A}_5$-representation $I \oplus W_4$ splits as a sum of five pairwise different one-dimensional representations after restriction to the subgroup $G_L \subset \mathfrak{A}_5$ by Corollary 5.2.3(iii). Therefore, any two-dimensional G_L-subrepresentation of $I \oplus W_4$ either contains I, or is contained in W_4. □

Lemma 5.3.7. Put $\mathbb{P}^4 = \mathbb{P}(W_5)$. Then the $\mathfrak{A}_5$-representation $H^0(\mathcal{O}_{\mathbb{P}^4}(2))$ does not contain three-dimensional $\mathfrak{A}_5$-invariant subspaces.

Proof. One has an identification of $\mathfrak{A}_5$-representations

$$H^0(\mathcal{O}_{\mathbb{P}^4}(2)) \cong \mathrm{Sym}^2(W_5).$$

Now the assertion follows from Corollary 5.2.1. □

5.4 Curves of low genera

In this section we describe some curves of low genera that admit a faithful action of the group $\mathfrak{A}_5$ (cf. Lemma 5.1.5).

Lemma 5.4.1. *There exists a unique smooth irreducible curve C of genus 4 with a non-trivial action of the group $\mathfrak{A}_5$. The curve C is non-hyperelliptic.*

Proof. To construct a curve of genus 4 with a non-trivial action of the group $\mathfrak{A}_5$ take an intersection of the unique $\mathfrak{A}_5$-invariant quadric and the unique $\mathfrak{A}_5$-invariant cubic in $\mathbb{P}^3 \cong \mathbb{P}(W_4)$ (see Lemma 5.3.3(viii)).

Now suppose that C is a curve of genus 4 with a non-trivial action of the group $\mathfrak{A}_5$. We claim that C is not hyperelliptic. Indeed, suppose that C is hyperelliptic. Then the double cover $\phi \colon C \to \mathbb{P}^1$ is given by the canonical linear system $|K_C|$, and thus ϕ is $\mathfrak{A}_5$-equivariant. In particular, ϕ is branched over an $\mathfrak{A}_5$-invariant set of 10 points in $\mathbb{P}^1$, which is impossible by Lemma 5.1.4.

We see that the curve C is not hyperelliptic. Consider the canonical embedding $\phi \colon C \hookrightarrow \mathbb{P}^3$. The action of the group $\mathfrak{A}_5$ lifts to the four-dimensional vector space $W = H^0(\mathcal{O}_C(K_C))$. As $\mathfrak{A}_5$-representations, one has either $W \cong I \oplus W_3$, or $W \cong I \oplus W_3'$, or $W \cong W_4$ (see Table 5.4). In each of the former two cases there exists an $\mathfrak{A}_5$-invariant hyperplane

$$H \subset \mathbb{P}^3 \cong \mathbb{P}(W^\vee).$$

The intersection of H with $\phi(C)$ is an $\mathfrak{A}_5$-invariant subset of a smooth curve $\phi(C)$ that consists of at most $\deg(\phi(C)) = 6$ points, which is impossible by Lemma 5.1.4. Therefore, one has $W \cong W_4$.

By [53, Example IV.5.2.2], the curve $\phi(C)$ is contained in a unique quadric $Q \subset \mathbb{P}^3$. The quadric Q is $\mathfrak{A}_5$-invariant, and thus smooth (see Lemma 5.3.3(v)).

By Lemma 5.3.3(vii), there exists a unique $\mathfrak{A}_5$-invariant cubic surface $R \subset \mathbb{P}^3$. We claim that $\phi(C)$ coincides with the curve $C' = Q \cap R$. Indeed, suppose that $C' \ne \phi(C)$. Note that $\phi(C)$ is irreducible, and both $\phi(C)$ and C' are curves of bi-degree $(3,3)$ on $Q \cong \mathbb{P}^1 \times \mathbb{P}^1$. Therefore, the intersection $\phi(C) \cap C'$ is an $\mathfrak{A}_5$-invariant set that consists of $\phi(C) \cdot C' = 18$ points (counted with multiplicities). The latter is impossible by Lemma 5.1.4. The obtained contradiction shows that the curve C is unique. □

Remark 5.4.2. The curve described in Lemma 5.4.1 has been studied by many people (see, e. g., [12], [37], [38], [50], [82], and [108]). It was called the *Bring's curve* by Felix Klein in [71, p. 157] after Erland Samuel Bring who studied this curve back in 1786. This was brought to our attention by Harry Braden in Edinburgh in May 2013.

Lemma 5.4.3. There exists a unique smooth irreducible curve C of genus 5 with a non-trivial action of the group $\mathfrak{A}_5$. The curve C is hyperelliptic.

Proof. To construct a hyperelliptic curve of genus 5 take a double cover Z of $\mathbb{P}^1$ branched over the unique $\mathfrak{A}_5$-orbit in $\mathbb{P}^1$ of length 12 (see Lemma 5.1.5). This shows the existence assertion.

To prove the uniqueness assertion suppose that C be a smooth irreducible curve of genus 5 that admits a faithful action of the group $\mathfrak{A}_5$.

We claim that C is hyperelliptic. Indeed, suppose that it is not. Then the linear system $|K_C|$ gives an $\mathfrak{A}_5$-equivariant embedding $C \hookrightarrow \mathbb{P}^4$. Using information given in Table 5.4 and arguing as in the proof of Lemma 5.4.1, we see that there is an $\mathfrak{A}_5$-equivariant identification $\mathbb{P}^4 \cong \mathbb{P}(W_5)$.

To simplify the notation, we identify the curve C with its canonical image $\phi(C) \subset \mathbb{P}^4$. Let $\mathcal{I}_C \subset \mathcal{O}_{\mathbb{P}^4}$ be the ideal sheaf of the curve $C \subset \mathbb{P}^4$. Then $h^0(\mathcal{I}_C \otimes \mathcal{O}_{\mathbb{P}^4}(2)) \geqslant 3$ (cf. [53, Exercise V.5.5]). Indeed, this follows from the exact sequence of $\mathfrak{A}_5$-representations

$$0 \longrightarrow H^0\big(\mathcal{I}_C \otimes \mathcal{O}_{\mathbb{P}^4}(2)\big) \longrightarrow H^0\big(\mathcal{O}_{\mathbb{P}^4}(2)\big) \longrightarrow H^0\big(\mathcal{O}_C \otimes \mathcal{O}_{\mathbb{P}^4}(2)\big),$$

because $h^0(\mathcal{O}_{\mathbb{P}^4}(2)) = 15$, and $h^0(\mathcal{O}_C \otimes \mathcal{O}_{\mathbb{P}^4}(2)) = 12$ by the Riemann–Roch formula and Kodaira vanishing. In fact, it follows from Max Noether's theorem (see [3, p. 117]) that $C \subset \mathbb{P}^4$ is projectively normal, which implies that

$$h^0(\mathcal{I}_C \otimes \mathcal{O}_{\mathbb{P}^4}(2)) = 3.$$

On the other hand, $H^0(\mathcal{O}_{\mathbb{P}^4}(2))$ does not contain three-dimensional $\mathfrak{A}_5$-invariant subspaces by Lemma 5.3.7. The obtained contradiction shows that C is hyperelliptic.

Since C is hyperelliptic, it is a double cover of $\mathbb{P}^1$ branched over 12 points, which must be the unique $\mathfrak{A}_5$-orbit in $\mathbb{P}^1$ of length 12 (see Lemma 5.1.5). Thus, the curve C coincides with the curve Z constructed above. □

The curve described in Lemma 5.4.3 can be given by the equation

$$w^2 = yx^{11} - 11x^6 y^6 - xy^{11}$$

in the weighted projective space $\mathbb{P}(1,1,6)$ (see [122, p. 41]). Similar to the Bring's curve, it has appeared many times in the literature (see, e. g., [76, Proposition 5], and also a paper [80] by Joseph McKelvey).

Lemma 5.4.4. There exists a unique smooth irreducible hyperelliptic curve C of genus 9 with a non-trivial action of the group $\mathfrak{A}_5$.

Proof. The curve C is a double cover of $\mathbb{P}^1$ branched over the unique $\mathfrak{A}_5$-orbit in $\mathbb{P}^1$ of length 20 (see Lemma 5.1.5). $\square$

The curve described in Lemma 5.4.4 is not the only smooth irreducible curve of genus 9 with a non-trivial action of the group $\mathfrak{A}_5$. Another (non-hyperelliptic) example is described in Lemma 5.3.3(xi).

5.5 SL$_2(\mathbb{C})$ and PSL$_2(\mathbb{C})$

Recall that the group $\mathrm{SL}_2(\mathbb{C})$ has a unique d-dimensional irreducible representation $\mathbb{W}_d$ for any positive integer d, and $\mathbb{W}_d \cong \mathrm{Sym}^{d-1}(W_2)$ (in particular, $\mathbb{W}_1$ is the trivial representation). All these representations are self-dual. The center of $\mathrm{SL}_2(\mathbb{C})$ acts trivially on $\mathbb{W}_d$ with odd d, so that these are actually representations of the group $\mathrm{PSL}_2(\mathbb{C})$. All finite dimensional irreducible representations of $\mathrm{SL}_2(\mathbb{C})$ arise in this way (see, e. g., [46, §11.1]).

Lemma 5.5.1. One has

$$\mathrm{Sym}^2(\mathbb{W}_3) \cong \mathbb{W}_1 \oplus \mathbb{W}_5,$$
$$\mathrm{Sym}^2(\mathbb{W}_7) \cong \mathbb{W}_1 \oplus \mathbb{W}_5 \oplus \mathbb{W}_9 \oplus \mathbb{W}_{13},$$
$$\Lambda^2(\mathbb{W}_5) \cong \mathbb{W}_3 \oplus \mathbb{W}_7,$$
$$\mathbb{W}_2 \otimes \mathbb{W}_6 \cong \mathbb{W}_5 \oplus \mathbb{W}_7$$

and

$$\mathrm{Sym}^3(\mathbb{W}_7) \cong \mathbb{W}_3 \oplus \mathbb{W}_7^{\oplus 2} \oplus \mathbb{W}_9 \oplus \mathbb{W}_{11} \oplus \mathbb{W}_{13} \oplus \mathbb{W}_{15} \oplus \mathbb{W}_{19}.$$

Proof. One has

$$\mathrm{Sym}^2(\mathbb{W}_{d+1}) \cong \mathrm{Sym}^{d+1}(\mathbb{W}_2) \cong \bigoplus_{k=0}^{\lfloor \frac{d}{2} \rfloor} \mathbb{W}_{2d-4k+1}, \qquad (5.5.2)$$

see, e.g., Exercises 11.14 and 11.34 in [46]. In particular, the isomorphism (5.5.2) implies the first two assertions of the lemma. Also, using [46, Exercise 11.35] one derives from (5.5.2) that

$$\Lambda^2(\mathbb{W}_{d+1}) \cong \bigoplus_{k=0}^{\lfloor \frac{d}{2}-1 \rfloor} \mathbb{W}_{2d-4k-1},$$

which gives the third assertion of the lemma. The last assertion of the lemma can be obtained by a direct computation (we used Lie software [78] to perform it). □

The following computation will be used in the proof of Lemma 12.1.1.

Lemma 5.5.3. Let $U \subset \operatorname{Sym}^3(\mathbb{W}_7)$ be a 54-dimensional subrepresentation of the group $\operatorname{SL}_2(\mathbb{C})$. Suppose that U contains subrepresentations $\mathbb{W}_{19}$ and $\mathbb{W}_7$. Then
$$U \cong \mathbb{W}_7 \oplus \mathbb{W}_{13} \oplus \mathbb{W}_{15} \oplus \mathbb{W}_{19}.$$

Proof. By Lemma 5.5.1 one has

$$\operatorname{Sym}^3(\mathbb{W}_7) \cong \mathbb{W}_3 \oplus \mathbb{W}_7^{\oplus 2} \oplus \mathbb{W}_9 \oplus \mathbb{W}_{11} \oplus \mathbb{W}_{13} \oplus \mathbb{W}_{15} \oplus \mathbb{W}_{19}.$$

Since $\dim(U) = 54$, we have one of the following three possible cases:

(i) either $U \cong \mathbb{W}_3 \oplus \mathbb{W}_7^{\oplus 2} \oplus \mathbb{W}_9 \oplus \mathbb{W}_{13} \oplus \mathbb{W}_{15}$,

(ii) or $U \cong \mathbb{W}_9 \oplus \mathbb{W}_{11} \oplus \mathbb{W}_{15} \oplus \mathbb{W}_{19}$,

(iii) or $U \cong \mathbb{W}_7 \oplus \mathbb{W}_{13} \oplus \mathbb{W}_{15} \oplus \mathbb{W}_{19}$.

The assumption that $\mathbb{W}_{19}$ is a subrepresentation of U excludes case (i), while the assumption that $\mathbb{W}_7$ is a subrepresentation of U excludes case (ii). Thus we are left with case (iii), which gives the assertion of the lemma. □

The group $\operatorname{PSL}_2(\mathbb{C})$ acts faithfully on the projective space $\mathbb{P}^d \cong \mathbb{P}(\mathbb{W}_{d+1})$. In particular, the standard action of the group $\operatorname{PSL}_2(\mathbb{C})$ on $\mathbb{P}^1$ is given by an identification $\mathbb{P}^1 \cong \mathbb{P}(\mathbb{W}_2)$.

Remark 5.5.4. Let C be a curve with a non-trivial action of the group $\operatorname{SL}_2(\mathbb{C})$. Then $C \cong \mathbb{P}^1$, and the action of $\operatorname{SL}_2(\mathbb{C})$ on C is the standard action of $\operatorname{PSL}_2(\mathbb{C})$ on $\mathbb{P}^1$.

Lemma 5.5.5. Put $\mathbb{P}^d = \mathbb{P}(\mathbb{W}_{d+1})$. Then there exists a unique one-dimensional $\operatorname{PSL}_2(\mathbb{C})$-orbit C in $\mathbb{P}^d$. The curve C is a rational normal curve of degree d.

Proof. It is easy to produce a rational normal curve C_d of degree d in $\mathbb{P}^d$ that is a $\mathrm{PSL}_2(\mathbb{C})$-orbit. On the other hand, there is a unique closed $\mathrm{PSL}_2(\mathbb{C})$-orbit in $\mathbb{P}^d$ (see, e.g., [46, Claim 23.52]). □

Lemma 5.5.6. Let $C \subset \mathbb{P}^d = \mathbb{P}(\mathbb{W}_{d+1})$ be a $\mathrm{PSL}_2(\mathbb{C})$-invariant rational normal curve of degree d, and H be a hyperplane section in $\mathbb{P}^d$. Then for any non-negative integer m one has an isomorphism of $\mathrm{SL}_2(\mathbb{C})$-representations

$$H^0(\mathcal{O}_C(mH|_C)) \cong \mathbb{W}_{md+1}.$$

Proof. One has

$$H^0(\mathcal{O}_C(mH|_C)) = H^0(\mathcal{O}_{\mathbb{P}^1}(md)) \cong \mathrm{Sym}^{md}(\mathbb{W}_2) \cong \mathbb{W}_{md+1}.$$

□

Lemma 5.5.7. Let W be a non-trivial representation of the group $\mathrm{SL}_2(\mathbb{C})$ of dimension $n+1$, and let C be an $\mathrm{SL}_2(\mathbb{C})$-invariant curve in $\mathbb{P}^n = \mathbb{P}(W)$ of degree d such that the action of $\mathrm{SL}_2(\mathbb{C})$ on C is non-trivial. Then there is an $\mathrm{SL}_2(\mathbb{C})$-invariant subspace $\mathbb{P}^d \subset \mathbb{P}^n$ containing C, and C is a rational normal curve.

Proof. Let $\mathbb{P} \subset \mathbb{P}^n$ be the linear span of the curve C, and let $W' \subset W$ be the $\mathrm{SL}_2(\mathbb{C})$-invariant subspace such that $\mathbb{P} = \mathbb{P}(W')$. By Remark 5.5.4 one has $C \cong \mathbb{P}^1$, and $\mathrm{SL}_2(\mathbb{C})$ acts on C via the standard action of $\mathrm{PSL}_2(\mathbb{C})$. In particular, the action of $\mathrm{SL}_2(\mathbb{C})$ on $\mathbb{P}$ factors through the action of $\mathrm{PSL}_2(\mathbb{C})$. Replacing W with W', we may assume that $\mathbb{P} = \mathbb{P}^n$ and C is not contained in any hyperplane.

Suppose that there is a non-trivial decomposition $W = W_1 \oplus W_2$ into subrepresentations of $\mathrm{SL}_2(\mathbb{C})$, and let $\mathbb{P}^{n_1}$ and $\mathbb{P}^{n_2}$ be the corresponding $\mathrm{SL}_2(\mathbb{C})$-invariant linear subspaces of $\mathbb{P}^n$. Then $\mathbb{P}^{n_1}$ does not intersect the curve C since the action of $\mathrm{SL}_2(\mathbb{C})$ on C is transitive. Let $T \subset \mathbb{P}^n$ be a general linear subspace of $\mathbb{P}^n$ of dimension $n_1 + 1$ containing $\mathbb{P}^{n_1}$ such that $T \cap C \neq \varnothing$. Then the intersection $T \cap C$ consists of one point (indeed, otherwise one would have $C \subset T$ since the action of $\mathrm{PSL}_2(\mathbb{C})$ on pairs of different points on C is transitive). Thus the projection of C from $\mathbb{P}^{n_1}$ is an isomorphism, so that there is a $\mathrm{PSL}_2(\mathbb{C})$-equivariant embedding of C into $\mathbb{P}^{n_2}$ as a curve of degree d. Replacing $\mathbb{P}^n$ by $\mathbb{P}^{n_2}$ and iterating this process, we may assume that W is an irreducible representation of $\mathrm{SL}_2(\mathbb{C})$. Now the assertion follows from Lemma 5.5.5. □

The classification of finite subgroups in $\mathrm{SL}_2(\mathbb{C})$ and $\mathrm{PSL}_2(\mathbb{C})$ is well known (see, e.g., [115, §4.4]). Let $G \subset \mathrm{PSL}_2(\mathbb{C})$ be a finite subgroup and $\tilde{G} \subset \mathrm{SL}_2(\mathbb{C})$ be a subgroup of minimal possible order that is mapped surjectively onto G under the natural projection $\mathrm{SL}_2(\mathbb{C}) \to \mathrm{PSL}_2(\mathbb{C})$. Then G is either a cyclic group $\boldsymbol{\mu}_n$, $n \geqslant 1$, or a dihedral group $\mathfrak{D}_{2n}$, $n \geqslant 2$ (this includes the groups $\mathfrak{D}_4 \cong \boldsymbol{\mu}_2 \times \boldsymbol{\mu}_2$ and $\mathfrak{D}_6 \cong \mathfrak{S}_3$), or the tetrahedral group $\mathfrak{A}_4$, or the octahedral group $\mathfrak{S}_4$, or the icosahedral group $\mathfrak{A}_5$. The group $\tilde{G}$ in these cases is isomorphic to $\boldsymbol{\mu}_n$, $2.\mathfrak{D}_{2n}$, $2.\mathfrak{A}_4$, $2.\mathfrak{S}_4$, or $2.\mathfrak{A}_5$, respectively. A subgroup of each of these isomorphism classes is unique up to conjugation in $\mathrm{PSL}_2(\mathbb{C})$ (respectively, in $\mathrm{SL}_2(\mathbb{C})$).

Lemma 5.5.8. For any subgroup $G \subset \mathrm{PSL}_2(\mathbb{C})$ let $N(G) \subset \mathrm{PSL}_2(\mathbb{C})$ be the normalizer of G. Then one has

$$N(\mathfrak{A}_5) \cong \mathfrak{A}_5, \ N(\mathfrak{S}_3) \cong \mathfrak{D}_{12},$$
$$N(\boldsymbol{\mu}_2 \times \boldsymbol{\mu}_2) \cong N(\mathfrak{A}_4) \cong \mathfrak{S}_4.$$

Proof. Let $G \subset \mathrm{PSL}_2(\mathbb{C})$ be a finite subgroup, and let $\tilde{G} \subset \mathrm{SL}_2(\mathbb{C})$ be a subgroup of minimal possible order that is mapped surjectively onto G under the natural projection $\mathrm{SL}_2(\mathbb{C}) \to \mathrm{PSL}_2(\mathbb{C})$. Suppose that G is not a cyclic group. Then the natural two-dimensional representation of $\tilde{G}$ is irreducible. In particular, by Schur's lemma the centralizer $C(\tilde{G})$ of $\tilde{G}$ in $\mathrm{SL}_2(\mathbb{C})$ consists of scalar matrices and thus is finite. Hence the normalizer

$$N(\tilde{G}) \subset \mathrm{SL}_2(\mathbb{C})$$

is also finite, which in turn implies that the normalizer

$$N(G) \subset \mathrm{PSL}_2(\mathbb{C})$$

is finite as well. The rest follows by classification of finite subgroups of $\mathrm{PSL}_2(\mathbb{C})$. □

The following description of orbits of finite subgroups of $\mathrm{PSL}_2(\mathbb{C})$ acting on $\mathbb{P}^1$ can be found, for example, in [115, §4.4].

Lemma 5.5.9. Let $G \subset \mathrm{PSL}_2(\mathbb{C})$ be a finite subgroup. Consider the natural action of G on $\mathbb{P}^1$. Let $\Sigma(G)$ be the set of all G-orbits of length less than $|G|$ on $\mathbb{P}^1$. One has the following possibilities:

(i) if $G \cong \boldsymbol{\mu}_n$, then $\Sigma(G)$ consists of two G-fixed points;

(ii) if $G \cong \mathfrak{D}_{2n}$, then $\Sigma(G)$ consists of one G-orbit of length 2 and two G-orbits of length n;

CREMONA GROUPS AND THE ICOSAHEDRON 83

(iii) if $G \cong \mathfrak{A}_4$, then $\Sigma(G)$ consists of two G-orbits of length 4 and one G-orbit of length 6;

(iv) if $G \cong \mathfrak{S}_4$, then $\Sigma(G)$ consists of one G-orbit of length 6, one G-orbit of length 8 and one G-orbit of length 12;

(v) if $G \cong \mathfrak{A}_5$, then $\Sigma(G)$ consists of one G-orbit of length 12, one G-orbit of length 20, and one G-orbit of length 30 (cf. Lemma 5.1.5).

Another way to approach representations of the group $\mathrm{SL}_2(\mathbb{C})$ is to consider the action of the Lie algebra $\mathfrak{sl}_2(\mathbb{C})$. Recall (see, e. g., [46, §11]) that $\mathbb{W}_2$ can be identified with the vector space of linear forms in two variables, say, x and y, and $\mathbb{W}_n$ can be identified with the vector space of homogeneous polynomials of degree $n-1$ in x and y. The action $\mathfrak{sl}_2(\mathbb{C})$ is given by the operators

$$\mathrm{e} = x\frac{\partial}{\partial y}, \quad \mathrm{f} = y\frac{\partial}{\partial x}$$

and

$$\mathrm{h} = x\frac{\partial}{\partial x} - y\frac{\partial}{\partial y}.$$

Denote by w_i, $i = -n+1, -n+3, \ldots, n-3, n-1$, the monomial

$$x^{\frac{n-1+i}{2}} y^{\frac{n-1-i}{2}} \in \mathbb{W}_n.$$

Then

$$\mathrm{e}\colon w_i \mapsto \frac{n-1-i}{2} w_{i+2}, \quad \mathrm{f}\colon w_i \mapsto \frac{n-1+i}{2} w_{i-2}$$

and

$$\mathrm{h}\colon w_i \mapsto iw_i.$$

In particular, w_{n-1} is a highest weight vector in $\mathbb{W}_n$. This basis gives rise to the basis

$$\left\{w_i \wedge w_j \mid i > j, \quad i,j \in \{-n+1, -n+3, \ldots, n-3, n-1\}\right\}$$

in $\Lambda^2(\mathbb{W}_n)$. Note that

$$\mathrm{h}\colon w_i \wedge w_j \mapsto (i+j) w_i \wedge w_j.$$

Let us consider the $\mathfrak{sl}_2(\mathbb{C})$-representation $\mathbb{W}_5$ in more detail. By Lemma 5.5.1 one has $\Lambda^2(\mathbb{W}_5) \cong \mathbb{W}_3 \oplus \mathbb{W}_7$. Let $U' \subset \Lambda^2(\mathbb{W}_5)$ be the unique three-dimensional $\mathfrak{sl}_2(\mathbb{C})$-subrepresentation, and let w' be a highest weight vector in U.

Lemma 5.5.10. One has
$$w' = w_4 \wedge w_{-2} - 3 w_2 \wedge w_0$$
up to scaling. The bi-vector w is not decomposable.

Proof. One must have $hw' = 2w'$, so that
$$w' = \alpha w_4 \wedge w_{-2} + \beta w_2 \wedge w_0$$
for some complex numbers α and β. On the other hand, we have
$$0 = ew' = (3\alpha + \beta) w_4 \wedge w_0,$$
so that $\beta = -3\alpha$.

If the bi-vector w' was decomposable, then one would have
$$0 = w' \wedge w' = -3 w_4 \wedge w_2 \wedge w_0 \wedge w_{-2},$$
which is a contradiction. $\square$

Corollary 5.5.11. The vector space U' is generated by the vectors
$$w' = w_4 \wedge w_{-2} - 3 w_2 \wedge w_0, \quad fw' = w_4 \wedge w_{-4} - 2 w_2 \wedge w_{-2}$$
and
$$\frac{f^2}{2} w' = w_2 \wedge w_{-4} - 3 w_0 \wedge w_{-2}.$$

Similarly, let $U \subset \Lambda^2(\mathbb{W}_5)$ be the unique seven-dimensional $\mathfrak{sl}_2(\mathbb{C})$-subrepresentation, and let w be a highest weight vector in U. Clearly, one has $w = w_4 \wedge w_2$. Therefore, we obtain

Lemma 5.5.12. The vector space U is generated by the vectors
$$u_6 = w = w_4 \wedge w_2, \quad u_4 = \frac{f}{3} w = w_4 \wedge w_0,$$
$$u_2 = \frac{f^2}{6} w = 2 w_2 \wedge w_0 + w_4 \wedge w_{-2},$$
$$u_0 = \frac{f^3}{6} w = 8 w_2 \wedge w_{-2} + w_4 \wedge w_{-4},$$
$$u_{-2} = \frac{f^4}{72} w = 2 w_0 \wedge w_{-2} + w_2 \wedge w_{-4},$$
$$u_{-4} = \frac{f^5}{360} w = w_0 \wedge w_{-4}$$
and
$$u_{-6} = \frac{f^6}{360} w = w_{-2} \wedge w_{-4}.$$

CREMONA GROUPS AND THE ICOSAHEDRON 85

In Chapter 7 we will study the Grassmannian $\mathrm{Gr}(2,\mathbb{W}_5)$ embedded into the projectivization of $\Lambda^2(\mathbb{W}_5)$. Lemma 5.5.12 allows one to deal with fixed points of (some) subgroups of $\mathrm{SL}_2(\mathbb{C})$ on $\mathrm{Gr}(2,\mathbb{W}_5)$. More precisely, denote by V the intersection of $\mathrm{Gr}(2,\mathbb{W}_5) \subset \Lambda^2(\mathbb{W}_5)$ with the projectivization of the subrepresentation U.

Lemma 5.5.13. *Fix a subgroup $G \cong \mu_3$ in $\mathrm{PSL}_2(\mathbb{C})$. Then the set of fixed points of G on V is a union of a conic and two points.*

Proof. Assume the notation of Lemma 5.5.12. Since G is unique up to conjugation, one may assume that G is the subgroup that acts trivially on the subspace U_0 of U generated by u_6, u_0 and u_{-6}, acts by a scalar equal to a non-trivial cubic root ω of 1 on the subspace U_1 of U generated by u_4 and u_{-2}, and acts by a scalar equal to ω^2 on the subspace U_2 of U generated by u_2 and u_{-4}. The fixed points of G in $\mathbb{P}(U)$ are the points of the projectivizations of the subspaces U_0, U_1, and U_2, while the fixed points of G on V correspond to the decomposable bi-vectors in the latter subspaces.

Let w be a decomposable bi-vector in U_0. Then

$$w = \alpha w_4 \wedge w_2 + \beta(8 w_2 \wedge w_{-2} + w_4 \wedge w_{-4}) + \gamma w_{-2} \wedge w_{-4}$$

for some complex numbers α, β, and γ. One has

$$0 = w \wedge w = (\alpha\gamma + 8\beta^2) w_4 \wedge w_2 \wedge w_{-2} \wedge w_{-4},$$

which is equivalent to $\alpha\gamma + 8\beta^2 = 0$. Thus, the intersection $V \cap \mathbb{P}(U_0)$ is a conic.

Let w be a decomposable bi-vector in U_1. Then

$$w = \alpha w_4 \wedge w_0 + \beta(2 w_0 \wedge w_{-2} + w_2 \wedge w_{-4})$$

for some complex numbers α and β. One has

$$0 = w \wedge w = -\alpha\beta w_4 \wedge w_2 \wedge w_0 \wedge w_{-4} + 2\beta^2 w_2 \wedge w_0 \wedge w_{-2} \wedge w_{-4},$$

which is equivalent to $\beta = 0$. Thus, the intersection $V \cap \mathbb{P}(U_1)$ is a point.

Let w be a decomposable bi-vector in U_2. Then

$$w = \alpha(2 w_2 \wedge w_0 + w_4 \wedge w_{-2}) + \beta w_0 \wedge w_{-4}$$

for some complex numbers α and β. One has

$$0 = w \wedge w = 2\alpha^2 w_4 \wedge w_2 \wedge w_0 \wedge w_{-2} - \alpha\beta w_4 \wedge w_0 \wedge w_{-2} \wedge w_{-4},$$

which is equivalent to $\alpha = 0$. Thus, the intersection $V \cap \mathbb{P}(U_2)$ is a point. $\square$

In a similar way we prove

Lemma 5.5.14. Fix a subgroup $G \cong \boldsymbol{\mu}_2$ in $\mathrm{PSL}_2(\mathbb{C})$. Then the set of fixed points of G on V is a disjoint union of a line and a twisted cubic.

Proof. Assume the notation of Lemma 5.5.12. Since G is unique up to conjugation, one may assume that G is the subgroup that acts trivially on the subspace U_+ of U generated by u_4, u_0, and u_{-4}, and acts by a scalar equal to a -1 on the subspace U_- of U generated by u_6, u_2, u_{-2}, and u_{-6}. The fixed points of G in $\mathbb{P}(U)$ are the points of the projectivizations of the subspaces U_+ and U_-, while the fixed points of G on V correspond to the decomposable bi-vectors in the latter subspaces.

Let w be a decomposable bi-vector in U_+. Then

$$w = \alpha w_4 \wedge w_0 + \beta(8w_2 \wedge w_{-2} + w_4 \wedge w_{-4}) + \gamma w_0 \wedge w_{-4}$$

for some complex numbers α, β and γ. One has

$$0 = w \wedge w = -8\alpha\beta w_4 \wedge w_2 \wedge w_0 \wedge w_{-2} + \\ + 8\beta^2 w_4 \wedge w_2 \wedge w_{-2} \wedge w_{-4} - 8\beta\gamma w_2 \wedge w_0 \wedge w_{-2} \wedge w_{-4},$$

which is equivalent to $\beta = 0$. Thus, the intersection $V \cap \mathbb{P}(U_+)$ is a line.

Let w be a decomposable bi-vector in U_-. Then

$$w = \alpha w_4 \wedge w_2 + \beta(2w_2 \wedge w_0 + w_4 \wedge w_{-2}) + \\ + \gamma(2w_0 \wedge w_{-2} + w_2 \wedge w_{-4}) + \delta w_{-2} \wedge w_{-4}$$

for some complex numbers α, β, γ, and δ. One has

$$0 = w \wedge w = 2(\alpha\gamma + \beta^2) w_4 \wedge w_2 \wedge w_0 \wedge w_{-2} + \\ + (\alpha\delta - \beta\gamma) w_4 \wedge w_2 \wedge w_{-2} \wedge w_{-4} + 2(\beta\delta + \gamma^2) w_2 \wedge w_0 \wedge w_{-2} \wedge w_{-4},$$

which is equivalent to

$$\alpha\gamma + \beta^2 = \alpha\delta - \beta\gamma = \beta\delta + \gamma^2 = 0.$$

Thus, the intersection $V \cap \mathbb{P}(U_-)$ is a twisted cubic. □

5.6 Binary icosahedral group

As was already mentioned in §5.5, there is a subgroup isomorphic to $\mathfrak{A}_5$ in the group $\mathrm{PSL}_2(\mathbb{C})$. All such subgroups are conjugate in $\mathrm{PSL}_2(\mathbb{C})$, but there are two *embeddings*

$$\mathfrak{A}_5 \hookrightarrow \mathrm{PSL}_2(\mathbb{C})$$

up to conjugation. They are related by an outer automorphism of the group $\mathfrak{A}_5$ that is provided by (any) odd element of the group $\mathrm{Aut}(\mathfrak{A}_5) \cong \mathfrak{S}_5$. The fact that there are twice as many embeddings $\mathfrak{A}_5 \hookrightarrow \mathrm{PSL}_2(\mathbb{C})$ as subgroups of $\mathrm{PSL}_2(\mathbb{C})$ isomorphic to $\mathfrak{A}_5$ is due to the absence of a subgroup isomorphic to $\mathfrak{S}_5$ in $\mathrm{PSL}_2(\mathbb{C})$.

Two non-conjugate embeddings of $\mathfrak{A}_5$ to $\mathrm{PSL}_2(\mathbb{C})$ correspond to two non-conjugate embeddings

$$2.\mathfrak{A}_5 \hookrightarrow \mathrm{SL}_2(\mathbb{C})$$

of the non-trivial central extension $2.\mathfrak{A}_5$ of the group $\mathfrak{A}_5$, i.e., to two non-isomorphic two-dimensional faithful representations U_2 and U_2' of the group $2.\mathfrak{A}_5$. Note that the *images* of the latter two embeddings are conjugate subgroups of $\mathrm{SL}_2(\mathbb{C})$.

Any two-dimensional representation of the group $2.\mathfrak{A}_5$ different from U_2 and U_2' is trivial (see, e.g., [27, p. 2]). In particular, for any (not necessarily non-trivial) action of the group $\mathfrak{A}_5$ on $\mathbb{P}^1$ there exists a unique two-dimensional representation U of the group $2.\mathfrak{A}_5$ such that there is an $\mathfrak{A}_5$-equivariant identification $\mathbb{P}^1 \cong \mathbb{P}(U)$.

An inclusion $j\colon 2.\mathfrak{A}_5 \hookrightarrow \mathrm{SL}_2(\mathbb{C})$ gives a structure of $2.\mathfrak{A}_5$-representation on the space $\mathbb{W}_d$ for any $d \geqslant 1$, and also a structure of $\mathfrak{A}_5$-representation on $\mathbb{W}_d$ with odd d. One can check that $\mathbb{W}_3$ is identified with either W_3 or W_3' as a representation of the group $\mathfrak{A}_5$, depending on the choice of j. Therefore, we adopt the following convention that we will use throughout the next chapters.

Convention 5.6.1. If the converse is not specified, when speaking of an embedding of $\mathfrak{A}_5$ to $\mathrm{PSL}_2(\mathbb{C})$ and of an embedding of $2.\mathfrak{A}_5$ to $\mathrm{SL}_2(\mathbb{C})$, we always choose one that gives an $\mathfrak{A}_5$-equivariant identification $\mathbb{W}_3 \cong W_3'$.

By a natural abuse of notation, we will sometimes denote by I, W_3, W_3', W_4, and W_5 the corresponding representations of the group $2.\mathfrak{A}_5$.

Lemma 5.6.2. One has the following isomorphisms of $2.\mathfrak{A}_5$-representations

$$\operatorname{Res}_{2.\mathfrak{A}_5}^{\operatorname{SL}_2(\mathbb{C})} \mathbb{W}_3 \cong W_3',$$
$$\operatorname{Res}_{2.\mathfrak{A}_5}^{\operatorname{SL}_2(\mathbb{C})} \mathbb{W}_5 \cong W_5,$$
$$\operatorname{Res}_{2.\mathfrak{A}_5}^{\operatorname{SL}_2(\mathbb{C})} \mathbb{W}_7 \cong W_3 \oplus W_4,$$
$$\operatorname{Res}_{2.\mathfrak{A}_5}^{\operatorname{SL}_2(\mathbb{C})} \mathbb{W}_9 \cong W_4 \oplus W_5,$$
$$\operatorname{Res}_{2.\mathfrak{A}_5}^{\operatorname{SL}_2(\mathbb{C})} \mathbb{W}_{13} \cong I \oplus W_3' \oplus W_4 \oplus W_5,$$
$$\operatorname{Res}_{2.\mathfrak{A}_5}^{\operatorname{SL}_2(\mathbb{C})} \mathbb{W}_{15} \cong W_3 \oplus W_3' \oplus W_4 \oplus W_5,$$

and

$$\operatorname{Res}_{2.\mathfrak{A}_5}^{\operatorname{SL}_2(\mathbb{C})} \mathbb{W}_{19} \cong W_3 \oplus W_3' \oplus W_4^{\oplus 2} \oplus W_5.$$

Proof. The first isomorphism is just Convention 5.6.1. The rest is mostly a direct computation (we used GAP software [48] to perform it). □

The group $2.\mathfrak{A}_5$ has a unique faithful four-dimensional irreducible representation U_4 (see [27, p. 2]).

Lemma 5.6.3. Let U be an irreducible two-dimensional representation of the group $2.\mathfrak{A}_5$. Then

(o) the representation U is self-dual;

(i) one has $U_4 \cong \operatorname{Sym}^3(U)$;

(ii) the space $\operatorname{Sym}^r(U)$ does not contain one-dimensional $2.\mathfrak{A}_5$-invariant subspaces for $1 \leqslant r \leqslant 11$;

(iii) the space $U \otimes \operatorname{Sym}^7(U)$ does not contain one-dimensional $2.\mathfrak{A}_5$-invariant subspaces;

(iv) the space $U \otimes \operatorname{Sym}^{11}(U)$ contains a unique one-dimensional $2.\mathfrak{A}_5$-invariant subspace;

(v) the space $\operatorname{Sym}^2(U) \otimes \operatorname{Sym}^{10}(U)$ contains a unique one-dimensional $2.\mathfrak{A}_5$-invariant subspace;

(vi) the space $\operatorname{Sym}^2(U) \otimes \operatorname{Sym}^{18}(U)$ contains a unique one-dimensional $2.\mathfrak{A}_5$-invariant subspace;

(vii) the space $\operatorname{Sym}^3(U) \otimes \operatorname{Sym}^9(U)$ contains a unique one-dimensional $2.\mathfrak{A}_5$-invariant subspace;

CREMONA GROUPS AND THE ICOSAHEDRON

(viii) the space $\operatorname{Sym}^4(U) \otimes \operatorname{Sym}^8(U)$ contains a unique one-dimensional $2.\mathfrak{A}_5$-invariant subspace.

Proof. Assertion (o) is basic representation theory of $2.\mathfrak{A}_5$ (see, e.g., [27, p. 2]). The rest is obtained by a direct computation (we used GAP software [48] to perform it). Note also that one can derive assertion (ii) from the fact that the group $\mathfrak{A}_5$ acting on $\mathbb{P}^1 \cong \mathbb{P}(U)$ does not have orbits of length less than 12 (see Lemma 5.1.4). □

Lemma 5.6.4. *The following assertions hold:*

(i) *the space $U_2 \otimes \operatorname{Sym}^k(U_2')$ does not contain one-dimensional $2.\mathfrak{A}_5$-invariant subspaces for $1 \leqslant k \leqslant 6$;*

(ii) *the space $U_2 \otimes \operatorname{Sym}^7(U_2')$ contains a unique one-dimensional $2.\mathfrak{A}_5$-invariant subspace;*

(iii) *the space $\operatorname{Sym}^2(U_2) \otimes \operatorname{Sym}^6(U_2')$ contains a unique one-dimensional $2.\mathfrak{A}_5$-invariant subspace;*

(iv) *the space $U_2 \otimes \operatorname{Sym}^{13}(U_2')$ contains a unique one-dimensional $2.\mathfrak{A}_5$-invariant subspace.*

Proof. This is a direct computation (we used GAP software [48] to perform it). □

5.7 Symmetric group

In this section we use the computations involving representations of the symmetric group $\mathfrak{S}_5$ to derive some information about a couple of particular representations of the group $\mathfrak{A}_5$ that will be used in §6.2.

Recall (see, e.g., [46, §3.1]) that the group $\mathfrak{S}_5$ has two four-dimensional irreducible representations U_4 and U_4', two five-dimensional irreducible representations U_5 and U_5' (the first of them is a representation of $\mathfrak{S}_5$ in $\operatorname{SL}_5(\mathbb{C})$, and the second one is a representation of $\mathfrak{S}_5$ in $\operatorname{GL}_5(\mathbb{C})$ whose image is not contained in $\operatorname{SL}_5(\mathbb{C})$), and a unique six-dimensional irreducible representation U_6. Moreover, one has

$$\operatorname{Res}_{\mathfrak{A}_5}^{\mathfrak{S}_5} U_4 \cong \operatorname{Res}_{\mathfrak{A}_5}^{\mathfrak{S}_5} U_4' \cong W_4, \quad \operatorname{Res}_{\mathfrak{A}_5}^{\mathfrak{S}_5} U_5 \cong \operatorname{Res}_{\mathfrak{A}_5}^{\mathfrak{S}_5} U_5' \cong W_5 \quad (5.7.1)$$

and

$$\operatorname{Res}_{\mathfrak{A}_5}^{\mathfrak{S}_5} U_6 \cong W_3 \oplus W_3'. \quad (5.7.2)$$

The remaining irreducible representations of $\mathfrak{S}_5$ are the trivial representation $I_{\mathfrak{S}_5}$ and the sign representation Sgn. Clearly, they both restrict to the trivial representation of the subgroup $\mathfrak{A}_5$ of $\mathfrak{S}_5$.

For the sake of the following two lemmas we denote by $I_{\mathfrak{A}_4}$ the trivial representation of the group $\mathfrak{A}_4$, and by J and J' its non-trivial one-dimensional representations. The following computation will be used in the proof of Lemma 6.2.8.

Lemma 5.7.3. Let $U \subset \operatorname{Sym}^2(U_6)$ be a 16-dimensional subrepresentation of the group $\mathfrak{S}_5$ that contains two one-dimensional subrepresentations. Then
$$\operatorname{Res}_{\mathfrak{A}_5}^{\mathfrak{S}_5} U \cong I^{\oplus 2} \oplus W_4 \oplus W_5^{\oplus 2}.$$
One has
$$\dim \operatorname{Hom}\bigl(\operatorname{Res}_{\mathfrak{A}_4}^{\mathfrak{S}_5} U, J\bigr) = \dim \operatorname{Hom}\bigl(\operatorname{Res}_{\mathfrak{A}_4}^{\mathfrak{S}_5} U, J'\bigr) = 2.$$
and
$$\dim \operatorname{Hom}\bigl(\operatorname{Res}_{\mathfrak{A}_4}^{\mathfrak{S}_5} U, I_{\mathfrak{A}_4}\bigr) = 3.$$

Proof. A direct computation shows that
$$\operatorname{Sym}^2(U_6) \cong I_{\mathfrak{S}_5} \oplus \operatorname{Sgn} \oplus U_4 \oplus U_5^{\oplus 2} \oplus U_5'.$$
Since U contains two one-dimensional subrepresentations, it must split as a sum of $\mathfrak{S}_5$-representations I, Sgn, U_4, and two five-dimensional irreducible $\mathfrak{S}_5$-representations (the latter two may both be isomorphic to U_5, or they may be different representations U_5 and U_5'). Now the first assertion of the lemma follows from (5.7.1). The remaining assertions follow from Corollaries 5.2.3(vi) and 5.2.4(ii). □

The following computation will be used in the proof of Lemma 6.2.6.

Lemma 5.7.4. Let $U \subset \operatorname{Sym}^3(U_6)$ be a 31-dimensional subrepresentation of the group $\mathfrak{S}_5$. Then
$$\operatorname{Res}_{\mathfrak{A}_5}^{\mathfrak{S}_5} U \cong W_3^{\oplus 3} \oplus W_3'^{\oplus 3} \oplus W_4^{\oplus 2} \oplus W_5.$$
Furthermore, the restriction of U to a subgroup $\mathfrak{A}_4 \subset \mathfrak{S}_5$ contains a unique subrepresentation isomorphic to J, a unique subrepresentation isomorphic to J', and
$$\dim \operatorname{Hom}\bigl(\operatorname{Res}_{\mathfrak{A}_4}^{\mathfrak{S}_5} U, I_{\mathfrak{A}_4}\bigr) = 2.$$

Proof. One computes
$$\mathrm{Sym}^3(U_6) \cong U_6^{\oplus 5} \oplus U_4^{\oplus 2} \oplus U_4'^{\oplus 2} \oplus U_5 \oplus U_5'.$$
It is straightforward to check that the 31-subrepresentation U must split as a sum of three $\mathfrak{S}_5$-representations U_6, one five-dimensional irreducible $\mathfrak{S}_5$-representation (which might be U_5 or U_5'), and two four-dimensional irreducible $\mathfrak{S}_5$-representations (each of them might be U_4 or U_4'). Now the first assertion of the lemma follows from (5.7.1) and (5.7.2). The remaining assertions follow from Corollaries 5.2.2(vi), 5.2.3(vi), and 5.2.4(ii). □

5.8 Dihedral group

In this section we make a couple of observations concerning the dihedral groups contained in $\mathfrak{A}_5$.

The group $\mathrm{PSL}_2(\mathbb{C})$ contains a unique dihedral subgroup $\mathfrak{D}_{10}$ up to conjugation, and there are two embeddings
$$\mathfrak{D}_{10} \hookrightarrow \mathrm{PSL}_2(\mathbb{C})$$
up to conjugation (cf. §5.6). There are two ways to lift this subgroup to $\mathrm{GL}_2(\mathbb{C})$. One of them appears from an irreducible two-dimensional representation of the group $\mathfrak{D}_{10}$. There are two representations like this, so that there are two embeddings
$$\mathfrak{D}_{10} \hookrightarrow \mathrm{GL}_2(\mathbb{C})$$
up to conjugation, but there is only one subgroup isomorphic to $\mathfrak{D}_{10}$ in $\mathrm{GL}_2(\mathbb{C})$ up to conjugation. This subgroup is not contained in $\mathrm{SL}_2(\mathbb{C}) \subset \mathrm{GL}_2(\mathbb{C})$. Also, there is a non-trivial central extension $2.\mathfrak{D}_{10}$ of the group $\mathfrak{D}_{10}$. There are two embeddings
$$2.\mathfrak{D}_{10} \hookrightarrow \mathrm{GL}_2(\mathbb{C})$$
up to conjugation, but again only one subgroup isomorphic to $2.\mathfrak{D}_{10}$ in $\mathrm{GL}_2(\mathbb{C})$ up to conjugation. This subgroup is contained in $\mathrm{SL}_2(\mathbb{C})$. For each embedding $\mathfrak{D}_{10} \hookrightarrow \mathrm{PSL}_2(\mathbb{C})$ there is one lift to $\mathrm{GL}_2(\mathbb{C})$ of each of these two types.

Note also that there is a unique embedding of the group $\mathfrak{S}_3$ into $\mathrm{PSL}_2(\mathbb{C})$ up to conjugation, and two lifts of $\mathfrak{S}_3$ to the group $\mathrm{GL}_2(\mathbb{C})$; one of them comes from the two-dimensional irreducible representation of $\mathfrak{S}_3$, and its image is not contained in $\mathrm{SL}_2(\mathbb{C})$, while the other comes from the two-dimensional irreducible representation of a non-trivial central extension $2.\mathfrak{S}_3$, and its image is contained in $\mathrm{SL}_2(\mathbb{C})$.

Lemma 5.8.1. Let U_1 and U_2 be two different irreducible two-dimensional representation of the group $\mathfrak{D}_{10}$, and let $I_{\mathfrak{D}_{10}}$ be the trivial representation of the group $\mathfrak{D}_{10}$. Then

(i) the space $U_1 \otimes \mathrm{Sym}^2(U_1)$ does not contain one-dimensional $\mathfrak{D}_{10}$-invariant subspaces;

(ii) the space $U_1 \otimes \mathrm{Sym}^2(U_2)$ contains a unique one-dimensional $\mathfrak{D}_{10}$-invariant subspace;

(iii) the space $U_1 \otimes \mathrm{Sym}^3(U_2)$ contains a unique one-dimensional $\mathfrak{D}_{10}$-invariant subspace.

(iv) one has
$$\mathrm{Sym}^6(U_1) \cong I_{\mathfrak{D}_{10}} \oplus U_1^{\oplus 2} \oplus U_2.$$

Proof. This is a direct computation (we used GAP software [48] to perform it). □

Lemma 5.8.2. There exists a projective space $\mathbb{P}^3$ with a faithful action of the group $G \cong \mathfrak{D}_{10}$ and a smooth G-invariant rational curve $C \subset \mathbb{P}^3$ of degree 4 such that C is contained in a smooth G-invariant quadric $R \subset \mathbb{P}^3$.

Proof. Consider the projective space $\mathbb{P}^4 = \mathbb{P}(\mathbb{W}_5)$ acted on by the group $\mathrm{PSL}_2(\mathbb{C})$. There is a $\mathrm{PSL}_2(\mathbb{C})$-invariant rational normal curve $C_4 \subset \mathbb{P}^4$ of degree 4 by Lemma 5.5.5. As a representation of the group $2.\mathfrak{D}_{10} \subset \mathrm{SL}_2(\mathbb{C})$ the vector space $\mathbb{W}_5$ splits as a sum of one one-dimensional and two two-dimensional representations by Lemma 5.6.2 and Corollary 5.2.4(i). In particular, there exists a unique point $P \in \mathbb{P}^4$ that is invariant with respect to the action of the group G. Note that the point P is not contained in the curve C_4, since $C_4 \cong \mathbb{P}^1$ has no G-fixed points by Lemma 5.5.9(ii).

Let $\beta \colon \mathbb{P}^4 \dashrightarrow \mathbb{P}^3$ be a linear projection from the point P. Then $\mathbb{P}^3$ carries an action of G such that β is G-equivariant, and $\mathbb{P}^3$ has no G-fixed points. The restriction of β to C_4 is a morphism since $P \notin C_4$.

Put $C = \beta(C_4)$. Then $C \subset \mathbb{P}^3$ is a rational curve of degree 4 that is not contained in a plane. In particular, C has at most one singular point, and hence C is smooth since there are no G-fixed points in $\mathbb{P}^3$.

Note that a smooth rational curve of degree 4 in $\mathbb{P}^3$ is always contained in a unique (irreducible) quadric. Let $R \subset \mathbb{P}^3$ be the quadric that contains C. Then R is G-invariant. Moreover, R is smooth since there are no G-fixed points in $\mathbb{P}^3$. □

CREMONA GROUPS AND THE ICOSAHEDRON 93

We will also need information about some representations of the group $\mathfrak{A}_5$ induced from its subgroups $\mathfrak{S}_3$ and $\mathfrak{D}_{10}$.

Lemma 5.8.3. Let $I_{\mathfrak{S}_3}$ be the trivial representation of the group $\mathfrak{S}_3$, and J be the non-trivial one-dimensional representation of $\mathfrak{S}_3$. Then

(o) one has $\mathrm{Sym}^3(J) \cong J$;

(i) the $\mathfrak{A}_5$-representation $\mathrm{Ind}_{\mathfrak{S}_3}^{\mathfrak{A}_5} I_{\mathfrak{S}_3}$ does not contain a subrepresentation $W_3 \oplus W_4$;

(ii) the $\mathfrak{A}_5$-representation $\mathrm{Ind}_{\mathfrak{S}_3}^{\mathfrak{A}_5} J$ does not contain a trivial subrepresentation I.

Proof. Assertion (o) is obvious.

Suppose that one has an embedding

$$W_3 \oplus W_4 \hookrightarrow \mathrm{Ind}_{\mathfrak{S}_3}^{\mathfrak{A}_5} I_{\mathfrak{S}_3}.$$

Since the restriction of the trivial $\mathfrak{A}_5$-representation I to a subgroup $\mathfrak{S}_3$ is the trivial $\mathfrak{S}_3$-representation $I_{\mathfrak{S}_3}$, it follows from Frobenius reciprocity (see, e.g., [46, Proposition 3.17]) that there is an embedding

$$I \hookrightarrow \mathrm{Ind}_{\mathfrak{S}_3}^{\mathfrak{A}_5} I_{\mathfrak{S}_3}.$$

This shows that there is an isomorphism of $\mathfrak{A}_5$-representations

$$H^0(\mathcal{O}_\Sigma \otimes \mathcal{O}_{V_5}(H)) \cong I \oplus W_3 \oplus W_4 \oplus U,$$

where U is a non-trivial two-dimensional $\mathfrak{A}_5$-representation. The latter is a contradiction (see Table 5.4), which gives assertion (i).

Finally, Frobenius reciprocity implies

$$\dim \mathrm{Hom}_{\mathfrak{A}_5}\left(\mathrm{Ind}_{\mathfrak{S}_3}^{\mathfrak{A}_5} J, I\right) = \dim \mathrm{Hom}_{\mathfrak{S}_3}(J, I_{\mathfrak{S}_3}) = 0,$$

which gives assertion (ii). □

Lemma 5.8.4. Let U_1 and U_2 be two different irreducible two-dimensional representations of the group $\mathfrak{D}_{10}$, and let $I_{\mathfrak{D}_{10}}$ be the trivial representation of $\mathfrak{D}_{10}$. Put $U = \mathrm{Sym}^3(U_1)$. Then

$$\dim \mathrm{Hom}_{\mathfrak{A}_5}\left(\mathrm{Ind}_{\mathfrak{D}_{10}}^{\mathfrak{A}_5} U, W_5\right) = 4.$$

Proof. Put
$$U' = \operatorname{Res}^{\mathfrak{A}_5}_{\mathfrak{D}_{10}} W_5.$$
Then one has
$$U' \cong I_{\mathfrak{D}_{10}} \oplus U_1 \oplus U_2$$
by Corollary 5.2.4(i). By Lemma 5.8.1(iv) we know that
$$U \cong I_{\mathfrak{D}_{10}} \oplus U_1^{\oplus 2} \oplus U_2.$$
This gives
$$\dim \operatorname{Hom}_{\mathfrak{D}_{10}}(U, U') = 4,$$
and the assertion follows by Frobenius reciprocity. □

Remark 5.8.5. Let R be an irreducible two-dimensional representation of the group $2.\mathfrak{D}_{10}$. Put $U = \operatorname{Sym}^6(R)$. Then U is a representation of the group $\mathfrak{D}_{10}$, and U does not contain a trivial subrepresentation by [115, Exercise 4.5.2(2)]. Therefore, the $\mathfrak{A}_5$-representation $\operatorname{Ind}^{\mathfrak{A}_5}_{\mathfrak{D}_{10}} U$ does not contain a trivial subrepresentation by Frobenius reciprocity.

Let us conclude this section with the following analog of Lemma 5.1.5 that will be used in the proof of Lemma 15.3.5.

Lemma 5.8.6. *Let C be a smooth irreducible curve of genus $g \leqslant 5$ that is faithfully acted on by the group $\mathfrak{D}_{10}$. Suppose that C does not contain $\mathfrak{D}_{10}$-orbits of length 2. Then $g = 1$.*

Proof. It follows from Lemma 4.4.1 that any $\mathfrak{D}_{10}$-orbit on C is of length 5 or 10. Put $\bar{C} = C/\mathfrak{D}_{10}$. Then $\bar{C}$ is a smooth curve of genus $\bar{g}$. Moreover, the Riemann–Hurwitz formula gives
$$2g - 2 = 10(2\bar{g} - 2) + 5a_5,$$
where a_5 is the number of $\mathfrak{D}_{10}$-orbits in C of length 5. This shows that $g = 1$. □

Chapter 6

Surfaces with icosahedral symmetry

In this chapter we study the action of the group $\mathfrak{A}_5$ on rational surfaces and $K3$ surfaces.

6.1 Projective plane

In this section we study the action of the group $\mathfrak{A}_5$ on the projective plane and extend the results obtained in Lemma 5.3.1. Everything discussed here is classical and well known.

Let us fix a faithful action of the group $\mathfrak{A}_5$ on $\mathbb{P}^2$.

Remark 6.1.1. There is an $\mathfrak{A}_5$-equivariant identification $\mathbb{P}^2 \cong \mathbb{P}(W_3)$ or $\mathbb{P}^2 \cong \mathbb{P}(W_3')$. Indeed, let G be a subgroup in $\mathrm{SL}_3(\mathbb{C})$ such that the image of G in $\mathrm{PGL}_3(\mathbb{C})$ is our subgroup $\mathfrak{A}_5$. Then we may assume that $G \cong \mathfrak{A}_5$, since there are no non-trivial extensions of $\mathfrak{A}_5$ by $\boldsymbol{\mu}_3$. Now everything follows from the representation theory of $\mathfrak{A}_5$ (see Table 5.4). In particular, there is a unique subgroup $\mathfrak{A}_5 \subset \mathrm{PGL}_3(\mathbb{C})$ up to conjugation. Note that there is no faithful action of the group $\mathfrak{S}_5$ on $\mathbb{P}^2$ (see, for example, [83]).

Keeping in mind that $\mathbb{P}^2 \cong \mathbb{P}(W_3)$ or $\mathbb{P}^2 \cong \mathbb{P}(W_3')$, we assume the notation of Lemma 5.3.1. We denote by Σ_6, Σ_{10}, Σ_{12}, Σ_{15}, and Σ_{20} the (unique) $\mathfrak{A}_5$-orbits of length 6, 10, 12, 15, and 20 in $\mathbb{P}^2$, respectively. By $\mathcal{L}_6$, $\mathcal{L}_{10}$, $\mathcal{L}_{12}$, $\mathcal{L}_{15}$, and $\mathcal{L}_{20}$ we denote the (unique) $\mathfrak{A}_5$-invariant curves in $\mathbb{P}^2$ that are unions of 6, 10, 12, 15, and 20 lines, respectively. By a small abuse of terminology we will refer to irreducible components of any of the latter curves $\mathcal{L}_r$ as *lines of $\mathcal{L}_r$* (although $\mathcal{L}_r$ is not a set of lines, but a union of lines).

Theorem 6.1.2. The following assertions hold:

(i) the points of pairwise intersections of the lines of $\mathcal{L}_6$ are the points of the $\mathfrak{A}_5$-orbit Σ_{15};

(ii) the lines passing through pairs of points of Σ_6 are the lines of $\mathcal{L}_{15}$;

(iii) there are exactly 5 points of Σ_{15} lying on each line of $\mathcal{L}_6$;

(iv) there are exactly 5 lines of $\mathcal{L}_{15}$ passing through each point of Σ_6;

(v) there are exactly 2 points of Σ_6 lying on each line of $\mathcal{L}_{15}$;

(vi) there are exactly 2 lines of $\mathcal{L}_6$ passing through each point of Σ_{15};

(vii) the $\mathfrak{A}_5$-orbit Σ_6 is disjoint from the curve $\mathcal{L}_6$;

(viii) the $\mathfrak{A}_5$-orbit Σ_{10} is disjoint from the curve $\mathcal{L}_6$;

(ix) the $\mathfrak{A}_5$-orbit Σ_6 is disjoint from the curve $\mathcal{L}_{10}$;

(x) there are exactly 2 points of Σ_{10} lying on each line of $\mathcal{L}_{15}$;

(xi) there are exactly 2 lines of $\mathcal{L}_{10}$ passing through each point of Σ_{15};

(xii) there are exactly 3 points of Σ_{15} lying on each line of $\mathcal{L}_{10}$;

(xiii) there are exactly 3 lines of $\mathcal{L}_{15}$ passing through each point of Σ_{10};

(xiv) there are exactly 2 points of Σ_{15} lying on each line of $\mathcal{L}_{15}$;

(xv) there are exactly 2 lines of $\mathcal{L}_{15}$ passing through each point of Σ_{15};

(xvi) the points of pairwise intersections of the lines of $\mathcal{L}_{15}$ are the points of the $\mathfrak{A}_5$-orbits Σ_6, Σ_{10}, and Σ_{15};

(xvii) the $\mathfrak{A}_5$-orbit Σ_{10} is disjoint from the curve $\mathcal{L}_{10}$;

(xviii) the points of pairwise intersections of the lines of $\mathcal{L}_{10}$ are the points of the $\mathfrak{A}_5$-orbit Σ_{15} and some $\mathfrak{A}_5$-orbit of length 30;

(xix) any point that is contained in a line of $\mathcal{L}_{15}$ and is not contained in the $\mathfrak{A}_5$-orbits Σ_6, Σ_{10}, and Σ_{15} has an $\mathfrak{A}_5$-orbit of length 30;

(xx) any point of $\mathbb{P}^2$ that has an $\mathfrak{A}_5$-orbit of length 30 is contained in the curve $\mathcal{L}_{15}$.

Proof. Let L be a line of $\mathcal{L}_6$. Then the stabilizer $G_L \subset \mathfrak{A}_5$ of L is isomorphic to $\mathfrak{D}_{10}$ by Lemma 5.1.1. Moreover, G_L acts on the line L faithfully by Corollary 5.2.2(v). The intersection of L with all other lines of $\mathcal{L}_6$ contains at most five points, and at the same time it is a union of G_L-orbits. Since the length of any G_L-orbit is either 2, or 5, or 10 by Lemma 5.5.9(ii), we conclude that the line L can intersect the union of all other lines of $\mathcal{L}_6$ either by two points or by five points. The former case is impossible by Lemma 5.1.3(ii). In the latter case, we see that no three lines of $\mathcal{L}_6$ intersect in one point, which implies assertions (i) and (iii) by Lemma 5.3.1(i). Assertions (ii) and (iv) follow from assertions (i) and (iii) by duality.

Note that no three points of Σ_6 lie on one line (cf. [58, §7]). This easily implies assertion (v), which in turn gives assertion (vi) by duality.

Suppose that some point $P \in \Sigma_6$ is contained in some line L of $\mathcal{L}_6$. By assertion (i) we conclude that P is the only point of Σ_6 contained in L. Thus P is a fixed point of the stabilizer $G_L \cong \mathfrak{D}_{10}$ of L, which is impossible by Lemma 5.5.9(ii). The obtained contradiction proves assertion (vii).

Suppose that some point $P \in \Sigma_{10}$ is contained in some line L of $\mathcal{L}_6$. By assertion (i) we conclude that P is the only point of Σ_{10} contained in L. This is clearly impossible. The obtained contradiction proves assertion (viii), which in turn gives assertion (ix) by duality.

Choose a subgroup $G \cong \boldsymbol{\mu}_2 \times \boldsymbol{\mu}_2$ in $\mathfrak{A}_5$. We know from Lemma 5.3.1(ii) that any line stabilized by G is one of the lines of $\mathcal{L}_{15}$. On the other hand, by Lemma 5.1.3(vii) there are two points among the points of Σ_{10} that form a G-orbit of length 2. The line passing through these two points is G-invariant, which means that some (and thus all) points of Σ_{10} are contained in the lines of $\mathcal{L}_{15}$.

Let p be the number of points of Σ_{10} contained in the line L (and thus any other line of $\mathcal{L}_{15}$), and c be the number of lines of $\mathcal{L}_{15}$ passing through the point P_1 (and thus any other point of Σ_{10}). Then

$$15p = 10c.$$

Since we already know that $p \geqslant 2$, this gives $c \geqslant 3$.

Note that if there are r lines in $\mathbb{P}^2$ passing through a point P, then there are $\frac{r(r-1)}{2}$ pairs of these lines such that the intersection point of the lines in a pair is P. Since there are $\frac{15 \cdot 14}{2} = 105$ pairs of lines of $\mathcal{L}_{15}$ in total, and there are 5 lines of $\mathcal{L}_{15}$ passing through each of the 6 points of Σ_6 by assertion (iv), we have an estimate

$$105 \geqslant \frac{6 \cdot 5 \cdot 4}{2} + \frac{10c(c-1)}{2} = 60 + \frac{10c(c-1)}{2}.$$

We conclude that $c \leqslant 3$, so that $c = 3$ and $p = 2$. This proves assertions (x) and (xiii), which in turn imply assertions (xi) and (xii) by duality. Counting the number of intersection points of various lines of $\mathcal{L}_{15}$ not contained in the $\mathfrak{A}_5$-orbits Σ_6 and Σ_{10}, we also obtain assertions (xiv), (xv), and (xvi).

Assertion (xvii) is checked in a straightforward way (the reader is encouraged to have a look at the regular icosahedron, which actually gives an alternative way to check some other assertions of the theorem).

In total there are $\frac{10 \cdot 9}{2} = 45$ pairs of lines of $\mathcal{L}_{10}$, and there are 2 lines of $\mathcal{L}_{10}$ passing through each of the 15 points of Σ_{15} by assertion (xi). Also, we know that the points of the $\mathfrak{A}_5$-orbit Σ_6 are not contained in $\mathcal{L}_{10}$ by assertion (ix). Applying Lemma 5.3.1(i), we see that the intersection points of the remaining 30 pairs of lines of $\mathcal{L}_{10}$ either form a union of an $\mathfrak{A}_5$-orbit of length 20 and the $\mathfrak{A}_5$-orbit Σ_{10}, or form the $\mathfrak{A}_5$-orbit Σ_{10}, or form an $\mathfrak{A}_5$-orbit of length 30. The first case is excluded by Lemma 5.1.3(vii). The second case is impossible by assertion (xvii). This gives the proof of assertion (xviii).

Let L be a line of $\mathcal{L}_{15}$, and let P be a point that is contained in L but not contained in the $\mathfrak{A}_5$-orbits Σ_6, Σ_{10}, and Σ_{15}. The stabilizer $G_L \subset \mathfrak{A}_5$ of the line L is isomorphic to $\boldsymbol{\mu}_2 \times \boldsymbol{\mu}_2$ by Lemma 5.1.1, and the action of G_L on L has a kernel isomorphic to $\boldsymbol{\mu}_2$. Thus, the group that actually acts on L faithfully is $\bar{G}_L \cong \boldsymbol{\mu}_2$. By Lemma 5.5.9(i) the point P has an $\mathfrak{A}_5$-orbit of length 30, provided that it is not contained in some other line of $\mathcal{L}_{15}$, and that it is not a fixed point of $\bar{G}_L$. The former case does not take place by assertion (xvi). In the latter case the length r of the $\mathfrak{A}_5$-orbit of P is at most 15, and r divides 30. By Lemma 5.3.1(i) this means that P is a point of one of the $\mathfrak{A}_5$-orbits Σ_6, Σ_{10}, or Σ_{15}, which is not the case by assumption. Therefore, we obtain assertion (xix).

Finally, if $P \in \mathbb{P}^2$ is a point such that the $\mathfrak{A}_5$-orbit of P has length 30, then the stabilizer of P in $\mathfrak{A}_5$ is isomorphic to $\boldsymbol{\mu}_2$. Choose a subgroup $G \cong \boldsymbol{\mu}_2$ in $\mathfrak{A}_5$. By Corollary 5.2.2(o) the fixed points of G in $\mathbb{P}^2$ are contained in a union of a line L_G and a point P_G outside L_G. Note that there is a subgroup $\tilde{G} \cong \boldsymbol{\mu}_2 \times \boldsymbol{\mu}_2$ in $\mathfrak{A}_5$ such that $\tilde{G}$ contains G, and both the stabilizer of the point P_G and the stabilizer of the line L_G in $\mathfrak{A}_5$ coincide with $\tilde{G}$. Hence the point P_G is contained in the $\mathfrak{A}_5$-orbit Σ_{15}, and the line L_G is a line of $\mathcal{L}_{15}$. This gives assertion (xx) and completes the proof of Theorem 6.1.2. □

Corollary 6.1.3. *No three points of Σ_6 lie on one line in $\mathbb{P}^2$. Moreover, Σ_6 is not contained in a conic.*

Proof. No three points of Σ_6 lie on one line in $\mathbb{P}^2$ by Theorem 6.1.2(ii),(v). If C is a conic in $\mathbb{P}^2$ that contains Σ_6, then C must be unique and, thus, $\mathfrak{A}_5$-invariant. The latter is impossible by Lemma 5.1.4. □

Remark 6.1.4. The configuration of 16 points of the $\mathfrak{A}_5$-orbits Σ_6 and Σ_{10} and 15 lines of $\mathcal{L}_{15}$ is included in a series of configurations of n^2 points and $3n+3$ lines each containing n points as the case with $n=4$ (see [42]). This configuration enjoys unexpectedly many symmetries compared to other configurations in the series.

Remark 6.1.5. By Lemma 5.3.1(vi), all $\mathfrak{A}_5$-invariant curves in $|\mathcal{O}_{\mathbb{P}^2}(6)|$ form a pencil. Let us denote it by $\mathcal{P}$. The properties of this pencil were described in [36, Theorems 1 and 2]. They all easily follow from Lemma 5.3.1. Let us mention some of them. The only non-reduced curve in $\mathcal{P}$ is a multiple conic $3\mathfrak{C}$, and the only reducible curve in $\mathcal{P}$ is the curve $\mathcal{L}_6$. Note that $\mathcal{L}_6$ has exactly 15 singular points; they are points of Σ_{15} and they are all nodes by Theorem 6.1.2(i),(vi). Keeping in mind that reduced and irreducible sextic curve in $\mathbb{P}^2$ may have at most 10 singular points, one can show that the only singular irreducible curves in $\mathcal{P}$ are the unique curve in $\mathcal{P}$ with 6 nodes in the points of Σ_6 and the unique curve in $\mathcal{P}$ with 10 nodes in the points of Σ_{10}. The normalization of the former curve is a unique smooth curve of genus 4 that admits a faithful action of $\mathfrak{A}_5$ (cf. Lemma 5.4.1). The normalization of the latter curve is a rational curve.

The following observation will be used in the proof of Lemma 7.6.7.

Lemma 6.1.6. Let L be a line of $\mathcal{L}_{15}$, and $G \cong \boldsymbol{\mu}_2 \times \boldsymbol{\mu}_2$ be the stabilizer of L in $\mathfrak{A}_5$. Then the following assertions hold:

(i) there exists a G-invariant pair of points $\{P_1, P_2\}$ in L such that P_1 and P_2 are interchanged by G, and the polar line of P_1 with respect to the $\mathfrak{A}_5$-invariant conic $\mathfrak{C} \subset \mathbb{P}^2$ contains P_2;

(ii) such G-invariant pair of points $\{P_1, P_2\}$ in L is unique;

(iii) the $\mathfrak{A}_5$-orbit of the points P_1 and P_2 is of length 30;

(iv) the points P_1 and P_2 are disjoint from the curve $\mathcal{L}_{12}$.

Proof. Recall that the $\mathfrak{A}_5$-action on $\mathbb{P}^2$ is given by one of the three-dimensional irreducible $\mathfrak{A}_5$-representations W_3 or W_3'. So, let us fix a subgroup $\mathfrak{A}_5 \subset \mathrm{GL}_3(\mathbb{C})$. Since W_3 and W_3' are defined over $\mathbb{R}$, we may assume that $\mathfrak{A}_5$ is the subgroup of $\mathrm{SO}_3(\mathbb{R}) \subset \mathrm{GL}_3(\mathbb{C})$, and the $\mathfrak{A}_5$-invariant conic $\mathfrak{C} \subset \mathbb{P}^2$ is given by
$$x^2 + y^2 + z^2 = 0$$
in the homogenous coordinates x, y, z on $\mathbb{P}^2$ that arise from an appropriate orthonormal basis in $\mathbb{R}^3$. Thus, the polar line of a point $[x_1 : y_1 : z_1]$ with

respect to the conic $\mathfrak{C}$ contains a point $[x_2 : y_2 : z_2]$ if and only if

$$x_1 x_2 + y_1 y_2 + z_1 z_2 = 0.$$

The subgroup $\mathfrak{A}_5 \subset \mathrm{SO}_3(\mathbb{R})$ consists of orientation preserving symmetries of a regular icosahedron, whose center is in the origin. Without loss of generality, we may assume that its vertices are $(\pm 1, 0, 0)$ and

$$\left(\pm \frac{1}{\sqrt{5}}, \pm \frac{2}{\sqrt{5}} \cos\left(\frac{2\pi n}{5}\right), \pm \frac{2}{\sqrt{5}} \sin\left(\frac{2\pi n}{5}\right) \right)$$

for $n = 0, 1, 2, 3, 4$.

The group $\mathfrak{A}_5$ is generated by the permutations $\sigma_1 = (1\ 2\ 3\ 4\ 5)$ and $\sigma_2 = (1\ 2)(3\ 4)$. The image of $\mathfrak{A}_5$ in $\mathrm{SO}_3(\mathbb{R})$ is generated by the matrices

$$M_1 = \begin{pmatrix} 1 & 0 & 0 \\ 0 & \cos\left(\frac{2\pi}{5}\right) & -\sin\left(\frac{2\pi}{5}\right) \\ 0 & \sin\left(\frac{2\pi}{5}\right) & \cos\left(\frac{2\pi}{5}\right) \end{pmatrix}$$

and

$$M_2 = \begin{pmatrix} \frac{1}{\sqrt{5}} & \frac{2}{\sqrt{5}} & 0 \\ \frac{2}{\sqrt{5}} & -\frac{1}{\sqrt{5}} & 0 \\ 0 & 0 & -1 \end{pmatrix}$$

corresponding to σ_1 and σ_2, respectively.

We may assume that the subgroup $G \subset \mathfrak{A}_5$ is generated by the permutations σ_2 and $\sigma_3 = (1\ 3)(2\ 4)$. One has

$$\sigma_3 = \sigma_1 \sigma_2 \sigma_1^3 \sigma_2 \sigma_1^3 \sigma_2 \sigma_1 \sigma_2,$$

so that the matrix in $\mathrm{SO}_3(\mathbb{R})$ corresponding to σ_3 is

$$M_3 = \begin{pmatrix} -1 & 0 & 0 \\ 0 & -1 & 0 \\ 0 & 0 & 1 \end{pmatrix}.$$

By Corollary 5.2.2(ii), there are exactly three lines in $\mathbb{P}^2$ that are G-invariant. Hence, we may assume that L is given by $z = 0$. Note that σ_3 fixes every point in L. Similarly, the line L contains exactly two points that are fixed by σ_2. These points are $Q_1 = [-2 : 1 - \sqrt{5} : 0]$ and $Q_2 = [-2 : 1 + \sqrt{5} : 0]$. By Lemma 5.3.1(i), they are contained in Σ_{15}.

The polar line of the point $\sigma_2([x:y:0])$ contains the point $[x:y:0]$ if and only if

$$x\left(\frac{x}{\sqrt{5}} + \frac{2y}{\sqrt{5}}\right) + y\left(\frac{2x}{\sqrt{5}} - \frac{y}{\sqrt{5}}\right) = 0. \tag{6.1.7}$$

This equation does have solutions with non-zero x and y, which proves assertion (i). In fact, x and y satisfy (6.1.7) if and only if either $x = y = 0$, or $[x:y:0]$ is one of the points $P_1 = [-2+\sqrt{5}:1:0]$ or $P_2 = [-2-\sqrt{5}:1:0]$. This proves assertion (ii).

The $\mathfrak{A}_5$-orbit Σ_6 consists of the point $[1:0:0]$ and five points

$$\left[1 : 2\cos\left(\frac{2\pi n}{5}\right) : 2\sin\left(\frac{2\pi n}{5}\right)\right]$$

for $n = 0, 1, 2, 3, 4$. In particular, the points $[1:0:0]$ and $[1:2:0]$ are the only points of Σ_6 that are contained in L by Theorem 6.1.2(v). This shows that P_1 and P_2 are not contained in Σ_6.

The $\mathfrak{A}_5$-orbit Σ_{10} consists of five points

$$\left[\frac{2+\sqrt{5}}{2} : \cos\left(\frac{2\pi n}{5}\right) + \cos\left(\frac{2\pi(n+1)}{5}\right) : \sin\left(\frac{2\pi n}{5}\right) + \sin\left(\frac{2\pi(n+1)}{5}\right)\right]$$

for $n = 0, 1, 2, 3, 4$, and five points

$$\left[\frac{1}{2} : \cos\left(\frac{2\pi n}{5}\right) + \cos\left(\frac{2\pi(n+1)}{5}\right) - \cos\left(\frac{2\pi(n+3)}{5}\right) : \right.$$
$$\left. \sin\left(\frac{2\pi n}{5}\right) + \sin\left(\frac{2\pi(n+1)}{5}\right) - \sin\left(\frac{2\pi(n+3)}{5}\right)\right]$$

for $n = 0, 1, 2, 3, 4$. In particular, the points

$$\left[\frac{2+\sqrt{5}}{2} : -\frac{1+\sqrt{5}}{2} : 0\right]$$

and

$$\left[\frac{1}{2} : -\frac{3+\sqrt{5}}{2} : 0\right]$$

are the only points of Σ_{10} that are contained in L by Theorem 6.1.2(x). This shows that P_1 and P_2 are not contained in Σ_{10}.

The points P_1 and P_2 are not contained in the $\mathfrak{A}_5$-orbit Σ_{15}, because Q_1 and Q_2 are the only points in Σ_{15} that are contained in L by Theorem 6.1.2(xiv). Thus, assertion (iii) follows from Theorem 6.1.2(xix).

The polar line of the point $[1:0:0]$ with respect to the conic $\mathfrak{C}$ is given by $x = 0$. This line intersects the conic $\mathfrak{C}$ by two points $[0:1:\sqrt{-1}]$

and $[0:1:-\sqrt{-1}]$, which are contained in Σ_{12}. Their polar lines are given by $y + \sqrt{-1}z = 0$ and $y - \sqrt{-1}z = 0$, respectively. These lines are irreducible components of $\mathcal{L}_{12}$ by Remark 5.3.2. Thus, if the point $[x:y:z]$ with real x, y, and z is contained in $\mathcal{L}_{12}$, then $[x:y:z] \in \Sigma_6$. In particular, the points P_1 and P_2 are not contained in $\mathcal{L}_{12}$ by assertion (iii). This proves assertion (iv). □

One can show that the $\mathfrak{A}_5$-orbit of the pair of points in Lemma 6.1.6 does not depend on the choice of the line of $\mathcal{L}_{15}$.

Corollary 6.1.8. Let P be a point of the $\mathfrak{A}_5$-orbit Σ_{15}, and let P_1 and P_2 be points of $\mathbb{P}^2$ not contained in Σ_{15}. Suppose that P_1 and P_2 are contained in the polar line of P with respect to the $\mathfrak{A}_5$-invariant conic $\mathfrak{C} \subset \mathbb{P}^2$, and that P_2 is contained in the polar line of P_1 with respect to $\mathfrak{C}$. Then a polar line of P_1 with respect to $\mathfrak{C}$ does not contain points of the $\mathfrak{A}_5$-orbit Σ_{12}.

Proof. Apply Lemma 6.1.6 together with Remark 5.3.2. □

Similarly to Lemma 6.1.6, we prove the following fact that will be used in the proof of Lemma 15.3.5.

Lemma 6.1.9. The intersection $\mathcal{L}_6 \cap \mathcal{L}_{15}$ consists of Σ_{15} and an $\mathfrak{A}_5$-orbit of length 30, that is disjoint from the curve $\mathcal{L}_{12}$.

Proof. Since $\mathcal{L}_6 \cdot \mathcal{L}_{15} = 90$, it follows from Theorem 6.1.2(vi),(xv) that the intersection $\mathcal{L}_6 \cap \mathcal{L}_{15}$ is a disjoint union of Σ_{15} and an $\mathfrak{A}_5$-invariant set Σ that consists of 30 points (counted with multiplicities). Since $\Sigma_6 \not\subset \mathcal{L}_6$ by Theorem 6.1.2(vii) and $\Sigma_{10} \not\subset \mathcal{L}_6$ by Theorem 6.1.2(viii), we see that Σ is an $\mathfrak{A}_5$-orbit of length 30 by Theorem 6.1.2(xix).

Let us use assumptions and notation of Lemma 6.1.6. Since the line $x = 0$ is one of the lines of $\mathcal{L}_6$, and $z = 0$ is one of the lines of $\mathcal{L}_{15}$, we see that the point $P = [0:1:0]$ is contained in the intersection $\mathcal{L}_6 \cap \mathcal{L}_{15}$. Since $P \notin \Sigma_{15}$, we conclude that $P \in \Sigma$. Note also that P is not contained in Σ_6. Arguing as in the proof of Lemma 6.1.6(iv), we see that P is not contained in $\mathcal{L}_{12}$. Thus, the $\mathfrak{A}_5$-orbit Σ is disjoint from $\mathcal{L}_{12}$. □

Now we present some facts about the action of the group $\mathfrak{A}_4$ on $\mathbb{P}^2$. Recall that $\mathfrak{A}_4$ has a unique faithful three-dimensional representation, which is also its unique irreducible three-dimensional representation U_3. One has

$$\mathrm{Res}_{\mathfrak{A}_4}^{\mathfrak{A}_5} W_3 \cong \mathrm{Res}_{\mathfrak{A}_4}^{\mathfrak{A}_5} W_3' \cong U_3$$

by Corollary 5.2.2(vi).

CREMONA GROUPS AND THE ICOSAHEDRON

Proposition 6.1.10. Put $\mathbb{P}^2 = \mathbb{P}(U_3)$. Then

(i) if an $\mathfrak{A}_4$-orbit in $\mathbb{P}^2$ has length $r < 12$, then $r \in \{3, 4, 6\}$; there is a unique $\mathfrak{A}_4$-orbit Ω_3 of length 3 in $\mathbb{P}^2$, and there are exactly three orbits Ω_4, Ω_4', and Ω_4'' of length 4 in $\mathbb{P}^2$;

(ii) if $\mathcal{A}$ is an $\mathfrak{A}_4$-irreducible curve that is a union of $r < 12$ lines in $\mathbb{P}^2$, then $r \in \{3, 4, 6\}$; there is a unique $\mathfrak{A}_4$-irreducible curve $\mathcal{A}_3$ that is a union of 3 lines in $\mathbb{P}^2$, and there are exactly three $\mathfrak{A}_4$-irreducible curves $\mathcal{A}_4$, $\mathcal{A}_4'$, and $\mathcal{A}_4''$ that are unions of 4 lines in $\mathbb{P}^2$ (as before, we will refer to the lines that are irreducible components of the curves $\mathcal{A}_3$, $\mathcal{A}_4$, $\mathcal{A}_4'$, and $\mathcal{A}_4''$ as the lines of $\mathcal{A}_3$, $\mathcal{A}_4$, $\mathcal{A}_4'$, and $\mathcal{A}_4''$, respectively);

(iii) the intersection points of the lines of $\mathcal{A}_3$ are the points of the $\mathfrak{A}_4$-orbit Ω_3, and the lines passing through pairs of points of Ω_3 are the lines of $\mathcal{A}_3$;

(iv) the $\mathfrak{A}_4$-orbits Ω_4, Ω_4', and Ω_4'' are disjoint from the curve $\mathcal{A}_3$;

(v) the $\mathfrak{A}_4$-orbit Ω_3 is disjoint from the curves $\mathcal{A}_4$, $\mathcal{A}_4'$, and $\mathcal{A}_4''$;

(vi) one can choose the $\mathfrak{A}_4$-orbits Ω_4, Ω_4', and Ω_4'' and the curves $\mathcal{A}_4$, $\mathcal{A}_4'$, and $\mathcal{A}_4''$ so that Ω_4 (resp., Ω_4', Ω_4'') is disjoint from $\mathcal{A}_4$ (resp., $\mathcal{A}_4'$, $\mathcal{A}_4''$);

(vii) if the $\mathfrak{A}_4$-orbits Ω_4, Ω_4', and Ω_4'' and the curves $\mathcal{A}_4$, $\mathcal{A}_4'$, and $\mathcal{A}_4''$ are chosen as in assertion (vi), then any line of $\mathcal{A}_4$ (resp., $\mathcal{A}_4'$, $\mathcal{A}_4''$) contains exactly one point of Ω' and one point of Ω'' (resp., one point of Ω and one point of Ω'', one point of Ω and one point of Ω'), and any point of Ω_4 (resp., Ω_4', Ω_4'') is contained in exactly one line of $\mathcal{A}_4'$ and one line of $\mathcal{A}_4''$ (resp., one line of $\mathcal{A}_4$ and one line of $\mathcal{A}_4''$, one line of $\mathcal{A}_4$ and one line of $\mathcal{A}_4'$);

(viii) the intersection points of the lines passing through pairs of points of the $\mathfrak{A}_4$-orbit Ω_4 (resp., Ω_4', Ω_4'') are the points of the $\mathfrak{A}_4$-orbits Ω_4 (resp., Ω_4', Ω_4'') and Ω_3;

(ix) any point that is contained in a line of $\mathcal{A}_3$ and is not contained in Ω_3 has an $\mathfrak{A}_4$-orbit of length 6;

(x) any point of $\mathbb{P}^2$ that has an $\mathfrak{A}_4$-orbit of length 6 is contained in the curve $\mathcal{A}_3$.

Proof. Assertions (i) and (ii) follow from Corollary 5.2.2. Assertion (iii) follows from uniqueness of Ω_3 and $\mathcal{A}_3$.

Suppose that the $\mathfrak{A}_4$-orbit Ω_4 is contained in the curve $\mathcal{A}_3$. Then each line of $\mathcal{A}_3$ contains at most one point of Ω_4 by assertion (iii), which is clearly impossible. The same argument applies to the $\mathfrak{A}_4$-orbits Ω_4' and Ω_4'', which gives assertion (iv). Assertion (v) is obtained from assertion (iv) by duality.

A stabilizer in $\mathfrak{A}_4$ of any point of Ω_4, Ω_4', and Ω_4'', and of any line of $\mathcal{A}_4$, $\mathcal{A}_4'$, and $\mathcal{A}_4''$ is isomorphic to μ_3. Choose a subgroup $G \cong \mu_3$ in $\mathfrak{A}_4$. By Corollary 5.2.2(i) there are exactly three points P, P', and P'' in $\mathbb{P}^2$ stabilized by G. Also, there are exactly three lines in $\mathbb{P}^2$ stabilized by G. These are the line L passing through the points P' and P'', the line L' passing through the points P and P'', and the line L'' passing through the points P and P'. Choose the $\mathfrak{A}_4$-orbits Ω_4, Ω_4', and Ω_4'' so that they contain the points P, P', and P'', respectively. Choose the curves $\mathcal{A}_4$, $\mathcal{A}_4'$, and $\mathcal{A}_4''$ so that among their irreducible components there are the lines L, L', and L'', respectively. Now it is straightforward to check assertions (vi) and (vii).

Let S be the set of lines passing through pairs of points of the $\mathfrak{A}_4$-orbit Ω_4. In total there are $\frac{6 \cdot 5}{2} = 15$ pairs of lines of S, and the lines of

$$\frac{4 \cdot 3 \cdot 2}{2} = 12$$

of these pairs intersect in the points of Ω_4. Thus the remaining 3 intersection points are the points of Ω_3 by assertion (i). The same argument applies to the sets of lines passing through pairs of points of the $\mathfrak{A}_4$-orbits Ω_4' or Ω_4'', which gives assertion (viii).

To prove assertions (ix) and (x) we argue literally like in the proofs of assertions (xix) and (xx) of Theorem 6.1.2. Let L be a line of $\mathcal{A}_3$, and let P be a point that is contained in L but not contained in the $\mathfrak{A}_4$-orbit Ω_3. The stabilizer $G_L \subset \mathfrak{A}_4$ of the line L is isomorphic to $\mu_2 \times \mu_2$, and the action of G_L on L has a kernel isomorphic to μ_2. Thus, the group that actually acts on L faithfully is $\bar{G}_L \cong \mu_2$. By Lemma 5.5.9(i) the point P has an $\mathfrak{A}_4$-orbit of length 6, provided that it is not contained in some other line of $\mathcal{A}_3$, and that it is not a fixed point of $\bar{G}_L$. The former case does not take place by assertion (iii). In the latter case the length r of the $\mathfrak{A}_4$-orbit of P is at most 3. By assertion (i) this means that P is a point of the $\mathfrak{A}_4$-orbit Ω_3, which is not the case by assumption. Therefore, we obtain assertion (ix).

Finally, if $P \in \mathbb{P}^2$ is a point such that the $\mathfrak{A}_4$-orbit of P has length 6, then the stabilizer of P in $\mathfrak{A}_4$ is isomorphic to μ_2. Choose a subgroup $G \cong \mu_2$ in $\mathfrak{A}_4$. By Corollary 5.2.2(o) the fixed points of G in $\mathbb{P}^2$ are contained in a union of a line L_G and a point P_G outside L_G. Note that there is a subgroup $\tilde{G} \cong \boldsymbol{\mu}_2 \times \boldsymbol{\mu}_2$ in $\mathfrak{A}_4$ such that $\tilde{G}$ contains G, and both the stabilizer of the point P_G and the stabilizer of the line L_G in $\mathfrak{A}_4$ coincide with $\tilde{G}$. Hence the point P_G is contained in the $\mathfrak{A}_4$-orbit Ω_3, and the line L_G is a line of $\mathcal{A}_3$. This gives assertion (x) and completes the proof of Proposition 6.1.10. □

We conclude this section by an observation that is valid not only for $\mathbb{P}^2$, but for any surface with an action of the group $\mathfrak{A}_5$.

Lemma 6.1.11 (cf. Lemma 5.3.1(i)). Let S be a surface faithfully acted on by $\mathfrak{A}_5$, and let P be a smooth point of S. Then the $\mathfrak{A}_5$-orbit of P consists of at least 6 points.

Proof. Let $G_P \subset \mathfrak{A}_5$ be the stabilizer of the point P. Suppose that the $\mathfrak{A}_5$-orbit of P consists of $l \leqslant 5$ points. Then either $l = 1$ or $l = 5$ by Corollary 5.1.2. Thus, Lemma 5.1.1 implies that either $G_P \cong \mathfrak{A}_5$ or $G_P \cong \mathfrak{A}_4$. On the other hand, G_P acts faithfully on the Zariski tangent space $T_P(S) \cong \mathbb{C}^2$ by Lemma 4.4.1. This is impossible, since neither $\mathfrak{A}_5$ nor $\mathfrak{A}_4$ has faithful two-dimensional representations (cf. Table 5.4). □

6.2 Quintic del Pezzo surface

Let S_5 be a smooth del Pezzo surface such that $K_{S_5}^2 = 5$. Then S_5 is unique up to isomorphism and
$$\mathrm{Aut}(S_5) \cong \mathfrak{S}_5.$$
This is well known (see, for example, [102] or [34, §8.5.4]). In particular, the surface S_5 is faithfully acted on by the groups $\mathfrak{A}_5$ and $\mathfrak{A}_4$. Clearly, the former action is unique. Since all subgroups in $\mathfrak{A}_5$ isomorphic to $\mathfrak{A}_4$ are conjugate in $\mathfrak{A}_5$, the latter action is unique up to conjugation. In this section, we study basic properties of the actions of these two groups on S_5. So, for simplicity, we fix a subgroup in $\mathfrak{A}_5$ that is isomorphic to $\mathfrak{A}_4$. Let us start with:

Lemma 6.2.1. There is an identification of $\mathfrak{A}_5$-representations
$$H^0(\mathcal{O}_{S_5}(-K_{S_5})) \cong W_3 \oplus W_3'.$$
As an $\mathfrak{A}_4$-representation, $H^0(\mathcal{O}_{S_5}(-K_{S_5}))$ splits as a direct sum of two irreducible three-dimensional representations.

Proof. The first assertion is well known (see, for example, [102] or [111, Lemma 1]). The second assertion follows from the first one by Corollary 5.2.2(vi). □

The divisor $-K_{S_5}$ is very ample and gives an $\mathfrak{A}_5$-equivariant embedding

$$S_5 \hookrightarrow \mathbb{P}^5 \cong \mathbb{P}(W_3 \oplus W_3').$$

Its image is known to be an intersection of quadrics (see, for example, [34, Corollary 8.5.2]). In the rest of this section, we will identify S_5 with its anticanonical image in $\mathbb{P}^5$.

Lemma 6.2.2. The following assertions hold:

(i) one has $\mathrm{Pic}(S_5)^{\mathfrak{A}_5} = \mathbb{Z}[-K_{S_5}]$ and $\mathrm{rk}\,\mathrm{Pic}\,(S_5)^{\mathfrak{A}_4} = 2$;

(ii) the surface S_5 contains 10 lines, and the group $\mathfrak{A}_5$ acts transitively on them;

(iii) there exists an $\mathfrak{A}_4$-equivariant birational morphism $\varpi\colon S_5 \to \mathbb{P}^2$ that contracts 4 lines in S_5, say L_1, L_2, L_3, and L_4, to an $\mathfrak{A}_4$-orbit in $\mathbb{P}^2$ of length 4;

(iv) the set of lines in S_5 splits into two $\mathfrak{A}_4$-orbits: $\{L_1, L_2, L_3, L_4\}$ and

$$\left\{L_{12}, L_{13}, L_{14}, L_{23}, L_{24}, L_{23}\right\},$$

where L_{ij} is the proper transform via ϖ of the line in $\mathbb{P}^2$ that passes through the points $\varpi(L_i)$ and $\varpi(L_j)$;

(v) the linear system $|\varpi^*(\mathcal{O}_{\mathbb{P}^2}(2)) - (L_1 + L_2 + L_3 + L_4)|$ is a base point free pencil that gives an $\mathfrak{A}_4$-equivariant morphism $\phi\colon S_5 \to \mathbb{P}^1$;

(vi) the morphism ϕ is a conic bundle (i.e., its fibers are isomorphic to reduced conics in $\mathbb{P}^2$), a general fiber of ϕ is a conic in S_5, and ϕ has exactly 3 singular fibers, which are $L_{12}+L_{34}$, $L_{13}+L_{24}$, and $L_{14}+L_{23}$;

(vii) the $\mathfrak{A}_4$-representation

$$H^0\Big(\mathcal{O}_{S_5}\big(\varpi^*(\mathcal{O}_{\mathbb{P}^2}(2)) - (L_1 + L_2 + L_3 + L_4)\big)\Big)$$

splits as a sum of two different non-trivial one-dimensional $\mathfrak{A}_4$-representations;

(viii) the conic bundle ϕ contains exactly two $\mathfrak{A}_4$-invariant fibers, say C_1 and C_1', which are both smooth conics; the action of the group $\mathfrak{A}_4$ on C_1 and C_1' is faithful;

(ix) the surface S_5 does not contain $\mathfrak{A}_4$-invariant conics that are different from C_1 and C_1';

(x) the surface S_5 does not contain $\mathfrak{A}_5$-invariant conics.

Proof. By Proposition 6.1.10(i), the plane $\mathbb{P}^2$ contains three $\mathfrak{A}_4$-orbits of length 4, and any three points in each of these orbits are not collinear. Blowing up one such $\mathfrak{A}_4$-orbit, we obtain a smooth del Pezzo surface of degree 5. Since all subgroups in $\mathfrak{A}_5$ isomorphic to $\mathfrak{A}_4$ are conjugate, this gives assertion (iii).

The birational morphism ϖ is not $\mathfrak{A}_5$-equivariant, because $\mathbb{P}^2$ does not have $\mathfrak{A}_5$-orbits of length 4 by Lemma 5.3.1(i). Hence, assertion (iii) implies that $\operatorname{rk}\operatorname{Pic}(S_5)^{\mathfrak{A}_5} = 1$ and $\operatorname{rk}\operatorname{Pic}(S_5)^{\mathfrak{A}_4} = 2$. Since $\operatorname{Pic}(S_5)$ has no torsion and $K_{S_5}^2 = 5$ is a square free integer, we get assertion (i).

Assertion (ii) is well known (see, for example, [34, §8.5.1]). Assertions (iv), (v), and (vi) are obvious.

Every conic in S_5 lies either in the pencil $|\varpi^*(\mathcal{O}_{\mathbb{P}^2}(2)) - (L_1 + L_2 + L_3 + L_4)|$ or in one of the four pencils $|\varpi^*(\mathcal{O}_{\mathbb{P}^2}(1)) - L_j|$. The latter four pencils are interchanged by the group $\mathfrak{A}_4$. Hence, every $\mathfrak{A}_4$-invariant conic in S_5 must lie in the pencil $|\varpi^*(\mathcal{O}_{\mathbb{P}^2}(2)) - (L_1 + L_2 + L_3 + L_4)|$. The $\mathfrak{A}_4$-action on $\mathbb{P}^1$ induced by the morphism ϕ is not faithful by Lemma 5.5.9(iii), because the points $\phi(L_{12})$, $\phi(L_{13})$, and $\phi(L_{14})$ form one $\mathfrak{A}_4$-orbit. Moreover, this $\mathfrak{A}_4$-action on $\mathbb{P}^1$ cannot be trivial, since otherwise there will be an infinite number of $\mathfrak{A}_4$-orbits of length 4 on $\mathbb{P}^2$ by Lemma 5.5.9(iii), which is not the case by Proposition 6.1.10(i). Hence, the image of the induced homomorphism

$$\mathfrak{A}_4 \to \operatorname{Aut}(\mathbb{P}^1)$$

is $\boldsymbol{\mu}_3$. This implies assertion (vii). In particular, ϕ has exactly two $\mathfrak{A}_4$-invariant fibers, which must be smooth conics by assertion (vi). This gives assertion (ix), and since these fibers are mapped to conics in $\mathbb{P}^2$ by the morphism ϖ, we also obtain assertion (viii). Assertion (x) immediately follows from assertion (i). □

Remark 6.2.3. Recall that $\mathbb{P}^2$ is faithfully acted on by $\mathfrak{A}_5$ (see §6.1). In the proof of Lemma 6.2.2(i), we have seen that the $\mathfrak{A}_4$-birational morphism $\varpi \colon S_5 \to \mathbb{P}^2$ is not an $\mathfrak{A}_5$-morphism. In particular, we will not make attempts to unify the notation for orbits and curves on S_5 and $\mathbb{P}^2$ (cf. §6.1).

Recall that the group $\mathfrak{A}_5$ has exactly five subgroups isomorphic to $\mathfrak{A}_4$. For each of them, we can construct an $\mathfrak{A}_4$-birational morphism $S_5 \to \mathbb{P}^2$ that contracts four disjoint lines among $L_1, L_2, L_3, L_4, L_{12}, L_{13}, L_{14}, L_{23},$

L_{24}, and L_{23}. This gives us five birational morphisms from S_5 to $\mathbb{P}^2$, and one of them is ϖ.

Remark 6.2.4. Let $\varpi_1\colon S_5 \to \mathbb{P}^2$ be a contraction of the lines L_1, L_{23}, L_{24}, and L_{34}, let $\varpi_2\colon S_5 \to \mathbb{P}^2$ be a contraction of the lines L_2, L_{13}, L_{34}, and L_{14}, let $\varpi_3\colon S_5 \to \mathbb{P}^2$ be a contraction of the lines L_3, L_{12}, L_{14}, and L_{24}, let $\varpi_4\colon S_5 \to \mathbb{P}^2$ be a contraction of the lines L_4, L_{12}, L_{23}, and L_{13}. Then they fit into a commutative diagram

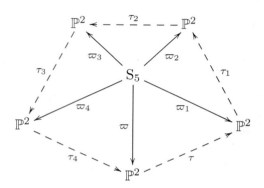

Here the map τ is a standard quadratic transformation with base points $\varpi(L_2)$, $\varpi(L_3)$, and $\varpi(L_4)$, the map τ_1 is a standard quadratic transformation with base points $\varpi_1(L_1)$, $\varpi_1(L_{23})$, and $\varpi_1(L_{24})$, the map τ_2 is a standard quadratic transformation with base points $\varpi_2(L_2)$, $\varpi_2(L_{13})$, and $\varpi_2(L_{34})$, the map τ_3 is a standard quadratic transformation with base points $\varpi_3(L_3)$, $\varpi_3(L_{14})$, and $\varpi_3(L_{24})$, while the map τ_4 is a standard quadratic transformation with base points $\varpi_4(L_{12})$, $\varpi_4(L_{13})$, and $\varpi_4(L_{23})$. Note that the dual graph of the lines on S_5 is the famous Petersen graph (see [34, Figure 8.4]). One can see from it that there are exactly five quadruples of pairwise disjoint lines in S_5. Therefore, the above $\mathfrak{A}_4$-birational morphisms are actually all possible $\mathfrak{A}_4$-birational morphisms from S_5 to $\mathbb{P}^2$.

Recall from Proposition 6.1.10(ii) that $\mathbb{P}^2$ contains a unique $\mathfrak{A}_4$-invariant triple of lines. Denote by T_1, T_2, and T_3 the proper transforms of these lines on S_5 via ϖ. It follows from Proposition 6.1.10(iv) that T_1, T_2, and T_3 are smooth rational cubic curves that are disjoint from the lines L_1, L_2, L_3, and L_4. By construction, each intersection $T_1 \cap T_2$, $T_1 \cap T_3$, and $T_2 \cap T_3$ consists of a single point. Together, these three points form an $\mathfrak{A}_4$-orbit of length 3.

Lemma 6.2.5. *The following assertions hold:*

(i) *if an $\mathfrak{A}_4$-orbit on S_5 has length r, then $r \in \{3, 4, 6, 12\}$;*

CREMONA GROUPS AND THE ICOSAHEDRON   109

(ii) there is a unique $\mathfrak{A}_4$-orbit of length 3 on S_5, and it consists of the points $T_1 \cap T_2$, $T_1 \cap T_3$, and $T_2 \cap T_3$;

(iii) the $\mathfrak{A}_4$-orbit of every point in $T_1 \cup T_2 \cup T_3$ that is not of length 3 is of length 6, and all $\mathfrak{A}_4$-orbits of length 6 on S_5 are contained in $T_1 \cup T_2 \cup T_3$;

(iv) the surface S_5 contains exactly four $\mathfrak{A}_4$-orbits of length 4, which are all contained in $C_1 \cup C_1'$;

(v) the conic C_1 (resp., C_1') contains exactly two $\mathfrak{A}_4$-orbit of length 4, and one of them consists of the points $L_1 \cap C_1$, $L_2 \cap C_1$, $L_3 \cap C_1$, and $L_4 \cap C_1$ (resp., $L_1 \cap C_1'$, $L_2 \cap C_1'$, $L_3 \cap C_1'$, and $L_4 \cap C_1'$).

Proof. Assertions (i), (ii), and (iii) follow from Proposition 6.1.10. Assertions (iv) and (v) follow from Lemma 5.5.9(iii) and Proposition 6.1.10. □

Now let us consider one technical result that will be used in the proof of Lemma 15.3.4.

Lemma 6.2.6. *Let D be a divisor in $|-3K_{S_5}|$. Then D is $\mathfrak{A}_4$-invariant if and only if one of the following possibilities holds:*

- *either $D = T_1 + T_2 + T_3 + 2C_1 + C_1'$,*
- *or $D = T_1 + T_2 + T_3 + C_1 + 2C_1'$,*
- *or $D = T_1 + T_2 + T_3 + Z$ for some divisor Z in the pencil generated by $3C_1$ and $3C_1'$.*

Proof. The divisors

$$T_1 + T_2 + T_3 + 3C_1, \; T_1 + T_2 + T_3 + 3C_1', \; T_1 + T_2 + T_3 + 2C_1 + C_1'$$

and $T_1 + T_2 + T_3 + C_1 + 2C_1'$ are $\mathfrak{A}_4$-invariant elements of $|-3K_{S_5}|$. Let $\mathcal{Q}$ be the pencil in the linear system $|-3K_{S_5}|$ that is generated by the divisors $T_1 + T_2 + T_3 + 3C_1$ and $T_1 + T_2 + T_3 + 3C_1'$. This pencil corresponds to an $\mathfrak{A}_4$-subrepresentation in $H^0(\mathcal{O}_{S_5}(-3K_{S_5}))$ that splits as a sum of two trivial one-dimensional $\mathfrak{A}_4$-representations by Lemma 6.2.2(vii). On the other hand, $T_1 + T_2 + T_3 + 2C_1 + C_1'$ and $T_1 + T_2 + T_3 + C_1 + 2C_1'$ correspond to two non-isomorphic non-trivial one-dimensional $\mathfrak{A}_4$-subrepresentations in $H^0(\mathcal{O}_{S_5}(-3K_{S_5}))$. This describes all one-dimensional $\mathfrak{A}_4$-subrepresentations in $H^0(\mathcal{O}_{S_5}(-3K_{S_5}))$ by Lemma 5.7.4. Hence, every $\mathfrak{A}_4$-invariant divisor in the linear system $|-3K_{S_5}|$ that is different from $T_1 + T_2 + T_3 + 2C_1 + C_1'$ and $T_1 + T_2 + T_3 + C_1 + 2C_1'$ is contained in $\mathcal{Q}$. □

Remark 6.2.7. Each curve in the pencil $\mathcal{Q}$ considered in the proof of Lemma 6.2.6 is $\mathfrak{A}_4$-invariant. Moreover, one has

$$T_1 + T_2 + T_3 + L_{12} + L_{13} + L_{14} + L_{23} + L_{24} + L_{34} \in \mathcal{Q}.$$

Furthermore, the curve $L_{12} + L_{13} + L_{14} + L_{23} + L_{24} + L_{34}$ is the unique curve of the pencil generated by $3C_1$ and $3C_1'$ that contains the $\mathfrak{A}_4$-orbit of length 3 in S_5, because the latter pencil is free from base points and the curve $L_{12} + L_{13} + L_{14} + L_{23} + L_{24} + L_{34}$ contains the $\mathfrak{A}_4$-orbit of length 3 (cf. Lemma 6.2.5(ii)).

Now we proceed to study the action of the group $\mathfrak{A}_5$ on the surface S_5. By Lemma 6.2.2(x), the $\mathfrak{A}_5$-orbit of the conic C_1 (resp., C_1') consists of 5 different conics. Denote by $\mathcal{G}_5$ (resp., $\mathcal{G}_5'$) the $\mathfrak{A}_5$-orbit of the conic C_1 (resp., C_1'). Then

$$\mathcal{G}_5 \sim \mathcal{G}_5' \sim -2K_{S_5}$$

by Lemma 6.2.2(i). Moreover, one has $\mathcal{G}_5 \neq \mathcal{G}_5'$, since the stabilizer $\mathfrak{A}_4$ of the conic C_1 does not stabilize any other conic in the same orbit by Lemma 5.1.3(i). Therefore, the curves $\mathcal{G}_5$ and $\mathcal{G}_5'$ generate a pencil that is free from base curves. Denote this pencil by $\mathcal{P}$. Let $\mathcal{L}_{10}$ be the curve in S_5 whose irreducible components are the lines L_1, L_2, L_3, L_4, L_{12}, L_{13}, L_{14}, L_{23}, L_{24}, and L_{23}. The topology of $S_5 \setminus \mathcal{L}_{10}$ has been studied by Jiro Sekiguchi in [114].

Lemma 6.2.8. The following assertions hold:

(i) there is an identification of $\mathfrak{A}_5$-representations

$$H^0\Big(\mathcal{O}_{S_5}(-2K_{S_5})\Big) \cong I^{\oplus 2} \oplus W_4 \oplus W_5^{\oplus 2};$$

(ii) every curve in $\mathcal{P}$ is $\mathfrak{A}_5$-invariant;

(iii) every $\mathfrak{A}_5$-invariant curve in $|-2K_{S_5}|$ is contained in $\mathcal{P}$;

(iv) one has $\mathcal{L}_{10} \in \mathcal{P}$.

Proof. Assertion (i) is well known (see, for example, [111, Corollary 2] or [61, §3.22]). It immediately follows from Lemma 5.7.3 and the existence of the pencil $\mathcal{P}$, because S_5 is an intersection of quadrics and

$$h^0\Big(\mathcal{O}_{S_5}(-2K_{S_5})\Big) = 16.$$

Assertions (ii) and (iii) follow from assertion (i). Finally, assertion (iv) follows from assertion (iii), since $\mathcal{L}_{10}$ is obviously $\mathfrak{A}_5$-invariant and $\mathcal{L}_{10} \sim -2K_{S_5}$ by Lemma 6.2.2(i). □

CREMONA GROUPS AND THE ICOSAHEDRON   111

The pencil $\mathcal{P}$ has been studied by William Edge in [39] and by Naoki Inoue and Fumiharu Kato in [61]. For every curve $C \in \mathcal{P}$, Edge explicitly described the equation of its image $\varpi(C) \subset \mathbb{P}^2$. In particular, he proved:

Theorem 6.2.9. The following assertions hold:

(i) the base locus of the pencil $\mathcal{P}$ consists of an $\mathfrak{A}_5$-orbit, say Σ_{20}, of length 20;

(ii) every curve in $\mathcal{P}$ is smooth at the points of Σ_{20};

(iii) any two different curves in $\mathcal{P}$ intersect at the points of Σ_{20} transversally;

(iv) the only reducible curves in $\mathcal{P}$ are $\mathcal{L}_{10}$, $\mathcal{G}_5$, and $\mathcal{G}'_5$;

(v) the pencil $\mathcal{P}$ contains two irreducible singular rational curves, say $\mathcal{R}$ and $\mathcal{R}'$, whose singular loci consist of $\mathfrak{A}_5$-orbits of length 6, say Σ_6 and Σ'_6, respectively;

(vi) one has $\Sigma_6 \neq \Sigma'_6$, and the $\mathfrak{A}_5$-orbits Σ_6 and Σ'_6 are the only $\mathfrak{A}_5$-orbits of length 6 on S_5;

(vii) the points of pairwise intersections of the lines of $\mathcal{L}_{10}$ form a single $\mathfrak{A}_5$-orbit of length 15, say Σ_{15};

(viii) the points of pairwise intersections of the conics of $\mathcal{G}_5$ (resp., $\mathcal{G}'_5$) form a single $\mathfrak{A}_5$-orbit of length 10, say Σ_{10} (resp., Σ'_{10}), and one has $\Sigma_{10} \neq \Sigma'_{10}$;

(ix) the $\mathfrak{A}_5$-orbits Σ_{10} and Σ'_{10} are the only $\mathfrak{A}_5$-orbits of length 10 on S_5, the $\mathfrak{A}_5$-orbits Σ_{15} and Σ_{20} are the only $\mathfrak{A}_5$-orbits of lengths 15 and 20 on S_5, respectively;

(x) the only singular curves in $\mathcal{P}$ are $\mathcal{L}_{10}$, $\mathcal{G}_5$, $\mathcal{G}'_5$, $\mathcal{R}$, and $\mathcal{R}'$;

(xi) every smooth curve in $\mathcal{P}$ is an $\mathfrak{A}_5$-invariant curve of genus 6.

Proof. The proofs of all these assertions can be found in [39] and [61]. For the reader's convenience, we recover them here using a different approach.

Put $\Sigma_{20} = \mathcal{G}_5 \cap \mathcal{G}'_5$. Then the base locus of the pencil $\mathcal{P}$ is the set Σ_{20}. Moreover, it follows from Lemma 6.2.8(iv) that

$$\Sigma_{20} = \mathcal{L}_{10} \cap \mathcal{G}_5 = \mathcal{L}_{10} \cap \mathcal{G}'_5.$$

On the other hand, we have

$$|\Sigma_{20}| \leqslant \mathcal{G}_5 \cdot \mathcal{G}_5' = (-2K_{S_5})^2 = 20,$$

which implies that the set Σ_{20} consists of 20 points (counted with multiplicities). Since $|\Sigma_{20}| \geqslant 6$ by Lemma 6.1.11, it follows from Lemma 5.1.1 that either Σ_{20} is an $\mathfrak{A}_5$-orbit of length 10 or $|\Sigma_{20}| = 20$.

Let Z and Z' be two sufficiently general curves in $\mathcal{P}$. Then both Z and Z' are irreducible and smooth away from Σ_{20} by Bertini's theorem. On the other hand, we have

$$20 = Z \cdot Z' \geqslant \sum_{P \in \Sigma_{20}} \mathrm{mult}_P(Z \cdot Z') \geqslant$$

$$\geqslant \sum_{P \in \Sigma_{20}} \mathrm{mult}_P(Z) \mathrm{mult}_P(Z') = \sum_{P \in \Sigma_{20}} \mathrm{mult}_P^2(\mathcal{P}),$$

which implies, in particular, that both Z and Z' are smooth at the points of Σ_{20}. Thus, the curves Z and Z' are smooth everywhere. Since we already know that either $|\Sigma_{20}| = 10$ or $|\Sigma_{20}| = 20$, the set Σ_{20} is an $\mathfrak{A}_5$-orbit of length 20 by Lemma 5.1.4. This proves assertion (i).

Let C and C' be any two different curves in $\mathcal{P}$. Then

$$20 = |\Sigma_{20}| = |C \cap C'| = C \cdot C' \geqslant$$

$$\geqslant \sum_{P \in \Sigma_{20}} \mathrm{mult}_P(C \cdot C') \geqslant \sum_{P \in \Sigma_{20}} \mathrm{mult}_P(C) \mathrm{mult}_P(C') =$$

$$= |\Sigma_{20}| \mathrm{mult}_{\Sigma_{20}}(C) \mathrm{mult}_{\Sigma_{20}}(C') \geqslant |\Sigma_{20}| = 20,$$

which implies assertions (ii) and (iii).

Suppose that C is a reducible divisor in the pencil $\mathcal{P}$. Since $\mathrm{rk}\,\mathrm{Pic}\,(S_5)^{\mathfrak{A}_5} = 1$ by Lemma 6.2.2(i), and the linear system $|-K_{S_5}|$ does not contain $\mathfrak{A}_5$-invariant curves by Lemma 6.2.1, we see that C is an $\mathfrak{A}_5$-irreducible curve. Denote by $C_1, \ldots, C_r$ its irreducible components. Then

$$10 = -K_{S_5} \cdot C = \sum_{i=1}^{r}(-K_{S_5}) \cdot C_i = r\deg(C_1).$$

Since $r > 1$, it follows from Corollary 5.1.2 that either $r = 10$ and C_1 is a line, or $r = 5$ and C_1 is a conic. By construction, all lines in S_5 are irreducible components of the curve $\mathcal{L}_{10}$, and the curve $\mathcal{L}_{10}$ is $\mathfrak{A}_5$-irreducible by Lemma 6.2.2(ii). Similarly, it follows from Lemma 6.2.2(ix) that $\mathcal{G}_5$ and $\mathcal{G}_5'$

CREMONA GROUPS AND THE ICOSAHEDRON 113

are the only $\mathfrak{A}_5$-irreducible curves on S_5 that are unions of five conics. Therefore, if C is reducible, then either $C = \mathcal{L}_{10}$ or $C = \mathcal{G}_5$ or $C = \mathcal{G}'_5$. This proves assertion (iv).

It is easy to see that there are 15 points of pairwise intersections of the irreducible components of $\mathcal{L}_{10}$ (cf. [34, Figure 8.4]). By Lemma 6.1.11, they form a single $\mathfrak{A}_5$-orbit, which we denote by Σ_{15}. This proves assertion (vii) and implies that the lines of $\mathcal{L}_{10}$ are disjoint away from Σ_{15}. In particular, $\mathcal{L}_{10}$ does not contain other $\mathfrak{A}_5$-orbits of length 15, and $\mathcal{L}_{10}$ does not contain $\mathfrak{A}_5$-orbits of lengths 6 and 12.

Let L be a line on S_5, and let G_L be its stabilizer in $\mathfrak{A}_5$. Then $G_L \cong \mathfrak{S}_3$ by Lemma 5.1.1. The subgroup $\boldsymbol{\mu}_3 \subset G_L$ permutes the three lines of $\mathcal{L}_{10}$ that intersect L. Hence, the action of $\boldsymbol{\mu}_3$ on L is non-trivial, which means that G_L acts on L faithfully. Therefore, $\mathcal{L}_{10}$ does not contain $\mathfrak{A}_5$-orbits of length 10 by Lemma 5.5.9(ii).

Let $C_1, \ldots, C_5$ (resp., $C'_1, \ldots, C'_5$) be the irreducible components of the curve $\mathcal{G}_5$ (resp., $\mathcal{G}'_5$). Then $C_i \cdot C_j = C_1 \cdot C_2$ for every $i \neq j$ by Lemma 5.1.3(i). Moreover, one has $C_i^2 = 0$ for any i by the adjunction formula. Hence, we have

$$20 = (-2K_{S_5}) \cdot (-2K_{S_5}) = \left(C_1 + \ldots + C_5\right)^2 = 5C_1^2 + 20 C_1 \cdot C_2 = 20 C_1 \cdot C_2.$$

Therefore, the intersection $C_i \cap C_j$ consists of a single point for every $i \neq j$. These intersection points give us at most 10 points in $\mathcal{G}_5$. Thus, by Lemma 6.1.11 the pairwise intersections of the curves $C_1, \ldots, C_5$ form a single $\mathfrak{A}_5$-orbit of length 10. Let us denote it by Σ_{10}. Similarly, the pairwise intersections of the curves $C'_1, \ldots, C'_5$ form a single $\mathfrak{A}_5$-orbit of length 10. Let us denote it by Σ'_{10}. By assertion (i), we know that $\Sigma_{10} \neq \Sigma'_{10}$.

By construction, the irreducible components of the curve $\mathcal{G}_5$ (resp., $\mathcal{G}'_5$) are disjoint away from Σ_{10} (resp., Σ'_{10}). This proves assertion (viii) and implies that neither $\mathcal{G}_5$ nor $\mathcal{G}'_5$ contains $\mathfrak{A}_5$-orbits of lengths 6 and 12. Moreover, the stabilizer of each irreducible component of the curve $\mathcal{G}_5$ (resp., $\mathcal{G}'_5$) acts faithfully on it by Lemma 6.2.2(viii). Hence, it follows from Lemma 5.5.9(iii) that $\mathcal{G}_5$ (resp., $\mathcal{G}'_5$) does not contain $\mathfrak{A}_5$-orbits of length 10 that are different from Σ_{10} (resp., Σ'_{10}), and does not contain $\mathfrak{A}_5$-orbits of length 15. Similarly, we see that the only $\mathfrak{A}_5$-orbit of length 20 contained in $\mathcal{G}_5 \cup \mathcal{G}'_5$ is Σ_{20}.

Let G be a subgroup in $\mathfrak{A}_5$ that is isomorphic to $\boldsymbol{\mu}_5$. By Remark 1.1.2, the surface S_5 contains a G-invariant point. Denote this point by P, and denote its $\mathfrak{A}_5$-orbit by Σ_6. Then either $|\Sigma_6| = 6$ or $|\Sigma_6| = 12$ by Lemma 5.1.1. This implies that $P \notin \Sigma_{20}$, $P \notin \mathcal{L}_{10}$, $P \notin \mathcal{G}_5$ and $P \notin \mathcal{G}'_5$. In particular, there exists a unique curve $\mathcal{R}$ in the pencil $\mathcal{P}$ that contains Σ_6, and $\mathcal{R}$ is different from $\mathcal{L}_{10}$, $\mathcal{G}_5$, and $\mathcal{G}'_5$. By assertion (iv), this curve is irreducible.

Since the arithmetic genus of $\mathcal{R}$ is 6 by the adjunction formula, it follows from Lemmas 4.4.6 and 5.1.5 that $|\Sigma_6| = 6$, that the curve $\mathcal{R}$ is singular at every point of the $\mathfrak{A}_5$-orbit Σ_6, that $\mathcal{R}$ is smooth away from Σ_6, and that $\mathcal{R}$ is rational. Furthermore, Lemma 4.4.6 also implies that either $\mathcal{R}$ has nodes in the points of Σ_6, or $\mathcal{R}$ has ordinary cusps in the points of Σ_6. The latter case is impossible, since the normalization of the curve $\mathcal{R}$ cannot contain $\mathfrak{A}_5$-orbits of length 6 by Lemma 5.1.4. Thus, each point in Σ_6 is a node of the curve $\mathcal{R}$.

Recall that $\mathrm{Aut}(S_5) \cong \mathfrak{S}_5$. Since $\mathfrak{A}_5$ is a normal subgroup in $\mathfrak{S}_5$, the pencil $\mathcal{P}$ is $\mathfrak{S}_5$-invariant. Note that the curve $\mathcal{L}_{10}$ is $\mathfrak{S}_5$-invariant by construction. On the other hand, the curve $\mathcal{R}$ is not $\mathfrak{S}_5$-invariant, because $\mathfrak{S}_5$ cannot act faithfully on a rational curve by Lemma 5.5.9. In particular, the pencil $\mathcal{P}$ contains another rational curve, say $\mathcal{R}'$, that has 6 singular points. This proves assertion (v).

Denote by Σ'_6 the singular locus of the curve $\mathcal{R}'$. Since the base locus of the pencil $\mathcal{P}$ is an $\mathfrak{A}_5$-orbit of length 20 by assertion (i), and $\mathcal{R} \neq \mathcal{R}'$ by construction, we conclude that $\Sigma_6 \neq \Sigma'_6$.

Then the stabilizer in $\mathfrak{A}_5$ of every point in $\Sigma_6 \cup \Sigma'_6$ is isomorphic to $\mathfrak{D}_{10}$. On the other hand, the group $\mathfrak{A}_5$ contains exactly 6 subgroups isomorphic to $\mathfrak{D}_{10}$ (see §5.1). Using Lemma 6.2.1, Corollary 5.2.2(v) and the fact that S_5 is an intersection of quadrics, one can deduce that each such subgroup in $\mathfrak{A}_5$ fixes at most two points in S_5. This shows that Σ_6 and Σ'_6 are the only $\mathfrak{A}_5$-orbits of length 6 in S_5, and proves assertion (vi).

Let C be an irreducible singular curve in the pencil $\mathcal{P}$. Then its arithmetic genus is 6 by the adjunction formula. Hence, C has at most 6 singular points by Lemma 4.4.6. Thus, the singular locus of C must be an $\mathfrak{A}_5$-orbit of length 6 by Lemma 6.1.11. Therefore, it follows from assertion (vi) that either $C = \mathcal{R}$ or $C = \mathcal{R}'$. This together with assertion (iv) proves assertion (x).

We see that every curve in $\mathcal{P}$ that is different from $\mathcal{L}_{10}$, $\mathcal{G}_5$, $\mathcal{G}'_5$, $\mathcal{R}$, and $\mathcal{R}'$ is smooth. By the adjunction formula, its genus is equal to 6. This proves assertion (xi). Hence, it follows from Lemma 5.1.5 that every $\mathfrak{A}_5$-orbit on S_5 of length less than 20 is contained in the union

$$\mathcal{L}_{10} \cup \mathcal{G}_5 \cup \mathcal{G}'_5 \cup \mathcal{R} \cup \mathcal{R}'.$$

By Lemmas 5.1.1 and 6.1.11, such $\mathfrak{A}_5$-orbits can be of lengths 6, 10, and 15. Since $\mathfrak{A}_5$-orbits of length 6 are singular loci of the curves $\mathcal{R}$ and $\mathcal{R}'$, and the curves $\mathcal{R}$ and $\mathcal{R}'$ do not contain $\mathfrak{A}_5$-orbits of lengths 10 and 15 by

Lemma 5.1.4, we see that all $\mathfrak{A}_5$-orbits of lengths 10 and 15 are contained in $\mathcal{L}_{10} \cup \mathcal{G}_5 \cup \mathcal{G}_5'$. However, we already described all $\mathfrak{A}_5$-orbits of lengths 10 and 15 that are contained in $\mathcal{L}_{10} \cup \mathcal{G}_5 \cup \mathcal{G}_5'$. This proves assertion (ix) and completes the proof of Theorem 6.2.9. □

Remark 6.2.10. As we saw in the proof of Theorem 6.2.9, the pencil $\mathcal{P}$ is $\mathfrak{S}_5$-invariant, and the action of the group $\mathfrak{S}_5$ on $\mathcal{P}$ arises from the non-trivial action of the group $\mathfrak{S}_5/\mathfrak{A}_5 \cong \boldsymbol{\mu}_2$ on $\mathbb{P}^1$. In particular, $\mathcal{P}$ has two $\mathfrak{S}_5$-invariant curves (cf. [111, Corollary 2]). One of these $\mathfrak{S}_5$-invariant curves in $\mathcal{P}$ is, of course, $\mathcal{L}_{10}$. Another curve is a smooth curve of genus 6 known as Wiman's sextic curve (see [123], [39], [61]). By construction, it admits a faithful action of the group $\mathfrak{S}_5$, and one can show that its full automorphism group is $\mathfrak{S}_5$. Its image on $\mathbb{P}^2$ via ϖ is a sextic curve that is singular at the points $\varpi(L_1)$, $\varpi(L_2)$, $\varpi(L_3)$ and $\varpi(L_4)$, and can be given by the equation

$$x^6 + y^6 + z^6 + (x^2 + y^2 + z^2)(x^4 + y^4 + z^4) = 12x^2y^2z^2$$

in appropriate homogeneous coordinates x, y, z.

Using properties of the pencil $\mathcal{P}$ described in Theorem 6.2.9, we can easily describe all $\mathfrak{A}_5$-orbits on S_5 similarly to what we did in Lemma 6.2.5 for $\mathfrak{A}_4$-orbits. To do this, we need one auxiliary result.

Lemma 6.2.11. Let $\mathcal{T}_{15}$ be the $\mathfrak{A}_5$-orbit of the curves T_1, T_2, and T_3. Then $\mathcal{T}_{15}$ is a union of 15 smooth rational cubic curves, and $\mathcal{T}_{15} \sim -3K_{S_5}$.

Proof. By construction, the stabilizer in $\mathfrak{A}_4$ of T_1 is $\boldsymbol{\mu}_2 \times \boldsymbol{\mu}_2$. By Lemma 5.1.1, this means that the stabilizer in $\mathfrak{A}_5$ of T_1 is also $\boldsymbol{\mu}_2 \times \boldsymbol{\mu}_2$. Hence, $\mathcal{T}_{15}$ is a union of 15 cubic curves. Therefore, $\mathcal{T}_{15} \sim -3K_{S_5}$ by Lemma 6.2.2(i). □

Now we are ready to describe $\mathfrak{A}_5$-orbits in S_5.

Theorem 6.2.12 (cf. [102]). The following assertions hold:

(i) if an $\mathfrak{A}_5$-orbit on S_5 has length r, then $r \in \{6, 10, 15, 20, 30, 60\}$;

(ii) the only $\mathfrak{A}_5$-orbits of length 6 on S_5 are Σ_6 and Σ_6';

(iii) the only $\mathfrak{A}_5$-orbits of length 10 on S_5 are Σ_{10} and Σ_{10}';

(iv) the only $\mathfrak{A}_5$-orbit of length 15 on S_5 is Σ_{15};

(v) the only $\mathfrak{A}_5$-orbit of length 20 on S_5 is Σ_{20};

(vi) the $\mathfrak{A}_5$-orbits Σ_6, Σ_6', Σ_{10}, Σ_{10}', and Σ_{15} are contained in the curve $\mathcal{T}_{15}$;

(vii) the $\mathfrak{A}_5$-orbit of every point in $\mathcal{T}_{15} \setminus (\Sigma_6 \cup \Sigma_6' \cup \Sigma_{10} \cup \Sigma_{10}' \cup \Sigma_{15})$ has length 30;

(viii) every $\mathfrak{A}_5$-orbit of length 30 in S_5 is contained in the curve $\mathcal{T}_{15}$;

(ix) the irreducible components of the curve $\mathcal{T}_{15}$ are disjoint away from
$$\Sigma_6 \cup \Sigma_6' \cup \Sigma_{10} \cup \Sigma_{10}' \cup \Sigma_{15};$$

(x) exactly 5 (resp., 5, 3, 3, 2) irreducible components of the curve $\mathcal{T}_{15}$ pass through one point in Σ_6 (resp., Σ_6', Σ_{10}, Σ_{10}', Σ_{15}), and each irreducible component of the curve $\mathcal{T}_{15}$ contains exactly one point of each of the $\mathfrak{A}_5$-orbits Σ_6, Σ_6', Σ_{10}, Σ_{10}', and Σ_{15}.

Proof. By Theorem 6.2.9, we have assertions (i), (ii), (iii), (iv), and (v). Assertion (vi) easily follows from Theorem 6.2.9 as well. Assertions (vii) and (viii) follow from Lemma 6.2.5. Assertions (ix) and (x) follow from assertions (vi) and (vii). □

Remark 6.2.13. It follows from the construction of the curves $\mathcal{L}_{10}$ and $\mathcal{T}_{15}$ that the intersection $\mathcal{L}_{10} \cap \mathcal{T}_{15}$ consists of Σ_{15} and one $\mathfrak{A}_5$-orbit of length 30. Similarly, it follows from Theorem 6.2.12 that the intersection $\mathcal{G}_5 \cap \mathcal{T}_{15}$ (resp., $\mathcal{G}_5' \cap \mathcal{T}_{15}$) consists of Σ_{10} and one $\mathfrak{A}_5$-orbit of length 30, and the intersection $\mathcal{R} \cap \mathcal{T}_{15}$ (resp., $\mathcal{R}' \cap \mathcal{T}_{15}$) consists of Σ_6 (resp., Σ_6') and one $\mathfrak{A}_5$-orbit of length 30. Moreover, for every smooth curve $C \in \mathcal{P}$, the intersection $C \cap \mathcal{T}_{15}$ consists of three distinct $\mathfrak{A}_5$-orbits of length 30 by Lemma 5.1.5.

By Lemma 6.2.1, we have two $\mathfrak{A}_5$-equivariant linear projections
$$\theta \colon S_5 \dashrightarrow \mathbb{P}^2 \cong \mathbb{P}(W_3)$$
and
$$\theta' \colon S_5 \dashrightarrow \mathbb{P}^2 \cong \mathbb{P}(W_3').$$

Lemma 6.2.14 ([102, Theorem 5]). *Both maps θ and θ' are finite morphisms of degree 5.*

CREMONA GROUPS AND THE ICOSAHEDRON   117

Proof. Let $\Pi_2 \cong \mathbb{P}^2$ be one of the two $\mathfrak{A}_5$-invariant planes in $\mathbb{P}^5$. It is enough to show that Π_2 is disjoint from S_5. Suppose it is not. Since S_5 is an intersection of quadrics, the set $\Pi_2 \cap S_5$ is also an intersection of quadrics. Hence, the intersection $\Pi_2 \cap S_5$ cannot be a finite number of points by Lemma 6.1.11. Thus, it contains either a line or a conic, which must be $\mathfrak{A}_5$-invariant. The latter is impossible by Lemma 6.2.2(i). □

Note that $\mathfrak{A}_5$-orbits on S_5 are mapped by θ and θ' to $\mathfrak{A}_5$-orbits on $\mathbb{P}^2$, and lines in S_5 are mapped by θ and θ' to lines in $\mathbb{P}^2$. This can be used to match Theorem 6.2.12 with Lemma 5.3.1 and Theorem 6.1.2. For example, all $\mathfrak{A}_5$-orbits of length 6 on S_5 are mapped by θ and θ' to the unique $\mathfrak{A}_5$-orbit on $\mathbb{P}^2$ of length 6. Similarly, the curve $\mathcal{L}_{10}$ is mapped by θ and θ' to the unique $\mathfrak{A}_5$-irreducible curve in $\mathbb{P}^2$ that is a union of 10 lines.

6.3 Clebsch cubic surface

By Lemma 5.3.3(vii), there exists a unique $\mathfrak{A}_5$-invariant cubic surface S_3 in $\mathbb{P}^3 \cong \mathbb{P}(W_4)$. The surface S_3 is smooth by Lemma 5.3.3(viii). We may identify S_3 with a surface in the projective space $\mathbb{P}^4$ with homogeneous coordinates $x_0, \ldots, x_4$ equipped with the standard action of the group $\mathfrak{S}_5$ that is given by the equations

$$\begin{cases} x_0 + \ldots + x_4 = 0, \\ x_0^3 + \ldots + x_4^3 = 0. \end{cases} \quad (6.3.1)$$

Remark 6.3.2. The surface S_3 has been described by Alfred Clebsch in [25]. Because of this, it is usually called the *Clebsch diagonal cubic surface*, or simply the Clebsch cubic. The group of automorphisms of the Clebsch cubic is $\mathfrak{S}_5$ (see, for example, [34, §9.5.4]).

Recall that every smooth cubic surface in $\mathbb{P}^2$ contains exactly 27 lines.

Lemma 6.3.3. *The following assertions hold:*

(i) *the 27 lines contained in S_3 are irreducible components of three $\mathfrak{A}_5$-irreducible curves $\mathcal{L}_6$, $\mathcal{L}_6'$, and $\mathcal{L}_{15}$, having 6, 6, and 15 irreducible components, respectively (as in §6.1, we will refer to the lines that are irreducible components of the curves $\mathcal{L}_6$, $\mathcal{L}_6'$, and $\mathcal{L}_{15}$ as the lines of $\mathcal{L}_6$, $\mathcal{L}_6'$, and $\mathcal{L}_{15}$, respectively);*

(ii) *the lines of $\mathcal{L}_6$ (resp., $\mathcal{L}_6'$) are pairwise disjoint;*

(iii) there exists a commutative diagram

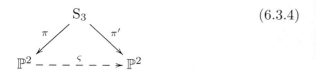 (6.3.4)

such that π (resp., π') is an $\mathfrak{A}_5$-birational morphism that contracts the lines of $\mathcal{L}_6$ (resp., $\mathcal{L}_6'$) to the unique $\mathfrak{A}_5$-orbit of length 6 in $\mathbb{P}^2$;

(iv) each line of $\mathcal{L}_6'$ (resp., of $\mathcal{L}_6$) is mapped by π (resp., by π') to a conic that passes through exactly 5 points of the unique $\mathfrak{A}_5$-orbit of length 6 in $\mathbb{P}^2$;

(v) each line of $\mathcal{L}_{15}$ is mapped by π and π' to a line that passes through exactly 2 points of the unique $\mathfrak{A}_5$-orbit of length 6 in $\mathbb{P}^2$.

Proof. By Lemma 5.3.1(i), the projective plane $\mathbb{P}^2$ contains a unique $\mathfrak{A}_5$-orbit of length 6. Let $\pi\colon S \to \mathbb{P}^2$ be a blow-up of this orbit. Then S is a smooth cubic surface by Corollary 6.1.3 (cf. [58, §7]). Moreover, S is acted on by $\mathfrak{A}_5$. In particular, S is isomorphic to S_3 by Lemma 5.3.3(vii). This easily implies all required assertions. □

Remark 6.3.5 (cf. Remark 6.2.3). Although there are $\mathfrak{A}_5$-birational morphisms from S_3 to $\mathbb{P}^2$, and thus some facts about the $\mathfrak{A}_5$-orbits and $\mathfrak{A}_5$-invariant curves on S_3 can be recovered from the corresponding facts about $\mathbb{P}^2$, in some cases we will prefer to take a more straightforward way that does not rely much on the latter contractions. In particular, we will not make attempts to unify the notation for orbits and curves on S_3 and $\mathbb{P}^2$ (cf. §6.1).

Lemma 6.3.3(iii) immediately implies the following

Corollary 6.3.6. One has $\operatorname{rk}\operatorname{Pic}(S_3)^{\mathfrak{A}_5} = 2$. Moreover, the curves $\mathcal{L}_6$ and $\mathcal{L}_6'$ generate the $\mathfrak{A}_5$-invariant Mori cone of S_3 over $\mathbb{Q}$.

Corollary 6.3.7. One has $\mathcal{L}_6 + \mathcal{L}_6' \sim -4K_{S_3}$ and $\mathcal{L}_{15} \sim -5K_{S_3}$.

Proof. Apply Lemma 6.3.3(iv),(v). □

Corollary 6.3.8. Let θ be any odd permutation in $\operatorname{Aut}(S_3) \cong \mathfrak{S}_5$. Then θ does not act biregularly on $\mathbb{P}^2 \cong \mathbb{P}(W_3)$, and $\theta(\mathcal{L}_6) = \mathcal{L}_6'$.

Proof. The first assertion follows from the fact that there is no regular action of $\mathfrak{S}_5$ on $\mathbb{P}^2$ (see Remark 6.1.1). If $\theta(\mathcal{L}_6) \neq \mathcal{L}_6'$, then $\theta(\mathcal{L}_6) = \mathcal{L}_6$ and θ acts biregularly on $\mathbb{P}^2$, which is a contradiction. □

Remark 6.3.9. Note that there is a unique subgroup of $\mathrm{PSL}_3(\mathbb{C})$ isomorphic to $\mathfrak{A}_5$ up to conjugation. On the other hand, there are two non-conjugate embeddings

$$\mathfrak{A}_5 \hookrightarrow \mathrm{PSL}_3(\mathbb{C})$$

that identify $\mathbb{P}^2$ either with $\mathbb{P}(W_3)$ or $\mathbb{P}(W_3')$, see Remark 6.1.1. These embeddings are related by an outer automorphism of the group $\mathfrak{A}_5$ that is provided by an odd element θ of the group $\mathrm{Aut}(\mathfrak{A}_5) \cong \mathfrak{S}_5$. Therefore, the commutative diagram (6.3.4) and the fact that $\mathrm{Aut}(S_3) \cong \mathfrak{S}_5$ show that the two embeddings of $\mathfrak{A}_5$ to $\mathrm{PSL}_3(\mathbb{C})$ that are non-conjugate in $\mathrm{PSL}_3(\mathbb{C})$ become conjugate in $\mathrm{Cr}_2(\mathbb{C})$.

Corollary 6.3.10. *Let θ be any odd permutation in $\mathrm{Aut}(S_3) \cong \mathfrak{S}_5$. Then*

$$\begin{cases} \theta^*\left(\pi^*(\mathcal{O}_{\mathbb{P}^2}(1))\right) \sim \pi^*(\mathcal{O}_{\mathbb{P}^2}(5)) - 2\mathcal{L}_6, \\ \theta^*(\mathcal{L}_6) \sim \pi^*(\mathcal{O}_{\mathbb{P}^2}(12)) - 5\mathcal{L}_6. \end{cases}$$

Proof. Since $\theta^*(K_{S_3}) \sim K_{S_3}$ and $-K_{S_3} \sim \pi^*(3L) - \mathcal{L}_6$, the required assertion follows from Corollaries 6.3.7 and 6.3.8. □

Corollary 6.3.11. *Let $\mathcal{M}$ be a linear system on $\mathbb{P}^2$ that does not have fixed curves, let $\mathcal{M}'$ be its proper transform on $\mathbb{P}^2$ via ς in (6.3.4), let n and n' be positive integers such that $\mathcal{M} \sim \mathcal{O}_{\mathbb{P}^2}(n)$ and $\mathcal{M}' \sim \mathcal{O}_{\mathbb{P}^2}(n')$. Denote by m and m' the multiplicities in the unique $\mathfrak{A}_5$-orbit of length 6 on $\mathbb{P}^2$ of the linear systems $\mathcal{M}$ and $\mathcal{M}'$, respectively. Then $m > \frac{n}{3}$ if and only if $m' < \frac{n'}{3}$.*

Proof. This follows from Corollary 6.3.10. □

Now it is time to describe $\mathfrak{A}_5$-orbits on S_3.

Lemma 6.3.12. *The following assertions hold:*

(i) *if an $\mathfrak{A}_5$-orbit on S_3 has length r, then $r \in \{10, 12, 15, 20, 30, 60\}$;*

(ii) *for any $r \in \{10, 15, 20\}$ there is a unique $\mathfrak{A}_5$-orbit Σ_r on S_3 of length r;*

(iii) *the curve $\mathcal{L}_6$ (resp., $\mathcal{L}_6'$) contains a unique $\mathfrak{A}_5$-orbit Σ_{12} (resp., Σ_{12}') of length 12;*

(iv) the $\mathfrak{A}_5$-orbits Σ_{12} and Σ'_{12} are the only $\mathfrak{A}_5$-orbits on S_3 of length 12;

(v) the $\mathfrak{A}_5$-orbits Σ_{10} and Σ_{15} are contained in $\mathcal{L}_{15}$, and every other $\mathfrak{A}_5$-orbit contained in $\mathcal{L}_{15}$ is of length 30;

(vi) there is a unique $\mathfrak{A}_5$-orbit Σ_{30} on S_3 of length 30 that is not contained in $\mathcal{L}_{15}$;

(vii) one has $\Sigma_{30} = \mathcal{L}_6 \cap \mathcal{L}'_6$, while $\Sigma_{10} \not\subset \mathcal{L}_6 \cup \mathcal{L}'_6$ and $\Sigma_{15} \not\subset \mathcal{L}_6 \cup \mathcal{L}'_6$.

Proof. All required assertions very easily follow from Lemma 5.3.1 using (6.3.4). We leave details to the reader. □

The combinatorics of the 27 lines on the Clebsch cubic surface S_3 is given by:

Lemma 6.3.13 (cf. [34, §9.5.4]). *The following assertions hold:*

(i) one has $\mathcal{L}_6 \cap \mathcal{L}'_6 = \Sigma_{30}$ and every point in Σ_{30} lies on exactly one line of $\mathcal{L}_6$ and on exactly one line in $\mathcal{L}'_6$;

(ii) the intersection $\mathcal{L}_6 \cap \mathcal{L}_{15}$ (resp., $\mathcal{L}'_6 \cap \mathcal{L}_{15}$) is an $\mathfrak{A}_5$-orbit of length 30, and every point in this $\mathfrak{A}_5$-orbit lies on exactly one line of $\mathcal{L}_6$ (resp., $\mathcal{L}'_6$) and on exactly one line of $\mathcal{L}_{15}$;

(iii) every point in Σ_{10} lies on exactly 3 lines of $\mathcal{L}_{15}$, and every line of $\mathcal{L}_{15}$ contains exactly two points in Σ_{10};

(iv) every point in Σ_{15} lies on exactly 2 lines of $\mathcal{L}_{15}$, and every line of $\mathcal{L}_{15}$ contains exactly two points in Σ_{15};

(v) the lines of $\mathcal{L}_{15}$ are disjoint outside of $\Sigma_{10} \cup \Sigma_{15}$.

Proof. All required assertions follow from Lemmas 5.3.1 and 6.3.12 using (6.3.4). □

Note that by Lemma 6.3.13 the $\mathfrak{A}_5$-orbit Σ_{10} is the set of Eckardt points of the cubic surface S_3.

Now let us prove two technical results that will be used in the proofs of Lemmas 15.3.1 and 15.3.4, respectively.

Lemma 6.3.14. *Let C be an irreducible conic in S_3. Then its $\mathfrak{A}_5$-orbit consists of at least 10 conics.*

CREMONA GROUPS AND THE ICOSAHEDRON

Proof. Suppose that the $\mathfrak{A}_5$-orbit of C consists of $r < 10$ conics. Let G be the stabilizer of C in $\mathfrak{A}_5$. By Lemma 5.1.1, we have either $r = 1$ and $G \cong \mathfrak{A}_5$, or $r = 5$ and $G \cong \mathfrak{A}_4$, or $r = 6$ and $G \cong \mathfrak{D}_{10}$. The first and the third cases are impossible by Corollary 5.2.3. Hence, $r = 5$, and the residual lines of the conics in the $\mathfrak{A}_5$-orbit of C form an $\mathfrak{A}_5$-invariant set of at most 5 lines, which is impossible by Lemma 6.3.3(i). □

Lemma 6.3.15. *Let G be a subgroup in $\mathfrak{A}_5$ that is isomorphic to $\mathfrak{A}_4$. Then there exists a unique G-invariant hyperplane section of S_3. Moreover, this hyperplane section consists of three coplanar lines of $\mathcal{L}_{15}$ that do not intersect in one point.*

Proof. Recall that we identified S_3 with a surface in $\mathbb{P}^4$ given by (6.3.1). Since all subgroups in $\mathfrak{A}_5$ isomorphic to $\mathfrak{A}_4$ are conjugate, we may assume that G permutes x_0, x_1, x_2, x_3. Then there exists a unique G-invariant hyperplane section of S_3 that is given by

$$x_4 = x_0 + x_1 + x_2 + x_3 = 0.$$

Hence, this hyperplane section is a cubic curve in $\mathbb{P}^2$ that is given by

$$x_0^3 + x_1^3 + x_2^3 = (x_0 + x_1 + x_2)^3.$$

Computing derivatives, we see that the singular points of this curve are $[1:1:-1]$, $[1:-1:1]$, and $[-1:1:1]$. Thus, it is a union of three lines that do not pass through one point. These must be lines of $\mathcal{L}_{15}$ by Lemma 6.3.3. □

By Lemma 5.3.3(viii), there exists a smooth irreducible $\mathfrak{A}_5$-invariant curve $\mathcal{B}_6 \subset S_3$ that is cut out on S_3 by the unique $\mathfrak{A}_5$-invariant quadric in $\mathbb{P}^3$. The curve $\mathcal{B}_6$, known as the Bring's curve (see Remark 5.4.2), is of degree 6 and is given by

$$\begin{cases} x_0 + \ldots + x_4 = 0, \\ x_0^2 + \ldots + x_4^3 = 0, \\ x_0^3 + \ldots + x_4^3 = 0. \end{cases}$$

Lemma 6.3.16. *One has $\mathcal{L}_6 \cap \mathcal{B}_6 = \Sigma_{12}$ and $\mathcal{L}_6' \cap \mathcal{B}_6 = \Sigma_{12}'$. Moreover, the intersection $\mathcal{L}_{15} \cap \mathcal{B}_6$ is an $\mathfrak{A}_5$-orbit of length 30 that is different from Σ_{30}. Furthermore, one has $\Sigma_{10} \not\subset \mathcal{B}_6$, $\Sigma_{15} \not\subset \mathcal{B}_6$, and $\Sigma_{20} \subset \mathcal{B}_6$.*

Proof. All required assertions follow from Lemmas 5.1.5 and 6.3.12. □

Let $\mathscr{C}$ (resp., $\mathscr{C}'$) be a smooth rational curve in S_3 that is a proper transform of the unique $\mathfrak{A}_5$-invariant conic in $\mathbb{P}^2$ (see Lemma 5.3.1(iii)) via π (resp., π').

Lemma 6.3.17. The curves $\mathscr{C}$ and $\mathscr{C}'$ have degree 6. One has

$$\mathcal{L}_6 \cap \mathscr{C} = \mathcal{L}_6' \cap \mathscr{C}' = \varnothing, \ \mathcal{L}_6 \cap \mathscr{C}' = \Sigma_{12}, \ \mathcal{L}_6' \cap \mathscr{C} = \Sigma_{12}'$$

and $\mathscr{C} \cap \mathscr{C}' = \Sigma_{20}$. Moreover, one has

$$\mathscr{C} + \mathscr{C}' \sim -4K_{S_3},$$
$$3\mathscr{C} \sim -2K_{S_3} + \mathcal{L}_6 \sim -10K_{S_3} - 2\mathcal{L}_6',$$
$$3\mathscr{C}' \sim -2K_{S_3} + \mathcal{L}_6' \sim -10K_{S_3} - 2\mathcal{L}_6,$$
$$\mathscr{C} \sim_\mathbb{Q} \frac{3}{2}\mathcal{L}_6 + \frac{1}{2}\mathcal{L}_6',$$
$$\mathscr{C}' \sim_\mathbb{Q} \frac{1}{2}\mathcal{L}_6 + \frac{3}{2}\mathcal{L}_6'.$$

Proof. This is obvious. □

The surface S_3 contains infinitely many $\mathfrak{A}_5$-irreducible curves of degree 12. Indeed, every curve in the pencil generated by $\mathcal{L}_6 + \mathcal{L}_6'$ and $2\mathcal{B}_6$ is $\mathfrak{A}_5$-invariant, and a general curve in this pencil is, in fact, irreducible. On the other hand, there are just finitely many $\mathfrak{A}_5$-irreducible curves in S_3 of degree less than 12 by:

Theorem 6.3.18. Let C be an $\mathfrak{A}_5$-irreducible curve in S_3 of degree less than 12. Then C is one of the curves $\mathcal{L}_6$, $\mathcal{L}_6'$, $\mathscr{C}$, $\mathscr{C}'$, or $\mathcal{B}_6$.

Proof. Assume that C is different from the curves $\mathcal{L}_6$, $\mathcal{L}_6'$, and $\mathcal{B}_6$. Then

$$\mathcal{B}_6 \cdot C = 2\deg(C) \leqslant 22,$$

which implies that $\mathcal{B}_6 \cdot C = 12$, because the only possible lengths of $\mathfrak{A}_5$-orbits in $\mathcal{B}_6$ are 12, 30, and 60 by Lemma 5.1.5. Thus, we have $\deg(C) = 6$.

It follows from Corollary 6.3.6 that

$$C \sim_\mathbb{Q} a\mathcal{L}_6 + b\mathcal{L}_6'$$

for some positive rational numbers a and b such that $\deg(C) = 6(a+b)$. Hence $\pi(C)$ is a curve of degree $12b$, and $\pi'(C)$ is a curve of degree $12a$.

Note that $\mathcal{L}_6^2 = -6$ and $\mathcal{L}_6 \cdot \mathcal{L}_6' = 30$ (see Corollary 6.3.7). Thus, we have

$$C \cdot \mathcal{L}_6 = \left(a\mathcal{L}_6 + b\mathcal{L}_6'\right) \cdot \mathcal{L}_6 = -6a + 30b = 5\deg(C) - 36a. \qquad (6.3.19)$$

Without loss of generality, we may assume that $a \leqslant b$. Let us show that $C = \mathscr{C}$. (The case $a \geqslant b$ is treated similarly, and gives $C = \mathscr{C}'$.)

One has
$$2a \leqslant a + b = \frac{\deg(C)}{6} = 1,$$
which implies that $\pi(C)$ is an $\mathfrak{A}_5$-invariant curve of degree $12a \leqslant 6$. By Lemma 5.3.1(iii), we see that either $a = \frac{1}{6}$ or $a = \frac{1}{2}$. If $a = \frac{1}{6}$, then $\pi(C)$ is the $\mathfrak{A}_5$-invariant conic by Lemma 5.3.1(iv), so that $C = \mathscr{C}$ by construction. If $a = \frac{1}{2}$, then $\pi(C)$ is contained in the pencil generated by $3\pi(\mathscr{C})$ and $\pi(\mathcal{B}_6)$ by Lemma 5.3.1(vi). By (6.3.19), we have $C \cdot \mathcal{L}_6 = 6$. In particular, the curve $\pi(C)$ passes through the $\mathfrak{A}_5$-orbit $\pi(\mathcal{L}_6)$. On the other hand, the curve $\pi(\mathcal{B}_6)$ is the only curve in this pencil that passes through $\pi(\mathcal{L}_6)$ (see Remark 6.1.5). Thus, we have $C = \mathscr{C}$, since we assumed that $C \neq \mathcal{B}_6$. □

6.4 Two-dimensional quadric

In this section we will consider the actions of the group $\mathfrak{A}_5$ on a two-dimensional quadric $\mathbb{P}^1 \times \mathbb{P}^1$, and study $\mathfrak{A}_5$-orbits and $\mathfrak{A}_5$-invariant curves of low degree on $\mathbb{P}^1 \times \mathbb{P}^1$. We will also make some remarks about the action of the group $\mathfrak{D}_{10}$ on $\mathbb{P}^1 \times \mathbb{P}^1$. Since these results will be applied later to $\mathbb{P}^1 \times \mathbb{P}^1$ embedded into particular threefolds, we introduce a bit awkward-looking notation for $\mathfrak{A}_5$-orbits and $\mathfrak{A}_5$-invariant curves for the sake of consistency with further chapters. We hope that the reader will not be confused by them.

We will pay special attention to two types of action. If U is any of the two irreducible two-dimensional representations of the group $2.\mathfrak{A}_5$ (see §5.6), then we will refer to the action of $\mathfrak{A}_5$ on

$$\mathbb{P}^1 \times \mathbb{P}^1 \cong \mathbb{P}(U) \times \mathbb{P}(U)$$

as a *diagonal* action. If U_2 and U_2' are two different irreducible two-dimensional representation of the group $2.\mathfrak{A}_5$, then we will refer to the action of $\mathfrak{A}_5$ on

$$\mathbb{P}^1 \times \mathbb{P}^1 \cong \mathbb{P}(U_2) \times \mathbb{P}(U_2')$$

as a *twisted diagonal* action. Note that there are two diagonal actions up to conjugation although the corresponding subgroup of $\mathrm{Aut}(\mathbb{P}^1 \times \mathbb{P}^1)$ is the same up to conjugation in these cases. On the other hand, there is a unique twisted diagonal action of $\mathfrak{A}_5$ on $\mathbb{P}^1 \times \mathbb{P}^1$ up to conjugation in $\mathrm{Aut}(\mathbb{P}^1 \times \mathbb{P}^1)$, because such action can be extended to an embedding of the group $\mathrm{Aut}(\mathfrak{A}_5) \cong \mathfrak{S}_5$ into $\mathrm{Aut}(\mathbb{P}^1 \times \mathbb{P}^1)$ (cf. Remark 6.3.9). There are

two more embeddings of $\mathfrak{A}_5$ into $\operatorname{Aut}(\mathbb{P}^1 \times \mathbb{P}^1)$ up to conjugation, given by identifications
$$\mathbb{P}^1 \times \mathbb{P}^1 \cong \mathbb{P}(I \oplus I) \times \mathbb{P}(U_2)$$
and
$$\mathbb{P}^1 \times \mathbb{P}^1 \cong \mathbb{P}(I \oplus I) \times \mathbb{P}(U_2').$$
The corresponding subgroups $\mathfrak{A}_5 \subset \operatorname{Aut}(\mathbb{P}^1 \times \mathbb{P}^1)$ are conjugate to each other in the latter two cases, but are not conjugate to subgroups corresponding to the former three actions.

Five embeddings described above are all possible embeddings of $\mathfrak{A}_5$ to $\operatorname{Aut}(\mathbb{P}^1 \times \mathbb{P}^1)$ up to conjugation. In particular, it is easy to check that the embedding corresponding to an identification
$$\mathbb{P}^1 \times \mathbb{P}^1 \cong \mathbb{P}(U) \times \mathbb{P}(I \oplus I)$$
is conjugate to the embedding corresponding to the identification
$$\mathbb{P}^1 \times \mathbb{P}^1 \cong \mathbb{P}(I \oplus I) \times \mathbb{P}(U),$$
where U is any of the two irreducible two-dimensional representations of the group $2.\mathfrak{A}_5$.

Now suppose that $\mathbb{P}^1 \times \mathbb{P}^1$ is equipped by some faithful action of the group $\mathfrak{A}_5$. Note that by Corollary 5.1.2 the group $\mathfrak{A}_5$ preserves the projections of $\mathbb{P}^1 \times \mathbb{P}^1$ onto both factors, and thus acts on both copies of $\mathbb{P}^1$. These actions lift to two two-dimensional representations U_2 and U_2' of the group $2.\mathfrak{A}_5$ (see §5.6), one of which may be trivial. Thus, we see that the actions described above are the only possible faithful actions of $\mathfrak{A}_5$ on $\mathbb{P}^1 \times \mathbb{P}^1$.

We can describe $\mathfrak{A}_5$-orbits on $\mathbb{P}^1 \times \mathbb{P}^1$.

Lemma 6.4.1. Let $\mathbb{P}^1 \times \mathbb{P}^1$ be equipped either with a diagonal or with a twisted diagonal action of the group $\mathfrak{A}_5$. Then for any $r \in \{12, 20, 30\}$ there are exactly two $\mathfrak{A}_5$-orbits of length r contained in $\mathbb{P}^1 \times \mathbb{P}^1$. Moreover, if Σ is an $\mathfrak{A}_5$-orbit in $\mathbb{P}^1 \times \mathbb{P}^1$, then $|\Sigma| \in \{12, 20, 30, 60\}$.

Proof. Let $\Sigma \subset \mathbb{P}^1 \times \mathbb{P}^1$ be an $\mathfrak{A}_5$-orbit of length $r < 60$. Consider the projection $p_1 \colon \mathbb{P}^1 \times \mathbb{P}^1 \to \mathbb{P}^1$ to the first factor. Since p_1 is $\mathfrak{A}_5$-equivariant, the image $p_1(\Sigma) \subset \mathbb{P}^1$ is an $\mathfrak{A}_5$-orbit. Thus
$$r = |\Sigma| = |p_1(\Sigma)| \in \{12, 20, 30\}$$

by Lemma 5.1.4. Moreover, the value of r uniquely defines $p_1(\Sigma)$. The curve $p_1^{-1}(p_1(\Sigma))$ is a disjoint union of r fibers of the $\mathbb{P}^1$-fibration

$$p_1 \colon \mathbb{P}^1 \times \mathbb{P}^1 \to \mathbb{P}^1,$$

and each of these r fibers contains exactly one point of Σ. Let

$$L \subset p_1^{-1}(p_1(\Sigma))$$

be one of these fibers, and $G_L \subset \mathfrak{A}_5$ be the stabilizer of L. By Lemma 5.1.1 the group G_L is a (non-trivial) cyclic group. Moreover, G_L acts faithfully on L by the assumption about the $\mathfrak{A}_5$-action on $\mathbb{P}^1 \times \mathbb{P}^1$. Therefore, G_L has exactly two fixed points on L by Lemma 5.5.9(i), which give rise to two orbits of length r on $\mathbb{P}^1 \times \mathbb{P}^1$. □

Lemma 6.4.2. Let $\mathbb{P}^1 \times \mathbb{P}^1$ be equipped either with a diagonal or with a twisted diagonal action of the group $\mathfrak{A}_5$. Then

(i) if there is a reduced $\mathfrak{A}_5$-invariant effective divisor of bi-degree $(0, r)$ on $\mathbb{P}^1 \times \mathbb{P}^1$, then either $r \in \{12, 20\}$, or $r \geqslant 30$;

(ii) there is a unique $\mathfrak{A}_5$-invariant effective divisor $\hat{\mathcal{L}}_{12}$ on $\mathbb{P}^1 \times \mathbb{P}^1$ of bi-degree $(0, 12)$, and $\hat{\mathcal{L}}_{12}$ is a union of 12 different curves of bi-degree $(0, 1)$; similarly, there is a unique $\mathfrak{A}_5$-invariant effective divisor $\hat{\mathcal{L}}_{20}$ on $\mathbb{P}^1 \times \mathbb{P}^1$ of bi-degree $(0, 20)$, and $\hat{\mathcal{L}}_{20}$ is a union of 20 different curves of bi-degree $(0, 1)$;

(iii) any $\mathfrak{A}_5$-orbit of length 12 in $\mathbb{P}^1 \times \mathbb{P}^1$ is contained in the curve $\hat{\mathcal{L}}_{12}$; similarly, any $\mathfrak{A}_5$-orbit of length 20 in $\mathbb{P}^1 \times \mathbb{P}^1$ is contained in the curve $\hat{\mathcal{L}}_{20}$.

Proof. The support of any $\mathfrak{A}_5$-invariant effective divisor D of bi-degree $(0, r)$ on $\mathbb{P}^1 \times \mathbb{P}^1$ is a union of at most r curves of bi-degree $(0, 1)$. Thus, D gives rise to an $\mathfrak{A}_5$-invariant subset of $\mathbb{P}^1$ consisting of at most r points via the projection $p_2 \colon \mathbb{P}^1 \times \mathbb{P}^1 \to \mathbb{P}^1$ to the second factor. Thus assertion (i) follows from Lemma 5.1.4. The curves $\hat{\mathcal{L}}_{12}$ and $\hat{\mathcal{L}}_{20}$ are preimages with respect to p_2 of the unique $\mathfrak{A}_5$-orbits on $\mathbb{P}^1$ of length 12 and 20, respectively (see Lemma 5.5.9(v)). This gives assertion (ii). Assertion (iii) follows from Lemma 6.4.1. □

It appears that one can tell one action of $\mathfrak{A}_5$ on $\mathbb{P}^1 \times \mathbb{P}^1$ from the other in terms of $\mathfrak{A}_5$-invariant curves of low degree.

Lemma 6.4.3. Let $\mathbb{P}^1 \times \mathbb{P}^1$ be equipped with a faithful action of the group $\mathfrak{A}_5$. Then

(i) the action of $\mathfrak{A}_5$ on $\mathbb{P}^1 \times \mathbb{P}^1$ is diagonal if and only if there is an $\mathfrak{A}_5$-invariant effective divisor on $\mathbb{P}^1 \times \mathbb{P}^1$ of bi-degree $(1,1)$;

(ii) the action of $\mathfrak{A}_5$ on $\mathbb{P}^1 \times \mathbb{P}^1$ is twisted diagonal if and only if there is an $\mathfrak{A}_5$-invariant effective divisor on $\mathbb{P}^1 \times \mathbb{P}^1$ of bi-degree $(1,7)$;

(iii) if the action of $\mathfrak{A}_5$ on $\mathbb{P}^1 \times \mathbb{P}^1$ is diagonal, then an $\mathfrak{A}_5$-invariant effective divisor $\hat{\mathscr{C}}$ on $\mathbb{P}^1 \times \mathbb{P}^1$ of bi-degree $(1,1)$ is unique, and $\hat{\mathscr{C}}$ is a smooth rational curve;

(iv) if the action of $\mathfrak{A}_5$ on $\mathbb{P}^1 \times \mathbb{P}^1$ is twisted diagonal, then an $\mathfrak{A}_5$-invariant effective divisor $C_{1,7}$ on $\mathbb{P}^1 \times \mathbb{P}^1$ of bi-degree $(1,7)$ is unique, and $C_{1,7}$ is a smooth rational curve.

Proof. The actions of the group $\mathfrak{A}_5$ on two copies of $\mathbb{P}^1$ that are two factors of $\mathbb{P}^1 \times \mathbb{P}^1$ give rise to two (not necessarily non-trivial) two-dimensional representations U and V of the group $2.\mathfrak{A}_5$ (see §5.6). The following possibilities may occur:

(D) $U \cong V$ is an irreducible representation of the group $2.\mathfrak{A}_5$, in which case the action of $\mathfrak{A}_5$ on $\mathbb{P}^1 \times \mathbb{P}^1$ is diagonal;

(TD) U and V are different irreducible representations of the group $2.\mathfrak{A}_5$, in which case the action of $\mathfrak{A}_5$ on $\mathbb{P}^1 \times \mathbb{P}^1$ is twisted diagonal;

(Tr) U is an irreducible representation and V is a trivial two-dimensional representation of the group $2.\mathfrak{A}_5$, or V is an irreducible representation and U is a trivial two-dimensional representation of the group $2.\mathfrak{A}_5$.

Note that an $\mathfrak{A}_5$-invariant effective divisor on $\mathbb{P}^1 \times \mathbb{P}^1$ of bi-degree (a, b) corresponds to a one-dimensional $2.\mathfrak{A}_5$-invariant subspace of $\mathrm{Sym}^a(U^\vee) \otimes \mathrm{Sym}^b(V^\vee)$. On the other hand, a one-dimensional $2.\mathfrak{A}_5$-invariant subspace does exist and is unique in $U^\vee \otimes V^\vee$ in case (D) by Lemma 5.6.3(o), does not exist in case (TD) by Lemma 5.6.4(i), and does not exist in case (Tr) by definition. Similarly, a one-dimensional $2.\mathfrak{A}_5$-invariant subspace does exist and is unique in $U^\vee \otimes \mathrm{Sym}^7(V^\vee)$ in case (TD) by Lemma 5.6.4(ii), does not exist in case (D) by Lemma 5.6.3(iii), and does not exist in case (Tr) by Lemma 5.6.3(ii). In particular, this gives assertions (i) and (ii) of the lemma.

Suppose that the action of $\mathfrak{A}_5$ on $\mathbb{P}^1 \times \mathbb{P}^1$ is diagonal, and let $\hat{\mathscr{C}}$ be the unique $\mathfrak{A}_5$-invariant effective divisor of bi-degree $(1,1)$ on $\mathbb{P}^1 \times \mathbb{P}^1$. Obviously, the divisor $\hat{\mathscr{C}}$ is reduced. It is also irreducible, since otherwise by Corollary 5.1.2 there would exist an $\mathfrak{A}_5$-invariant curve of bi-degree $(0,1)$ on $\mathbb{P}^1 \times \mathbb{P}^1$, which is impossible by Lemma 6.4.2(i). Now the isomorphism $\hat{\mathscr{C}} \xrightarrow{\sim} \mathbb{P}^1$ provided by the projection to the second factor shows that the curve $\hat{\mathscr{C}}$ is smooth and rational, and thus proves assertion (iii).

Suppose that the action of $\mathfrak{A}_5$ on $\mathbb{P}^1 \times \mathbb{P}^1$ is twisted diagonal, and let $C_{1,7}$ be the unique $\mathfrak{A}_5$-invariant effective divisor of bi-degree $(1,7)$ on $\mathbb{P}^1 \times \mathbb{P}^1$. If $C_{1,7}$ is reducible or not reduced, then there exists an $\mathfrak{A}_5$-invariant curve of bi-degree $(0,r)$ on $\mathbb{P}^1 \times \mathbb{P}^1$ for some $r \leqslant 7$. This is impossible by Lemma 6.4.2(i), which shows that $C_{1,7}$ is actually an irreducible curve. As before, the isomorphism $C_{1,7} \xrightarrow{\sim} \mathbb{P}^1$ provided by the projection to the second factor shows that the curve $C_{1,7}$ is smooth and rational, and thus proves assertion (iv). □

Lemma 6.4.4. Let $\mathbb{P}^1 \times \mathbb{P}^1$ be equipped with a diagonal action of the group $\mathfrak{A}_5$. Let $\hat{\mathscr{C}} \subset \mathbb{P}^1 \times \mathbb{P}^1$ be the unique $\mathfrak{A}_5$-invariant curve of bi-degree $(1,1)$ (see Lemma 6.4.3(iii)) and $\hat{\mathcal{L}}_{12} \subset \mathbb{P}^1 \times \mathbb{P}^1$ be the unique $\mathfrak{A}_5$-invariant curve of bi-degree $(0,12)$ (see Lemma 6.4.2(ii)). Let $\hat{Z}$ be an $\mathfrak{A}_5$-invariant effective divisor on $\mathbb{P}^1 \times \mathbb{P}^1$ of bi-degree (a,b) such that $\hat{\mathscr{C}} \not\subset \mathrm{Supp}(\hat{Z})$. Then

  (i) one has $a + b \geqslant 12$;

  (ii) if $\hat{Z} \neq \hat{\mathcal{L}}_{12}$, then $5a + b \geqslant 16$;

  (iii) one has $5a + b \neq 18$.

Proof. The degree of the zero-cycle $\hat{Z}|_{\hat{\mathscr{C}}}$ is $\hat{Z} \cdot \hat{\mathscr{C}} = a + b$. On the other hand, the curve $\hat{\mathscr{C}} \cong \mathbb{P}^1$ does not contain $\mathfrak{A}_5$-orbits of length less than 12 by Lemma 5.1.4. Thus, we have $a + b \geqslant 12$, which proves assertion (i).

Suppose that $5a + b < 16$. Then one has $0 \leqslant a \leqslant 3$. If $a = 0$, then $b \leqslant 15$. Since the zero-cycle $\hat{Z}|_{\hat{\mathscr{C}}}$ has degree b in this case, we have $b = 12$ by Lemma 5.1.4, so that $\hat{Z} = \hat{\mathcal{L}}_{12}$ by Lemma 6.4.2(ii). If $a = 1$, then by assertion (i) one has
$$5a + b \geqslant 4a + 12 \geqslant 16.$$
This proves assertion (ii).

Suppose that $5a + b = 18$. It follows from Lemma 5.1.4 that
$$18 - 4a = a + b = \hat{Z} \cdot \hat{\mathscr{C}} = 12k_1 + 20k_2 + 30k_3 + 60k_4$$

for some non-negative integers k_1, k_2, k_3, and k_4. This immediately leads to a contradiction and proves assertion (iii). □

Lemma 6.4.5. Let $\mathbb{P}^1 \times \mathbb{P}^1$ be equipped with a diagonal action of the group $\mathfrak{A}_5$. Let $\hat{Z} \subset \mathbb{P}^1 \times \mathbb{P}^1$ be an $\mathfrak{A}_5$-invariant effective divisor of bi-degree (a, b) such that $5a + b = 16$. Then $a = 1$ and $b = 11$.

Proof. Let $\hat{\mathscr{C}} \subset \mathbb{P}^1 \times \mathbb{P}^1$ be the unique $\mathfrak{A}_5$-invariant curve of bi-degree $(1,1)$ (see Lemma 6.4.3(iii)). Then $\hat{Z} = n\hat{\mathscr{C}} + \hat{T}$ for some non-negative integer n and an $\mathfrak{A}_5$-invariant effective divisor $\hat{T}$ whose support does not contain $\hat{\mathscr{C}}$. Note that $\hat{T}$ is a divisor of bi-degree $(a-n, b-n)$. In particular, $a \geqslant n$. On the other hand, it follows from Lemma 5.1.4 that

$$16 - 4a - 2n = a + b - 2n = \hat{T} \cdot \hat{\mathscr{C}} = 12k_1 + 20k_2 + 30k_3 + 60k_4$$

for some non-negative integers k_1, k_2, k_3 and k_4. This implies that $n = 0$, $a = 1$ and $b = 11$. □

Lemma 6.4.6. Let $\mathbb{P}^1 \times \mathbb{P}^1$ be equipped with a diagonal action of the group $\mathfrak{A}_5$. Then

(i) there exists a unique $\mathfrak{A}_5$-invariant effective divisor $C_{1,11}$ on $\mathbb{P}^1 \times \mathbb{P}^1$ of bi-degree $(1, 11)$;

(ii) the divisor $C_{1,11}$ is a smooth rational curve.

Proof. Write $\mathbb{P}^1 \times \mathbb{P}^1 \cong \mathbb{P}(U) \times \mathbb{P}(U)$ for some irreducible two-dimensional representation U of the group $2.\mathfrak{A}_5$. An $\mathfrak{A}_5$-invariant effective divisor on $\mathbb{P}^1 \times \mathbb{P}^1$ of bi-degree (a, b) corresponds to a one-dimensional $2.\mathfrak{A}_5$-invariant subspace of $\mathrm{Sym}^a(U^\vee) \otimes \mathrm{Sym}^b(U^\vee)$. Thus assertion (i) is implied by Lemma 5.6.3(iv).

We claim that the divisor $C_{1,11}$ is an irreducible curve. Indeed, suppose that it is reducible or non-reduced. Then a union of some of the irreducible components of $C_{1,11}$ is an $\mathfrak{A}_5$-invariant curve of bi-degree $(0, b)$ for some $1 \leqslant b \leqslant 10$. The latter is impossible by Lemma 6.4.2(i).

Therefore, $C_{1,11}$ is an irreducible curve. It is also smooth and rational, since the projection to the second factor $\mathbb{P}^1 \times \mathbb{P}^1 \to \mathbb{P}^1$ induces an isomorphism of $C_{1,11}$ with $\mathbb{P}^1$. This proves assertion (ii). □

Lemma 6.4.7. Let $\mathbb{P}^1 \times \mathbb{P}^1$ be equipped with a diagonal action of the group $\mathfrak{A}_5$. Let $\hat{\mathscr{C}} \subset \mathbb{P}^1 \times \mathbb{P}^1$ be the unique $\mathfrak{A}_5$-invariant curve of bi-degree $(1, 1)$ (see Lemma 6.4.3(iii)) and $\hat{\mathcal{L}}_{12} \subset \mathbb{P}^1 \times \mathbb{P}^1$ be the unique $\mathfrak{A}_5$-invariant curve of bi-degree $(0, 12)$ (see Lemma 6.4.2(ii)). Then the

CREMONA GROUPS AND THE ICOSAHEDRON   129

unique $\mathfrak{A}_5$-invariant effective divisor of bi-degree $(1, 13)$ on $\mathbb{P}^1 \times \mathbb{P}^1$ is the curve $\hat{\mathscr{C}} + \hat{\mathcal{L}}_{12}$.

Proof. Let $\hat{Z}$ be an $\mathfrak{A}_5$-invariant effective divisor on $\mathbb{P}^1 \times \mathbb{P}^1$ of bi-degree $(1, 13)$. If $\hat{\mathscr{C}}$ is an irreducible component of $\hat{Z}$, then $\hat{Z} = \hat{\mathscr{C}} + \hat{\mathcal{L}}_{12}$ by Lemma 6.4.2(ii).

Suppose that $\hat{\mathscr{C}}$ is not contained in $\mathrm{Supp}(\hat{Z})$. Then $\hat{Z}|_{\hat{\mathscr{C}}}$ is an effective $\mathfrak{A}_5$-invariant divisor on $\hat{\mathscr{C}} \cong \mathbb{P}^1$ of degree $\hat{\mathscr{C}} \cdot \hat{Z} = 14$. This is impossible by Lemma 5.1.4. $\square$

Lemma 6.4.8. *Let $\mathbb{P}^1 \times \mathbb{P}^1$ be equipped with a diagonal action of the group $\mathfrak{A}_5$. Then*

(i) *there exists a unique $\mathfrak{A}_5$-invariant effective divisor $C_{2,10}$ on $\mathbb{P}^1 \times \mathbb{P}^1$ of bi-degree $(2, 10)$;*

(ii) *the divisor $C_{2,10}$ is an irreducible smooth hyperelliptic curve of genus 9.*

Proof. Write $\mathbb{P}^1 \times \mathbb{P}^1 \cong \mathbb{P}(U) \times \mathbb{P}(U)$ for some irreducible two-dimensional representation U of the group $2.\mathfrak{A}_5$. An $\mathfrak{A}_5$-invariant effective divisor on $\mathbb{P}^1 \times \mathbb{P}^1$ of bi-degree (a, b) corresponds to a one-dimensional $2.\mathfrak{A}_5$-invariant subspace of $\mathrm{Sym}^a(U^\vee) \otimes \mathrm{Sym}^b(U^\vee)$. Thus assertion (i) is implied by Lemma 5.6.3(v).

Let us show that $C_{2,10}$ is $\mathfrak{A}_5$-irreducible and reduced. Let $\hat{\mathscr{C}} \subset \mathbb{P}^1 \times \mathbb{P}^1$ be the unique $\mathfrak{A}_5$-invariant curve of bi-degree $(1, 1)$ (see Lemma 6.4.3(iii)). Put

$$C_{2,10} = n\hat{\mathscr{C}} + \sum_{i=1}^{r} \hat{Z}_i,$$

where n is a positive integer, and each $\hat{Z}_i$ is an $\mathfrak{A}_5$-irreducible curve that is different from $\hat{\mathscr{C}}$ (note that we do not assume that all these curves are different from each other). Then $\hat{Z}_i$ is a curve of bi-degree (a_i, b_i) for some non-negative integers a_i and b_i. Since $\hat{Z}_i \neq \hat{\mathscr{C}}$ for every $\hat{Z}_i$, we have

$$a_i + b_i \geqslant 12$$

by Lemma 6.4.4(i). On the other hand, one has

$$\sum_{i=1}^{r} a_i = 2 - n$$

and

$$\sum_{i=1}^{r} b_i = 10 - n,$$

since $C_{2,10}$ is a divisor of bi-degree $(2,10)$. Thus, we have $n=0$ and $r=1$, which simply means that the divisor $C_{2,10}$ is an $\mathfrak{A}_5$-irreducible curve.

Let us show that the curve $C_{2,10}$ is irreducible. Put

$$C_{2,10} = \sum_{i=1}^{s} \hat{C}_i,$$

where each $\hat{C}_i$ is an irreducible curve. Since $C_{2,10}$ is a curve of bi-degree $(2,10)$, we see that either $s=1$ or $s=2$. But $s \neq 2$ by Corollary 5.1.2. Thus, we see that $s=1$, which means that $C_{2,10}$ is irreducible.

We claim that the curve $C_{2,10}$ is smooth. Indeed, if $C_{2,10}$ has a singular point, then it has at least 12 singular points by Lemma 6.4.1. The latter is impossible since the number of singular points of $C_{2,10}$ cannot exceed $p_a(C_{2,10})$ by Lemma 4.4.6, and $p_a(C_{2,10}) = 9$ by the adjunction formula. Thus, the curve $C_{2,10}$ is a smooth curve of genus 9. Since the bi-degree of $C_{2,10}$ is $(2,10)$, the curve $C_{2,10}$ is hyperelliptic. □

The curve $C_{2,10}$ constructed in Lemma 6.4.8 is the unique smooth hyperelliptic curve of genus 9 that admits a faithful action of $\mathfrak{A}_5$ (see Lemma 5.4.4).

Lemma 6.4.9. Let $\mathbb{P}^1 \times \mathbb{P}^1$ be equipped with a diagonal action of the group $\mathfrak{A}_5$. Then

(i) there exists a unique $\mathfrak{A}_5$-invariant effective divisor on $\mathbb{P}^1 \times \mathbb{P}^1$ of bi-degree $(1,18)$;

(ii) there exists a unique $\mathfrak{A}_5$-invariant effective divisor on $\mathbb{P}^1 \times \mathbb{P}^1$ of bi-degree $(3,9)$;

(iii) there exists a unique $\mathfrak{A}_5$-invariant effective divisor on $\mathbb{P}^1 \times \mathbb{P}^1$ of bi-degree $(4,8)$.

Proof. Write $\mathbb{P}^1 \times \mathbb{P}^1 \cong \mathbb{P}(U) \times \mathbb{P}(U)$ for some irreducible two-dimensional representation U of the group $2.\mathfrak{A}_5$. An $\mathfrak{A}_5$-invariant effective divisor on $\mathbb{P}^1 \times \mathbb{P}^1$ of bi-degree (a,b) corresponds to a one-dimensional $2.\mathfrak{A}_5$-invariant subspace of $\operatorname{Sym}^a(U^\vee) \otimes \operatorname{Sym}^b(U^\vee)$. Now everything is implied by assertions (vi), (vii), and (viii) of Lemma 5.6.3. □

Lemma 6.4.10. Let $\mathbb{P}^1 \times \mathbb{P}^1$ be equipped with a diagonal action of the group $\mathfrak{A}_5$. Then $\mathbb{P}^1 \times \mathbb{P}^1$ contains a finite number of $\mathfrak{A}_5$-irreducible curves of bi-degree (a,b) with $5a+b < 30$. Moreover, $\mathbb{P}^1 \times \mathbb{P}^1$ contains exactly two $\mathfrak{A}_5$-irreducible curves of bi-degree (a,b) with $5a+b = 20$, and no $\mathfrak{A}_5$-irreducible curves of bi-degree (a,b) with $5a+b = 22$.

Proof. Let Z be an $\mathfrak{A}_5$-irreducible curve of bi-degree (a,b) on $\mathbb{P}^1 \times \mathbb{P}^1$ with $5a+b < 30$. One has $0 \leqslant a \leqslant 5$. If $a = 0$, then the assertion follows from Lemma 6.4.2(i),(ii). If $(a,b) = (1,1)$, then Z coincides with the unique $\mathfrak{A}_5$-invariant curve $\hat{\mathscr{C}} \subset \mathbb{P}^1 \times \mathbb{P}^1$ of bi-degree $(1,1)$ by Lemma 6.4.3(iii). Therefore, we will assume that $(a,b) \neq (1,1)$, which means that $\hat{\mathscr{C}}$ is not an irreducible component of Z (recall that $\hat{\mathscr{C}}$ is a smooth rational curve by Lemma 6.4.3(iii)). In particular, the intersection $Z \cap \hat{\mathscr{C}}$ is an $\mathfrak{A}_5$-invariant set that consists of $a+b$ points (counted with multiplicities). Since $a+b \leqslant 5a+b < 30$, we see that

$$a + b \in \{12, 20, 24\}$$

by Lemma 5.1.4. In particular, if $a = 5$, then $b \geqslant 7$, which contradicts the inequality $5a+b < 30$. Therefore, we are left with the possibilities $1 \leqslant a \leqslant 4$.

Suppose that $a = 1$. Then $0 \leqslant b \leqslant 24$. Since we assumed that $(a,b) \neq (1,1)$, this implies that $b \in \{11, 19, 23\}$. Assume that in one of these cases Z is not unique. Then there is a pencil $\mathcal{P}_{a,b}$ of $\mathfrak{A}_5$-invariant curves of bi-degree (a,b) on $\mathbb{P}^1 \times \mathbb{P}^1$, and at least one of them (namely, Z itself) is $\mathfrak{A}_5$-irreducible. This means that the base locus of the pencil $\mathcal{P}_{a,b}$ is finite. Let Z' be a general curve in the pencil $\mathcal{P}_{a,b}$. Then $Z \cap Z'$ is an $\mathfrak{A}_5$-invariant set that consists of $r = Z \cdot Z' = 2ab$ points (counted with multiplicities). In particular, if $(a,b) = (1,11)$, then $r = 22$; if $(a,b) = (1,19)$, then $r = 38$; if $(a,b) = (1,23)$, then $r = 46$. Neither of these cases is possible by Lemma 6.4.1, so that for any $b \in \{11, 19, 23\}$ there is at most one $\mathfrak{A}_5$-irreducible curve of bi-degree $(1,b)$ on $\mathbb{P}^1 \times \mathbb{P}^1$. Note also that in the case $(a,b) = (1,11)$ one can apply Lemma 6.4.6(i) to get this result.

Suppose that $a = 2$. Then $0 \leqslant b \leqslant 19$. This implies that $b \in \{10, 18\}$. If $b = 10$, then Z is unique by Lemma 6.4.8(i). If $b = 18$, then Z is unique by Lemma 6.4.9(i).

Suppose that $a = 3$. Then $0 \leqslant b \leqslant 14$. This implies that $b = 9$, so that Z is unique by Lemma 6.4.9(ii).

Suppose that $a = 4$. Then $0 \leqslant b \leqslant 9$. This implies that $b = 8$, so that Z is unique by Lemma 6.4.9(iii).

Finally, checking the above possibilities case by case, we see that one has $5a+b = 20$ only for $(a,b) = (0, 20)$ and $(a,b) = (2, 10)$. These two cases indeed correspond to $\mathfrak{A}_5$-irreducible curves by Lemmas 6.4.2(ii) and 6.4.8(ii). Similarly, we check that the case $5a+b = 22$ is impossible. □

Lemma 6.4.11. *Let $\mathbb{P}^1 \times \mathbb{P}^1$ be equipped with a twisted diagonal action of the group $\mathfrak{A}_5$. Then*

(o) there are no $\mathfrak{A}_5$-invariant effective divisors on $\mathbb{P}^1 \times \mathbb{P}^1$ of bi-degree $(1, r)$ for $r \leqslant 6$ (cf. Lemma 6.4.3(ii));

(i) there exists a unique $\mathfrak{A}_5$-invariant effective divisor $C_{2,6}$ on $\mathbb{P}^1 \times \mathbb{P}^1$ of bi-degree $(2, 6)$;

(ii) the divisor $C_{2,6}$ is a smooth hyperelliptic curve of genus 5;

(iii) there exists a unique $\mathfrak{A}_5$-invariant effective divisor $C_{1,13}$ on $\mathbb{P}^1 \times \mathbb{P}^1$ of bi-degree $(1, 13)$;

(iv) the divisor $C_{1,13}$ is a smooth rational curve (cf. Lemma 6.4.7).

Proof. Write $\mathbb{P}^1 \times \mathbb{P}^1 \cong \mathbb{P}(U_2) \times \mathbb{P}(U_2')$ for the two different irreducible two-dimensional representations U_2 and U_2' of the group $2.\mathfrak{A}_5$. An $\mathfrak{A}_5$-invariant effective divisor on $\mathbb{P}^1 \times \mathbb{P}^1$ of bi-degree (a, b) corresponds to a one-dimensional $2.\mathfrak{A}_5$-invariant subspace of $\mathrm{Sym}^a(U_2^\vee) \otimes \mathrm{Sym}^b(U_2'^\vee)$. Thus, assertion (o) is implies by Lemma 5.6.4(i), assertion (i) is implied by Lemma 5.6.4(iii), and assertion (iii) is implied by Lemma 5.6.4(iv).

Let us prove assertion (ii). Suppose that the divisor $C_{2,6}$ is reducible or not reduced. Then either a union of some of the irreducible components of $C_{2,6}$ is an $\mathfrak{A}_5$-invariant curve of bi-degree $(1, b)$ for some $0 \leqslant b \leqslant 3$, or a union of some of the irreducible components of $C_{2,6}$ is an $\mathfrak{A}_5$-invariant curve of bi-degree $(0, b)$ for some $1 \leqslant b \leqslant 6$. The former case is impossible by assertion (o), and the latter case is impossible by Lemma 6.4.2(i). The obtained contradiction shows that $C_{2,6}$ is an irreducible curve.

Suppose that the curve $C_{2,6}$ is singular. Then $C_{2,6}$ has at least 12 singular points by Lemma 6.4.1. On the other hand, one has $p_a(C_{2,6}) = 5$ by adjunction. This gives a contradiction with Lemma 4.4.6.

Therefore, the curve $C_{2,6}$ is smooth. In particular, its genus equals 5. The hyperelliptic structure on $C_{2,6}$ is given by the projection onto the second factor of $\mathbb{P}^1 \times \mathbb{P}^1$ (actually, we already know from Lemma 5.4.3 that any smooth irreducible curve of genus 5 is hyperelliptic). This completes the proof of assertion (ii).

Let us prove assertion (iv). Suppose that the divisor $C_{1,13}$ is reducible or not reduced. Then a union of some of the irreducible components of $C_{1,13}$ is an $\mathfrak{A}_5$-invariant curve Z of bi-degree $(1, b)$ for some $0 \leqslant b \leqslant 12$, so that the union of the remaining irreducible components is an $\mathfrak{A}_5$-invariant curve of bi-degree $(0, c)$ for some $1 \leqslant c \leqslant 13 - b$. By Lemma 6.4.2(i) this implies that $c = 12$ and $b = 1$. Thus Z has bi-degree $(1, 1)$, which contradicts Lemma 6.4.3(i).

Therefore, the divisor $C_{1,13}$ is an irreducible curve. By the (bi-)degree reason it is also smooth and rational. This proves assertion (iv). □

The following two lemmas describe some properties of actions of the group $\mathfrak{D}_{10}$ on $\mathbb{P}^1 \times \mathbb{P}^1$. As in the case of $\mathfrak{A}_5$, we say that the action of the group $\mathfrak{D}_{10}$ on the surface $\mathbb{P}^1 \times \mathbb{P}^1$ is *twisted diagonal*, if there is a $\mathfrak{D}_{10}$-equivariant identification $\mathbb{P}^1 \times \mathbb{P}^1 \cong \mathbb{P}(U) \times \mathbb{P}(V)$ for some *irreducible* non-isomorphic representations U and V of the group $\mathfrak{D}_{10}$. In particular, we require that the twisted diagonal action of $\mathfrak{D}_{10}$ preserves the projections of $\mathbb{P}^1 \times \mathbb{P}^1$ to its first and second factors.

Lemma 6.4.12 (cf. Lemma 6.4.3). Let $\mathbb{P}^1 \times \mathbb{P}^1$ be equipped with an action of the group $\mathfrak{D}_{10}$ that preserves the projections to the factors and that is faithful on the second factor. Suppose that there is an irreducible $\mathfrak{D}_{10}$-invariant divisor on $\mathbb{P}^1 \times \mathbb{P}^1$ of bi-degree $(1,2)$. Then the action of $\mathfrak{D}_{10}$ on $\mathbb{P}^1 \times \mathbb{P}^1$ is twisted diagonal.

Proof. Recall that an embedding $\mathfrak{D}_{10} \hookrightarrow \mathrm{PSL}_2(\mathbb{C})$ corresponds to an embedding $\mathfrak{D}_{10} \hookrightarrow \mathrm{GL}_2(\mathbb{C})$ (see §5.8). Write $\mathbb{P}^1 \times \mathbb{P}^1 \cong \mathbb{P}(U) \times \mathbb{P}(V)$ for two two-dimensional representations U and V of the group $\mathfrak{D}_{10}$. Note that U is a faithful representation of $\mathfrak{D}_{10}$, while V may be not faithful (or even trivial).

A $\mathfrak{D}_{10}$-invariant effective divisor on $\mathbb{P}^1 \times \mathbb{P}^1$ of bi-degree (a,b) corresponds to a one-dimensional $\mathfrak{D}_{10}$-invariant subspace of $\mathrm{Sym}^a(U^\vee) \otimes \mathrm{Sym}^b(V^\vee)$. Since there is an irreducible $\mathfrak{D}_{10}$-invariant divisor $C \subset \mathbb{P}^1 \times \mathbb{P}^1$ of bi-degree $(1,2)$, we see that U is not isomorphic to V by Lemma 5.8.1(i).

Suppose that the action of $\mathfrak{D}_{10}$ on $\mathbb{P}^1 \times \mathbb{P}^1$ is not twisted diagonal. Since we already know that U and V are non-isomorphic representations of the group $\mathfrak{D}_{10}$, we conclude that U splits as a sum of two one-dimensional $\mathfrak{D}_{10}$-representations. In particular, there are two disjoint $\mathfrak{D}_{10}$-invariant curves L and L' of bi-degree $(1,0)$ on $\mathbb{P}^1 \times \mathbb{P}^1$. Let C be an irreducible $\mathfrak{D}_{10}$-invariant curve of bi-degree $(1,2)$ on $\mathbb{P}^1 \times \mathbb{P}^1$ (which exists by assumption). Then the projection to the second factor of $\mathbb{P}^1 \times \mathbb{P}^1$ establishes an isomorphism

$$C \cong \mathbb{P}^1 = \mathbb{P}(V).$$

Neither L nor L' is an irreducible component of the curve C, so that the intersections $C \cap L$ and $C \cap L'$ are disjoint $\mathfrak{D}_{10}$-invariant subsets of C that consist of at most two points. The latter is impossible by Lemma 5.5.9(ii). □

Lemma 6.4.13. Let $\mathbb{P}^1 \times \mathbb{P}^1$ be equipped with a twisted diagonal action of the group $\mathfrak{D}_{10}$. Then

(i) there are no $\mathfrak{D}_{10}$-invariant divisors on $\mathbb{P}^1 \times \mathbb{P}^1$ of bi-degree $(0,1)$, $(1,0)$, and $(1,1)$;

(ii) there is a unique $\mathfrak{D}_{10}$-invariant divisor $C_{1,2} \subset \mathbb{P}^1 \times \mathbb{P}^1$ of bi-degree $(1,2)$;

(iii) there is a unique $\mathfrak{D}_{10}$-invariant divisor $C_{1,3} \subset \mathbb{P}^1 \times \mathbb{P}^1$ of bi-degree $(1,3)$;

(iv) the divisors $C_{1,2}$ and $C_{1,3}$ are smooth rational curves.

Proof. Write $\mathbb{P}^1 \times \mathbb{P}^1 \cong \mathbb{P}(U) \times \mathbb{P}(V)$ for the two different irreducible two-dimensional representations U and V of the group $\mathfrak{D}_{10}$. A $\mathfrak{D}_{10}$-invariant effective divisor on $\mathbb{P}^1 \times \mathbb{P}^1$ of bi-degree (a,b) corresponds to a one-dimensional $\mathfrak{D}_{10}$-invariant subspace of $\mathrm{Sym}^a(U^\vee) \otimes \mathrm{Sym}^b(V^\vee)$. Thus, assertion (i) follows from the definition of the twisted diagonal action, assertion (ii) follows from Lemma 5.8.1(ii), and assertion (iii) follows from Lemma 5.8.1(iii).

Let $C_{1,2}$ be the unique $\mathfrak{D}_{10}$-invariant divisor on $\mathbb{P}^1 \times \mathbb{P}^1$ of bi-degree $(1,2)$. If $C_{1,2}$ is reducible or non-reduced, then there is a $\mathfrak{D}_{10}$-invariant curve of bi-degree $(1,0)$, or $(1,1)$ on $\mathbb{P}^1 \times \mathbb{P}^1$, which is not the case by assertion (i). Similarly, let $C_{1,3}$ be the unique $\mathfrak{D}_{10}$-invariant divisor on $\mathbb{P}^1 \times \mathbb{P}^1$ of bi-degree $(1,3)$. If $C_{1,3}$ is reducible or non-reduced, then there is a $\mathfrak{D}_{10}$-invariant curve of bi-degree $(1,0)$, $(0,1)$ or $(1,1)$ on $\mathbb{P}^1 \times \mathbb{P}^1$, which is impossible by assertion (i). Therefore, we see that $C_{1,2}$ and $C_{1,3}$ are irreducible curves. Then the projection of $C_{1,2}$ and $C_{1,3}$ to the second factor of $\mathbb{P}^1 \times \mathbb{P}^1$ establishes isomorphisms between these curves and $\mathbb{P}^1$, which completes the proof of assertion (iv). □

The following technical lemma will be used in the proof of Proposition 17.3.1.

Lemma 6.4.14. *Let $\mathbb{P}^1 \times \mathbb{P}^1$ be equipped with a diagonal action of the group $\mathfrak{A}_5$, and let $\hat{\mathscr{C}}$ be the unique $\mathfrak{A}_5$-invariant curve in $\mathbb{P}^1 \times \mathbb{P}^1$ of bi-degree $(1,1)$ (see Lemma 6.4.3(iii)). Let Σ be an $\mathfrak{A}_5$-orbit on $\mathbb{P}^1 \times \mathbb{P}^1$ such that*

$$|\Sigma| \in \{20, 30, 60\}.$$

Put

$$n = \begin{cases} 3 \text{ if } |\Sigma| = 20, \\ 5 \text{ if } |\Sigma| = 30, \\ 11 \text{ if } |\Sigma| = 60. \end{cases}$$

Let $\mathcal{M}$ be a linear system on $\mathbb{P}^1 \times \mathbb{P}^1$ consisting of all curves of bi-degree $(n-2, 5n-2)$ that pass through Σ. Then $\mathcal{M}$ does not have base curves except possibly the curve $\hat{\mathscr{C}}$.

CREMONA GROUPS AND THE ICOSAHEDRON

Proof. Suppose that the base locus of the linear system $\mathcal{M}$ contains a base curve C that is different from $\hat{\mathscr{C}}$. Since $\mathcal{M}$ is $\mathfrak{A}_5$-invariant, we may assume that C is $\mathfrak{A}_5$-invariant as well. Let $\mathcal{I}_C \subset \mathcal{O}_{\mathbb{P}^1 \times \mathbb{P}^1}$ be the ideal sheaf of C, let $\mathcal{I}_\Sigma \subset \mathcal{O}_{\mathbb{P}^1 \times \mathbb{P}^1}$ be the ideal sheaf of the subset Σ, and let $\mathcal{I}_{C \cup \Sigma} \subset \mathcal{O}_{\mathbb{P}^1 \times \mathbb{P}^1}$ be the ideal sheaf of the subset $C \cup \Sigma$. Then

$$h^0\Big(\mathcal{I}_\Sigma \otimes \mathcal{O}_{\mathbb{P}^1 \times \mathbb{P}^1}(n-2, 5n-2)\Big) = h^0\Big(\mathcal{I}_{C \cup \Sigma} \otimes \mathcal{O}_{\mathbb{P}^1 \times \mathbb{P}^1}(n-2, 5n-2)\Big),$$

since C is contained in the base locus of $\mathcal{M}$. But

$$h^0\Big(\mathcal{I}_{C \cup \Sigma} \otimes \mathcal{O}_{\mathbb{P}^1 \times \mathbb{P}^1}(n-2, 5n-2)\Big) \leqslant h^0\Big(\mathcal{I}_C \otimes \mathcal{O}_{\mathbb{P}^1 \times \mathbb{P}^1}(n-2, 5n-2)\Big).$$

On the other hand, we have

$$h^0\Big(\mathcal{I}_\Sigma \otimes \mathcal{O}_{\mathbb{P}^1 \times \mathbb{P}^1}(n-2, 5n-2)\Big) \geqslant h^0\Big(\mathcal{O}_{\mathbb{P}^1 \times \mathbb{P}^1}(n-2, 5n-2)\Big) - |\Sigma|.$$

The curve C is a divisor on $\mathbb{P}^1 \times \mathbb{P}^1$ of bi-degree (a, b) for some non-negative integers a and b. In particular, one has

$$\begin{cases} a \leqslant n-2, \\ b \leqslant 5n-2. \end{cases} \tag{6.4.15}$$

We also have

$$h^0\Big(\mathcal{O}_{\mathbb{P}^1 \times \mathbb{P}^1}(n-2-a, 5n-2-b)\Big) =$$
$$= h^0\Big(\mathcal{I}_C \otimes \mathcal{O}_{\mathbb{P}^1 \times \mathbb{P}^1}(n-2, 5n-2)\Big) \geqslant h^0\Big(\mathcal{I}_{C \cup \Sigma} \otimes \mathcal{O}_{\mathbb{P}^1 \times \mathbb{P}^1}(n-2, 5n-2)\Big) =$$
$$= h^0\Big(\mathcal{I}_\Sigma \otimes \mathcal{O}_{\mathbb{P}^1 \times \mathbb{P}^1}(n-2, 5n-2)\Big) \geqslant h^0\Big(\mathcal{O}_{\mathbb{P}^1 \times \mathbb{P}^1}(n-2, 5n-2)\Big) - |\Sigma|,$$

which implies that

$$|\Sigma| \geqslant h^0\Big(\mathcal{O}_{\mathbb{P}^1 \times \mathbb{P}^1}(n-2, 5n-2)\Big) - h^0\Big(\mathcal{O}_{\mathbb{P}^1 \times \mathbb{P}^1}(n-2-a, 5n-2-b)\Big) =$$
$$= (n-1)(5n-1) - (n-1-a)(5n-1-b) = a(5n-1) + b(n-1) - ab.$$

Thus, we have

$$ab + b + a + |\Sigma| \geqslant bn + 5na. \tag{6.4.16}$$

Note that (6.4.16) implies that $a \leqslant 1$. Indeed, suppose that $a \geqslant 2$. Recall that $a \leqslant n-2$ by (6.4.15). Hence

$$(n-2)b + b + n - 2 + |\Sigma| \geqslant ab + b + a + |\Sigma| \geqslant bn + 5na \geqslant bn + 10n$$

by (6.4.16). This gives

$$|\Sigma| \geqslant 2 + 9n + b \geqslant 2 + 9n = \begin{cases} 29 \text{ if } |\Sigma| = 20, \\ 47 \text{ if } |\Sigma| = 30, \\ 101 \text{ if } |\Sigma| = 60, \end{cases}$$

which is absurd.

We see that either $a = 0$ or $a = 1$. Plugging each possibility for a into (6.4.16), we conclude that either $a = 0$ and $b \leqslant 10$, or $a = 1$ and $b \leqslant 6$. Both of these cases are impossible by Lemma 6.4.4(i), since $C \neq \hat{\mathscr{C}}$. □

6.5 Hirzebruch surfaces

Recall that the Hirzebruch surface $\mathbb{F}_n$, $n \geqslant 0$, is defined as the projectivization of the vector bundle $\mathcal{O}_{\mathbb{P}^1} \oplus \mathcal{O}_{\mathbb{P}^1}(n)$ on $\mathbb{P}^1$. In particular, one has

$$\mathbb{F}_0 \cong \mathbb{P}^1 \times \mathbb{P}^1,$$

and $\mathbb{F}_1$ is the blow-up of $\mathbb{P}^2$ in one point. In this section, we show that the group $\mathfrak{A}_5$ does not act on the surfaces $\mathbb{F}_n$ with odd n, and all $\mathfrak{A}_5$-actions on the surfaces $\mathbb{F}_n$ with even n are related by $\mathfrak{A}_5$-birational maps. All results here are well known to experts (see, for example, [18]).

Lemma 6.5.1. *Let S be a surface faithfully acted on by the group $\mathfrak{A}_5$. Suppose that $S \cong \mathbb{F}_n$ for some $n > 0$. Then n is even. Moreover, there exists an $\mathfrak{A}_5$-birational map $\rho \colon \mathbb{F}_n \dashrightarrow \mathbb{P}^1 \times \mathbb{P}^1$ such that the diagram*

$$\begin{array}{ccc} \mathbb{F}_n & \overset{\rho}{\dashrightarrow} & \mathbb{P}^1 \times \mathbb{P}^1 \\ {\scriptstyle \pi} \downarrow & & \downarrow {\scriptstyle \pi_2} \\ \mathbb{P}^1 & = & \mathbb{P}^1 \end{array}$$

commutes, where π is a natural projection, π_2 is a projection to the second factor, and $\mathfrak{A}_5$-acts trivially on the first factor of $\mathbb{P}^1 \times \mathbb{P}^1$.

Proof. Let s be the section of π such that $s^2 = -n$. Then s is $\mathfrak{A}_5$-invariant. We claim that $n \geqslant 2$. Indeed, if $n = 1$, then we can contract s and obtain $\mathbb{P}^2$ faithfully acted on by $\mathfrak{A}_5$ that admits an $\mathfrak{A}_5$-invariant point. The latter is impossible by Remark 6.1.1.

CREMONA GROUPS AND THE ICOSAHEDRON 137

Let Ξ_{30} be the $\mathfrak{A}_5$-orbit in $s \cong \mathbb{P}^1$ that consists of 30 points (it exists and it is unique by Lemma 5.5.9(v)). Then there is a commutative diagram

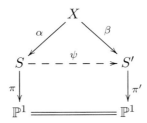

where ψ is an $\mathfrak{A}_5$-birational map, α is the blow-up of the set Ξ_{30}, and β is the blow down of the proper transforms of the fibers of π that pass through the points of the set Ξ_{30}.

We claim that $S' \cong \mathbb{F}_{n'}$, where $n' = n + 30$. Indeed, put $s' = \psi(s)$. Then s' is a section of π' and

$$s' \cdot s' = s \cdot s - 30 = -n - 30 < 0,$$

which implies that $S' \cong \mathbb{F}_{n+30}$. Note that s' is $\mathfrak{A}_5$-invariant by construction.

Let l' be a fiber of π'. Then there exists an $\mathfrak{A}_5$-invariant irreducible curve in the linear system $|s' + n'l'|$. Indeed, contracting s', we obtain a weighted projective plane $\mathbb{P}(1, 1, n')$, which we can identify with a cone in $\mathbb{P}^{n'+1}$ over a rational normal curve in $\mathbb{P}^{n'}$ of degree n'. Since the vertex of this cone is $\mathfrak{A}_5$-invariant, there is an $\mathfrak{A}_5$-invariant hyperplane section of $\mathbb{P}(1, 1, n')$ that does not pass through its vertex. This gives us an $\mathfrak{A}_5$-invariant irreducible curve in $|s' + n'l'|$. Denote this curve by C'. Then C' is a section of π' that is disjoint from s'.

Let Ξ'_{32} be the $\mathfrak{A}_5$-invariant set in C' that consists of 32 points (it exists and it is unique by Lemma 5.5.9(v); it is a union of an $\mathfrak{A}_5$-orbit of length 20 and an $\mathfrak{A}_5$-orbit of length 12). Then there is a commutative diagram

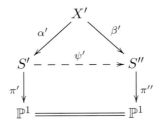

where ψ' is an $\mathfrak{A}_5$-birational map, α' is the blow-up of the set Ξ'_{32}, and β' is the blow-down of the proper transforms of the fibers of π' that pass through the points of the set Ξ'_{32}.

We claim that $S'' \cong \mathbb{F}_{n-2}$. Indeed, put $s'' = \psi'(s')$. Then s'' is a section of π'' and

$$s'' \cdot s'' = s' \cdot s' + 32 = s \cdot s - 30 + 32 = -n + 2 \leqslant 0,$$

which implies that $S'' \cong \mathbb{F}_{n-2}$. Note that s'' is $\mathfrak{A}_5$-invariant by construction.

Suppose that $n = 2$. Then $S'' \cong \mathbb{P}^1 \times \mathbb{P}^1$ and π'' is a projection to one of its factors. We may assume that π'' is a projection to the second factor. Then s'' is a fiber of the projection to the first factor, since $s'' \cdot s'' = 0$. Therefore, $\mathfrak{A}_5$ acts on the first factor of $S'' \cong \mathbb{P}^1 \times \mathbb{P}^1$ with a fixed point, which implies that $\mathfrak{A}_5$ acts trivially on it (see Lemma 5.5.9(v)) and we are done.

Thus, we may assume that $n \neq 2$. Since we already proved that $\mathfrak{A}_5$ does not act faithfully on $\mathbb{F}_1$, we see that $n \neq 3$. Hence $n \geqslant 4$. Now we can iterate the above construction $\lfloor \frac{n}{2} \rfloor$ times to obtain the required assertion. $\square$

Lemma 6.5.1 shows that a Hirzebruch surface $\mathbb{F}_n$ with $n > 0$ equipped with an action of the group $\mathfrak{A}_5$ is $\mathfrak{A}_5$-birational to the surface $\mathbb{P}^1 \times \mathbb{P}^1$ equipped with an $\mathfrak{A}_5$-action that is trivial on one of its factors. Now we are going to show that the same holds for $\mathbb{F}_0 \cong \mathbb{P}^1 \times \mathbb{P}^1$. The idea of the proof is to produce an $\mathfrak{A}_5$-birational map from $\mathbb{P}^1 \times \mathbb{P}^1$ to $\mathbb{F}_n$ with $n > 0$, and then return back to $\mathbb{P}^1 \times \mathbb{P}^1$ using Lemma 6.5.1.

Lemma 6.5.2. Let S be a surface isomorphic to $\mathbb{P}^1 \times \mathbb{P}^1$ equipped with an action of $\mathfrak{A}_5$ that is non-trivial on the second factor. Then there exists a commutative diagram

$$\begin{array}{ccc} S & \dashrightarrow^{\psi} & S' \\ \pi_2 \downarrow & & \downarrow \pi'_2 \\ \mathbb{P}^1 & = & \mathbb{P}^1 \end{array}$$

where $S' \cong \mathbb{P}^1 \times \mathbb{P}^1$ with an action of $\mathfrak{A}_5$ that is trivial on the first factor, the morphisms π_2 and π'_2 are projections to the second factors, and ψ is an $\mathfrak{A}_5$-birational map.

Proof. If the $\mathfrak{A}_5$-action on S is already trivial on the first factor, then we are done. So, we assume that this is not the case. There exists an $\mathfrak{A}_5$-invariant section s of π_2 such that $s^2 < 20$ by Lemma 6.4.3(i),(ii). Indeed, if $\mathfrak{A}_5$ acts on S diagonally, then such s can be chosen as a divisor of bi-degree $(1, 1)$, and if the $\mathfrak{A}_5$-action on S is twisted diagonal, then such s can be chosen as a divisor of bi-degree $(1, 7)$.

CREMONA GROUPS AND THE ICOSAHEDRON   139

Let Ξ_{20} be the $\mathfrak{A}_5$-orbit in $s \cong \mathbb{P}^1$ that consists of 20 points (it exists and it is unique by Lemma 5.5.9(v)). Then there is a commutative diagram

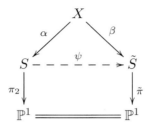

where ψ is an $\mathfrak{A}_5$-birational map, α is the blow-up of the set Ξ_{20}, and β is the blow-down of the proper transforms of the fibers of π_2 that pass through the points of the set Ξ_{20}. Put $\tilde{s} = \psi(s)$. Then $\tilde{s}$ is a section of $\tilde{\pi}$ and

$$\tilde{s}^2 = s^2 - 20 = -20.$$

This implies that $\tilde{S} \cong \mathbb{F}_{20}$. Now the assertion follows from Lemma 6.5.1. □

Lemma 6.5.3. Let $S \cong \mathbb{P}^1 \times \mathbb{P}^1$ and $S' \cong \mathbb{P}^1 \times \mathbb{P}^1$ be equipped with faithful actions of the group $\mathfrak{A}_5$. Then there exists an $\mathfrak{A}_5$-birational map $S \dashrightarrow S'$.

Proof. Twisting (if necessary) the actions of $\mathfrak{A}_5$ on S and S' by an element of $\mathrm{Aut}(\mathbb{P}^1 \times \mathbb{P}^1)$ that interchanges the factors of $\mathbb{P}^1 \times \mathbb{P}^1$, we may assume that both actions are non-trivial on the second factors. Now applying Lemma 6.5.2, we reduce the problem to the case when both actions are trivial on the first factors. Therefore, there are two-dimensional irreducible representations U and U' of the group $2.\mathfrak{A}_5$ such that there are identifications

$$S \cong \mathbb{P}(I \oplus I) \times \mathbb{P}(U)$$

and

$$S' \cong \mathbb{P}(I \oplus I) \times \mathbb{P}(U').$$

If $U \cong U'$, then there is nothing to prove. So, we assume that U and U' are different representations of the group $2.\mathfrak{A}_5$.

Applying Lemma 6.5.2 to the surface $\mathbb{P}(U') \times \mathbb{P}(U)$, we obtain an $\mathfrak{A}_5$-birational map

$$\psi_1 \colon \mathbb{P}(U') \times \mathbb{P}(U) \dashrightarrow S.$$

Applying Lemma 6.5.2 to the surface $\mathbb{P}(U) \times \mathbb{P}(U')$, we obtain an $\mathfrak{A}_5$-birational map

$$\psi_2 \colon \mathbb{P}(U) \times \mathbb{P}(U') \dashrightarrow S'.$$

Since the actions of the group $\mathfrak{A}_5$ on $\mathbb{P}(U') \times \mathbb{P}(U)$ and $\mathbb{P}(U) \times \mathbb{P}(U')$ are conjugate in $\mathrm{Aut}(\mathbb{P}^1 \times \mathbb{P}^1)$ (see §6.4), we obtain the required result. □

6.6 Icosahedral subgroups of $\mathrm{Cr}_2(\mathbb{C})$

In this section, we describe all subgroups in $\mathrm{Cr}_2(\mathbb{C})$ isomorphic to $\mathfrak{A}_5$ up to conjugation. To be precise, we show that there are exactly three non-conjugate subgroups isomorphic to $\mathfrak{A}_5$ in $\mathrm{Cr}_2(\mathbb{C})$. The fact that there are at least three of them has been proved by Shinzo Bannai and Hiro-o Tokunaga (see [119], [4]). We will give an alternative proof of their result. The fact that $\mathrm{Cr}_2(\mathbb{C})$ contains at most three non-conjugate subgroups isomorphic to $\mathfrak{A}_5$ is also well known (see, for example, [18]). We will present the proof of this result as well.

Up to conjugation, the group $\mathrm{Aut}(\mathbb{P}^2) \cong \mathrm{PGL}_3(\mathbb{C})$ contains a unique subgroup isomorphic to $\mathfrak{A}_5$, so that $\mathbb{P}^2$ is identified with $\mathbb{P}(W_3)$ or $\mathbb{P}(W_3')$ (see Remark 6.1.1). Recall that a smooth del Pezzo surface S of degree 5 is unique up to isomorphism and $\mathrm{Aut}(S) \cong \mathfrak{S}_5$ (see §6.2). Various actions of the group $\mathfrak{A}_5$ on $\mathbb{P}^1 \times \mathbb{P}^1$ are described in §6.4.

The main result of this section is:

Theorem 6.6.1 ([18, Theorem B.10]). *Let S be a rational surface that is faithfully acted on by the group $\mathfrak{A}_5$. Then the following assertions hold:*

(i) *the surface S is $\mathfrak{A}_5$-equivariantly birational to one of the following three surfaces: the projective plane $\mathbb{P}^2$, the smooth del Pezzo surface of degree 5, or $\mathbb{P}^1 \times \mathbb{P}^1$ acted on diagonally by $\mathfrak{A}_5$;*

(ii) *the projective plane $\mathbb{P}^2$ is $\mathfrak{A}_5$-birationally rigid and $\mathrm{Bir}^{\mathfrak{A}_5}(\mathbb{P}^2) \cong \mathfrak{S}_5$;*

(iii) *the smooth del Pezzo surface of degree 5 is $\mathfrak{A}_5$-birationally superrigid.*

Corollary 6.6.2 ([4, Corollary 0.1], [18, Theorem B.2]). *Up to conjugation, the group $\mathrm{Cr}_2(\mathbb{C})$ contains exactly 3 subgroups isomorphic to $\mathfrak{A}_5$.*

Let us prove Theorem 6.6.1 (our proof is a simplified version of the arguments from [18]). We start with:

Lemma 6.6.3. *Let S be a rational surface that is faithfully acted on by the group $\mathfrak{A}_5$. Then S is $\mathfrak{A}_5$-equivariantly birational to one of the following surfaces: the projective plane $\mathbb{P}^2$, the smooth del Pezzo surface of degree 5, or a Hirzebruch surface $\mathbb{F}_n$ for some $n \geqslant 0$.*

Proof. Applying $\mathfrak{A}_5$-equivariant Minimal Model Program, we may assume that S is smooth and there exists an $\mathfrak{A}_5$-equivariant morphism $\pi \colon S \to Z$ such that we have a dichotomy:

(DP) either Z is a point and $\operatorname{rk}\operatorname{Pic}(S)^{\mathfrak{A}_5} = 1$;

(CB) or Z is a smooth rational curve, $\operatorname{rk}\operatorname{Pic}(S)^{\mathfrak{A}_5} = 2$ and a general fiber of π is isomorphic to $\mathbb{P}^1$.

In the case (DP), either $S \cong \mathbb{P}^2$ or S is the smooth del Pezzo surface of degree 5. This follows from the classification of automorphism groups of smooth del Pezzo surfaces (see [35] or [34]). Thus, to complete the proof, we may assume that we are in the case (CB). We must prove that $S \cong \mathbb{F}_n$ for some $n \geqslant 0$. Actually, this follows from a much more general result, namely [35, Lemma 5.6], that is applicable to many other finite groups (not just for $\mathfrak{A}_5$). We give an independent proof in the case of $\mathfrak{A}_5$ for the reader's convenience.

To prove that $S \cong \mathbb{F}_n$, it is enough to show that every fiber of π is isomorphic to $\mathbb{P}^1$. Suppose that this is not the case. Then there exists a reducible fiber F of the morphism π. Since $\operatorname{rk}\operatorname{Pic}(S)^{\mathfrak{A}_5} = 2$, the fiber F consists of two irreducible components, F_1 and F_2, such that both of them are isomorphic to $\mathbb{P}^1$ and they intersect transversally at one point $Q \in F$. Moreover, there exists an element $g \in \mathfrak{A}_5$ that interchanges F_1 and F_2 (this also follows from the equality $\operatorname{rk}\operatorname{Pic}(S)^{\mathfrak{A}_5} = 2$). In particular, g is of even order.

Note that the action of $\mathfrak{A}_5$ on the curve $Z \cong \mathbb{P}^1$ is non-trivial and hence faithful, since otherwise $\mathfrak{A}_5$ would fix the point $\pi(F)$ on Z, and thus act on the set $\{F_1, F_2\}$ non-trivially. The latter is impossible, because $\mathfrak{A}_5$ is simple (cf. Corollary 5.1.2).

Put $P = \pi(F)$, and let Ω be the $\mathfrak{A}_5$-orbit of P. One has

$$|\Omega| \in \{12, 20, 30, 60\}$$

by Lemma 5.1.4. Since g lies in the stabilizer $G_P \subset \mathfrak{A}_5$ of the point P, the group G_P must have an even order. This is only possible if $|\Omega| = 30$. Thus, one has $G_P \cong \boldsymbol{\mu}_2$.

The group G_P fixes the point $Q \in S$. Thus, it faithfully acts on the Zariski tangent space $T_Q(S) \cong \mathbb{C}^2$ by Lemma 4.4.1. If it acts by reflection, i.e., there is a trivial subrepresentation of the group G_P in $T_Q(S)$, then Lemma 4.4.3 implies that there exists a curve $C \subset S$ passing through Q that is fixed pointwise by G_P. Note that Q is the unique G_P-invariant point in F, since $g(F_1) = F_2$. Thus, the curve C is not contained in F, so that it is mapped surjectively to Z by π. The latter is impossible, since G_P acts non-trivially on Z.

Hence, the non-trivial element of G_P acts on $T_Q(S) \cong \mathbb{C}^2$ by a multiplication by -1. Therefore, it preserves the one-dimensional Zariski tangent spaces to the curves F_1 and F_2. On the other hand, G_P must interchange them, since it interchanges F_1 and F_2 (which intersect transversally at Q). The obtained contradiction completes the proof of Lemma 6.6.3. □

By Lemma 6.6.3, the assertion of Theorem 6.6.1 follows from Lemmas 6.5.1 and 6.5.3 and the following two lemmas. The first of them deals with the del Pezzo surface of degree 5.

Lemma 6.6.4 ([4, Theorem 0.1], [18, Lemma B.12]). *Let S be the smooth del Pezzo surface of degree 5. Then S is $\mathfrak{A}_5$-birationally superrigid, and*

$$\mathrm{Bir}^{\mathfrak{A}_5}(S) = \mathrm{Aut}^{\mathfrak{A}_5}(S) \cong \mathfrak{S}_5.$$

Proof. The surface S does not contain $\mathfrak{A}_5$-orbits of length less than 5 by Lemma 6.1.11. Thus, S is $\mathfrak{A}_5$-birationally superrigid by Lemma 3.3.8. Hence

$$\mathrm{Bir}^{\mathfrak{A}_5}(S) = \mathrm{Aut}^{\mathfrak{A}_5}(S).$$

On the other hand, one has $\mathrm{Aut}^{\mathfrak{A}_5}(S) \cong \mathfrak{S}_5$, since $\mathrm{Aut}(S) \cong \mathfrak{S}_5$. □

Now we pass to the case of the projective plane.

Lemma 6.6.5 ([18, Lemma B.13]). *Let $S = \mathbb{P}^2$. Then S is $\mathfrak{A}_5$-birationally rigid, one has*

$$\mathrm{Bir}^{\mathfrak{A}_5}(S) \cong \mathfrak{S}_5,$$

and S is not $\mathfrak{A}_5$-birationally superrigid.

Proof. Let S_3 be the Clebsch cubic surface in $\mathbb{P}^3$ (see §6.3). By Lemma 6.3.3(iii), there exists a commutative diagram

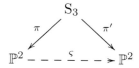

such that π (resp., π') is an $\mathfrak{A}_5$-birational morphism that contracts six disjoint lines to the unique $\mathfrak{A}_5$-orbit Σ_6 of length 6 in $\mathbb{P}^2$ (cf. [57, Proposition 1]).

Recall that $\mathrm{Aut}(S_3) \cong \mathfrak{S}_5$. Take any odd permutation $\theta \in \mathrm{Aut}(S_3)$. Put

$$\varsigma = \pi \circ \theta \circ \pi^{-1} \colon \mathbb{P}^2 \dashrightarrow \mathbb{P}^2.$$

CREMONA GROUPS AND THE ICOSAHEDRON   143

Then $\varsigma \notin \mathrm{Aut}(\mathbb{P}^2)$, because $\mathfrak{S}_5$ does not act faithfully on $\mathbb{P}^2$ (see Remark 6.1.1). In particular, $\mathbb{P}^2$ is not $\mathfrak{A}_5$-birationally superrigid. Note that $\mathfrak{A}_5$ and θ generate the group $\mathfrak{S}_5$ by construction.

Now we are ready to use Theorem 3.3.1. Let $\mathcal{M}$ be a mobile linear system on $\mathbb{P}^2$. Choose a positive rational number μ such that

$$K_{\mathbb{P}^2} + \mu\mathcal{M} \sim_{\mathbb{Q}} 0.$$

Denote by $\mathcal{M}'$ the proper transform of $\mathcal{M}$ via ς. Let μ' be a positive rational number such that

$$K_{\mathbb{P}^2} + \mu'\mathcal{M}' \sim_{\mathbb{Q}} 0.$$

We claim that either $(\mathbb{P}^2, \mu\mathcal{M})$ or $(\mathbb{P}^2, \mu'\mathcal{M}')$ has canonical singularities. Indeed, let P be a point in $\mathbb{P}^2$, and let Σ be its $\mathfrak{A}_5$-orbit. Take two general curves M_1 and M_2 in $\mathcal{M}$. If $\Sigma \neq \Sigma_6$, then $|\Sigma| \geqslant 10$ by Lemma 5.3.1(i), so that

$$\frac{9}{\mu^2} = \frac{K_{\mathbb{P}^2}^2}{\mu^2} = M_1 \cdot M_2 \geqslant |\Sigma| \cdot \mathrm{mult}_P(M_1) \cdot \mathrm{mult}_P(M_2) \geqslant$$
$$\geqslant |\Sigma| \cdot \mathrm{mult}_P^2(\mathcal{M}) > 10\mathrm{mult}_P^2(\mathcal{M}),$$

which implies, in particular, that $\mathrm{mult}_P(\mathcal{M}) < \frac{1}{\mu}$. Thus, it follows from Lemma 2.4.4(ii) that $(\mathbb{P}^2, \mu\mathcal{M})$ has canonical singularities outside of Σ_6. Similarly, we see that $(\mathbb{P}^2, \mu'\mathcal{M}')$ has canonical singularities outside of Σ_6. On the other hand, either $(\mathbb{P}^2, \mu\mathcal{M})$ or $(\mathbb{P}^2, \mu'\mathcal{M}')$ has canonical singularities at the points of Σ_6 by Corollary 6.3.11 and Lemma 2.2.7.

We see that either $(\mathbb{P}^2, \mu\mathcal{M})$ or $(\mathbb{P}^2, \mu'\mathcal{M}')$ has canonical singularities. Thus, it follows from Theorem 3.3.1 that $\mathbb{P}^2$ is $\mathfrak{A}_5$-birationally rigid, and the group $\mathrm{Bir}^{\mathfrak{A}_5}(\mathbb{P}^2)$ is generated by ς and $\mathrm{Aut}^{\mathfrak{A}_5}(\mathbb{P}^2)$. □

Arguing as in the proof of Lemmas 6.6.4 and 6.6.5 and using Theorem 3.4.2, we see that neither the del Pezzo surface of degree 5 nor $\mathbb{P}^2$ contains an $\mathfrak{A}_5$-invariant Halphen pencil. This is not very surprising, since there exist no rational surface that is acted on by $\mathfrak{A}_5$ and admits an $\mathfrak{A}_5$-equivariant elliptic fibration. The latter can be shown using $\mathfrak{A}_5$-equivariant Minimal Model Program.

6.7 $K3$ surfaces

In this section we describe properties of $K3$ surfaces acted on by the group $\mathfrak{A}_5$. Recall that a $K3$ surface S is a normal projective surface with mild singularities such that $K_S \sim 0$ and $h^1(\mathcal{O}_S) = 0$. Usually, $K3$ surfaces are assumed to be smooth or having at most du Val singularities. In the latter case, the minimal resolution of S is again a smooth $K3$ surface. The basic properties of $K3$ surfaces are described in [109] and [6].

Our main tool here will be the following lemma implied by the results of Gang Xiao.

Lemma 6.7.1. Let $\tilde{S}$ be a smooth $K3$ surface with a faithful action of the group $\mathfrak{A}_5$. Then

(i) either $\operatorname{Pic}(\tilde{S})^{\mathfrak{A}_5} \cong \mathbb{Z}$, or $\operatorname{Pic}(\tilde{S})^{\mathfrak{A}_5} \cong \mathbb{Z}^2$;

(ii) any $\mathfrak{A}_5$-orbit $\Sigma \subset \tilde{S}$ has length $|\Sigma| \in \{12, 20, 30, 60\}$;

(iii) there are exactly two $\mathfrak{A}_5$-orbits of length 12 in $\tilde{S}$;

(iv) there are exactly three $\mathfrak{A}_5$-orbits of length 20 in $\tilde{S}$;

(v) there are exactly four $\mathfrak{A}_5$-orbits of length 30 in $\tilde{S}$.

Proof. Put $\bar{S} = \tilde{S}/\mathfrak{A}_5$. Then $\operatorname{rk}\operatorname{Pic}(\tilde{S})^{\mathfrak{A}_5} = \operatorname{rk}\operatorname{Pic}(\bar{S})$.

Since $\mathfrak{A}_5$ is a simple non-abelian group, the action of $\mathfrak{A}_5$ on $\tilde{S}$ is symplectic. Therefore, it follows from [124, Theorem 3] that $\bar{S}$ is a $K3$ surface with exactly 9 du Val singular points, namely, 4 ordinary double points, 3 singular points of type A_2, and 2 singular points of type A_4. It is well known that the natural morphism $\varsigma \colon \tilde{S} \to \bar{S}$ is unramified away from the singular points of the surface $\bar{S}$ (cf. [124, §1]).

Recall that a du Val singular point of type A_n is analytically isomorphic to a quotient of $\mathbb{C}^2$ by the group $\boldsymbol{\mu}_{n+1}$. In particular, the fiber of the morphism ς over a point $P \in \bar{S}$ of type A_n consists of a single $\mathfrak{A}_5$-orbit of length $\frac{60}{n+1}$. This implies assertions (ii), (iii), (iv), and (v).

Let $\psi \colon \hat{S} \to \bar{S}$ be the minimal resolution of singularities. Then

$$\operatorname{rk}\operatorname{Pic}(\hat{S}) = \operatorname{rk}\operatorname{Pic}(\bar{S}) + 4 \cdot 1 + 3 \cdot 2 + 2 \cdot 4 = \operatorname{rk}\operatorname{Pic}(\bar{S}) + 18,$$

which implies that $\operatorname{rk}\operatorname{Pic}(\bar{S}) \leqslant 2$, since $\operatorname{rk}\operatorname{Pic}(\hat{S}) \leqslant 20$ (see [6, Proposition VIII.3.3]). Thus, we see that

$$\operatorname{rk}\operatorname{Pic}(\tilde{S})^{\mathfrak{A}_5} = \operatorname{rk}\operatorname{Pic}(\bar{S}) \leqslant 2.$$

Since $\operatorname{Pic}(\tilde{S})$ has no torsion (see, for example, [6, Proposition VIII.3.1]), this implies assertion (i) and completes the proof of the lemma. □

Remark 6.7.2. If $\tilde{S}$ is a smooth $K3$ surface with a faithful action of the group $\mathfrak{A}_5$, then one has

$$\operatorname{rk}\operatorname{Pic}(\tilde{S}) = \operatorname{rk}\operatorname{Pic}(\tilde{S})^{\mathfrak{A}_5} + 18.$$

This follows from the fact that the coinvariant sublattice in $H^2(\tilde{S}, \mathbb{Z})$ has rank 18 (see, e. g., [88, §6]). In particular, this provides an alternative way to prove assertion (i) of Lemma 6.7.1.

Lemma 6.7.3. Let S be a $K3$ surface with at most du Val singularities that is faithfully acted on by the group $\mathfrak{A}_5$. Then

(i) either $\operatorname{Pic}(S)^{\mathfrak{A}_5} \cong \mathbb{Z}$, or $\operatorname{Pic}(S)^{\mathfrak{A}_5} \cong \mathbb{Z}^2$;

(ii) if $\operatorname{rk}\operatorname{Pic}(S)^{\mathfrak{A}_5} = 2$, then S is smooth;

(iii) if S is singular, then $\operatorname{Sing}(S)$ consists of one $\mathfrak{A}_5$-orbit, every singular point of the surface S is an ordinary double point, and

$$|\operatorname{Sing}(S)| \in \{1, 5, 6, 10, 12, 15\};$$

(iv) if $|\operatorname{Sing}(S)| = 12$, then $\operatorname{Sing}(S)$ is the only $\mathfrak{A}_5$-orbit in S that consists of 12 points.

Proof. Let $\theta \colon \tilde{S} \to S$ be the minimal resolution of singularities. Then the action of the group $\mathfrak{A}_5$ on S lifts to the faithful action on $\tilde{S}$. By Lemma 6.7.1(i) we have $\operatorname{rk}\operatorname{Pic}(\tilde{S})^{\mathfrak{A}_5} \leqslant 2$, so that

$$\operatorname{rk}\operatorname{Pic}(S)^{\mathfrak{A}_5} \leqslant \operatorname{rk}\operatorname{Pic}(\tilde{S})^{\mathfrak{A}_5} \leqslant 2.$$

Since the morphism θ is birational, and $\operatorname{Pic}(\tilde{S})$ has no torsion (see, for example, [6, Proposition VIII.3.1]), we see that $\operatorname{Pic}(S)$ has no torsion as well. This implies that either $\operatorname{Pic}(S)^{\mathfrak{A}_5} \cong \mathbb{Z}$, or $\operatorname{Pic}(S)^{\mathfrak{A}_5} \cong \mathbb{Z}^2$. The latter proves assertion (i) and also implies that $\operatorname{Sing}(S)$ consists of at most one $\mathfrak{A}_5$-orbit. Furthermore, θ must be an isomorphism if $\operatorname{rk}\operatorname{Pic}(S)^{\mathfrak{A}_5} = 2$, which gives assertion (ii).

To complete the proof, we assume that the surface S is singular. Then $\operatorname{rk}\operatorname{Pic}(S)^{\mathfrak{A}_5} = 1$ and $\operatorname{Sing}(S)$ is an $\mathfrak{A}_5$-orbit. If S has a singular point of type $\mathbb{D}_k$ for some $k \geqslant 4$ or $\mathbb{E}_k$ for some $k \in \{6, 7, 8\}$, then

$$\operatorname{rk}\operatorname{Pic}(\tilde{S})^{\mathfrak{A}_5} - \operatorname{rk}\operatorname{Pic}(S)^{\mathfrak{A}_5} \geqslant 2,$$

because the $\mathfrak{A}_5$-orbit of the θ-exceptional curve that corresponds to the "fork" in the dual graph of the singularity $\mathbb{D}_k$ or $\mathbb{E}_k$ generates a proper

$\mathfrak{A}_5$-invariant linear subspace in $\operatorname{Pic}(\tilde{S}) \otimes \mathbb{Q}$. Since $\operatorname{rk}\operatorname{Pic}(\tilde{S})^{\mathfrak{A}_5} \leqslant 2$, we see that every singular point of the surface S is a singular point of type A_k. Similarly, we see that k must be even if $k \geqslant 2$ (consider the "central" curve in the dual graph of the singularity A_k). Since

$$\operatorname{rk}\operatorname{Pic}(\tilde{S}) = \operatorname{rk}\operatorname{Pic}(S) + k|\operatorname{Sing}(S)|,$$

we see that $k|\operatorname{Sing}(S)| \leqslant 19$, because $\operatorname{rk}\operatorname{Pic}(\tilde{S}) \leqslant 20$ (see [6, Proposition VIII.3.3]).

Let P be a singular point of the surface S, and let G_P be its stabilizer in $\mathfrak{A}_5$. Note that G_P acts on the dual graph of the singularity A_k.

Let us show that $k = 1$. Suppose that $k \geqslant 2$. Since k is even, the group G_P must fix a point $\tilde{P}$ in $\tilde{S}$ such that $\theta(\tilde{P}) = P$. In particular, by Lemmas 6.7.1(ii) and 5.1.1 this implies that $|G_P| \in \{1, 2, 3, 5\}$. One has

$$19 \geqslant k|\operatorname{Sing}(S)| = \frac{k|\mathfrak{A}_5|}{|G|} = \frac{60k}{|G|} \geqslant 12k \geqslant 24,$$

which is a contradiction. Thus, we see that $k = 1$, i.e., the surface S has only isolated ordinary double points.

Since $\operatorname{Sing}(S)$ is an $\mathfrak{A}_5$-orbit and $|\operatorname{Sing}(S)| \leqslant 19$, we have

$$|\operatorname{Sing}(S)| \in \{1, 5, 6, 10, 12, 15\}$$

by Corollary 5.1.2. This completes the proof of assertion (iii).

Suppose that $|\operatorname{Sing}(S)| = 12$. Then $G \cong \boldsymbol{\mu}_5$ by Lemma 5.1.1. Let E be the θ-exceptional curve such that $\theta(E) = P$. Then G acts on $E \cong \mathbb{P}^1$. Since $G \cong \boldsymbol{\mu}_5$, we see that G fixes at least two points in E. Thus the $\mathfrak{A}_5$-orbits of each of these two points have length 12. By Lemma 6.7.1(iii), the surface $\tilde{S}$ has exactly two $\mathfrak{A}_5$-orbits of length 12. The image of each of these orbits with respect to θ is $\operatorname{Sing}(S)$. Therefore, the surface S does not contain other $\mathfrak{A}_5$-orbits of length 12 except $\operatorname{Sing}(S)$. This gives assertion (iv) and completes the proof of Lemma 6.7.3. □

Lemma 6.7.4. Let S be a $K3$ surface with at most du Val singularities that is faithfully acted on by the group $\mathfrak{A}_5$. Suppose that

$$|\operatorname{Sing}(S)| \in \{5, 10, 15\}.$$

Let H_S be an ample Cartier divisor on S such that the class of H_S in the Picard group $\operatorname{Pic}(S)$ is $\mathfrak{A}_5$-invariant and is not divisible. Assume that the number H_S^2 is not divisible by 4. Then

$$\operatorname{Cl}(S)^{\mathfrak{A}_5} = \operatorname{Pic}(S)^{\mathfrak{A}_5},$$

and this group is generated by H_S.

Proof. Suppose that the first assertion of the lemma is wrong. Then there exists a Weil divisor R on S such that the class of R in $\mathrm{Cl}(S)$ is $\mathfrak{A}_5$-invariant and R is not a Cartier divisor on S. In particular, the surface S is singular. By Lemma 6.7.3(ii),(iii), the surface S has isolated ordinary double points and $\mathrm{rk}\,\mathrm{Pic}\,(S)^{\mathfrak{A}_5} = 1$. Thus, the divisor $2R$ is a Cartier divisor, and H_S generates $\mathrm{Pic}(S)^{\mathfrak{A}_5}$, since $\mathrm{Pic}(S)^{\mathfrak{A}_5}$ does not have torsion (see Lemma 6.7.3(i)). We have
$$2R \sim lH_S$$
for some positive integer l. Hence, we have $4R^2 = H_S^2 l^2$, so that l is even.

Let $\theta\colon \tilde{S} \to S$ be the minimal resolution of singularities of the surface S. Denote by $\tilde{R}$ the proper transform of the curve R on the surface $\tilde{S}$. Put
$$r = |\mathrm{Sing}(S)|.$$
Denote by $E_1, \ldots, E_r$ the exceptional curves of θ. Then
$$2\tilde{R} \sim \theta^*(lH_S) - m\sum_{i=1}^r E_i,$$
for some positive integer m. Moreover, m is odd, because R is not a Cartier divisor. We have
$$4\tilde{R}^2 = H_S^2 l^2 - 2m^2 r,$$
which implies that r is even. This shows that $r = 10$, so that
$$4\tilde{R}^2 = H_S^2 l^2 - 20m^2.$$
Since $\tilde{R}^2$ and l are even numbers, we see that m is even as well.

The obtained contradiction shows that every $\mathfrak{A}_5$-invariant Weil divisor on S is actually a Cartier divisor, so that $\mathrm{Cl}(S)^{\mathfrak{A}_5} = \mathrm{Pic}(S)^{\mathfrak{A}_5}$. Recall that one has $\mathrm{Pic}(S)^{\mathfrak{A}_5} \cong \mathbb{Z}$ by Lemma 6.7.3(i),(ii). Since H_S is not divisible in $\mathrm{Pic}(S)^{\mathfrak{A}_5}$ by assumption, we see that $\mathrm{Pic}(S)^{\mathfrak{A}_5}$ is generated by H_S. □

Lemma 6.7.5. *Let S be a singular $K3$ surface with at most du Val singularities that is faithfully acted on by the group $\mathfrak{A}_5$. Suppose that there exists an $\mathfrak{A}_5$-equivariant double cover $\eta\colon S \to \mathbb{P}^2$ that is branched over a reduced sextic curve $B \subset \mathbb{P}^2$. Then*

(i) *there is an $\mathfrak{A}_5$-equivariant identification $\mathbb{P}^2 \cong \mathbb{P}(W_3)$ or $\mathbb{P}^2 \cong \mathbb{P}(W_3')$;*

(ii) *the set $\mathrm{Sing}(S)$ consists of one $\mathfrak{A}_5$-orbit, every singular point of the surface S is an ordinary double point, and $|\mathrm{Sing}(S)| \in \{6, 10, 15\}$;*

(iii) if $|\operatorname{Sing}(S)| = 6$, then B is a unique irreducible $\mathfrak{A}_5$-invariant sextic curve whose normalization has genus 4 (cf. Remark 5.4.2);

(iv) if $|\operatorname{Sing}(S)| = 10$, then B is a unique irreducible $\mathfrak{A}_5$-invariant rational sextic curve;

(v) if $|\operatorname{Sing}(S)| = 15$, then the curve B is a unique $\mathfrak{A}_5$-invariant union of six lines.

Proof. Assertion (i) holds by Remark 6.1.1. Since B is $\mathfrak{A}_5$-invariant, all remaining assertions follow from Remark 6.1.5. □

Lemma 6.7.6. Let S be a smooth $K3$ surface that is faithfully acted on by the group $\mathfrak{A}_5$. Let H_S be a Cartier divisor on S, and let D be a non-ample divisor on S such that $D^2 > 0$ and the linear system $|D|$ is base point free. Suppose that the classes of H_S and D in $\operatorname{Pic}(S)$ are $\mathfrak{A}_5$-invariant. Then there exists a (not necessarily $\mathfrak{A}_5$-invariant) smooth rational curve F on the surface S such that $D \cdot F = 0$ and

$$D^2 \left(2H_S^2 + r(H_S \cdot F)^2 \right) = 2(H_S \cdot D)^2$$

for some $r \in \{1, 5, 6, 10, 12, 15\}$. Furthermore, the $\mathfrak{A}_5$-orbit of the curve F consists of r disjoint irreducible components (and F is one of them). Moreover, if F' is an irreducible curve on S such that $D \cdot F' = 0$, then F' is contained in the $\mathfrak{A}_5$-orbit of F. Finally, if $D^2 = 2$, then $r \in \{6, 10, 15\}$.

Proof. Since $|D|$ is base point free and $D^2 > 0$, the linear system $|D|$ gives an $\mathfrak{A}_5$-equivariant morphism $\phi \colon S \to \mathbb{P}^n$. By [109, §4.1], either ϕ is birational onto its image and $\phi(S)$ is a $K3$ surface with at most du Val singularities, or $\phi(S)$ is a normal surface of minimal degree (see [40]) and ϕ is a generically two-to-one cover. Taking the Stein factorization of ϕ, we obtain a commutative diagram

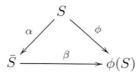

such that α is a birational morphism, and β is either an isomorphism or a finite morphism of degree 2. Note that both α and β are $\mathfrak{A}_5$-equivariant. Furthermore, since S is a smooth $K3$ surface, we conclude that $\bar{S}$ is a $K3$ surface with at most du Val singularities.

CREMONA GROUPS AND THE ICOSAHEDRON 149

The surface $\bar{S}$ is faithfully acted on by $\mathfrak{A}_5$. Since D is not ample, $\bar{S}$ is indeed singular. By Lemma 6.7.3(iii), the set $\mathrm{Sing}(\bar{S})$ consists of one $\mathfrak{A}_5$-orbit, every singular point of the surface $\bar{S}$ is an ordinary double point, and
$$|\mathrm{Sing}(\bar{S})| \in \{1, 5, 6, 10, 12, 15\}.$$

Put $r = |\mathrm{Sing}(\bar{S})|$. Let $F_1, \ldots, F_r$ be α-exceptional curves. Then each F_i is a smooth rational curve such that $F_i^2 = -2$. Moreover, the curves $F_1, \ldots, F_r$ are disjoint. Put $F_S = \sum_{i=1}^{r} F_i$ and $F = F_1$. Since $\mathrm{rk}\,\mathrm{Pic}\,(S)^{\mathfrak{A}_5} \leqslant 2$ by Lemma 6.7.3(i), we have

$$0 = \det \begin{pmatrix} D^2 & D \cdot F_S & D \cdot H_S \\ D \cdot F_S & F_S^2 & F_S \cdot H_S \\ D \cdot H_S & F_S \cdot H_S & H_S^2 \end{pmatrix} = \det \begin{pmatrix} D^2 & 0 & D \cdot H_S \\ 0 & -2r & rF \cdot H_S \\ D \cdot H_S & rF \cdot H_S & H_S^2 \end{pmatrix},$$

which implies that

$$D^2 \left(2H_S^2 + r(H_S \cdot F)^2 \right) = 2 \left(H_S \cdot D \right)^2.$$

Since the set $\mathrm{Sing}(\bar{S})$ consists of one $\mathfrak{A}_5$-orbit, every irreducible curve on S that has trivial intersection with D must be one of the curves $F_1, \ldots, F_r$.

If $D^2 = 2$, then $\phi(S) \cong \mathbb{P}^2$ and β is a double cover branched over a reduced sextic curve $B \subset \phi(S)$, which implies that $r \in \{6, 10, 15\}$ by Lemma 6.7.5(ii). $\square$

Lemma 6.7.7. *Let S be a smooth $K3$ surface that is faithfully acted on by the group $\mathfrak{A}_5$. Let H_S be an ample divisor on S, and let D be an ample divisor on S such that $D^2 \notin \{2, 8\}$, the linear system $|D|$ is base point free and D is not very ample. Suppose that the classes of H_S and D in $\mathrm{Pic}(S)$ are $\mathfrak{A}_5$-invariant. Then*

$$k^2 D^2 - 4kD \cdot H_S + 4H_S^2 = 0$$

for some positive integer k.

Proof. Since D is ample, one has

$$h^1(\mathcal{O}_S(D)) = h^2(\mathcal{O}_S(D)) = 0$$

by Kodaira vanishing (cf. Theorem 2.3.5), so that

$$h^0(\mathcal{O}_S(D)) = \frac{D^2}{2} + 2$$

by the Riemann–Roch formula. Put $n = \frac{D^2}{2} + 1$. Since $|D|$ is base point free, the linear system $|D|$ gives an $\mathfrak{A}_5$-equivariant morphism $\phi \colon S \to \mathbb{P}^n$. Since D is ample, it follows from [109, §4.1] that either ϕ is an isomorphism on its image, or $\phi(S)$ is a normal surface of minimal degree (see [40]) and ϕ is a two-to-one cover. The former case is impossible, since D is not very ample by assumption.

By [40], one of the following possibilities occurs:

(i) $n = 2$ and $\phi(S) = \mathbb{P}^2$;

(ii) $n = 5$ and $\phi(S) \cong \mathbb{P}^2$ is a Veronese surface in $\mathbb{P}^5$;

(iii) $\phi(S)$ is a cone over a rational normal curve of degree $n - 1$;

(iv) $\phi(S)$ is a rational normal scroll.

Since $D^2 \ne 2$, the case (i) is impossible. Since $D^2 \ne 8$, the case (ii) is impossible. Since S does not contain $\mathfrak{A}_5$-invariant points by Lemma 6.7.1(ii), the case (iii) is impossible as well. Therefore, we see that $\phi(S) \cong \mathbb{F}_m$ is a rational normal scroll. In fact, it follows from [33, Theorem 2.3] or [104, Corollary 2.4] (cf. [109, Proposition 5.7]) that $m \in \{0, 1, 2, 3, 4\}$. Moreover, since $\mathfrak{A}_5$ does not act faithfully on $\mathbb{F}_m$ with odd m by Lemma 6.5.1, we see that $m \in \{0, 2, 4\}$. Nevertheless, we will not use the bounds on m to obtain a contradiction.

One has
$$D \sim \phi^*(s + al)$$
for some integer $a > m$, where s and l are curves on $\mathbb{F}_m$ such that $s^2 = -m$, $l^2 = 0$ and $l \cdot s = 1$.

Since the morphism ϕ is $\mathfrak{A}_5$-equivariant, the class of the divisor $\phi^*(l)$ is $\mathfrak{A}_5$-invariant. Since $\operatorname{rk} \operatorname{Pic}(S)^{\mathfrak{A}_5} \leqslant 2$ by Lemma 6.7.3(i), the determinant of the matrix
$$\begin{pmatrix} D^2 & D \cdot H_S & D \cdot \phi^*(l) \\ D \cdot H_S & H_S^2 & H_S \cdot \phi^*(l) \\ D \cdot \phi^*(l) & H_S \cdot \phi^*(l) & \phi^*(l) \cdot \phi^*(l) \end{pmatrix}$$
must vanish. Put $k = H_S \cdot \phi^*(l)$. Then $k > 0$, because H_S is ample. Since
$$D \cdot \phi^*(l) = 2\Big(s + al\Big) \cdot l = 2$$
and $\phi^*(l) \cdot \phi^*(l) = l^2 = 0$, we have
$$0 = \det(M) = -k^2 D^2 + 4k D \cdot H_S - 4H_S^2,$$
and the assertion of Lemma 6.7.7 follows. $\qquad\square$

Part III
Quintic del Pezzo threefold

Chapter 7

Quintic del Pezzo threefold

In this chapter we recall a construction and basic properties of the main character of the remaining part of the book, the quintic del Pezzo threefold V_5. This is a rational smooth Fano threefold with many remarkable features. It was studied from various points of view, including its biregular properties (see, e.g., [86] and [47]), birational maps (see, e.g., [66, §4.3]), Kähler–Einstein metrics (see [22]), exceptional collections (see [91]), instanton bundles (see [110]) and others.

7.1 Construction and basic properties

Fix a five-dimensional vector space $\mathscr{W}_5$. There is a unique way to endow $\mathscr{W}_5$ with a structure of an irreducible representation of the group $\mathrm{SL}_2(\mathbb{C})$, so that $\mathscr{W}_5 \cong \mathbb{W}_5$ in the notation of §5.5. The Grassmannian

$$\mathrm{Gr}(2, \mathscr{W}_5) \cong \mathrm{Gr}(2, 5)$$

of two-dimensional planes in $\mathscr{W}_5$ has an $\mathrm{SL}_2(\mathbb{C})$-equivariant Plücker embedding into the projective space

$$\mathbb{P}(\Lambda^2(\mathscr{W}_5)) \cong \mathbb{P}^9.$$

By Lemma 5.5.1 one has

$$\Lambda^2(\mathscr{W}_5) \cong \mathbb{W}_3 \oplus \mathbb{W}_7,$$

so that there is a unique $\mathrm{SL}_2(\mathbb{C})$-invariant section of $\mathrm{Gr}(2, \mathscr{W}_5)$ by a subspace $\mathcal{R} \subset \mathbb{P}^9$ of codimension 3. Throughout this book the latter variety is referred to as V_5, and the corresponding seven-dimensional $\mathrm{SL}_2(\mathbb{C})$-invariant subspace in $\Lambda^2(\mathscr{W}_5)$ is referred to as $\mathscr{W}_7$, so that

$$V_5 \subset \mathcal{R} = \mathbb{P}(\mathscr{W}_7) \cong \mathbb{P}^6 \subset \mathbb{P}(\Lambda^2(\mathscr{W}_5)) \cong \mathbb{P}^9.$$

We will also keep the notation $\mathscr{W}_5$ to refer to the initial five-dimensional vector space.

The following assertion is well known (see [66, Theorem 3.4.8], [110, Lemma 2.8]), but we include its proof for the reader's convenience.

Lemma 7.1.1. The variety V_5 is a smooth threefold.

Proof. Suppose that this is not the case, i.e., either $\dim(V_5) \geqslant 4$, or V_5 is singular at some point x. This means that for some point

$$x \in V_5 \subset \mathrm{Gr}(2, \mathscr{W}_5)$$

there is a hyperplane $\mathcal{H} \subset \mathbb{P}^9$ containing $\mathcal{R}$ such that $\mathrm{Gr}(2, \mathscr{W}_5) \cap \mathcal{H}$ is singular at x. In terms of the projectively dual variety

$$\mathrm{Gr}(2, \mathscr{W}_5)^\vee \cong \mathrm{Gr}(2, \mathscr{W}_5^\vee)$$

this means that there is a point

$$h = \mathcal{H}^\perp \in \mathrm{Gr}(2, \mathscr{W}_5^\vee)$$

that is contained in the two-dimensional subspace

$$\mathcal{R}^\perp \subset (\mathbb{P}^9)^\vee \cong \mathbb{P}(\Lambda^2(\mathscr{W}_5^\vee)).$$

By Lemma 5.5.1 there is a unique three-dimensional $\mathrm{SL}_2(\mathbb{C})$-subrepresentation U' in $\Lambda^2(\mathscr{W}_5^\vee)$, so that $\mathcal{R}^\perp$ is a unique $\mathrm{SL}_2(\mathbb{C})$-invariant two-dimensional subspace in $(\mathbb{P}^9)^\vee$, and $\mathcal{R}^\perp = \mathbb{P}(U')$. Let w' be a highest weight vector in U'. The bi-vector w' is not decomposable by Lemma 5.5.10.

We claim that, contrary to our previous conclusions, $\mathcal{R}^\perp$ does not intersect the Grassmannian $\mathrm{Gr}(2, \mathscr{W}_5^\vee) \subset (\mathbb{P}^9)^\vee$. Indeed, suppose that it does. Then U' contains some decomposable bi-vector. Thus there exists a closed $\mathrm{SL}_2(\mathbb{C})$-orbit in $\mathcal{R}^\perp$ that consists of points corresponding to decomposable bi-vectors. Since the unique closed $\mathrm{SL}_2(\mathbb{C})$-orbit in $\mathcal{R}^\perp$ is the $\mathrm{SL}_2(\mathbb{C})$-orbit corresponding to the bi-vector w' (see, e.g., [46, Claim 23.52]), we conclude that w' is a decomposable bi-vector. The obtained contradiction shows that

$$\mathcal{R}^\perp \cap \mathrm{Gr}(2, \mathscr{W}_5^\vee) = \varnothing$$

and completes the proof. □

An approach used in the proof of Lemma 7.1.1 allows one to write down explicitly the equations defining the variety V_5. Namely, let

$$\{w_4, w_2, w_0, w_{-2}, w_{-4}\}$$

be the weight basis in the $\mathrm{SL}_2(\mathbb{C})$-representation $\mathscr{W}_5 \cong \mathbb{W}_5$, and let

$$\left\{w_i \wedge w_j \mid i > j, \ i, j \in \{-4, -2, 0, 2, 4\}\right\}$$

be the corresponding basis in $\Lambda^2(\mathscr{W}_5)$. For a bi-vector $w \in \Lambda^2(\mathscr{W}_5)$ define

$$\xi_{i,j}(w), \quad i > j, \ i, j \in \{-4, -2, 0, 2, 4\},$$

as the coefficient at $w_i \wedge w_j$ in the expansion of w with respect to the latter basis. The functions $\xi_{i,j}$ are interpreted as homogeneous coordinates on the projective space $\mathbb{P}^9 \cong \mathbb{P}(\Lambda^2(\mathscr{W}_5))$. Corollary 5.5.11 implies (cf. [110, §2.1]) that the variety $V_5 \subset \mathbb{P}^9$ is given by usual Plücker equations of $\mathrm{Gr}(2, \mathscr{W}_5)$ together with

$$\xi_{4,-2} - 3\xi_{2,0} = \xi_{4,-4} - 2\xi_{2,-2} = \xi_{2,-4} - 3\xi_{0,-2} = 0. \tag{7.1.2}$$

In principle, this can be used to obtain various information about the variety V_5, but we will usually avoid this and favor more geometric arguments.

Keeping in mind Lemma 7.1.1, it is easy to check that V_5 is a Fano threefold of index 2 and anticanonical degree 40 with $\mathrm{Pic}(V_5) = \mathbb{Z}[H]$, where H is the class of a hyperplane section of $V_5 \subset \mathbb{P}^6$. Moreover, any non-singular Fano threefold of Picard rank 1, index 2, and anticanonical degree 40 is isomorphic to V_5 (see [86, Lemma 3.3(i)], [66, Corollary 3.4.2]). The threefold $V_5 \subset \mathbb{P}^6$ is an intersection of quadrics, because the Grassmannian $\mathrm{Gr}(2, 5) \subset \mathbb{P}^9$ is an intersection of quadrics. Also, using projective normality of $\mathrm{Gr}(2, 5) \subset \mathbb{P}^9$ together with Kodaira vanishing, one can see that $V_5 \subset \mathbb{P}^6$ is projectively normal as well.

In the rest of the book we always denote by H the class of the hyperplane section of $V_5 \subset \mathbb{P}^6$. By degree of any subvariety of V_5 we mean its degree with respect to H. By a line we mean a curve $C \subset V_5$ such that $H \cdot C = 1$. Similarly, by a conic we mean an irreducible smooth curve $C \subset V_5$ such that $H \cdot C = 2$.

The Riemann–Roch formula and Kodaira vanishing (cf. Theorem 2.3.5) imply that

$$h^0(\mathcal{O}_{V_5}(lH)) = \frac{5l(l+1)(l+2)}{6} + l + 1 \tag{7.1.3}$$

for every non-negative integer l. In particular, we have
$$h^0(\mathcal{O}_{V_5}(H)) = 7, \quad h^0(\mathcal{O}_{V_5}(2H)) = 23$$
and $h^0(\mathcal{O}_{V_5}(3H)) = 54$.

The automorphism group of V_5 contains a subgroup isomorphic to $\mathrm{PSL}_2(\mathbb{C})$ by construction; actually, we will see below in Proposition 7.1.10 that $\mathrm{Aut}(V_5) \cong \mathrm{PSL}_2(\mathbb{C})$.

The variety V_5 has a natural $\mathrm{PSL}_2(\mathbb{C})$-invariant stratification.

Theorem 7.1.4 ([86, Lemma 1.5], [66, Remark 3.4.9 and p. 61], [110, Proposition 2.13]). *The variety V_5 is a union of three $\mathrm{PSL}_2(\mathbb{C})$-orbits. The unique one-dimensional orbit is a rational normal curve $\mathscr{C} \subset V_5$ of degree 6. The unique two-dimensional orbit is of the form $\mathscr{S} \setminus \mathscr{C}$, where $\mathscr{S}$ is some surface in the linear system $|2H|$. There is a $\mathrm{PSL}_2(\mathbb{C})$-equivariant identification*
$$V_5 \setminus \mathscr{S} \cong \mathrm{PSL}_2(\mathbb{C})/\mathfrak{S}_4.$$

Remark 7.1.5. One can interpret $\mathscr{W}_7$ as the vector space of homogeneous polynomials of degree 6 in two variables, say x and y (cf. §5.5). Then the open $\mathrm{PSL}_2(\mathbb{C})$-orbit contained in V_5 can be constructed as the $\mathrm{PSL}_2(\mathbb{C})$-orbit of the polynomial
$$F_3 = xy(x^4 - y^4),$$
the two-dimensional $\mathrm{PSL}_2(\mathbb{C})$-orbit $\mathscr{S} \subset V_5$ can be constructed as the $\mathrm{PSL}_2(\mathbb{C})$-orbit of the polynomial
$$F_2 = xy^5,$$
while the one-dimensional $\mathrm{PSL}_2(\mathbb{C})$-orbit $\mathscr{C} \subset V_5$ can be constructed as the $\mathrm{PSL}_2(\mathbb{C})$-orbit of the polynomial $F_1 = y^6$. We refer the reader to [86, Lemma 1.5 and §3] for details.

The group $\mathrm{Aut}(V_5)$ contains a subgroup isomorphic to $\mathfrak{A}_5$.

Lemma 7.1.6. *One has*
$$H^0(\mathcal{O}_{V_5}(H)) \cong \mathscr{W}_7 \cong \mathbb{W}_7$$
as $\mathrm{SL}_2(\mathbb{C})$-representations (or, more precisely, $\mathrm{PSL}_2(\mathbb{C})$-representations). Furthermore, one has
$$H^0(\mathcal{O}_{V_5}(H)) \cong \mathscr{W}_7 \cong W_3 \oplus W_4$$
as $\mathfrak{A}_5$-representations.

Proof. The first assertion follows from the construction of the variety V_5. To obtain the second assertion keep in mind Convention 5.6.1 and apply Lemma 5.6.2. □

Our next tool will be the description of lines contained in V_5. Let $\mathfrak{H}_\ell$ denote the Hilbert scheme of lines on V_5.

Theorem 7.1.7. *There is a $\mathrm{PSL}_2(\mathbb{C})$-equivariant identification of $\mathfrak{H}_\ell$ with the projective plane $\mathbb{P}_\ell^2 = \mathbb{P}(\mathbb{W}_3)$. In particular, there is an $\mathfrak{A}_5$-equivariant identification of $\mathfrak{H}_\ell$ with $\mathbb{P}(W_3')$.*

Proof. For the $\mathrm{PSL}_2(\mathbb{C})$-equivariant identification see [110, Proposition 2.20] (cf. [47, Theorem I]). The $\mathfrak{A}_5$-equivariant identification follows from the $\mathrm{PSL}_2(\mathbb{C})$-equivariant one and Lemma 5.6.2 (or rather Convention 5.6.1). □

It appears that the intersection of lines in V_5 can be detected on their parameter space $\mathbb{P}_\ell^2$ (see [110, Remark 2.28]).

Theorem 7.1.8. *Let $L \subset V_5$ be a line. Consider the set $\ell(L) \subset \mathbb{P}^2$ that consists of all lines $L' \subset V_5$ such that $L' \cap L \neq \varnothing$ and $L' \neq L$. Let $L_{\mathfrak{C}}$ be the line in $\mathbb{P}_\ell^2$ that is polar to the point $L \in \mathbb{P}_\ell^2$ with respect to the conic $\mathfrak{C}$. Then $\ell(L)$ coincides with $L_{\mathfrak{C}}$ if L is not contained in the conic $\mathfrak{C}$, and $\ell(L)$ coincides with $L_{\mathfrak{C}} \setminus \{L\}$ if L is contained in $\mathfrak{C}$.*

We have seen in Lemma 5.5.5 that there is a unique $\mathrm{PSL}_2(\mathbb{C})$-invariant conic $\mathfrak{C} \subset \mathbb{P}_\ell^2$. On the other hand, V_5 is a union of three $\mathrm{PSL}_2(\mathbb{C})$-orbits: the one-dimensional orbit $\mathscr{C}$, the two-dimensional orbit $\mathscr{S} \setminus \mathscr{C}$, and the open orbit $V_5 \setminus \mathscr{S}$ (see Theorem 7.1.4). There is a nice correspondence between these two stratifications.

Theorem 7.1.9. *The following assertions hold:*

(i) *for any point $P \in V_5$ there are exactly three (resp., two, one) distinct lines through P on V_5 if and only if P is a point of the three-dimensional $\mathrm{PSL}_2(\mathbb{C})$-orbit $V_5 \setminus \mathscr{S}$ (resp., if and only if P is a point of the two-dimensional $\mathrm{PSL}_2(\mathbb{C})$-orbit $\mathscr{S} \setminus \mathscr{C}$, if and only if P is a point of the one-dimensional $\mathrm{PSL}_2(\mathbb{C})$-orbit $\mathscr{C}$);*

(ii) *the lines on V_5 that are contained in the surface $\mathscr{S}$ are those that correspond to the points of the conic $\mathfrak{C} \subset \mathbb{P}_\ell^2$, and they are exactly the tangent lines to the curve $\mathscr{C}$;*

(iii) *any two lines contained in $\mathscr{S}$ are disjoint;*

(iv) any line on V_5 that intersects the curve $\mathscr{C}$ is contained in the surface $\mathscr{S}$;

(v) any line in V_5 that is disjoint from the curve $\mathscr{C}$ intersects the surface $\mathscr{S}$ transversally in two smooth points of $\mathscr{S}$.

Proof. For assertion (i) see [110, Corollary 2.24] or [60, 1.2.1(3)]. The first part of assertion (ii) follows from [60, 1.2.1(1)] and the fact that $\mathfrak{C}$ is the unique one-dimensional $\mathrm{PSL}_2(\mathbb{C})$-orbit in $\mathbb{P}_\ell^2$ (see Lemma 5.5.5). The second part of assertion (ii) follows from an explicit description of the surface $\mathscr{S}$ (see [66, Remark 3.4.9(b),(c)] or [60, 1.2.1(3)], cf. the proof of [86, Lemma 3.4]). Assertion (iii) holds since any two lines tangent to a given rational normal curve are disjoint. Assertion (iv) is implied by assertion (i).

Let L be a line in V_5 that is disjoint from the curve $\mathscr{C}$. By assertion (ii) this means that L is not contained in the surface $\mathscr{S}$, and the corresponding point of $\mathbb{P}_\ell^2$ is not contained in the conic $\mathfrak{C}$. Thus, by assertion (ii) and Theorem 7.1.8 there are exactly two lines contained in $\mathscr{S}$ that intersect L. By assertion (iii) this means that L intersects $\mathscr{S}$ in two points. Now assertion (v) is implied by the fact that $\mathscr{S} \cdot L = 2$. □

Now we are ready to give a complete description of the automorphism group of V_5. The following assertion is well known (see, e. g., [85, Proposition 4.4]).

Proposition 7.1.10. One has
$$\mathrm{Aut}(V_5) \cong \mathrm{PSL}_2(\mathbb{C}).$$
In particular, a subgroup $\mathfrak{A}_5 \subset \mathrm{Aut}(V_5)$ is unique up to conjugation, and the subgroup $\mathrm{Aut}^{\mathfrak{A}_5}(V_5) \subset \mathrm{Aut}(V_5)$ of biregular $\mathfrak{A}_5$-automorphisms of V_5 coincides with $\mathfrak{A}_5$.

Proof. We claim that the group $\mathrm{Aut}(V_5)$ acts faithfully on $\mathbb{P}_\ell^2$. Indeed, suppose that some element $g \in \mathrm{Aut}(V_5)$ acts on $\mathbb{P}_\ell^2$ trivially. Let P be a general point of $V_5 \setminus \mathscr{C}$. By Theorem 7.1.9(i) there are at least two lines in V_5 that pass through P, and P is the unique point of intersection of these lines. Thus $g(P) = P$, so that g acts trivially on V_5.

By Theorem 7.1.9(i) the surface $\mathscr{S} \subset V_5$ is $\mathrm{Aut}(V_5)$-invariant. Thus Theorem 7.1.9(ii) implies that the conic $\mathfrak{C} \subset \mathbb{P}_\ell^2$ is $\mathrm{Aut}(V_5)$-invariant. Therefore, the action of $\mathrm{Aut}(V_5)$ on $\mathfrak{C}$ is faithful, so that
$$\mathrm{Aut}(V_5) \subset \mathrm{Aut}(\mathfrak{C}) \cong \mathrm{PSL}_2(\mathbb{C}).$$
The latter means that $\mathrm{Aut}(V_5) \cong \mathrm{PSL}_2(\mathbb{C})$.

The remaining assertions follow from the classification of finite subgroups of $\mathrm{PSL}_2(\mathbb{C})$ and Lemma 5.5.8. □

7.2 $\mathrm{PSL}_2(\mathbb{C})$-invariant anticanonical surface

Lemma 7.2.1. One has an identification of $\mathrm{SL}_2(\mathbb{C})$-representations
$$H^0(\mathcal{O}_{V_5}(2H)) \cong \mathbb{W}_1 \oplus \mathbb{W}_9 \oplus \mathbb{W}_{13}.$$

Proof. Recall that by Lemma 7.1.6 one has an identification of $\mathrm{SL}_2(\mathbb{C})$-representations
$$H^0(\mathcal{O}_{V_5}(H)) \cong H^0(\mathcal{O}_{\mathbb{P}^6}(H)) \cong \mathbb{W}_7.$$
Therefore by Lemma 5.5.1 one has an identification of $\mathrm{SL}_2(\mathbb{C})$-representations
$$H^0(\mathcal{O}_{\mathbb{P}^6}(2H)) \cong \mathbb{W}_1 \oplus \mathbb{W}_5 \oplus \mathbb{W}_9 \oplus \mathbb{W}_{13}.$$
Since $V_5 \subset \mathbb{P}^6$ is projectively normal, the restriction map
$$H^0\left(\mathcal{O}_{\mathbb{P}^6}(2H)\right) \to H^0\left(\mathcal{O}_{V_5}(2H)\right)$$
is surjective. Moreover, $h^0(\mathcal{O}_{V_5}(2H)) = 23$ by (7.1.3) and the assertion follows. □

Recall that $\mathscr{S}$ is an $\mathrm{SL}_2(\mathbb{C})$-invariant surface in $|2H|$ by Theorem 7.1.4. Lemma 7.2.1 shows that it is a unique surface with this property.

Let $\nu\colon \mathcal{Y} \to V_5$ be a blow-up of the curve $\mathscr{C}$, and let $E_\mathcal{Y}$ be the ν-exceptional surface. Since $\mathscr{C}$ is an intersection of quadrics and
$$-K_\mathcal{Y} \sim \nu^*(2H) - E_\mathcal{Y},$$
the linear system $|-K_\mathcal{Y}|$ is free from base points. In particular, $-K_\mathcal{Y}$ is nef. Moreover, since $-K_\mathcal{Y}^3 = 14$ (cf. (11.5.3) below), the divisor $-K_\mathcal{Y}$ is also big. Denote by $\hat{\mathscr{S}}$ the proper transform of the surface $\mathscr{S}$ on the threefold $\mathcal{Y}$.

Lemma 7.2.2. The following assertions hold:

(i) one has $\mathrm{Sing}(\mathscr{S}) = \mathscr{C}$;

(ii) the surface $\mathscr{S}$ is swept out by the tangent lines to the curve $\mathscr{C}$;

(iii) the surface $\mathscr{S}$ has an ordinary cusp along the curve $\mathscr{C}$;

(iv) the surface $\hat{\mathscr{S}}$ is smooth, and the morphism $\nu|_{\hat{\mathscr{S}}}\colon \hat{\mathscr{S}} \to \mathscr{S}$ is the normalization of the surface $\mathscr{S}$;

(v) the morphism $\nu|_{\hat{\mathscr{S}}}\colon \hat{\mathscr{S}} \to \mathscr{S}$ is bijective;

(vi) one has $\hat{\mathscr{S}} \cong \mathbb{P}^1 \times \mathbb{P}^1$.

Proof. Assertion (ii) is just a reformulation of Theorem 7.1.9(ii).

It follows from assertion (ii) that the surface $\mathscr{S}$ has a singularity of type A_r with $r \geqslant 3$ along the curve $\mathscr{C}$. Let C be a general hyperplane section of the surface $\mathscr{S}$. Then C is an irreducible curve, and its arithmetic genus equals 6 by the adjunction formula. On the other hand, C has singularities of type A_r with $r \geqslant 3$ at 6 points of the intersection $C \cap \mathscr{C}$. Hence, these are ordinary cusps by Lemma 4.4.6. This implies that the surface $\mathscr{S}$ has an ordinary cusp along the curve $\mathscr{C}$ and proves assertion (iii).

Assertion (i) is a combination of assertions (ii) and (iii). Assertions (iv) and (v) are implied by assertion (iii). For assertion (vi) see [66, p. 61] or [86, Lemma 1.6]. □

Since the morphism
$$\nu|_{\hat{\mathscr{S}}}\colon \hat{\mathscr{S}} \to \mathscr{S}$$
is the normalization of the surface $\mathscr{S}$ by Lemma 7.2.2(iv), the $\mathfrak{A}_5$-action on $\mathscr{S}$ lifts to $\hat{\mathscr{S}}$. Denote by $\hat{\mathscr{C}}$ the set-theoretic preimage of the curve $\mathscr{C}$ on $\hat{\mathscr{S}}$, i.e., the curve $\left(\nu|_{\hat{\mathscr{S}}}^{*}(\mathscr{C})\right)_{red}$. Then the induced morphism
$$\nu|_{\hat{\mathscr{C}}}\colon \hat{\mathscr{C}} \to \mathscr{C}$$
is an isomorphism.

Lemma 7.2.3. *The following assertions hold:*

(i) *one has $\hat{\mathscr{S}} \sim \nu^*(2H) - 2E_Y$;*

(ii) *the divisor $\nu^*(H)|_{\hat{\mathscr{S}}}$ is a divisor of bi-degree $(1,5)$ on $\hat{\mathscr{S}} \cong \mathbb{P}^1 \times \mathbb{P}^1$;*

(iii) *the curve $\hat{\mathscr{C}}$ is a divisor of bi-degree $(1,1)$, and $E_Y|_{\hat{\mathscr{S}}} = 2\hat{\mathscr{C}}$.*

Proof. Assertion (i) is implied by the fact that the multiplicity of $\mathscr{S}$ along the curve $\mathscr{C}$ is 2 by Lemma 7.2.2(iii). For assertion (ii) see [86, Lemma 1.6] or [66, p. 61]. Since the surface $\mathscr{S}$ has an ordinary cusp along the curve $\mathscr{C}$ by Lemma 7.2.2(iii), we obtain the equality $E_Y|_{\hat{\mathscr{S}}} = 2\hat{\mathscr{C}}$. Moreover, if $\hat{\mathscr{C}}$ is a curve of bi-degree (a,b) on $\hat{\mathscr{S}}$, then by assertion (ii) one has
$$5a + b = \hat{\mathscr{C}} \cdot \nu^*(H) = \mathscr{C} \cdot H = 6.$$
This shows that $(a,b) = (1,1)$ and completes the proof of assertion (iii). □

CREMONA GROUPS AND THE ICOSAHEDRON

In particular, Lemma 7.2.3(ii),(iii) implies that $-K_{\mathcal{Y}}|_{\hat{\mathscr{S}}}$ is a divisor of bi-degree $(0,8)$ on $\hat{\mathscr{S}} \cong \mathbb{P}^1 \times \mathbb{P}^1$. Thus, the divisor $-K_{\mathcal{Y}}$ is not ample. In fact, we can say more.

Proposition 7.2.4. *The linear system $|-K_{\mathcal{Y}}|$ gives a $\mathrm{PSL}_2(\mathbb{C})$-equivariant birational morphism*
$$\zeta \colon \mathcal{Y} \to \mathbb{P}^9$$
whose image X_{14} is a Fano variety with canonical Gorenstein singularities. The exceptional locus of ζ is the surface $\hat{\mathscr{S}}$. Moreover, $\zeta(\hat{\mathscr{S}})$ is a rational normal curve of degree 8, the threefold X_{14} is singular along the curve $\zeta(\hat{\mathscr{S}})$, and $\zeta(E_{\mathcal{Y}})$ is the unique hyperplane section of X_{14} that passes through the curve $\zeta(\hat{\mathscr{S}})$.

Proof. The construction of X_{14} is done by Priska Jahnke, Thomas Peternell, and Ivo Radloff (see their construction No. 12 on [67, pp. 615–616]). Let us briefly recall it. Since $|-K_{\mathcal{Y}}|$ is base point free and
$$h^0(\mathcal{O}_{\mathcal{Y}}(-K_{\mathcal{Y}})) = 10$$
by the Riemann–Roch formula and Theorem 2.3.5, the linear system $|-K_{\mathcal{Y}}|$ gives a morphism $\zeta \colon \mathcal{Y} \to \mathbb{P}^9$. Denote its image by X_{14}. By [67, Propositions 1.3 and 1.6], the variety X_{14} is a Fano threefold with canonical Gorenstein singularities. Since $-K_{\mathcal{Y}}|_{\hat{\mathscr{S}}}$ is a divisor of bi-degree $(0,8)$ on $\hat{\mathscr{S}} \cong \mathbb{P}^1 \times \mathbb{P}^1$, the image $\zeta(\hat{\mathscr{S}})$ is a rational curve of degree 8. By Lemma 5.5.7, this curve is a rational normal curve of degree 8. In particular, it is contained in a unique hyperplane section in $\mathbb{P}^9$. This hyperplane section is $\zeta(E_{\mathcal{Y}})$, because
$$\hat{\mathscr{S}} \sim -K_{\mathcal{Y}} - E_{\mathcal{Y}}$$
by Lemma 7.2.3(i). Since the induced morphism
$$\zeta \colon \mathcal{Y} \to X_{14}$$
is crepant, the threefold X_{14} is singular along the curve $\zeta(\hat{\mathscr{S}})$. Note also that $\hat{\mathscr{S}}$ is the whole exceptional locus of ζ because $\mathrm{rk}\,\mathrm{Pic}\,(\mathcal{Y}) = 2$. □

From Lemma 7.2.3(iii), we also obtain the following.

Corollary 7.2.5 (cf. [86, Lemma 1.6]). *The action of the group $\mathfrak{A}_5$ on the surface $\hat{\mathscr{S}} \cong \mathbb{P}^1 \times \mathbb{P}^1$ is diagonal (see §6.4).*

Proof. The divisor $E_{\mathcal{Y}}|_{\hat{\mathscr{S}}}$ is $\mathfrak{A}_5$-invariant. By Lemma 7.2.3(iii) this means that there is an $\mathfrak{A}_5$-invariant curve of bi-degree $(1,1)$ on $\hat{\mathscr{S}} \cong \mathbb{P}^1 \times \mathbb{P}^1$. Now the assertion follows from Lemma 6.4.3(i). □

Remark 7.2.6. The assertion of Corollary 7.2.5 can be checked in a more direct way (which is actually what happens in [86, Lemma 1.6]). Indeed, the group $\mathrm{PSL}_2(\mathbb{C})$ acts on the surface $\mathscr{S}$ so that it is identified with $\mathbb{P}(\mathbb{W}_2) \times \mathbb{P}(\mathbb{W}_2)$, which restricts to a diagonal action of the group $\mathfrak{A}_5$. Still we prefer to prove Corollary 7.2.5 as we did it, since we will have to use a similar method later in other situations (cf. Lemmas 11.1.9 and 11.4.8 below).

Remark 7.2.7. The description of the surface $\mathscr{S}$ given in Remark 7.2.6 also allows one to describe the $\mathrm{PSL}_2(\mathbb{C})$-equivariant morphism of $\hat{\mathscr{S}}$ to $\mathbb{P}^6$ given by the morphism ν. Namely, we have an embedding

$$\hat{\mathscr{S}} \cong \mathbb{P}(\mathbb{W}_2) \times \mathbb{P}(\mathbb{W}_2) \to \mathbb{P}\big(\mathbb{W}_2 \otimes \mathrm{Sym}^5(\mathbb{W}_2)\big) \cong \mathbb{P}(\mathbb{W}_2 \otimes \mathbb{W}_6).$$

By Lemma 5.5.1 one has

$$\mathbb{W}_2 \otimes \mathbb{W}_6 \cong \mathbb{W}_5 \oplus \mathbb{W}_7,$$

which gives a $\mathrm{PSL}_2(\mathbb{C})$-equivariant projection

$$\mathbb{P}(\mathbb{W}_2 \otimes \mathbb{W}_6) \dashrightarrow \mathbb{P}(\mathbb{W}_7) \cong \mathbb{P}^6.$$

Now we can determine the normal bundle of the curve $\mathscr{C}$ in V_5.

Lemma 7.2.8. *The normal bundle of the curve $\mathscr{C} \cong \mathbb{P}^1$ in V_5 is*

$$\mathcal{N}_{\mathscr{C}/V_5} \cong \mathcal{O}_{\mathbb{P}^1}(5) \oplus \mathcal{O}_{\mathbb{P}^1}(5).$$

In particular, one has $E_{\mathcal{Y}} \cong \mathbb{P}^1 \times \mathbb{P}^1$.

Proof. The normal bundle of the curve $\mathscr{C}$ in V_5 is

$$\mathcal{N} = \mathcal{N}_{\mathscr{C}/V_5} \cong \mathcal{O}_{\mathbb{P}^1}(a) \oplus \mathcal{O}_{\mathbb{P}^1}(b)$$

for some integers a and b such that $a \leqslant b$ and

$$a + b = c_1(\mathcal{N}) = K_{V_5} \cdot \mathscr{C} + 2g(\mathscr{C}) - 2 = 10.$$

We must prove that $a = b = 5$. Put $n = b - a$. Then $E_{\mathcal{Y}} \cong \mathbb{F}_n$. We need to show that $n = 0$.

Let l be a fiber of the natural projection $E_{\mathcal{Y}} \to \mathscr{C}$, and let s be the section of this projection such that $s^2 = -n$. Then

$$E_{\mathcal{Y}}|_{E_{\mathcal{Y}}} \sim -s + \frac{10-n}{2}l$$

since
$$E_\mathcal{Y}^3 = -c_1(\mathcal{N}) = -10.$$
On the other hand, the surface $E_\mathcal{Y}$ contains the curve $\hat{\mathscr{C}}$ and
$$2\hat{\mathscr{C}} = \mathscr{S}|_{E_\mathcal{Y}} \sim \left(\nu^*(2H) - 2E_\mathcal{Y}\right)|_{E_\mathcal{Y}} \sim 2s + (n+2)l$$
by Lemma 7.2.3(iii), so that $\hat{\mathscr{C}} \sim s + \frac{n+2}{2}l$ on $E_\mathcal{Y}$. Since $\hat{\mathscr{C}}$ is a smooth rational $\mathrm{PSL}_2(\mathbb{C})$-invariant curve different from s, we have
$$|s \cap \hat{\mathscr{C}}| \leqslant s \cdot \hat{\mathscr{C}} = \left(s + \frac{n+2}{2}l\right) \cdot s = \frac{2-n}{2}.$$

Suppose that $n \neq 0$. Then s is also a $\mathrm{PSL}_2(\mathbb{C})$-invariant curve. Thus, the intersection $s \cap \hat{\mathscr{C}}$ must be empty, so that $n = 2$. Then
$$-K_\mathcal{Y} \cdot s = \left(\nu^*(2H) - E_\mathcal{Y}\right)|_{E_\mathcal{Y}} \cdot s = \left(s + \frac{n+14}{2}l\right) \cdot s = \left(s + 8l\right) \cdot s = 6.$$

By Proposition 7.2.4, the linear system $|-K_\mathcal{Y}|$ gives a $\mathrm{PSL}_2(\mathbb{C})$-equivariant birational morphism $\zeta \colon \mathcal{Y} \to \mathbb{P}^9$. Since $-K_\mathcal{Y} \cdot s = 6$ and s is $\mathrm{PSL}_2(\mathbb{C})$-invariant, the image $\zeta(s)$ is a rational normal sextic curve by Lemma 5.5.7. In particular, $\zeta(s)$ spans a $\mathrm{PSL}_2(\mathbb{C})$-invariant subspace of $\mathbb{P}^9$ of dimension 6. On the other hand, one has a $\mathrm{PSL}_2(\mathbb{C})$-equivariant identification
$$\mathbb{P}^9 \cong \mathbb{P}\left(\mathbb{W}_1 \oplus \mathbb{W}_9\right)$$
by Lemma 7.2.1. The obtained contradiction shows that $n = 0$. $\square$

Using Lemmas 7.2.3(iii) and 7.2.8, we obtain

Corollary 7.2.9. *The curve $\hat{\mathscr{C}}$ has bi-degree $(1,1)$ on $E_\mathcal{Y} \cong \mathbb{P}^1 \times \mathbb{P}^1$, and the action of the group $\mathfrak{A}_5$ on the surface $E_\mathcal{Y}$ is diagonal (see §6.4).*

Proof. Let l be a fiber of the natural projection $E_\mathcal{Y} \to \mathscr{C}$, and let s be the section of this projection such that $s^2 = 0$. Then
$$E_\mathcal{Y}|_{E_\mathcal{Y}} \sim -s + 5l$$
since $E_\mathcal{Y}^3 = -10$. On the other hand, the surface $E_\mathcal{Y}$ contains the curve $\hat{\mathscr{C}}$ and
$$2\hat{\mathscr{C}} = \mathscr{S}|_{E_\mathcal{Y}} \sim \left(\nu^*(2H) - 2E_\mathcal{Y}\right)|_{E_\mathcal{Y}} \sim 2s + 2l$$
by Lemma 7.2.3(iii), so that $\hat{\mathscr{C}} \sim s + l$ on $E_\mathcal{Y}$. Since $\hat{\mathscr{C}}$ is $\mathfrak{A}_5$-invariant, the required assertion follows from Lemma 6.4.3(i). $\square$

Corollary 7.2.5 allows us to describe $\mathfrak{A}_5$-orbits contained in the surface $\mathscr{S}$.

Lemma 7.2.10. For any $r \in \{12, 20, 30\}$ there exists a unique $\mathfrak{A}_5$-orbit $\Sigma_r \subset \mathscr{C}$, and a unique $\mathfrak{A}_5$-orbit $\Sigma'_r \subset \mathscr{S} \setminus \mathscr{C}$. Moreover, if $\Sigma \subset \mathscr{S}$ is an $\mathfrak{A}_5$-orbit of length $r < 60$, then $r \in \{12, 20, 30\}$, and Σ coincides either with Σ_r, or with Σ'_r.

Proof. The first assertion follows by Lemma 5.5.9(v) applied to the rational curve $\mathscr{C} \cong \mathbb{P}^1$. The rest follows from Lemma 6.4.1 since the normalization map $\nu \colon \hat{\mathscr{S}} \to \mathscr{S}$ is bijective by Lemma 7.2.2(v), and the action of the group $\mathfrak{A}_5$ on the surface $\hat{\mathscr{S}} \cong \mathbb{P}^1 \times \mathbb{P}^1$ is diagonal by Corollary 7.2.5. □

Corollary 7.2.11. The closure of the union of $\mathfrak{A}_5$-orbits of length less than 60 in V_5 is at most one-dimensional.

Proof. Choose a positive integer $r < 60$. Let P be a point in V_5 that lies in a closure of the union of $\mathfrak{A}_5$-orbits of length r. Then the length of its $\mathfrak{A}_5$-orbit divides r and, in particular, it is less that 60. Indeed, let $G_P \subset \mathfrak{A}_5$ be the stabilizer of the point P. Then G_P contains the stabilizer of a point in an $\mathfrak{A}_5$-orbit of length r, that is a subgroup in $\mathfrak{A}_5$ of order $\frac{60}{r}$. Hence, the $\mathfrak{A}_5$-orbit of P has length $\frac{60}{|G_P|}$ that must divide r.

We see that the closure of the union of $\mathfrak{A}_5$-orbits of length r in V_5 is contained in the union of $\mathfrak{A}_5$-orbits of length less than 60. On the other hand, we know from Lemma 7.2.10 that the union of $\mathfrak{A}_5$-orbits of length less than 60 intersects the surface $\mathscr{S}$ by a finite set. Hence, it is at most one-dimensional. □

We will also need some information about curves contained in $\mathscr{S}$.

Lemma 7.2.12. The following assertions hold:

(i) there exists a unique $\mathfrak{A}_5$-invariant curve $\mathcal{L}_{12} \subset \mathscr{S}$ that is a union of 12 lines; similarly, there exists a unique $\mathfrak{A}_5$-invariant curve $\mathcal{L}_{20} \subset \mathscr{S}$ that is a union of 20 lines;

(ii) the irreducible components of each of the curves $\mathcal{L}_{12}$ and $\mathcal{L}_{20}$ are disjoint;

(iii) each of the irreducible components of $\mathcal{L}_{12}$ contains one point of the $\mathfrak{A}_5$-orbit Σ_{12} and one point of the $\mathfrak{A}_5$-orbit Σ'_{12}; similarly, each of the irreducible components of $\mathcal{L}_{20}$ contains one point of the $\mathfrak{A}_5$-orbit Σ_{20} and one point of the $\mathfrak{A}_5$-orbit Σ'_{20};

(iv) the irreducible components of $\mathcal{L}_{12}$ are tangent to the curve $\mathscr{C}$ at the points of the $\mathfrak{A}_5$-orbit Σ_{12}; similarly, the irreducible components of $\mathcal{L}_{20}$ are tangent to the curve $\mathscr{C}$ at the points of the $\mathfrak{A}_5$-orbit Σ_{20};

(v) one has $\mathcal{L}_{12} \cap \mathscr{C} = \Sigma_{12}$ and $\mathcal{L}_{20} \cap \mathscr{C} = \Sigma_{20}$;

(vi) the curves $\mathcal{L}_{12}$ and $\mathcal{L}_{20}$ are disjoint.

Proof. By Lemma 7.2.3(ii) any line on $\mathscr{S}$ is an image of a curve of bi-degree $(0, 1)$ on $\hat{\mathscr{S}} \cong \mathbb{P}^1 \times \mathbb{P}^1$. By Corollary 7.2.5 the action of the group $\mathfrak{A}_5$ on $\hat{\mathscr{S}}$ is diagonal. Therefore, by Lemma 6.4.2(ii) there is a unique $\mathfrak{A}_5$-invariant curve $\hat{\mathcal{L}}_{12}$ that is a union of 12 curves of bi-degree $(0, 1)$ in $\hat{\mathscr{S}}$, and there is a unique $\mathfrak{A}_5$-invariant curve $\hat{\mathcal{L}}_{20}$ that is a union of 20 curves of bi-degree $(0, 1)$ in $\hat{\mathscr{S}}$. Since the morphism $\nu \colon \hat{\mathscr{S}} \to \mathscr{S}$ is bijective by Lemma 7.2.2(v), this implies assertion (i). Since the tangent lines to the rational normal curve $\mathscr{C}$ are disjoint from each other (cf. Theorem 7.1.9(iii)), we also obtain assertions (ii) and (vi). Lemmas 6.4.2(iii) and 7.2.10 imply assertion (iii). Assertion (iv) follows from Theorem 7.1.9(ii). Assertion (v) follows from assertion (iv) together with the fact that any tangent line to the rational normal curve $\mathscr{C}$ has a single common point with $\mathscr{C}$. □

Keeping in mind Lemma 7.2.12(i), in the rest of the book we denote by $\mathcal{L}_{12}$ and $\mathcal{L}_{20}$ the unique $\mathfrak{A}_5$-invariant curves in $\mathscr{S}$ that are unions of 12 and 20 lines, respectively.

Lemma 7.2.13. *Let Z be an $\mathfrak{A}_5$-invariant curve in $\mathscr{S}$ such that $\mathscr{C}$ is not its irreducible component. Then*

(i) *one has $\deg(Z) \geqslant 12$;*

(ii) *if Z is different from $\mathcal{L}_{12}$, then $\deg(Z) \geqslant 16$;*

(iii) *one has $\deg(Z) \neq 18$.*

Proof. Let $\hat{Z} \subset \hat{\mathscr{S}}$ be the proper transform of the curve Z. Then $\hat{Z}$ is a curve of bi-degree (a, b) in $\hat{\mathscr{S}} \cong \mathbb{P}^1 \times \mathbb{P}^1$ for some non-negative integers a and b. Note that $\hat{\mathscr{C}}$ is not an irreducible component of the curve $\hat{Z}$ by assumption. By Lemma 7.2.3(ii) we have

$$\deg(Z) = H \cdot Z = \nu^*(H) \cdot \hat{Z} = 5a + b \geqslant a + b.$$

Moreover, the action of $\mathfrak{A}_5$ on $\hat{\mathscr{S}} \cong \mathbb{P}^1 \times \mathbb{P}^1$ is diagonal by Corollary 7.2.5. Therefore, everything follows from Lemma 6.4.4. □

An easy application of Lemma 7.2.13 is a lower bound on degrees of $\mathfrak{A}_5$-irreducible curves in V_5.

Lemma 7.2.14. There are no $\mathfrak{A}_5$-irreducible curves of degree less than 6 in V_5.

Proof. Suppose that $C \subset V_5$ is an $\mathfrak{A}_5$-irreducible curve of degree $d \leqslant 5$. By Lemma 7.2.13(i) one has $C \not\subset \mathscr{S}$. Hence the intersection $\mathscr{S} \cap C$ consists of

$$\mathscr{S} \cdot C = 2d \leqslant 10$$

points (counted with multiplicities). On the other hand, $\mathscr{S} \cap C$ must be a union of several $\mathfrak{A}_5$-orbits contained in $\mathscr{S}$. This gives a contradiction with Lemma 7.2.10. □

The next lemma shows that $\mathfrak{A}_5$-invariant curves of low degree in the surface $\mathscr{S}$ are rather sparse. This is a particular case of a more general assertion that we will establish later in Theorem 8.3.1.

Lemma 7.2.15. The surface $\mathscr{S}$ contains only a finite number of $\mathfrak{A}_5$-invariant curves of degree less than 30. Moreover, $\mathscr{S}$ contains exactly two $\mathfrak{A}_5$-irreducible curves of degree 20, and no $\mathfrak{A}_5$-irreducible curves of degree 22.

Proof. Let $Z \subset \mathscr{S}$ be an $\mathfrak{A}_5$-irreducible curve of degree $d < 30$. We may assume that $Z \neq \mathscr{C}$. Let $\hat{Z}$ be the proper transform of Z on the surface $\hat{\mathscr{S}} \cong \mathbb{P}^1 \times \mathbb{P}^1$. Then $\hat{Z}$ is a curve of bi-degree (a, b) for some non-negative integers a and b. By Lemma 7.2.3(ii) we have

$$\deg(Z) = H \cdot Z = \nu^*(H) \cdot \hat{Z} = 5a + b.$$

Moreover, the action of $\mathfrak{A}_5$ on $\hat{\mathscr{S}} \cong \mathbb{P}^1 \times \mathbb{P}^1$ is diagonal by Corollary 7.2.5. Therefore, everything follows from Lemma 6.4.10. □

7.3 Small orbits

Lemma 7.3.1. Let Γ be a group, and $G, B \subset \Gamma$ be its finite subgroups. Consider the action of G on the set Γ/B of left cosets that is given by

$$g(\gamma B) = (g\gamma)B, \quad g \in G, \gamma \in \Gamma.$$

Let $N(G)$ be the normalizer of G in Γ, and let $k(G)$ be the number of subgroups of the form $\gamma^{-1}G\gamma$, $\gamma \in \Gamma$, contained in B. Then the number $f(G)$ of the cosets of Γ/B fixed by G equals

$$f(G) = \frac{k(G) \cdot |N(G)|}{|B|}$$

provided that $k(G) \neq 0$. If $k(G) = 0$, then G does not fix any coset of Γ/B.

Proof. Suppose that the coset γB is fixed by G. This means that one has

$$G\gamma B = \gamma B,$$

so that $(\gamma^{-1}G\gamma)B = B$, and thus $\gamma^{-1}G\gamma \subset B$. Therefore, if G has at least one fixed coset, then $k(G) > 0$, and we may assume that $G \subset B$. Let $F(G) \subset \Gamma$ be the set of all elements $\gamma \in \Gamma$ such that $\gamma^{-1}G\gamma \subset B$. Then

$$|F(G)| = k(G) \cdot |N(G)|.$$

Moreover, the group B acts freely on $F(G)$, and the cosets of Γ/B fixed by G are in one-to-one correspondence with the elements of the quotient set $F(G)/B$, so that

$$f(G) = \frac{|F(G)|}{|B|} = \frac{k(G) \cdot |N(G)|}{|B|}.$$

$\square$

Corollary 7.3.2. *In the notation of Lemma 7.3.1, let $\Gamma \cong \mathrm{PSL}_2(\mathbb{C})$ and $B \cong \mathfrak{S}_4$. Then*

(i) $f(G) = 0$, if $G \cong \mathfrak{A}_5$, $G \cong \mathfrak{D}_{10}$ or $G \cong \boldsymbol{\mu}_5$;

(ii) $f(G) = 1$, if $G \cong \mathfrak{A}_4$ or $G \cong \boldsymbol{\mu}_2 \times \boldsymbol{\mu}_2$;

(iii) $f(G) = 2$, if $G \cong \mathfrak{S}_3$.

Proof. If $G \cong \mathfrak{A}_5$, $G \cong \mathfrak{D}_{10}$, or $G \cong \boldsymbol{\mu}_5$, then $k(G) = 0$. For other groups we use the information on the normalizers given by Lemma 5.5.8. If $G \cong \mathfrak{A}_4$, then $k(G) = 1$ and $N(G) \cong \mathfrak{S}_4$. If $G \cong \boldsymbol{\mu}_2 \times \boldsymbol{\mu}_2$, then $k(G) = 4$ and $N(G) \cong \mathfrak{S}_4$. If $G \cong \mathfrak{S}_3$, then $k(G) = 4$ and $N(G) \cong \mathfrak{D}_{12}$. Now everything follows by Lemma 7.3.1. $\square$

Remark 7.3.3. Let A be a finite group acting on a set S, and G be a subgroup of A. Denote by $f(G)$ the number of points in S fixed by G. Then the number of A-orbits in S containing at least one point with a given stabilizer G equals
$$n(G) = f(G) \cdot \frac{c(G) \cdot |G|}{|A|},$$
where $c(G)$ is the number of subgroups in A conjugate to G.

Now we are able to classify $\mathfrak{A}_5$-orbits of small length contained in the open $\mathrm{PSL}_2(\mathbb{C})$-orbit $V_5 \setminus \mathscr{S}$.

Lemma 7.3.4. *Let $\Sigma \subset V_5 \setminus \mathscr{S}$ be an $\mathfrak{A}_5$-orbit of length $r < 20$. Then*
$$r \in \{5, 10, 15\}.$$

Moreover, if $r = 5$ or $r = 15$, then there is a unique $\mathfrak{A}_5$-orbit of length r contained in $V_5 \setminus \mathscr{S}$, and there are exactly two orbits of length 10 contained in $V_5 \setminus \mathscr{S}$.

Proof. Let P be a point of Σ, and $G = G_P \subset \mathfrak{A}_5$ be the stabilizer of P. By Lemma 5.1.1 the group G is isomorphic either to $\mathfrak{A}_5$, or to $\mathfrak{A}_4$, or to $\mathfrak{D}_{10}$, or to $\mathfrak{S}_3$, or to $\boldsymbol{\mu}_5$, or to $\boldsymbol{\mu}_2 \times \boldsymbol{\mu}_2$. For each of these subgroups there is a unique conjugacy class in $\mathfrak{A}_5$. By Theorem 7.1.4 there is a $\mathrm{PSL}_2(\mathbb{C})$-equivariant identification
$$V_5 \setminus \mathscr{S} \cong \mathrm{PSL}_2(\mathbb{C})/\mathfrak{S}_4.$$
Therefore, the number $f(G)$ of points on $V_5 \setminus \mathscr{S}$ fixed by G is given by Corollary 7.3.2. In particular, we see that there are no $\mathfrak{A}_5$-orbits containing points with stabilizers isomorphic to $\mathfrak{A}_5$, $\mathfrak{D}_{10}$ or $\boldsymbol{\mu}_5$, i.e., $\mathfrak{A}_5$-orbits of length 1, 6, or 12 on $V_5 \setminus \mathscr{S}$. To compute the numbers of orbits that contain points with stabilizers isomorphic to $\mathfrak{A}_4$, $\mathfrak{S}_3$ or $\boldsymbol{\mu}_2 \times \boldsymbol{\mu}_2$, i.e., orbits of length 5, 10, or 15 on $V_5 \setminus \mathscr{S}$, we apply Remark 7.3.3 and use information contained in Table 5.1. □

Keeping in mind Lemma 7.3.4, in the rest of the book we denote by Σ_r for $r \in \{5, 15\}$ the unique $\mathfrak{A}_5$-orbit of length r in $V_5 \setminus \mathscr{S}$, and by Σ_{10} and Σ'_{10} the two $\mathfrak{A}_5$-orbits of length 10 in $V_5 \setminus \mathscr{S}$.

We also keep the notation Σ_{12} and Σ'_{12} for the two $\mathfrak{A}_5$-orbits of length 12 contained in the surface $\mathscr{S}$, and the notation Σ_{20} and Σ'_{20} for the two $\mathfrak{A}_5$-orbits of length 20 contained in $\mathscr{S}$ (cf. Lemma 7.2.10).

Theorem 7.3.5. Let $\Sigma \subset V_5$ be an $\mathfrak{A}_5$-orbit of length r. If $r \geqslant 20$, then
$$r \in \{20, 30, 60\}.$$
If $r < 20$, then
$$r \in \{5, 10, 12, 15\}.$$
Moreover, if $r = 12$, then Σ coincides either with $\Sigma_{12} \subset \mathscr{C}$, or with $\Sigma'_{12} \subset \mathscr{S} \setminus \mathscr{C}$. If $r \in \{5, 10, 15\}$, then $\Sigma \subset V_5 \setminus \mathscr{S}$, and
$$\Sigma \in \{\Sigma_5, \Sigma_{10}, \Sigma'_{10}, \Sigma_{15}\}.$$

Proof. The first assertion follows by Corollary 5.1.2. The rest follows by Lemmas 7.2.10 and 7.3.4. □

As a first application of Theorem 7.3.5 we can obtain restrictions on singular $\mathfrak{A}_5$-invariant curves contained in V_5.

Corollary 7.3.6. Let $C \subset V_5$ be an irreducible $\mathfrak{A}_5$-invariant curve. Suppose that C is singular. Then $p_a(C) \geqslant 5$. Moreover, if C does not contain the $\mathfrak{A}_5$-orbit Σ_5, then $p_a(C) \geqslant 10$.

Proof. Recall that we have
$$p_a(C) - |\mathrm{Sing}(C)| \geqslant 0$$
by Lemma 4.4.6. On the other hand, by Theorem 7.3.5 one has $|\mathrm{Sing}(C)| \geqslant 5$, and also $|\mathrm{Sing}(C)| \geqslant 10$ provided that Σ_5 is disjoint from C. □

Also, Theorem 7.3.5 allows to show that smooth curves of some particular genera cannot be embedded into V_5 as $\mathfrak{A}_5$-invariant curves, even if they admit a non-trivial action of $\mathfrak{A}_5$. An example of such a situation is provided by the following:

Corollary 7.3.7. There are no smooth irreducible $\mathfrak{A}_5$-invariant curves of genus 13 in V_5.

Proof. A smooth irreducible curve of genus 13 contains three $\mathfrak{A}_5$-orbits of length 12 by Lemma 5.1.5, while V_5 contains only two by Theorem 7.3.5. □

Finally, we make an observation about stabilizers of general points of $\mathfrak{A}_5$-invariant irreducible curves in V_5.

Lemma 7.3.8. If an irreducible curve $C \subset V_5$ is $\mathfrak{A}_5$-invariant, then a stabilizer of a general point of C is trivial.

Proof. Suppose that a general point $P \in C$ has a non-trivial stabilizer $G_P \subset \mathfrak{A}_5$. Then G_P does not depend on P, because C is irreducible. Thus G_P acts trivially on the whole curve C. Since $\mathfrak{A}_5$ is a simple group, this implies that the action of $\mathfrak{A}_5$ on C is trivial, and thus $\mathfrak{A}_5$ has fixed points in V_5. This contradicts Theorem 7.3.5. □

7.4 Lines

Theorem 7.1.7 enables us to list $\mathfrak{A}_5$-invariant unions of small numbers of lines in V_5. This is our first step toward a classification of $\mathfrak{A}_5$-invariant curves of low degree in V_5.

Proposition 7.4.1. Let $\mathcal{L}$ be an $\mathfrak{A}_5$-irreducible curve on V_5 whose irreducible components are r lines. Suppose that $r \leqslant 20$. Then

$$r \in \{6, 10, 12, 15, 20\}.$$

Moreover, $\mathfrak{A}_5$-irreducible curves $\mathcal{L}_r$ whose irreducible components are r lines on V_5 are unique for each $r \in \{6, 10, 12, 15, 20\}$.

Proof. This follows from Theorem 7.1.7 and Lemma 5.3.1(i). □

Keeping in mind Proposition 7.4.1, in the rest of the book we denote by $\mathcal{L}_r$ for $r \in \{6, 10, 12, 15, 20\}$ the unique $\mathfrak{A}_5$-irreducible curve on V_5 that is a union of r lines (cf. Lemma 7.2.12(i)). Let us point out that we do not pursue a goal to unify this notation with the notation of §6.1, §6.2, and §6.3. As in Chapter 6, we will abuse terminology a little bit and refer to the lines that are irreducible components of the curve $\mathcal{L}_r$ as the lines of $\mathcal{L}_r$.

Remark 7.4.2. By Theorems 7.1.7 and 6.1.2(xix) there is an infinite number of $\mathfrak{A}_5$-irreducible curves that are unions of 30 lines on V_5. By Theorems 6.1.2(xx) and 7.1.8 and Remark 5.3.2 the corresponding lines are exactly those that intersect the curve $\mathcal{L}_{15}$. We will see later in Corollary 12.4.7 that there is an infinite number of irreducible $\mathfrak{A}_5$-invariant curves of degree 30 on V_5.

Lemma 7.4.3. One has

$$\mathscr{S} \cap \mathcal{L}_6 = \Sigma'_{12}.$$

In particular, the curve $\mathcal{L}_6$ contains the $\mathfrak{A}_5$-orbit Σ'_{12} and is disjoint from the $\mathfrak{A}_5$-orbits Σ_{12}, Σ_{20}, and Σ'_{20}.

Proof. Recall that the $\mathfrak{A}_5$-orbit Σ_{12} is contained in the curve $\mathscr{C}$ by construction, and the surface $\mathscr{S}$ is singular along the curve $\mathscr{C}$ (see Lemma 7.2.2(i)). Note also that $\mathcal{L}_6 \not\subset \mathscr{S}$ by Lemma 7.2.13(i). Thus if $\Sigma_{12} \subset \mathcal{L}_6$, then

$$12 = \mathscr{S} \cdot \mathcal{L}_6 \geqslant \sum_{P \in \Sigma_{12}} \mathrm{mult}_P(\mathscr{S}) = 24,$$

which is absurd. (Another way to obtain a contradiction here is to apply Theorem 7.1.9(iv).) Therefore, $\Sigma_{12} \not\subset \mathcal{L}_6$. On the other hand, the intersection $\mathscr{S} \cap \mathcal{L}_6$ consists of $\mathscr{S} \cdot \mathcal{L}_6 = 12$ points (counted with multiplicities), so that the remaining assertions are implied by Lemma 7.2.10. □

Theorem 7.1.9(i) implies the following.

Corollary 7.4.4. *One has $\Sigma_5 \subset \mathcal{L}_{15}$. The lines of $\mathcal{L}_{15}$ split into five triples of non-coplanar lines passing through points of Σ_5.*

Proof. Note that $\Sigma_5 \not\subset \mathscr{S}$ by Lemma 7.2.10. Thus by Theorem 7.1.9(i) there are exactly three lines passing through each of the points of Σ_5. We claim that each of these lines contains exactly one point of Σ_5. Indeed, suppose that there exists a line $L \subset V_5$ that contains at least 2 points in Σ_5, say, P_1 and P_2. If $\Sigma_5 \subset L$, then L is $\mathfrak{A}_5$-invariant, which is impossible by Lemma 7.1.6. Thus there is a point in Σ_5, say P_3, that is not contained in L. On the other hand, the action of $\mathfrak{A}_5$ on Σ_5 is doubly transitive by Lemma 5.1.3(i). Hence there exist two lines L' and L'' in V_5 such that $P_1, P_3 \in L'$ and $P_2, P_3 \in L''$. This means that the three lines L, L', and L'' are contained in a plane spanned by the points P_1, P_2, and P_3. The latter is impossible, since V_5 is an intersection of quadrics.

We see that each line in V_5 passing through a point in Σ_5 contains a unique point in Σ_5. Thus, there are exactly 15 lines passing through the points of Σ_5. This set of 15 lines is $\mathfrak{A}_5$-invariant, so that by Proposition 7.4.1 their union coincides with the curve $\mathcal{L}_{15}$. Moreover, the lines of $\mathcal{L}_{15}$ split into five triples of lines passing through points of Σ_5. Since V_5 is an intersection of quadrics, none of these triples is contained in a plane. □

In the sequel we will use the following agreement. If Z is a possibly reducible subvariety of V_5, by a general point of Z we will mean a point in some open dense subset of Z. Note that even if Z is $\mathfrak{A}_5$-invariant there is no notion of a stabilizer of a general point of Z in $\mathfrak{A}_5$, but one can speak about triviality of a stabilizer of a general point of an $\mathfrak{A}_5$-irreducible subvariety Z.

Now we are going to find the stabilizers of general points of the curves $\mathcal{L}_r$ introduced above.

Lemma 7.4.5. Let $Z \subset V_5$ be an $\mathfrak{A}_5$-irreducible curve, and let $C_1, \ldots, C_r$ be irreducible components of Z. Suppose that $r \in \{5, 6, 12\}$. Then the stabilizer of a general point of the curve Z is trivial.

Proof. It is enough to show that the kernel of the action of the stabilizer G_1 of the curve C_1 on C_1 is trivial. Suppose that G_1 acts on C_1 with a non-trivial kernel $K \subset G_1$. Note that the length $l(P)$ of the $\mathfrak{A}_5$-orbit of a general point $P \in C_1$ equals $r \cdot [G_1 : K]$.

Suppose that $r = 5$. By Lemma 5.1.1 one has $G_1 \cong \mathfrak{A}_4$. Then K contains the only non-trivial normal subgroup of G_1 which is isomorphic to $\boldsymbol{\mu}_2 \times \boldsymbol{\mu}_2$. Thus $l(P) \leqslant 15$, which is impossible since there is only a finite number of points in V_5 whose orbit has length less than 20 by Theorem 7.3.5.

Suppose that $r = 6$. By Lemma 5.1.1 one has $G_1 \cong \mathfrak{D}_{10}$. Then K contains the only non-trivial normal subgroup of G_1 which is isomorphic to $\boldsymbol{\mu}_5$. Thus $l(P) \leqslant 12$, which is impossible by Theorem 7.3.5.

Suppose that $r = 12$. Then $G_1 \cong K \cong \boldsymbol{\mu}_5$ and $l(P) \leqslant 12$, which is again impossible by Theorem 7.3.5. □

Lemma 7.4.6. The stabilizer of a general point of the curve $\mathcal{L}_r$ is trivial, if $r \in \{6, 10, 12\}$, and is isomorphic to $\boldsymbol{\mu}_2$, if $r = 15$.

Proof. The cases $r = 6$ and $r = 12$ are implied by Lemma 7.4.5. In the other cases we need (as in the proof of Lemma 7.4.5) to find the kernel of the action of the stabilizer G_L of a line L of $\mathcal{L}_r$ on L.

Suppose that $r = 10$. By Lemma 5.1.1 one has $G_L \cong \mathfrak{S}_3$. Suppose that G_L acts on L with a non-trivial kernel $K \subset G_L$. Then either $K \cong \mathfrak{S}_3$, or $K \cong \boldsymbol{\mu}_3$. In the former case each point of L is contained in an $\mathfrak{A}_5$-orbit of length at most 10, which is impossible by Theorem 7.3.5.

Thus $K \cong \boldsymbol{\mu}_3$. This gives a contradiction with Lemma 5.5.13. In spite of this, we will give an alternative approach to this case that avoids the computations required to prove Lemma 5.5.13.

Let $P \in L$ be a general point. Note that $L \not\subset \mathscr{S}$ by Lemma 7.2.13(i), so that by Theorem 7.1.9(i) there are exactly two more lines in V_5 except for L that pass through P. Denote these lines by L' and L''. Since the point P and the line L are stabilized by K, the set $\{L', L''\}$ must also be K-invariant. Since $K \cong \boldsymbol{\mu}_3$, this implies that each of the lines L' and L'' is stabilized by K.

Recall that the lines that are contained in V_5 and intersect L are parameterized by a line $\ell(L) \subset \mathbb{P}_\ell^2$ (see Theorem 7.1.8). We have shown that all such lines are fixed by K. This means that each point of $\ell(L)$ is fixed

CREMONA GROUPS AND THE ICOSAHEDRON   173

by K. On the other hand, by Theorem 7.1.7 and Corollary 5.2.2(i) the group $K \cong \boldsymbol{\mu}_3$ has only three fixed points on $\mathbb{P}^2_\ell$, which is a contradiction.

Suppose that $r = 15$. By Lemma 5.1.1 one has $G_L \cong \boldsymbol{\mu}_2 \times \boldsymbol{\mu}_2$. The lines of $\mathcal{L}_{15}$ split into five triples of lines passing through points of the $\mathfrak{A}_5$-orbit Σ_5 (see Corollary 7.4.4). Choose a point $P \in \Sigma_5$. The stabilizer $G_P \subset \mathfrak{A}_5$ of the point P is isomorphic to $\mathfrak{A}_4$ by Lemma 5.1.1.

The group $G_L \subset G_P$ acts on L with a fixed point P. Therefore, the action of G_L on L is not faithful, because the group

$$\boldsymbol{\mu}_2 \times \boldsymbol{\mu}_2 \subset \mathrm{PSL}_2(\mathbb{C})$$

does not have fixed points on $\mathbb{P}^1$ by Lemma 5.5.9(ii). Note also that since P is a fixed point of G_L and G_L is the unique subgroup of G_P isomorphic to $\boldsymbol{\mu}_2 \times \boldsymbol{\mu}_2$, each of the two other lines passing through P are also stabilized by G_L. Denote the latter lines by L' and L''.

Assume that the action of G_L on the line L is trivial. Recall that the lines L, L', and L'' are interchanged by G_P. Since G_L is a normal subgroup of G_P, this implies that the action of G_L on L' and L'' is also trivial. On the other hand, the tangent vectors to L, L', and L'' at P generate the Zariski tangent space $T_P(V_5)$, because the lines L, L', and L'' are not coplanar by Corollary 7.4.4. Thus G_L acts trivially on $T_P(V_5)$, which is impossible by Lemma 4.4.1. The obtained contradiction shows that the kernel of the action of G_L on L is isomorphic to $\boldsymbol{\mu}_2$. □

As an application of Lemma 7.4.6, we can show the following fact about the curve $\mathcal{L}_6$.

Lemma 7.4.7. *The lines of $\mathcal{L}_6$ are pairwise disjoint.*

Proof. Suppose that the lines of $\mathcal{L}_6$ are not disjoint. Recall that there are at most three lines passing through a given point on V_5 (see Theorem 7.1.9(i)). Let k_2 be the number of points that are contained in exactly two lines of $\mathcal{L}_6$, and k_3 be the number of points that are contained in exactly three lines of $\mathcal{L}_6$. Let m be the number of these points contained in a line L_1 of $\mathcal{L}_6$ (this number is the same for any line of $\mathcal{L}_6$). Then $m \leqslant 5$ and

$$2k_2 + 3k_3 = 6m.$$

By Lemma 5.1.1 the stabilizer $G_{L_1} \subset \mathfrak{A}_5$ of the line L_1 is isomorphic to $\mathfrak{D}_{10}$, and G_{L_1} acts faithfully on L_1 by Lemma 7.4.6. Since m is a cardinality of some G_{L_1}-invariant subset of L_1, this implies that either $m = 5$ or $m = 2$ by Lemma 5.5.9(ii). In the former case each line of $\mathcal{L}_6$ intersects any other line of $\mathcal{L}_6$, which is impossible since V_5 is an intersection of quadrics. In the latter case one has

$$(k_2, k_3) \in \{(6,0), (3,2), (0,4)\},$$

which is impossible by Theorem 7.3.5. □

Remark 7.4.8. Another proof of Lemma 7.4.7 can be obtained by applying Theorems 6.1.2(vii) and 7.1.8 together with Remark 5.3.2.

Lemma 7.4.7 immediately implies the following

Corollary 7.4.9. *The curve $\mathcal{L}_6$ does not contain $\mathfrak{A}_5$-orbits of length different from 12 and 60.*

7.5 Orbit of length five

In this section we study the unique $\mathfrak{A}_5$-orbit of length 5 in V_5, and the rational maps of V_5 related to this orbit.

Remark 7.5.1. Since $\mathrm{Pic}(V_5) = \mathbb{Z}[H]$, the degree of every surface contained in V_5 is a multiple of 5. In particular, there are no surfaces of degree $d \leqslant 4$ contained in V_5. Moreover, any surface $S \subset V_5$ of degree $\deg(S) = 5$ is a hyperplane section of V_5. For any four-dimensional linear subspace $\Pi \subset \mathbb{P}^6$ the intersection $\Pi \cap V_5$ is a curve. Indeed, if $\Pi \cap V_5$ contains a surface S, then S is contained in an intersection of quadrics in Π (since V_5 is an intersection of quadrics itself), and so one has $\deg(S) \leqslant 4$. Furthermore, Remark 4.4.7 implies that the degree of the curve $\Pi \cap V_5$ is at most 5.

Let $\Pi_2 \cong \mathbb{P}^2$ and $\Pi_3 \cong \mathbb{P}^3$ be the $\mathfrak{A}_5$-invariant subspaces of $\mathbb{P}^6 = \mathbb{P}(\mathscr{W}_7)$ corresponding to the splitting of the vector space $\mathscr{W}_7$ from Lemma 7.1.6. The following lemmas describe the base loci of linear subsystems of $|\mathcal{O}_{V_5}(H)|$.

Lemma 7.5.2. *The plane Π_2 is disjoint from V_5.*

Proof. Note that $\Pi_2 \not\subset V_5$ by Remark 7.5.1. Assume that the intersection $T = \Pi_2 \cap V_5$ is not empty. If T is a curve, then T is either a conic, or a pair of lines, or a line, since V_5 is an intersection of quadrics. Neither of these cases is possible by Lemma 7.2.14. Thus T is a finite collection of points. By Remark 4.4.7 one has $|T| \leqslant 4$, which is impossible by Theorem 7.3.5. □

By Lemma 7.5.2, the linear projection from the two-dimensional $\mathfrak{A}_5$-invariant subspace $\Pi_2 \subset \mathbb{P}^6$ induces an $\mathfrak{A}_5$-equivariant surjective morphism

$$\theta_{\Pi_2} : V_5 \to \mathbb{P}^3 \cong \mathbb{P}(W_4),$$

which is a finite cover of degree 5. This projection will play an important role in §12.2 and §15.3.

Lemma 7.5.3. The subspace Π_3 intersects V_5 by the $\mathfrak{A}_5$-orbit Σ_5.

Proof. Note that the intersection $\Pi_3 \cap V_5$ does not contain any surface by Remark 7.5.1. If $\Pi_3 \cap V_5$ contains a curve C, then $\deg(C) \leqslant 4$ by Remark 4.4.7, which is impossible by Lemma 7.2.14. Hence $T = \Pi_3 \cap V_5$ is a collection of $t > 0$ points. One has $t \geqslant 5$ by Theorem 7.3.5 and $t \leqslant 5$ by degree reasons, so that $t = 5$. Thus $T = \Sigma_5$ by Theorem 7.3.5. □

Corollary 7.5.4. Let Σ be a non-empty $\mathfrak{A}_5$-invariant subset in V_5 such that $\Sigma \neq \Sigma_5$. Then its linear span in $\mathbb{P}^6$ is the whole $\mathbb{P}^6$.

Proof. Let Π be the linear span of Σ in $\mathbb{P}^6$. Then Π must be an $\mathfrak{A}_5$-invariant subspace of $\mathbb{P}^6$. Since $\Pi \neq \Pi_2$ by Lemma 7.5.2, one has either $\Pi = \Pi_3$ or $\Pi = \mathbb{P}^6$. The former case is impossible by Lemma 7.5.3, because $\Sigma \neq \Sigma_5$. □

Corollary 7.5.5. If $C \subset V_5$ is an irreducible $\mathfrak{A}_5$-invariant curve of degree 6, then C is a rational normal curve.

Proof. If C is not a rational normal sextic curve, then the linear span of C is a proper $\mathfrak{A}_5$-invariant subspace of $\mathbb{P}^6$, which is impossible by Corollary 7.5.4. □

A consequence of Corollary 7.5.5 is the following preliminary classification of $\mathfrak{A}_5$-invariant curves of degree 6 (which is the lowest possible degree of an $\mathfrak{A}_5$-invariant curve by Lemma 7.2.14).

Lemma 7.5.6. Suppose that Z is an $\mathfrak{A}_5$-invariant curve of degree $\deg(Z) = 6$. Then one of the following possibilities occurs:

(i) $Z = \mathcal{L}_6$ is the union of the 6 lines;

(ii) Z is a rational normal sextic curve.

Moreover, there exists at most two $\mathfrak{A}_5$-invariant sextic rational normal curves contained in V_5, and each of them contains exactly one of the $\mathfrak{A}_5$-orbits Σ_{12} and Σ'_{12}. Furthermore, if there are two $\mathfrak{A}_5$-invariant sextic rational normal curves contained in V_5, then they are disjoint.

Proof. If Z is reducible, Corollary 5.1.2 implies that Z is a union of 6 lines, so that Z coincides with $\mathcal{L}_6$ by Proposition 7.4.1. Therefore, we may suppose that Z is irreducible, so that Z is a sextic rational normal curve by Corollary 7.5.5.

By Lemma 5.5.9(v) the curve Z contains exactly one $\mathfrak{A}_5$-orbit of length 12 and does not contain $\mathfrak{A}_5$-orbits of smaller length. Recall that by Theorem 7.3.5 there are exactly two $\mathfrak{A}_5$-orbits of length 12 contained in V_5, namely, Σ_{12} and Σ'_{12}. On the other hand, two distinct rational normal curves of degree m in $\mathbb{P}^m$ cannot have more than $m+2$ common points (see [52, Theorem 1.18], or rather the proof of the latter theorem). In particular, two distinct sextic rational normal curves in $\mathbb{P}^6$ cannot have 12 common points, and the last two assertions of the lemma follow. $\square$

Remark 7.5.7. We will see later in Lemma 11.3.2 that there are indeed exactly two $\mathfrak{A}_5$-invariant sextic rational normal curves contained in V_5.

Consider the rational $\mathfrak{A}_5$-invariant map

$$\theta_{\Pi_3}\colon V_5 \dashrightarrow \mathbb{P}^2 \cong \mathbb{P}(W_3)$$

given by the projection from the linear span $\Pi_3 \cong \mathbb{P}^3$ of Σ_5.

Lemma 7.5.8. *Let $f\colon \bar{V}_5 \to V_5$ be the blow-up of the points of Σ_5. Then there is a commutative diagram*

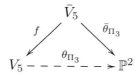

where $\bar{\theta}_{\Pi_3}$ is a morphism whose general fiber is a smooth elliptic curve. Moreover, the f-exceptional divisors are sections of $\bar{\theta}_{\Pi_3}$.

Proof. Recall from Lemma 7.5.3 that Π_3 intersects V_5 by the $\mathfrak{A}_5$-orbit Σ_5. Since Σ_5 consists of $5 = \deg(V_5)$ points, we see that Σ_5 is cut out on V_5 by hyperplane sections in a scheme-theoretic sense. This together with the adjunction formula implies all required assertions. $\square$

Lemma 7.5.9. *Let P be a point of the orbit Σ_5, and $T = T_P(V_5)$ be the Zariski tangent space to V_5 at the point P. Then T is an irreducible representation of the stabilizer $G_P \subset \mathfrak{A}_5$ of the point P. In particular, the length of any G_P-orbit in the projectivization $\mathbb{P}(T)$ is at least 3.*

Proof. By Lemma 5.1.1 one has $G_P \cong \mathfrak{A}_4$. The group $\mathfrak{A}_4$ does not have two-dimensional irreducible representations. Moreover, the normal subgroup $\boldsymbol{\mu}_2 \times \boldsymbol{\mu}_2 \subset \mathfrak{A}_4$ acts trivially in any one-dimensional representation of $\mathfrak{A}_4$. On the other hand, the three-dimensional space T is a faithful representation of G_P by Lemma 4.4.1. Therefore, T must be an irreducible representation. □

Remark 7.5.10. The Zariski tangent space $T_P(V_5)$ at a point $P \in \Sigma_5$ is identified with a fiber of a normal bundle to $\Pi_3 \subset \mathbb{P}^6$. Thus, Lemma 7.1.6 and Corollary 5.2.2(vi) provide another proof of Lemma 7.5.9.

Lemma 7.5.9 has the following useful application.

Lemma 7.5.11. Let $Z \subset V_5$ be an $\mathfrak{A}_5$-invariant curve such that $\Sigma_5 \subset Z$. Then for any $P \in \Sigma_5$ one has $\mathrm{mult}_P(Z) \geqslant 3$. If Z is irreducible, then for any $P \in \Sigma_5$ one has $\mathrm{mult}_P(Z) \geqslant 4$.

Proof. Let $f' \colon \tilde{V} \to V_5$ be a blow-up of the point P. Denote by B its exceptional divisor, and denote by $\tilde{Z}$ the proper transform of Z via f'. Then
$$\mathrm{mult}_P(Z) = E \cdot \tilde{Z} \geqslant |E \cap \tilde{Z}|.$$
On the other hand, we can identify $E \cong \mathbb{P}^2$ with a projectivization of the Zariski tangent space $T_P(V_5)$. By Lemma 7.5.9, the surface E does not contain G_P-orbits of length 1 or 2. Therefore, one has
$$|E \cap \tilde{Z}| \geqslant 3,$$
which implies that $\mathrm{mult}_P(Z) \geqslant 3$.

Suppose that Z is irreducible. Let us prove that $\mathrm{mult}_P(Z) \geqslant 4$. Let
$$\mu \colon \bar{Z} \to Z$$
be the normalization of the curve Z. Then $\bar{Z}$ admits a faithful action of $\mathfrak{A}_5$ arising from its action on Z. On the other hand, one has
$$\mathrm{mult}_P(Z) \geqslant |\mu^{-1}(P)|.$$
We claim that $|\mu^{-1}(P)| \geqslant 4$. Indeed, suppose that $|\mu^{-1}(P)| \leqslant 3$. Then $\mu^{-1}(\Sigma_5) \subset \bar{Z}$ is an $\mathfrak{A}_5$-invariant subset that consists of
$$|\mu^{-1}(P)| \cdot |\Sigma_5| \in \{5, 10, 15\}$$
points. The latter contradicts Lemma 5.1.4. □

Remark 7.5.12. Let $S \subset V_5$ be an arbitrary surface containing the curve $\mathcal{L}_{15}$. Then S is singular at the points of Σ_5. Indeed, let $P \in \Sigma_5$. By Corollary 7.4.4 there are three non-coplanar lines of $\mathcal{L}_{15}$ that pass through P, say, L_1, L_2, and L_3. Since the tangent vectors to the lines L_1, L_2, and L_3 are contained in the Zariski tangent space $T_P(S)$, it must be at least three-dimensional.

Lemma 7.5.13. *The curve $\mathcal{L}_{15}$ is not contained in any surface in the linear system $|2H|$.*

Proof. Suppose that this is not the case. Let $\mathcal{M}$ be the linear subsystem in $|2H|$ that consists of all surfaces passing through $\mathcal{L}_{15}$. Then $\mathcal{M}$ is not empty. Let S be a general surface in $\mathcal{M}$. Let $m = \mathrm{mult}_P(S)$ for some (and thus any) point $P \in \Sigma_5$. Since $\mathcal{L}_{15} \subset S$ by assumption, we see that the surface S is singular at the points of Σ_5 by Remark 7.5.12. In particular, one has $m \geqslant 2$.

Let us use the notation of Lemma 7.5.8. Let $N \subset \bar{V}_5$ be a general fiber of $\bar{\theta}_{\Pi_3}$, and $\bar{S}$ be the proper transform of the surface S with respect to f. Since the f-exceptional divisors B_i, $1 \leqslant i \leqslant 5$, are sections of the fibration $\bar{\theta}_{\Pi_3}$ by Lemma 7.5.8, one has

$$0 \leqslant \bar{S} \cdot N = f^*(S) \cdot N - m \sum_{i=1}^{5} B_i \cdot N = 10 - 5m.$$

Hence $m = 2$, and $\bar{S} \cdot N = 0$, so that S is contracted by the projection θ_{Π_3}. Therefore, the image of S is a conic, because $\bar{\theta}_{\Pi_3}$ is given by the linear system $|f^*(H) - \sum_{i=1}^{5} B_i|$.

The conic $\theta_{\Pi_3}(S)$ contains the set $\theta_{\Pi_3}(\mathcal{L}_{15})$. On the other hand, θ_{Π_3} maps $\mathcal{L}_{15}$ to the unique $\mathfrak{A}_5$-orbit of length 15 in $\mathbb{P}^2$ (cf. Lemma 5.3.1(i)), since θ_{Π_3} is a linear projection. This contradicts Lemma 5.3.1(v). □

7.6 Five hyperplane sections

The group $\mathfrak{A}_5$ contains exactly five subgroups isomorphic to $\mathfrak{A}_4$. Each such subgroup leaves invariant a unique point in Σ_5. By Lemma 7.1.6 and Corollaries 5.2.2(vi) and 5.2.3(vi), it also leaves invariant a unique surface in $|H|$. This gives us five hyperplane sections of V_5 whose stabilizers are subgroups in $\mathfrak{A}_5$ isomorphic to $\mathfrak{A}_4$. Denote these hypersurfaces by $H_1, \ldots, H_5$, and denote their stabilizers by $G_1, \ldots, G_5$, respectively.

Lemma 7.6.1. The surfaces $H_1, \ldots, H_5$ are disjoint from the $\mathfrak{A}_5$-orbits Σ_5, Σ_{12}, and Σ'_{12}.

Proof. By Lemmas 7.1.6, 7.5.3, and 5.3.4(iii), the surfaces $H_1, \ldots, H_5$ are disjoint from Σ_5. By Lemma 5.1.3(vi), each group G_i acts transitively on the points of the $\mathfrak{A}_5$-orbit Σ_{12} (resp., Σ'_{12}). Thus Corollary 7.5.4 implies that the surfaces $H_1, \ldots, H_5$ are disjoint from Σ_{12} (resp., Σ'_{12}). □

Note that each H_i is an irreducible surface, since $\mathrm{Pic}(V_5)$ is generated by H. In fact, each surface H_i is smooth. To prove this, we need:

Lemma 7.6.2. Let S be a hyperplane section of V_5. Then either S has du Val singularities or S is singular along some line in V_5 and is smooth away from it.

Proof. If S has isolated singularities, then S is normal and $-K_S$ is a Cartier divisor by the adjunction formula. In the latter case either S has du Val singularities or S is a cone over a smooth elliptic curve by [13, Theorem 1] or [56, Theorem 2.2], which can also be shown arguing as in the proof of [15, Theorem 1]. Since at most three lines on V_5 can pass through one point by Theorem 7.1.9(i), the surface S is not a cone. Thus, if S has isolated singularities, its singularities are du Val.

To complete the proof, we may assume that S is singular along a curve F. Let T be a general hyperplane section of V_5. Then T is a smooth surface, and $H_1|_T$ is an irreducible curve by Bertini's theorem, because H_1 is an irreducible surface. On the other hand, one has

$$H_1|_T \in |-K_T|,$$

which implies that the arithmetic genus of $H_1|_T$ is 1 by the adjunction formula. In particular, the curve $H_1|_T$ has at most one singular point. Since $H_1|_T$ is singular at every point of the intersection $T \cap F$, we see that F is a line. Moreover, this also shows that S has isolated singularities away from F. Thus, the surface S is smooth away from F by [106, Theorem 1.1], which completes the proof of the lemma. □

Remark 7.6.3. Let S be a hyperplane section of V_5 such that S has at most isolated singularities. Then S cannot contain more than 7 lines. This can be shown using classification of del Pezzo surfaces of degree 5 with du Val singularities (see, for example, [125]). Alternatively, this follows from the formula for the number of lines on S that was found in [90] by Yasuhiro Ohshima. Namely, he proved that the number of lines on S is

$$\frac{1}{2}\Big(5-\mu\Big)\Big(4-\mu\Big) + |\mathrm{Sing}(S)|,$$

where μ is a sum of Milnor numbers of singularities of S. Going through the list of all possible singularities of S (see, for example, [93]), we see that S cannot contain more than 7 lines.

Now we are ready to prove that each surface H_i is smooth.

Lemma 7.6.4. *The surfaces $H_1, \ldots, H_5$ are smooth. Each of them is a del Pezzo surface of degree 5 that is anticanonically embedded in the corresponding hyperplane in $\mathbb{P}^6$.*

Proof. If H_i is normal, then $-K_{H_i} \sim H|_{H_i}$ by the adjunction formula, which implies the second assertion of the lemma. Thus, to complete the proof, it is enough to show that H_1 is smooth.

Recall that V_5 does not contain $\mathfrak{A}_5$-invariant curves that are unions of at most 5 lines by Proposition 7.4.1. Thus the singularities of H_1 are du Val by Lemma 7.6.2. Let us show that H_1 is smooth. Suppose that H_1 is singular. Then H_1 may have at most 2 singular points (cf. [93] or [125]). Since $G_1 \cong \mathfrak{A}_4$ cannot act transitively on the set of two points, we see that H_1 contains a G_1-invariant point. Recall that there is a unique G_1-invariant point P_1 in V_5 by Theorem 7.3.5. By Lemma 7.6.1, one cannot have $P_1 \in H_1$. The obtained contradiction shows that H_1 is smooth. □

Let us consider the surface H_1 in more detail.

Remark 7.6.5. Recall that H_1, which is a smooth del Pezzo surface of degree 5 by Lemma 7.6.4, is faithfully acted on by the group $\mathfrak{A}_5$ (see §6.2). It is important to understand that this action is not compatible with our action of $\mathfrak{A}_5$ on V_5. These two actions agree only on a subgroup $G_1 \cong \mathfrak{A}_4$. In particular, the notation used in §6.2 for $\mathfrak{A}_5$-invariant curves and $\mathfrak{A}_5$-orbits (e.g., $\mathcal{L}_{10}$, $\mathcal{G}_5$, $\mathcal{G}_5'$, Σ_{10} or Σ_{10}') has nothing to do with the similar notation we use here.

By Lemma 6.2.2(iii), there exists an $\mathfrak{A}_4$-equivariant birational morphism $\varpi \colon H_1 \to \mathbb{P}^2$ that contracts 4 lines contained in H_1. Denote these lines by L_1, L_2, L_3, and L_4.

Lemma 7.6.6. *The lines L_1, L_2, L_3, and L_4 are irreducible components of the curve $\mathcal{L}_{10}$.*

Proof. Let $\mathcal{L}$ be the $\mathfrak{A}_5$-orbit of the line L_1. Then we have either $\mathcal{L} = \mathcal{L}_{10}$ or $\mathcal{L} = \mathcal{L}_{20}$ by Proposition 7.4.1. We observe that actually $\mathcal{L} = \mathcal{L}_{10}$. Indeed, if $\mathcal{L} = \mathcal{L}_{20}$, then the linear span of the hypersurface H_1 contains four lines L_1, L_2, L_3, and L_4 that are tangent to the normal rational sextic curve $\mathscr{C}$ by Lemma 7.2.12(iv), which gives a contradiction. □

By Lemma 6.2.2(iv), the surface H_1 contains six more lines L_{12}, L_{13}, L_{14}, L_{23}, L_{24}, L_{23} such that L_{ij} is the proper transform via ϖ of the line in $\mathbb{P}^2$ that passes through the points $\varpi(L_i)$ and $\varpi(L_j)$, and the group G_1 acts transitively on them. Denote by $\mathcal{L}_{30}$ their $\mathfrak{A}_5$-orbit.

Lemma 7.6.7. *The curve $\mathcal{L}_{30}$ is a union of 30 lines. One has $\mathcal{L}_{30} \not\subset \mathscr{S}$. The curve $\mathcal{L}_{30}$ is disjoint from the curves $\mathscr{C}$ and $\mathcal{L}_{12}$.*

Proof. Observe that $\mathcal{L}_{30} \neq \mathcal{L}_6$. This follows, for example, from Lemmas 7.4.3 and 7.6.1. By Lemma 7.6.6, one has $\mathcal{L}_{30} \neq \mathcal{L}_{10}$. Similarly, one has $\mathcal{L}_{30} \neq \mathcal{L}_{15}$ by Corollary 7.4.4 and Lemma 7.6.1. Thus, the curve $\mathcal{L}_{30}$ is a union of 30 lines by Proposition 7.4.1.

Note that $\mathcal{L}_{30}$ is not a disjoint union of lines by construction. Thus $\mathcal{L}_{30}$ is not contained in the surface $\mathscr{S}$ by Theorem 7.1.9(iii). Moreover, we have

$$\mathcal{L}_{30} \cap \mathscr{S} \not\subset \mathscr{C},$$

because every point in $\mathscr{C}$ is contained in exactly one line in V_5 and this line is contained in $\mathscr{S}$ by Theorem 7.1.9(i),(ii). This means that $\mathcal{L}_{30}$ is disjoint from $\mathscr{C}$. Finally, $\mathcal{L}_{30}$ does not intersect the curve $\mathcal{L}_{12}$ by Theorem 7.1.8 and Corollary 6.1.8. □

By Lemma 6.2.2(viii), the surface contains two G_1-invariant conics, which are both proper transforms of the conics in $\mathbb{P}^2$ that pass through the points $\varpi(L_1)$, $\varpi(L_2)$, $\varpi(L_3)$, $\varpi(L_4)$. Denote them by C_1 and C_1'. Note that the conics C_1 and C_1' are disjoint, and

$$C_1 \sim C_1' \sim L_{12} + L_{34} \sim L_{13} + L_{24} \sim L_{14} + L_{23}.$$

By Lemma 6.2.2(ix), the surface H_1 does not contain other $\mathfrak{A}_4$-invariant conics.

Denote by $\mathcal{G}_5$ (resp., $\mathcal{G}_5'$) the $\mathfrak{A}_5$-orbit of the conic C_1 (resp., C_1') on V_5.

Lemma 7.6.8. *The curves $\mathcal{G}_5$ and $\mathcal{G}_5'$ are unions of 5 conics, and $\mathcal{G}_5 \neq \mathcal{G}_5'$.*

Proof. Neither C_1 nor C_1' is $\mathfrak{A}_5$-invariant by Lemma 7.2.14. This shows that the $\mathfrak{A}_5$-orbit of C_1 and the $\mathfrak{A}_5$-orbit of C_1' both consist of exactly five conics. Moreover, these $\mathfrak{A}_5$-orbits are different, since the stabilizer $G_1 \cong \mathfrak{A}_4$ of the conic C_1 does not stabilize any other conic in the same $\mathfrak{A}_5$-orbit by Lemma 5.1.3(i). □

Remark 7.6.9. Suppose that Σ is an $\mathfrak{A}_5$-orbit in V_5 such that the intersection $\Sigma \cap H_1$ consists of 4 points. Then Σ is contained in one of the curves $\mathcal{G}_5$ or $\mathcal{G}_5'$, because all $\mathfrak{A}_4$-orbits in H_1 are contained in $C_1 \cup C_1'$ by Lemma 6.2.5(iv).

We will need the following auxiliary result to work with the $\mathfrak{A}_5$-orbits Σ_{10} and Σ_{10}'.

Lemma 7.6.10. *Let Σ be one of the $\mathfrak{A}_5$-orbits Σ_{10} or Σ_{10}'. Let P be a point of Σ, and let $G_P \cong \mathfrak{S}_3$ be a stabilizer of the point P in $\mathfrak{A}_5$ (see Lemma 5.1.1). Then $H^0(\mathcal{O}_P \otimes \mathcal{O}_{V_5}(H))$ is isomorphic to the (unique) nontrivial representation of the group G_P.*

Proof. Let $\mathcal{I}_\Sigma \subset \mathcal{O}_{V_5}$ be the ideal sheaf of Σ. Then there is a usual exact sequence

$$0 \to \mathcal{I}_\Sigma \otimes \mathcal{O}_{V_5}(H) \to \mathcal{O}_{V_5}(H) \to \mathcal{O}_\Sigma \otimes \mathcal{O}_{V_5}(H) \to 0.$$

Since the $\mathfrak{A}_5$-orbit Σ is not contained in any hyperplane in $\mathbb{P}^6$ by Corollary 7.5.4, we have

$$H^0\big(\mathcal{I}_\Sigma \otimes \mathcal{O}_{V_5}(H)\big) = 0.$$

By Lemma 7.1.6 this gives an injective morphism of $\mathfrak{A}_5$-representations

$$W_3 \oplus W_4 \cong H^0\big(\mathcal{O}_{V_5}(H)\big) \hookrightarrow H^0\big(\mathcal{O}_\Sigma \otimes \mathcal{O}_{V_5}(H)\big).$$

Now the assertion follows from Remark 4.4.5 and Lemma 5.8.3(ii). □

Lemma 7.6.11. *Let D be an $\mathfrak{A}_5$-invariant section of the line bundle $\mathcal{O}_{V_5}(nH)$ for some odd integer $n > 0$. Then D vanishes at the points of the $\mathfrak{A}_5$-orbits Σ_{10} and Σ_{10}'.*

Proof. Let Σ be one of the $\mathfrak{A}_5$-orbits Σ_{10} or Σ_{10}'. Suppose that D does not vanish on Σ. Then it defines an $\mathfrak{A}_5$-invariant section of the line bundle $\mathcal{O}_{V_5}(nH)|_\Sigma$. Keep in mind that a stabilizer in $\mathfrak{A}_5$ of a point of the $\mathfrak{A}_5$-orbit Σ is isomorphic to $\mathfrak{S}_3$ by Lemma 5.1.1. Now we obtain a contradiction applying Remark 4.4.5 together with Lemmas 7.6.10 and 5.8.3(o),(ii). □

CREMONA GROUPS AND THE ICOSAHEDRON

Corollary 7.6.12. Both $\mathfrak{A}_5$-orbits Σ_{10} and Σ'_{10} are contained in the union

$$H_1 \cup H_2 \cup H_3 \cup H_4 \cup H_5.$$

7.7 Projection from a line

Let L be a line in V_5.

Theorem 7.7.1. If L is not contained in the surface $\mathscr{S}$, then the normal bundle

$$\mathcal{N}_{L/V_5} \cong \mathcal{O}_{\mathbb{P}^1} \oplus \mathcal{O}_{\mathbb{P}^1}.$$

If L is contained in the surface $\mathscr{S}$, then the normal bundle

$$\mathcal{N}_{L/V_5} \cong \mathcal{O}_{\mathbb{P}^1}(1) \oplus \mathcal{O}_{\mathbb{P}^1}(-1).$$

Proof. See [110, Proposition 2.27] and Theorem 7.1.9(ii). □

Now let us assume that L is not contained in $\mathscr{S}$. Let $v\colon \check{V}_5 \to V_5$ be a blow-up of the line L. Then there exists a commutative diagram

$$\begin{array}{c} \check{V}_5 \\ {}^u\swarrow \quad \searrow{}^v \\ V_5 \dashrightarrow{w} Q \end{array} \qquad (7.7.2)$$

where Q is a smooth quadric threefold in $\mathbb{P}^4$, the morphism u is a blow-up of the line L, the morphism v is a blow-up of smooth rational cubic curve in Q, and w is a linear projection from L. The existence of (7.7.2) is well known (see, for example, [84, p. 117], [66, Proposition 3.4.1]).

Let F be the v-exceptional divisor, and let E be the u-exceptional divisor. Then $E \cong \mathbb{P}^1 \times \mathbb{P}^1$, since the normal bundle of $L \cong \mathbb{P}^1$ in V_5 is isomorphic to $\mathcal{O}_{\mathbb{P}^1} \oplus \mathcal{O}_{\mathbb{P}^1}$ by Theorem 7.7.1. This implies that $v(E)$ is a smooth hyperplane section of the quadric Q that passes through the curve $v(F)$ (see [66, Proposition 3.4.1(iii)]).

Lemma 7.7.3. One has $F \cong \mathbb{F}_1$ and $F \sim u^*(H) - 2E$.

Proof. One has $F \cong \mathbb{F}_1$ by Lemma 4.4.8, and the linear equivalence $F \sim u^*(H) - 2E$ follows from

$$v^*(-K_Q) - F \sim -K_{\check{V}_5} \sim u^*(2H) - E,$$

since $v(E)$ is a hyperplane section of the quadric Q that passes through $v(F)$. □

Corollary 7.7.4. *One has*

$$F|_E \sim 2s + l,$$

where l is a fiber of the natural projection $E \to L$, and s is a section of this projection such that $s^2 = 0$.

Hence, $u(F)$ is a hyperplane section of V_5 that is singular along the line L. Note that $u(F)$ is an irreducible surface, since $\mathrm{Pic}(V_5)$ is generated by H. Moreover, Theorem 7.1.9(i) implies

Corollary 7.7.5. *The surface $u(F)$ is swept out by lines in V_5 that intersect L.*

Keeping in mind that w is a linear projection and $v(F)$ is a rational cubic curve, we also get:

Corollary 7.7.6. *Every line contained in $u(F)$ intersects the line L.*

Remark 7.7.7. For every subgroup $G \subset \mathfrak{A}_5$, the commutative diagram (7.7.2) is G-equivariant provided that L is G-invariant.

Let us consider the commutative diagram (7.7.2) in the case when L is a line of $\mathcal{L}_6$. Recall from §5.1 that the group $\mathfrak{A}_5$ contains exactly six subgroups isomorphic to $\mathfrak{D}_{10}$. By Lemma 7.1.6 and Corollaries 5.2.2(v) and 5.2.3(v), each such subgroup leaves invariant a unique surface in $|H|$. This gives us six hyperplane sections of V_5 whose stabilizers are subgroups in $\mathfrak{A}_5$ isomorphic to $\mathfrak{D}_{10}$. Denote them by $H_1, \ldots, H_6$, and denote their stabilizers by $G_1, \ldots, G_6$, respectively.

Lemma 7.7.8. *For every hyperplane section H_i, there is a unique line of $\mathcal{L}_6$ that is contained in H_i.*

Proof. By Proposition 7.4.1, there is a unique line in V_5 that is G_1-invariant (it is one of the lines of $\mathcal{L}_6$). Denote this line by L_1. Then $L_1 \subset H_1$. Indeed, if $L_1 \not\subset H_1$, then the intersection $L_1 \cap H_1$ consists of a single point that must be G_1-invariant, which is impossible by Theorem 7.3.5.

If H_1 contains another line of $\mathcal{L}_6$, then it contains all lines of $\mathcal{L}_6$ by Lemma 5.1.3(ii). The latter is impossible, since the intersection $H_1 \cap H_2$ cannot contain a curve of degree 6. □

Denote by $L_1, \ldots, L_6$ the lines of $\mathcal{L}_6$ that are contained in $H_1, \ldots, H_6$, respectively. Then the lines L_i are not contained in the surface $\mathscr{S}$ by Lemma 7.2.13(i). Thus, if we put $L = L_1$, then the diagram (7.7.2) exists and is G_1-equivariant by Remark 7.7.7. We will use this G_1-equivariant diagram several times in the sequel, and denote the v-exceptional invariant divisor by F_1 and the u-exceptional invariant divisor by E_1 in this case.

Corollary 7.7.9. *The surface H_1 is singular along L_1 and is swept out by lines in V_5 that intersect L_1.*

Remark 7.7.10. By Lemmas 7.4.6 and 5.5.9(ii), there is a unique G_1-invariant pair of points in L_1. By Theorem 7.3.5 and Lemma 7.4.3, they are contained in the $\mathfrak{A}_5$-orbit Σ'_{12}. Note that by Lemma 5.1.3(v) this pair of points is the unique G_1-orbit of length 2 contained in Σ'_{12}.

Remark 7.7.11. One has
$$\Sigma_5 = H_1 \cap H_2 \cap H_3 \cap H_4 \cap H_5 \cap H_6.$$
This follows from Lemma 7.1.6 and Corollaries 5.2.2(v) and 5.2.3(v). Thus, the hyperplane sections $H_1, \ldots, H_6$ are contracted by the rational map θ_{Π_3} to the six lines in $\mathbb{P}^2 \cong \mathbb{P}(W_3)$. By Lemma 7.7.8, the lines $\theta_{\Pi_3}(H_1), \ldots, \theta_{\Pi_3}(H_6)$ are contained in the degeneration locus of the elliptic fibration $\bar{\theta}_{\Pi_3}$. These six lines form a unique $\mathfrak{A}_5$-invariant sixtuple of lines in $\mathbb{P}^2$ (see §6.1).

7.8 Conics

In this section we will proceed to describe $\mathfrak{A}_5$-invariant curves of low degree in V_5. More precisely, we are going to classify all $\mathfrak{A}_5$-irreducible curves that are unions of at most 6 conics. One approach to this problem is to use a description of the Hilbert scheme of conics on V_5 similar to one given for lines in Theorem 7.1.7 (see Remark 7.8.5 below). In this section we will demonstrate a more straightforward approach instead of this.

By Lemma 7.2.14, there are no $\mathfrak{A}_5$-invariant conics in V_5. In §7.6, we have constructed two $\mathfrak{A}_5$-irreducible curves $\mathcal{G}_5$ and $\mathcal{G}'_5$ on V_5 that are both unions of 5 conics (see Lemma 7.6.8).

Lemma 7.8.1. *The curves $\mathcal{G}_5$ and $\mathcal{G}'_5$ are the only $\mathfrak{A}_5$-irreducible curves on V_5 that are unions of 5 conics.*

Proof. Let Z be an $\mathfrak{A}_5$-irreducible curve on V_5 that is a union of 5 conics. We must show that either $Z = \mathcal{G}_5$ or $Z = \mathcal{G}_5'$. Let us use the notation of §7.6. The conic C_1 (resp., C_1') is an irreducible component of $\mathcal{G}_5$ (resp., $\mathcal{G}_5'$). To complete the proof, we must show that either C_1 or C_1' is an irreducible component of Z.

By Lemma 5.1.1, one of the irreducible components of Z, say Z_1, must be G_1-invariant. If $Z_1 \not\subset H_1$, then the G_1-invariant subset $Z_1 \cap H_1$ consists of at most 2 points, which is impossible by Lemma 6.2.5(i), since H_1 is a smooth del Pezzo surface. Thus, we see that $Z_1 \subset H_1$. Therefore, either $Z_1 = C_1$ or $Z_1 = C_1'$ by Lemma 6.2.2(viii). □

Lemma 7.8.2. *There exists a unique $\mathfrak{A}_5$-irreducible curve on V_5 that is a union of 6 conics.*

Proof. By Lemma 5.8.2 there exists a faithful action of G_1 on $\mathbb{P}^3$ such that G_1 leaves invariant a smooth rational quartic curve C contained in a smooth G_1-invariant quadric $R \subset \mathbb{P}^3$. Thus, there exists a commutative diagram

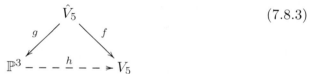
(7.8.3)

where g is a blow-up of the curve C, the morphism f is a contraction of a proper transform of the quadric R on the threefold $\hat{V}_5$, and h is the map given by the linear system of all cubic surfaces that contain C. The existence of (7.8.3) is well known (see [84, p. 117], [9, Example 3.4]). Furthermore, the morphism f is a blow-up of a smooth conic $C_1 \subset V_5$.

By construction, the commutative diagram (7.8.3) is G_1-equivariant. However, the action of G_1 on V_5 obtained from (7.8.3) could a priori disagree with our initial action of G_1 on V_5. Fortunately, all subgroups of $\mathrm{Aut}(V_5) \cong \mathrm{PSL}_2(\mathbb{C})$ isomorphic to $G_1 \cong \mathfrak{D}_{10}$ are conjugate. Thus, we may assume that the action of G_1 on V_5 agrees with the action of $\mathfrak{A}_5$ on V_5.

Note that the conic C_1 is G_1-invariant. Its stabilizer $\tilde{G}_1 \subset \mathfrak{A}_5$ contains the group G_1. By Lemma 5.1.1 one has either $\tilde{G}_1 = G_1$, or $\tilde{G}_1 = \mathfrak{A}_5$. The latter is impossible by Lemma 7.2.14, so that the stabilizer of C_1 coincides with G_1. Thus C_1 is an irreducible component of an $\mathfrak{A}_5$-invariant curve that is a union of 6 conics.

Let us prove that there are no other curves like this. Let Z be an $\mathfrak{A}_5$-invariant curve that is a union of 6 conics. By Lemma 5.1.1, one irreducible component of Z must be G_1-invariant. Denote it by C'_1. Let us show that $C'_1 = C_1$. To do this, we will use notation of §7.7.

By Lemma 7.1.6 and Corollaries 5.2.2(v) and 5.2.3(v), the linear span of H_1 is a projectivization of a sum of three two-dimensional irreducible G_1-representations. This shows that $C_1 \not\subset H_1$ and $C'_1 \not\subset H_1$. Since

$$C'_1 \cdot H_1 = C_1 \cdot H_1 = 2,$$

we conclude that each set $C'_1 \cap H_1$ and $C_1 \cap H_1$ consists of two points, because there are no G_1-invariant points in V_5 by Theorem 7.3.5.

Denote by Σ (resp., Σ') the $\mathfrak{A}_5$-orbit containing the set $C_1 \cap H_1$ (resp., $C'_1 \cap H_1$). Then

$$|\Sigma| = |\Sigma'| = 12$$

by Theorem 7.3.5. Note that $C'_1 \cap H_1 \not\subset L_1$ and $C_1 \cap H_1 \not\subset L_1$, because H_1 is singular along L_1 by Corollary 7.7.9. Thus, it follows from Remark 7.7.10 that both Σ and Σ' are different from Σ'_{12}. Therefore, we have

$$\Sigma = \Sigma' = \Sigma_{12}$$

by Theorem 7.3.5. Since Σ_{12} contains a unique G_1-orbit of length 2 by Lemma 5.1.3(v), we see that $C'_1 \cap H_1 = C_1 \cap H_1$.

Let Π be the two-dimensional linear subspace in $\mathbb{P}^2$ that contains C_1, let Π' be the two-dimensional linear subspace in $\mathbb{P}^2$ that contains C'_1, and let P_1 be the unique G_1-invariant point in $\mathbb{P}^6$. Recall that $P_1 \not\in V_5$ by Theorem 7.3.5. Moreover, the point P_1 is contained in both planes Π and Π'. Furthermore, the point P_1 is not contained in the line spanned by $C_1 \cap H_1 = C'_1 \cap H_1$. Thus, we have $\Pi = \Pi'$. But $V_5 \cap \Pi = C_1$ and $V_5 \cap \Pi' = C'_1$ since V_5 is an intersection of quadrics. Thus, we see that $C_1 = C'_1$. This shows the uniqueness of an $\mathfrak{A}_5$-irreducible curve on V_5 that is a union of 6 conics and completes the proof of Lemma 7.8.2. □

Proposition 7.8.4. *Let $\mathcal{G}$ be an $\mathfrak{A}_5$-irreducible curve on V_5 whose irreducible components are r conics. Then either $|r| = 5$, or $|r| = 6$, or $|r| \geqslant 10$. Furthermore, there are exactly two $\mathfrak{A}_5$-irreducible curves that are unions of 5 conics, and a unique $\mathfrak{A}_5$-irreducible curve that is a union of 6 conics on V_5.*

Proof. By Lemma 7.2.14, one has $r \neq 1$. Thus, using Corollary 5.1.2 we obtain that either $r \in \{5, 6\}$, or $r \geqslant 10$. The cases $r = 5$ and $r = 6$ are described by Lemmas 7.8.1 and 7.8.2, respectively. □

Keeping in mind Proposition 7.8.4, in the rest of the book we denote by $\mathcal{G}_5$ and $\mathcal{G}'_5$ the two $\mathfrak{A}_5$-irreducible curves that are unions of 5 conics, and by $\mathcal{G}_6$ we denote the $\mathfrak{A}_5$-irreducible curve that is a union of 6 conics on V_5. Similarly with the situation with lines, we do not pursue a goal to unify this notation with the notation of §6.2. Also, we will abuse terminology and refer to the conics that are irreducible components of the curves $\mathcal{G}_5$, $\mathcal{G}'_5$, and $\mathcal{G}_6$ as the conics of $\mathcal{G}_5$, $\mathcal{G}'_5$, and $\mathcal{G}_6$, respectively.

Remark 7.8.5. One can prove Lemmas 7.8.1 and 7.8.2 and Proposition 7.8.4 using the fact that there is a $\mathrm{PSL}_2(\mathbb{C})$-equivariant identification of a certain compactification of the variety of conics contained in V_5 with the projective space $\mathbb{P}^4 \cong \mathbb{P}(\mathscr{W}_5)$ (see [110, Proposition 2.32] or [60, Proposition 1.22]). Still we prefer to take a more straightforward way to avoid the involved constructions of [60].

Chapter 8

Anticanonical linear system

The purpose of this chapter is to study linear subsystems of the linear system $|2H|$ on V_5 and to derive some immediate corollaries. In particular, we describe the geometry of $\mathfrak{A}_5$-invariant surfaces in $|2H|$. This enables us to advance our knowledge about $\mathfrak{A}_5$-invariant curves of low degree on V_5.

8.1 Invariant anticanonical surfaces

Lemma 8.1.1. One has an identification of $\mathfrak{A}_5$-representations
$$H^0\bigl(\mathcal{O}_{V_5}(2H)\bigr) \cong I^{\oplus 2} \oplus W_3' \oplus W_4^{\oplus 2} \oplus W_5^{\oplus 2}.$$

Proof. By Lemma 7.2.1, one has an $\mathrm{SL}_2(\mathbb{C})$-equivariant identification
$$H^0\bigl(\mathcal{O}_{V_5}(2H)\bigr) \cong \mathbb{W}_1 \oplus \mathbb{W}_9 \oplus \mathbb{W}_{13},$$
and the assertion follows by Lemma 5.6.2. □

Corollary 8.1.2. Any 13-dimensional $\mathfrak{A}_5$-invariant subspace in $H^0(\mathcal{O}_{V_5}(2H))$ that contains a one-dimensional $\mathfrak{A}_5$-subrepresentation must contain the three-dimensional $\mathfrak{A}_5$-subrepresentation W_3'. Any 10-dimensional $\mathfrak{A}_5$-invariant subspace in $H^0(\mathcal{O}_{V_5}(2H))$ that contains W_3' must contain all one-dimensional $\mathfrak{A}_5$-invariant subspaces of $H^0(\mathcal{O}_{V_5}(2H))$.

Denote by $\mathcal{Q}_2$ and $\mathcal{Q}_3$ the unique $\mathfrak{A}_5$-invariant linear subsystems in $|2H|$ of dimensions 1 and 2, respectively, i.e., the linear subsystems in
$$|2H| = \mathbb{P}\bigl(H^0(\mathcal{O}_{V_5}(2H))\bigr)$$

that correspond to the $\mathfrak{A}_5$-subrepresentations $I^{\oplus 2}$ and W_3' in $H^0(\mathcal{O}_{V_5}(2H))$. We will use this notation throughout the rest of the book.

Lemma 8.1.3. Let Σ be a non-empty $\mathfrak{A}_5$-invariant subset in V_5, and let $\mathcal{I}_\Sigma \subset \mathcal{O}_{V_5}$ be its ideal sheaf. Suppose that
$$h^0\Big(\mathcal{I}_\Sigma \otimes \mathcal{O}_{V_5}(2H)\Big) \geqslant 2.$$
Let $\mathcal{Q}$ be the linear subsystem in $|2H|$ consisting of surfaces passing through Σ. Then $\mathcal{Q}$ does not have fixed components.

Proof. Suppose that $\mathcal{Q}$ has fixed components. Denote their union by S. Then S is $\mathfrak{A}_5$-invariant. Moreover, one has either $S \sim 2H$, or $S \sim H$. If $S \sim 2H$, then
$$h^0\Big(\mathcal{I}_\Sigma \otimes \mathcal{O}_{V_5}(2H)\Big) = 1,$$
which is not the case by assumption. The case $S \sim H$ is impossible by Lemma 7.1.6. □

We proceed by studying the properties of the pencil $\mathcal{Q}_2$. Note that the surface $\mathscr{S}$ is contained in $\mathcal{Q}_2$.

Remark 8.1.4. Any divisor $S \in \mathcal{Q}_2$ is an irreducible (reduced) surface. Indeed, any irreducible component $S' \subset S$ different from S itself is contained in the linear system $|H|$, since $\mathrm{Pic}(V_5) = \mathbb{Z}[H]$. Therefore, if S is reducible, then $S = S_1 \cup S_2$, where both components S_1 and S_2 lie in $|H|$. By Corollary 5.1.2, both S_1 and S_2 must be $\mathfrak{A}_5$-invariant. Similarly, if S is not reduced, then $\mathrm{Supp}(S)$ is an $\mathfrak{A}_5$-invariant surface in $|H|$. The latter is impossible by Lemma 7.1.6.

The next observation is that many $\mathfrak{A}_5$-irreducible curves of low degree are contained in surfaces from $\mathcal{Q}_2$.

Lemma 8.1.5. Let C be an $\mathfrak{A}_5$-irreducible curve of degree $d < 30$ in V_5. Suppose that a stabilizer of a general point of C (i.e., a point in some open dense subset of C) is trivial. Then there exists a surface $S \in \mathcal{Q}_2$ that contains C.

Proof. By assumption C contains an $\mathfrak{A}_5$-orbit Σ of length $|\mathfrak{A}_5| = 60$. Take a surface $S \in \mathcal{Q}_2$ passing through one of the points of Σ. Then S contains the whole orbit Σ because S is $\mathfrak{A}_5$-invariant. On the other hand, if the curve C is not contained in S, then the intersection $C \cap S$ consists of just
$$C \cdot S = 2d < 60$$
points (counted with multiplicities). Therefore, C is contained in S. □

Corollary 8.1.6. Let C be an irreducible $\mathfrak{A}_5$-invariant curve of degree less than 30 in V_5. Then there exists a surface in the pencil $\mathcal{Q}_2$ that contains C.

Proof. Apply Lemmas 7.3.8 and 8.1.5. □

We can replace the assumption of irreducibility in Corollary 8.1.6 by a weaker assumption of $\mathfrak{A}_5$-irreducibility, at the cost of decreasing the degree of the curves to which it is applicable.

Corollary 8.1.7. Let C be an $\mathfrak{A}_5$-irreducible curve of degree less than 20 in V_5. Suppose that $C \neq \mathcal{L}_{15}$. Then there exists a surface in the pencil $\mathcal{Q}_2$ that contains C.

Proof. By Corollary 8.1.6 we may assume that C is reducible. Let r be the number of its irreducible components. Then

$$r \in \{5, 6, 10, 12, 15\}$$

by Corollary 5.1.2. If $r \in \{5, 6, 12\}$, then the assertion follows from Lemmas 7.4.5 and 8.1.5. If $r = 10$, then C is a union of 10 lines, so that one has $C = \mathcal{L}_{10}$ by Proposition 7.4.1. In this case the assertion follows from Lemmas 7.4.6 and 8.1.5. Finally, if $r = 15$, then C is a union of 15 lines, so that $C = \mathcal{L}_{15}$ by Proposition 7.4.1. □

We proceed by describing the base locus of the pencil $\mathcal{Q}_2$.

Theorem 8.1.8. *The following assertions hold:*

(i) *the base locus of the pencil $\mathcal{Q}_2$ is an irreducible curve C_{20} of degree 20;*

(ii) *the normalization of the curve C_{20} is a hyperelliptic curve of genus 9;*

(iii) *the curve C_{20} contains the $\mathfrak{A}_5$-orbits Σ_{12} and Σ'_{12}, and one has $\mathscr{C} \cap C_{20} = \Sigma_{12}$;*

(iv) *the curve C_{20} is smooth outside of Σ_{12};*

(v) *the curve C_{20} has ordinary cusps at every point of Σ_{12};*

(vi) *the curve C_{20} contains the $\mathfrak{A}_5$-orbit $\Sigma'_{20} \subset \mathscr{S}$ of length 20 that is disjoint from the curve $\mathscr{C}$;*

(vii) *any $\mathfrak{A}_5$-orbit $\Sigma \subset C_{20}$ different from Σ_{12}, Σ'_{12}, and Σ'_{20} has length $|\Sigma| = 60$;*

(viii) one has $C_{20} \cap \mathcal{L}_{12} = \Sigma_{12} \cup \Sigma'_{12}$;

(ix) one has $C_{20} \cap \mathcal{L}_{20} = \Sigma'_{20}$.

Proof. Let us use notation of §7.2. Let S be a general surface in $\mathcal{Q}_2$, and let $\hat{S}$ be its proper transform on $\mathcal{Y}$. Put $\hat{C}_{20} = \hat{S}|_{\hat{\mathscr{S}}}$. By Lemma 7.2.3(ii) the divisor $\nu^*(2H)|_{\hat{\mathscr{S}}}$ has bi-degree $(2, 10)$, and by Lemma 7.2.3(iii) the curve $\hat{\mathscr{C}}$ is a divisor of bi-degree $(1, 1)$. Furthermore, one has $E_{\mathcal{Y}}|_{\hat{\mathscr{S}}} = 2\hat{\mathscr{C}}$ by Lemma 7.2.3(i). On the other hand, we have

$$\hat{S} \sim \nu^*(2H) - \mathrm{mult}_{\mathscr{C}}(S) E_{\mathcal{Y}}.$$

Thus, $\hat{C}_{20}$ is an effective divisor of bi-degree

$$(2 - 2\mathrm{mult}_{\mathscr{C}}(S), 10 - 2\mathrm{mult}_{\mathscr{C}}(S))$$

on the surface $\hat{\mathscr{S}} \cong \mathbb{P}^1 \times \mathbb{P}^1$.

Let us show that $\mathrm{mult}_{\mathscr{C}}(S) = 0$ and $\mathrm{Supp}(\hat{C}_{20})$ does not contain the curve $\hat{\mathscr{C}}$. Put $\hat{C}_{20} = a\hat{\mathscr{C}} + \Omega$, where a is a non-negative integer and Ω is an effective divisor on $\hat{\mathscr{S}}$ whose support does not contain the curve $\hat{\mathscr{C}}$. Then Ω is an $\mathfrak{A}_5$-invariant divisor of bi-degree

$$\Big(2 - 2\mathrm{mult}_{\mathscr{C}}(S) - a, 10 - 2\mathrm{mult}_{\mathscr{C}}(S) - a\Big).$$

In particular, one has $\Omega \neq 0$, so that $\Omega|_{\hat{\mathscr{C}}}$ is a divisor on $\hat{\mathscr{C}}$ of degree

$$\Omega \cdot \hat{\mathscr{C}} = 12 - 4\mathrm{mult}_{\mathscr{C}}(S) - 2a.$$

Since a smooth curve $\hat{\mathscr{C}}$ does not contain $\mathfrak{A}_5$-orbits of length less than 12 by Lemma 5.1.4, we conclude that $a = \mathrm{mult}_{\mathscr{C}}(S) = 0$. In particular, we see that S does not contain the curve $\mathscr{C}$, and $\hat{C}_{20}$ is a divisor of bi-degree $(2, 10)$. Therefore, by Corollary 7.2.5 and Lemma 6.4.8 the divisor $\hat{C}_{20}$ is a smooth hyperelliptic curve of genus 9 (cf. Lemma 5.4.4).

Put $C_{20} = \nu(\hat{C}_{20})$. We claim that C_{20} is the base locus of the pencil $\mathcal{Q}_2$. Indeed, the base locus of $\mathcal{Q}_2$ consists of the curve C_{20} and possibly of the curve $\mathscr{C}$, because $\hat{C}_{20} = \hat{S} \cap \hat{\mathscr{S}}$. However, we already proved that $\mathrm{mult}_{\mathscr{C}}(S) = 0$. This shows that $\mathscr{C} \not\subset \mathrm{Bs}(\mathcal{Q}_2)$. Thus, one has $\mathrm{Bs}(\mathcal{Q}_2) = C_{20}$. Note that

$$\deg(C_{20}) = C_{20} \cdot H = \hat{C}_{20} \cdot \nu^*(H) = \hat{C}_{20} \cdot \nu^*(H)|_{\hat{\mathscr{S}}} = 20,$$

since $\nu^*(H)|_{\hat{\mathscr{S}}}$ is a divisor of bi-degree $(1, 5)$ on the surface $\hat{\mathscr{S}}$. This completes the proof of assertion (i). Moreover, since $\nu|_{\hat{C}_{20}} \colon \hat{C}_{20} \to C_{20}$ is the normalization of the curve C_{20}, we also obtain assertion (ii).

Recall that the surface $\mathscr{S}$ is singular along $\mathscr{C}$ (see Lemma 7.2.2(i)). Thus the curve C_{20} must be singular at every point of $S \cap \mathscr{C}$, because C_{20} is a complete intersection of the surfaces S and $\mathscr{S}$. Moreover, since $S \cdot \mathscr{C} = 12$, we see that $S \cap \mathscr{C} = \Sigma_{12}$ by Lemma 7.2.10.

Since $\hat{C}_{20} \cdot \hat{\mathscr{C}} = 12$ and $\mathscr{C}$ is a smooth curve, the intersection $\hat{C}_{20} \cap \hat{\mathscr{C}}$ consists of exactly 12 points by Lemma 5.1.4. Since

$$\nu|_{\hat{\mathscr{S}}} \colon \hat{\mathscr{S}} \to \mathscr{S}$$

is a bijection by Lemma 7.2.2(v), we see that

$$\nu|_{\hat{C}_{20}} \colon \hat{C}_{20} \to C_{20}$$

is also a bijection. In particular, singular points of $\hat{C}_{20}$ are singularities of type A_r with $r \geqslant 3$. Since the curve C_{20} is a complete intersection in V_5 of two surfaces from the linear system $|2H|$, we have $p_a(C_{20}) = 21$; this can be seen, for example, by taking a smooth deformation (not-invariant with respect to the action of $\mathfrak{A}_5$) of the curve C_{20} and the surfaces in question (cf. [53, Corollary III.9.10]), and then applying the adjunction formula twice. Therefore, the singular points of C_{20} are ordinary cusps by Lemma 4.4.6, which proves assertion (v). Since

$$\nu|_{\hat{\mathscr{S}}} \colon \hat{\mathscr{S}} \to \mathscr{S}$$

induces an isomorphism $\hat{\mathscr{S}} \setminus \hat{\mathscr{C}} \to \mathscr{S} \setminus \mathscr{C}$, we see that C_{20} is smooth outside of Σ_{12}, which proves assertion (iv).

By Lemma 5.1.5, the curve $\hat{C}_{20}$ contains exactly two $\mathfrak{A}_5$-orbits of length 12 and one $\mathfrak{A}_5$-orbit of length 20 (and no other $\mathfrak{A}_5$-orbits of length less than 60). By Lemma 7.2.10, this shows that C_{20} contains the $\mathfrak{A}_5$-orbit Σ'_{12} and the unique $\mathfrak{A}_5$-orbit Σ'_{20} of length 20 that is contained in $\mathscr{S} \setminus \mathscr{C}$. This proves assertions (iii) and (vi). Moreover, since

$$\nu|_{\hat{C}_{20}} \colon \hat{C}_{20} \to C_{20}$$

is bijective, we get assertion (vii).

Let $\hat{\mathcal{L}}_{12}$ and $\hat{\mathcal{L}}_{20}$ be proper transforms of the curves $\mathcal{L}_{12}$ and $\mathcal{L}_{20}$ on the surface $\hat{\mathscr{S}} \cong \mathbb{P}^1 \times \mathbb{P}^1$. Then each of the irreducible components of $\hat{\mathcal{L}}_{12}$ and $\hat{\mathcal{L}}_{20}$ is a curve of bi-degree $(0, 1)$ on $\hat{\mathscr{S}}$. Therefore, each of these irreducible components intersects the curve $\hat{C}_{20}$ in two points (counted with multiplicities). Since the morphism $\nu|_{\hat{\mathscr{S}}}$ is bijective by Lemma 7.2.2(v), we see that the intersection $C_{20} \cap \mathcal{L}_{12}$ consists of at most 24 points, while the intersection $C_{20} \cap \mathcal{L}_{20}$ consists of at most 40 points. We already know that

$$\Sigma_{12} \cup \Sigma'_{12} \subset C_{20} \cap \mathcal{L}_{12}$$

by assertion (iii) and Lemma 7.2.12(iii). This gives assertion (viii).

We know that
$$\Sigma'_{20} \subset C_{20} \cap \mathcal{L}_{20}$$

by assertion (vi) and Lemma 7.2.12(iii). Let L be an irreducible component of $\mathcal{L}_{20}$, and $G_L \subset \mathfrak{A}_5$ be the stabilizer of L. Then $G_L \cong \boldsymbol{\mu}_3$ by Lemma 5.1.1, and the action of G_L on L is faithful by Corollary 7.2.5. Therefore, there are exactly two points on L that are fixed by G_L by Lemma 5.5.9(i), and all other points of L are contained in G_L-orbits of length 3. One of the G_L-fixed points on L is contained in the $\mathfrak{A}_5$-orbit Σ_{20}, and the other is contained in the $\mathfrak{A}_5$-orbit Σ'_{20}. Since Σ_{20} is disjoint from the curve C_{20} by assertion (vii), we conclude that L intersects C_{20} in a single point contained in the $\mathfrak{A}_5$-orbit Σ'_{20}. This implies assertion (ix) and completes the proof of Theorem 8.1.8. □

Keeping in mind Theorem 8.1.8, we denote the base curve of the pencil $\mathcal{Q}_2$ by C_{20} throughout the rest of the book.

Corollary 8.1.9. *Let C be a curve of degree less than 20 that is contained in a surface $S \in \mathcal{Q}_2$. Then S is the only surface in $\mathcal{Q}_2$ containing C.*

Corollary 8.1.10. *The only $\mathfrak{A}_5$-irreducible curves of degree 20 contained in the surface $\mathscr{S}$ are C_{20} and $\mathcal{L}_{20}$.*

Proof. There are exactly two $\mathfrak{A}_5$-irreducible curves of degree 20 in $\mathscr{S}$ by Lemma 7.2.15, so that there is no room for anything except these two examples. □

Corollary 8.1.11. *Let C be an $\mathfrak{A}_5$-irreducible curve of degree $d < 30$ in V_5 such that C is contained in some surface S of the pencil $\mathcal{Q}_2$. Choose a surface $S' \in \mathcal{Q}_2$ different from S. Then the intersection $C \cap S'$ coincides with the intersection $C \cap C_{20}$, and can be described as a union of those of the $\mathfrak{A}_5$-orbits Σ_{12}, Σ'_{12}, and Σ'_{20} that are contained in C.*

Proof. The intersection $C \cap S'$ consists of $2d < 60$ points (counted with multiplicities). One has

$$C \cap S' \subset S \cap S' = C_{20}.$$

Now the assertion follows from Theorem 8.1.8(vii). □

Corollary 8.1.12. Let C be an $\mathfrak{A}_5$-irreducible curve of degree less than 30 in V_5 such that C is contained in some surface $S \in \mathcal{Q}_2$. Let C' be an $\mathfrak{A}_5$-irreducible curve in V_5 such that C' is contained in some surface $S' \in \mathcal{Q}_2$ different from S. Then the intersection $C \cap C'$ is a union of those of the $\mathfrak{A}_5$-orbits Σ_{12}, Σ'_{12}, and Σ'_{20} that are contained in both C and C'.

Proof. One has
$$C \cap C' \subset C \cap S',$$
so that the assertion follows from Corollary 8.1.11. □

Corollary 8.1.13. Let S be a surface of the pencil $\mathcal{Q}_2$ different from $\mathscr{S}$. Then S is smooth at the points of the $\mathfrak{A}_5$-orbits Σ_{12} and Σ'_{12}. Moreover, the curve $\mathscr{C}$ intersects S transversally at the points of Σ_{12}, and the curve $\mathcal{L}_{12}$ intersects S transversally at the points of Σ_{12} and Σ'_{12}.

Proof. If S was singular at the points of Σ'_{12}, the curve $C_{20} = S \cap \mathscr{S}$ would be singular at the points of Σ'_{12}; this is not the case by Theorem 8.1.8(iv). If S was singular at the points of Σ_{12}, then the curve C_{20} would have singularities of multiplicity at least 4 at the points of Σ_{12}, because the surface $\mathscr{S}$ is singular at these points by Lemma 7.2.2(i); this is not the case by Theorem 8.1.8(v). Therefore, the surface S is smooth at the points of Σ_{12} and Σ'_{12}. (Later in Theorem 8.2.1 we will obtain more detailed information about the singularities of the surfaces of the pencil $\mathcal{Q}_2$.)

Note that the curves $\mathscr{C}$ and $\mathcal{L}_{12}$ are contained in the surface $\mathscr{S}$, so that they are not contained in the surface S by Corollary 8.1.9. Thus the intersection $S \cap \mathscr{C}$ consists of $S \cdot \mathscr{C} = 12$ points (counted with multiplicities), and the intersection $S \cap \mathcal{L}_{12}$ consists of $S \cdot \mathcal{L}_{12} = 24$ points (counted with multiplicities). On the other hand, the $\mathfrak{A}_5$-orbit Σ_{12} is contained in $\mathscr{C}$ by construction, and the $\mathfrak{A}_5$-orbits Σ_{12} and Σ'_{12} are contained in $\mathcal{L}_{12}$ by Lemma 7.2.12(iii). Therefore, the assertion follows from Corollary 8.1.11. □

As an application of Theorem 8.1.8 we obtain restrictions on degrees of $\mathfrak{A}_5$-invariant curves in V_5.

Lemma 8.1.14. Let C be an $\mathfrak{A}_5$-irreducible curve of degree $d < 20$ in V_5. Then
$$d \in \{6, 10, 12, 15, 16, 18\}.$$

Proof. Note that $C \not\subset \mathrm{Bs}(\mathcal{Q}_2)$, since $\mathrm{Bs}(\mathcal{Q}_2) = C_{20}$ is an irreducible curve of degree 20 by Theorem 8.1.8(i). Thus, a general $\mathfrak{A}_5$-invariant surface $S \in \mathcal{Q}_2$ intersects C by $2d$ points (counted with multiplicities). It means that the number $2d$ equals a sum of lengths of several $\mathfrak{A}_5$-orbits in V_5 (also possibly with multiplicities). Therefore, Theorem 7.3.5 immediately implies that

$$d \in \{5, 6, 10, 11, 12, 15, 16, 17, 18\}.$$

Moreover, the case $d = 5$ is impossible by Lemma 7.2.14.

Suppose that $d = 11$. Then Theorem 7.3.5 implies that there is an $\mathfrak{A}_5$-orbit of length 10 contained in $S \cap C$, and C is non-singular in the points of this orbit. Thus C must be reducible by Lemma 5.1.4. Therefore, Corollary 5.1.2 implies that C cannot be $\mathfrak{A}_5$-irreducible, which is a contradiction.

Suppose that $d = 17$. Then Theorem 7.3.5 implies that there is an $\mathfrak{A}_5$-orbit of length 10 contained in $S \cap C$, and C is non-singular in the point of this orbit. By Lemma 5.1.4 we see that C must be reducible. Now Corollary 5.1.2 implies that C cannot be $\mathfrak{A}_5$-irreducible, which is a contradiction. □

Corollary 8.1.15. Let C be a (possibly $\mathfrak{A}_5$-reducible) $\mathfrak{A}_5$-invariant curve of degree $d < 20$ in V_5. Then

$$d \in \{6, 10, 12, 15, 16, 18\}.$$

Using Theorem 8.1.8, we can also advance our knowledge about the curves contained in the surface $\mathscr{S}$.

Lemma 8.1.16. The following assertions hold:

(i) there exists a unique $\mathfrak{A}_5$-invariant curve $C_{16} \subset \mathscr{S}$ of degree 16;

(ii) the curve C_{16} is smooth and rational;

(iii) the $\mathfrak{A}_5$-orbit Σ_{12} is contained in C_{16}, while the $\mathfrak{A}_5$-orbit Σ'_{12} is disjoint from C_{16};

(iv) the $\mathfrak{A}_5$-orbits Σ_{20} and Σ_{30} are disjoint from C_{16}, while the $\mathfrak{A}_5$-orbits Σ'_{20} and Σ'_{30} are contained in C_{16};

(v) one has $C_{16} \cap \mathscr{C} = \Sigma_{12}$;

(vi) one has $C_{16} \cap \mathcal{L}_{12} = \Sigma_{12}$;

(vii) one has $C_{16} \cap \mathcal{L}_{20} = \Sigma'_{20}$;

(viii) one has $C_{16} \cap C_{20} = \Sigma_{12} \cup \Sigma'_{20}$.

Proof. Let us use notation of §7.2. Suppose that Z is an $\mathfrak{A}_5$-invariant curve of degree 16 contained in $\mathscr{S}$, and let $\hat{Z}$ be its proper transform on $\mathcal{Y}$. Then $\hat{Z}$ is a curve of bi-degree (a, b) in $\hat{\mathscr{S}} \cong \mathbb{P}^1 \times \mathbb{P}^1$ for some non-negative integers a and b. By Lemma 7.2.3(ii), we have $5a + b = 16$. Recall that the action of the group $\mathfrak{A}_5$ on the surface $\hat{\mathscr{S}} \cong \mathbb{P}^1 \times \mathbb{P}^1$ is diagonal by Corollary 7.2.5. This implies that $a = 1$ and $b = 11$ by Lemma 6.4.5. Similarly, if $\hat{Z} \subset \hat{\mathscr{S}}$ is a curve of bi-degree $(1, 11)$, then the curve $Z = \nu(\hat{Z})$ is a curve of degree 16 on $\mathscr{S}$. By Lemma 6.4.6(i) there exists a unique $\mathfrak{A}_5$-invariant curve $\hat{C}_{16} \subset \hat{\mathscr{S}}$ of bi-degree $(1, 11)$. In particular, this proves assertion (i).

By Lemma 6.4.6(ii) the curve $\hat{C}_{16}$ is smooth and rational. In particular, the curve $C_{16} = \nu(\hat{C}_{16})$ is also rational.

Let us show that the curve C_{16} is smooth. Suppose that this is not the case. Then $\mathrm{Sing}(C_{16}) = C_{16} \cap \mathscr{C}$ since

$$\nu|_{\hat{\mathscr{S}}} \colon \hat{\mathscr{S}} \to \mathscr{S}$$

induces an isomorphism $\hat{\mathscr{S}} \setminus \hat{\mathscr{C}} \to \mathscr{S} \setminus \mathscr{C}$. Note that $\hat{C}_{16}$ intersects $\hat{\mathscr{C}}$ in 12 points, because $\hat{C}_{16} \cdot \hat{\mathscr{C}} = 12$ and a smooth curve $\hat{\mathscr{C}}$ does not contain $\mathfrak{A}_5$-orbits of length less than 12 by Lemma 5.1.4. Since the normalization

$$\nu|_{\hat{\mathscr{S}}} \colon \hat{\mathscr{S}} \to \mathscr{S}$$

is bijective by Lemma 7.2.2(v), we see that $C_{16} \cap \mathscr{C} = \Sigma_{12}$ by Lemma 7.2.10. Let $S \in \mathcal{Q}_2$ be a general surface. Then $\Sigma_{12} \subset S$ by Theorem 8.1.8(iii) and $C_{16} \not\subset S$ by Theorem 8.1.8(i). The intersection $C_{16} \cap S$ is an $\mathfrak{A}_5$-invariant subset that consists of $2\deg(C_{16}) = 32$ points (counted with multiplicities). Thus $(C_{16} \cap S) \setminus \Sigma_{12}$ is an $\mathfrak{A}_5$-invariant set that consists of

$$2\deg(C_{16}) - n|\Sigma_{12}| = 32 - 12n$$

points (counted with multiplicities) for some integer $n \geqslant \mathrm{mult}_{\Sigma_{12}}(C_{16}) \geqslant 2$. A contradiction with Lemma 7.2.10 completes the proof of assertion (ii).

We already know that $\Sigma_{12} \subset C_{16}$. Since the curve C_{16} is smooth and rational, it contains a unique $\mathfrak{A}_5$-orbit of length 12 by Lemma 5.5.9(v). Thus C_{16} does not contain Σ'_{12}. This proves assertion (iii).

Since the intersection $\hat{C}_{16} \cap \hat{\mathscr{C}}$ consists of 12 points and

$$\nu|_{\hat{\mathscr{C}}} \colon \hat{\mathscr{C}} \to \mathscr{C}$$

is bijective by Lemma 7.2.2(v), we see that $C_{16} \cap \mathscr{C} = \Sigma_{12}$. This gives assertion (v). In particular, the curve C_{16} is disjoint from the $\mathfrak{A}_5$-orbits Σ_{20} and Σ_{30} contained in $\mathscr{C}$. On the other hand, by Lemma 5.5.9(v) the curve C_{16} contains an $\mathfrak{A}_5$-orbit of length 20 and an $\mathfrak{A}_5$-orbit of length 30. Hence it follows from Lemma 7.2.10 that the $\mathfrak{A}_5$-orbits Σ'_{20} and Σ'_{30} are contained in C_{16}, which proves assertion (iv).

Let $\hat{\mathcal{L}}_{12}$ and $\hat{\mathcal{L}}_{20}$ be proper transforms of the curves $\mathcal{L}_{12}$ and $\mathcal{L}_{20}$ on the surface $\hat{\mathscr{S}} \cong \mathbb{P}^1 \times \mathbb{P}^1$. Then each of the irreducible components of $\hat{\mathcal{L}}_{12}$ and $\hat{\mathcal{L}}_{20}$ is a curve of bi-degree $(0, 1)$ on $\hat{\mathscr{S}}$. Therefore, each of these irreducible components intersects the curve $\hat{C}_{16}$ in a single point. Since the morphism $\nu|_{\hat{\mathscr{S}}}$ is bijective by Lemma 7.2.2(v), we see that the intersection $C_{16} \cap \mathcal{L}_{12}$ consists of 12 points, while the intersection $C_{16} \cap \mathcal{L}_{20}$ consists of 20 points. Now assertions (vi) and (vii) are implied by assertions (iii) and (iv).

Finally, note that the intersection $C_{16} \cap C_{20}$ consists of $2H \cdot C_{16} = 32$ points (counted with multiplicities). Therefore, assertion (viii) is implied by assertions (iii) and (vi) and Theorem 8.1.8(iii),(vi),(vii). □

Keeping in mind Lemma 8.1.16, we will denote the unique $\mathfrak{A}_5$-invariant curve of degree 16 contained in the surface $\mathscr{S}$ by C_{16} throughout the rest of the book.

8.2 Singularities of invariant anticanonical surfaces

Now we will describe the singular surfaces in the pencil $\mathcal{Q}_2$. The proof of the following theorem is similar to the proof of [96, Lemma 4.7].

Theorem 8.2.1. Let S_5, S_{10}, S'_{10}, and S_{15} be the surfaces in the pencil $\mathcal{Q}_2$ that contain the orbits Σ_5, Σ_{10}, Σ'_{10}, and Σ_{15}, respectively. Then

(i) the surfaces S_5, S_{10}, S'_{10}, S_{15}, and $\mathscr{S}$ are pairwise different;

(ii) the surfaces S_5, S_{10}, S'_{10}, S_{15}, and $\mathscr{S}$ are the only singular surfaces of the pencil $\mathcal{Q}_2$;

(iii) the surfaces S_5, S_{10}, S'_{10}, and S_{15} are smooth outside of Σ_5, Σ_{10}, Σ'_{10}, and Σ_{15}, respectively;

(iv) the surfaces S_5, S_{10}, S'_{10}, S_{15} have isolated ordinary double points at every point of the sets Σ_5, Σ_{10}, Σ'_{10}, and Σ_{15}, respectively;

(v) any surface $S \in \mathcal{Q}_2$ except for $\mathscr{S}$ is a $K3$ surface (with at most ordinary double points).

Proof. Recall that the surface $\mathscr{S}$ does not contain the $\mathfrak{A}_5$-orbits Σ_5, Σ_{10}, Σ'_{10}, and Σ_{15} by Lemma 7.2.10, so that $\mathscr{S}$ is different from the surfaces S_5, S_{10}, S'_{10}, and S_{15}.

Let S be a surface in the pencil $\mathcal{Q}_2$ such that $S \neq \mathscr{S}$. Note that S is irreducible by Remark 8.1.4. Let us show that S has at most du Val singularities. Suppose this is not true. Then the singularities of S are not Kawamata log terminal, since S is a Cartier divisor on V_5.

Let μ be the largest (positive rational) number such that the pair $(V_5, \mu S)$ is log canonical, i.e., μ is the log canonical threshold of the log pair (V_5, S) along V_5 (see [73, Definition 8.1]). Note that $\mu \leqslant 1$. Let F be a minimal center in $\mathbb{LCS}(V_5, \mu S)$ (see Definition 2.4.7). Then $F \neq S$ by Theorem 2.4.8, because the singularities of S are worse than Kawamata log terminal. Hence, F is either a curve or a point. Let Z be the $\mathfrak{A}_5$-orbit of the subvariety F. Then $Z \subset S$.

Choose any $\epsilon \in \mathbb{Q}$ such that $1 \gg \epsilon > 0$. By Lemma 2.4.10, we can find an $\mathfrak{A}_5$-invariant effective $\mathbb{Q}$-divisor D such that $D \sim_{\mathbb{Q}} (1+\epsilon)S$, the log pair (V_5, D) is log canonical, and every minimal center in $\mathbb{LCS}(V_5, D)$ is an irreducible component of the subvariety Z.

Let $\mathcal{I}_Z \subset \mathcal{O}_{V_5}$ be the ideal sheaf of the subvariety Z. Then $\mathcal{I}_Z = \mathcal{I}(V_5, D)$ by Remark 2.3.4, so that there is an exact sequence of $\mathfrak{A}_5$-representations

$$0 \to H^0\Big(\mathcal{I}_Z \otimes \mathcal{O}_{V_5}(H)\Big) \to H^0\Big(\mathcal{O}_{V_5}(H)\Big) \to H^0\Big(\mathcal{O}_Z \otimes \mathcal{O}_{V_5}(H)\Big) \to 0$$

by Theorem 2.3.5. Put

$$q = h^0\big(\mathcal{I}_Z \otimes \mathcal{O}_{V_5}(H)\big).$$

Then it follows from Corollary 7.5.4 that $q \in \{0, 3\}$, and if $q = 3$, then $Z = \Sigma_5$. Thus, we have

$$h^0\Big(\mathcal{O}_Z \otimes \mathcal{O}_{V_5}(H)\Big) = 7 - q = \begin{cases} 7 \text{ if } Z \neq \Sigma_5, \\ 4 \text{ if } Z = \Sigma_5. \end{cases}$$

Since

$$h^0\Big(\mathcal{O}_{\Sigma_5} \otimes \mathcal{O}_{V_5}(H)\Big) = |\Sigma_5| = 5,$$

we see that $Z \neq \Sigma_5$, and hence $q = 0$. In particular, if F is a point, then $|Z| = 7$, which is impossible by Theorem 7.3.5. Thus, we see that F is a curve. Moreover, F is a smooth rational curve by Theorem 2.4.8. By Remark 2.4.9, the curve Z is a disjoint union of its irreducible components (that are smooth irreducible curves isomorphic to F).

Put $d = H \cdot F$. Let r be the number of irreducible components of the curve Z. Then

$$r(d+1) = h^0\Big(\mathcal{O}_Z \otimes \mathcal{O}_{V_5}(H)\Big) = 7$$

by Theorem 2.3.5, which implies that either $r = 1$ or $r = 7$. If $r = 7$, then $d = 0$, which is a contradiction. Hence we have $r = 1$ and $d = 6$.

We see that F is a rational normal sextic curve by Corollary 7.5.5. Since $S \neq \mathscr{S}$, one has $\mathscr{C} \not\subset S$ by Corollary 8.1.9. Hence $F \neq \mathscr{C}$. By Lemma 5.5.9(v) there is an $\mathfrak{A}_5$-orbit Σ of length 12 contained in F. By Lemma 7.5.6 one has $\Sigma = \Sigma'_{12}$. Recall that the curve

$$C_{20} = \mathrm{Bs}(\mathcal{Q}_2) = S \cap \mathscr{S}$$

is smooth at the points of the $\mathfrak{A}_5$-orbit Σ'_{12} by Theorem 8.1.8(iv). The latter gives a contradiction with our conclusion that the surface S is singular at the points of the curve F and, in particular, at the points of Σ'_{12}.

Therefore, the surface S has only du Val singularities. In particular, S is a $K3$ surface by the adjunction formula. Using Lemma 6.7.3(i),(ii),(iii), we conclude that either S is smooth, or the set $\mathrm{Sing}(S)$ consists of one $\mathfrak{A}_5$-orbit, every singular point of the surface S is an ordinary double point, and

$$|\mathrm{Sing}(S)| \in \{1, 5, 6, 10, 12, 15\}.$$

In particular, this proves assertion (v).

Since $C_{20} \subset S$, we see that S contains two $\mathfrak{A}_5$-orbits of length 12 by Theorem 8.1.8(iii). In particular, $|\mathrm{Sing}(S)| \neq 12$ by Lemma 6.7.3(iv). Thus, if S is singular, then

$$\mathrm{Sing}(S) \in \{\Sigma_5, \Sigma_{10}, \Sigma'_{10}, \Sigma_{15}\}$$

by Theorem 7.3.5. This proves assertions (ii) and (iii). Vice versa, if S is one of the surfaces S_5, S_{10}, S'_{10}, or S_{15}, then it follows from Lemma 6.7.1(ii) that S must be singular at every point of the $\mathfrak{A}_5$-orbit Σ_5, Σ_{10}, Σ'_{10}, or Σ_{15}, respectively. This proves assertions (i) and (iv), since we already showed that S has at most ordinary double points, and completes the proof of Theorem 8.2.1. $\square$

Corollary 8.2.2. Let S be a surface in the pencil $\mathcal{Q}_2$. Then S contains only a finite number of $\mathfrak{A}_5$-orbits of length less than 60.

Proof. If $S = \mathscr{S}$, then the assertion follows from Lemma 7.2.10. Thus, we assume that $S \neq \mathscr{S}$, so that S is a $K3$ surface with at most ordinary double points by Theorem 8.2.1(v). Hence, its minimal resolution of singularities $\tilde{S}$ is a smooth $K3$ surface, and the action of $\mathfrak{A}_5$ on S lifts to an action of $\mathfrak{A}_5$ on $\tilde{S}$. Now everything follows from Lemma 6.7.1 applied to $\tilde{S}$. □

Remark 8.2.3. It would be interesting to know how the pencil $\mathcal{Q}_2$ fits into the moduli space of $K3$ surfaces with an action of the group $\mathfrak{A}_5$ and with $\mathfrak{A}_5$-invariant polarization of degree 10 (cf. [55]). Although we do not know a precise answer to this question, we can observe that there is an infinite countable number of $K3$ surfaces $S \in \mathcal{Q}_2$ such that $\operatorname{rk}\operatorname{Pic}(S)^{\mathfrak{A}_5} > 1$. Indeed, one has $\operatorname{rk}\operatorname{Pic}(S)^{\mathfrak{A}_5} > 1$ if and only if $\operatorname{rk}\operatorname{Pic}(S) = 20$ by Remark 6.7.2. On the other hand, the surfaces of the pencil $\mathcal{Q}_2$ indeed varies in moduli, since $\mathcal{Q}_2$ contains both smooth and singular $K3$ surfaces. Therefore, it follows from [89, Theorem 1.1] that there is an infinite countable number of $K3$ surfaces in $\mathcal{Q}_2$ with Picard rank 20. Note that the structure of the pencil $\mathcal{Q}_2$ is very similar to that of the pencil of $\mathfrak{A}_5$-invariant quartic $K3$ surfaces in $\mathbb{P}^3$ mentioned in Example 1.3.8 (see [54] for details).

Remark 8.2.4. Let $S \in \mathcal{Q}_2$ be a $K3$ surface such that $\operatorname{rk}\operatorname{Pic}(S)^{\mathfrak{A}_5} = 1$. Then the group $\operatorname{Cl}(S)^{\mathfrak{A}_5}$ coincides with its subgroup $\operatorname{Pic}(S)^{\mathfrak{A}_5}$ and is generated by the class $H_S = H|_S$. Indeed, if S is smooth, then

$$\operatorname{Cl}(S)^{\mathfrak{A}_5} = \operatorname{Pic}(S)^{\mathfrak{A}_5},$$

and the latter group is isomorphic to $\mathbb{Z}$ by Lemma 6.7.1(i). Furthermore, since $H_S^2 = 10$ is square-free, we see that H_S is not divisible in $\operatorname{Pic}(S)^{\mathfrak{A}_5}$, so that H_S generates $\operatorname{Pic}(S)^{\mathfrak{A}_5}$. If S is singular, then

$$|\operatorname{Sing}(S)| \in \{5, 10, 15\}$$

by Theorem 8.2.1(ii),(iii), so that the required result holds by Lemma 6.7.4.

We conclude this section with two general results about the surfaces in $\mathcal{Q}_2$.

Lemma 8.2.5. *Let S be a surface in $\mathcal{Q}_2$ different from $\mathscr{S}$. Put $H_S = H|_S$. Then for any $n \geqslant 1$ the restriction map*

$$H^0\big(\mathcal{O}_{V_5}(nH)\big) \to H^0\big(\mathcal{O}_S(nH_S)\big)$$

is surjective.

Proof. Note that by Theorem 8.2.1(v) the surface S is normal. By Kodaira vanishing one has
$$h^1\Big(\mathcal{O}_{V_5}\big((n-2)H\big)\Big) = 0.$$
Therefore, the exact sequence of sheaves
$$0 \to \mathcal{O}_{V_5}\big((n-2)H\big) \to \mathcal{O}_{V_5}(nH) \to \mathcal{O}_S(nH_S) \to 0$$
gives the exact sequence of cohomology groups
$$0 \to H^0\Big(\mathcal{O}_{V_5}\big((n-2)H\big)\Big) \to H^0\Big(\mathcal{O}_{V_5}(nH)\Big) \to H^0\Big(\mathcal{O}_S(nH_S)\Big) \to 0.$$
$\square$

Lemma 8.2.6. *Let S be a surface in $\mathcal{Q}_2$ different from $\mathscr{S}$. Put $H_S = H|_S$. Then*

(i) *the linear system $|H_S|$ does not contain $\mathfrak{A}_5$-invariant curves;*

(ii) *the only $\mathfrak{A}_5$-invariant curve in $|2H_S|$ is C_{20};*

(iii) *the $\mathfrak{A}_5$-representation $H^0(\mathcal{O}_S(2H_S))$ does not contain two-dimensional $\mathfrak{A}_5$-subrepresentations.*

Proof. Note that by Theorem 8.2.1(v) the surface S is normal. By Lemma 8.2.5 there is a surjective morphism of $\mathfrak{A}_5$-representations
$$H^0(\mathcal{O}_{V_5}(H)) \to H^0(\mathcal{O}_S(H_S)).$$
Thus, $H^0(\mathcal{O}_S(H_S))$ does not have one-dimensional $\mathfrak{A}_5$-subrepresentations by Lemma 7.1.6, which implies assertion (i).

Similarly, by Lemma 8.2.5 there is a surjective morphism of $\mathfrak{A}_5$-representations
$$H^0(\mathcal{O}_{V_5}(2H)) \to H^0(\mathcal{O}_S(2H_S)).$$
By Lemma 8.1.1, one has
$$H^0(\mathcal{O}_{V_5}(2H)) \cong I^{\oplus 2} \oplus W_3' \oplus W_4^{\oplus 2} \oplus W_5^{\oplus 2}$$
as a representation of the group $\mathfrak{A}_5$. Thus, we have
$$H^0(\mathcal{O}_S(2H_S)) \cong I \oplus W_3' \oplus W_4^{\oplus 2} \oplus W_5^{\oplus 2}$$
as a representation of $\mathfrak{A}_5$. In particular, $H^0(\mathcal{O}_S(2H_S))$ does not contain two-dimensional $\mathfrak{A}_5$-subrepresentations, which implies assertion (iii). Moreover, since $C_{20} \subset S$ and $C_{20} \sim 2H_S$, we see that the unique one-dimensional

$\mathfrak{A}_5$-subrepresentation in $H^0(\mathcal{O}_S(2H_S))$ is given by the section whose set of zeroes is C_{20}, i.e., C_{20} is the only $\mathfrak{A}_5$-invariant curve in the linear system $|2H_S|$. This gives assertion (ii). □

As it is stated, Lemma 8.2.6 applies to any surface in $\mathcal{Q}_2$ except for $\mathscr{S}$. Note however that there are only two $\mathfrak{A}_5$-irreducible curves of degree 20 in the surface $\mathscr{S}$ by Corollary 8.1.10, which may be regarded as an analog of Lemma 8.2.6 for this case. The same result can be obtained even in a more straightforward manner if one considers linear systems on the non-normal surface $\mathscr{S}$ (which we would like to avoid here). The same applies to Lemma 8.2.5.

8.3 Curves in invariant anticanonical surfaces

By Remark 7.4.2 there is an infinite number of reducible $\mathfrak{A}_5$-invariant curves of degree 30 in V_5. We will see later that V_5 also contains infinitely many irreducible $\mathfrak{A}_5$-invariant curves of degree 30 (see Corollary 12.4.7). Quite surprisingly, Theorem 8.2.1 can be used to show that V_5 contains at most finitely many $\mathfrak{A}_5$-invariant curves of degree less than 30. This fact somehow justifies our further attempts to give a classification of low degree $\mathfrak{A}_5$-invariant curves on V_5 (see Proposition 10.2.5 and also §13.6 below).

Theorem 8.3.1. *There is only a finite number of $\mathfrak{A}_5$-invariant curves of degree less than 30 on V_5.*

Proof. Let C be an $\mathfrak{A}_5$-invariant curve of degree $d < 30$ on V_5. We may assume that C is $\mathfrak{A}_5$-irreducible.

By Corollary 7.2.11, we may assume that the stabilizer of a general point of C is trivial. Then C is contained in some surface $S \in \mathcal{Q}_2$ by Lemma 8.1.5. By Lemma 7.2.15, we may assume that $S \neq \mathscr{S}$, so that S is a $K3$ surface by Theorem 8.2.1(v).

Put $H_S = H|_S$. If $\operatorname{rk}\operatorname{Pic}(S)^{\mathfrak{A}_5} = 1$, then $\operatorname{Cl}(S)^{\mathfrak{A}_5} = \operatorname{Pic}(S)^{\mathfrak{A}_5}$ is generated by the divisor H_S by Remark 8.2.4. In particular, C is a Cartier divisor on S. Since
$$d < 30 = 3H_S^2,$$
we see that one has either $C \sim H_S$, or $C \sim 2H_S$. The former case is impossible by Lemma 8.2.6(i), while in the latter case there is a unique possibility $C = C_{20}$ by Lemma 8.2.6(ii). Hence we may assume that $\operatorname{rk}\operatorname{Pic}(S)^{\mathfrak{A}_5} > 1$, so that $\operatorname{rk}\operatorname{Pic}(S)^{\mathfrak{A}_5} = 2$ and S is smooth by Lemma 6.7.3(i),(ii). Note that there is a countable number of $K3$ surfaces

in the pencil $\mathcal{Q}_2$ with $\mathfrak{A}_5$-equivariant Picard rank equal to 2 by Remark 8.2.3. Since the Hilbert scheme of $\mathfrak{A}_5$-invariant curves of degree d in V_5 is of finite type, we conclude that there is only a finite number of surfaces in $\mathcal{Q}_2$ that contain $\mathfrak{A}_5$-invariant curves of degree d. Thus it is enough to show that a given smooth $K3$ surface $S \in \mathcal{Q}_2$ contains a finite number of $\mathfrak{A}_5$-irreducible curves of degree d.

Suppose that a smooth $K3$ surface $S \in \mathcal{Q}_2$ contains an infinite number of $\mathfrak{A}_5$-invariant curves of degree d. We may assume that there is a one-dimensional linear system $\mathcal{R}$ that consists of $\mathfrak{A}_5$-invariant divisors and contains our curve C. The divisors of the linear system $\mathcal{R}$ cover the whole surface S. Let P be a general point of the curve $C_{20} \subset S$, so that the $\mathfrak{A}_5$-orbit Σ of P has length 60. There is a divisor $C' \in \mathcal{R}$ such that $P \in \mathrm{Supp}(C')$. The curve $\mathrm{Supp}(C')$ contains C_{20} as an irreducible component, since otherwise one would have
$$60 > 2d = C_{20} \cdot C' \geqslant |\Sigma| = 60.$$

We have
$$C' = mC_{20} + Z,$$
where $m \geqslant 1$ and Z is an $\mathfrak{A}_5$-invariant one-cycle of degree at most 9. By Lemma 8.1.14 we conclude that $m = 1$, and either $C' = C_{20}$, or Z is an $\mathfrak{A}_5$-irreducible curve of degree 6. In the former case one has $\mathcal{R} \subset |2H_S|$, and we get a contradiction with Lemma 8.2.6(iii). In the latter case $d = 26$, and we either have $Z = \mathcal{L}_6$, or Z is a rational normal sextic curve by Lemma 7.5.6.

Suppose that $Z = \mathcal{L}_6$. Then $\mathcal{L}_6^2 = -12$ by Lemma 7.4.7, so that
$$C \cdot \mathcal{L}_6 = (C_{20} + \mathcal{L}_6) \cdot \mathcal{L}_6 = 0.$$
The intersection $\mathscr{S} \cap C$ consists of
$$\mathscr{S} \cdot C = 2d = 52$$
points (counted with multiplicities). By Lemma 7.2.10 we see that C contains one of the $\mathfrak{A}_5$-orbits Σ_{12} or Σ'_{12}. Since the surface $\mathscr{S}$ is singular at the points of Σ_{12} by Lemma 7.2.2(i), we conclude that C contains the $\mathfrak{A}_5$-orbit Σ'_{12}. The latter contradicts the observation that $C \cdot \mathcal{L}_6 = 0$. Indeed, $\mathcal{L}_6$ and C have no common irreducible components since they are both $\mathfrak{A}_5$-irreducible, but $\Sigma'_{12} \subset \mathcal{L}_6$ by Lemma 7.4.3.

Suppose that Z is a rational normal sextic curve. Then $Z^2 = -2$, so that
$$C \cdot Z = (C_{20} + Z) \cdot Z = 10.$$

Since the curve C is $\mathfrak{A}_5$-irreducible, it does not contain Z as an irreducible component, and the intersection $C \cap Z$ is an $\mathfrak{A}_5$-invariant set that consists of 10 points (counted with multiplicities). This is impossible by Lemma 5.1.4. □

The following easy computation will be used frequently throughout the book.

Lemma 8.3.2. Let $C_1 \subset V_5$ and $C_2 \subset V_5$ be $\mathfrak{A}_5$-invariant curves of degrees d_1 and d_2, respectively. Suppose that C_1 and C_2 are contained in a surface $S \in \mathcal{Q}_2$ such that $S \neq \mathscr{S}$. Put $s_i = C_i^2$ and $k = C_1 \cdot C_2$. Then

$$10k^2 - 2d_1 d_2 k + (d_1^2 s_2 + d_2^2 s_1 - 10 s_1 s_2) = 0.$$

Proof. The surface S is a $K3$ surface with at most ordinary double points by Theorem 8.2.1(v).

Put $H_S = H|_S$. Since $\operatorname{rk} \operatorname{Pic}(S)^{\mathfrak{A}_5} \leqslant 2$ by Lemma 6.7.3(i), the determinant of the matrix

$$M = \begin{pmatrix} H_S^2 & H_S \cdot C_1 & H_S \cdot C_2 \\ H_S \cdot C_1 & C_1^2 & C_1 \cdot C_2 \\ H_S \cdot C_2 & C_1 \cdot C_2 & C_2^2 \end{pmatrix} = \begin{pmatrix} 10 & d_1 & d_2 \\ d_1 & s_1 & k \\ d_2 & k & s_2 \end{pmatrix}$$

must vanish. One has

$$0 = \det(M) = -10k^2 + 2d_1 d_2 k - (d_1^2 s_2 + d_2^2 s_1 - 10 s_1 s_2),$$

and the assertion follows. □

Knowing the structure of the singular surfaces in $\mathcal{Q}_2$ enables us to obtain restrictions on the curves contained in them. First of all, we have the following easy consequence of the fact that the singular surfaces in the pencil $\mathcal{Q}_2$ that are different from $\mathscr{S}$ have only ordinary double points.

Lemma 8.3.3. Let S be a surface in $\mathcal{Q}_2$, and let Z be an $\mathfrak{A}_5$-invariant curve contained in S. Suppose that $S \neq \mathscr{S}$ and S is singular. Then $\deg(Z)$ is divisible by 5.

Proof. By Theorem 8.2.1(v), the surface S is a $K3$ surface with at most ordinary double points. In particular, the intersection form on S has values in $\frac{1}{2}\mathbb{Z}$. On the other hand, it follows from Lemma 6.7.3(i),(ii) that $\operatorname{rk} \operatorname{Pic}(S)^{\mathfrak{A}_5} = 1$ since S is singular.

Put $H_S = H|_S$. Then $H_S^2 = 10$ and $H_S \cdot Z = \deg(Z)$. One has

$$H_S^2 Z^2 - (H_S \cdot Z)^2 = 10Z^2 - \deg(Z)^2 = 0$$

because H_S and Z must be numerically proportional. Since $2Z^2 \in \mathbb{Z}$, we see that $\deg(Z)^2$ is divisible by 5, and thus $\deg(Z)$ is also divisible by 5. □

Corollary 8.3.4. Let Z be an $\mathfrak{A}_5$-irreducible curve of degree $\deg(Z) < 30$ such that the stabilizer of a general point of Z is trivial. Suppose that Z contains one of the orbits Σ_5, Σ_{10}, Σ'_{10}, or Σ_{15}. Then $\deg(Z)$ is divisible by 5.

Proof. By Lemma 8.1.5 there is a surface $S \in \mathcal{Q}_2$ such that $Z \subset S$. Since S contains one of the orbits Σ_5, Σ_{10}, Σ'_{10}, or Σ_{15}, Theorem 8.2.1(iv) implies that S is singular, and we have $S \neq \mathscr{S}$ by Theorem 8.2.1(i). Now the assertion follows by Lemma 8.3.3. □

Corollary 8.3.5 (cf. Corollary 7.4.9). Each of the curves $\mathcal{L}_6$, $\mathcal{L}_{12}$, and $\mathcal{G}_6$ is disjoint from each of the orbits Σ_5, Σ_{10}, Σ'_{10}, and Σ_{15}.

Proof. This follows from Lemma 7.4.5 and Corollary 8.3.4. □

Remark 8.3.6. In the case of curve $\mathcal{L}_6$ Corollary 8.3.5 is covered by Corollary 7.4.9. In the case of curve $\mathcal{L}_{12}$ the assertion of Corollary 8.3.5 is also easily implied by the fact that $\mathcal{L}_{12}$ is a union of 12 disjoint lines (see Lemma 7.2.12(ii)). Moreover, this argument shows that $\mathcal{L}_{12}$ does not contain $\mathfrak{A}_5$-orbits of length different from 12 and 60.

Chapter 9

Combinatorics of lines and conics

In this chapter we study combinatorics of $\mathfrak{A}_5$-invariant curves in V_5 of small degree that are unions of lines or conics (cf. Propositions 7.4.1 and 7.8.4).

9.1 Lines

In this section we study $\mathfrak{A}_5$-invariant unions of lines on V_5 (cf. Lemma 7.4.7). Note that in many cases the assertions of this section can be proved using Theorems 7.1.7 and 7.1.8 together with the well-known facts about the action of the group $\mathfrak{A}_5$, like those that are listed in Theorem 6.1.2 (cf. Remark 7.4.8). Still we prefer to give proofs that involve more three-dimensional geometry in most cases.

Lemma 9.1.1. *The orbit Σ_5 is disjoint from the curve $\mathcal{L}_{10}$.*

Proof. Suppose that $\Sigma_5 \subset \mathcal{L}_{10}$. Let L be a line of $\mathcal{L}_{10}$, and let p be the number of points of Σ_5 contained in L. The stabilizer $G_L \cong \mathfrak{S}_3$ of L acts on L faithfully by Lemma 7.4.6. In particular, each G_L-invariant subset of L contains at least two points by Lemma 5.5.9(ii), so that $p \geqslant 2$. Hence L is contained in the linear span Π_3 of the $\mathfrak{A}_5$-orbit Σ_5, which contradicts Lemma 7.5.3. □

Lemma 9.1.2. *The lines of $\mathcal{L}_{10}$ are pairwise disjoint.*

Proof. Suppose that the lines of $\mathcal{L}_{10}$ are not disjoint. Recall that there are at most three lines passing through a given point on V_5 (see Theorem 7.1.9(i)). Let k_2 be the number of points that are contained in exactly two lines of $\mathcal{L}_{10}$, and k_3 be the number of points that are contained in exactly three lines of $\mathcal{L}_{10}$. Let m be the number of these points contained in some line L_1 of $\mathcal{L}_{10}$ (this number is the same for any line of $\mathcal{L}_{10}$).

Suppose that $m \geqslant 6$. Let $L_2, \ldots, L_7$ be (some of) the lines of $\mathcal{L}_{10}$ that intersect L_1. Apart from L_1 there are at least five lines of $\mathcal{L}_{10}$ that intersect L_2. Thus one of them must intersect both L_1 and L_2. This means that there exist three lines of $\mathcal{L}_{10}$ that are coplanar, which is impossible since V_5 is an intersection of quadrics.

Therefore $m \leqslant 5$, and
$$2k_2 + 3k_3 = 10m \leqslant 50.$$

By Theorem 7.3.5 and Lemma 9.1.1 for each i one either has $k_i = 0$, or $k_i \geqslant 10$. Hence
$$(k_2, k_3, m) \in \{(10, 10, 5), (0, 10, 3), (10, 0, 2)\}.$$

In particular, the curve $\mathcal{L}_{10}$ contains either the orbit Σ_{10}, or the orbit Σ'_{10}, or both of them (see Theorem 7.3.5). We may assume that $\Sigma_{10} \subset \mathcal{L}_{10}$.

By Corollary 8.1.7 there is a surface $S \in \mathcal{Q}_2$ such that $\mathcal{L}_{10} \subset S$. Note that $S \neq \mathscr{S}$ by Lemma 7.2.13(i). Hence S is a $K3$ surface that has at most ordinary double points by Theorem 8.2.1(v). Since $S \supset \mathcal{L}_{10} \supset \Sigma_{10}$ by our assumptions, it follows from Theorem 8.2.1(iii),(iv) that $\mathrm{Sing}(S) = \Sigma_{10}$. Futhermore, one has $\Sigma'_{10} \not\subset S$ by Theorem 8.2.1(i). In particular,
$$(k_2, k_3, m) \neq (10, 10, 5).$$

Moreover, $\mathrm{rk}\,\mathrm{Pic}\,(S)^{\mathfrak{A}_5} = 1$ by Lemma 6.7.3(i),(ii).

Put $H_S = H|_S$. Let $L_1, \ldots, L_{10}$ be the lines of $\mathcal{L}_{10}$. Then
$$L_i^2 = -2 + \frac{m}{2},$$
and $L_i \cdot L_j = \frac{1}{2}$ if $i \neq j$ and $L_i \cap L_j \neq \varnothing$. Thus $\mathcal{L}_{10}^2 = 25$ if
$$(k_2, k_3, m) = (0, 10, 3),$$
and $\mathcal{L}_{10}^2 = 0$ if
$$(k_2, k_3, m) = (10, 0, 2).$$

On the other hand, the matrix
$$\begin{pmatrix} H_S^2 & H_S \cdot \mathcal{L}_{10} \\ H_S \cdot \mathcal{L}_{10} & \mathcal{L}_{10}^2 \end{pmatrix} = \begin{pmatrix} 10 & 10 \\ 10 & \mathcal{L}_{10}^2 \end{pmatrix}$$
must be degenerate since $\mathrm{rk}\,\mathrm{Pic}\,(S)^{\mathfrak{A}_5} = 1$. Hence $\mathcal{L}_{10}^2 = 10$, which is a contradiction. $\square$

Lemma 9.1.2 immediately implies the following:

Corollary 9.1.3. *Let* $\Sigma \in \{\Sigma_{12}, \Sigma'_{12}, \Sigma_{15}\}$. *Then* $\Sigma \not\subset \mathcal{L}_{10}$.

Proof. The lines of $\mathcal{L}_{10}$ are pairwise disjoint by Lemma 9.1.2, and each of them contains one and the same number of points of Σ. □

Remark 9.1.4. Another proof of Lemma 9.1.2 can be obtained by applying Theorems 6.1.2(xvii) and 7.1.8 together with Remark 5.3.2. In particular, this approach does not require proving Lemma 9.1.1, but allows to derive it from Lemma 9.1.2 alongside Corollary 9.1.3.

Another consequence of Lemma 9.1.2 is the following:

Corollary 9.1.5. *Let* $\Sigma \in \{\Sigma_{10}, \Sigma'_{10}\}$. *Then* $\Sigma \not\subset \mathcal{L}_{10}$.

Proof. Suppose that $\Sigma \subset \mathcal{L}_{10}$. Let L be a line of $\mathcal{L}_{10}$, and let p be the number of points of Σ contained in L. The stabilizer $G_L \cong \mathfrak{S}_3$ of L acts on L faithfully by Lemma 7.4.6. In particular, each G_L-invariant subset of L contains at least two points, so that $p \geqslant 2$ by Lemma 5.5.9(ii). On the other hand, the lines of $\mathcal{L}_{10}$ are pairwise disjoint by Lemma 9.1.2, so that

$$10 = |\Sigma| = 10p \geqslant 20,$$

which is a contradiction. □

Remark 9.1.6. Since $\mathcal{L}_{10}$ is not contained in the surface $\mathscr{S}$ by Lemma 7.2.13(i), and the lines of $\mathcal{L}_{10}$ are pairwise disjoint by Lemma 9.1.2, we conclude that the intersection $\mathcal{L}_{10} \cap \mathscr{S}$ is either a union of two $\mathfrak{A}_5$-orbits of length 10, or a single $\mathfrak{A}_5$-orbit Σ of length $|\Sigma| \in \{10, 20\}$. Keeping in mind Lemmas 7.2.10 and 7.2.2(i), we see that $\mathcal{L}_{10} \cap \mathscr{S}$ coincides with the $\mathfrak{A}_5$-orbit Σ'_{20} contained in $\mathscr{S} \setminus \mathscr{C}$. In particular, $\mathcal{L}_{10}$ is disjoint from the curve $\mathscr{C}$ and from the $\mathfrak{A}_5$-orbit $\Sigma_{20} \subset \mathscr{C}$. Note that by Theorem 7.1.9(i) for every point $P \in \mathscr{S}$ that is not contained in $\mathscr{C}$, there are exactly two lines on V_5 passing through P. Hence, the only lines that have non-empty intersection with the $\mathfrak{A}_5$-orbit Σ'_{20} and are not contained in $\mathscr{S}$ are lines of $\mathcal{L}_{10}$.

Lemma 9.1.7. *The curves* $\mathcal{L}_6$ *and* $\mathcal{L}_{10}$ *are disjoint.*

Proof. By Corollary 8.1.7 there is a surface $S \in \mathcal{Q}_2$ such that $\mathcal{L}_{10} \subset S$. Note that $S \neq \mathscr{S}$ by Lemma 7.2.13(i). Hence S is a $K3$ surface that has at most ordinary double points by Theorem 8.2.1(v). Moreover, it follows from Theorem 8.2.1(ii),(iii) that the set of singular points of S is contained in $\Sigma_5 \cup \Sigma_{10} \cup \Sigma'_{10} \cup \Sigma_{15}$. Hence $\mathrm{Sing}(S) \cap \mathcal{L}_{10} = \varnothing$ by Lemma 9.1.1 and Corollaries 9.1.3 and 9.1.5. In particular, $\mathcal{L}_{10}^2 = -20$ by Lemma 9.1.2.

Let $H_S = H|_S$. Then H_S is not numerically proportional to $\mathcal{L}_{10}$ because $\mathcal{L}_{10}^2 < 0$, and thus $\mathrm{rk}\,\mathrm{Pic}\,(S)^{\mathfrak{A}_5} \geqslant 2$. Therefore, $\mathrm{rk}\,\mathrm{Pic}\,(S)^{\mathfrak{A}_5} = 2$ by Lemma 6.7.3(i), which implies that S is smooth by Lemma 6.7.3(ii).

Suppose that $\mathcal{L}_6 \subset S$. Put $k = \mathcal{L}_6 \cdot \mathcal{L}_{10}$. By Lemma 7.4.7 one has

$$\mathcal{L}_6^2 = -12.$$

Applying Lemma 8.3.2, we conclude that

$$10k^2 - 120k - 4320 = 0.$$

The latter equation does not have integer solutions, which is a contradiction.

We see that $\mathcal{L}_6$ intersects S in a finite number of points. Hence the intersection $\mathcal{L}_6 \cap S$ consists of 12 points (counted with multiplicities). By Theorem 7.3.5 this means that $|\mathcal{L}_6 \cap S| = 12$, and $\mathcal{L}_6 \cap S$ is a single $\mathfrak{A}_5$-orbit (in fact, Lemma 7.4.3 implies that $\mathcal{L}_6 \cap S = \Sigma'_{12}$). Therefore $\mathcal{L}_6 \cap \mathcal{L}_{10} = \varnothing$ by Corollary 9.1.3. □

Remark 9.1.8. Another proof of Lemma 9.1.7 can be obtained by applying Theorems 6.1.2(viii) and 7.1.8 together with Remark 5.3.2.

Lemma 9.1.9. One has $\mathcal{L}_6 \cap \mathcal{L}_{12} = \Sigma'_{12}$.

Proof. Recall that $\Sigma'_{12} \subset \mathcal{L}_6$ by Lemma 7.4.3 and $\Sigma'_{12} \subset \mathcal{L}_{12}$ by Lemma 7.2.12(iii). In particular, each line of $\mathcal{L}_6$ intersect the curve $\mathcal{L}_{12}$ in at least two points. Suppose that some (and thus any) line L of $\mathcal{L}_6$ intersects $\mathcal{L}_{12}$ in at least three points. Then L intersects the surface $\mathscr{S} \supset \mathcal{L}_{12}$ in at least three points, so that $L \subset \mathscr{S}$ since $\mathscr{S} \cdot L = 2$. Thus $\mathcal{L}_6 \subset \mathscr{S}$, which contradicts Lemma 7.2.13(i). □

Lemma 9.1.10. The following assertions hold:

(i) the lines of $\mathcal{L}_{15}$ are pairwise disjoint outside of Σ_5 (cf. Corollary 7.4.4);

(ii) one has $\Sigma_{15} \subset \mathcal{L}_{15}$;

(iii) each line of $\mathcal{L}_{15}$ contains a unique point of Σ_{15};

(iv) a stabilizer $G_P \subset \mathfrak{A}_5$ of a point $P \in \mathcal{L}_{15}$ is isomorphic to $\mathfrak{A}_4$ if $P \in \Sigma_5$, to $\boldsymbol{\mu}_2 \times \boldsymbol{\mu}_2$ if $P \in \Sigma_{15}$, or to $\boldsymbol{\mu}_2$ if $P \notin \Sigma_5 \cup \Sigma_{15}$.

Proof. Let us show that the lines of $\mathcal{L}_{15}$ are disjoint outside of Σ_5. Suppose that there are two lines of $\mathcal{L}_{15}$, say, L_1 and L_2, such that L_1 intersects L_2 outside Σ_5. We are going to show that this is impossible. Let P_1 be the point of Σ_5 contained in L_1, and P_2 be the point of Σ_5 contained in L_2. In particular, one has $P_1 \neq P_2$.

Let $G_{P_1,P_2} \subset \mathfrak{A}_5$ be the stabilizer of the (ordered) pair of points (P_1, P_2). Then $G_{P_1,P_2} \cong \boldsymbol{\mu}_3$. Since a stabilizer $\boldsymbol{\mu}_2 \times \boldsymbol{\mu}_2 \subset \mathfrak{A}_5$ of any line of $\mathcal{L}_{15}$ does not contain a subgroup $\boldsymbol{\mu}_3$, we see that G_{P_1,P_2} acts transitively both on the three lines of $\mathcal{L}_{15}$ passing through P_1 and on the three lines of $\mathcal{L}_{15}$ passing through P_2. In particular, each of the lines of $\mathcal{L}_{15}$ passing through P_2 intersects some of the lines of $\mathcal{L}_{15}$ passing through P_1, and vice versa. Since $\mathfrak{A}_5$ acts transitively on the pairs of points of Σ_5 by Lemma 5.1.3(i), we conclude that the same holds after replacing P_1 and P_2 by two arbitrary points $P_i, P_j \in \Sigma_5$.

Let $T_i \subset \mathbb{P}^6$ be the linear span of the three lines of $\mathcal{L}_{15}$ passing through the point $P_i \in \Sigma_5$. Then $T_i \cong \mathbb{P}^3$ by Corollary 7.4.4. Since any line of $\mathcal{L}_{15}$ passing through P_i, $i = 1, 2$, intersects T_3, we see that the linear span $T_{1,2,3}$ of the union $T_1 \cup T_2 \cup T_3$ coincides with the linear span of the union $P_1 \cup P_2 \cup T_3$, so that $\dim(T_{1,2,3}) \leqslant 5$. Let L be an arbitrary line of $\mathcal{L}_{15}$. Then L intersects each of the subspaces T_1, T_2, and T_3 in the points lying on the lines that pass through P_1, P_2, and P_3, respectively. Moreover, the points of intersection of L with all these three subspaces cannot coincide, since there are at most 3 lines passing through a point on V_5 by Theorem 7.1.9(i). Thus $L \subset T_{1,2,3}$, so that $T_{1,2,3}$ contains the curve $\mathcal{L}_{15}$. This is impossible by Corollary 7.5.4. The obtained contradiction shows that the lines of $\mathcal{L}_{15}$ are pairwise disjoint outside of Σ_5, i.e., proves assertion (i).

Now let us show the remaining assertions of the lemma. Let L_1 be a line of $\mathcal{L}_{15}$, and $G_1 \subset \mathfrak{A}_5$ be the stabilizer of the line L_1. Then $G_1 \cong \boldsymbol{\mu}_2 \times \boldsymbol{\mu}_2$ acts on L_1 with a kernel $G_P \cong \boldsymbol{\mu}_2$ by Lemma 7.4.6. Hence L_1 contains exactly two G_1-invariant points by Lemma 5.5.9(i), say, Q_1 and Q_1'. Since the point $L_1 \cap \Sigma_5$ is unique and thus G_1-invariant, one can assume that

$$Q_1 = L_1 \cap \Sigma_5.$$

Now it is easy to see that the $\mathfrak{A}_5$-orbit of the point Q_1' has length 15, since the lines of $\mathcal{L}_{15}$ are disjoint outside of Σ_5; in particular, we see that each line of $\mathcal{L}_{15}$ contains a unique point of this orbit. By Theorem 7.3.5 we conclude that the $\mathfrak{A}_5$-orbit of Q_1' is Σ_{15}, which gives assertions (ii) and (iii). Finally, it follows from by Lemma 5.5.9(i) that any point $P \in L_1$ different from Q_1 and Q_1' is contained in a G_1-orbit of length two, which implies assertion (iv). □

Recall that by Lemma 7.2.10 the surface $\mathscr{S}$ contains a unique $\mathfrak{A}_5$-orbit Σ'_{30} outside the curve $\mathscr{C}$. An easy consequence of Lemma 9.1.10 is the following:

Corollary 9.1.11. One has
$$\mathcal{L}_{15} \cap \mathscr{S} = \mathcal{L}_{15} \cap C_{16} = \Sigma'_{30}.$$
In particular, the curve $\mathcal{L}_{15}$ is disjoint from the curves $\mathscr{C}$, $\mathcal{L}_{20}$, and C_{20}.

Proof. Note that the $\mathfrak{A}_5$-orbits Σ_5 and Σ_{15} are not contained in the surface $\mathscr{S}$ by Lemma 7.2.10. Thus by Lemma 9.1.10 the intersection $\mathcal{L}_{15} \cap \mathscr{S}$ is a single $\mathfrak{A}_5$-orbit Σ of length $\mathcal{L}_{15} \cdot \mathscr{S} = 30$. Thus each line of $\mathcal{L}_{15}$ intersects $\mathscr{S}$ in two points. If some line L intersects the curve $\mathscr{C}$, then $L \subset \mathscr{S}$ since $\mathscr{S}$ is singular along $\mathscr{C}$ by Lemma 7.2.2(i). This would imply $\mathcal{L}_{15} \subset \mathscr{S}$, which is not the case by Lemma 7.2.13(ii). Now Lemma 7.2.10 implies that $\Sigma = \Sigma'_{30}$. Since Σ'_{30} is contained in the curve C_{16} by Lemma 8.1.16(iv), this gives
$$\mathcal{L}_{15} \cap C_{16} = \Sigma'_{30}.$$
On the other hand, the curve $\mathscr{C}$ is disjoint from the $\mathfrak{A}_5$-orbit Σ'_{30} by definition of Σ'_{30}, the curve $\mathcal{L}_{20}$ is disjoint from Σ'_{30} by Lemma 7.2.12(ii), and the curve C_{20} is disjoint from Σ'_{30} by Theorem 8.1.8(vii). This gives the remaining assertions of the corollary. □

Remark 9.1.12. It is also easy to describe the intersection of the curve $\mathcal{L}_{15}$ with other surfaces of the pencil $\mathcal{Q}_2$. Namely, we have $\mathcal{L}_{15} \cap S_5 = \Sigma_5$ because the surface S_5 has singularities at the points of Σ_5 by Theorem 8.2.1(iv), and a similar argument shows that $\mathcal{L}_{15} \cap S_{15} = \Sigma_{15}$. Since a surface S of the pencil $\mathcal{Q}_2$ that is different from S_5 and S_{15} does not contain the $\mathfrak{A}_5$-orbits Σ_5 and Σ_{15} by Theorem 8.1.8(vii), we conclude using Lemma 9.1.10(iv) that the intersection of $\mathcal{L}_{15}$ with S is a single $\mathfrak{A}_5$-orbit of length 30. Note also that the latter orbit is not contained in $\mathscr{S}$ provided that $S \neq \mathscr{S}$ by Theorem 8.1.8(vii).

Remark 9.1.13. The lines of $\mathcal{L}_{15}$ are contracted by the $\mathfrak{A}_5$-equivariant projection
$$\theta_{\Pi_3} \colon V_5 \dashrightarrow \mathbb{P}^2 \cong \mathbb{P}(W_3)$$
from the subspace $\Pi_3 \subset \mathbb{P}^6$ (see §7.5). The image
$$\theta_{\Pi_3}(\mathcal{L}_{15}) \subset \mathbb{P}^2$$
is the unique $\mathfrak{A}_5$-orbit of length 15 in $\mathbb{P}^2$ (cf. Lemma 5.3.1(i)).

Lemma 9.1.14. Each line of $\mathcal{L}_6$ intersects exactly five lines of $\mathcal{L}_{15}$, and each line of $\mathcal{L}_{15}$ intersects exactly two lines of $\mathcal{L}_6$. The intersection of the curves $\mathcal{L}_{15}$ and $\mathcal{L}_6$ is a single $\mathfrak{A}_5$ orbit of length 30 different from Σ_{30} and Σ'_{30}.

Proof. By Theorem 7.1.8 it is enough to show that each point $P \in \mathbb{P}^2_\ell$ that represents a line of $\mathcal{L}_6$ is polar with respect to the conic $\mathfrak{C} \subset \mathbb{P}^2_\ell$ to a line $l_P \subset \mathbb{P}^2_\ell$ containing exactly 5 points that represent the lines of $\mathcal{L}_{15}$, and each point $Q \in \mathbb{P}^2_\ell$ that represents a line of $\mathcal{L}_{15}$ is polar with respect to the conic $\mathfrak{C}$ to a line $l_Q \subset \mathbb{P}^2_\ell$ containing exactly 2 points that represent the lines of $\mathcal{L}_6$. Therefore, the first two assertions of the lemma follow from Theorems 6.1.2(iii),(v) and 7.1.8 together with Remark 5.3.2.

To prove the remaining assertion choose a point $P \in \mathcal{L}_6 \cap \mathcal{L}_{15}$, and let Σ be the $\mathfrak{A}_5$-orbit of the point P. We know from Lemma 9.1.10(iv) that either $\Sigma = \Sigma_5$, or $\Sigma = \Sigma_{15}$, or $|\Sigma| = 30$. This means that $|\Sigma| = 30$ by Corollary 7.4.9, and $\Sigma = \mathcal{L}_6 \cap \mathcal{L}_{15}$. Finally, note that Σ is not contained in the surface $\mathscr{S}$ by Lemma 7.4.3. □

Lemma 9.1.15. Each line of $\mathcal{L}_{15}$ intersects exactly two lines of $\mathcal{L}_{10}$, and each line of $\mathcal{L}_{10}$ intersects exactly three lines of $\mathcal{L}_{15}$. The intersection of the curves $\mathcal{L}_{15}$ and $\mathcal{L}_{10}$ is a single $\mathfrak{A}_5$ orbit of length 30 different from Σ_{30} and Σ'_{30}.

Proof. By Theorem 7.1.8 it is enough to show that each point $P \in \mathbb{P}^2_\ell$ that represents a line of $\mathcal{L}_{10}$ is polar with respect to the conic $\mathfrak{C} \subset \mathbb{P}^2_\ell$ to a line $l_P \subset \mathbb{P}^2_\ell$ containing exactly 3 points that represent the lines of $\mathcal{L}_{15}$, and each point $Q \in \mathbb{P}^2_\ell$ that represents a line of $\mathcal{L}_{15}$ is polar with respect to the conic $\mathfrak{C}$ to a line $l_Q \subset \mathbb{P}^2_\ell$ containing exactly 2 points that represent the lines of $\mathcal{L}_{10}$. Therefore, the first two assertions of the lemma follow from Theorems 6.1.2(x),(xii) and 7.1.8 together with Remark 5.3.2.

Now choose a point $P \in \mathcal{L}_{10} \cap \mathcal{L}_{15}$, and let Σ be the $\mathfrak{A}_5$-orbit of the point P. We know from Lemma 9.1.10(iv) that either $\Sigma = \Sigma_5$, or $\Sigma = \Sigma_{15}$, or $|\Sigma| = 30$. This gives $|\Sigma| = 30$ by Lemma 9.1.1 and Corollary 9.1.3. It remains to note that Σ is not contained in the surface $\mathscr{S}$ by Remark 9.1.6. □

9.2 Conics

In this section we study $\mathfrak{A}_5$-invariant unions of conics in V_5.

Lemma 9.2.1. The orbit Σ_5 is disjoint from the curve $\mathcal{G}_5$ (resp., $\mathcal{G}'_5$).

Proof. It is enough to prove the assertion for the curve $\mathcal{G}_5$. Suppose that $\Sigma_5 \subset \mathcal{G}_5$. Consider the rational $\mathfrak{A}_5$-invariant map

$$\theta \colon V_5 \dashrightarrow \mathbb{P}^2 \cong \mathbb{P}(W_3)$$

given by the projection from the linear span $\Pi_3 \subset \mathbb{P}^6$ of the $\mathfrak{A}_5$-orbit Σ_5 (see §7.5). Since $\Sigma_5 \subset \mathcal{G}_5$, we see that the image $\theta_{\Pi_3}(\mathcal{G}_5) \subset \mathbb{P}^2$ is either a union of at most 5 lines, or a union of at most 5 points. Both of these cases are impossible by Lemma 5.3.1(i),(ii). □

Lemma 9.2.2. *Let $\Sigma \in \{\Sigma_{12}, \Sigma'_{12}\}$. Then $\Sigma \not\subset \mathcal{G}_5$ and $\Sigma \not\subset \mathcal{G}'_5$.*

Proof. It is enough to prove the assertion for the curve $\mathcal{G}_5$. Choose a general surface $S \in \mathcal{Q}_2$. By Theorem 8.1.8(i), one has $\mathcal{G}_5 \not\subset S$, so that the intersection $S \cap \mathcal{G}_5$ consists of 20 points (counted with multiplicities). By Theorem 7.3.5 this means that $S \cap \mathcal{G}_5$ cannot contain an $\mathfrak{A}_5$-orbit of length 12. On the other hand, the surface S contains both $\mathfrak{A}_5$-orbits Σ_{12} and Σ'_{12} by Theorem 8.1.8(iii). □

Lemma 9.2.3. *The conics of $\mathcal{G}_5$ (resp., of $\mathcal{G}'_5$) are pairwise disjoint.*

Proof. It is enough to prove the assertion for the conics of $\mathcal{G}_5$. Suppose that there are two conics C_1 and C_2 of $\mathcal{G}_5$ that are not disjoint. Let k_i, $2 \leqslant i \leqslant 5$, be the number of points that are contained in exactly i conics of $\mathcal{G}_5$. Let m be the number of these points contained in C_1 (this number is the same for any conic of $\mathcal{G}_5$). Since V_5 is an intersection of quadrics, any two conics in V_5 intersect by at most two points. In particular, $m \leqslant 8$.

Suppose that $m = 8$. Then each of the conics of $\mathcal{G}_5$ intersects each other conic of $\mathcal{G}_5$ in two points. Consider the three-dimensional linear span Π of $C_1 \cup C_2$. Since $m = 8$ implies that $k_i = 0$ for $i > 2$, any other conic of $\mathcal{G}_5$ intersects $C_1 \cup C_2$ in 4 points, and thus is contained in Π. This means that $\Pi \subset \mathbb{P}^6$ is an $\mathfrak{A}_5$-invariant subspace, which contradicts Corollary 7.5.4.

We see that $m < 8$. Moreover, by Lemma 7.4.5 the stabilizer $G_{C_1} \cong \mathfrak{A}_4$ of the conic C_1 acts on C_1 faithfully, so that $m \neq 7$ by Lemma 5.5.9(iii). Thus $m \leqslant 6$.

One has

$$2k_2 + 3k_3 + 4k_4 + 5k_5 = 5m \leqslant 30.$$

By Theorem 7.3.5 and Lemmas 9.2.1 and 9.2.2, for each i one has either $k_i = 0$, or $k_i = 10$, or $k_i \geqslant 15$. In particular, $k_4 = k_5 = 0$.

Suppose that $k_3 > 0$. Then $k_3 = 10$, $k_2 = 0$, and $m = 6$. Let k be the number of points of intersection of C_1 with some (and thus any, see Lemma 5.1.3(i)) conic of $\mathcal{G}_5$ different from C_1. Then $4k = 2m = 12$. Hence $k = 3$, which is impossible since V_5 is an intersection of quadrics.

We see that $k_3 = k_4 = k_5 = 0$. Thus either $k_2 = 15$ and $m = 6$, or $k_2 = 10$ and $m = 4$. Since the action of $\mathfrak{A}_5$ on the conics of $\mathcal{G}_5$ is doubly transitive by Lemma 5.1.3(i), we see that m must be divisible by 4. Thus, $k_2 = 10$ and $m = 4$. In particular, $\mathcal{G}_5$ contains either the orbit Σ_{10} or the orbit Σ'_{10} (see Theorem 7.3.5). We may assume that $\Sigma_{10} \subset \mathcal{G}_5$.

By Corollary 8.1.7 there is a surface $S \in \mathcal{Q}_2$ such that $\mathcal{G}_5 \subset S$. Note that $S \neq \mathscr{S}$ by Lemma 7.2.13(i). Hence S is a $K3$ surface that has at most ordinary double points by Theorem 8.2.1(v). Since by our assumptions we have $S \supset \mathcal{G}_5 \supset \Sigma_{10}$, it follows from Theorem 8.2.1(iii),(iv) that $S = S_{10}$, i.e., $\mathrm{Sing}(S) = \Sigma_{10}$. Thus $\mathrm{rk}\,\mathrm{Pic}\,(S)^{\mathfrak{A}_5} = 1$ by Lemma 6.7.3(i),(ii).

Put $H_S = H|_S$. Then $\mathcal{G}_5 \sim_{\mathbb{Q}} \lambda H_S$ for some $\lambda \in \mathbb{Q}$. Since

$$10 = H_S \cdot \mathcal{G}_5 = \lambda H_S^2 = 10\lambda,$$

we see that $\mathcal{G}_5 \sim_{\mathbb{Q}} H_S$. Note that $\mathcal{G}_5$ is a Cartier divisor on S and $\mathcal{G}_5 \sim H_S$ by Remark 8.2.4. (Actually, one can obtain the first of these assertions in a straightforward way, using the fact that the singularities of S are ordinary double points, and there are exactly two components of $\mathcal{G}_5$ passing through a given singular point of S.) The latter means that the linear system $|H_S|$ contains an $\mathfrak{A}_5$-invariant curve, which contradicts Lemma 8.2.6(i). □

Corollary 9.2.4. Let $\Sigma \in \{\Sigma_{10}, \Sigma'_{10}, \Sigma_{15}\}$. Then $\Sigma \not\subset \mathcal{G}_5$ (resp., $\Sigma \not\subset \mathcal{G}'_5$).

Proof. It is enough to prove the assertion for the curve $\mathcal{G}_5$. The conics of $\mathcal{G}_5$ are pairwise disjoint by Lemma 9.2.3, and each of them contains one and the same number of points of Σ. Therefore, it is enough to show that $\Sigma_{10} \not\subset \mathcal{G}_5$.

Suppose that $\Sigma_{10} \subset \mathcal{G}_5$, so that each conic of $\mathcal{G}_5$ contains exactly two points of Σ_{10}. Then the stabilizer $G_C \cong \mathfrak{A}_4$ of a conic C of $\mathcal{G}_5$ has an orbit of length at most 2 on C. On the other hand, G_C acts on C faithfully by Lemma 7.4.5. This contradicts Lemma 5.5.9(iii). □

Remark 9.2.5. Since $\mathcal{G}_5$ is not contained in the surface $\mathscr{S}$ by Lemma 7.2.13(i) and the conics of $\mathcal{G}_5$ are pairwise disjoint by Lemma 9.2.3, it follows from Theorem 7.3.5 that the intersection $\mathcal{G}_5 \cap \mathscr{S}$ either contains an $\mathfrak{A}_5$-orbit of length 5 or 10, or is a single $\mathfrak{A}_5$-orbit of length 20. Keeping in mind Lemmas 7.2.10 and 7.2.2(i), we see that $\mathcal{G}_5 \cap \mathscr{S}$ coincides with the unique $\mathfrak{A}_5$-orbit Σ'_{20} contained in $\mathscr{S} \setminus \mathscr{C}$. Similarly, one has

$$\mathcal{G}'_5 \cap \mathscr{S} = \Sigma'_{20}.$$

In particular, the curves $\mathcal{G}_5$ and $\mathcal{G}'_5$ are not disjoint, but each of them is disjoint from the curve $\mathscr{C}$ and from the $\mathfrak{A}_5$-orbit $\Sigma_{20} \subset \mathscr{C}$. Recall that one also has
$$\mathcal{L}_{10} \cap \mathscr{S} = \Sigma'_{20}$$
by Remark 9.1.6. In particular, the curves $\mathcal{G}_5$ and $\mathcal{L}_{10}$ (resp., $\mathcal{G}'_5$ and $\mathcal{L}_{10}$) are not disjoint.

Lemma 9.2.6. *The curves $\mathcal{L}_6$ and $\mathcal{G}_5$ (resp., $\mathcal{G}'_5$) are disjoint.*

Proof. It is enough to prove the assertion for the curve $\mathcal{G}_5$. Suppose that some line L of $\mathcal{L}_6$ intersects some conic C of $\mathcal{G}_5$. Then L intersects any conic of $\mathcal{G}_5$, because the action of the stabilizer $G_L \cong \mathfrak{D}_{10}$ of L on the conics of $\mathcal{G}_5$ is transitive by Lemma 5.1.3(iii). This means that each line of $\mathcal{L}_6$ intersects each conic of $\mathcal{G}_5$.

Let C_1 and C_2 be two conics of $\mathcal{G}_5$, and let $\Pi \subsetneq \mathbb{P}^6$ be the linear span of $C_1 \cup C_2$. Then any line L of $\mathcal{L}_6$ intersects Π by at least two points, and thus is contained in Π. This means that the linear span of $\mathcal{L}_6$ is a proper subspace of $\mathbb{P}^6$, which contradicts Corollary 7.5.4. □

Lemma 9.2.7. *The curves $\mathcal{L}_{15}$ and $\mathcal{G}_5$ (resp., $\mathcal{G}'_5$) are disjoint.*

Proof. It is enough to prove the assertion for the curve $\mathcal{G}_5$. Suppose that some line L of $\mathcal{L}_{15}$ intersects some conic C of $\mathcal{G}_5$. Let L' and L'' be the lines of $\mathcal{L}_{15}$ that have a common point in Σ_5 with L (see Corollary 7.4.4). Let $G \subset \mathfrak{A}_5$ be the stabilizer of the curve $L \cup L' \cup L''$, and $G_C \subset \mathfrak{A}_5$ be the stabilizer of the conic C. Then $G \cong G_C \cong \mathfrak{A}_4$ (see Lemma 5.1.1).

Suppose that the subgroups G and G_C coincide. Then the intersection
$$\Sigma = C \cap (L \cup L' \cup L'')$$
is a G-invariant subset of $C \cong \mathbb{P}^1$. Thus $|\Sigma| \geqslant 4$ by Lemma 5.5.9(iii). Hence at least one (and thus any) of the lines L, L', and L'' intersects the conic C in two points. This is impossible since V_5 is an intersection of quadrics.

We see that G and G_C are different subgroups of $\mathfrak{A}_5$. Denote by $C_1 = C, C_2, \ldots, C_5$ the conics of $\mathcal{G}_5$; we can label them so that the stabilizer of C_5 in $\mathfrak{A}_5$ coincides with G. As we have already checked, the conic C_5 is disjoint from the line L (and from the lines L' and L'' as well). Note that G acts transitively on the set $\{L, L', L''\}$ by definition, and also acts transitively on the set $\{C_1, \ldots, C_4\}$ by Lemma 5.1.3(i).

Let c be the number of conics among $C_1, \ldots, C_4$ that intersect some (and thus any) of the lines L, L', and L'', and let l be the number of lines among

L, L', and L'' that intersect some (and thus any) of the conics $C_1, \ldots, C_4$. One has $3c = 4l$. This gives $c = 4$.

By Corollary 8.1.7 there is a surface $S \in \mathcal{Q}_2$ that contains the curve $\mathcal{G}_5$. Since $c = 4$, we also have $L \subset S$, so that $\mathcal{L}_{15} \subset S$. The latter gives a contradiction with Lemma 7.5.13. □

Lemma 9.2.8. *The conics of $\mathcal{G}_6$ are pairwise disjoint.*

Proof. Suppose that there are two conics C_1 and C_2 of $\mathcal{G}_6$ that are not disjoint. Let k_i, $2 \leqslant i \leqslant 6$, be the number of points that are contained in exactly i conics of $\mathcal{G}_6$. Let m be the number of these points contained in C_1 (this number is the same for any conic of $\mathcal{G}_6$). Since V_5 is an intersection of quadrics, any two conics on V_5 intersect each other in at most two points, so that $m \leqslant 10$.

Suppose that $m = 10$. Then each of the conics of $\mathcal{G}_6$ intersects each other conic of $\mathcal{G}_6$ in two points, and all these points are pairwise distinct. Hence any conic of $\mathcal{G}_6$ is contained in the linear span $\Pi \subsetneq \mathbb{P}^6$ of $C_1 \cup C_2$, which is impossible by Corollary 7.5.4.

We see that $m < 10$. Since the stabilizer $G_{C_1} \cong \mathfrak{D}_{10}$ of the conic C_1 acts on C_1 faithfully by Lemma 7.4.5, this means that $m \in \{2, 5, 7\}$ (see Lemma 5.5.9(ii)). One has

$$2k_2 + 3k_3 + 4k_4 + 5k_5 + 6k_6 = 6m \leqslant 42.$$

By Theorem 7.3.5 and Corollary 8.3.5 for each i one has either $k_i = 0$, or $k_i = 12$, or $k_i \geqslant 20$. In particular, $k_4 = k_5 = k_6 = 0$, and $m \neq 2$.

Suppose that $k_3 > 0$. Then either

$$2k_2 + 3k_3 = 6m = 30,$$

or

$$2k_2 + 3k_3 = 6m = 42.$$

Thus $k_3 < 20$, so that $k_3 = 12$ and $k_2 = 3$, which is a contradiction.

Therefore, one has $k_3 = 0$. Hence

$$k_2 = 3m \in \{15, 21\},$$

which means that $k_2 = 21$. This is impossible by Theorem 7.3.5. □

Remark 9.2.9. In particular, Lemma 9.2.8 implies that $\mathcal{G}_6$ does not contain $\mathfrak{A}_5$-orbits of length 20. Also, it implies that $\mathcal{G}_6$ does not contain the $\mathfrak{A}_5$-orbits Σ_5, Σ_{10}, Σ'_{10}, and Σ_{15}, but this fact (namely, Corollary 8.3.5) was already used in its proof.

Lemma 9.2.10. The curves $\mathcal{L}_6$ and $\mathcal{G}_6$ are disjoint.

Proof. By Corollary 8.1.7 there is a surface $S \in \mathcal{Q}_2$ such that $\mathcal{L}_6 \subset S$. Note that $S \neq \mathscr{S}$ by Lemma 7.2.13(i). Hence S is smooth by Lemma 8.3.3. In particular, $\mathcal{L}_6^2 = -12$, since the lines of $\mathcal{L}_6$ are disjoint by Lemma 7.4.7.

Suppose that $\mathcal{G}_6 \subset S$. Put $k = \mathcal{G}_6 \cdot \mathcal{L}_6$. By Lemma 9.2.8 one has $\mathcal{G}_6^2 = -12$. Applying Lemma 8.3.2, we conclude that

$$10k^2 - 144k - 3600 = 0.$$

The latter equation does not have integer solutions, which is a contradiction.

Thus $\mathcal{G}_6$ intersects S in a finite number of points. Suppose that there exists a conic C of $\mathcal{G}_6$ and a line L of $\mathcal{L}_6$ such that $C \cap L \neq \varnothing$. Since $C \not\subset S$, we see that C intersects S in at most $C \cdot S = 4$ points. By Lemma 7.4.5, the stabilizer $G_C \cong \mathfrak{D}_{10}$ of C acts on C faithfully.

Note that the set $C \cap L$ consists of a single point P since V_5 is an intersection of quadrics. Hence L is not stabilized by G_C, because G_C acts on C without fixed points by Lemma 5.5.9(ii). Recall that the action of $\mathfrak{A}_5$ on the lines of $\mathcal{L}_6$ is doubly transitive by Lemma 5.1.3(ii) so that the G_C-orbit of L consists of 5 lines. Since the lines of $\mathcal{L}_6$ are disjoint, the G_C-orbit of the point P has length at least 5. On the other hand, all points in G_C-orbit of the point P are intersection points of the conic C and the surface S, which contradicts the fact that $|C \cap S| \leqslant 4$. $\square$

Corollary 9.2.11. One has $\Sigma'_{12} \not\subset \mathcal{G}_6$ and $\Sigma_{12} \subset \mathcal{G}_6$. Moreover,

$$\mathcal{L}_{12} \cap \mathcal{G}_6 = \mathscr{S} \cap \mathcal{G}_6 = \Sigma_{12}.$$

Proof. One has $\Sigma'_{12} \subset \mathcal{L}_6$ by Lemma 7.4.3 and $\mathcal{L}_6 \cap \mathcal{G}_6 = \varnothing$ by Lemma 9.2.10, which implies that $\Sigma'_{12} \not\subset \mathcal{G}_6$. Recall that the curve $\mathcal{G}_6$ is not contained in the surface $\mathscr{S}$ by Lemma 7.2.13(ii). Thus the intersection $\mathscr{S} \cap \mathcal{G}_6$ consists of

$$\mathscr{S} \cdot \mathcal{G}_6 = 24$$

points (counted with multiplicities). Since $\Sigma'_{12} \not\subset \mathcal{G}_6$, we see that

$$\mathscr{S} \cap \mathcal{G}_6 = \Sigma_{12}$$

by Lemma 7.2.10. In particular, one has $\mathcal{L}_{12} \cap \mathcal{G}_6 = \Sigma_{12}$ by Lemma 7.2.12(iii). $\square$

Lemma 9.2.12. The curves $\mathcal{L}_{10}$ and $\mathcal{G}_6$ are disjoint.

Proof. Suppose that there is a point $P \in \mathcal{L}_{10} \cap \mathcal{G}_6$, and let Σ be the $\mathfrak{A}_5$-orbit of P. Then $|\Sigma|$ is divisible by 10 due to Lemma 9.1.2, and $|\Sigma|$ is divisible by 6 due to Lemma 9.2.8. Thus $|\Sigma| \geqslant 30$.

By Corollary 8.1.7 there exists a surface $S \in \mathcal{Q}_2$ that contains the curve $\mathcal{G}_6$. Note that $S \neq \mathscr{S}$ by Lemma 7.2.13(ii). Hence S is smooth by Lemma 8.3.3. In particular, $\mathcal{G}_6^2 = -12$, since the conics of $\mathcal{G}_6$ are disjoint by Lemma 9.2.8.

Suppose that $\mathcal{L}_{10} \subset S$. Put $k = \mathcal{G}_6 \cdot \mathcal{L}_{10}$. By Lemma 9.1.2 one has $\mathcal{L}_{10}^2 = -20$. Applying Lemma 8.3.2, we conclude that

$$k^2 - 12k - 432 = 0.$$

The latter equation does not have integer solutions, which is a contradiction.

Therefore, the curve $\mathcal{L}_{10}$ is not contained in S. Thus the intersection $S \cap \mathcal{L}_{10}$ consists of $S \cdot \mathcal{L}_{10} = 20$ points (counted with multiplicities). This gives a contradiction since $\Sigma \subset S \cap \mathcal{L}_{10}$. □

Chapter 10

Special invariant curves

The purpose of this chapter is to prove that there are no $\mathfrak{A}_5$-invariant irreducible curves on V_5 with certain properties, and to classify $\mathfrak{A}_5$-invariant curves of degree at most 15.

10.1 Irreducible curves

We start with an easy observation concerning curves on $K3$ surfaces.

Lemma 10.1.1. Let S be a $K3$ surface with at most du Val singularities, and let F be an effective $\mathbb{Q}$-divisor on S. Let H_S be an ample Cartier divisor on S. Put $d = F \cdot H_S$. Then $F^2 \geqslant -2d^2$. Moreover, if $F^2 = -2d^2$, then $F = dL$ for some smooth rational curve L, such that L is contained in a smooth locus of S and $L \cdot H_S = 1$.

Proof. Write $F = \sum_{i=1}^r a_i F_i$, where F_i are (different) irreducible curves, and a_i are positive integers. Put $d_i = F_i \cdot H_S$. Then $d_i \geqslant 1$, and

$$d = \sum_{i=1}^r a_i d_i.$$

One has $F_i^2 \geqslant -2$, and $F_i^2 = -2$ if and only if F_i is a smooth rational curve contained in the smooth locus of S. Therefore

$$F^2 = \left(\sum_{i=1}^r a_i F_i\right)^2 \geqslant \sum_{i=1}^r a_i^2 F_i^2 \geqslant -2\sum_{i=1}^r a_i^2 \geqslant$$

$$\geqslant -2\sum_{i=1}^r a_i^2 d_i^2 \geqslant -2\left(\sum_{i=1}^r a_i d_i\right)^2 = -2d^2.$$

Moreover, if all inequalities in the above computation turn into equalities, then $r = 1$, $d_1 = 1$, and $F_1^2 = -2$. □

Lemma 10.1.2. Let C be a smooth irreducible $\mathfrak{A}_5$-invariant curve of genus 4 and degree d in V_5. Then $d \neq 12$.

Proof. Suppose that $d = 12$. Then the curve C contains two $\mathfrak{A}_5$-orbits of length 12 by Lemma 5.1.5. Thus both orbits Σ_{12} and Σ'_{12} are contained in C, since these are the only $\mathfrak{A}_5$-orbits of length 12 on V_5 by Theorem 7.3.5.

We claim that $C \subset \mathscr{S}$. Suppose that $C \not\subset \mathscr{S}$. Then

$$24 = \mathscr{S} \cdot C \geqslant 2|\Sigma_{12}| + |\Sigma'_{12}| = 36,$$

since the $\mathfrak{A}_5$-orbits Σ_{12} and Σ'_{12} are contained in the surface $\mathscr{S}$ by construction, and $\mathscr{S}$ is singular at the points of Σ_{12} by Lemma 7.2.2(i). The obtained contradiction shows that $C \subset \mathscr{S}$. Now the assertion follows by Lemma 7.2.13(ii). □

Lemma 10.1.3. Let C be a smooth irreducible $\mathfrak{A}_5$-invariant curve of genus 5 and degree d in V_5. Then $d \neq 12$.

Proof. Suppose that $d = 12$. By Corollary 8.1.6 there exists a surface $S \in \mathcal{Q}_2$ that contains C. By Lemma 7.2.13(ii) one has $S \neq \mathscr{S}$, and S is non-singular by Lemma 8.3.3.

Put $H_S = H|_S$. Then $H_S^2 = 10$ and $C^2 = 8$. Moreover, we have

$$H_S \cdot C = d = 12.$$

One has $(2H_S - C)^2 = 0$ and $(2H_S - C) \cdot H_S = 8$, which implies that

$$h^2(\mathcal{O}_S(2H_S - C)) = h^0(\mathcal{O}_S(C - 2H_S)) = 0$$

by Serre duality. Hence

$$h^0(\mathcal{O}_S(2H_S - C)) = h^1(\mathcal{O}_S(2H_S - C)) + 2 \geqslant 2$$

by the Riemann–Roch formula.

Let us show that the linear system $|2H_S - C|$ does not have base components. Indeed, suppose that it does. Then

$$2H_S - C \sim F + M$$

for some effective divisors F and M on S such that the linear system $|M|$ is free from base components, and F is the fixed part of the linear system $|2H_S - C|$. In particular, F is an $\mathfrak{A}_5$-invariant divisor. Note that $|M|$ is non-trivial since
$$h^0(\mathcal{O}_S(2H_S - C)) > 1.$$
Moreover, one has $\deg(M) \geqslant 3$ since S is not uniruled. Hence
$$\deg(F) = 20 - \deg(C) - \deg(M) \leqslant 5,$$
which is impossible by Corollary 8.1.15.

Thus, we see that $|2H_S - C|$ does not have base components. Since
$$(2H_S - C)^2 = 0,$$
we see that $|2H_S - C|$ does not have base points. Moreover, it follows from [109, Proposition 2.6] that there exists a smooth elliptic curve E on the surface S such that $|E|$ is a pencil that does not have base points, and
$$2H_S - C \sim rE$$
for some positive integer r. Keeping in mind that
$$8 = (2H_S - C) \cdot H_S = r\deg(E),$$
we see that either $r = 1$ or $r = 2$. One has
$$h^0(\mathcal{O}_S(2H_S - C)) \neq 2$$
by Lemma 8.2.6(iii). Hence $r \neq 1$, because
$$h^0(\mathcal{O}_S(2H_S - C)) = r + 1.$$
We see that $r = 2$ and $2H_S - C \sim 2E$.

Note that $(H_S - E)^2 = 2$ and $(H_S - E) \cdot H_S = 6$, which implies that
$$h^2(\mathcal{O}_S(H_S - E)) = h^0(\mathcal{O}_S(E - H_S)) = 0$$
by Serre duality. Thus
$$h^0(\mathcal{O}_S(H_S - E)) = h^1(\mathcal{O}_S(H_S - E)) + 3 \geqslant 3$$
by the Riemann–Roch formula.

Let us prove that the linear system $|H_S - E|$ is free from base components. Suppose that it is not. Then

$$H_S - E \sim F + M,$$

for some effective divisors F and M on S such that the linear system $|M|$ is free from base components, and F is the fixed part of the linear system $|H_S - E|$. In particular, F is an $\mathfrak{A}_5$-invariant divisor. Note that $|M|$ is non-trivial since

$$h^0(\mathcal{O}_S(H_S - E)) > 1.$$

Hence

$$\deg(F) = 10 - \deg(E) - \deg(M) \leqslant 5,$$

which contradicts Corollary 8.1.15.

We see that $|H_S - E|$ is free from base components. Since $(H_S - E)^2 = 2$, it follows from Theorem 7.3.5 that $|H_S - E|$ is free from base points. Actually, the same result follows from [109, Corollary 3.1] without using Theorem 7.3.5.

We see that $|H_S - E|$ is free from base points. We claim that the divisor $H_S - E$ is ample. Indeed, suppose that $H_S - E$ is not ample. Since

$$(H_S - E)^2 = 2,$$

it follows from Lemma 6.7.6 that S contains an irreducible curve F such that

$$20 + r(H_S \cdot F) = 2H_S^2 + r(H_S \cdot F) = ((H_S - E) \cdot H_S)^2 = 36$$

for some $r \in \{6, 10, 15\}$. This gives

$$(H_S \cdot F)^2 = \frac{16}{r} \notin \mathbb{Z},$$

which is a contradiction.

Thus, we see that $H_S - E$ is ample, so that

$$h^1(\mathcal{O}_S(H_S - E)) = 0$$

by Kodaira vanishing (cf. Theorem 2.3.5). Hence

$$h^0(\mathcal{O}_S(H_S - E)) = 3$$

and the linear system $|H_S - E|$ gives an $\mathfrak{A}_5$-equivariant morphism $\eta \colon S \to \mathbb{P}^2$. Since the divisor $H_S - E$ is ample, the morphism η is a finite morphism of degree 2 that is branched along a smooth sextic curve $B \subset \mathbb{P}^2$.

CREMONA GROUPS AND THE ICOSAHEDRON

Since $(H_S - E) \cdot E = 4$, we conclude that either $\eta(E)$ is a quartic curve and the induced map $E \to \eta(E)$ is an isomorphism, or $\eta(E)$ is a conic and the induced map $E \to \eta(E)$ is a double cover. The former case is impossible, since E is a smooth elliptic curve, while a smooth plane quartic has genus 3. The latter case is impossible since otherwise one would have

$$0 = E^2 = 2\eta(E)^2 = 8.$$

The obtained contradiction completes the proof of Lemma 10.1.3. □

Lemma 10.1.4. Let $C \subset V_5$ be a smooth rational $\mathfrak{A}_5$-invariant curve of degree d. Then $d \neq 12$.

Proof. Suppose that $d = 12$. By Corollary 8.1.6 there exists a surface $S \in \mathcal{Q}_2$ such that $C \subset S$. Moreover, $S \neq \mathscr{S}$ by Lemma 7.2.13(ii), and S is non-singular by Lemma 8.3.3. Put $H_S = H|_S$. Then $H_S^2 = 10$ and $H_S \cdot C = 12$. One also has $C^2 = -2$.

Let us show that the linear system $|5H_S - 2C|$ is free from base components. Indeed, suppose that it is not. Then

$$5H_S - 2C \sim F + M$$

for some effective divisors F and M on S such that the linear system $|M|$ is free from base components, and F is the fixed part of the linear system $|5H_S - 2C|$. In particular, F is an $\mathfrak{A}_5$-invariant divisor. We have $M^2 \geq 0$ and $M \cdot F \geq 0$. Moreover, we must have $F^2 \leq -2$, because

$$1 = h^0(\mathcal{O}_S(F)) \geq 2 + \frac{F^2}{2}$$

by the Rieman–Roch theorem. One has

$$(F + M) \cdot H_S = (5H_S - 2C) \cdot H_S = 26.$$

Therefore $H_S \cdot M \leq 20$, because $H \cdot F \geq 6$ by Corollary 8.1.15. Furthermore, one has $\deg(M) \geq 3$ since the $K3$ surface S is not uniruled. Thus we obtain $H_S \cdot F \leq 23$. One has

$$F^2 \geq -2 \cdot 23^2 = -1058$$

by Lemma 10.1.1. Applying the Hodge index theorem to H_S and M, we get

$$H_S^2 M^2 - (M \cdot H_S)^2 \leq 0,$$

which implies that
$$M^2 \leqslant \frac{(M \cdot H_S)^2}{10} \leqslant 40.$$
On the other hand, by Lemma 8.3.2 we must have
$$10(M \cdot F)^2 - 2(H_S \cdot M)(H_S \cdot F)(M \cdot F)+$$
$$+ (H_S \cdot M)^2 F^2 + (H_S \cdot F)^2 M^2 - 10 M^2 F^2 = 0.$$
Keeping in mind that
$$M^2 + 2M \cdot F + F^2 = (M+F)^2 = (5H_S - 2C)^2 = 2,$$
and checking the possible values of M^2, $M \cdot H$, $M \cdot F$, and F^2 satisfying the above conditions, we get
$$M^2 = 10, \ F^2 = -40, \ H_S \cdot M = 10, \ H_S \cdot F = 16$$
and $M \cdot F = 16$.

We are going to exclude the latter possibility. Let us show first that the divisor F must be reduced. Indeed, suppose that it is not. If F is irreducible, then Corollary 8.1.15 implies that F is reduced since
$$\deg(F) = H_S \cdot F = 16.$$
If F is reducible, then Corollary 8.1.15 implies that $F = F_1 \cup F_2$, where F_1 and F_2 are $\mathfrak{A}_5$-irreducible curves such that $\deg(F_1) = 10$ and $\deg(F_2) = 6$. Applying Corollary 8.1.15 to the curves F_1 and F_2, we see that the curves F_1 and F_2 (and so F itself) are reduced.

Denote by s the number of irreducible components of F. Since F is reduced, we have
$$-40 = F^2 \geqslant -2s \geqslant -2\deg(F) = -32,$$
which is a contradiction.

Therefore, the linear system $|5H_S - 2C|$ is free from base components. We claim that the divisor $5H_S - 2C$ is ample. Indeed, suppose that $5H_S - 2C$ is not ample. Since $(5H_S - 2C)^2 = 2$, it follows from Lemma 6.7.6 that S contains an irreducible curve F such that
$$20 + r(H_S \cdot F) = 2H_S^2 + r(H_S \cdot F) = ((5H_S - 2C) \cdot H_S)^2 = 776$$
for some $r \in \{6, 10, 15\}$. This gives
$$(H_S \cdot F)^2 = \frac{656}{r} \notin \mathbb{Z},$$

which is a contradiction.

Thus, we see that $5H_S - 2C$ is ample, so that

$$h^1(\mathcal{O}_S(5H_S - 2C)) = 0$$

by Kodaira vanishing. Hence

$$h^0(\mathcal{O}_S(5H_S - 2C)) = 3$$

and the linear system $|5H_S - 2C|$ gives an $\mathfrak{A}_5$-equivariant morphism $\eta\colon S \to \mathbb{P}^2$. Since $5H_S - 2C$ is ample, the morphism η is a finite morphism of degree 2 that is branched over a smooth sextic curve $B \subset \mathbb{P}^2$.

Let ς be the biregular involution of S that arises from the double cover η. Put $C' = \varsigma(C)$. Then $H_S \cdot C' = 12$ and $(C')^2 = -2$. Put $k = C \cdot C'$. Then $k \geqslant 0$ provided that $C \neq C'$. Applying Lemma 8.3.2 we conclude that

$$10k^2 - 288k - 616 = 0.$$

This implies that either $k = -2$ or $k = \frac{154}{5}$. But k is an integer. Thus, we have $k = -2$, which implies that $C = C'$.

Since $\varsigma(C) = C$ and $(5H_S - 2C) \cdot C = 64$, we see that $\eta(C)$ is an irreducible curve of degree 32. Thus, we have

$$-2 = C^2 = 2\eta(C)^2 = 2048,$$

which is absurd. The obtained contradiction completes the proof of Lemma 10.1.4. □

Remark 10.1.5. One can give another proof of Lemma 10.1.4 using the lattices associated to $K3$ surfaces. Suppose that there is a smooth rational curve $C \subset V_5$ of degree 12. By Corollary 8.1.6 there exists a surface $S \in \mathcal{Q}_2$ such that $C \subset S$. Moreover, $S \neq \mathscr{S}$ by Lemma 7.2.13(ii), and S is non-singular by Lemma 8.3.3. Put $H_S = H|_S$. Then $H_S^2 = 10$, $H_S \cdot C = 12$ and $C^2 = -2$.

From [55, §10.3] we know that the $\mathfrak{A}_5$-invariant lattice

$$\Lambda = H^2(S, \mathbb{Z})^{\mathfrak{A}_5}$$

is a lattice of rank 4 with intersection form

$$\mathrm{M} = \begin{pmatrix} 20 & -10 & 0 & 0 \\ -10 & 20 & 0 & 0 \\ 0 & 0 & 0 & 1 \\ 0 & 0 & 1 & 0 \end{pmatrix}$$

All vectors $v \in \Lambda$ with $v^2 = -2$ are contained in one $O(\Lambda)$-orbit. One of them is $v_C = (0, 0, 1, -1)$. Thus there exists a vector

$$v_H = (x, y, z, t) \in \Lambda \cong \mathbb{Z}^4$$

such that $v_H \cdot v_C = 12$ and $v_H \cdot v_H = 10$ (here the scalar products are taken with respect to the bilinear form M). The former equality yields $t - z = 12$. The latter equality gives

$$12 = 20x^2 - 20xy + 20y^2 + 2zt = 20(x^2 - xy + y^2) + 2z(z + 12). \quad (10.1.6)$$

Note that $x^2 - xy + y^2 \geqslant 0$ for all $x, y \in \mathbb{Z}$, while the function $\alpha(z) = z(z+12)$ attains its minimum at $z = -6$, and $\alpha(-6) = -36$. Thus we conclude from (10.1.6) that $x^2 - xy + y^2 = r$ and

$$z^2 + 12z + (10r - 6) = 0 \quad (10.1.7)$$

for some $r \in \{0, 1, 2\}$. It is easy to check that for such r equation (10.1.7) has no integer solutions in z, which gives a contradiction.

Lemma 10.1.8. Let C be a smooth irreducible $\mathfrak{A}_5$-invariant curve of genus 5 and degree d in V_5. Then $d \neq 16$.

Proof. The proof is similar to that of Lemma 10.1.4 (although the computations are slightly different). Suppose that $d = 16$. By Corollary 8.1.6 there exists a surface $S \in \mathcal{Q}_2$ that contains C. Moreover, $S \neq \mathscr{S}$ by Lemma 8.1.16(i),(ii), and S is non-singular by Lemma 8.3.3. Put $H_S = H|_S$. Then $H_S^2 = 10$ and $H_S \cdot C = 16$. One also has $C^2 = 8$ by the adjunction formula. Thus we have $(3H_S - C)^2 = 2$.

Let us show that the linear system $|3H_S - C|$ is free from base components. Indeed, suppose that it is not. Then

$$3H_S - C \sim F + M$$

for some effective divisors F and M on S such that the linear system $|M|$ is free from base components, and F is the fixed part of the linear system $|3H_S - C|$. In particular, F is an $\mathfrak{A}_5$-invariant divisor. We have $M^2 \geqslant 0$ and $M \cdot F \geqslant 0$. Moreover, we must have $F^2 \leqslant -2$, because

$$1 = h^0(\mathcal{O}_S(F)) \geqslant 2 + \frac{F^2}{2}$$

by the Riemann–Roch formula. One has

$$(F + M) \cdot H_S = (3H_S - C) \cdot H_S = 14.$$

Therefore $H_S \cdot M \leqslant 8$ since $H \cdot F \geqslant 6$ by Corollary 8.1.15. Furthermore, one has $H_S \cdot M \geqslant 3$ since S is not uniruled. Thus $H_S \cdot F \leqslant 11$. Hence

$$F^2 \geqslant -2 \cdot 11^2 = -242$$

by Lemma 10.1.1. Applying the Hodge index theorem to H_S and M, we get

$$H_S^2 M^2 - (M \cdot H_S)^2 \leqslant 0,$$

which implies that

$$M^2 \leqslant \frac{(M \cdot H_S)^2}{10} \leqslant \frac{64}{10}.$$

Thus we have $M^2 \leqslant 6$. On the other hand, by Lemma 8.3.2 we must have

$$10(M \cdot F)^2 - 2(H_S \cdot M)(H_S \cdot F)(M \cdot F) + \\ + (H_S \cdot M)^2 F^2 + (H_S \cdot F)^2 M^2 - 10 M^2 F^2 = 0.$$

Keeping in mind that

$$M^2 + 2M \cdot F + F^2 = (M + F)^2 = (3H_S - C)^2 = 2,$$

and checking the possible values of M^2, $M \cdot H_S$, $M \cdot F$, and F^2 satisfying the above conditions, we obtain a contradiction (cf. the proof of Lemma 10.1.4).

Therefore, the linear system $|3H_S - C|$ is free from base components. Since $(3H_S - C)^2 = 2$, Theorem 7.3.5 implies that $|3H_S - C|$ is free from base points.

We claim that the divisor $3H_S - C$ is ample. Indeed, suppose that it is not ample. Since $(3H_S - C)^2 = 2$, it follows from Lemma 6.7.6 that S contains an irreducible curve F such that

$$20 + r(H_S \cdot F) = 2H_S^2 + r(H_S \cdot F) = \big((3H_S - C) \cdot H_S\big)^2 = 196$$

for some $r \in \{6, 10, 15\}$. This gives

$$(H_S \cdot F)^2 = \frac{176}{r} \notin \mathbb{Z},$$

which is a contradiction.

Thus, we see that $3H_S - C$ is ample, so that

$$h^1\big(\mathcal{O}_S(3H_S - C)\big) = 0$$

by Kodaira vanishing. Hence

$$h^0\big(\mathcal{O}_S(3H_S - C)\big) = 3$$

and the linear system $|3H_S - C|$ gives an $\mathfrak{A}_5$-equivariant morphism $\eta\colon S \to \mathbb{P}^2$. Since $3H_S - C$ is ample, the morphism η is a finite morphism of degree 2 that is branched over a smooth sextic curve $B \subset \mathbb{P}^2$.

Let ς be the biregular involution of S that arises from the double cover η. Put $C' = \varsigma(C)$. Then $H_S \cdot C' = 16$ and $(C')^2 = 8$. Put $k = C \cdot C'$. Applying Lemma 8.3.2, we conclude that

$$10k^2 - 512k + 3456 = 0.$$

This implies that either $k = 8$ or $k = \frac{216}{5}$. But k is an integer. Thus, we have $k = 8$. Since $(3H_S - C) \cdot C = 40$, we have $\eta(C)^2 = 1600$, and

$$32 = (C + C')^2 = 2\eta(C + C')^2 = 2\eta(C)^2 = 3200,$$

which is absurd. The obtained contradiction completes the proof of Lemma 10.1.8. $\square$

10.2 Preliminary classification of low degree curves

The assertions proved in Chapter 8 and §10.1 allow us to advance slightly further in the explicit description of low degree $\mathfrak{A}_5$-irreducible curves on V_5. We start with describing $\mathfrak{A}_5$-irreducible curves of degree 10.

Lemma 10.2.1. *Let Z be an $\mathfrak{A}_5$-irreducible curve of degree 10 in V_5. Suppose that Z has r irreducible components of degree d. Then one of the following possibilities occurs:*

(i) *$r = 10$ and $d = 1$, so that $Z = \mathcal{L}_{10}$;*

(ii) *$r = 5$ and $d = 2$, so that $Z = \mathcal{G}_5$ or $Z = \mathcal{G}'_5$.*

Proof. Suppose that $r > 1$. Since $\deg(Z) = 10$, either $r = 10$ and $d = 1$, or $r = 5$ and $d = 2$ by Corollary 5.1.2. The assertion in these cases follows from Propositions 7.4.1 and 7.8.4.

Therefore, we may suppose that $r = 1$, so that Z is an irreducible curve of degree $d = 10$. We are going to show that this is impossible. By Corollary 8.1.6 there exists a surface $S \in \mathcal{Q}_2$ such that $Z \subset S$. By Lemma 7.2.13(i) one has $S \neq \mathscr{S}$.

Let us show that S is smooth. Indeed, suppose that S is singular. Then S is a $K3$ surface with isolated ordinary double points by Theorem 8.2.1(v).

Moreover, it follows from Lemma 6.7.3(i),(ii) that $\operatorname{rk}\operatorname{Pic}(S)^{\mathfrak{A}_5} = 1$. Since Z is a curve of degree 10, we have $Z \sim_{\mathbb{Q}} H_S$, where $H_S = H|_S$. In particular, we have $Z^2 = H_S^2 = 10$.

Denote by $O_1, O_2, \ldots, O_r$ the singular points of the surface S. Then

$$r \in \{5, 10, 15\},$$

and $\{O_1, \ldots, O_r\}$ is a single $\mathfrak{A}_5$-orbit (see Theorem 8.2.1(ii),(iii)). Let $\theta \colon \tilde{S} \to S$ be the minimal resolution of singularities of the surface S, i.e., the blow-up of all points $O_1, O_2, \ldots, O_r$, let $F_1, F_2, \ldots, F_r$ be θ-exceptional divisors such that $\theta(F_i) = O_i$, and let $\tilde{Z}$ be the proper transform of the curve Z on the surface $\tilde{S}$. Then

$$\tilde{Z} \sim_{\mathbb{Q}} \theta^*(Z) - m \sum_{i=1}^{r} F_i$$

for some rational number m. Since

$$\tilde{Z} \cdot F_i = m F_i^2 = -2m,$$

we see that m is a non-negative half-integer that is strictly positive if Z passes through singular points of the surface S. Moreover,

$$\tilde{Z}^2 = 10 - m^2 \sum_{i=1}^{r} F_i^2 = 10 - 2rm^2,$$

and $\tilde{Z}^2$ is an even integer that is greater or equal to -2, since $\tilde{S}$ is a smooth $K3$ surface. If $m \geqslant 3/2$, then

$$-2 \leqslant \tilde{Z}^2 = 10 - 2rm^2 \leqslant 10 - \frac{9r}{2} \leqslant 10 - \frac{45}{2} = -\frac{25}{2},$$

which is absurd. Thus, we see that

$$m \in \{0, \frac{1}{2}, 1\}.$$

If $m = 1$, then

$$-2 \leqslant \tilde{Z}^2 = 10 - 2rm^2 = 10 - 2r,$$

which implies that $r = 5$ and $\tilde{Z}^2 = 0$. In this case, one has $p_a(\tilde{Z}) = 1$, which implies that $\tilde{Z}$ is smooth by Lemma 4.4.6, because $\tilde{Z} \subset \tilde{S}$ does not contain

$\mathfrak{A}_5$-invariant points by Theorem 7.3.5. Since $\mathfrak{A}_5$ cannot act on a smooth elliptic curve in a non-trivial way (see Lemma 5.1.5), we see that $m \neq 1$.

If $m = 1/2$, then
$$-2 \leqslant \tilde{Z}^2 = 10 - \frac{r}{2},$$
which implies that r must be even, so that $r = 10$ and $\tilde{Z}^2 = 5$. The latter is impossible, since $\tilde{Z}^2$ is an even number.

Thus, we see that $m = 0$, i.e., Z does not pass through singular points of the surface S. In this case $\tilde{Z} \sim_{\mathbb{Q}} \theta^*(H_S)$, so that $\tilde{Z} \sim \theta^*(H_S)$, because $\operatorname{Pic}(\tilde{S})^{\mathfrak{A}_5}$ has no torsion (see Lemma 6.7.3(i)). Thus $Z \sim H_S$. The latter contradicts Lemma 8.2.6(i).

Therefore, we see that S is smooth, and it follows from the Hodge index theorem that $Z^2 \leqslant 10$. Thus, one has $p_a(Z) \leqslant 6$ by the adjunction formula. Since S does not contain the $\mathfrak{A}_5$-orbit Σ_5 by Theorem 8.2.1(iv), we see that the curve Z must be smooth by Corollary 7.3.6. Hence the genus g of the curve Z does not exceed 4 by the Castelnuovo bound (see [3, §III.2]). Therefore, one has $g \in \{0, 4\}$ by Lemma 5.1.5.

Recall that $Z \not\subset \mathscr{S}$ by Lemma 7.2.13(i). Thus the intersection $Z \cap \mathscr{S}$ consists of 20 points (counted with multiplicities). By Lemma 5.1.4 these points are distinct, so that $Z \cap \mathscr{S}$ is a single $\mathfrak{A}_5$-orbit of length 20. In particular, one has $g \neq 4$ by Lemma 5.1.5. Moreover, all $\mathfrak{A}_5$-orbits of length 12 contained in V_5 are contained in $\mathscr{S}$ by Theorems 7.3.5. Hence the case $g = 0$ is also impossible, since otherwise Z would contain an $\mathfrak{A}_5$-orbit of length 12 by Lemma 5.5.9(v), and thus $Z \cap \mathscr{S}$ would also contain an orbit of length 12, which is a contradiction. □

The next step is to describe $\mathfrak{A}_5$-irreducible curves of degree 12.

Lemma 10.2.2. Let Z be an $\mathfrak{A}_5$-irreducible curve of degree 12 in V_5. Suppose that Z has r irreducible components of degree d. Then one of the following possibilities occurs:

(i) $r = 12$ and $d = 1$, so that $Z = \mathcal{L}_{12}$;

(ii) $r = 6$ and $d = 2$, so that $Z = \mathcal{G}_6$.

Proof. Suppose that $r > 1$. One has $rd = 12$ and $r \in \{5, 6, 10, 12\}$ by Corollary 5.1.2. Hence either $r = 12$ and $d = 1$, or $r = 6$ and $d = 2$. In the former case the assertion follows from Proposition 7.4.1, and in the latter case it follows by Proposition 7.8.4.

Therefore, we may suppose that $r = 1$, so that Z is an irreducible curve. We are going to show that this is impossible. By Corollary 8.1.6 there exists a surface $S \in \mathcal{Q}_2$ such that $Z \subset S$. Moreover, $S \neq \mathscr{S}$ by Lemma 7.2.13(ii), and thus S is non-singular by Lemma 8.3.3. In particular, Z does not pass through the $\mathfrak{A}_5$-orbit Σ_5, since $\Sigma_5 \not\subset S$ by Theorem 8.2.1(iv).

It follows from the Hodge index theorem that $Z^2 \leqslant 14$, which implies that $p_a(Z) \leqslant 8$ by the adjunction formula. Therefore, Z is smooth by Corollary 7.3.6. Let g be the genus of the curve Z. Then Lemma 5.1.5 implies that $g \in \{0, 4, 5, 6\}$. Note that $g \neq 4$ by Lemma 10.1.2 and $g \neq 5$ by Lemma 10.1.3. Also $g \neq 0$ by Lemma 10.1.4.

Thus, we see that $g = 6$. Recall that the base locus of $\mathcal{Q}_2$ is the irreducible curve C_{20} by Theorem 8.1.8(i). Thus, the curve Z is not contained in C_{20}. Moreover, the intersection $Z \cap C_{20}$ is an $\mathfrak{A}_5$-invariant set that consists of
$$2 \deg(Z) = 24$$
points (counted with multiplicities). Lemma 5.1.4 implies that Z contains an $\mathfrak{A}_5$-orbit of length 12, which is impossible by Lemma 5.1.5. □

Now we will describe $\mathfrak{A}_5$-irreducible curves of degree 15.

Lemma 10.2.3. *Let Z be an $\mathfrak{A}_5$-irreducible curve of degree 15 in V_5. Then Z is reducible.*

Proof. Assume that Z is irreducible. Then there exists a surface $S \in \mathcal{Q}_2$ that contains Z by Corollary 8.1.6. Let S' be a general surface in $\mathcal{Q}_2$. By Theorem 8.1.8(i), the intersection $S \cap S'$ is an irreducible curve C_{20}. Note that $Z \neq C_{20}$. Hence,
$$Z \cap S' = Z \cap C_{20} \subset C_{20}.$$

On the other hand, any $\mathfrak{A}_5$-orbit contained in the curve C_{20} is of length 12, 20, or 60 by Theorem 8.1.8(vii). This is impossible, since $Z \cdot S'$ is an effective $\mathfrak{A}_5$-invariant zero-cycle on V_5 of degree 30 whose support is contained in C_{20}. □

Lemma 10.2.4. *Let Z be an $\mathfrak{A}_5$-irreducible curve of degree 15 in V_5. Suppose that Z has r irreducible components of degree d. Then $r = 15$ and $d = 1$, so that $Z = \mathcal{L}_{15}$.*

Proof. One has $r > 1$ by Lemma 10.2.3. Thus Corollary 5.1.2 implies that either $r = 15$ and $d = 1$, or $r = 5$ and $d = 3$. In the former case Z is the curve $\mathcal{L}_{15}$ by Proposition 7.4.1. In the latter case each irreducible component C_i, $1 \leqslant i \leqslant 5$, of Z is a twisted cubic, since V_5 is an intersection of quadrics. Let us assume the latter case and derive a contradiction to show that it is impossible.

Consider the rational $\mathfrak{A}_5$-invariant map

$$\theta_{\Pi_3} : V_5 \dashrightarrow \mathbb{P}^2 \cong \mathbb{P}(W_3)$$

given by the projection from the subspace $\Pi_3 \subset \mathbb{P}^6$ (see §7.5). If $\Sigma_5 \subset Z$, then for any point $P \in \Sigma_5$ there are at least 3 irreducible components of Z passing through P by Lemma 7.5.11. In this case the image $\theta_{\Pi_3}(Z) \subset \mathbb{P}^2$ is a union of at most 5 points, which is impossible by Lemma 5.3.1(i). Hence one has $\Sigma_5 \not\subset Z$, and the image $\theta_{\Pi_3}(Z) \subset \mathbb{P}^2$ is either a union of at most 5 lines, or a union of at most 5 rational plane cubic curves. The former case is impossible by Lemma 5.3.1(ii). In the latter case each of the curves

$$\theta_{\Pi_3}(C_i) \subset \mathbb{P}^2, \quad 1 \leqslant i \leqslant 5,$$

has a unique singular point. Thus there exists an $\mathfrak{A}_5$-invariant subset of $\mathbb{P}^2$ that consists of at most 5 points, which contradicts Lemma 5.3.1(i). □

We summarize the information on the low degree $\mathfrak{A}_5$-invariant curves on V_5 that is currently available to us in the following lemma (cf. Lemmas 7.5.6 and 8.1.14, and also Theorem 13.6.1 below).

Proposition 10.2.5. Let Z be an $\mathfrak{A}_5$-irreducible curve of degree $\deg(Z) \leqslant 15$ in V_5. Suppose that Z has r irreducible components of degree d. Then one of the following possibilities occurs:

(i) $\deg(Z) = 6$, $r = 1$, and $d = 6$, so that Z is a sextic rational normal curve (cf. Remark 7.5.7);

(ii) $\deg(Z) = 6$, $r = 6$, and $d = 1$, so that $Z = \mathcal{L}_6$;

(iii) $\deg(Z) = 10$, $r = 10$, and $d = 1$, so that $Z = \mathcal{L}_{10}$;

(iv) $\deg(Z) = 10$, $r = 5$, and $d = 2$, so that $Z = \mathcal{G}_5$ or $Z = \mathcal{G}'_5$;

(v) $\deg(Z) = 12$, $r = 12$, and $d = 1$, so that $Z = \mathcal{L}_{12}$;

(vi) $\deg(Z) = 12$, $r = 6$, and $d = 2$, so that $Z = \mathcal{G}_6$;

(vii) $\deg(Z) = 15$, $r = 15$, and $d = 1$, so that $Z = \mathcal{L}_{15}$.

Proof. By Lemma 8.1.14 one has $\deg(Z) \in \{6, 10, 12, 15\}$. If $\deg(Z) = 6$, the assertion follows from Lemma 7.5.6. If $\deg(Z) = 10$, the assertion follows from Lemma 10.2.1. If $\deg(Z) = 12$, the assertion follows from Lemma 10.2.2. Finally, if $\deg(Z) = 15$, the assertion follows from Lemma 10.2.4. $\square$

Lemma 10.2.6. *The curve $\mathcal{L}_{15}$ is not contained in any surface of the pencil $\mathcal{Q}_2$. Each of the curves $\mathscr{C}$ and $\mathcal{L}_{12}$ is contained in $\mathscr{S}$ and is not contained in any surface of the pencil $\mathcal{Q}_2$ different from $\mathscr{S}$.*

Proof. The first assertion follows from Lemma 7.5.13. The curves $\mathscr{C}$ and $\mathcal{L}_{12}$ are contained in the surface $\mathscr{S}$ by construction. Thus the second assertion of the lemma follows from Corollary 8.1.9. $\square$

Lemma 10.2.7. *Let Z be an $\mathfrak{A}_5$-irreducible curve of degree $\deg(Z) \leqslant 15$ in V_5, and suppose that Z is different from $\mathscr{C}$, $\mathcal{L}_{12}$, and $\mathcal{L}_{15}$. Then there exists a unique surface $S \in \mathcal{Q}_2$ that contains Z. Moreover, the surface S is smooth and does not contain any other $\mathfrak{A}_5$-invariant curve of degree at most 15.*

Proof. Corollary 8.1.7 implies that there is a surface $S \in \mathcal{Q}_2$ that contains Z. Furthermore, Lemma 7.2.13(ii) implies that $S \neq \mathscr{S}$. Moreover, S is unique by Corollary 8.1.9.

We claim that S is smooth. Indeed, suppose that S is not smooth. Then $\operatorname{rk}\operatorname{Pic}(S)^{\mathfrak{A}_5} = 1$ by Lemma 6.7.3(i),(ii), and $\deg(Z)$ is divisible by 5 by Lemma 8.3.3. Thus, it follows from Proposition 10.2.5 that either $Z = \mathcal{L}_{10}$, or $Z = \mathcal{G}_5$, or $Z = \mathcal{G}'_5$. By Lemma 9.1.1, Corollaries 9.1.3 and 9.1.5, Lemma 9.2.1, and Corollary 9.2.4, the curve Z does not contain $\mathfrak{A}_5$-orbits Σ_5, Σ_{10}, Σ'_{10}, and Σ_{15}. Thus, Z must be contained in a smooth locus of the surface S by Theorem 8.2.1(ii),(iii). This implies that $Z^2 < 0$, because Z is a disjoint union of smooth rational curves by Lemmas 9.1.2 and 9.2.3. The latter contradicts the fact that $\operatorname{rk}\operatorname{Pic}(S)^{\mathfrak{A}_5} = 1$. Thus, we see that S is smooth.

Recall that by Lemma 7.5.6 there exists at most one $\mathfrak{A}_5$-invariant irreducible curve of degree 6 different from $\mathscr{C}$. By a small abuse of notation we denote this curve by $\mathscr{C}'$ assuming that it exists. We will see in Lemma 11.3.2 that it does exist, but this is not important here because the remaining part of the proof works under the assumption that $\mathscr{C}'$ does not exist as well.

By Lemmas 7.4.7, 9.1.2, 9.2.3, and 9.2.8, the curve Z is a union of 6 (respectively, 10, 5, and 6) smooth rational curves provided that Z is as in case (ii) of Proposition 10.2.5 (respectively, case (iii), (iv), and (vi)).

Note that $Z^2 = 2p_a(Z) - 2$ by the adjunction formula. Using this together with the classification from Proposition 10.2.5, we obtain the list of possible values of degrees and self-intersection numbers for the curve $Z \subset S$ given in Table 10.1.

Table 10.1: Intersection numbers for low degree curves

	$\mathscr{C}'$	$\mathcal{L}_6$	$\mathcal{L}_{10}$	$\mathcal{G}_5$	$\mathcal{G}_5'$	$\mathcal{G}_6$
$\deg(Z)$	6	6	10	10	10	12
Z^2	-2	-12	-20	-10	-10	-12

Suppose that S contains another $\mathfrak{A}_5$-irreducible curve Z' of degree at most 15. Note that Z' is different from $\mathscr{C}$, $\mathcal{L}_{12}$, and $\mathcal{L}_{15}$ by Lemma 10.2.6. Thus Z' is also one of the curves listed in Table 10.1, and the degree of Z' and the self-intersection of Z' are as in Table 10.1.

Going through all possible cases listed in Table 10.1 and applying Lemma 8.3.2, we obtain the only a priori possible case (up to permutation): $Z = \mathcal{G}_5$, $Z' = \mathcal{G}_5'$, and $Z \cdot Z' = 30$. Let us show that this case is also impossible.

Suppose that $Z = \mathcal{G}_5$ and $Z' = \mathcal{G}_5'$. Put $H_S = H|_S$. Then $Z + Z'$ is numerically equivalent to $2H_S$. Indeed, they have the same intersection numbers with H_S and Z, and the latter two classes generate $\mathrm{Pic}(S)^{\mathfrak{A}_5} \otimes \mathbb{Q}$ by Lemma 6.7.3(i). Since $\mathrm{Pic}(S)^{\mathfrak{A}_5}$ does not have torsion (see Lemma 6.7.3(i)), this implies that $Z + Z' \sim 2H_S$. Thus, $Z + Z' \subset S$ and $C_{20} \subset S$ are both contained in $|2H_S|$, which is impossible by Lemma 8.2.6(ii). The obtained contradiction completes the proof. $\square$

A first application of Lemma 10.2.7 is the following information about intersections among the curves $\mathcal{L}_{10}$, $\mathcal{G}_5$, and $\mathcal{G}_5'$.

Corollary 10.2.8. One has

$$\mathcal{G}_5 \cap \mathcal{G}_5' = \mathcal{G}_5 \cap \mathcal{L}_{10} = \mathcal{G}_5' \cap \mathcal{L}_{10} = \Sigma_{20}'.$$

Proof. By Lemma 10.2.7 there is a surface $S_{\mathcal{G}_5} \in \mathcal{Q}_2$ such that $\mathcal{G}_5 \subset S_{\mathcal{G}_5}$, while the curves $\mathcal{L}_{10}$ and $\mathcal{G}_5'$ are not contained in $S_{\mathcal{G}_5}$. Similarly, there is a surface $S_{\mathcal{G}_5'} \in \mathcal{Q}_2$ such that $\mathcal{G}_5' \subset S_{\mathcal{G}_5'}$, while the curves $\mathcal{L}_{10}$ and $\mathcal{G}_5$ are not contained in $S_{\mathcal{G}_5'}$. Each of the intersections $S_{\mathcal{G}_5} \cap \mathcal{G}_5'$, $S_{\mathcal{G}_5} \cap \mathcal{L}_{10}$, and $S_{\mathcal{G}_5'} \cap \mathcal{L}_{10}$

consists of 20 points (counted with multiplicities). On the other hand, we have

$$\Sigma'_{20} \subset \mathcal{G}_5 \cap \mathcal{L}_{10} \subset S_{\mathcal{G}_5} \cap \mathcal{L}_{10}, \quad \Sigma'_{20} \subset \mathcal{G}'_5 \cap \mathcal{L}_{10} \subset S_{\mathcal{G}'_5} \cap \mathcal{L}_{10}$$

and

$$\Sigma'_{20} \subset \mathcal{G}_5 \cap \mathcal{G}'_5 \subset S_{\mathcal{G}_5} \cap \mathcal{G}'_5$$

by Remarks 9.1.6 and 9.2.5. □

Chapter 11
Two Sarkisov links

Recall that $\mathrm{Aut}(V_5) \cong \mathrm{PSL}_2(\mathbb{C})$ and $\mathrm{Aut}^{\mathfrak{A}_5}(V_5) \cong \mathfrak{A}_5$ (see Proposition 7.1.10). The purpose of this chapter is to describe two $\mathfrak{A}_5$-Sarkisov links that both start and end with V_5 (see §11.4). The first link and $\mathrm{Aut}^{\mathfrak{A}_5}(V_5)$ generate a subgroup in $\mathrm{Bir}^{\mathfrak{A}_5}(V_5)$ isomorphic to $\mathfrak{S}_5$. The second link and $\mathrm{Aut}^{\mathfrak{A}_5}(V_5)$ generate a subgroup in $\mathrm{Bir}^{\mathfrak{A}_5}(V_5)$ isomorphic to $\mathfrak{A}_5 \times \boldsymbol{\mu}_2$. Together these two subgroups generate a subgroup in $\mathrm{Bir}^{\mathfrak{A}_5}(V_5)$ isomorphic to $\mathfrak{S}_5 \times \boldsymbol{\mu}_2$. We will see later in Chapter 18 that $\mathrm{Bir}^{\mathfrak{A}_5}(V_5)$ coincides with this subgroup.

We will use partial information about the first $\mathfrak{A}_5$-Sarkisov link to prove that there exists an $\mathfrak{A}_5$-invariant rational normal sextic curve $\mathscr{C}' \subset V_5$ such that $\mathscr{C}' \neq \mathscr{C}$ (see §11.3). Then we will use the curve $\mathscr{C}'$ in §11.4 to construct the second link. We will also prove that the sextic curve $\mathscr{C}'$, the sextic curve $\mathscr{C}$, and the reducible sextic curve $\mathcal{L}_6$ that is a union of six lines are the only $\mathfrak{A}_5$-invariant curves of degree 6 in V_5 (see Lemma 11.3.2).

11.1 Anticanonical divisors through the curve $\mathcal{L}_6$

Let $\mathcal{Q}(\mathcal{L}_6)$ be the linear subsystem in $|2H|$ consisting of all surfaces that contain the curve $\mathcal{L}_6$. Let $L_1, \ldots, L_6$ be the lines of $\mathcal{L}_6$.

Lemma 11.1.1. *The linear system $\mathcal{Q}(\mathcal{L}_6)$ does not have fixed components.*

Proof. For each line L_i of $\mathcal{L}_6$, let us fix 3 sufficiently general points P_i, O_i, and Q_i in L_i. Put

$$\Xi = \{P_1, O_1, Q_1, P_2, O_2, Q_2, \ldots, P_6, O_6, Q_6, \}.$$

Then $|\Xi| = 18$. On the other hand, $h^0(\mathcal{O}_{V_5}(2H)) = 23$ by (7.1.3). Thus the linear subsystem in $|2H|$ that consists of all surfaces passing through Ξ is at least four-dimensional. Every surface $M \in |2H|$ that passes through Ξ must contain $\mathcal{L}_6$, since otherwise we would have

$$2 = M \cdot L_i \geqslant \mathrm{mult}_{P_i}(M) + \mathrm{mult}_{O_i}(M) + \mathrm{mult}_{Q_i}(M) \geqslant 3.$$

Thus, the dimension of the linear system $\mathcal{Q}(\mathcal{L}_6)$ is at least 4, which implies that the linear system $\mathcal{Q}(\mathcal{L}_6)$ does not have fixed components by Lemma 8.1.3. □

The linear system $\mathcal{Q}(\mathcal{L}_6)$ contains an $\mathfrak{A}_5$-invariant surface by Corollary 8.1.7. Denote it by $S_{\mathcal{L}_6}$. By Lemmas 7.2.13(i) and 8.3.3, the surface $S_{\mathcal{L}_6}$ is a smooth $K3$ surface.

Lemma 11.1.2. *The divisor $2H|_{S_{\mathcal{L}_6}} - \mathcal{L}_6$ is very ample on $S_{\mathcal{L}_6}$.*

Proof. Put $H_{S_{\mathcal{L}_6}} = H|_{S_{\mathcal{L}_6}}$ and $D = 2H_{S_{\mathcal{L}_6}} - \mathcal{L}_6$. Then $\mathcal{L}_6^2 = -12$ by Lemma 7.4.7, so that

$$D^2 = (2H_{S_{\mathcal{L}_6}} - \mathcal{L}_6)^2 = 4H_{S_{\mathcal{L}_6}}^2 - 4\deg(\mathcal{L}_6) + \mathcal{L}_6^2 = 40 - 24 - 12 = 4.$$

Since $S_{\mathcal{L}_6}$ is a smooth $K3$ surface, we have

$$h^2(\mathcal{O}_{S_{\mathcal{L}_6}}(D)) = h^0(\mathcal{O}_{S_{\mathcal{L}_6}}(-D)) = 0$$

by Serre duality, and the Riemann–Roch formula implies that

$$h^0(\mathcal{O}_{S_{\mathcal{L}_6}}(D)) = \frac{D^2}{2} + 2 + h^1(\mathcal{O}_{S_{\mathcal{L}_6}}(D)) \geqslant \frac{D^2}{2} + 2 = 4. \tag{11.1.3}$$

In particular, the mobile part of the linear system $|D|$ is non-trivial.

We claim that the linear system $|D|$ does not have fixed components. Indeed, suppose that it does. Let us denote by Z the fixed part of $|D|$. Then Z must be $\mathfrak{A}_5$-invariant. On the other hand, we have

$$Z \cdot H_{S_{\mathcal{L}_6}} \leqslant D \cdot H_{S_{\mathcal{L}_6}} = 14.$$

Thus, Lemma 10.2.7 implies that $\mathrm{Supp}(Z) = \mathcal{L}_6$. Therefore, we see that the linear system $|2H_{S_{\mathcal{L}_6}} - m\mathcal{L}_6|$ is free from base curves for some integer $m \geqslant 2$. Then the divisor $2H_{S_{\mathcal{L}_6}} - m\mathcal{L}_6$ is nef. On the other hand, we have

$$\left(2H_{S_{\mathcal{L}_6}} - m\mathcal{L}_6\right)^2 = 40 - 24m - 12m^2 < 0,$$

because $m \geqslant 2$. The obtained contradiction shows that the linear system $|D|$ does not have fixed components.

Since $D^2 = 4$ and the linear system $|D|$ does not have fixed components, Theorem 7.3.5 implies that $|D|$ is free from base points (this also follows from [109, Corollary 3.2]).

Let us show that D is ample. Suppose it is not. By Lemma 6.7.6, there exists a smooth rational curve F on the surface $S_{\mathcal{L}_6}$ such that

$$D^2\left(2H_{S_{\mathcal{L}_6}}^2 + r(H_{S_{\mathcal{L}_6}} \cdot F)^2\right) = 2(H_{S_{\mathcal{L}_6}} \cdot D)^2$$

for some $r \in \{1, 5, 6, 10, 12, 15\}$. This gives

$$4\big(20 + r(H_{S_{\mathcal{L}_6}} \cdot F)^2\big) = 392.$$

Hence, one has $r(H_{S_{\mathcal{L}_6}} \cdot F)^2 = 78$ for some $r \in \{1, 5, 6, 10, 12, 15\}$, which is absurd.

Thus, we see that D is ample and $|D|$ is free from base points, so that $h^1(\mathcal{O}_{S_{\mathcal{L}_6}}(D)) = 0$ by Kodaira vanishing. Hence, (11.1.3) implies $h^0(\mathcal{O}_{S_{\mathcal{L}_6}}(D)) = 4$.

We must show that D is very ample. Suppose it is not. Then it follows from Lemma 6.7.7 that

$$4k^2 - 56k + 40 = k^2 D^2 - 4kD \cdot H_S + 4H_S^2 = 0$$

for some positive integer k. However, the latter equation does not have integer solutions. Thus, D must be very ample. $\square$

Let $\pi\colon \mathcal{W} \to V_5$ be a blow-up of the lines $L_1, \ldots, L_6$, and let $E_1, \ldots, E_6$ be the exceptional divisors of π such that

$$\pi(E_1) = L_1, \ldots, \pi(E_6) = L_6.$$

Denote by $\check{S}_{\mathcal{L}_6}$ the proper transform of the surface $S_{\mathcal{L}_6}$ on the threefold $\mathcal{W}$. Then

$$-K_\mathcal{W} \sim \pi^*(2H) - \sum_{i=1}^{6} E_i \sim \check{S}_{\mathcal{L}_6},$$

because $S_{\mathcal{L}_6}$ is smooth. This implies that $|-K_\mathcal{W}|$ is the proper transform of the linear system $\mathcal{Q}(\mathcal{L}_6)$ on the threefold $\mathcal{W}$ via π. Since $\mathcal{Q}(\mathcal{L}_6)$ does not have fixed components by Lemma 11.1.1, the linear system $|-K_\mathcal{W}|$ does not have fixed components either.

The induced morphism

$$\pi|_{\check{S}_{\mathcal{L}_6}}\colon \check{S}_{\mathcal{L}_6} \to S_{\mathcal{L}_6}$$

is an isomorphism. Thus, Lemma 11.1.2 implies:

Corollary 11.1.4. The surface $\check{S}_{\mathcal{L}_6}$ is a smooth $K3$ surface, and the divisor $-K_\mathcal{W}|_{\check{S}_{\mathcal{L}_6}}$ is very ample.

Proof. Since $\pi|_{\check{S}_{\mathcal{L}_6}} \colon \check{S}_{\mathcal{L}_6} \to S_{\mathcal{L}_6}$ is an isomorphism, the first assertion follows from the fact that $S_{\mathcal{L}_6}$ is a smooth $K3$ surface. Moreover, one has

$$-K_\mathcal{W}|_{\check{S}_{\mathcal{L}_6}} \sim (\pi|_{\check{S}_{\mathcal{L}_6}})^*\Big(2H|_{S_{\mathcal{L}_6}} - \mathcal{L}_6\Big),$$

so that the second assertion is implied by Lemma 11.1.2. □

In particular, Corollary 11.1.4 implies that the linear system $|-K_\mathcal{W}|_{\check{S}_{\mathcal{L}_6}}|$ does not have base points. Moreover, we have

$$h^0\big(\mathcal{O}_{\check{S}_{\mathcal{L}_6}}(-K_\mathcal{W}|_{\check{S}_{\mathcal{L}_6}})\big) = 4.$$

The latter follows from the Riemann–Roch formula and Kodaira vanishing, since the divisor $-K_\mathcal{W}|_{\check{S}_{\mathcal{L}_6}}$ is very ample by Corollary 11.1.4. On the other hand, the exact sequence of sheaves

$$0 \to \mathcal{O}_\mathcal{W} \to \mathcal{O}_\mathcal{W}(-K_\mathcal{W}) \to \mathcal{O}_{\check{S}_{\mathcal{L}_6}} \otimes \mathcal{O}_\mathcal{W}(-K_\mathcal{W}) \to 0$$

implies the exact sequence of cohomology groups

$$0 \to H^0(\mathcal{O}_\mathcal{W}) \to H^0\big(\mathcal{O}_\mathcal{W}(-K_\mathcal{W})\big) \to H^0\Big(\mathcal{O}_{\check{S}_{\mathcal{L}_6}}\big(-K_\mathcal{W}|_{\check{S}_{\mathcal{L}_6}}\big)\Big) \to 0,$$
(11.1.5)

because $h^1(\mathcal{O}_\mathcal{W}) = h^1(\mathcal{O}_{V_5}) = 0$ by Kodaira vanishing. Since $h^0(\mathcal{O}_\mathcal{W}) = 1$, the exact sequence (11.1.5) and Corollary 11.1.4 imply:

Corollary 11.1.6. The linear system $|-K_\mathcal{W}|$ is free from base points, and $h^0(\mathcal{O}_\mathcal{W}(-K_\mathcal{W})) = 5$.

Recall that the normal bundle

$$\mathcal{N}_{L_i/V_5} \cong \mathcal{O}_{L_i} \oplus \mathcal{O}_{L_i}$$

by Theorem 7.7.1, since $\mathcal{L}_6 \not\subset \mathscr{S}$ by Lemma 7.2.13(i). Therefore, one has

$$E_i \cong \mathbb{P}^1 \times \mathbb{P}^1.$$

Let $l_i \subset E_i$ be a fiber of the projection

$$\pi_{E_i} \colon E_i \to L_i,$$

and s_i be a section of this projection such that $s_i^2 = 0$. Then
$$-E_i|_{E_i} \sim s_i, \quad E_i \cdot E_j = 0 \text{ for } i \neq j.$$
Furthermore, one has the following intersection numbers on $\mathcal{W}$:
$$\pi^*(H)^3 = 5, \quad \pi^*(H)^2 \cdot E_i = 0, \quad \pi^*(H) \cdot E_i^2 = -1, \quad E_i^3 = 0. \quad (11.1.7)$$
In particular, we compute
$$-K_\mathcal{W}^3 = \left(\pi^*(2H) - \sum_{i=1}^{6} E_i\right)^3 = 4.$$

Corollary 11.1.8. The divisor $-K_\mathcal{W}$ is nef and big. Moreover, the divisor
$$-K_\mathcal{W}|_{E_i} \sim s_i + 2l_i$$
is ample.

Let $G_{L_i} \cong \mathfrak{D}_{10}$ be the stabilizer of the line L_i in $\mathfrak{A}_5$ (see Lemma 5.1.1). We can determine the action of G_{L_i} on the exceptional divisor E_i.

Lemma 11.1.9. The action of G_{L_i} on $E_i \cong \mathbb{P}^1 \times \mathbb{P}^1$ is twisted diagonal (see §6.4).

Proof. Recall that G_{L_i} acts on L_i faithfully by Lemma 7.4.5. Thus, it acts on E_i faithfully as well.

Put $C = \check{S}_{\mathcal{L}_6} \cap E_i$. Then $C \sim -K_\mathcal{W}|_{E_i}$ is a divisor of bi-degree $(1,2)$ (see Corollary 11.1.8). Moreover, since the surface $S_{\mathcal{L}_6} \subset V_5$ is smooth, we see that C is actually an irreducible curve. Now the assertion follows from Lemma 6.4.12. □

Let $\tilde{\mathcal{Q}}_\mathcal{W} \subset H^0(\mathcal{O}_{V_5}(2H))$ be the linear subspace of sections vanishing along $\mathcal{L}_6$, i.e., the linear subspace that corresponds to the linear system $\mathcal{Q}(\mathcal{L}_6)$.

Lemma 11.1.10. There is an isomorphism of $\mathfrak{A}_5$-representations
$$\tilde{\mathcal{Q}}_\mathcal{W} \cong I \oplus W_4.$$

Proof. By Corollary 11.1.6, one has $\dim(\tilde{\mathcal{Q}}_\mathcal{W}) = 5$. By Lemma 8.1.1, the $\mathfrak{A}_5$-representation $\tilde{\mathcal{Q}}_\mathcal{W}$ may be isomorphic either to W_5, or to $I^{\oplus 2} \oplus W_3'$, or to $I \oplus W_4$. The former two cases are impossible since there exists a surface $S_{\mathcal{L}_6}$ in $\mathcal{Q}_2$ containing $\mathcal{L}_6$, and moreover there exists only one such surface by Corollary 8.1.9. □

11.2 Rational map to $\mathbb{P}^4$

Let $\phi\colon \mathcal{W} \to \mathbb{P}^4 \cong \mathbb{P}(I \oplus W_4)$ be the morphism given by the linear system $|-K_{\mathcal{W}}|$, and let us denote by $\varphi\colon \mathcal{W} \to \phi(\mathcal{W})$ the same morphism viewed as a morphism onto its image. Thus, there exists a commutative diagram

where Φ is the map given by the linear system $\mathcal{Q}(\mathcal{L}_6)$ of all surfaces in $|2H|$ passing through $\mathcal{L}_6$.

Lemma 11.2.1. *The image $\phi(\mathcal{W})$ is a quartic threefold, and the morphism φ is birational.*

Proof. Since $-K_{\mathcal{W}}^3 = 4$ and ϕ is a morphism, we see that either $\phi(\mathcal{W})$ is a quartic threefold and φ is a birational morphism, or $\phi(\mathcal{W})$ is a quadric threefold and φ is a composition of a birational morphism and double cover of the quadric $\phi(\mathcal{W})$. Since $\check{S}_{\mathcal{L}_6} \in |-K_{\mathcal{W}}|$, the latter case is impossible. Indeed, if $\phi(\mathcal{W})$ is a quadric threefold, then $\phi(\check{S}_{\mathcal{L}_6})$ must be its hyperplane section, which is impossible, because the induced morphism

$$\phi|_{\check{S}_{\mathcal{L}_6}} \colon \check{S}_{\mathcal{L}_6} \to \phi(\check{S}_{\mathcal{L}_6})$$

is an isomorphism by (11.1.5) and Corollary 11.1.4. □

Put $X_4 = \phi(\mathcal{W})$. The map $\phi \circ \pi^{-1}$ induces a biregular action of $\mathfrak{A}_5$ on $\mathbb{P}^4$. By Lemma 11.1.10, this action is given by the standard reducible five-dimensional representation $I \oplus W_4$ of the group $\mathfrak{A}_5$. In particular, there exists a unique $\mathfrak{A}_5$-invariant hyperplane $\mathcal{H} \subset \mathbb{P}^4$.

Lemma 11.2.2. *The surface $\mathcal{H} \cap X_4$ is smooth.*

Proof. The surface $\mathcal{H} \cap X_4$ is the image under φ of the surface $\check{S}_{\mathcal{L}_6}$ (cf. Lemma 11.1.10). Using Corollary 11.1.4, we see that

$$\mathcal{H} \cap X_4 \cong \check{S}_{\mathcal{L}_6}.$$

The latter surface is smooth by Corollary 11.1.4. □

CREMONA GROUPS AND THE ICOSAHEDRON 245

Corollary 11.2.3. The threefold X_4 is smooth at every point of $\mathcal{H} \cap X_4$. In particular, X_4 has at most isolated singularities. Moreover, the birational morphism $\varphi \colon \mathcal{W} \to X_4$ is an isomorphism in a neighborhood of the surface $\check{S}_{\mathcal{L}_6}$, since every point of $\check{S}_{\mathcal{L}_6}$ is mapped to a smooth point of X_4, and the morphism φ is crepant.

Consider the $\mathfrak{A}_5$-irreducible curve $\mathcal{L}_{12}$ that is a union of 12 lines.

Lemma 11.2.4. The curve $\phi \circ \pi^{-1}(\mathcal{L}_{12}) \subset \mathbb{P}^4$ is a reducible curve that is a union of 12 lines.

Proof. One has $\mathcal{L}_{12} \cdot S_{\mathcal{L}_6} = 24$. Moreover, $\mathcal{L}_{12} \not\subset S_{\mathcal{L}_6}$ by Lemma 10.2.6. Furthermore, the two $\mathfrak{A}_5$-orbits Σ_{12} and Σ'_{12} are contained in $\mathcal{L}_{12}$ by Lemma 7.2.12(iii). These $\mathfrak{A}_5$-orbits are also contained in $S_{\mathcal{L}_6}$ by Theorem 8.1.8(iii). Thus,
$$\mathcal{L}_{12} \cap S_{\mathcal{L}_6} = \Sigma_{12} \cup \Sigma'_{12}.$$

Let $R_1, \ldots, R_{12}$ be irreducible components of the curve $\mathcal{L}_{12}$. By Lemma 9.1.9, we have $\mathcal{L}_{12} \cap \mathcal{L}_6 = \Sigma'_{12}$, so that each line R_i intersects $\mathcal{L}_6$ at one point. Denote by $\check{R}_1, \ldots, \check{R}_{12}$ the proper transforms of the lines $R_1, \ldots, R_{12}$ on the threefold $\mathcal{W}$. Then
$$-K_\mathcal{W} \cdot \check{R}_i = \left(\pi^*(2H) - \sum_{i=1}^{6} E_i\right) \cdot \check{R}_i = 1,$$
which implies that each $\phi(\check{R}_i)$ is a line in $\mathbb{P}^4$.

We claim that all lines $\phi(\check{R}_1), \ldots, \phi(\check{R}_{12})$ are different. Indeed, denote by $\check{\mathcal{L}}_{12}$ the proper transform of the curve $\mathcal{L}_{12}$ on the threefold $\mathcal{W}$, and denote by $\check{\Sigma}_{12}$ the proper transform of the $\mathfrak{A}_5$-orbit Σ_{12} on the threefold $\mathcal{W}$ (recall that $\pi \colon \mathcal{W} \to V_5$ is an isomorphism over Σ_{12}, since $\Sigma_{12} \not\subset \mathcal{L}_6$). Then
$$\check{\mathcal{L}}_{12} \cap \check{S}_{\mathcal{L}_6} = \check{\Sigma}_{12},$$
because $\check{\Sigma}_{12} \subset \check{\mathcal{L}}_{12}$ and $\check{\mathcal{L}}_{12} \cdot \check{S}_{\mathcal{L}_6} = 12$. Since $\varphi \colon \mathcal{W} \to X_4$ is an isomorphism in a neighborhood of the surface $\check{\Sigma}_{12}$ by Corollary 11.2.3, and each of the curves $\check{R}_1, \ldots, \check{R}_{12}$ contains exactly one point of $\check{\Sigma}_{12}$ by Lemma 7.2.12(iii), we see that all lines $\phi(\check{R}_1), \ldots, \phi(\check{R}_{12})$ are indeed different. □

Note that $\mathbb{P}^4$ contains a unique $\mathfrak{A}_5$-invariant point. Denote it by P. Then $P \notin \mathcal{H}$.

Lemma 11.2.5. The quartic X_4 contains the point P.

Proof. Recall that $\mathcal{L}_{12} \not\subset S_{\mathcal{L}_6}$ by Lemma 10.2.6. Thus, the curve $\phi \circ \pi^{-1}(\mathcal{L}_{12})$ is not contained in $\mathcal{H}$. The curve $\phi \circ \pi^{-1}(\mathcal{L}_{12})$ is a reducible curve that is a union of 12 lines by Lemma 11.2.4. Now it follows from Lemma 5.3.6 that each irreducible component of the curve $\phi \circ \pi^{-1}(\mathcal{L}_{12})$ contains P. In particular, the threefold $\phi(\mathcal{W})$ contains the point P as well. $\square$

Lemma 11.2.6. *The morphism φ contracts finitely many curves.*

Proof. Suppose that φ contracts an irreducible surface. Denote it by E. Then $\varphi(E)$ is either a curve or a point. We claim that $E \cap \check{S}_{\mathcal{L}_6} = \varnothing$. Indeed, if $E \cap \check{S}_{\mathcal{L}_6} \neq \varnothing$, then $E|_{\check{S}_{\mathcal{L}_6}}$ is an effective divisor on $\check{S}_{\mathcal{L}_6}$ such that

$$E|_{\check{S}_{\mathcal{L}_6}} \cdot (-K_{\mathcal{W}}|_{\check{S}_{\mathcal{L}_6}}) = E \cdot \check{S}_{\mathcal{L}_6} \cdot (-K_{\mathcal{W}}) = E \cdot (-K_{\mathcal{W}})^2 = 0,$$

which implies that $-K_{\mathcal{W}}|_{\check{S}_{\mathcal{L}_6}}$ is not ample. The latter contradicts Corollary 11.1.4.

Since $E \cap \check{S}_{\mathcal{L}_6} = \varnothing$, and the intersection $E_i \cap \check{S}_{\mathcal{L}_6}$ is not empty for every i, we have $E \neq E_i$ for every i, so that the surface E is not π-exceptional. Thus, $\pi(E)$ is a surface in V_5. Moreover, we have

$$S_{\mathcal{L}_6} \cap \pi(E) \subset \mathcal{L}_6,$$

since E is disjoint from $\check{S}_{\mathcal{L}_6}$. In particular, we have

$$\pi(E)|_{S_{\mathcal{L}_6}} = \sum_{i=1}^{6} b_i L_i$$

for some non-negative integers $b_1, \ldots, b_6$ (note that the numbers $b_1, \ldots, b_6$ may not be the same since we do not assume E to be $\mathfrak{A}_5$-invariant). On the other hand, we have $\pi(E) \sim aH$ for some positive integer a, because the group $\mathrm{Pic}(V_5)$ is generated by H. Thus, we have

$$0 < 10a^2 = a^2 H^2 \cdot S_{\mathcal{L}_6} = \pi(E)|_{S_{\mathcal{L}_6}} \cdot \pi(E)|_{S_{\mathcal{L}_6}} = \left(\sum_{i=1}^{6} b_i L_i\right)^2 = -2\sum_{i=1}^{6} b_i^2 \leqslant 0,$$

since the lines of $\mathcal{L}_6$ are pairwise disjoint by Lemma 7.4.7. The obtained contradiction shows that E does not exist. $\square$

Corollary 11.2.7. *The threefold X_4 has terminal singularities.*

Choose homogeneous coordinates x, y, z, t, and w in $\mathbb{P}^4$ so that the hyperplane $\mathcal{H}$ is given by the equation $w = 0$, and x, y, z, t correspond to coordinates in the vector space $W_4 \subset \tilde{\mathcal{Q}}_\mathcal{W}$ (so that x, y, z, t are naturally interpreted as homogeneous coordinates on $\mathcal{H}$). In these coordinates we have $P = [0:0:0:0:1]$. Since $P \in X_4$ by Lemma 11.2.5 and W_4 is an irreducible $\mathfrak{A}_5$-representation, we can write the equation of the quartic X_4 in the form

$$w^2 q_2(x,y,z,t) + w q_3(x,y,z,t) + q_4(x,y,z,t) = 0, \qquad (11.2.8)$$

where q_2, q_3, and q_4 are $\mathfrak{A}_5$-invariant forms of degrees 2, 3, and 4, respectively. In particular, X_4 is singular at P. Since X_4 has a terminal singularity at P by Corollary 11.2.7, the form q_2 in (11.2.8) is not a zero polynomial. In fact, we can say more.

Lemma 11.2.9. *The form $q_2(x,y,z,t)$ is of maximal rank and, in particular, P is an ordinary double point of X_4.*

Proof. The coordinates x, y, z, t are naturally interpreted as homogeneous coordinates on $\mathcal{H} \cong \mathbb{P}^3$. Since the polynomial q_2 is not a zero-polynomial, equation $q_2 = 0$ defines a smooth quadric in $\mathcal{H}$ by Lemma 5.3.3(v). In particular, the point P is an isolated ordinary double point on X_4. □

Remark 11.2.10. The point P is not the only singularity of the quartic X_4. Indeed, a three-dimensional quartic with a unique isolated ordinary double point is $\mathbb{Q}$-factorial by [17, Theorem 2] (see also [113]). But X_4 is not $\mathbb{Q}$-factorial, since $\mathcal{W}$ is projective and $\varphi \colon \mathcal{W} \to X_4$ is a small resolution by Lemma 11.2.6. Alternatively, one can show that $\mathrm{Sing}(X_4) \neq \{P\}$ using the fact that V_5 is rational, while $\mathbb{Q}$-factorial three-dimensional nodal quartics are non-rational (see [29], [81]).

Lemma 11.2.11. *The form $q_3(x,y,z,t)$ is not a zero polynomial.*

Proof. Suppose that q_3 is a zero polynomial. Then $X_4 \subset \mathbb{P}^4$ is defined by the equation
$$w^2 q_2(x,y,z,t) + q_4(x,y,z,t) = 0,$$
where q_2 and q_4 are $\mathfrak{A}_5$-invariant polynomials of degree 2 and 4, respectively. Let us find singular points of X_4. Let $O = [x_0 : y_0 : z_0 : t_0 : w_0]$ be a

singular point of X_4. Then

$$\begin{cases} w_0^2 \dfrac{\partial q_2}{\partial x}(x_0, y_0, z_0, t_0) + \dfrac{\partial q_4}{\partial x}(x_0, y_0, z_0, t_0) = 0, \\ w_0^2 \dfrac{\partial q_2}{\partial y}(x_0, y_0, z_0, t_0) + \dfrac{\partial q_4}{\partial y}(x_0, y_0, z_0, t_0) = 0, \\ w_0^2 \dfrac{\partial q_2}{\partial z}(x_0, y_0, z_0, t_0) + \dfrac{\partial q_4}{\partial z}(x_0, y_0, z_0, t_0) = 0, \\ w_0^2 \dfrac{\partial q_2}{\partial t}(x_0, y_0, z_0, t_0) + \dfrac{\partial q_4}{\partial t}(x_0, y_0, z_0, t_0) = 0. \end{cases}$$

Multiplying these equations by x_0, y_0, z_0, and t_0, respectively, and adding them up together, we get

$$\begin{aligned} 0 = x_0 w_0^2 \dfrac{\partial q_2}{\partial x}(x_0, y_0, z_0, t_0) + x_0 \dfrac{\partial q_4}{\partial x}(x_0, y_0, z_0, t_0) + \\ + y_0 w_0^2 \dfrac{\partial q_2}{\partial y}(x_0, y_0, z_0, t_0) + y_0 \dfrac{\partial q_4}{\partial y}(x_0, y_0, z_0, t_0) + \\ + z_0 w_0^2 \dfrac{\partial q_2}{\partial z}(x_0, y_0, z_0, t_0) + z_0 \dfrac{\partial q_4}{\partial z}(x_0, y_0, z_0, t_0) + \\ + t_0 w_0^2 \dfrac{\partial q_2}{\partial t}(x_0, y_0, z_0, t_0) + t_0 \dfrac{\partial q_4}{\partial t}(x_0, y_0, z_0, t_0) = \\ = 2 w_0^2 q_2(x_0, y_0, z_0, t_0) + 4 q_4(x_0, y_0, z_0, t_0) \end{aligned}$$

by Euler's formula. This implies that

$$q_4(x_0, y_0, z_0, t_0) = w_0 q_2(x_0, y_0, z_0, t_0) = 0,$$

because

$$2wq_2 = \dfrac{\partial}{\partial w}\left(w^2 q_2 + q_4\right) = 0$$

at the point O.

Thus, we see that either

$$q_4(x_0, y_0, z_0, t_0) = q_2(x_0, y_0, z_0, t_0) = 0$$

or

$$q_4(x_0, y_0, z_0, t_0) = w_0 = 0.$$

In the latter case, one would have $O \in \mathcal{H} \cap X_4$, because the equations

$$q_4(x, y, z, t) = w = 0$$

define exactly the surface $\mathcal{H} \cap X_4$. On the other hand, O is a singular point of X_4, while X_4 is smooth at every point of $\mathcal{H} \cap X_4$ by Corollary 11.2.3, which gives a contradiction.

Therefore, we see that $q_4(x_0, y_0, z_0, t_0) = q_2(x_0, y_0, z_0, t_0) = 0$. Equations

$$q_4(x, y, z, t) = q_2(x, y, z, t) = 0.$$

define a surface $T \subset X_4$. This surface is a cone with vertex

$$P = [0:0:0:0:1]$$

over a curve in $\mathcal{H}$ given by the same equations. By Lemma 5.3.3(xi), the latter curve is smooth. Hence, the surface T is smooth outside of the point P. In particular, the quartic X_4 must be smooth in every point of the surface T different from P.

We see that $\mathrm{Sing}(X_4) = \{P\}$, which is impossible by Remark 11.2.10. The obtained contradiction shows that q_3 is a non-zero polynomial. □

The following proposition is not essential for the proof of the main results of the book, but it helps to understand the geometry of X_4 better.

Proposition 11.2.12. There are exactly 24 lines on X_4 that pass through the point P.

Proof. Note that the union of all lines in X_4 passing through P are defined in $\mathbb{P}^4$ by equations $q_2 = q_3 = q_4 = 0$. In particular, to prove the lemma it is enough to show that the equations $w = q_2 = q_3 = q_4 = 0$ define 24 points in $\mathbb{P}^4$. This is what we are going to do.

By Lemma 11.2.11, the polynomial q_3 in (11.2.8) is not a zero polynomial. By Lemma 11.2.9, the form q_2 is of maximal rank. By Lemma 5.3.3(v),(vii), the forms q_2 and q_3 are unique up to scaling. By Lemma 5.3.3(x), the form q_4 modulo q_2^2 is unique up to scaling.

We claim that q_4 is not divisible by q_2. Indeed, if q_2 divides q_4, then one has $q_4 = \lambda q_2^2$ for some $\lambda \in \mathbb{C}$, which implies that X_4 is singular along the curve $w = q_2 = q_3 = 0$. The latter contradicts Corollary 11.2.3.

Since q_4 is not divisible by q_2, the curve $w = q_2 = q_4 = 0$ is a smooth irreducible curve of degree 8 and genus 9 by Lemma 5.3.3(xi). Since q_3 is not a zero polynomial, the curve $w = q_2 = q_3 = 0$ is a smooth irreducible curve of degree 6 and genus 4 by Lemma 5.3.3(viii). In particular, the latter two curves are different, so that the system of equations

$$w = q_2 = q_3 = q_4 = 0$$

defines a finite subset in $\mathbb{P}^4$. Moreover, since the curve $w = q_2 = q_3 = 0$ is a smooth irreducible curve of degree 6, we see that equations

$$w = q_2 = q_3 = q_4 = 0$$

define a set that consists either of exactly 12 points or of exactly 24 points. This follows from Lemma 5.1.4.

The curve $w = q_2 = q_3 = 0$, being a smooth curve of genus 4, contains exactly two $\mathfrak{A}_5$-orbits of length 12 by Lemma 5.1.5. Similarly, the curve $w = q_2 = q_4 = 0$, being a smooth curve of genus 9, contains exactly two $\mathfrak{A}_5$-orbits of length 12 by Lemma 5.1.5. Moreover, the surface given by $w = q_2 = 0$ contains exactly two $\mathfrak{A}_5$-orbits of length 12 by Lemma 5.3.3(vi). This shows that equations

$$w = q_2 = q_3 = q_4 = 0$$

cut out exactly these two $\mathfrak{A}_5$-orbits of length 12. This means that there are exactly 24 lines on X_4 passing through the point P. □

Remark 11.2.13. By Proposition 11.2.12, there are exactly 24 lines on X_4 that pass through P. These lines split into two $\mathfrak{A}_5$-orbits, each consisting of 12 lines. One of these two $\mathfrak{A}_5$-orbits is mapped by $\pi \circ \varphi^{-1}$ to the curve $\mathcal{L}_{12}$ on V_5 (cf. Lemma 11.2.4). Another $\mathfrak{A}_5$-orbit is contracted by $\pi \circ \varphi^{-1}$ to the $\mathfrak{A}_5$-orbit $\Sigma'_{12} \subset \mathcal{L}_6$.

By Remark 11.2.10, the point P is not the only singularity of the quartic X_4. In fact, one can see this in a more straightforward way. Recall that the curve $\mathcal{L}_{15}$ has five connected components by Corollary 7.4.4 and Lemma 9.1.10(i).

Lemma 11.2.14. *Five connected components of the curve $\mathcal{L}_{15}$ are contracted by the birational map $\varphi \circ \pi^{-1}$ to five singular points of X_4 that form an $\mathfrak{A}_5$-orbit, and that are not ordinary double points.*

Proof. Let $\check{\mathcal{L}}_{15}$ be the proper transform of the curve $\mathcal{L}_{15}$ on the threefold $\mathcal{W}$. Then $\check{\mathcal{L}}_{15}$ also has five connected components, since the $\mathfrak{A}_5$-orbit Σ_5 is not contained in the curve $\mathcal{L}_6$ by Corollary 7.4.9 (cf. Corollary 7.4.4). On the other hand, the curve $\check{\mathcal{L}}_{15}$ is contracted by φ (see Lemma 9.1.14). Thus, $\varphi(\check{\mathcal{L}}_{15})$ is either an $\mathfrak{A}_5$-orbit of length 5 or a single $\mathfrak{A}_5$-invariant point in X_4. In the latter case, a proper transform via $\varphi \circ \pi^{-1}$ of a general hyperplane section of X_4 that passes through $\varphi(\check{\mathcal{L}}_{15})$ is a surface in $|2H|$ that contains $\mathcal{L}_{15}$, which is impossible by Lemma 7.5.13.

CREMONA GROUPS AND THE ICOSAHEDRON

Thus, we see that the curve $\check{\mathcal{L}}_{15}$ is contracted by φ to five singular points of the threefold X_4. These points cannot be ordinary double points by Lemma 4.2.2. □

Now we are ready to explicitly determine the equation of the quartic X_4.

Proposition 11.2.15. One can choose homogeneous coordinates x, y, z, t, and w in $\mathbb{P}^4$ so that the hyperplane $\mathcal{H}$ is given by $w = 0$, and the quartic X_4 is given by the equation

$$w^2\left(x^2 + y^2 + z^2 + t^2 + (x+y+z+t)^2\right) +$$
$$+ w\left(x^3 + y^3 + z^3 + t^3 - (x+y+z+t)^3\right) +$$
$$+ \frac{9}{16}\left(x^4 + y^4 + z^4 + t^4 + (x+y+z+t)^4\right) -$$
$$- \frac{81}{320}\left(x^2 + y^2 + z^2 + t^2 + (x+y+z+t)^2\right)^2 = 0.$$

Proof. The proof is organized as follows. First we fix some homogeneous coordinates x', y', z', t', and w' on $\mathbb{P}^4$ and find explicitly the $\mathfrak{A}_5$-invariant forms of low degree. Then we express the equation of the quartic X_4 via these forms with indeterminate coefficients, and use geometrical information about X_4 that we already know to find the latter coefficients.

Recall that by Lemma 11.1.10 there is an $\mathfrak{A}_5$-equivariant identification of $\mathbb{P}^4$ with a projectivization of the standard reducible five-dimensional representation $I \oplus W_4$ of the group $\mathfrak{A}_5$. Let x', y', z', t', and w' be homogenous coordinates on $\mathbb{P}^4$ such that $\mathfrak{A}_5$ permutes them. Then

$$P = [1:1:1:1:1]$$

in these coordinates, and the hyperplane $\mathcal{H}$ is given by

$$x' + y' + z' + t' + w' = 0.$$

Moreover,

$$\begin{aligned} &x' + y' + z' + t' + w', \\ &(x')^2 + (y')^2 + (z')^2 + (t')^2 + (w')^2, \\ &(x')^3 + (y')^3 + (z')^3 + (t')^3 + (w')^3, \\ &(x')^4 + (y')^4 + (z')^4 + (t')^4 + (w')^4 \end{aligned} \qquad (11.2.16)$$

are $\mathfrak{A}_5$-invariant forms, and any $\mathfrak{A}_5$-invariant form of degree at most 4 can be expressed through them, see Lemma 5.3.3(v),(vii),(x). Put

$$\begin{cases} x = \dfrac{4}{5}x' - \dfrac{y'}{5} - \dfrac{z'}{5} - \dfrac{t'}{5} - \dfrac{w'}{5}, \\ y = -\dfrac{x'}{5} + \dfrac{4}{5}y' - \dfrac{z'}{5} - \dfrac{t'}{5} - \dfrac{w'}{5}, \\ z = -\dfrac{x'}{5} - \dfrac{y'}{5} + \dfrac{4}{5}z' - \dfrac{t'}{5} - \dfrac{w'}{5}, \\ t = -\dfrac{x'}{5} - \dfrac{y'}{5} - \dfrac{z'}{5} + \dfrac{4t'}{5} - \dfrac{w'}{5}, \\ w = x' + y' + z' + t' + w'. \end{cases}$$

Then $P = [0 : 0 : 0 : 0 : 1]$ in these new coordinates, and the hyperplane $\mathcal{H}$ is given by $w = 0$. (Note that this choice of coordinates agrees with the one made to obtain equation (11.2.8) and further used in the proof of Lemma 11.2.11.) We have

$$\begin{cases} x' = x + \dfrac{w}{5}, \\ y' = y + \dfrac{w}{5}, \\ z' = z + \dfrac{w}{5}, \\ t' = t + \dfrac{w}{5}, \\ w' = \dfrac{w}{5} - x - y - z - t. \end{cases}$$

Plugging the latter expressions into the forms (11.2.16), we get the $\mathfrak{A}_5$-invariant forms of low degree expressed in the new coordinates: these are w,

$$x^2 + y^2 + z^2 + t^2 + (x+y+z+t)^2, \quad x^3 + y^3 + z^3 + t^3 - (x+y+z+t)^3$$

and
$$x^4 + y^4 + z^4 + t^4 + (x+y+z+t)^4,$$

respectively. Denote the last three of them by f_2, f_3, and f_4, respectively.

The equation of the $\mathfrak{A}_5$-invariant quartic X_4 can be written as

$$w^2 q_2(x,y,z,t) + w q_3(x,y,z,t) + q_4(x,y,z,t) = 0,$$

where $q_2 = \lambda f_2$, $q_3 = \alpha f_3$, and $q_4 = \beta f_4 + \gamma f_2^2$ for some complex numbers λ, α, β, and γ. Since q_2 is not a zero polynomial by Lemma 11.2.9, we may assume after rescaling that $\lambda = 1$. Also $\alpha \neq 0$, since q_3 is not a zero polynomial by Lemma 11.2.11. (Actually, the computations below applied to the case $\alpha = 0$ imply that $\beta = \gamma = 0$, which is clearly impossible, so we could avoid using Lemma 11.2.11 here.) Thus, by rescaling w and (x, y, z, t) we can assume that $\alpha = 1$.

Let $\chi \colon X_4 \dashrightarrow \mathbb{P}^3$ be a linear projection given by

$$[x : y : z : t : w] \mapsto [x : y : z : t].$$

Then χ is a projection from $P \in X_4$, and χ is $\mathfrak{A}_5$-equivariant. The map χ contracts 24 lines on X_4 passing through P to 24 different points in $\mathbb{P}^3$. Outside of these 24 lines in X_4, the map χ is a double cover that is branched along a sextic surface. Denote this surface by S_6. Then S_6 is given by

$$f_3^2(x, y, z, t) = 4 f_2(x, y, z, t) \Big(\beta f_4(x, y, z, t) + \gamma f_2^2(x, y, z, t) \Big). \quad (11.2.17)$$

By Lemma 11.2.14, the curve $\mathcal{L}_{15}$ is contracted by $\varphi \circ \pi^{-1}$ to 5 singular points of X_4 that form a single $\mathfrak{A}_5$-orbit, and that are not ordinary double points. Denote these points by O_1, O_2, O_3, O_4, and O_5. Then $\chi(O_i)$ are singular points of the surface S_6, because O_i are singular points of X_4 that are not contained in the 24 lines on X_4 passing through P (cf. Remark 11.2.13). Moreover, the singularities of the surface S_6 at these 5 points must be worse than ordinary double points, because O_1, O_2, O_3, O_4, and O_5 are not ordinary double points of the threefold X_4.

Note that $\mathbb{P}^3$ contains a unique $\mathfrak{A}_5$-orbit of length 5 by Lemma 5.3.3(ii). Denote it by Θ_5. Then $[1 : 1 : 1 : 1] \in \Theta_5$. Since $\mathbb{P}^3$ does not contain $\mathfrak{A}_5$-orbits of length less than 5 (i.e., of length 1) by Lemma 5.3.3(i), we have

$$\Theta_5 = \{O_1, O_2, O_3, O_4, O_5\}.$$

In particular, we have $[1 : 1 : 1 : 1] \in S_6$. Plugging $[1 : 1 : 1 : 1]$ into (11.2.17), we get

$$\gamma = \frac{9 - 52\beta}{80}.$$

The fact that S_6 is singular at $[1 : 1 : 1 : 1]$ does not impose any extra constraints on α, β, and γ. This is not surprising, because surfaces acted on faithfully by $\mathfrak{A}_5$ do not have $\mathfrak{A}_5$-orbits of length 5 consisting of smooth points by Lemma 6.1.11. However, the fact that the singularity of S_6 at $[1 : 1 : 1 : 1]$ is worse than an ordinary double point implies that $\beta = \frac{9}{16}$ and S_6 has a triple point in $[1 : 1 : 1 : 1]$. Thus, we have $\gamma = -\frac{81}{320}$, which completes the proof of Proposition 11.2.15. □

Corollary 11.2.18. The threefold X_4 has exactly 6 singular points, which are the ordinary double point P and five singular points of type cD_4 with Tjurina number 10 (cf. [70, Theorem 2.3(iii)]). These five singular points are images of the connected components of $\mathcal{L}_{15}$ via $\varphi \circ \pi^{-1}$. The morphism

$$\varphi \colon \mathcal{W} \to X_4$$

is a small resolution of singularities of the quartic threefold X_4.

The following lemma answers a question of Michael Wemyss.

Lemma 11.2.19. Let L be a line of $\mathcal{L}_{15}$. Denote by $\check{L}$ its proper transform on $\mathcal{W}$. Then

$$\mathcal{N}_{\check{L}/\mathcal{W}} \cong \mathcal{O}_{\mathbb{P}^1}(-1) \oplus \mathcal{O}_{\mathbb{P}^1}(-1).$$

Proof. By Lemma 7.2.13(ii), one has $L \not\subset \mathscr{S}$. Thus, we have

$$\mathcal{N}_{L/V_5} \cong \mathcal{O}_{\mathbb{P}^1} \oplus \mathcal{O}_{\mathbb{P}^1}$$

by Theorem 7.7.1. Let us use notation of §7.7. We have $E \cong \mathbb{P}^1 \times \mathbb{P}^1$.

Recall that L intersects exactly two lines of $\mathcal{L}_6$ by Lemma 9.1.14. Denote these two lines by L_1 and L_2. Denote by $\tilde{L}_1$, $\tilde{L}_2$, and $\tilde{\mathcal{C}}_6$ the proper transforms of the curves L_1, L_2, and $\mathcal{C}_6$ on the threefold $\tilde{V}$, respectively. Put $\tilde{P}_1 = \tilde{L}_1 \cap E$ and $\tilde{P}_2 = \tilde{L}_2 \cap E$. By Lemma 4.2.6, we have

$$\mathcal{N}_{\check{L}/\mathcal{W}} \cong \mathcal{O}_{\mathbb{P}^1}(-1) \oplus \mathcal{O}_{\mathbb{P}^1}(-1)$$

if and only if the points $\tilde{P}_1$ and $\tilde{P}_2$ are not contained in one section of the natural projection $E \to L$ that has self-intersection 0. Let us show that this is the case.

Note that $v(E)$ is a hyperplane section of the quadric Q that passes through the curve $v(F)$, and F contains $\tilde{L}_1$ and $\tilde{L}_2$ by Corollary 7.7.5. In particular, $F|_E$ is a smooth curve that contains the points $\tilde{P}_1$ and $\tilde{P}_2$. By Corollary 7.7.4 one has

$$F|_E \sim 2\tilde{s} + \tilde{l},$$

where $\tilde{l}$ is a fiber of the natural projection $E \to L$, and $\tilde{s}$ is a section of this projection that has self-intersection 0.

Suppose that $\tilde{P}_1$ and $\tilde{P}_2$ are contained in one curve $\tilde{s}_{1,2} \in |\tilde{s}|$. Then $\tilde{s}_{1,2}$ must be an irreducible component of the curve $F|_E$, because

$$F|_E \cdot \tilde{s}_{1,2} = \left(2\tilde{s} + \tilde{l}\right) \cdot \tilde{s} = 1,$$

11.3 A remarkable sextic curve

Recall that X_4 has an isolated ordinary double point in P by Lemma 11.2.9, and φ is a small resolution of singularities of X_4 by Lemma 11.2.6. In particular, $\varphi^{-1}(P)$ is a (possibly reducible) curve. Denote this curve by $\check{\mathscr{C}}'$. Put $\mathscr{C}' = \pi(\check{\mathscr{C}}')$. In fact, Lemma 4.2.2 implies that $\check{\mathscr{C}}'$ and $\mathscr{C}'$ are rational curves. Still we prefer not to take this approach, but instead describe the curve $\mathscr{C}'$ using elementary methods.

Lemma 11.3.1. *The curve $\mathscr{C}'$ is a smooth rational $\mathfrak{A}_5$-invariant curve of degree 6 such that $\Sigma'_{12} \subset \mathscr{C}'$. The curve $\check{\mathscr{C}}'$ is the proper transform of $\mathscr{C}'$ via π. Moreover, one has $\check{\mathscr{C}}' \cong \mathscr{C}'$ and $\mathscr{C}' \neq \mathscr{C}$.*

Proof. The linear system

$$\mathcal{Q} = |2H - \mathcal{L}_6 - \mathscr{C}'|$$

has dimension 3 by Lemma 11.1.10, so that a general surface $M \in \mathcal{Q}$ is irreducible by Lemma 8.1.3. Let M_1 and M_2 be general surfaces in $\mathcal{Q}$. Then

$$M_1 \cdot M_2 = \mathcal{L}_6 + \mathscr{C}' + Z,$$

where Z is an effective one-cycle on V_5. Thus

$$20 = H \cdot M_1 \cdot M_2 = H \cdot \mathcal{L}_6 + H \cdot \mathscr{C}' + H \cdot Z =$$
$$= 6 + \deg(\mathscr{C}') + H \cdot Z \geqslant 6 + \deg(\mathscr{C}'),$$

which implies that $\deg(\mathscr{C}') \leqslant 14$.

By Proposition 10.2.5, the curve $\mathscr{C}'$ is either one of the curves $\mathcal{L}_6$, $\mathcal{L}_{10}$, $\mathcal{G}_5$, $\mathcal{G}'_5$, $\mathcal{L}_{12}$, or $\mathcal{G}_6$, or a rational sextic normal curve, or a union of a rational sextic normal curve and the curve $\mathcal{L}_6$, or a union of two rational sextic normal curves.

Recall that $E_1, \ldots, E_6$ denote the exceptional divisors of π (see §11.1). We know that $-K_\mathcal{W}|_{E_i}$ is ample by Corollary 11.1.8. Thus, no irreducible component of the curve $\check{\mathscr{C}}'$ is contained in E_i. In particular, one has $\check{\mathscr{C}}'$ is the proper transform of $\mathscr{C}'$ via π. Moreover, the curve $\check{\mathscr{C}}'$ must be connected by Zariski's main theorem (see [53, Corollary III.11.4]). In particular, the curve $\mathscr{C}'$ is connected, and no irreducible component of $\mathscr{C}'$ is contained in $\mathcal{L}_6$. On the other hand, the curves $\mathcal{L}_{10}$, $\mathcal{G}_5$, $\mathcal{G}'_5$, $\mathcal{L}_{12}$, and $\mathcal{G}_6$ are not connected by Lemmas 9.1.2, 9.2.3, 7.2.12(ii), and 9.2.8, while a union of two $\mathfrak{A}_5$-invariant rational sextic normal curves in V_5 is not connected by Lemma 7.5.6. Hence, $\mathscr{C}'$ is a sextic rational normal curve. In particular, the curve $\mathscr{C}'$ is smooth, which implies that $\check{\mathscr{C}}' \cong \mathscr{C}'$.

Note that the curve $\mathscr{C}'$ intersects the curve $\mathcal{L}_6$, because $\mathscr{C}'$ is contracted by the map φ. Since a line cannot intersect a sextic normal curve at more than two points, and there are no $\mathfrak{A}_5$-orbits on $\mathscr{C}'$ of length less than 12 by Lemma 5.1.4, we conclude that $\mathscr{C}' \cap \mathcal{L}_6$ is an $\mathfrak{A}_5$-orbit of length 12, which is Σ'_{12} by Theorem 7.3.5 and Lemma 7.4.3. On the other hand, the unique $\mathfrak{A}_5$-orbit of length 12 contained in the curve $\mathscr{C}$ is Σ_{12} (see Lemma 7.2.10). Therefore, one has $\mathscr{C}' \neq \mathscr{C}$. □

As a by-product of the above construction we are able now to describe all $\mathfrak{A}_5$-invariant curves of degree 6 on V_5.

Lemma 11.3.2. *The curves $\mathscr{C}'$, $\mathscr{C}$, and $\mathcal{L}_6$ are the only $\mathfrak{A}_5$-invariant curves of degree 6 in V_5.*

Proof. By Lemma 7.5.6 we know that if Z is an $\mathfrak{A}_5$-invariant curve of degree 6 in V_5, then Z is either $\mathcal{L}_6$, or a sextic rational normal curve, and moreover, there are at most two $\mathfrak{A}_5$-invariant sextic rational normal curves in V_5. On the other hand, Lemma 11.3.1 shows that we have already constructed two of them, namely, $\mathscr{C}$ and $\mathscr{C}'$. □

By Lemma 10.2.7, there exists a surface $S_{\mathscr{C}'} \in \mathcal{Q}_2$ that contains $\mathscr{C}'$. By Lemma 7.2.13(i), one has $S_{\mathscr{C}'} \neq \mathscr{S}$. Thus, Lemma 8.3.3 implies that $S_{\mathscr{C}'}$ is smooth.

Let $\psi \colon \mathcal{U} \to V_5$ be a blow-up of the curve $\mathscr{C}'$. Denote by $\mathcal{F}_\mathcal{U}$ the ψ-exceptional surface.

Lemma 11.3.3 (cf. Lemma 7.2.8). *The normal bundle of the curve $\mathscr{C}' \cong \mathbb{P}^1$ in V_5 is*

$$\mathcal{N}_{\mathscr{C}'/V_5} \cong \mathcal{O}_{\mathbb{P}^1}(5) \oplus \mathcal{O}_{\mathbb{P}^1}(5).$$

In particular, one has $\mathcal{F}_\mathcal{U} \cong \mathbb{P}^1 \times \mathbb{P}^1$.

Proof. The normal bundle of the curve $\mathscr{C}'$ in V_5 is

$$\mathcal{N} = \mathcal{N}_{\mathscr{C}'/V_5} \cong \mathcal{O}_{\mathbb{P}^1}(a) \oplus \mathcal{O}_{\mathbb{P}^1}(b)$$

Proof. Recall that by Lemma 9.1.9 one has $\mathcal{L}_6 \cap \mathcal{L}_{12} = \Sigma'_{12}$. Choose a point $Q \in \Sigma'_{12}$. Then Q lies on one irreducible component L of the curve $\mathcal{L}_6$, since the irreducible components of $\mathcal{L}_6$ are disjoint by Lemma 7.4.7. Similarly, Q lies on one irreducible component R of the curve $\mathcal{L}_{12}$, since the irreducible components of $\mathcal{L}_{12}$ are disjoint by Lemma 7.2.12(ii). This proves assertion (i).

Suppose that the tangent vectors to the curves L, R, and $\mathscr{C}'$ at the point Q are linearly independent in the Zariski tangent space to V_5 at Q. Let $\check{\mathscr{C}}'$ and $\check{R}$ be proper transforms of the curves $\mathscr{C}'$ and R on the threefold $\mathcal{W}$, respectively. Then
$$\check{R} \cap \check{\mathscr{C}}' = \emptyset.$$
The latter is impossible, since $\varphi(\check{R})$ is a line in X_4 that passes through the point $P = \varphi(\check{\mathscr{C}}')$ (see Lemma 11.2.4) and $\varphi^{-1}(P) = \check{\mathscr{C}}'$. This proves assertion (ii). $\square$

Lemma 11.3.6. *The curves $\mathscr{C}'$ and $\mathcal{G}_6$ are disjoint.*

Proof. Apply Corollary 8.1.12 and Lemma 10.2.7 together with Corollary 9.2.11 and Lemma 11.3.4. $\square$

Lemma 11.3.7. *The curves $\mathscr{C}'$ and $\mathcal{L}_{15}$ are disjoint.*

Proof. Suppose that $\mathscr{C}'$ and $\mathcal{L}_{15}$ are not disjoint. Let $f \colon \bar{V}_5 \to V_5$ be the blow-up of V_5 at Σ_5. Denote by $\bar{\mathscr{C}}'$ and $\bar{\mathcal{L}}_{15}$ the proper transforms of the curves $\mathscr{C}'$ and $\mathcal{L}_{15}$ on the threefold $\bar{V}_5$, respectively. Then the curves $\bar{\mathscr{C}}'$ and $\bar{\mathcal{L}}_{15}$ are not disjoint either, since $\Sigma_5 \not\subset \mathscr{C}'$ by Lemma 5.1.4. On the other hand, the curve $\bar{\mathcal{L}}_{15}$ is a disjoint union of 15 smooth rational curves (cf. Lemma 9.1.10(i)). Since $\mathscr{C}'$ does not contain $\mathfrak{A}_5$-orbits of length 15 (see Lemma 5.1.4), we see that $\bar{\mathscr{C}}'$ intersects every irreducible component of the curve $\bar{\mathcal{L}}_{15}$ in at least 2 points.

Let $B_1, \ldots, B_5$ be f-exceptional divisors that are contracted to the points of Σ_5 by f. By Lemma 7.5.3, the linear system $|f^*(H) - \sum_{i=1}^{5} B_i|$ is free from base points and gives an $\mathfrak{A}_5$-equivariant morphism
$$\bar{\theta}_{\Pi_3} \colon \bar{V}_5 \to \mathbb{P}^2$$
whose general fiber is an elliptic curve. By Corollary 7.4.4, the curve $\bar{\mathcal{L}}_{15}$ is contracted by the morphism $\bar{\theta}_{\Pi_3}$ to the unique $\mathfrak{A}_5$-orbit $\Omega \subset \mathbb{P}^2$ that consists of 15 points (see Lemma 5.3.1(i)). Moreover, Ω is contained in $\bar{\theta}_{\Pi_3}(\bar{\mathscr{C}}')$, because $\bar{\mathscr{C}}' \cap \bar{\mathcal{L}}_{15} \neq \emptyset$. We compute
$$\left(f^*(H) - \sum_{i=1}^{5} B_i \right) \cdot \bar{\mathscr{C}}' = f^*(H) \cdot \bar{\mathscr{C}}' = H \cdot \mathscr{C}' = 6.$$

Therefore, the curve $\bar{\theta}_{\Pi_3}(\bar{\mathscr{C}}')$ is an $\mathfrak{A}_5$-invariant sextic curve in $\mathbb{P}^2$, because there exist no $\mathfrak{A}_5$-invariant curves in $\mathbb{P}^2$ of degree 1 and 3 (see Lemma 5.3.1(iv)), and the only $\mathfrak{A}_5$-invariant conic in $\mathbb{P}^2$ does not contain the $\mathfrak{A}_5$-orbit Ω (see Lemma 5.3.1(v)).

Recall that by Lemma 5.3.1(vi) all $\mathfrak{A}_5$-invariant curves of degree 6 form a pencil (cf. the proof of Lemma 6.7.5). This pencil contains the curve $\bar{\theta}_{\Pi_3}(\bar{\mathscr{C}}')$. On the other hand, the curve $\bar{\theta}_{\Pi_3}(\bar{\mathscr{C}}')$ is singular at every point of the $\mathfrak{A}_5$-orbit Ω, because the curve $\bar{\mathcal{L}}_{15}$ is contracted by the morphism $\bar{\theta}_{\Pi_3}$ to Ω, and $\bar{\mathscr{C}}'$ intersects every irreducible component of the curve $\bar{\mathcal{L}}_{15}$ in at least 2 points. The latter is impossible by Lemma 6.7.5(v), since the curve $\bar{\theta}_{\Pi_3}(\bar{\mathscr{C}}')$ is irreducible (cf. Lemma 6.7.5(iv)). The obtained contradiction shows that $\bar{\mathscr{C}}' \cap \mathcal{L}_{15} = \varnothing$. □

11.4 Two Sarkisov links

The quartic X_4 is invariant with respect to the $\mathfrak{S}_5$-action on $\mathbb{P}^4$ that extends our $\mathfrak{A}_5$ action on $\mathbb{P}^4$. Let τ be an odd involution in the group $\mathfrak{S}_5 \subset \mathrm{Bir}(V_5)$. In fact, we only need that τ is an odd element in $\mathfrak{S}_5 \subset \mathrm{Bir}(V_5)$, but for the sake of additional symmetry we chose it to be an involution.

Remark 11.4.1. The birational involution τ does not act biregularly on V_5. Indeed, V_5 does not admit a biregular action of $\mathfrak{S}_5$ at all, because its automorphism group is isomorphic to $\mathrm{PSL}_2(\mathbb{C})$ by Proposition 7.1.10. Since τ does not act biregularly on V_5, it does not act biregularly on $\mathcal{W}$ either. Indeed, if τ acted biregularly on $\mathcal{W}$, it would preserve the contraction $\pi \colon \mathcal{W} \to V_5$ and thus would descend to V_5.

We have constructed an $\mathfrak{A}_5$-Sarkisov link of type II (cf. [28, Definition 3.4]). Namely, there exists a commutative diagram

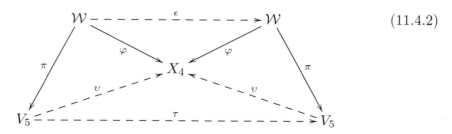
(11.4.2)

Here π is the blow-up of the six lines of $\mathcal{L}_6$, the rational map v is given by the linear system $\mathcal{Q}(\mathcal{L}_6)$, the morphism φ is the regularization of the map v. The birational map τ is a non-biregular $\mathfrak{A}_5$-birational map such that the subgroup $\mathrm{Aut}^{\mathfrak{A}_5}(V_5) \cong \mathfrak{A}_5$ and the element τ generate a subgroup

CREMONA GROUPS AND THE ICOSAHEDRON 261

in $\mathrm{Bir}^{\mathfrak{A}_5}(V_5)$ isomorphic to $\mathfrak{S}_5$. The birational map ϵ, which is just the involution τ considered as an element in $\mathrm{Bir}^{\mathfrak{A}_5}(\mathcal{W})$, is not biregular by Remark 11.4.1 and thus is a composition of flops in the curves contracted by φ (cf. Corollary 11.2.18).

Remark 11.4.3. Recall that X_4 contains exactly 24 lines that pass through the singular point P (see Proposition 11.2.12). These 24 lines form one $\mathfrak{S}_5$-orbit and split into two $\mathfrak{A}_5$-orbits consisting of 12 lines each. Thus, these two $\mathfrak{A}_5$-orbits are swapped by the involution τ (which acts biregularly on X_4 by construction). One of these two $\mathfrak{A}_5$-orbits arises from the curve $\mathcal{L}_{12} \subset V_5$ as described in Lemma 11.2.4. Another $\mathfrak{A}_5$-orbit arises from the curve $\mathcal{L}_{12}$ contained in the second copy of V_5 in a similar way.

The subgroup generated by $\mathrm{Aut}^{\mathfrak{A}_5}(V_5) \cong \mathfrak{A}_5$ and τ is not the whole group $\mathrm{Bir}^{\mathfrak{A}_5}(V_5)$, because the quartic X_4 has a natural non-biregular $\mathfrak{A}_5$-equivariant birational involution. Indeed, projecting from the point $P \in X_4$, we obtain a rational map $\chi \colon X_4 \dashrightarrow \mathbb{P}^3$ that is generically two-to-one. This map induces an $\mathfrak{S}_5$-birational involution $\iota \in \mathrm{Bir}^{\mathfrak{S}_5}(X_4)$. Note that ι is actually an $\mathfrak{S}_5$-equivariant birational involution, because ι commutes with every element of $\mathfrak{S}_5$; this can be seen using the structure of the double cover χ. Therefore, we get

Corollary 11.4.4. *One has* $\langle \mathfrak{S}_5, \iota \rangle \cong \mathfrak{S}_5 \times \boldsymbol{\mu}_2$.

Remark 11.4.5. The birational automorphisms ι and $\iota \circ \tau$ are not biregular on V_5. Indeed, the group $\mathrm{Aut}(V_5) \cong \mathrm{PSL}_2(\mathbb{C})$ does not contain subgroups isomorphic to $\mathfrak{S}_5$ or $\mathfrak{A}_5 \times \boldsymbol{\mu}_2$. Similarly to Remark 11.4.1, this implies that ι and $\iota \circ \tau$ do not act biregularly on $\mathcal{W}$.

It is well known that ι acts biregularly on X_4 if and only if the form q_3 in (11.2.8) is a zero polynomial. Since q_3 is not a zero polynomial by Lemma 11.2.11, we have:

Corollary 11.4.6. *The birational involution ι does not act biregularly on* X_4.

Let $\mathcal{Q}(\mathscr{C}')$ be the linear subsystem in $|2H|$ consisting of all surfaces that contain the curve $\mathscr{C}'$. Then $\mathcal{Q}(\mathscr{C}')$ does not have fixed points outside of the curve $\mathscr{C}'$, because the normal rational curve $\mathscr{C}'$ is a scheme-theoretic intersection of quadrics in $\mathbb{P}^6$. Moreover, the linear system $\mathcal{Q}(\mathscr{C}')$ contains the smooth $\mathfrak{A}_5$-invariant surface $S_{\mathscr{C}'}$ such that $\mathscr{C}' \subset S'_{\mathscr{C}}$ (see §11.3).

Lemma 11.4.7. *The divisor $2H|_{S_{\mathscr{C}'}} - \mathscr{C}'$ is very ample on $S_{\mathscr{C}'}$.*

Proof. Put $H_{S_{\mathscr{C}'}} = H|_{S_{\mathscr{C}'}}$ and $D = 2H_{S_{\mathscr{C}'}} - \mathscr{C}'$. Then

$$H_{S_{\mathscr{C}'}} \cdot D = D^2 = 14.$$

Since

$$h^2(\mathcal{O}_{S_{\mathscr{C}'}}(D)) = h^0(\mathcal{O}_{S_{\mathscr{C}'}}(-D)) = 0$$

by Serre duality, the Riemann–Roch formula implies that

$$h^0(\mathcal{O}_{S_{\mathscr{C}'}}(D)) \geq \frac{D^2}{2} + 2 = 9.$$

In particular, the mobile part of the linear system $|D|$ is non-trivial.

We claim that the linear system $|D|$ does not have fixed components. Indeed, suppose that it does. Let us denote by Z the fixed part of $|D|$. Then Z must be $\mathfrak{A}_5$-invariant. On the other hand, we have

$$Z \cdot H_{S_{\mathscr{C}'}} \leq D \cdot H_{S_{\mathscr{C}'}} = 14.$$

Thus, Lemma 10.2.7 implies that $\operatorname{Supp}(Z) = \mathscr{C}'$. Therefore, we see that the linear system $|2H_{S_{\mathscr{C}'}} - m\mathscr{C}'|$ is free from base curves for some integer $m \geq 2$. In particular, the divisor $2H_{S_{\mathscr{C}'}} - m\mathscr{C}'$ is nef. On the other hand, one has

$$\left(2H_{S_{\mathscr{C}'}} - m\mathscr{C}'\right)^2 = 40 - 24m - 2m^2 < 0,$$

because $m \geq 2$. The obtained contradiction shows that the linear system $|D|$ does not have fixed components.

Since $D^2 = 14$ and the linear system $|D|$ does not have fixed components, the linear system $|D|$ is free from base points by [109, Corollary 3.2]. Let us show that D is actually ample. Suppose it is not. By Lemma 6.7.6, there exists a smooth rational curve R on the surface $S_{\mathscr{C}'}$ such that

$$D^2\left(2H_{S_{\mathscr{C}'}}^2 + r(H_{S_{\mathscr{C}'}} \cdot R)^2\right) = 2(H_{S_{\mathscr{C}'}} \cdot D)^2$$

for some $r \in \{1, 5, 6, 10, 12, 15\}$. This gives

$$14(20 + r(H_{S_{\mathscr{C}'}} \cdot R)^2) = 392,$$

so that $r(H_{S_{\mathscr{C}'}} \cdot R)^2 = 8$ for some $r \in \{1, 5, 6, 10, 12, 15\}$, which is absurd.

Thus, we see that D is ample. Therefore, one has $h^0(\mathcal{O}_{S_{\mathscr{C}'}}(D)) = 9$ by Kodaira vanishing and the Riemann–Roch formula.

CREMONA GROUPS AND THE ICOSAHEDRON 263

We must show that D is very ample. Suppose it is not. Since the linear system $|D|$ is free from base points, it follows from Lemma 6.7.7 that

$$14k^2 - 56k + 40 = k^2 D^2 - 4kD \cdot H_S + 4H_S^2 = 0$$

for some positive integer k. However, the latter equation does not have integer solutions. Thus, the divisor D is very ample. □

Denote by $\bar{S}_{\mathscr{C}'}$ the proper transform of the surface $S_{\mathscr{C}'}$ on the threefold $\mathcal{U}$ (see §11.3). Then

$$\bar{S}_{\mathscr{C}'} \sim \psi^*(2H) - \mathcal{F}_\mathcal{U},$$

because the surface $S_{\mathscr{C}'}$ is smooth. Note that $\mathcal{F}_\mathcal{U} \cong \mathbb{P}^1 \times \mathbb{P}^1$ by Lemma 11.3.3.

Lemma 11.4.8 (cf. Lemma 7.2.9). *The action of $\mathfrak{A}_5$ on $\mathcal{F}_\mathcal{U}$ is twisted diagonal (see §6.4).*

Proof. Since the surface $S_{\mathscr{C}'}$ is smooth, we see that $\bar{S}_{\mathscr{C}'}|_{\mathcal{F}_\mathcal{U}}$ is a smooth $\mathfrak{A}_5$-invariant irreducible curve. On the other hand, $\bar{S}_{\mathscr{C}'}|_{\mathcal{F}_\mathcal{U}}$ is a divisor of bi-degree $(1,7)$ on $\mathcal{F}_\mathcal{U} \cong \mathbb{P}^1 \times \mathbb{P}^1$, since

$$\bar{S}_{\mathscr{C}'}|_{\mathcal{F}_\mathcal{U}} \sim \left(\psi^*(2H) - \mathcal{F}_\mathcal{U}\right)|_{\mathcal{F}_\mathcal{U}}.$$

This implies that the action of $\mathfrak{A}_5$ on $\mathcal{F}_\mathcal{U}$ is twisted diagonal by Lemma 6.4.3(ii). □

Note that

$$-K_\mathcal{U} \sim \psi^*(2H) - \mathcal{F}_\mathcal{U}.$$

Thus, the linear system $|-K_\mathcal{U}|$ is the proper transform of the linear system $\mathcal{Q}(\mathscr{C}')$ on the threefold $\mathcal{U}$ via ψ. We compute

$$\mathcal{F}_\mathcal{U}^3 = -c_1\left(\mathcal{N}_{\mathscr{C}'/V_5}\right) = K_{V_5} \cdot \mathscr{C}' - 2g(\mathscr{C}') + 2 = -10$$

and $\psi^*(H) \cdot \mathcal{F}_\mathcal{U}^2 = -6$. Hence, one has $-K_\mathcal{U}^3 = 14$. Since the normal rational sextic curve $\mathscr{C}'$ is a scheme-theoretic intersection of quadrics in $\mathbb{P}^6$, we obtain:

Corollary 11.4.9. *The linear system $|-K_\mathcal{U}|$ is free from base points, so that the divisor $-K_\mathcal{U}$ is nef and big.*

In particular, we have
$$h^0(\mathcal{O}_\mathcal{U}(-K_\mathcal{U})) = 10$$
by the Riemann–Roch formula and Theorem 2.3.5. Thus, the linear system $|-K_\mathcal{U}|$ defines a morphism $\mathcal{U} \to \mathbb{P}^9$. Denote its image by X'_{14}. Let
$$\zeta' : \mathcal{U} \to X'_{14}$$
be the induced morphism. The following result can be deduced from [67, Propositions 1.3 and 1.6].

Lemma 11.4.10 (cf. Proposition 7.2.4). *The morphism ζ' is birational, and X'_{14} is a Fano variety with at most canonical Gorenstein singularities such that $-K^3_{X'_{14}} = 14$.*

Proof. For $l \gg 0$, the linear system $|-lK_\mathcal{U}|$ gives a birational morphism
$$\hat{\zeta}' : \mathcal{U} \to \hat{X}_{14}$$
such that $\hat{X}'_{14}$ is a normal variety and
$$-K_\mathcal{U} \sim (\hat{\zeta}')^*(-K_{\hat{X}'_{14}}).$$
This follows from the Basepoint-free Theorem (see, for example, [74, Theorem 3.3]). Thus
$$-K^3_{\hat{X}'_{14}} = -K^3_\mathcal{U} = 14$$
and $\hat{X}'_{14}$ is a Fano threefold with canonical singularities.

By Lemma 11.4.7, the divisor $2H|_{S_{\mathscr{C}'}} - \mathscr{C}'$ is very ample on $S_{\mathscr{C}'}$. Thus, the divisor $-K_\mathcal{U}|_{\bar{S}_{\mathscr{C}'}}$ is very ample. Hence, it follows from [109, Theorem 6.1](ii) that the graded ring
$$\bigoplus_{i \geqslant 1} H^0\left(\mathcal{O}_{\bar{S}_{\mathscr{C}'}}(-iK_\mathcal{U}|_{\bar{S}_{\mathscr{C}'}})\right)$$
is generated by $H^0(\mathcal{O}_{\bar{S}_{\mathscr{C}'}}(-K_\mathcal{U}|_{\bar{S}_{\mathscr{C}'}}))$. On the other hand, the exact sequence of sheaves
$$0 \to \mathcal{O}_\mathcal{U} \to \mathcal{O}_\mathcal{U}(-K_\mathcal{U}) \to \mathcal{O}_{\bar{S}_{\mathscr{C}'}} \otimes \mathcal{O}_\mathcal{U}(-K_\mathcal{U}) \to 0$$
implies the exact sequence of cohomology groups
$$0 \longrightarrow H^0(\mathcal{O}_\mathcal{U}) \longrightarrow H^0\left(\mathcal{O}_\mathcal{U}(-K_\mathcal{U})\right) \longrightarrow H^0\left(\mathcal{O}_{\bar{S}_{\mathscr{C}'}}(-K_\mathcal{U}|_{\bar{S}_{\mathscr{C}'}})\right) \longrightarrow 0,$$
(11.4.11)

CREMONA GROUPS AND THE ICOSAHEDRON

since $h^1(\mathcal{O}_\mathcal{U}) = h^1(\mathcal{O}_{V_5}) = 0$ by Kodaira vanishing. Thus, it follows from [62, Lemma 2.9] that the graded ring

$$\bigoplus_{i \geqslant 1} H^0\big(\mathcal{O}_\mathcal{U}(-iK_\mathcal{U})\big)$$

is generated by $H^0(\mathcal{O}_\mathcal{U}(-K_\mathcal{U}))$. This implies that X'_{14} is projectively normal (in particular, it is normal by [53, Exercise I.3.18(a)]) and $X'_{14} \cong \hat{X}'_{14}$. Thus, X'_{14} is a Fano threefold with canonical singularities and

$$-K^3_{X'_{14}} = -K^3_\mathcal{U} = 14.$$

$\square$

We have the following diagram of maps between the varieties in question.

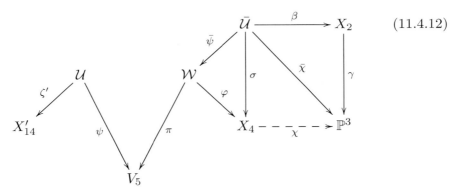
(11.4.12)

Here $\bar{\psi}$ is a blow-up of the proper transform $\check{\mathscr{C}}'$ of the curve $\mathscr{C}'$ on the variety $\mathcal{W}$, and the morphism σ is a blow-up of the singular point P on the quartic X_4. The morphism β is a birational morphism that contracts the proper transforms on $\bar{\mathcal{U}}$ of the lines on X_4 passing through the point P and the proper transforms of the curves contracted by φ except the curve $\check{\mathscr{C}}'$ (note that β is given by the linear system $|-2K_{\bar{\mathcal{U}}}|$). The morphism

$$\gamma \colon X_2 \to \mathbb{P}^3$$

is a double cover branched along a singular sextic surface (defined in the appropriate coordinates by the equation (11.2.17)), so that γ is given by the linear system $|-K_{X_2}|$.

Remark 11.4.13. The morphism β contracts the proper transforms of the 24 lines on X_4 that pass through the point P (see Proposition 11.2.12) to 24 isolated ordinary double points of the threefold X_2. Note that X_2 has exactly 29 singular points: 24 isolated ordinary double points just mentioned, and 5 non-ordinary isolated double points that are images of the singular points of X_4 of type cD_4 (see Corollary 11.2.18).

Note that the group
$$\langle \mathfrak{S}_5, \iota \rangle \cong \mathfrak{S}_5 \times \boldsymbol{\mu}_2$$
acts biregularly on X_2. Now we will show that $\iota \circ \tau$ acts biregularly on X'_{14}, but ι does not.

Lemma 11.4.14. *The element $\iota \circ \tau \in \mathrm{Bir}(V_5)$ acts on $\mathcal{U}$ biregularly in codimension 1. On the other hand, ι does not act biregularly in codimension 1 on $\mathcal{U}$.*

Proof. Let $E_1, \ldots, E_6$ denote the exceptional divisors of the morphism $\pi \colon \mathcal{W} \to V_5$ (see §11.1). Let $\tilde{E}_i \subset X_4$ be the image of the divisor E_i under the morphism φ, and put $\bar{E}_i = \sigma^{-1} \tilde{E}_i$, so that
$$\bar{E}_i = \bar{\psi}^{-1} E_i = \bar{\psi}^*(E_i).$$
Denote by $\bar{\mathcal{F}}_\mathcal{U}$ the surface in $\bar{\mathcal{U}}$ that is a proper transform of the surface $\mathcal{F}_\mathcal{U}$, so that $\bar{\mathcal{F}}_\mathcal{U}$ is the exceptional surface of the morphism $\bar{\psi}$. Then $\bar{E}_i \cdot \bar{E}_j = 0$ for $i \neq j$. Furthermore, one has the following intersection numbers on $\bar{\mathcal{U}}$ (cf. (11.1.7)):
$$\big((\pi \circ \bar{\psi})^*(H)\big)^2 \cdot \bar{E}_i = 0, \quad (\pi \circ \bar{\psi})^*(H) \cdot \bar{E}_i^2 = -1, \quad (\pi \circ \bar{\psi})^*(H) \cdot \bar{E}_i \cdot \bar{\mathcal{F}}_\mathcal{U} = 0,$$
$$\bar{E}_i^3 = 0, \quad \bar{E}_i^2 \cdot \bar{\mathcal{F}}_\mathcal{U} = 0, \quad \bar{E}_i \cdot \bar{\mathcal{F}}_\mathcal{U}^2 = -2. \quad (11.4.15)$$

Both τ and ι are biregular on X_2 and thus act on $\bar{\mathcal{U}}$ biregularly in codimension 1. Note that any birational automorphism ξ acting on $\bar{\mathcal{U}}$ biregularly in codimension 1 also acts on $\mathcal{U}$ biregularly in codimension 1 if and only if the divisor $\sum_{i=1}^{6} \bar{E}_i$ is invariant under ξ. We are going to show that this holds for $\iota \circ \tau$ and fails for ι.

Note that the regularization $\bar{\chi} \colon \bar{\mathcal{U}} \to \mathbb{P}^3$ of the rational map $\chi \colon X_4 \dashrightarrow \mathbb{P}^3$ is given by the linear system
$$|-K_{\bar{\mathcal{U}}}| = \left| 2(\pi \circ \bar{\psi})^*(H) - \sum_{i=1}^{6} \bar{E}_i - \bar{\mathcal{F}}_\mathcal{U} \right|.$$
Keeping in mind (11.4.15), we compute
$$\big(2(\pi \circ \bar{\psi})^*(H) - \sum_{i=1}^{6} \bar{E}_i - \bar{\mathcal{F}}_\mathcal{U}\big)^2 \cdot \bar{E}_i = 2.$$

Therefore, we see that $\bar{\chi}(\bar{E}_i) \subset \mathbb{P}^3$ is either a plane or a quadric. In the former case, the restriction of $\bar{\chi}$ to $\bar{E}_i$ has degree 2 at a general point of $\bar{E}_i$, so that $\bar{\chi}(\bar{E}_i) \neq \bar{\chi}(\bar{E}_j)$ for $i \neq j$, which is impossible by Lemma 5.3.5(i). Thus, $\bar{\chi}(\bar{E}_i)$ is a quadric, and the restriction of $\bar{\chi}$ to $\bar{E}_i$ is birational at a general point of $\bar{E}_i$. In particular, none of the surfaces $\bar{E}_1, \ldots, \bar{E}_6$ is ι-invariant, because ι is the involution of the double cover $\bar{\chi}$. Moreover, the surfaces $\bar{\chi}(\bar{E}_1), \ldots, \bar{\chi}(\bar{E}_6)$ are distinct by Lemma 5.3.5(iii). Therefore, the involution ι does not map any $\bar{E}_i$ to some $\bar{E}_j$. This implies that the action of ι on $\mathcal{U}$ is *not* biregular in codimension 1.

Let $\bar{E}'_i = \iota(\bar{E}_i)$ and $\bar{E}''_i = \tau(\bar{E}_i)$. By Lemma 5.3.5(v) one has

$$\left\{\bar{\chi}(\bar{E}_1), \ldots, \bar{\chi}(\bar{E}_6)\right\} = \left\{\bar{\chi}(\bar{E}''_1), \ldots, \bar{\chi}(\bar{E}''_6)\right\}.$$

On the other hand, we have already seen that

$$\left\{\bar{E}_1, \ldots, \bar{E}_6\right\} \neq \left\{\bar{E}'_1, \ldots, \bar{E}'_6\right\}.$$

We claim that $\{\bar{E}_1, \ldots, \bar{E}_6\} \neq \{\bar{E}''_1, \ldots, \bar{E}''_6\}$. Indeed, suppose that the latter is not true. Then τ preserves the exceptional divisor $E_1 \cup \ldots \cup E_6$ of the contraction $\pi: \mathcal{W} \to V_5$. Since τ acts biregularly in codimension 1 on $\mathcal{W}$ by Lemma 11.2.6, the involution τ must act biregularly in codimension 1 on V_5. The latter implies that τ acts biregularly on V_5 (see, for example, [28, Proposition 3.5]). However, τ does not act biregularly on V_5 by Remark 11.4.1.

Since $\{\bar{E}_1, \ldots, \bar{E}_6\} \neq \{\bar{E}''_1, \ldots, \bar{E}''_6\}$, we have

$$\left\{\bar{E}''_1, \ldots, \bar{E}''_6\right\} = \left\{\bar{E}'_1, \ldots, \bar{E}'_6\right\}.$$

Therefore,

$$\left\{\iota\bigl(\tau(\bar{E}_1)\bigr), \ldots, \iota\bigl(\tau(\bar{E}_6)\bigr)\right\} = \left\{\bar{E}_1, \ldots, \bar{E}_6\right\},$$

and the assertion of Lemma 11.4.14 follows. □

Lemma 11.4.16 (cf. Lemma 11.2.6). *The morphism $\zeta': \mathcal{U} \to X'_{14}$ contracts finitely many curves.*

Proof. Suppose that ζ' contracts an irreducible surface. Denote it by E. Then $\zeta'(E)$ is either a curve or a point. As above, consider the surface $S_{\mathscr{C}'} \in \mathcal{Q}_2$ that contains the curve $\mathscr{C}'$. The surface $S_{\mathscr{C}'}$ is smooth. Let $\bar{S}_{\mathscr{C}'}$ be the proper transform of the surface $S_{\mathscr{C}'}$ on $\mathcal{U}$ via ψ. Then the induced morphism

$$\psi|_{\bar{S}_{\mathscr{C}'}}: \bar{S}_{\mathscr{C}'} \to S_{\mathscr{C}'}$$

is an isomorphism.

We claim that $E \cap \bar{S}_{\mathscr{C}'} = \varnothing$. Indeed, suppose that $E \cap \bar{S}_{\mathscr{C}'} \neq \varnothing$. Then $E|_{\bar{S}_{\mathscr{C}'}}$ is an effective divisor on the surface $\bar{S}_{\mathscr{C}'}$. Put

$$\bar{D} = -K_{\mathcal{U}}|_{\bar{S}_{\mathscr{C}'}}.$$

Then $\bar{D}$ is very ample on $\bar{S}_{\mathscr{C}'}$ by Lemma 11.4.7. On the other hand, we have

$$E|_{\bar{S}_{\mathscr{C}'}} \cdot \bar{D} = E \cdot (-K_{\mathcal{U}})^2 = 0,$$

which implies that $\bar{D}$ is not ample. The latter is a contradiction.

Therefore, we have $E \cap \bar{S}_{\mathscr{C}'} = \varnothing$, so that $E \neq \mathcal{F}_{\mathcal{U}}$ since the intersection $\mathcal{F}_{\mathcal{U}} \cap \bar{S}_{\mathscr{C}'}$ is not empty by construction. Thus the surface E is not ψ-exceptional, which means that $\psi(E)$ is a surface in V_5. Since $\mathrm{Pic}(V_5) = \mathbb{Z}[H]$, we have

$$\psi(E) \sim aH$$

for some positive integer a. Since $E \cap \bar{S}_{\mathscr{C}'} = \varnothing$, we have $\psi(E) \cap S_{\mathscr{C}'} \subset \mathscr{C}'$. Therefore,

$$\psi(E)|_{S_{\mathscr{C}'}} = b\mathscr{C}'$$

for some positive integer b. Hence

$$0 < 10a^2 = a^2 H^2 \cdot S_{\mathscr{C}'} = \psi(E)|_{S_{\mathscr{C}'}} \cdot \psi(E)|_{S_{\mathscr{C}'}} = b^2 \mathscr{C}' \cdot \mathscr{C}' = -2b^2 < 0,$$

which is absurd. $\square$

Corollary 11.4.17. *The element $\iota \circ \tau \in \mathrm{Bir}(V_5)$ acts biregularly on X'_{14}. On the other hand, ι is not biregular on X'_{14}.*

Proof. The involution ι is not biregular on X'_{14}, since it does not act on $\mathcal{U}$ biregularly in codimension 1 by Lemma 11.4.14, while X'_{14} and $\mathcal{U}$ are isomorphic in codimension 1 by Lemma 11.4.16.

By Lemmas 11.4.14 and 11.4.16, the involution $\iota \circ \tau$ acts on X'_{14} biregularly in codimension 1. Since it preserves the class of the ample divisor $-K_{X'_{14}}$, it must be biregular (see, for example, [28, Proposition 3.5]). $\square$

CREMONA GROUPS AND THE ICOSAHEDRON

Thus, we have constructed a second $\mathfrak{A}_5$-Sarkisov link of type II (cf. [28, Definition 3.4]). Namely, there exists a commutative diagram

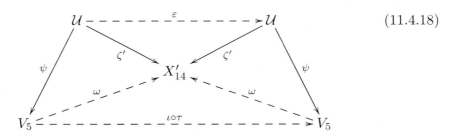
(11.4.18)

Here ψ is the blow-up of the sextic rational normal curve $\mathscr{C}'$, the morphism ζ' is the regularization of the rational map ω, the rational map ε is a simultaneous flop of all curves contracted by ζ'. The birational map ε, which is just the involution $\iota \circ \tau$ considered as an element in $\mathrm{Bir}^{\mathfrak{A}_5}(\mathcal{U})$, is a composition of flops in the curves contracted by ζ'. Later we will show in Lemma 11.5.6 that ε is a composition of Atiyah flops (see §4.2) in the irreducible components of the proper transform $\bar{\mathcal{L}}_6$ of the curve $\mathcal{L}_6$ on the variety $\mathcal{U}$.

Remark 11.4.19. The $\mathfrak{A}_5$-Sarkisov link (11.4.18) is the Sarkisov link number 107 in the classification obtained in [31]. The construction provided above is the only way we know to prove the existence of this link.

Recall from Lemma 11.3.2 that the threefold V_5 contains exactly two $\mathfrak{A}_5$-invariant rational normal sextic curves, namely $\mathscr{C}$ and $\mathscr{C}'$. By Proposition 7.2.4, a construction similar to the one above applied to the curve $\mathscr{C}$ provides what is sometimes called a *bad link*.

11.5 Action on the Picard group

Now we are going to compute the action of τ^* on the Picard group of the threefold $\mathcal{W}$ (note that τ acts on $\mathrm{Pic}(\mathcal{W})$ since it acts on $\mathcal{W}$ biregularly in codimension 1 by Lemma 11.2.6). Recall that the divisors $\pi^*(H)$ and E_i, $1 \leqslant i \leqslant 6$, generate the group $\mathrm{Pic}(\mathcal{W})$, while the divisors $\pi^*(H)$ and $\sum_{i=1}^{6} E_i$ generate the group $\mathrm{Pic}(\mathcal{W})^{\mathfrak{A}_5}$.

Lemma 11.5.1. One has

$$\tau^*(\pi^*(H)) \sim 13\pi^*(H) - 7\sum_{j=1}^{6} E_j$$

and
$$\tau^*(E_i) \sim 4\pi^*(H) - E_i - 2\sum_{i=j}^{6} E_j.$$

Proof. For some $a, b \in \mathbb{Z}$ one has
$$\tau^*(\pi^*(H)) \sim a\pi^*(H) - b\sum_{j=1}^{6} E_j.$$

Since the class
$$-K_\mathcal{W} \sim 2\pi^*(H) - \sum_{j=1}^{6} E_j$$

is invariant under τ, we also have
$$\tau^*\left(\sum_{j=1}^{6} E_j\right) \sim (2a-2)\pi^*(H) - (2b-1)\sum_{j=1}^{6} E_j. \tag{11.5.2}$$

Using (11.1.7), we compute the intersection numbers
$$10 = \left(2\pi^*(H) - \sum_{j=1}^{6} E_j\right) \cdot (\pi^*(H))^2 =$$
$$= \left(2\pi^*(H) - \sum_{j=1}^{6} E_j\right) \cdot \left(\tau^*(\pi^*(H))\right)^2 = 10a^2 - 12ab - 12b^2,$$

and
$$14 = \left(2\pi^*(H) - \sum_{j=1}^{6} E_j\right)^2 \cdot \pi^*(H) =$$
$$= \left(2\pi^*(H) - \sum_{j=1}^{6} E_j\right)^2 \cdot \left(\tau^*(\pi^*(H))\right) = 14a - 24b.$$

Keeping in mind that the action of τ is non-trivial (since τ does not act biregularly on $\mathcal{W}$ by Remark 11.4.1), we conclude that $a = 13$ and $b = 7$. This gives the first assertion of the lemma. The second is implied by (11.5.2) and symmetry arguments. □

Now we are going to compute the action of $(\iota\circ\tau)^*$ on the Picard group of the threefold $\mathcal{U}$ (note that $\iota\circ\tau$ acts on $\mathrm{Pic}(\mathcal{U})$ since it acts on $\mathcal{U}$ biregularly in codimension 1 by Lemma 11.4.14). The divisor $\psi^*(H)$ and the ψ-exceptional divisor $\mathcal{F}_\mathcal{U}$ generate the group $\mathrm{Pic}(\mathcal{U}) = \mathrm{Pic}(\mathcal{U})^{\mathfrak{A}_5}$. We have the following intersection numbers on $\mathcal{U}$ (cf. (11.4.15)):

$$\psi^*(H)^3 = 5, \quad \psi^*(H)^2 \cdot \mathcal{F}_\mathcal{U} = 0, \quad \psi^*(H) \cdot \mathcal{F}_\mathcal{U}^2 = -6, \quad \mathcal{F}_\mathcal{U}^3 = -10. \quad (11.5.3)$$

Lemma 11.5.4. One has

$$(\iota\circ\tau)^*(\psi^*(H)) \sim 3\psi^*(H) - 2\mathcal{F}_\mathcal{U}$$

and

$$(\iota\circ\tau)^*(\mathcal{F}_\mathcal{U}) \sim 4\psi^*(H) - 3\mathcal{F}_\mathcal{U}.$$

Proof. For some $a, b \in \mathbb{Z}$ one has

$$(\iota\circ\tau)^*(\psi^*(H)) \sim a\psi^*(H) - b\mathcal{F}_\mathcal{U}.$$

Since the class $-K_\mathcal{U} \sim 2\psi^*(H) - \mathcal{F}_\mathcal{U}$ is invariant under $\iota\circ\tau$, we also have

$$(\iota\circ\tau)^*(\mathcal{F}_\mathcal{U}) \sim (2a - 2)\psi^*(H) - (2b - 1)\mathcal{F}_\mathcal{U}.$$

Using (11.5.3), we compute the intersection numbers

$$10 = (2\psi^*(H) - \mathcal{F}_\mathcal{U}) \cdot (\psi^*(H))^2 =$$
$$= (2\psi^*(H) - \mathcal{F}_\mathcal{U}) \cdot ((\iota\circ\tau)^*(\psi^*(H)))^2 = 10a^2 - 12ab - 2b^2,$$

and

$$14 = (2\psi^*(H) - \mathcal{F}_\mathcal{U})^2 \cdot \psi^*(H) =$$
$$= (2\psi^*(H) - \mathcal{F}_\mathcal{U})^2 \cdot ((\iota\circ\tau)^*(\psi^*(H))) = 14a - 14b.$$

Keeping in mind that the action of $\iota\circ\tau$ is non-trivial (since $\iota\circ\tau$ does not act biregularly on $\mathcal{W}$ by Remark 11.4.5), we conclude that $a = 3$ and $b = 2$. □

Later we will show that the subgroup $\mathfrak{A}_5 \subset \mathrm{Aut}(V_5)$ and the elements τ and ι generate the group $\mathrm{Bir}^{\mathfrak{A}_5}(V_5)$ (see the proof of Theorem 1.4.1 in §18.1). For this we will need the following auxiliary result.

Lemma 11.5.5. Let $\mathcal{D}$ be an $\mathfrak{A}_5$-invariant mobile linear system on V_5. Then there exists an element
$$f \in \langle \mathfrak{A}_5, \tau, \iota \rangle$$
such that
$$\mathrm{mult}_{\mathscr{C}'}(f^{-1}\mathcal{D}) \leqslant \frac{1}{\lambda}$$
and
$$\mathrm{mult}_{\mathcal{L}_6}(f^{-1}\mathcal{D}) \leqslant \frac{1}{\lambda},$$
where λ is a positive rational number defined by the relation
$$\lambda f^{-1}\mathcal{D} \sim_{\mathbb{Q}} -K_{V_5}.$$

Proof. For every $g \in \langle \mathfrak{A}_5, \tau, \iota \rangle$, one has
$$g^{-1}\mathcal{D} \sim n(g)H \sim_{\mathbb{Q}} -\frac{n(g)}{2}K_{V_5}$$
for some natural number $n(g)$ (that depends on g of course). Thus, there exists $f \in \langle \mathfrak{A}_5, \tau, \iota \rangle$ such that $n(f)$ is the smallest one. We claim that this f is the one we are looking for. Indeed, we have
$$(f \circ \tau)^{-1}\mathcal{D} \sim \Big(13n(f) - 24\mathrm{mult}_{\mathcal{L}_6}(f^{-1}\mathcal{D})\Big)H$$
by Lemma 11.5.1. Similarly, we have
$$(f \circ \iota \circ \tau)^{-1}\mathcal{D} \sim \Big(3n(f) - 4\mathrm{mult}_{\mathscr{C}'}(f^{-1}\mathcal{D})\Big)H$$
by Lemma 11.5.4. Hence,
$$n(f \circ \tau) = 13n(f) - 24\mathrm{mult}_{\mathcal{L}_6}(f^{-1}\mathcal{D})$$
and
$$n(f \circ \iota \circ \tau) = 3n(f) - 4\mathrm{mult}_{\mathscr{C}'}(f^{-1}\mathcal{D}).$$
Since f minimizes $n(f)$, we see that
$$\mathrm{mult}_{\mathcal{L}_6}(f^{-1}\mathcal{D})) \leqslant \frac{n(f)}{2}$$
and
$$\mathrm{mult}_{\mathscr{C}'}(f^{-1}\mathcal{D})) \leqslant \frac{n(f)}{2},$$
which is exactly what we had to prove. $\square$

CREMONA GROUPS AND THE ICOSAHEDRON 273

Let $\bar{\mathcal{L}}_6$ be the proper transform of the curve $\mathcal{L}_6$ on the threefold $\mathcal{U}$. Note that $-K_\mathcal{U} \cdot \bar{\mathcal{L}}_6 = 0$ by Lemmas 7.4.3 and 11.3.1, which implies that every irreducible component of the curve $\bar{\mathcal{L}}_6$ is contracted by ζ'.

Lemma 11.5.6. The following assertions hold:

(i) the only irreducible curves contracted by the morphism $\zeta' \colon \mathcal{U} \to X'_{14}$ are irreducible components of $\bar{\mathcal{L}}_6$;

(ii) the normal bundle of every irreducible component of $\bar{\mathcal{L}}_6$ is isomorphic to $\mathcal{O}_{\mathbb{P}^1}(-1) \oplus \mathcal{O}_{\mathbb{P}^1}(-1)$;

(iii) the morphism ζ' contracts each irreducible component of the curve $\bar{\mathcal{L}}_6$ to an isolated ordinary double point of X'_{14};

(iv) the birational map ε is a composition of Atiyah flops in the irreducible components of the curve $\bar{\mathcal{L}}_6$.

Proof. By Lemma 11.4.16, the morphism ζ' contracts only finitely many curves. Suppose that ζ' contracts an irreducible curve that is not an irreducible component of $\bar{\mathcal{L}}_6$. Then there exists an $\mathfrak{A}_5$-irreducible curve $\bar{Z} \subset \mathcal{U}$ such that $\bar{Z} \neq \bar{\mathcal{L}}_6$ and $-K_\mathcal{U} \cdot \bar{Z} = 0$.

Since $\mathcal{F}_\mathcal{U} \cong \mathbb{P}^1 \times \mathbb{P}^1$ by Lemma 11.3.3, we have $\bar{Z} \not\subset \mathcal{F}_\mathcal{U}$. In particular, $\bar{Z}$ is not contracted by ψ. Put $Z = \psi(\bar{Z})$. Then Z is an $\mathfrak{A}_5$-irreducible curve on V_5 such that $Z \neq \mathscr{C}'$ and $Z \neq \mathcal{L}_6$.

The linear system $|3\psi^*(H) - 2\mathcal{F}_\mathcal{U}|$ does not have base components, since it is a proper transform of the linear system $|\psi^*(H)|$ via $\iota \circ \tau$ by Lemma 11.5.4, and $\iota \circ \tau$ acts on $\mathcal{U}$ biregularly in codimension 1 by Lemma 11.4.14. On the other hand, we have

$$\bar{Z} \cdot \left(3\psi^*(H) - 2\mathcal{F}_\mathcal{U}\right) = -\bar{Z} \cdot \psi^*(H) - 2K_\mathcal{U} \cdot \bar{Z} = -\bar{Z} \cdot \psi^*(H) < 0,$$

because $\bar{Z}$ is not contracted by ψ. Thus, we see that the curve $\bar{Z}$ is contained in the base locus of the linear system $|3\psi^*(H) - 2\mathcal{F}_\mathcal{U}|$. For the very same reason, the curve $\bar{\mathcal{L}}_6$ is contained in the base locus of the linear system $|3\psi^*(H) - 2\mathcal{F}_\mathcal{U}|$.

Let $\bar{M}$ and $\bar{M}'$ be two general surfaces in $|3\psi^*(H) - 2\mathcal{F}_\mathcal{U}|$. Then

$$\bar{M} \cdot \bar{M}' = \bar{Z} + \bar{\mathcal{L}}_6 + \Omega,$$

where Ω is an effective one-cycle on $\mathcal{U}$. Note that $\Omega \neq 0$, because the linear system $|3\psi^*(H) - 2\mathcal{F}_\mathcal{U}|$ is not composed of a pencil. Moreover, Lemma 4.3.1 implies that $\mathrm{Supp}(\Omega)$ is not contained in $\mathcal{F}_\mathcal{U}$. In particular, one has

$$\Omega \cdot \psi^*(H) > 0.$$

Thus, we have

$$21 = \left(3\psi^*(H) - 2\mathcal{F}_\mathcal{U}\right)^2 \cdot \psi^*(H) = \bar{M} \cdot \bar{M}' \cdot \psi^*(H) =$$
$$= \left(\bar{Z} + \bar{\mathcal{L}}_6 + \Omega\right) \cdot \psi^*(H) > \left(\bar{Z} + \bar{\mathcal{L}}_6\right) \cdot \psi^*(H) =$$
$$= Z \cdot H + \mathcal{L}_6 \cdot H = \deg(Z) + 6,$$

which implies that $\deg(Z) \leqslant 14$.

Note that $Z \cap \mathscr{C}' \neq \varnothing$, since $-K_\mathcal{U} \cdot \bar{Z} = 0$. In particular, we see that $Z \neq \mathscr{C}$ by Lemma 11.3.4. Since we already know that $Z \neq \mathscr{C}'$ and $Z \neq \mathcal{L}_6$, we have $\deg(Z) \geqslant 10$ by Proposition 10.2.5 and Lemma 11.3.2.

We claim that $Z \neq \mathcal{L}_{12}$. Indeed, the curve $\mathcal{L}_{12}$ intersects the curve $\mathscr{C}'$ transversally at points of Σ'_{12} (see Lemma 11.3.4). Thus, if $Z = \mathcal{L}_{12}$, then we have $\mathcal{F}_\mathcal{U} \cdot \bar{Z} = 12$, so that

$$0 = -K_\mathcal{U} \cdot \bar{Z} = \left(\psi^*(2H) - \mathcal{F}_\mathcal{U}\right) \cdot \bar{Z} = 12,$$

which is a contradiction.

By Lemma 10.2.7, there exists a smooth surface $S \in \mathcal{Q}_2$ that contains Z and does not contain $\mathscr{C}'$. One has $S \cdot \mathscr{C}' = 12$. By Theorem 7.3.5, this implies that S intersects $\mathscr{C}'$ transversally in 12 points. In particular, we see that $\bar{Z} \cdot \mathcal{F}_\mathcal{U} \leqslant 12$, which implies that

$$0 = -K_\mathcal{U} \cdot \bar{Z} = \left(2\psi^*(H) - \mathcal{F}_\mathcal{U}\right) \cdot \bar{Z} = 2\deg(Z) - \bar{Z} \cdot \mathcal{F}_\mathcal{U} \geqslant 8,$$

which is a contradiction. Thus, we see that the only irreducible curves contracted by the morphism $\zeta' \colon \mathcal{U} \to X'_{14}$ are irreducible components of $\bar{\mathcal{L}}_6$, which proves assertion (i).

Let L_1 be a line of $\mathcal{L}_6$. Recall that $L_1 \not\subset \mathscr{S}$ by Lemma 7.2.13(ii), so that

$$\mathcal{N}_{L_1/V_5} \cong \mathcal{O}_{\mathbb{P}^1} \oplus \mathcal{O}_{\mathbb{P}^1}$$

by Theorem 7.7.1. Denote by $\bar{L}_1$ the proper transform of the line L_1 on $\mathcal{U}$. Let us show that the normal bundle of $\bar{L}_1$ in $\mathcal{U}$ is $\mathcal{O}_{\mathbb{P}^1}(-1) \oplus \mathcal{O}_{\mathbb{P}^1}(-1)$.

Let us use the notation of §7.7. One has $E_1 \cong \mathbb{P}^1 \times \mathbb{P}^1$.

Denote by $\check{\mathscr{C}}'$ the proper transform of the curve $\mathscr{C}'$ on the threefold $\check{V}_5$. Since the intersection $L_1 \cap \mathscr{C}'$ consists of two points, the intersection $E_1 \cap \check{\mathscr{C}}'$ also consists of two points. Denote them by $\check{P}_1$ and $\check{P}_2$. By Lemma 4.2.6, we have

$$\mathcal{N}_{\bar{L}_1/\mathcal{U}} \cong \mathcal{O}_{\mathbb{P}^1}(-1) \oplus \mathcal{O}_{\mathbb{P}^1}(-1)$$

if and only if the points $\check{P}_1$ and $\check{P}_2$ are not contained in one section of the natural projection $E_1 \to L_1$ that has self-intersection 0. Let us show that this is the case.

Note that $F_1|_{E_1}$ is a smooth curve, since $v(E_1)$ is a hyperplane section of Q that contains $v(F_1)$. By Corollary 7.7.4 we have

$$F_1|_{E_1} \sim 2\check{s} + \check{l},$$

where $\check{l}$ is a fiber of the natural projection $E_1 \to L_1$, and $\check{s}$ is a section of this projection that has self-intersection 0.

Suppose that $\check{P}_1$ and $\check{P}_2$ are contained in one curve $\check{s}_{1,2} \in |\check{s}|$. Then $\check{s}_{1,2}$ is invariant with respect to the stabilizer $G_1 \cong \mathfrak{D}_{10}$ of the line L_1. Hence, the point $\check{s}_{1,2} \cap F_1 \cap E_1$ is G_1-invariant, and the point

$$\pi\left(\check{s}_{1,2} \cap F_1 \cap E_1\right) \in L_1$$

is G_1-invariant as well. Therefore, the group G_1 does not act faithfully on the line L_1 by Lemma 5.5.9(ii), which contradicts Lemma 7.4.6. The obtained contradiction shows that

$$\mathcal{N}_{\bar{L}_1/\mathcal{U}} \cong \mathcal{O}_{\mathbb{P}^1}(-1) \oplus \mathcal{O}_{\mathbb{P}^1}(-1),$$

which proves assertion (ii).

Since X'_{14} is a normal variety, we see that ζ' contracts each irreducible component of the curve $\bar{\mathcal{L}}_6$ to an ordinary double point of X'_{14} by Remark 4.2.1, which proves assertion (iii).

By Remark 4.2.5, the birational map ε is a composition of Atiyah flops in the irreducible components of the curve $\bar{\mathcal{L}}_6$, which proves assertion (iv) and completes the proof of Lemma 11.4.16. □

Corollary 11.5.7 (cf. Corollary 11.2.18). *The threefold X'_{14} has exactly 6 singular points, and each of these points is an ordinary double point.*

Let $\bar{\mathcal{L}}_{12}$ be the proper transform of the curve $\mathcal{L}_{12}$ on the threefold $\mathcal{U}$. Put $\bar{\mathcal{L}}'_{12} = \varepsilon(\bar{\mathcal{L}}_{12})$.

Lemma 11.5.8. *All irreducible components of the curve $\bar{\mathcal{L}}'_{12}$ are contracted by the morphism ψ, i.e., the irreducible components of the curve $\bar{\mathcal{L}}'_{12}$ are fibers of the natural projection $\mathcal{F}_\mathcal{U} \to \mathscr{C}'$.*

Proof. Let $\bar{R}$ be an irreducible component of the curve $\bar{\mathcal{L}}_{12}$. Put $\bar{R}' = \varepsilon(\bar{R})$. Suppose that $\bar{R}'$ is not contracted by the morphism ψ.

Pick a sufficiently general point $\bar{P}' \in \bar{R}'$. Then there exists a surface $\bar{M}' \in |\psi^*(H)|$ such that $\bar{P}' \in \bar{M}'$ and $\bar{R}' \not\subset \bar{M}'$. Let $\bar{M}$ be the proper transform of the surface $\bar{M}'$ on the threefold $\mathcal{U}$. Then
$$\bar{M} \sim 3\psi^*(H) - 2\mathcal{F}_\mathcal{U}$$
by Lemma 11.5.4. Since the choice of the point $\bar{P}'$ was generic, we know that the birational map ε^{-1} is biregular at P (cf. Lemma 11.5.6(iv)). Put
$$\bar{P} = \varepsilon^{-1}(\bar{P}').$$
Then $\bar{P} \in \bar{M}$ and $\bar{R} \not\subset \bar{M}$. On the other hand, we know that
$$\bar{R} \cdot \bar{M} = \bar{R} \cdot (3\psi^*(H) - 2\mathcal{F}_\mathcal{U}) = 1,$$
which implies that $\bar{R} \cap \bar{M} = \bar{P}$.

Put $R = \psi(\bar{R})$. Then R intersects exactly one irreducible component L of the curve $\mathcal{L}_6$ by Lemma 9.1.9. Let $\bar{L}$ be the proper transform of the curve L on the threefold $\mathcal{U}$. Since
$$\bar{L} \cdot \left(3\psi^*(H) - 2\mathcal{F}_\mathcal{U}\right) = -1,$$
we see that the curve $\bar{L}$ is contained in the base locus of the linear system $\bar{M}$. In particular, we obtain $\bar{L} \subset \bar{M}$. Since $\bar{P} \notin \bar{L}$ by construction, we see that
$$\bar{L} \cap \bar{R} = \varnothing.$$
This implies that the tangent vectors to the curves L, R, and $\mathscr{C}'$ at the point
$$Q = L \cap R \cap \mathscr{C}'$$
generate the Zariski tangent space to V_5 at Q. The latter contradicts Lemma 11.3.5(ii). □

Let $\bar{\mathcal{L}}_{15}$ be the proper transform of the curve $\mathcal{L}_{15}$ on the threefold $\mathcal{U}$.

Lemma 11.5.9. One has $\varepsilon(\bar{\mathcal{L}}_{15}) = \bar{\mathcal{L}}_{15}$.

Proof. Note that the birational map ε is well defined in a general point of the curve $\bar{\mathcal{L}}_{15}$ by Lemma 11.5.6(iv). Put
$$\bar{\mathcal{L}}'_{15} = \varepsilon(\bar{\mathcal{L}}_{15}).$$
Then $\bar{\mathcal{L}}'_{15}$ is an $\mathfrak{A}_5$-irreducible curve that has 15 irreducible components. If we prove that $\bar{\mathcal{L}}'_{15}$ is not contained in the ψ-exceptional surface $\mathcal{F}_\mathcal{U}$ and is

mapped by ψ to a curve of degree 15 in V_5, then we are done by Proposition 7.4.1.

In the notation of §11.4, there exists a commutative diagram

$$\begin{array}{ccc} \tilde{\mathcal{U}} & \xrightarrow{\tilde{\varepsilon}} & \tilde{\mathcal{U}} \\ r \downarrow & & \downarrow r \\ \mathcal{U} & \dashrightarrow{\varepsilon} & \mathcal{U} \end{array}$$

where r is the blow-up of the proper transforms $\bar{L}_1, \ldots, \bar{L}_6$ of the six lines of $\mathcal{L}_6$, and $\tilde{\varepsilon}$ is a biregular involution. Denote by $\tilde{\mathcal{L}}_{15}$ the proper transform of the curve $\mathcal{L}_{15}$ on the threefold $\tilde{\mathcal{U}}$. Put $\tilde{\mathcal{L}}'_{15} = \tilde{\varepsilon}(\tilde{\mathcal{L}}_{15})$. Then $\bar{\mathcal{L}}'_{15} = r(\tilde{\mathcal{L}}'_{15})$.

Let $\bar{H}$ be a sufficiently general surface in $|\psi^*(H)|$. Denote by $\tilde{H}$ its proper transform on $\tilde{\mathcal{U}}$ via r. Put $\tilde{M} = \tilde{\varepsilon}(\tilde{H})$ and $\bar{M} = r(\tilde{M})$. Then $\bar{M} \sim 3\bar{H} - 2\mathcal{F}_{\mathcal{U}}$ by Lemma 11.5.4. Thus, we have

$$\tilde{M} \sim r^*(3\bar{H} - 2\mathcal{F}_{\mathcal{U}}) - m\sum_{i=1}^{6} \tilde{E}_i,$$

where $\tilde{E}_1, \ldots, \tilde{E}_6$ are divisors that are contracted by the morphism r to the curves $\bar{L}_1, \ldots, \bar{L}_6$, respectively, and m is some non-negative integer.

Recall that each surface $\tilde{E}_i$ is isomorphic to $\mathbb{P}^1 \times \mathbb{P}^1$ and $\tilde{E}_i|_{\tilde{E}_i}$ is a divisor of bi-degree $(-1,-1)$ by Lemma 11.5.6(ii). Let $\tilde{s}_i$ be a section of the projection $\tilde{E}_i \to \bar{L}_i$ such that $\tilde{s}_i^2 = 0$. Then $\tilde{s}_i \cdot \tilde{M} \geqslant 0$, since $\bar{H}$ is general. Thus, we have

$$\tilde{M} \cdot \tilde{s}_i = \left(r^*(3\bar{H} - 2\mathcal{F}_{\mathcal{U}}) - m\sum_{i=1}^{6}\tilde{E}_i\right)\tilde{s}_i =$$
$$= r^*(3\bar{H} - 2\mathcal{F}_{\mathcal{U}})\tilde{s}_i + m = \left(3\bar{H} - 2\mathcal{F}_{\mathcal{U}}\right) \cdot r(\tilde{s}_i) + m = -1 + m,$$

which implies that $m \geqslant 1$.

Recall that each line of $\mathcal{L}_{15}$ intersects exactly two lines of $\mathcal{L}_6$ by Lemma 9.1.14, and the curves $\mathscr{C}'$ and $\mathcal{L}_{15}$ are disjoint by Lemma 11.3.7. Therefore, we have

$$\bar{H} \cdot \bar{\mathcal{L}}'_{15} = \tilde{H} \cdot \tilde{\mathcal{L}}'_{15} = \tilde{M} \cdot \tilde{\mathcal{L}}_{15} =$$
$$= r^*(3\bar{H} - 2\mathcal{F}_{\mathcal{U}}) \cdot \tilde{\mathcal{L}}_{15} - m\sum_{i=1}^{6}\tilde{E}_i \cdot \tilde{\mathcal{L}}_{15} =$$
$$= 3\bar{H} \cdot \bar{\mathcal{L}}_{15} - 2\mathcal{F}_{\mathcal{U}} \cdot \bar{\mathcal{L}}_{15} - 30m = 3H \cdot \mathcal{L}_{15} - 30m = 45 - 30m,$$

which implies that $m = 1$ and $\bar{H} \cdot \bar{\mathcal{L}}'_{15} = 15$. Thus, if $\bar{\mathcal{L}}'_{15} \not\subset \mathcal{F}_\mathcal{U}$, then one has $\psi(\bar{\mathcal{L}}'_{15}) = \mathcal{L}_{15}$ by Proposition 7.4.1.

To complete the proof, we must show that $\bar{\mathcal{L}}'_{15} \not\subset \mathcal{F}_\mathcal{U}$. Denote by $\bar{\Sigma}_5$ the $\mathfrak{A}_5$-orbit in $\mathcal{U}$ that is mapped to Σ_5 by ψ. Then $\bar{\Sigma}_5 \subset \bar{\mathcal{L}}_{15}$. Since Σ_5 is not contained in $\mathcal{L}_6$ by Corollary 7.4.9, we see that the birational map ε is biregular at the points of $\bar{\Sigma}_5$ by Lemma 11.5.6(iv). Therefore, if $\bar{\mathcal{L}}'_{15}$ is contained in $\mathcal{F}_\mathcal{U}$, then $\mathcal{F}_\mathcal{U}$ contains $\varepsilon(\bar{\Sigma}_5)$, which implies that $\mathscr{C}' = \psi(\mathcal{F}_\mathcal{U})$ contains an $\mathfrak{A}_5$-orbit consisting of at most 5 points. The latter is impossible by Lemma 5.5.9(v), since $\mathscr{C}' \cong \mathbb{P}^1$ and $\mathfrak{A}_5$ acts faithfully on $\mathscr{C}'$ by Lemma 7.3.8. $\square$

Part IV

Invariant subvarieties

Chapter 12

Invariant cubic hypersurface

In Chapter 8, we studied linear subsystems of the linear system $|2H|$ on V_5 and described in detail the geometry of its $\mathfrak{A}_5$-invariant members. The main purpose of this chapter is to study the unique $\mathfrak{A}_5$-invariant surface in the linear systems $|3H|$. We also make some remarks about one particular $\mathfrak{A}_5$-invariant surface in $|4H|$ that appears in the construction of one of the Sarkisov links found in Chapter 11.

12.1 Linear system of cubics

We already know the structure of $\mathfrak{A}_5$-representations $H^0(\mathcal{O}_{V_5}(H))$ and $H^0(\mathcal{O}_{V_5}(2H))$ (see Lemmas 7.1.6 and 8.1.1). Now we are ready to determine the structure of the $\mathfrak{A}_5$-representation $H^0(\mathcal{O}_{V_5}(3H))$.

Lemma 12.1.1. One has an identification of $\mathrm{SL}_2(\mathbb{C})$-representations
$$H^0(\mathcal{O}_{V_5}(3H)) \cong \mathbb{W}_7 \oplus \mathbb{W}_{13} \oplus \mathbb{W}_{15} \oplus \mathbb{W}_{19}$$
and an identification of $\mathfrak{A}_5$-representations
$$H^0(\mathcal{O}_{V_5}(3H)) \cong I \oplus W_3^{\oplus 3} \oplus W_3'^{\oplus 3} \oplus W_4^{\oplus 5} \oplus W_5^{\oplus 3}.$$

Proof. Recall that by Lemma 7.1.6 one has an identification of $\mathrm{SL}_2(\mathbb{C})$-representations
$$H^0(\mathcal{O}_{V_5}(H)) \cong H^0(\mathcal{O}_{\mathbb{P}^6}(H)) \cong \mathbb{W}_7.$$
Therefore, one has an identification of $\mathrm{SL}_2(\mathbb{C})$-representations
$$H^0(\mathcal{O}_{\mathbb{P}^6}(3H)) \cong \mathrm{Sym}^3(\mathbb{W}_7).$$

Since $V_5 \subset \mathbb{P}^6$ is projectively normal, there is an $\mathrm{SL}_2(\mathbb{C})$-equivariant surjective map
$$H^0(\mathcal{O}_{\mathbb{P}^6}(3H)) \to H^0(\mathcal{O}_{V_5}(3H)).$$
Note that $h^0(\mathcal{O}_{V_5}(3H)) = 54$ by (7.1.3).

Keeping in mind that $\mathscr{C}$ is a $\mathrm{PSL}_2(\mathbb{C})$-invariant rational normal sextic curve, we see that
$$H^0(\mathcal{O}_{\mathscr{C}}(3H|_{\mathscr{C}})) \cong \mathbb{W}_{19}$$
by Lemma 5.5.6. Thus we get a non-trivial homomorphism of $\mathrm{SL}_2(\mathbb{C})$-representations
$$H^0(\mathcal{O}_{V_5}(3H)) \to H^0(\mathcal{O}_{\mathscr{C}}(3H|_{\mathscr{C}})) \cong \mathbb{W}_{19}.$$

A $\mathrm{PSL}_2(\mathbb{C})$-invariant element $\mathscr{S}$ of the linear system $|2H|$ produces an injective homomorphism of $\mathrm{SL}_2(\mathbb{C})$-representations
$$\mathbb{W}_7 \cong H^0(\mathcal{O}_{V_5}(H)) \to H^0(\mathcal{O}_{V_5}(3H)).$$

Now the first assertion of the lemma follows from Lemma 5.5.3, and the second assertion follows from Lemma 5.6.2. $\square$

12.2 Curves in the invariant cubic

By Lemma 12.1.1 we know that the linear system $|3H|$ contains a unique $\mathfrak{A}_5$-invariant divisor.

Lemma 12.2.1 (cf. Remark 8.1.4). *The $\mathfrak{A}_5$-invariant divisor $S \in |3H|$ is an irreducible (reduced) surface.*

Proof. Any irreducible component $S' \subset S$ different from S itself is contained in one of the linear systems $|H|$ or $|2H|$, since $\mathrm{Pic}(V_5) = \mathbb{Z}[H]$. Thus, if S is reducible or non-reduced, then by Corollary 5.1.2 there exists an $\mathfrak{A}_5$-invariant irreducible component S' of S such that $S' \in |H|$. The latter is impossible by Lemma 7.1.6. $\square$

Remark 12.2.2. The $\mathfrak{A}_5$-invariant surface in $|3H|$ can be described very explicitly. Indeed, as we know from §7.5, the linear projection from the two-dimensional $\mathfrak{A}_5$-invariant subspace $\Pi_2 \subset \mathbb{P}^6$ induces an $\mathfrak{A}_5$-equivariant surjective morphism
$$\theta_{\Pi_2} \colon V_5 \to \mathbb{P}^3 \cong \mathbb{P}(W_4).$$
The unique $\mathfrak{A}_5$-invariant surface in $|3H|$ is mapped by the morphism θ_{Π_2} to the Clebsch cubic surface (see §6.3). Moreover, it is exactly the preimage of the Clebsch cubic surface with respect to θ_{Π_2}.

CREMONA GROUPS AND THE ICOSAHEDRON

In the sequel we will denote the unique $\mathfrak{A}_5$-invariant surface in $|3H|$ by S_{Cl} to emphasize its relation with the Clebsch cubic.

Lemma 12.2.3. *The surface S_{Cl} does not contain the $\mathfrak{A}_5$-orbit Σ_5. In particular, S_{Cl} does not contain the curve $\mathcal{L}_{15}$. On the other hand, S_{Cl} contains the $\mathfrak{A}_5$-orbit Σ_{15}.*

Proof. Note that the image of the $\mathfrak{A}_5$-orbit Σ_5 with respect to the morphism θ_{Π_2} cannot be contained in the Clebsch cubic surface in $\mathbb{P}^3$ by Lemma 6.1.11, which implies that Σ_5 is not contained in the surface S_{Cl} by Remark 12.2.2. In particular, one has $\mathcal{L}_{15} \not\subset S_{Cl}$, since $\mathcal{L}_{15}$ contains Σ_5 (see Corollary 7.4.4). Thus the intersection $S_{Cl} \cap \mathcal{L}_{15}$ is an $\mathfrak{A}_5$-invariant set that consists of
$$S_{Cl} \cdot \mathcal{L}_{15} = 45$$
points (counted with multiplicities), which implies $\Sigma_{15} \subset S_{Cl}$ by Theorem 7.3.5. □

Lemma 12.2.4. *The surface S_{Cl} contains the $\mathfrak{A}_5$-orbits Σ_{10} and Σ'_{10}.*

Proof. This follows from Lemma 7.6.11. □

We can find some other $\mathfrak{A}_5$-orbits and $\mathfrak{A}_5$-invariant curves of low degree that are contained in the surface S_{Cl}.

Lemma 12.2.5. *The following assertions hold:*

(i) *the curves $\mathscr{C}$, $\mathscr{C}'$, $\mathcal{L}_6$ and the $\mathfrak{A}_5$-orbits Σ_{12}, Σ'_{12}, and Σ_{20} are contained in the surface S_{Cl};*

(ii) *the curve $\mathcal{L}_{12}$ is contained in the surface S_{Cl};*

(iii) *the surface S_{Cl} is smooth at a general point of the curve $\mathscr{C}$;*

(iv) *the surface S_{Cl} is smooth at every point of $\mathcal{L}_{12} \setminus \Sigma_{12}$ and, in particular, at every point of Σ'_{12};*

(v) *one has $\mathscr{S}|_{S_{Cl}} = 3\mathscr{C} + \mathcal{L}_{12}$;*

(vi) *the curve $C_{20} = \mathrm{Bs}(\mathcal{Q}_2)$ is transversal to the surface S_{Cl} at the points of Σ'_{12}.*

Proof. If $\mathscr{C} \not\subset S_{Cl}$, then $S_{Cl} \cap \mathscr{C}$ is an $\mathfrak{A}_5$-invariant set that consists of 18 points (counted with multiplicities). The latter is impossible by Lemma 5.1.4. Thus, we see that $\mathscr{C} \subset S_{Cl}$. Similarly, we check that $\mathscr{C}'$ and $\mathcal{L}_6$ are contained in S_{Cl}. Also, recall that the $\mathfrak{A}_5$-orbits Σ_{12} and Σ_{20} are contained in the curve $\mathscr{C}$ by construction, while the $\mathfrak{A}_5$-orbit Σ'_{12} is contained in the curve $\mathscr{C}'$ by Lemma 11.3.1. Thus, we conclude that the $\mathfrak{A}_5$-orbits Σ_{12}, Σ'_{12}, and Σ_{20} are contained in S_{Cl}. This proves assertion (i).

To prove the remaining assertions, let us use notation of §7.2. Denote by $\hat{S}_{Cl}$ the proper transform of the surface S_{Cl} on $\mathcal{Y}$, and denote by $\hat{\mathcal{L}}_{12}$ the proper transform of the curve $\mathcal{L}_{12}$ on $\mathcal{Y}$. By Lemma 7.2.3(ii) the curve $\hat{S}_{Cl}|_{\hat{\mathscr{S}}}$ is a divisor on $\hat{\mathscr{S}} \cong \mathbb{P}^1 \times \mathbb{P}^1$ of bi-degree

$$\Big(3 - 2\mathrm{mult}_{\mathscr{C}}(S_{Cl}), 15 - 2\mathrm{mult}_{\mathscr{C}}(S_{Cl})\Big),$$

so that

$$3 - 2\mathrm{mult}_{\mathscr{C}}(S_{Cl}) \geqslant 0.$$

This shows that $\mathrm{mult}_{\mathscr{C}}(S_{Cl}) = 1$, i.e., the surface S_{Cl} is smooth at a general point of the curve $\mathscr{C}$, and in particular gives assertion (iii). Thus, $\hat{S}_{Cl}|_{\hat{\mathscr{S}}}$ is a divisor on $\hat{\mathscr{S}} \cong \mathbb{P}^1 \times \mathbb{P}^1$ of bi-degree $(1, 13)$. By Corollary 7.2.5 and Lemma 6.4.7 (cf. Lemma 6.4.11(iii),(iv)) we have

$$\hat{S}_{Cl}|_{\hat{\mathscr{S}}} = \hat{\mathscr{C}} + \hat{\mathcal{L}}_{12}. \tag{12.2.6}$$

In particular, $\mathcal{L}_{12}$ is contained in S_{Cl}, and we get assertion (ii). Moreover, since the curve $\hat{\mathscr{C}} + \hat{\mathcal{L}}_{12}$ is smooth away from $\hat{\mathscr{C}} \cap \hat{\mathcal{L}}_{12}$, the surface $\hat{S}_{Cl}$ is smooth at every point of the intersection $\hat{S}_{Cl} \cap \hat{\mathscr{S}}$ away from $\hat{\mathscr{C}}$. Since

$$\mathcal{L}_{12} \cap \mathscr{C} = \Sigma_{12}$$

by Lemma 7.2.12(v) and $\Sigma'_{12} \subset \mathcal{L}_{12}$ by Lemma 7.2.12(iii), this implies assertion (iv).

Put $m = \mathrm{mult}_{\mathscr{C}}(\mathscr{S} \cdot S_{Cl})$. Let us notice that

$$\mathscr{S}|_{S_{Cl}} = m\mathscr{C} + \mathcal{L}_{12},$$

by (12.2.6). Moreover, we have

$$m = \mathrm{mult}_{\mathscr{C}}(\mathscr{S}) \cdot \mathrm{mult}_{\mathscr{C}}(S_{Cl}) + \mathrm{mult}_{\hat{\mathscr{C}}}(\hat{\mathscr{S}} \cdot \hat{S}_{Cl}) = 3,$$

because $\hat{\mathscr{C}}$ is the only curve contained in $E_{\mathcal{Y}}$ that dominates $\mathscr{C}$ and is contained in the intersection $\hat{S}_{Cl} \cap \hat{\mathscr{S}}$. This proves assertion (v).

Denote by $\hat{C}_{20}$ the proper transform of the curve C_{20} on $\mathcal{Y}$, denote by $\hat{\Sigma}_{12}$ the unique $\mathfrak{A}_5$-orbit in $\hat{\mathscr{C}}$ of length 12, and denote by $\hat{\Sigma}'_{12}$ the $\mathfrak{A}_5$-orbit in $\hat{\mathscr{S}}$ that is mapped to Σ'_{12} by ν. Then $\hat{C}_{20}$ contains both $\mathfrak{A}_5$-orbits $\hat{\Sigma}_{12}$ and $\hat{\Sigma}'_{12}$ by Theorem 8.1.8(iii). Moreover, the curve $\hat{\mathcal{L}}_{12}$ also contains both $\mathfrak{A}_5$-orbits $\hat{\Sigma}_{12}$ and $\hat{\Sigma}'_{12}$. Since
$$\hat{C}_{20} \cdot \hat{\mathcal{L}}_{12} = 24$$
on the surface $\hat{\mathscr{S}}$, we see that $\hat{C}_{20}$ intersects the curve $\hat{S}_{Cl}|_{\hat{\mathscr{S}}} = \hat{\mathscr{C}} + \hat{\mathcal{L}}_{12}$ transversally at $\hat{\Sigma}'_{12}$. The latter implies that the curve C_{20} is transversal to the surface S_{Cl} at the points of Σ'_{12} and proves assertion (vi). □

Corollary 12.2.7. *The surface S_{Cl} does not contain the $\mathfrak{A}_5$-orbits Σ'_{20} and Σ'_{30}. Moreover, S_{Cl} does not contain the curves $\mathcal{L}_{10}$, $\mathcal{G}_5$, $\mathcal{G}'_5$, C_{16}, and C_{20}.*

Proof. By Lemma 12.2.5(v) we have
$$S_{Cl} \cap \mathscr{S} = \mathscr{C} \cup \mathcal{L}_{12},$$
while the $\mathfrak{A}_5$-orbits Σ'_{20} and Σ'_{30} are not contained in the curves $\mathscr{C}$ and $\mathcal{L}_{12}$. Thus, the latter $\mathfrak{A}_5$-orbits are not contained in S_{Cl}. On the other hand, the curve $\mathcal{L}_{10}$ contains Σ'_{20} by Remark 9.1.6, the curves $\mathcal{G}_5$ and $\mathcal{G}'_5$ contain Σ'_{20} by Remark 9.2.5, the curve C_{16} contains Σ'_{20} by Lemma 8.1.16, and the curve C_{20} contains Σ'_{20} by Theorem 8.1.8(vi). Therefore, neither of these curves is contained in the surface S_{Cl}. □

Corollary 12.2.8. *The intersection of the surface S_{Cl} with each of the curves $\mathcal{L}_{10}$, $\mathcal{G}_5$, and $\mathcal{G}'_5$ consists of a single $\mathfrak{A}_5$-orbit of length 30 that is disjoint from the surface $\mathscr{S}$.*

Proof. Let Z be one of the curves $\mathcal{L}_{10}$, $\mathcal{G}_5$, or $\mathcal{G}'_5$. We know that Z is not contained in the surface S_{Cl} by Corollary 12.2.7. Thus the intersection $\Sigma = Z \cap S_{Cl}$ consists of $Z \cdot S_{Cl} = 30$ points counted with multiplicities. Note that the $\mathfrak{A}_5$-orbits Σ_5, Σ_{10}, Σ'_{10}, and Σ_{15} are disjoint from Z; this follows from Lemma 9.1.1 and Corollary 9.1.3 if $Z = \mathcal{L}_{10}$, and from Lemma 9.2.1 and Corollaries 9.1.5 and 9.2.4 if $Z = \mathcal{G}_5$ or $Z = \mathcal{G}'_5$. Thus Theorem 7.3.5 implies that Σ is a single $\mathfrak{A}_5$-orbit of length 30.

We claim that Σ is not contained in the surface $\mathscr{S}$. Indeed, suppose that it is. Then $\Sigma \subset \mathscr{C}$ by Lemmas 7.2.10 and 12.2.5(v). On the other hand, the curve Z is disjoint from the curve $\mathscr{C}$ by Remarks 9.1.6 and 9.2.5, which gives a contradiction. □

Corollary 12.2.9. *One has $S_{Cl} \cap C_{16} = \Sigma_{12}$.*

Proof. We know that the curve C_{16} is not contained in the surface S_{Cl} by Corollary 12.2.7. Thus the intersection $\Sigma = S_{Cl} \cap C_{16}$ consists of

$$S_{Cl} \cdot C_{16} = 48$$

points (counted with multiplicities). Using this information together with Theorem 7.3.5 and Lemma 8.1.16(iii) we obtain $\Sigma = \Sigma_{12}$. □

Corollary 12.2.10. *One has $S_{Cl} \cap \mathcal{L}_{20} = \Sigma_{20}$.*

Proof. This follows from Lemma 12.2.5(v) together with Lemmas 7.2.10 and 7.2.12(iii). □

Corollary 12.2.11. *One has $S_{Cl} \cap C_{20} = \Sigma_{12} \cup \Sigma'_{12}$.*

Proof. This follows from Lemma 12.2.5(v) together with Lemma 7.2.10 and Theorem 8.1.8(iii). □

Recall from Lemma 7.2.10 that Σ_{30} and Σ'_{30} are the unique $\mathfrak{A}_5$-orbits of length 30 such that $\Sigma_{30} \subset \mathscr{C}$ and $\Sigma'_{30} \subset \mathscr{S} \setminus \mathscr{C}$.

Lemma 12.2.12. *One has*

$$S_{Cl} \cap \mathcal{L}_{15} = \Sigma_{15} \cup \Sigma,$$

where Σ is an $\mathfrak{A}_5$-orbit of length 30 that is contained in the curve $\mathcal{L}_6$ and is different from Σ_{30} and Σ'_{30}.

Proof. By Lemma 12.2.3 we know that the intersection $\Omega = S_{Cl} \cap \mathcal{L}_{15}$ consists of $S_{Cl} \cdot \mathcal{L}_{15} = 45$ points (counted with multiplicities), and that $\Sigma_{15} \subset \Omega$, while $\Sigma_5 \not\subset \Omega$. Also, we know that $\mathcal{L}_6 \subset S_{Cl}$ by Lemma 12.2.5(i), and that the intersection $\Sigma = \mathcal{L}_6 \cap \mathcal{L}_{15}$ is a single $\mathfrak{A}_5$-orbit of length 30 by Lemma 9.1.14. This means that

$$S_{Cl} \cap \mathcal{L}_{15} = \Sigma_{15} \cup \Sigma.$$

Finally, Σ is different from the $\mathfrak{A}_5$-orbits Σ_{30} and Σ'_{30} contained in the surface $\mathscr{S}$, because the curve $\mathcal{L}_6$ is disjoint from these orbits by Lemma 7.4.3. □

Note that the curve $\mathcal{L}_6$ is contained in the surface S_{Cl} by Lemma 12.2.5(i), while the Clebsch cubic surface $\theta_{\Pi_2}(S_{Cl})$ contains exactly two $\mathfrak{A}_5$-invariant sixtuples of lines (see the proof of Lemma 6.6.5). Thus, $\theta_{\Pi_2}(\mathcal{L}_6)$ is one of them. In §15.3, we will find the preimages under θ_{Π_2} of both of these sixtuples. It would be interesting to understand how the biregular $\mathfrak{S}_5$-action on the Clebsch cubic surface $\theta_{\Pi_2}(S_{Cl})$ in $\mathbb{P}^3$ interacts with the birational $\mathfrak{S}_5$-action on V_5.

Lemma 12.2.13. Let $\psi\colon \mathcal{U} \to V_5$ be a blow-up of the curve $\mathscr{C}'$, let $\mathcal{F}_\mathcal{U} \cong \mathbb{P}^1 \times \mathbb{P}^1$ be its exceptional divisor (see §11.3, and in particular Lemma 11.3.3), and let $\bar{S}_{Cl}$ be the proper transform of the surface S_{Cl} on the threefold $\mathcal{U}$. Then the following assertions hold:

(i) there exists a unique $\mathfrak{A}_5$-invariant effective divisor $\bar{Z}$ in $\mathcal{F}_\mathcal{U}$ that has bi-degree $(1, 13)$;

(ii) the divisor $\bar{Z}$ is a smooth rational curve, and $\bar{Z} = \bar{S}_{Cl}|_{\mathcal{F}_\mathcal{U}}$;

(iii) the surface S_{Cl} is smooth at every point of the curve $\mathscr{C}'$.

Proof. Assertion (i) follows from Lemmas 11.4.8 and 6.4.11(iii). By Lemma 6.4.11(iv), the divisor $\bar{Z}$ is a smooth rational curve.

We claim that $\bar{Z}$ is cut out on $\mathcal{F}_\mathcal{U}$ by the surface $\bar{S}_{Cl}$. Indeed, one has $\mathscr{C}' \subset S_{Cl}$ by Lemma 12.2.5(i), and S_{Cl} is smooth at the general point of the curve $\mathscr{C}'$ by Lemma 12.2.5(iv). This gives

$$\bar{S}_{Cl}|_{\mathcal{F}_\mathcal{U}} \sim \left(\psi^*(3H) - \mathcal{F}_\mathcal{U}\right)|_{\mathcal{F}_\mathcal{U}},$$

which implies that $\bar{S}_{Cl}|_{\mathcal{F}_\mathcal{U}}$ is a divisor of bi-degree $(1, 13)$ on $\mathcal{F}_\mathcal{U} \cong \mathbb{P}^1 \times \mathbb{P}^1$. Thus, we have $\bar{Z} = \bar{S}_{Cl}|_{\mathcal{F}_\mathcal{U}}$ by assertion (i). This completes the proof of assertion (ii).

Assertion (iii) follows from the fact that $\bar{Z} = \bar{S}_{Cl}|_{\mathcal{F}_\mathcal{U}}$ is a section of the natural projection $\mathcal{F}_\mathcal{U} \to \mathscr{C}'$. □

Lemma 12.2.14. The surface S_{Cl} is singular at the points of the $\mathfrak{A}_5$-orbit Σ_{12}, and is smooth at every other point of the curve $\mathscr{C}$. In the notation of §7.2, the proper transform $\hat{S}_{Cl}$ of the surface S_{Cl} on the threefold $\mathcal{Y}$ is smooth at every point of $\hat{S}_{Cl} \cap E_\mathcal{Y}$ except possibly the points of the unique $\mathfrak{A}_5$-orbit of length 12 contained in the curve $\hat{\mathscr{C}} \subset E_\mathcal{Y}$.

Proof. The proof is similar to that of Lemma 12.2.5. Let us use notation of §7.2. One has $E_\mathcal{Y} \cong \mathbb{P}^1 \times \mathbb{P}^1$ and $E_\mathcal{Y}^3 = -10$ by Lemma 7.2.8. Let l be a fiber of the natural projection $E_\mathcal{Y} \to \mathscr{C}$, and let s be the section of this projection such that $s^2 = 0$. Then

$$E_\mathcal{Y}|_{E_\mathcal{Y}} \sim -s + 5l.$$

Recall that $\hat{\mathscr{C}} \sim s + l$ and the action of the group $\mathfrak{A}_5$ on the surface $E_\mathcal{Y} \cong \mathbb{P}^1 \times \mathbb{P}^1$ is diagonal by Corollary 7.2.9.

Let $l_1, \ldots, l_{12}$ be fibers of the natural projection of $E_{\mathcal{Y}} \to \mathscr{C}$ that are mapped to the points of the $\mathfrak{A}_5$-orbit Σ_{12}. Then $\sum_{i=1}^{12} l_i$ is the unique $\mathfrak{A}_5$-invariant curve in the linear system $|12l|$ (see Lemma 6.4.2(ii)). Moreover, we know from Lemma 6.4.7 that the curve $\mathscr{C} + \sum_{i=1}^{12} l_i$ is the unique $\mathfrak{A}_5$-invariant curve in the linear system $|s + 13l|$.

Let $\hat{S}_{Cl}$ be the proper transform of the surface S_{Cl} on the threefold $\mathcal{Y}$. Then
$$\hat{S}_{Cl} \sim \nu^*(3H) - \mathrm{mult}_{\mathscr{C}}(S_{Cl}) E_{\mathcal{Y}} \sim \nu^*(3H) - E_{\mathcal{Y}}$$
by Lemma 12.2.5(iii). Thus
$$\hat{S}_{Cl}|_{E_{\mathcal{Y}}} \sim \big(\nu^*(3H) - E_{\mathcal{Y}}\big)\big|_{E_{\mathcal{Y}}} \sim s + 13l,$$
and one-cycle $\hat{S}_{Cl}|_{E_{\mathcal{Y}}}$ is $\mathfrak{A}_5$-invariant. Therefore, one has
$$\hat{S}_{Cl}|_{E_{\mathcal{Y}}} = \hat{\mathscr{C}} + \sum_{i=1}^{12} l_i.$$

This means that the surface $\hat{S}_{Cl}$ is smooth at every point in $\hat{S}_{Cl} \cap E_{\mathcal{Y}}$ except possibly for the 12 points of the intersection
$$\hat{\Sigma}_{12} = \hat{\mathscr{C}} \bigcap \big(l_1 \cup \ldots \cup l_{12}\big),$$
because the curve $\hat{\mathscr{C}} + \sum_{i=1}^{12} l_i$ is smooth outside of $\hat{\Sigma}_{12}$. Note that $\hat{\Sigma}_{12}$ is the unique $\mathfrak{A}_5$-orbit of length 12 in the curve $\hat{\mathscr{C}} \cong \mathbb{P}^1$ by Lemma 5.5.9(v). In particular, we conclude that the surface S_{Cl} is singular at the points of $\Sigma_{12} = \nu(\hat{\Sigma}_{12})$ and is smooth at every other point of the curve $\mathscr{C}$. □

In §15.2, we will prove that the singularities of the surface S_{Cl} are du Val. This implies that S_{Cl} is a surface of general type and, in particular, is not rational. Furthermore, we will prove that the surface S_{Cl} is smooth away from Σ_{12} and has isolated ordinary double points at the points of Σ_{12}.

12.3 Bring's curve in the invariant cubic

It is well known that there exists a unique smooth curve of genus 4 that admits a faithful action of $\mathfrak{A}_5$ (see Lemma 5.4.1). It is called *Bring's curve* (see Remark 5.4.2). The purpose of this section is to study its appearance in V_5.

We start with a preliminary result about the curve $\mathcal{G}_6$. By Corollary 8.1.7, there is a surface $S_{\mathcal{G}_6} \in \mathcal{Q}_2$ such that $\mathcal{G}_6 \subset S_{\mathcal{G}_6}$. By Lemma 7.2.13(iii), one has $S_{\mathcal{G}_6} \neq \mathscr{S}$. By Lemma 8.3.3 the surface $S_{\mathcal{G}_6}$ is smooth.

Lemma 12.3.1. The restriction morphism
$$H^0(\mathcal{O}_{V_5}(3H)) \to H^0(\mathcal{O}_{\mathcal{G}_6}(3H|_{\mathcal{G}_6}))$$
is surjective.

Proof. Put $H_{S_{\mathcal{G}_6}} = H|_{S_{\mathcal{G}_6}}$. By Lemma 8.2.5 the restriction map
$$H^0\big(\mathcal{O}_{V_5}(3H)\big) \to H^0\big(\mathcal{O}_{S_{\mathcal{G}_6}}(3H_{S_{\mathcal{G}_6}})\big) \qquad (12.3.2)$$
is surjective.

Since $\mathcal{G}_6$ is a disjoint union of six conics by Lemma 9.2.8, we have $\mathcal{G}_6^2 = -12$ on $S_{\mathcal{G}_6}$. Thus,
$$(3H_{S_{\mathcal{G}_6}} - \mathcal{G}_6)^2 = 6$$
and
$$(3H_{S_{\mathcal{G}_6}} - \mathcal{G}_6) \cdot H_{S_{\mathcal{G}_6}} = 18,$$
which implies that
$$h^2(\mathcal{O}_{S_{\mathcal{G}_6}}(3H_{S_{\mathcal{G}_6}} - \mathcal{G}_6)) = h^0(\mathcal{O}_{S_{\mathcal{G}_6}}(\mathcal{G}_6 - 3H_{S_{\mathcal{G}_6}})) = 0$$
by Serre duality. Hence
$$h^0(\mathcal{O}_{S_{\mathcal{G}_6}}(3H_{S_{\mathcal{G}_6}} - \mathcal{G}_6)) = h^1(\mathcal{O}_{S_{\mathcal{G}_6}}(3H_{S_{\mathcal{G}_6}} - \mathcal{G}_6)) + 5$$
by the Riemann–Roch formula.

The linear system $|3H_{S_{\mathcal{G}_6}} - \mathcal{G}_6|$ does not have fixed components. Indeed, suppose that it does. Then
$$3H_{S_{\mathcal{G}_6}} - \mathcal{G}_6 \sim Z + M$$
for some effective divisors Z and M on $S_{\mathcal{G}_6}$ such that the linear system $|M|$ is free from base components, and Z is the fixed part of the linear system $|3H_{S_{\mathcal{G}_6}} - \mathcal{G}_6|$. In particular, Z is an $\mathfrak{A}_5$-invariant divisor. Moreover, one has $\deg(M) \geqslant 3$ since the surface $S_{\mathcal{G}_6}$ is not uniruled. We compute
$$\deg(Z) = 3H_{S_{\mathcal{G}_6}}^2 - \deg(\mathcal{G}_6) - \deg(M) = 18 - \deg(M) \leqslant 15,$$

which is impossible by Lemma 10.2.7.

Since the linear system $|3H_{S_{\mathcal{G}_6}} - \mathcal{G}_6|$ does not have fixed components, we conclude that the divisor $3H_{S_{\mathcal{G}_6}} - \mathcal{G}_6$ is big and nef. Thus

$$h^1\big(\mathcal{O}_{S_{\mathcal{G}_6}}(3H_{S_{\mathcal{G}_6}} - \mathcal{G}_6)\big) = 0$$

by Theorem 2.3.5. In particular, we have an exact sequence

$$0 \to H^0\Big(\mathcal{O}_{S_{\mathcal{G}_6}}(3H_{S_{\mathcal{G}_6}} - \mathcal{G}_6)\Big) \to H^0\Big(\mathcal{O}_{S_{\mathcal{G}_6}}(3H_{S_{\mathcal{G}_6}})\Big) \to$$
$$\to H^0\Big(\mathcal{O}_{\mathcal{G}_6} \otimes \mathcal{O}_{S_{\mathcal{G}_6}}(3H_{S_{\mathcal{G}_6}})\Big) \to 0. \quad (12.3.3)$$

The restriction morphism

$$H^0(\mathcal{O}_{V_5}(3H)) \to H^0(\mathcal{O}_{\mathcal{G}_6}(3H|_{\mathcal{G}_6}))$$

is a composition of two surjective morphisms

$$H^0(\mathcal{O}_{V_5}(3H)) \to H^0(\mathcal{O}_{S_{\mathcal{G}_6}}(3H_{S_{\mathcal{G}_6}})) \to H^0(\mathcal{O}_{\mathcal{G}_6}(3H|_{\mathcal{G}_6}))$$

given by (12.3.2) and (12.3.3). □

Now we are ready to show that the curve $\mathcal{G}_6$ is contained in the surface S_{Cl}.

Lemma 12.3.4. *The surface S_{Cl} contains the curve $\mathcal{G}_6$.*

Proof. Let C be one of the conics of $\mathcal{G}_6$. Then the stabilizer $G_C \subset \mathfrak{A}_5$ of the conic C is isomorphic to $\mathfrak{D}_{10}$ (see Lemma 5.1.1). Moreover, there is an isomorphism of $\mathfrak{A}_5$-representations

$$H^0(\mathcal{O}_{\mathcal{G}_6}(3H|_{\mathcal{G}_6})) \cong \operatorname{Ind}_{\mathfrak{D}_{10}}^{\mathfrak{A}_5} H^0(\mathcal{O}_C(3H|_C)) \quad (12.3.5)$$

by Remark 4.4.5. On the other hand, if S_{Cl} does not contain $\mathcal{G}_6$, then S_{Cl} defines an $\mathfrak{A}_5$-invariant section of the line bundle $\mathcal{O}_{V_5}(3H|_{\mathcal{G}_6})$. Thus, it is enough to show that the $\mathfrak{A}_5$-representation $H^0(\mathcal{O}_{\mathcal{G}_6}(3H|_{\mathcal{G}_6}))$ does not contain trivial subrepresentations. Due to (12.3.5), for this we need to analyze the equivariant structure on $H^0(\mathcal{O}_C(3H|_C))$.

One has $C \cong \mathbb{P}^1$, and there are essentially two ways to choose a $2.\mathfrak{D}_{10}$-equivariant structure on the line bundle $\mathcal{O}_{\mathbb{P}^1}(1)$: one of them corresponds to a lifting of the group $\mathfrak{D}_{10}$ as its central extension $2.\mathfrak{D}_{10}$ to $\operatorname{SL}_2(\mathbb{C})$, and the other corresponds to an isomorphic lifting of $\mathfrak{D}_{10}$ to $\operatorname{GL}_2(\mathbb{C})$ (see §5.8 for a description of the actions of $\mathfrak{D}_{10}$ on $\mathbb{P}^1$ and relevant representations).

In the former case the two-dimensional space $H^0(\mathcal{O}_{\mathbb{P}^1}(1))$ is an irreducible representation of the group $2.\mathfrak{D}_{10}$, while in the latter case it is an irreducible two-dimensional representation of the group $\mathfrak{D}_{10}$. Both of these choices give a $\mathfrak{D}_{10}$-equivariant structure on the line bundles $\mathcal{O}_C(H|_C)$ and $\mathcal{O}_C(3H|_C)$ on the curve C, and we obtain two different actions of the group $\mathfrak{D}_{10}$ on the space

$$H^0(\mathcal{O}_C(3H|_C)) \cong \mathrm{Sym}^6(H^0(\mathcal{O}_{\mathbb{P}^1}(1))).$$

On the other hand, we have

$$\dim \mathrm{Hom}_{\mathfrak{A}_5}\left(H^0(\mathcal{O}_{\mathcal{G}_6}(3H|_{\mathcal{G}_6})), W_5\right) \leqslant 3$$

by Lemmas 12.1.1 and 12.3.1. Therefore, it follows from (12.3.5) and Lemma 5.8.4 that the latter choice of a $2.\mathfrak{D}_{10}$-equivariant structure on $\mathcal{O}_{\mathbb{P}^1}(1)$ does not agree with the $\mathfrak{D}_{10}$-equivariant structure on the line bundle $\mathcal{O}_C(3H|_C)$ that we inherit from the $\mathfrak{A}_5$-equivariant structure on the line bundle $\mathcal{O}_{\mathcal{G}_6}(3H|_{\mathcal{G}_6})$ on the curve $\mathcal{G}_6$. Hence, $H^0(\mathcal{O}_{\mathcal{G}_6}(3H|_{\mathcal{G}_6}))$ does not contain trivial $\mathfrak{A}_5$-subrepresentations by (12.3.5) and Remark 5.8.5, which implies, in particular, that the curve $\mathcal{G}_6$ is not contained in the surface S_{Cl}. □

Note that $S_{\mathcal{G}_6}$ is not an irreducible component of the unique $\mathfrak{A}_5$-invariant surface $S_{Cl} \in |3H|$ by Lemma 12.2.1. On the other hand, S_{Cl} contains the curve $\mathcal{G}_6$ by Lemma 12.3.4. This allows us to construct a new $\mathfrak{A}_5$-invariant curve contained in the intersection of the surfaces S_{Cl} and $S_{\mathcal{G}_6}$.

Lemma 12.3.6. One has

$$S_{Cl}|_{S_{\mathcal{G}_6}} = \mathcal{G}_6 + \mathcal{B}_{18},$$

where $\mathcal{B}_{18}$ is an irreducible smooth curve of genus 4. Moreover, one has

$$\mathcal{G}_6 \cap \mathcal{B}_{18} = \Sigma_{12},$$

and the curves $\mathcal{G}_6$ and $\mathcal{B}_{18}$ are tangent at the points of Σ_{12}. The surface S_{Cl} is smooth at every point of the intersection $S_{\mathcal{G}_6} \cap S_{Cl}$ that is not contained in the $\mathfrak{A}_5$-orbit Σ_{12}.

Proof. Recall that we have $\mathcal{G}_6^2 = -12$ by Lemma 9.2.8. One has

$$S_{Cl}|_{S_{\mathcal{G}_6}} = m\mathcal{G}_6 + \mathcal{B}_{18},$$

where m is a positive integer and $\mathcal{B}_{18}$ is an effective $\mathfrak{A}_5$-invariant divisor on $S_{\mathcal{G}_6}$. Put $H_{S_{\mathcal{G}_6}} = H|_{S_{\mathcal{G}_6}}$. Thus

$$30 = H \cdot S_{Cl} \cdot S_{\mathcal{G}_6} = H_{S_{\mathcal{G}_6}} \cdot S_{Cl}|_{S_{\mathcal{G}_6}} = 12m + H_{S_{\mathcal{G}_6}} \cdot \mathcal{B}_{18}.$$

By Lemma 8.1.14 this means that either $m = 1$ and the divisor $\mathcal{B}_{18}$ is reduced, or the surface $S_{\mathcal{G}_6}$ contains an $\mathfrak{A}_5$-invariant curve of degree 6. The latter is impossible by Lemma 10.2.7, so that $\mathcal{B}_{18}$ is a curve of degree

$$H \cdot \mathcal{B}_{18} = 30 - H \cdot \mathcal{G}_6 = 18.$$

Similarly, Lemma 10.2.7 implies that the curve $\mathcal{B}_{18}$ is $\mathfrak{A}_5$-irreducible. Moreover, we have

$$\mathcal{B}_{18}^2 = \left(3H_{S_{\mathcal{G}_6}} - \mathcal{G}_6\right)^2 = 6.$$

Suppose that $\mathcal{B}_{18}$ is reducible. Denote by $C_1, \ldots, C_r$ its irreducible components. Since $\mathcal{B}_{18}$ is an $\mathfrak{A}_5$-irreducible curve of degree 18, we have $r = 6$ by Corollary 5.1.2. Thus $C_i \cdot C_j = C_1 \cdot C_2$ for every $i \neq j$, since $\mathfrak{A}_5$ acts doubly transitively on the set $\{C_1, \ldots, C_6\}$ by Lemma 5.1.3(ii). Also, one has $C_i^2 = C_1^2$ for any i. Hence, we have

$$6 = \mathcal{B}_{18}^2 = \left(C_1 + \ldots + C_6\right)^2 = 6C_1^2 + 30 C_1 \cdot C_2,$$

which implies that $C_1^2 + 5C_1 \cdot C_2 = 1$. On the other hand, $C_1^2 \geqslant -2$ by the adjunction formula, since $S_{\mathcal{G}_6}$ is a smooth $K3$ surface. Thus, we have

$$1 = C_1^2 + 5C_1 \cdot C_2 \geqslant -2 + 5C_1 \cdot C_2,$$

which gives $C_1 \cdot C_2 = 0$ and $C_1^2 = 1$. On the other hand, the adjunction formula implies that C_1^2 is an even integer. The obtained contradiction shows that the curve $\mathcal{B}_{18}$ is irreducible.

Since $\mathcal{B}_{18}^2 = 6$, the arithmetic genus of the curve $\mathcal{B}_{18}$ is 4. This implies that $\mathcal{B}_{18}$ is smooth by Corollary 7.3.6.

We compute

$$90 = S_{Cl} \cdot S_{Cl} \cdot S_{\mathcal{G}_6} = (\mathcal{G}_6 + \mathcal{B}_{18})^2 = -6 + 2\mathcal{G}_6 \cdot \mathcal{B}_{18},$$

which gives $\mathcal{G}_6 \cdot \mathcal{B}_{18} = 48$. Thus the set $\mathcal{G}_6 \cap \mathcal{B}_{18}$ consists of 48 points (counted with multiplicities). By Theorem 7.3.5 and Corollary 9.2.11 this implies that $\mathcal{G}_6 \cap \mathcal{B}_{18} = \Sigma_{12}$, and the curves $\mathcal{G}_6$ and $\mathcal{B}_{18}$ are tangent at the points of Σ_{12}.

Finally, we see that the curve

$$S_{Cl}|_{S_{\mathcal{G}_6}} = \mathcal{G}_6 + \mathcal{B}_{18}$$

is smooth at every point except the points of Σ_{12}, and the surface S_{Cl} itself is smooth at the points of the latter intersection except the points of Σ_{12}. This completes the proof of Lemma 12.3.6. □

12.4 Intersecting invariant quadrics and cubic

Recall from §11.3 that there exists a unique $\mathfrak{A}_5$-invariant surface $S_{\mathscr{C}'}$ in the pencil $\mathcal{Q}_2$ that contains the curve $\mathscr{C}'$. Also, recall from §11.1 that there exists a unique $\mathfrak{A}_5$-invariant surface $S_{\mathcal{L}_6}$ in the pencil $\mathcal{Q}_2$ that contains the curve $\mathcal{L}_6$. Finally, recall from §12.3 that there exists a unique $\mathfrak{A}_5$-invariant surface $S_{\mathcal{G}_6}$ in the pencil $\mathcal{Q}_2$ that contains the curve $\mathcal{G}_6$. The surfaces $S_{\mathscr{C}'}$, $S_{\mathcal{L}_6}$, and $S_{\mathcal{G}_6}$ are smooth $K3$ surfaces.

Since the curves $\mathscr{C}'$ and $\mathcal{L}_6$ are contained in S_{Cl} by Lemma 12.2.5(i), we see that $S_{Cl}|_{S_{\mathscr{C}'}}$ and $S_{Cl}|_{S_{\mathcal{L}_6}}$ are reducible divisors on the surfaces $S_{\mathscr{C}'}$ and $S_{\mathcal{L}_6}$, respectively. Recall that we already know two instances of such behavior of intersections of S_{Cl} with surfaces of the pencil $\mathcal{Q}_2$. Namely, one has

$$S_{Cl} \cap \mathscr{S} = \mathscr{C} \cup \mathcal{L}_{12}$$

by Lemma 12.2.5(v), and

$$S_{Cl}|_{S_{\mathcal{G}_6}} = \mathcal{G}_6 + \mathcal{B}_{18}$$

by Lemma 12.3.6. Our next goal is to describe the restrictions $S_{Cl}|_{S_{\mathscr{C}'}}$ and $S_{Cl}|_{S_{\mathcal{L}_6}}$ in a similar way. We start with the surface $S_{\mathscr{C}'}$.

Lemma 12.4.1. One has

$$S_{Cl}|_{S_{\mathscr{C}'}} = \mathscr{C}' + Z_{\mathscr{C}'},$$

where $Z_{\mathscr{C}'}$ is an irreducible curve that is smooth away from the $\mathfrak{A}_5$-orbit Σ_{12}. Moreover, the curve $Z_{\mathscr{C}'}$ has ordinary cusps at every point of Σ_{12}, and the normalization of the curve $Z_{\mathscr{C}'}$ has genus 15. Furthermore, the intersection $\mathscr{C}' \cap Z_{\mathscr{C}'}$ is an $\mathfrak{A}_5$-orbit of length 20, and the curves $\mathscr{C}'$ and $Z_{\mathscr{C}'}$ intersect transversally at the points of $\mathscr{C}' \cap Z_{\mathscr{C}'}$. Also, the surface S_{Cl} is smooth at every point of the intersection $S_{Cl} \cap S_{\mathscr{C}'}$ that is not contained in Σ_{12}.

Proof. By Lemma 12.2.13(iii), the surface S_{Cl} is smooth at every point of the curve $\mathscr{C}'$. We have
$$S_{Cl}|_{S_{\mathscr{C}'}} = m\mathscr{C}' + Z_{\mathscr{C}'},$$
where m is a positive integer and $Z_{\mathscr{C}'}$ is an effective divisor whose support does not contain the curve $\mathscr{C}'$. In fact, $m = 1$. Indeed, let $\psi\colon \mathcal{U} \to V_5$ be a blow-up of the curve $\mathscr{C}'$, let $\mathcal{F}_\mathcal{U} \cong \mathbb{P}^1 \times \mathbb{P}^1$ be its exceptional divisor (see §11.3). Denote by $\bar{S}_{\mathscr{C}'}$ the proper transform of the surface $S_{\mathscr{C}'}$ on the threefold $\mathcal{U}$. Then
$$\bar{S}_{\mathscr{C}'}|_{\mathcal{F}_\mathcal{U}} \sim \Big(\psi^*(2H) - \mathcal{F}_\mathcal{U}\Big)\Big|_{\mathcal{F}_\mathcal{U}} \sim s + 7l,$$
and $\bar{S}_{\mathscr{C}'}|_{\mathcal{F}_\mathcal{U}}$ is a smooth $\mathfrak{A}_5$-invariant curve. If $m \geqslant 2$, then the divisors $\bar{S}_{Cl}|_{\mathcal{F}_\mathcal{U}}$ and $\bar{S}_{\mathscr{C}'}|_{\mathcal{F}_\mathcal{U}}$ must share an irreducible component that is mapped surjectively to the curve $\mathscr{C}'$ by ψ. The latter is impossible by Lemma 12.2.13(i),(ii), and we conclude that $m = 1$.

Put $H_{S_{\mathscr{C}'}} = H|_{S_{\mathscr{C}'}}$. We have
$$\mathscr{C}' + Z_{\mathscr{C}'} = S_{Cl}|_{S_{\mathscr{C}'}} \sim 3H_{S_{\mathscr{C}'}}.$$
Thus it follows from Proposition 10.2.5 and Lemmas 10.2.6 and 10.2.7 that $Z_{\mathscr{C}'}$ is an $\mathfrak{A}_5$-irreducible curve.

Recall that S_{Cl} is singular at every point of Σ_{12} by Lemma 12.2.14. Hence the curve $Z_{\mathscr{C}'}$ is also singular at every point of Σ_{12}, because $\Sigma_{12} \not\subset \mathscr{C}'$ by Lemma 11.3.1. On the other hand, we have
$$48 = \Big(3H_{S_{\mathscr{C}'}} - \mathscr{C}'\Big) \cdot 2H_{S_{\mathscr{C}'}} = Z_{\mathscr{C}'} \cdot C_{20} \geqslant$$
$$\geqslant \sum_{O \in \Sigma_{12}} \mathrm{mult}_O(Z_{\mathscr{C}'}) \cdot \mathrm{mult}_O(C_{20}) \geqslant \sum_{O \in \Sigma_{12}} 2\mathrm{mult}_O(Z_{\mathscr{C}'}) =$$
$$= 2|\Sigma_{12}|\mathrm{mult}_{\Sigma_{12}}(Z_{\mathscr{C}'}) = 24\mathrm{mult}_{\Sigma_{12}}(Z_{\mathscr{C}'}) \geqslant 48,$$
since C_{20} is singular at every point of Σ_{12} by Theorem 8.1.8(v) and C_{20} is not an irreducible component of $Z_{\mathscr{C}'}$. This shows that $\mathrm{mult}_{\Sigma_{12}}(Z_{\mathscr{C}'}) = 2$ and $Z_{\mathscr{C}'} \cap C_{20} = \Sigma_{12}$. In particular, one has $\Sigma'_{12} \not\subset Z_{\mathscr{C}'}$, since $\Sigma'_{12} \subset C_{20}$ by Theorem 8.1.8(iii).

Note that we have
$$Z_{\mathscr{C}'} \cdot \mathscr{C}' = \Big(S_{Cl}|_{S_{\mathscr{C}'}} - \mathscr{C}'\Big) \cdot \mathscr{C}' = S_{Cl} \cdot \mathscr{C}' + 2 = 20.$$
Since $\mathscr{C}' \cong \mathbb{P}^1$, it follows from Lemma 5.5.9(v) that the intersection $Z_{\mathscr{C}'} \cap \mathscr{C}'$ is the $\mathfrak{A}_5$-orbit of length 20, the curve $Z_{\mathscr{C}'}$ is smooth at every point of the

intersection $Z_{\mathscr{C}'} \cap \mathscr{C}'$, and the curves $\mathscr{C}'$ and $Z_{\mathscr{C}'}$ intersect transversally at these points.

Suppose that $Z_{\mathscr{C}'}$ is reducible. Denote by $C_1, \ldots, C_r$ its irreducible components. Since $Z_{\mathscr{C}'}$ is an $\mathfrak{A}_5$-irreducible curve of degree 24, we have either $r = 6$ or $r = 12$ by Corollary 5.1.2. On the other hand, we have

$$20 = Z_{\mathscr{C}'} \cdot \mathscr{C}' = \Big(C_1 + \ldots + C_r\Big) \cdot \mathscr{C}' = rC_1 \cdot \mathscr{C}',$$

which is absurd. The obtained contradiction shows that $Z_{\mathscr{C}'}$ is irreducible.

By the adjunction formula, the arithmetic genus of the curve $Z_{\mathscr{C}'}$ is 27, since $Z_{\mathscr{C}'}^2 = 52$. Hence, $Z_{\mathscr{C}'}$ cannot have more than 27 singular points by Lemma 4.4.6. This implies that $Z_{\mathscr{C}'}$ is smooth away from Σ_{12}. Indeed, the curve $Z_{\mathscr{C}'}$ is singular at every point of Σ_{12} and $Z_{\mathscr{C}'}$ does not contain Σ'_{12}. Therefore, if $Z_{\mathscr{C}'}$ is singular at some point $P \in Z_{\mathscr{C}'}$ such that $P \notin \Sigma_{12}$, then its orbit has length different from 12 by Theorem 7.3.5, so that this length is at least 20 points by Theorem 6.7.1(ii),(iii). We conclude that $Z_{\mathscr{C}'}$ has at least 32 singular points, which is a contradiction.

We see that $Z_{\mathscr{C}'}$ is an irreducible curve of arithmetic genus 27 that is smooth away from Σ_{12}. In particular, the surface S_{Cl} is smooth at every point of the intersection $S_{Cl} \cap S_{\mathscr{C}'}$ that is not contained in Σ_{12} by Lemma 12.2.13(iii). Moreover, it follows from Lemma 4.4.6 that the genus g of the normalization of the curve $Z_{\mathscr{C}'}$ is

$$g = 27 - n|\Sigma_{12}| = 27 - 12n$$

for some positive integer n. Now Lemma 5.1.5 implies that $n = 1$ and $g = 15$. By Lemma 4.4.6 the equality $n = 1$ means that the curve $Z_{\mathscr{C}'}$ has either a node or an ordinary cusp at each point of Σ_{12}. Also, we know from Lemma 5.1.5 that a smooth curve of genus 15 with an action of the group $\mathfrak{A}_5$ contains a single $\mathfrak{A}_5$-orbit of length 12. This implies that the singularities of $Z_{\mathscr{C}'}$ at the points of Σ_{12} are ordinary cusps and completes the proof of Lemma 12.4.1. □

Now we pass to the surface $S_{\mathcal{L}_6}$.

Lemma 12.4.2. One has

$$S_{Cl}|_{S_{\mathcal{L}_6}} = \mathcal{L}_6 + Z_{\mathcal{L}_6},$$

where $Z_{\mathcal{L}_6}$ is an irreducible curve that is smooth away from the $\mathfrak{A}_5$-orbit Σ_{12}. Moreover, the curve $Z_{\mathcal{L}_6}$ has ordinary cusps at every point of Σ_{12}, and the normalization of the curve $Z_{\mathcal{L}_6}$ has genus 10. Furthermore, the intersection $\mathcal{L}_6 \cap Z_{\mathcal{L}_6}$ is an $\mathfrak{A}_5$-orbit of length 30, and the curves $\mathcal{L}_6$ and $Z_{\mathcal{L}_6}$ intersect transversally at the points of $\mathcal{L}_6 \cap Z_{\mathcal{L}_6}$. Also, the surface S_{Cl} is smooth at every point of the intersection $S_{Cl} \cap S_{\mathcal{L}_6}$ that is not contained in Σ_{12}.

Proof. The proof is similar to that of Lemma 12.4.1. Let $L_1, \ldots, L_6$ be the lines of $\mathcal{L}_6$. Recall that the $\mathfrak{A}_5$-orbit Σ'_{12} is contained in $\mathcal{L}_6$ (see Lemma 7.4.3). Thus S_{Cl} is smooth at the general point of every line $L_1, \ldots, L_6$, since S_{Cl} is smooth at every point of Σ'_{12} by Lemma 12.2.5(iv).

Since S_{Cl} contains the lines $L_1, \ldots, L_6$ by Lemma 12.2.5(i), we have

$$S_{Cl}|_{S_{\mathcal{L}_6}} = m \sum_{i=1}^{6} L_i + Z_{\mathcal{L}_6},$$

where m is a positive integer and $Z_{\mathcal{L}_6}$ is an effective divisor whose support does not contain the lines $L_1, \ldots, L_6$.

Let $\pi \colon \mathcal{W} \to V_5$ be the blow-up of the lines $L_1, \ldots, L_6$. Denote by $E_1, \ldots, E_6$ the exceptional divisors of π that are mapped to $L_1, \ldots, L_6$, respectively; one has $E_i \cong \mathbb{P}^1 \times \mathbb{P}^1$ (see §11.1). The stabilizer $G_{L_i} \subset \mathfrak{A}_5$ of a line L_i is isomorphic to $\mathfrak{D}_{10}$ (see Lemma 5.1.1), and the action of G_{L_i} on E_i is twisted diagonal by Lemma 11.1.9.

Let $l_i \subset E_i$ be a fiber of the projection $\pi_{E_i} \colon E_i \to L_i$, and s_i be a section of π_{E_i} such that $s_i^2 = 0$. Denote by $\check{S}_{\mathcal{L}_6}$ and $\check{S}_{Cl}$ the proper transforms of the surfaces $S_{\mathcal{L}_6}$ and S_{Cl} on the threefold $\mathcal{W}$, respectively. Then

$$\check{S}_{Cl}|_{E_i} \sim s_i + 3l_i$$

is an irreducible curve by Lemma 6.4.13(iii). Since this curve is smooth, we see that S_{Cl} is smooth in every point of $\mathcal{L}_6$. Since the surface $S_{\mathcal{L}_6}$ is smooth, we know that

$$\check{S}_{\mathcal{L}_6}|_{E_i} \sim s_i + 2l_i$$

is also an irreducible curve. Therefore, we have $m = 1$.

Put $H_{S_{\mathcal{L}_6}} = H|_{S_{\mathcal{L}_6}}$. We have

$$\mathcal{L}_6 + Z_{\mathcal{L}_6} = S_{Cl}|_{S_{\mathcal{L}_6}} \sim 3H_{S_{\mathcal{L}_6}}.$$

Thus, it immediately follows from Proposition 10.2.5 and Lemmas 10.2.6 and 10.2.7 that $Z_{\mathcal{L}_6}$ is an $\mathfrak{A}_5$-irreducible curve.

The lines $L_1, \ldots, L_6$ are pairwise disjoint by Lemma 7.4.7. Therefore, we have $\mathcal{L}_6^2 = -12$ and

$$Z_{\mathcal{L}_6} \cdot \mathcal{L}_6 = 3H_{S_{\mathcal{L}_6}} \cdot \mathcal{L}_6 - \mathcal{L}_6^2 = 30.$$

Now Theorem 7.3.5 implies that the intersection $\mathcal{L}_6 \cap Z_{\mathcal{L}_6}$ is an $\mathfrak{A}_5$-orbit of length 30, the curve $Z_{\mathcal{L}_6}$ is smooth at the points of this orbit, and the curves $Z_{\mathcal{L}_6}$ and $\mathcal{L}_6$ intersect transversally at these points.

Recall that S_{Cl} is singular at every point of Σ_{12} by Lemma 12.2.14. Hence the curve $Z_{\mathcal{L}_6}$ is also singular at every point of Σ_{12}, because $\Sigma_{12} \not\subset \mathcal{L}_6$ by Lemma 7.4.3. On the other hand, we have

$$48 = (3H_{S_{\mathcal{L}_6}} - \mathcal{L}_6) \cdot 2H_{S_{\mathcal{L}_6}} = Z_{\mathcal{L}_6} \cdot C_{20} \geqslant$$
$$\geqslant \sum_{O \in \Sigma_{12}} \mathrm{mult}_O(Z_{\mathcal{L}_6}) \cdot \mathrm{mult}_O(C_{20}) \geqslant \sum_{O \in \Sigma_{12}} 2\mathrm{mult}_O(Z_{\mathcal{L}_6}) =$$
$$= 2|\Sigma_{12}|\mathrm{mult}_{\Sigma_{12}}(Z_{\mathcal{L}_6}) = 24\mathrm{mult}_{\Sigma_{12}}(Z_{\mathcal{L}_6}) \geqslant 48,$$

since C_{20} is singular at every point of Σ_{12} by Theorem 8.1.8(v) and C_{20} is not an irreducible component of $Z_{\mathcal{L}_6}$. We conclude that $\mathrm{mult}_{\Sigma_{12}}(Z_{\mathcal{L}_6}) = 2$ and $Z_{\mathcal{L}_6} \cap C_{20} = \Sigma_{12}$. In particular, one has $\Sigma'_{12} \not\subset Z_{\mathcal{L}_6}$, since $\Sigma'_{12} \subset C_{20}$ by Theorem 8.1.8(iii).

Suppose that $Z_{\mathcal{L}_6}$ is reducible. Denote by r the number of irreducible components of the curve $Z_{\mathcal{L}_6}$, and denote by $C_1, \ldots, C_r$ its irreducible components. Since $Z_{\mathcal{L}_6}$ is an $\mathfrak{A}_5$-irreducible curve of degree 24, we have either $r = 6$ or $r = 12$ by Corollary 5.1.2. If $r = 12$, then

$$30 = Z_{\mathcal{L}_6} \cdot \mathcal{L}_6 = \left(C_1 + \ldots + C_{12}\right) \cdot \mathcal{L}_6 = 12 C_1 \cdot \mathcal{L}_6,$$

which is absurd. Hence, one has $r = 6$. Thus $C_i \cdot C_j = C_1 \cdot C_2$ for every $i \neq j$, since $\mathfrak{A}_5$ acts doubly transitively on the set $\{C_1, \ldots, C_6\}$ by Lemma 5.1.3(ii). Also, we have $C_i^2 = C_1^2$ for every i. Hence, we have

$$42 = \left(C_1 + \ldots + C_6\right)^2 = 6C_1^2 + 30 C_1 \cdot C_2.$$

Thus, we obtain
$$C_1^2 + 5 C_1 \cdot C_2 = 7,$$

which implies that $C_1^2 = 2$ and $C_1 \cdot C_2 = 1$, because C_1^2 is even and $C_1^2 \geqslant -2$. In particular, the intersection $C_1 \cap C_2$ consists of a single point.

Let G_1 be the stabilizer of the curve C_1 in $\mathfrak{A}_5$. Then G_1 acts faithfully on C_1 by Lemma 7.4.5. Since G_1 acts transitively on the set $\{C_2, \ldots, C_6\}$ by Lemma 5.1.3(ii), we see that the intersection

$$C_1 \bigcap \left(C_2 \cup C_3 \cup C_4 \cup C_5 \cup C_6\right)$$

is a G_1-orbit of length 5. Thus, the intersection points

$$C_i \cap C_j, \quad 1 \leqslant i < j \leqslant 6,$$

form an $\mathfrak{A}_5$-invariant set that consists of $\frac{6 \cdot 5}{2} = 15$ points, which is impossible by Lemma 6.7.1(ii). The obtained contradiction shows that $Z_{\mathcal{L}_6}$ is irreducible.

Suppose that $Z_{\mathcal{L}_6}$ is singular at some point $P \in S_{\mathcal{L}_6}$ that is not contained in Σ_{12}. Denote its $\mathfrak{A}_5$-orbit by Σ. Since the $\mathfrak{A}_5$-orbit Σ'_{12} is not contained in $Z_{\mathcal{L}_6}$, we see that $|\Sigma| \neq 12$ by Theorem 7.3.5. This implies that $|\Sigma| \geqslant 20$ by Theorem 6.7.1(ii),(iii). On the other hand, we have

$$Z_{\mathcal{L}_6}^2 = \left(3H_{S_{\mathcal{L}_6}} - \mathcal{L}_6\right)^2 = 42,$$

which implies that the arithmetic genus of the curve $Z_{\mathcal{L}_6}$ is 22 by the adjunction formula. In particular, $Z_{\mathcal{L}_6}$ cannot have more than 22 singular points by Lemma 4.4.6. However, $Z_{\mathcal{L}_6}$ is singular at every point of the $\mathfrak{A}_5$-orbit Σ that consists of at least 20 points, and $Z_{\mathcal{L}_6}$ is also singular at every point of Σ_{12}, so that in total $Z_{\mathcal{L}_6}$ has at least 32 singular points. The obtained contradiction shows that $Z_{\mathcal{L}_6}$ is smooth away from Σ_{12}, which implies that S_{Cl} is smooth at every point of the intersection $S_{Cl} \cap S_{\mathcal{L}_6}$ that is not contained in Σ_{12}, because we already proved that S_{Cl} is smooth in every point of $\mathcal{L}_6$.

It follows from Lemma 4.4.6 that the genus g of the normalization of the curve $Z_{\mathcal{L}_6}$ is

$$g = 22 - n|\Sigma_{12}| = 22 - 12n$$

for some positive integer n. This immediately implies that $n = 1$ and $g = 10$. By Lemma 4.4.6 the equality $n = 1$ means that the curve $Z_{\mathcal{L}_6}$ has either a node or an ordinary cusp at each point of Σ_{12}. Also, we know from Lemma 5.1.5 that a smooth curve of genus 10 with an action of the group $\mathfrak{A}_5$ contains a single $\mathfrak{A}_5$-orbit of length 12. This implies that the singularities of $Z_{\mathcal{L}_6}$ at the points of Σ_{12} are ordinary cusps and completes the proof of Lemma 12.4.2. □

Lemmas 12.2.5(v), 12.3.6, 12.4.1, and 12.4.2 describe the intersections of the surface S_{Cl} with the surfaces $\mathscr{S}$, $S_{\mathcal{G}_6}$, $S_{\mathscr{C}'}$ and $S_{\mathcal{L}_6}$, respectively. Now we are going to describe the intersections of S_{Cl} with the remaining surfaces of the pencil $\mathcal{Q}_2$. First we need:

Lemma 12.4.3. *Let S be a surface in $\mathcal{Q}_2$ that is different from $\mathscr{S}$, $S_{\mathcal{G}_6}$, $S_{\mathscr{C}'}$ and $S_{\mathcal{L}_6}$. Put $Z = S_{Cl}|_S$. Then the following assertions hold:*

(i) the one-cycle Z is an $\mathfrak{A}_5$-irreducible curve;

(ii) one has $\mathrm{mult}_{\Sigma_{12}}(Z) = 2$ and $\mathrm{mult}_{\Sigma'_{12}}(Z) = 1$;

(iii) if Z is reducible, then it has 6 irreducible components and each of them is singular in exactly two points of the $\mathfrak{A}_5$-orbit Σ_{12}.

Proof. Suppose that Z is either non-reduced or not $\mathfrak{A}_5$-irreducible. Then its support contains an $\mathfrak{A}_5$-irreducible curve R of degree at most 15. By Lemma 10.2.6, we know that $R \ne \mathcal{L}_{15}$. Thus, it follows from Proposition 10.2.5 and Lemma 11.3.2 that R is one of the curves $\mathscr{C}$, $\mathscr{C}'$, $\mathcal{L}_6$, $\mathcal{L}_{10}$, $\mathcal{G}_5$, $\mathcal{G}'_5$, $\mathcal{L}_{12}$, or $\mathcal{G}_6$. By our assumption about the surface S and Lemma 10.2.7, we conclude that R is one of the curves $\mathcal{L}_{10}$, $\mathcal{G}_5$, or $\mathcal{G}'_5$. However, these curves are not contained in S_{Cl} by Corollary 12.2.7. This proves assertion (i).

Put $H_S = H|_S$. Recall that S is smooth at every point of the curve C_{20} by Theorems 8.2.1(ii),(iii) and 8.1.8(vii). On the other hand, we know that $\Sigma_{12} \subset Z$ and $\Sigma'_{12} \subset Z$ by Theorem 8.1.8(iii) and Lemma 12.2.5(i). Moreover, the curve Z is singular at every point of Σ_{12} by Lemma 12.2.14. Thus, we have

$$60 = 3H_S \cdot 2H_S = Z \cdot C_{20} \geqslant$$
$$\geqslant \sum_{O \in Z \cap C_{20}} \mathrm{mult}_O(Z \cdot C_{20}) \geqslant \sum_{O \in Z \cap C_{20}} \mathrm{mult}_O(Z) \cdot \mathrm{mult}_O(C_{20}) \geqslant$$
$$\geqslant \sum_{O \in \Sigma_{12}} 2\mathrm{mult}_O(Z) + \sum_{O \in \Sigma'_{12}} \mathrm{mult}_O(Z) =$$
$$= 24\mathrm{mult}_{\Sigma_{12}}(Z) + 12\mathrm{mult}_{\Sigma'_{12}}(Z) \geqslant 24\mathrm{mult}_{\Sigma_{12}}(Z) + 12 \geqslant 60,$$

since C_{20} is singular at every point of Σ_{12} by Theorem 8.1.8(v) and C_{20} is not an irreducible component of Z by assertion (i). This implies that $\mathrm{mult}_{\Sigma_{12}}(Z) = 2$ and $\mathrm{mult}_{\Sigma'_{12}}(Z) = 1$, which proves assertion (ii).

Let us prove assertion (iii). Suppose that Z is reducible. Denote by $C_1, \ldots, C_r$ its irreducible components. Since Z is a curve of degree 30, we have
$$r \in \{5, 6, 10, 15, 30\}$$
by Corollary 5.1.2.

Let c be the number of irreducible components of the curve Z that pass through some (and hence any) point in Σ_{12}, and let p be the number of points in Σ_{12} that are contained in some (and hence any) irreducible component of the curve Z. Then
$$rp = 12c,$$

which implies that $c \geqslant 5$ unless $r = 1$ or $r = 6$. On the other hand, we have

$$2 = \mathrm{mult}_{\Sigma_{12}}(Z) \geqslant c,$$

which implies that $r = 6$ and $p = 2c$.

Let $G_1 \subset \mathfrak{A}_5$ be the stabilizer of the curve C_1. Then $G_1 \cong \mathfrak{D}_{10}$ by Lemma 5.1.1. By Lemma 5.1.3(v), the $\mathfrak{A}_5$-orbit Σ_{12} splits into two G_1-orbits of lengths 2 and 10, respectively. Thus, we obtain $p = 2$ and $c = 1$, since we already proved that $c \leqslant 2$. In particular, each of the two points of $\Sigma_{12} \cap C_1$ must be a singular point of the curve C_1 by assertion (ii). This proves assertion (iii). $\square$

Recall from Theorem 8.2.1 that $\mathcal{Q}_2$ contains exactly five singular surfaces, which are the surfaces $\mathscr{S}$, S_5, S_{10}, S'_{10}, and S_{15}.

Lemma 12.4.4. Let S be a surface in $\mathcal{Q}_2$ that is different from $\mathscr{S}$, $S_{\mathcal{G}_6}$, $S_{\mathscr{C}'}$, $S_{\mathcal{L}_6}$, S_{10}, S'_{10}, and S_{15}. Put $Z = S_{Cl}|_S$. Then the following assertions hold:

(i) the divisor Z is an irreducible curve;

(ii) either Z is smooth away from Σ_{12}, or Z has 30 singular points away from Σ_{12}, which form one $\mathfrak{A}_5$-orbit consisting of nodes or of ordinary cusps of Z;

(iii) the curve Z has ordinary cusps at every point of Σ_{12};

(iv) if Z is smooth away from Σ_{12}, the normalization of the curve Z has genus 34;

(v) if Z is not smooth away from Σ_{12}, the normalization of the curve Z has genus 4.

Proof. By Lemma 12.4.3(i), the divisor Z is an $\mathfrak{A}_5$-irreducible curve. Moreover, we know that $\mathrm{mult}_{\Sigma_{12}}(Z) = 2$ and Z is smooth at every point of Σ'_{12} by Lemma 12.4.3(ii).

Let us prove assertion (i). Suppose that Z is reducible. Then it has six irreducible components by Lemma 12.4.3(iii). Denote them by $C_1, \ldots, C_6$. Since the group $\mathfrak{A}_5$ acts on the set $\{C_1, \ldots, C_6\}$ doubly transitively by Lemma 5.1.3(ii), we have $C_i \cdot C_j = C_1 \cdot C_2$ for every $i \neq j$. Also, we have $C_i^2 = C_1^2$ for every i. We compute

$$90 = \bigl(C_1 + \ldots + C_6\bigr)^2 = 6C_1^2 + 30 C_1 \cdot C_2. \tag{12.4.5}$$

Recall that either S is smooth or $S = S_5$. In any case, the curve Z is contained in the smooth locus of the surface S by Theorem 8.2.1(ii),(iii), since Z does not contain the $\mathfrak{A}_5$-orbit Σ_5 by Lemma 12.2.3. Thus, C_1^2 is an even integer, and $C_1^2 \geqslant -2$ by the adjunction formula. Therefore, it follows from (12.4.5) that either $C_1^2 = 0$ or $C_1^2 = 10$. If $C_1^2 = 0$, then the arithmetic genus of the curve C_i is 1 by the adjunction formula. On the other hand, each C_i has at least two singular points by Lemma 12.4.3(iii). Thus, Lemma 4.4.6 implies that $C_1^2 = 10$, and by (12.4.5) we obtain $C_1 \cdot C_2 = 1$. In particular, the intersection $C_1 \cap C_2$ consists of a single point.

Let G_1 be the stabilizer of the curve C_1 in $\mathfrak{A}_5$. Then $G_1 \cong \mathfrak{D}_{10}$ by Lemma 5.1.1, and G_1 acts faithfully on the curve C_1 by Lemma 7.4.5. Since $\mathfrak{A}_5$ acts doubly transitively on the set $\{C_1, \ldots, C_6\}$ by Lemma 5.1.3(ii), we see that the intersection

$$C_1 \cap \left(C_2 \cup C_3 \cup C_4 \cup C_5 \cup C_6 \right)$$

is a G_1-orbit of length 5. This implies that the intersection points

$$C_i \cap C_j, \quad 1 \leqslant i < j \leqslant 6,$$

form an $\mathfrak{A}_5$-invariant set that consists of $\frac{6 \cdot 5}{2} = 15$ points, which is impossible, because we assume that $S \neq S_{15}$. The obtained contradiction proves assertion (i).

We have $Z^2 = 90$. Thus, the arithmetic genus of the curve Z is 46 by the adjunction formula. Denote by g the genus of the normalization of the curve Z.

Let us prove assertion (ii). Suppose that Z is singular away from Σ_{12}. Denote by Σ the subset in Z consisting of all singular points of the curve Z that are not contained in Σ_{12}. We observe that $\Sigma'_{12} \not\subset \Sigma$, because Z is smooth at every point of Σ'_{12} by Lemma 12.4.3(ii). Thus $|\Sigma| \neq 12$ by Theorem 7.3.5, so that $|\Sigma| \geqslant 20$ by Lemma 6.7.1(ii),(iii). On the other hand, we have

$$0 \leqslant g \leqslant 46 - |\Sigma_{12}| - |\Sigma| \leqslant 34 - |\Sigma|,$$

by Lemma 4.4.6. Therefore, Lemma 6.7.1(ii),(iii) implies that either $|\Sigma| = 20$ or $|\Sigma| = 30$, and Σ is a single $\mathfrak{A}_5$-orbit. Thus it follows from Lemma 4.4.6 that

$$g = 46 - n_1|\Sigma_{12}| - n_2|\Sigma| = 46 - 12n_1 - n_2|\Sigma| \geqslant 0$$

for some positive integers n_1 and n_2. If $|\Sigma| = 20$, then either $n_1 = n_2 = 1$ and $g = 14$, or $n_1 = 2$, $n_2 = 1$ and $g = 2$. Both cases are impossible by Lemma 5.1.5. Hence, we must have $|\Sigma| = 30$. Then $n_1 = n_2 = 1$ and $g = 4$.

Thus, it follows from Lemma 4.4.6 that Z has either nodes or ordinary cusps at the points of Σ. This proves assertion (ii).

By Lemma 4.4.6, one has

$$g = \begin{cases} 46 - 12n, & \text{if } Z \text{ is smooth away from } \Sigma_{12}, \\ 16 - 12n, & \text{if } Z \text{ is singular away from } \Sigma_{12}. \end{cases}$$

for some positive integer n. Since

$$\mathrm{mult}_{\Sigma_{12}}(Z) = 2$$

and Z contains the $\mathfrak{A}_5$-orbit Σ'_{12}, we conclude that the normalization of the curve Z contains at least two $\mathfrak{A}_5$-orbits of length 12. Thus Lemma 5.1.5 implies that $n = 1$. This proves assertions (iv) and (v). Since $n = 1$, Lemma 4.4.6 implies that Z has either nodes or ordinary cusps at the points of Σ_{12}. Note that any smooth curve of genus 4 or 34 with an action of the group $\mathfrak{A}_5$ contains exactly two $\mathfrak{A}_5$-orbits of length 12 by Lemma 5.1.5. Therefore, the singularities of Z at the points of Σ_{12} are ordinary cusps. This proves assertion (iii) and completes the proof of Lemma 12.4.4. □

Remark 12.4.6. Later we will see that the curve Z in Lemma 12.4.4 is always smooth outside of the $\mathfrak{A}_5$-orbit Σ_{12} (see Remark 15.2.14), so that the second alternative in Lemma 12.4.4(ii) never holds.

By Remark 7.4.2 there is an infinite number of reducible $\mathfrak{A}_5$-invariant curves of degree 30 in V_5. Lemma 12.4.4(i) immediately implies the following:

Corollary 12.4.7. There is an infinite number of irreducible $\mathfrak{A}_5$-invariant curves of degree 30 in V_5.

Now we will describe the intersections of the remaining three surfaces S_{10}, S'_{10}, and S_{15} of the pencil $\mathcal{Q}_2$ with the surface S_{Cl}.

Lemma 12.4.8. Put $Z_{10} = S_{Cl}|_{S_{10}}$, $Z'_{10} = S_{Cl}|_{S'_{10}}$, and $Z_{15} = S_{Cl}|_{S_{15}}$. Then

(i) the divisors Z_{10}, Z'_{10}, and Z_{15} are irreducible curves;

(ii) the curves Z_{10}, Z'_{10}, and Z_{15} have ordinary cusps at every point of Σ_{12};

(iii) the curve Z_{10} (resp., Z'_{10}) is smooth away from $\Sigma_{10} \cup \Sigma_{12}$ (resp., $\Sigma'_{10} \cup \Sigma_{12}$), and it has nodes at every point of Σ_{10} (resp., Σ'_{10});

CREMONA GROUPS AND THE ICOSAHEDRON   303

(iv) the curve Z_{15} is smooth away from $\Sigma_{15} \cup \Sigma_{12}$, and it has nodes at every point of Σ_{15};

(v) the normalizations of the curves Z_{10} and Z'_{10} have genus 24;

(vi) the normalization of the curve Z_{15} has genus 19.

Proof. Let S be one of the surfaces S_{10}, S'_{10} or S_{15}. By Theorem 8.2.1(iii),(iv), the surface S is a $K3$ surface with ordinary double points, and its singular locus coincides with the $\mathfrak{A}_5$-orbit Σ_{10}, Σ'_{10}, or Σ_{15} that is contained in S.

Put $Z = S_{Cl}|_S$, so that Z is one of the divisors Z_{10}, Z'_{10}, or Z_{15}. Then the divisor Z is an $\mathfrak{A}_5$-irreducible curve by Lemma 12.4.3(i). Note that Z contains the singular locus of S, since the $\mathfrak{A}_5$-orbits Σ_{10}, Σ'_{10}, and Σ_{15} are contained in S_{Cl} by Lemmas 12.2.3 and 12.2.4. Moreover, we have $\mathrm{mult}_{\Sigma_{12}}(Z) = 2$ and Z is smooth at every point of Σ'_{12} by Lemma 12.4.3(ii).

Let $u \colon \check{S} \to S$ be the blow-up of $\mathrm{Sing}(S)$. Then $\check{S}$ is a smooth $K3$ surface. Put
$$t = |\mathrm{Sing}(S)| \in \{10, 15\}.$$
Denote by $N_1, \ldots, N_t$ the exceptional divisors of u, and denote by $\check{Z}$ the proper transform of the curve Z on the surface $\check{S}$. Then
$$\check{Z} \sim u^*(3H_S) - m \sum_{i=1}^{t} N_i$$
for some positive integer m, because Z is a Cartier divisor on the surface S by Lemma 6.7.4. Thus, we have
$$\check{Z}^2 = \left(u^*(3H_S) - m \sum_{i=1}^{t} N_i\right)^2 = 90 - 2tm^2. \qquad (12.4.9)$$

Let us prove assertion (i). Suppose that Z is reducible. Then it has six irreducible components by Lemma 12.4.3(iii). Denote them by $C_1, \ldots, C_6$, and denote their proper transforms on the surface $\check{S}$ by $\check{C}_1, \ldots, \check{C}_6$, respectively. Then $\check{C}_i \cdot \check{C}_j = \check{C}_1 \cdot \check{C}_2$ for every $i \neq j$, because $\mathfrak{A}_5$ acts on the set $\{\check{C}_1, \ldots, \check{C}_6\}$ doubly transitively by Lemma 5.1.3(ii). Similarly, we have $\check{C}_i^2 = \check{C}_1^2$. Moreover, $\check{C}_i^2 \geqslant -2$ by the adjunction formula. Thus
$$90 - 2tm^2 = \check{Z}^2 = \left(\sum_{i=1}^{6} \check{C}_i\right)^2 =$$
$$= 6\check{C}_1^2 + 30 \check{C}_1 \cdot \check{C}_2 \geqslant -12 + 30 \check{C}_1 \cdot \check{C}_2 \geqslant -12. \qquad (12.4.10)$$

In particular, this gives $m \leqslant 2$. Moreover, if $t = 10$, then it follows from (12.4.10) that m must be divisible by 3, which is inconsistent with the inequality $m \leqslant 2$. Thus, one has $t = 15$, so that $S = S_{15}$. Then (12.4.10) gives
$$15 - 5m^2 = \check{C}_1^2 + 5\check{C}_1 \cdot \check{C}_2 \geqslant -2 + 5\check{C}_1 \cdot \check{C}_2 \geqslant -2,$$
which implies that $m = 1$ and
$$\check{C}_1^2 + 5\check{C}_1 \cdot \check{C}_2 = 10.$$
Since $\check{C}_1^2$ is an even integer, we see that either $\check{C}_1^2 = 0$ or $\check{C}_1^2 = 10$. If $\check{C}_1^2 = 0$, then the arithmetic genus of the curve $\check{C}_1$ is 1 by the adjunction formula. On the other hand, C_1 has at least two singular points by Lemma 12.4.3(iii). A contradiction with Lemma 4.4.6 shows that $\check{C}_1^2 = 10$. This implies
$$\check{C}_1 \cdot \check{C}_2 = 0,$$
which is impossible by the Hodge index theorem, since $\check{C}_2^2 = \check{C}_1^2 = 10$. The obtained contradiction proves assertion (i).

The curve $\check{Z}$ is singular. Indeed, denote by $\check{\Sigma}_{12}$ the $\mathfrak{A}_5$-orbit in $\check{S}$ that is mapped to Σ_{12} by u. Then $\check{Z}$ is singular at every point of $\check{\Sigma}_{12}$, because Z is singular at every point of the $\mathfrak{A}_5$-orbit Σ_{12} and u is an isomorphism in a neighborhood of Σ_{12}. Hence
$$p_a(\check{Z}) \geqslant |\check{\Sigma}_{12}| = 12$$
by Lemma 4.4.6. On the other hand, we have
$$p_a(\check{Z}) = 46 - tm^2 \leqslant 46 - 10m^2$$
by (12.4.9) and the adjunction formula. This gives $m = 1$.

We claim that each point in $\mathrm{Sing}(S)$ is a node of the curve Z. Indeed, let O be a point of $\mathrm{Sing}(S)$. Without loss of generality, we may assume that $O = u(N_1)$. One has
$$2 = 2m = \check{Z} \cdot N_1 \geqslant |\check{Z} \cap N_1|,$$
which implies that the intersection $\check{Z} \cap N_1$ consists of either one or two points. In the latter case, the curve $\check{Z}$ is smooth at both points of $\check{Z} \cap N_1$, which implies that Z has a node at O. On the other hand, if the intersection $\check{Z} \cap N_1$ consists of a single point, then $\mathfrak{A}_5$-orbit of this point in $\check{S}$

consists of t points, which is impossible by Lemma 6.7.1(ii). This shows that each point in $\mathrm{Sing}(S)$ is a node of the curve Z.

The arithmetic genus of the curve $\check{Z}$ equals $46 - t$. We conclude that to prove assertions (iii) and (iv) it is enough to check that $\check{Z}$ is smooth away from $\check{\Sigma}_{12}$.

Suppose that $\check{Z}$ is singular at some point $P \in \check{Z}$ that is not contained in $\check{\Sigma}_{12}$. Denote by $\check{\Sigma}$ the $\mathfrak{A}_5$-orbit of the point P. Since we already know that Z is smooth at the points of the $\mathfrak{A}_5$-orbit Σ'_{12} by Lemma 12.4.3(ii), we see that $|\check{\Sigma}| \neq 12$ by Theorem 7.3.5. Therefore

$$|\check{\Sigma}| \in \{20, 30, 60\}$$

by Theorem 6.7.1(ii),(iii). It follows from Lemma 4.4.6 that the genus of the normalization of the curve $\check{Z}$ (which is the same as the normalization of the curve Z) does not exceed

$$46 - t - |\check{\Sigma}_{12}| - |\check{\Sigma}| \leqslant 34 - t - |\check{\Sigma}|,$$

which implies that $t = 10$ and $|\check{\Sigma}| = 20$. Applying Lemma 4.4.6 once again, we see that $\check{Z}$ is smooth away from $\check{\Sigma}_{12} \cup \check{\Sigma}$, the genus of the normalization of the curve Z is 4, and $\check{Z}$ has either nodes or ordinary cusps at the points of $\check{\Sigma}$. In particular, the normalization of the curve $\check{Z}$ is a smooth curve of genus 4 faithfully acted on by the group $\mathfrak{A}_5$ that has an $\mathfrak{A}_5$-orbit of length 20. The latter is impossible by Lemma 5.1.5. The obtained contradiction proves assertions (iii) and (iv).

The normalization of the curve $\check{Z}$ is a curve of genus

$$g = 46 - t - \check{n}|\check{\Sigma}_{12}| = 46 - t - 12\check{n}$$

for some positive integer $\check{n}$. Since

$$\mathrm{mult}_{\Sigma_{12}}(Z) = 2$$

and Z contains the $\mathfrak{A}_5$-orbit Σ'_{12}, we conclude that the normalization of the curve $\check{Z}$ contains at least two $\mathfrak{A}_5$-orbits of length 12. Thus Lemma 5.1.5 implies that $\check{n} = 1$. This proves assertions (v) and (vi).

Since $\check{n} = 1$ we see that the singularities of $\check{Z}$ at the points of $\check{\Sigma}_{12}$ (and thus also the singularities of Z at the points of Σ_{12}) are either nodes or ordinary cusps. Recall that each smooth curve of genus 19 or 24 faithfully acted on by the group $\mathfrak{A}_5$ contains exactly two $\mathfrak{A}_5$-orbits of length 12 by Lemma 5.1.5. Since $\Sigma'_{12} \subset Z$, we conclude that the singularities of Z at the points of Σ_{12} are ordinary cusps. This proves assertion (ii) and completes the proof of Lemma 12.4.8. $\square$

Corollary 12.4.11. The surface S_{Cl} is smooth at the points of the $\mathfrak{A}_5$-orbits Σ_{10}, Σ'_{10}, and Σ_{15}.

Proof. Let us use notation of Lemma 12.4.8. If S_{Cl} is singular at the points of Σ_{10} (resp., Σ'_{10}, Σ_{15}), then the curve Z_{10} (resp., Z'_{10}, Z_{15}) has multiplicity at least 4 at the points of Σ_{10} (resp., Σ'_{10}, Σ_{15}), because the surface S_{10} (resp., S'_{10}, S_{15}) is singular at every point of Σ_{10} (resp., Σ'_{10}, Σ_{15}) by Theorem 8.2.1(iv). This gives a contradiction with Lemma 12.4.8(iii),(iv). □

Corollary 12.4.12. Let S be a surface in $\mathcal{Q}_2$ that is different from $\mathscr{S}$, $S_{\mathscr{C}'}$, $S_{\mathcal{L}_6}$, and $S_{\mathcal{G}_6}$. Then either S_{Cl} is smooth at every point of the locus

$$\left(S \cap S_{Cl}\right) \setminus \Sigma_{12},$$

or S_{Cl} has double points at some $\mathfrak{A}_5$-orbit Σ of length 30 contained in $S \cap S_{Cl}$ and is smooth away from $\Sigma \cup \Sigma_{12}$.

Proof. The assertion follows from Lemmas 12.4.4(ii) and 12.4.8(iii),(iv) together with Corollary 12.4.11. □

We will see later that S_{Cl} is smooth away from Σ_{12} (see Theorem 15.2.11).

12.5 A remarkable rational surface

Let us recall the second $\mathfrak{A}_5$-Sarkisov link (11.4.18) constructed in §11.4. We constructed a commutative diagram

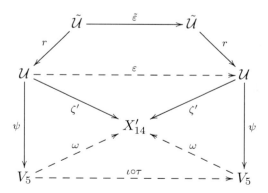

where ψ is a blow-up of the sextic rational normal curve $\mathscr{C}'$, the morphism ζ' is the map given by the linear system $|-K_\mathcal{U}|$ (it contracts the proper transforms of the six lines of $\mathcal{L}_6$ on $\mathcal{U}$ to six isolated ordinary double points),

CREMONA GROUPS AND THE ICOSAHEDRON   307

the map r is the blow-up of the proper transforms of the six lines of $\mathcal{L}_6$, the map ε is a simultaneous Atiyah flop of all curves contracted by ζ', and $\tilde{\varepsilon}$ is a biregular involution (see Corollary 11.4.17 and Lemma 11.5.6(i),(iv)).

Remark 12.5.1. By construction, the r-exceptional divisor is $\tilde{\varepsilon}$-invariant, although some of its irreducible components are not $\tilde{\varepsilon}$-invariant. The details of the action of $\tilde{\varepsilon}$ on the set of irreducible components of the r-exceptional divisor depend on the choice of the element $\tau \in \mathfrak{S}_5$.

Denote by $\mathcal{F}_\mathcal{U}$ the exceptional divisor of the morphism ψ, and denote by $\bar{\mathcal{L}}_6$ the proper transform of the curve $\mathcal{L}_6$ on the threefold $\mathcal{U}$ (cf. §11.3).

Outside of the curve $\bar{\mathcal{L}}_6$, the birational involution ε acts biregularly by Corollary 11.4.17 and Lemma 11.5.6(i),(iv). Thus, its acts on $\operatorname{Pic}(\mathcal{U})$. Furthermore, we know this action by Lemma 11.5.4. Namely, we have

$$\varepsilon^*(\psi^*(H)) \sim 3\psi^*(H) - 2\mathcal{F}_\mathcal{U},$$
$$\varepsilon^*(\mathcal{F}_\mathcal{U}) \sim 4\psi^*(H) - 3\mathcal{F}_\mathcal{U}.$$

In particular, $\psi \circ \varepsilon(\mathcal{F}_\mathcal{U})$ is an irreducible $\mathfrak{A}_5$-invariant surface in $|4H|$ that has multiplicity 3 at a general point of the curve $\mathscr{C}'$. Denote this surface by $\mathscr{F}$, and denote its proper transform on $\mathcal{U}$ by $\bar{\mathscr{F}}$.

Remark 12.5.2. By construction, the surface $\mathscr{F}$ is rational. Moreover, one has $\operatorname{mult}_{\mathscr{C}'}(\mathscr{F}) = 3$. Furthermore, the curve $\mathcal{L}_{12}$ is contained in $\mathscr{F}$ by Lemma 11.5.8. The latter implies that $\mathscr{F}$ does not contain the curve C_{20}. Indeed, if it does, then

$$\mathscr{F} \cdot \mathscr{S} = C_{20} + \mathcal{L}_{12} + \Delta$$

for some $\mathfrak{A}_5$-invariant effective one-cycle Δ on V_5 of degree

$$H \cdot \Delta = 8H^3 - \deg(C_{20}) - \deg(\mathcal{L}_{12}) = 8,$$

which is impossible by Corollary 8.1.15.

Lemma 12.5.3. *The multiplicity of the surface $\mathscr{F}$ at a general point of every line of $\mathcal{L}_6$ is 2.*

Proof. Denote by $L_1, \ldots, L_6$ the lines of $\mathcal{L}_6$. Put $m = \operatorname{mult}_{L_i}(\mathscr{F})$. Let us show that $m = 2$. Denote by $\bar{L}_i$ the proper transform of the line L_i on $\mathcal{U}$. Then $\bar{L}_i \cdot \mathcal{F}_\mathcal{U} = 2$, because $\mathcal{L}_6 \cap \mathscr{C}' = \Sigma'_{12}$ and the curve $\mathcal{L}_6$ intersects the curve $\mathscr{C}'$ transversally at the points of Σ'_{12} by Lemma 11.3.4. Since

$$\bar{\mathscr{F}} \sim 4\psi^*(H) - 3\mathcal{F}_\mathcal{U},$$

we have
$$\bar{L}_i \cdot \tilde{\mathcal{F}} = \bar{L}_i \cdot \left(4\psi^*(H) - 3\mathcal{F}_\mathcal{U}\right) = 4 - 6 = -2.$$

Denote by $\tilde{E}_i$ the r-exceptional divisor that is mapped to the curve $\bar{L}_i$. Then $\tilde{E}_i \cong \mathbb{P}^1 \times \mathbb{P}^1$ and $\tilde{E}_i|_{\tilde{E}_i}$ is a divisor of bi-degree $(-1,-1)$ on $\tilde{E}_i$, since the normal bundle of $\bar{L}_i \cong \mathbb{P}^1$ on $\mathcal{U}$ is $\mathcal{O}_{\mathbb{P}^1}(-1) \oplus \mathcal{O}_{\mathbb{P}^1}(-1)$ by Lemma 11.5.6(ii). Denote by $\tilde{\tilde{\mathcal{F}}}$ the proper transform of the surface $\tilde{\mathcal{F}}$ on the threefold $\tilde{\mathcal{U}}$. Then

$$\tilde{\tilde{\mathcal{F}}} \sim r^*(\tilde{\mathcal{F}}) - m\left(\sum_{i=1}^{6} \tilde{E}_i\right),$$

which implies that $\tilde{\tilde{\mathcal{F}}}|_{\tilde{E}_i}$ is a divisor on $\tilde{E}_i$ of bi-degree $(-2+m, m)$. In particular, one has $m \geqslant 2$.

Recall from §11.1 that there exists a unique $\mathfrak{A}_5$-invariant surface

$$S_{\mathcal{L}_6} \in |2H|$$

that contains the curve $\mathcal{L}_6$, and $S_{\mathcal{L}_6}$ is a smooth $K3$ surface. Furthermore, the divisor $2H|_{S_{\mathcal{L}_6}} - \mathcal{L}_6$ is very ample on $S_{\mathcal{L}_6}$ by Lemma 11.1.2. One has

$$\left(\mathcal{F}|_{S_{\mathcal{L}_6}} - 3\mathcal{L}_6\right) \cdot \left(2H|_{S_{\mathcal{L}_6}} - \mathcal{L}_6\right) = \left(4H|_{S_{\mathcal{L}_6}} - 3\mathcal{L}_6\right) \cdot \left(2H|_{S_{\mathcal{L}_6}} - \mathcal{L}_6\right) = -16,$$

which means that the divisor $\mathcal{F}|_{S_{\mathcal{L}_6}} - 3\mathcal{L}_6$ cannot be effective. This implies that $m < 3$, so that $m = 2$. $\square$

Lemma 12.5.4. *Let S be a surface in the pencil $\mathcal{Q}_2$ that is different from the surface $\mathscr{S}$. Then the divisors $2C_{20}$ and $\mathcal{F}|_S$ generate a pencil in the linear system $|4H|_S|$ that is free from fixed curves. Let Z be an $\mathfrak{A}_5$-irreducible curve in S of degree at most 20. Suppose that Z contains the $\mathfrak{A}_5$-orbit Σ'_{12}. Then there is a divisor in the pencil generated by $2C_{20}$ and $\mathcal{F}|_S$ whose support contains Z.*

Proof. By Remark 12.5.2, the surface $\mathcal{F}$ does not contain the curve C_{20}. Since
$$2C_{20} \sim 4H|_S \sim \mathcal{F}|_S,$$
we see that $2C_{20}$ and $\mathcal{F}|_S$ indeed generate some pencil in $|4H|_S|$ that is free from fixed curves. Denote this pencil by $\mathcal{P}$.

By Lemma 12.5.3, the surface $\mathcal{F}$ is singular in every point of $\mathcal{L}_6$. In particular, it is singular at the points of $\Sigma'_{12} \subset \mathcal{L}_6$ (see Lemma 7.4.3). Also, one has $\Sigma'_{12} \subset C_{20}$ by Theorem 8.1.8(iii). Hence, we have

$$\mathrm{mult}_{\Sigma'_{12}}(D) \geqslant 2$$

CREMONA GROUPS AND THE ICOSAHEDRON 309

for every divisor $D \in \mathcal{P}$.

Let P be a general point in Z, and let Σ_P be its $\mathfrak{A}_5$-orbit. Then $|\Sigma_P| = 60$ by Corollary 8.2.2. Let D be a divisor in $\mathcal{P}$ such that $P \in \mathrm{Supp}(D)$. We claim that $Z \subset \mathrm{Supp}(D)$. Indeed, if Z is not contained in $\mathrm{Supp}(D)$, then

$$80 = D \cdot Z \geqslant \sum_{O \in \Sigma_P} \mathrm{mult}_O(D \cdot Z) + \sum_{O \in \Sigma'_{12}} \mathrm{mult}_O(D \cdot Z) \geqslant$$

$$\geqslant |\Sigma_P| + \sum_{O \in \Sigma'_{12}} \mathrm{mult}_O(D \cdot Z) = |\Sigma_P| + \sum_{O \in \Sigma'_{12}} \mathrm{mult}_O(D) \mathrm{mult}_O(Z) \geqslant$$

$$\geqslant |\Sigma_P| + \sum_{O \in \Sigma'_{12}} \mathrm{mult}_O(D) \geqslant |\Sigma_P| + 2|\Sigma'_{12}| = 84,$$

which is absurd. □

Recall that we have $\mathcal{F}_\mathcal{U} \cong \mathbb{P}^1 \times \mathbb{P}^1$ and $\mathcal{F}_\mathcal{U}^3 = -10$ by Lemma 11.3.3 (see also (11.5.3)). This implies that $\mathcal{F}_\mathcal{U}|_{\mathcal{F}_\mathcal{U}}$ is a divisor of bi-degree $(-1, 5)$ on $\mathcal{F}_\mathcal{U}$.

Lemma 12.5.5. Let $\bar{Z}$ be the unique $\mathfrak{A}_5$-invariant curve in $\mathcal{F}_\mathcal{U} \cong \mathbb{P}^1 \times \mathbb{P}^1$ of bi-degree $(1, 13)$ (see Lemma 12.2.13). Then the following assertions hold:

(i) the birational map $\varepsilon \colon \mathcal{U} \dashrightarrow \mathcal{U}$ is well defined in a general point of $\bar{Z}$;

(ii) the image $\psi \circ \varepsilon(\bar{Z})$ is a rational curve C_{22} of degree 22 that is contained in the surface $\mathscr{F}$;

(iii) the curve C_{22} has ordinary cusps at Σ'_{12} and is smooth elsewhere.

Proof. By Lemma 12.2.5(i), the surface S_{Cl} contains the curve $\mathcal{L}_6$ and the $\mathfrak{A}_5$-orbit Σ'_{12}. Furthermore, by Lemma 12.2.5(iv), the surfaces S_{Cl} is smooth at the points of the $\mathfrak{A}_5$-orbit Σ'_{12}.

Denote by $\bar{\mathcal{L}}_6$ the proper transform of the curve $\mathcal{L}_6$ on the threefold $\mathcal{U}$, and denote by $\bar{L}_1, \ldots, \bar{L}_6$ the irreducible components of the curve $\bar{\mathcal{L}}_6$. Then ε is a morphism outside of the curves $\bar{L}_1, \ldots, \bar{L}_6$ by Lemma 11.5.6(i),(iv). In particular, the birational map ε is well defined in a general point of $\bar{Z}$, which gives assertion (i).

Put $\bar{C} = \varepsilon(\bar{Z})$. Denote by $\bar{S}_{Cl}$ the proper transform of the surface S_{Cl} on the threefold $\mathcal{U}$. Since $\mathcal{L}_6 \subset S_{Cl}$, we have $\bar{\mathcal{L}}_6 \subset \bar{S}_{Cl}$. Recall that

$$\bar{Z} = \mathcal{F}_\mathcal{U} \cap \bar{S}_{Cl}$$

by Lemma 12.2.13(ii). Note that $\mathcal{L}_6 \cap \mathscr{C}' = \Sigma'_{12}$ by Lemma 11.3.4. Since $\bar{\mathcal{L}}_6 \subset \bar{S}_{Cl}$ and $\bar{\mathcal{L}}_6 \cdot \mathcal{F}_\mathcal{U} = 12$, we see that the intersection $\bar{\mathcal{L}}_6 \cap \bar{Z}$ consists of 12 points, and the curve $\bar{Z}$ intersects the curve $\bar{\mathcal{L}}_6$ transversally in 12 points.

Denote by $\tilde{Z}$ the proper transform of the curve $\bar{Z}$ on $\tilde{\mathcal{U}}$, denote by $\tilde{C}$ the proper transform of the curve $\bar{C}$ on $\tilde{\mathcal{U}}$, and denote by $\tilde{E}_1, \ldots, \tilde{E}_6$ the r-exceptional divisors that are mapped to the curves $\bar{L}_1, \ldots, \bar{L}_6$, respectively. Then

$$\tilde{Z} \cdot \tilde{E}_1 = \tilde{Z} \cdot \tilde{E}_2 = \tilde{Z} \cdot \tilde{E}_3 = \tilde{Z} \cdot \tilde{E}_4 = \tilde{Z} \cdot \tilde{E}_5 = \tilde{Z} \cdot \tilde{E}_6 = 2,$$

because $\bar{\mathcal{L}}_6$ intersects the curve $\bar{Z}$ transversally in 12 points. Thus

$$\tilde{C} \cdot \tilde{E}_1 = \tilde{C} \cdot \tilde{E}_2 = \tilde{C} \cdot \tilde{E}_3 = \tilde{C} \cdot \tilde{E}_4 = \tilde{C} \cdot \tilde{E}_5 = \tilde{C} \cdot \tilde{E}_6 = 2$$

as well, because the curve $\tilde{\varepsilon}(\tilde{Z}) = \tilde{C}$ and the set $\{\tilde{E}_1, \ldots, \tilde{E}_6\}$ are $\tilde{\varepsilon}$-invariant by Remark 12.5.1.

Denote by $\tilde{\bar{\mathscr{F}}}$ and $\tilde{\mathcal{F}}_\mathcal{U}$ the proper transforms of the surfaces $\bar{\mathscr{F}}$ and $\mathcal{F}_\mathcal{U}$ on the threefold $\tilde{\mathcal{U}}$, respectively. Then

$$\tilde{\varepsilon}(\tilde{\bar{\mathscr{F}}}) = \tilde{\mathcal{F}}_\mathcal{U}$$

by construction. Moreover, we have $\tilde{\mathcal{F}}_\mathcal{U} \sim r^*(\mathcal{F}_\mathcal{U})$, because $\mathcal{F}_\mathcal{U}$ does not contain the curves $\bar{L}_1, \ldots, \bar{L}_6$. Furthermore, we have

$$\tilde{\bar{\mathscr{F}}} \sim r^*(\bar{\mathscr{F}}) - 2\sum_{i=1}^{6} \tilde{E}_i,$$

because the multiplicity of the surface $\bar{\mathscr{F}}$ at a general point of every line of $\mathcal{L}_6$ is 2 by Lemma 12.5.3. Thus, we have

$$\bar{Z} \cdot \bar{\mathscr{F}} = \tilde{Z} \cdot r^*(\bar{\mathscr{F}}) =$$

$$= \tilde{Z} \cdot \left(\tilde{\bar{\mathscr{F}}} + 2\sum_{i=1}^{6} \tilde{E}_i\right) = \tilde{C} \cdot \left(\tilde{\mathcal{F}}_\mathcal{U} + 2\sum_{i=1}^{6} \tilde{E}_i\right) =$$

$$= \tilde{C} \cdot \tilde{\mathcal{F}}_\mathcal{U} + 24 = \tilde{C} \cdot r^*(\mathcal{F}_\mathcal{U}) + 24 = \bar{C} \cdot \mathcal{F}_\mathcal{U} + 24. \quad (12.5.6)$$

We have

$$\bar{\mathscr{F}}|_{\mathcal{F}_\mathcal{U}} \sim \left(\psi^*(4H) - 3\mathcal{F}_\mathcal{U}\right)|_{\mathcal{F}_\mathcal{U}},$$

which implies that $\bar{\mathscr{F}}|_{\mathcal{F}_\mathcal{U}}$ is a curve of bi-degree $(3, 9)$ on $\mathcal{F}_\mathcal{U} \cong \mathbb{P}^1 \times \mathbb{P}^1$. Thus, one has $\bar{Z} \cdot \bar{\mathscr{F}} = 48$, because $\bar{Z}$ is a curve of bi-degree $(1, 13)$ on $\mathcal{F}_\mathcal{U}$.

Put $C_{22} = \psi(\bar{C})$. Since
$$\bar{Z} \cdot \bar{\mathscr{F}} = \bar{C} \cdot \mathcal{F}_{\mathcal{U}} + 24$$
and $\bar{Z} \cdot \bar{\mathscr{F}} = 48$, we see that $\bar{C} \cdot \mathcal{F}_{\mathcal{U}} = 24$. We have
$$\begin{aligned}20 = -K_{\mathcal{U}} \cdot \bar{Z} &= (\zeta')^*(-K_{X'_{14}}) \cdot \bar{Z} = \\ &= -K_{X'_{14}} \cdot \zeta'(\bar{Z}) = -K_{X'_{14}} \cdot \zeta'(\bar{C}) = \\ &= (\zeta')^*(-K_{X'_{14}}) \cdot \bar{C} = -K_{\mathcal{U}} \cdot \bar{C}, \quad (12.5.7)\end{aligned}$$
because $\iota \circ \tau \in \mathrm{Bir}(V_5)$ acts biregularly on X'_{14} by Corollary 11.4.17. Since we know that $\bar{C} \cdot \mathcal{F}_{\mathcal{U}} = 24$ and
$$(\psi^*(2H) - \mathcal{F}_{\mathcal{U}}) \cdot \bar{C} = -K_{\mathcal{U}} \cdot \bar{C} = 20,$$
we obtain $\psi^*(H) \cdot \bar{C} = 22$. Thus, we have
$$22 = \psi^*(H) \cdot \bar{C} = H \cdot C_{22},$$
which implies that C_{22} is a curve of degree 22 in V_5. By construction, the curve C_{22} is contained in $\mathscr{F}$. This proves assertion (ii).

Let us show that C_{22} is singular at every point of Σ'_{12}. Since $\mathcal{F}_{\mathcal{U}} \cdot \bar{C} = 24$ and $\bar{C}$ is a smooth rational curve, the intersection $\mathcal{F}_{\mathcal{U}} \cap \bar{C}$ is a single $\mathfrak{A}_5$-orbit of length 12 by Lemma 5.1.4. Hence, the intersection $\mathscr{C}' \cap C_{22}$ coincides with the $\mathfrak{A}_5$-orbit Σ'_{12}, because $\psi(\mathcal{F}_{\mathcal{U}}) = \mathscr{C}'$, and Σ'_{12} is the unique $\mathfrak{A}_5$-orbit of length 12 in $\mathscr{C}'$ by Lemmas 5.5.9(v) and 11.3.1.

Recall that the curve C_{20} (which is the base curve of the pencil $\mathcal{Q}_2$) contains the $\mathfrak{A}_5$-orbits Σ_{12}, Σ'_{12}, and Σ'_{20} by Theorem 8.1.8(iii),(vi). Moreover, it does not contain any other $\mathfrak{A}_5$-orbits of length less than 60 by Theorem 8.1.8(vii). Furthermore, the curve C_{20} is singular at the points of Σ_{12} by Theorem 8.1.8(v).

By Corollary 8.1.6, the curve C_{22} is contained in a surface $S \in \mathcal{Q}_2$. The surface S is smooth by Lemma 8.3.3. Then $C_{20} \cdot C_{22} = 44$ on the surface S. This implies that
$$C_{20} \cap C_{22} = \Sigma'_{12} \cup \Sigma'_{20}.$$
Moreover, this shows that either C_{22} is singular at every point of Σ'_{12}, or C_{22} is tangent to C_{20} at each of these 12 points. In the latter case, C_{22} is transversal to $\mathscr{C}'$ at every point of $\Sigma'_{12} = \mathscr{C}' \cap C_{22}$, because $\mathscr{C}'$ and C_{20} intersect transversally at these 12 points by Lemma 11.3.4. On the other hand, if C_{22} intersects the curve $\mathscr{C}'$ transversally at every point of
$$\Sigma'_{12} = \mathscr{C}' \cap C_{22},$$

then $\bar{C} \cdot \mathcal{F}_\mathcal{U} = 12$. The latter is impossible, since we already proved that $\bar{C} \cdot \mathcal{F}_\mathcal{U} = 24$. Thus, we see that the curve C_{22} is singular at every point of Σ'_{12}.

Let us show that S does not contain the curve $\mathscr{C}'$. Suppose it does. Then we have $S = S_{\mathscr{C}'}$ (see §11.3). In particular, the divisor $2H|_S - \mathscr{C}'$ is very ample by Lemma 11.4.7. Thus,

$$0 < \left(2H|_S - \mathscr{C}'\right) \cdot C_{22} = 44 - \mathscr{C}' \cdot C_{22}.$$

This implies that either $\mathscr{C}' \cdot C_{22} = 24$ or $\mathscr{C}' \cdot C_{22} = 36$, because $\mathscr{C}' \cap C_{22} = \Sigma'_{12}$ and C_{22} is singular at the points of Σ'_{12}. Then Lemma 8.3.2 implies that either $C_{22}^2 = \frac{193}{7}$ or $C_{22}^2 = -\frac{311}{7}$, which is absurd. This shows that S does not contain the curve $\mathscr{C}'$.

Since $\mathscr{C}' \cdot S = 12$ and $\mathscr{C}' \not\subset S$, the intersection $\mathscr{C}' \cap S$ consists of the $\mathfrak{A}_5$-orbit Σ'_{12}, and the curve $\mathscr{C}'$ intersects the surface S transversally in these 12 points.

Since $\psi(\mathcal{F}_\mathcal{U} \cap \bar{C}) = \Sigma'_{12}$ and the intersection $\mathcal{F}_\mathcal{U} \cap \bar{C}$ consists of a single $\mathfrak{A}_5$-orbit of length 12, we see that C_{22} is smooth outside of Σ'_{12}, and each singular point of the curve C_{22} is a singularity of type A_r with $r \geqslant 3$.

Let us show that each singular point of C_{22} is an ordinary cusp. Denote by $\bar{S}$ the proper transform of the surface S on the threefold $\mathcal{U}$. The restriction $\psi|_{\bar{S}} \colon \bar{S} \to S$ is a blow-up of the $\mathfrak{A}_5$-orbit Σ'_{12}. Moreover, we have

$$\mathcal{F}_\mathcal{U}|_{\bar{S}} = F_1 + \ldots + F_{12},$$

where $F_1, \ldots, F_{12}$ are exceptional curves of $\psi|_{\bar{S}}$. Since $\bar{C} \cdot \mathcal{F}_\mathcal{U} = 24$, we see that $\bar{C} \cdot F_i = 2$ and the intersection $\bar{C} \cap F_i$ consists of a single point for every i. We conclude that each singular point of C_{22} is an ordinary cusp. This proves assertion (iii) and completes the proof of Lemma 12.5.5. □

Chapter 13

Curves of low degree

In this chapter we classify $\mathfrak{A}_5$-irreducible curves of low degree contained in V_5 (cf. Lemma 8.1.14 and Proposition 10.2.5) and study $\mathfrak{A}_5$-invariant anticanonical divisors passing through them.

13.1 Curves of degree 16

Recall from Lemma 8.1.16(i) that the surface $\mathscr{S} \subset V_5$ contains a unique $\mathfrak{A}_5$-irreducible curve C_{16} of degree 16. This curve is smooth and rational by Lemma 8.1.16(ii). The curve C_{16} is not the only $\mathfrak{A}_5$-irreducible curve of degree 16 contained in V_5. In this section we construct an $\mathfrak{A}_5$-irreducible curve C'_{16} of degree 16 in V_5 different from C_{16}, and prove that V_5 does not contain other $\mathfrak{A}_5$-irreducible curves of degree 16. We will use notation of §12.5.

Lemma 13.1.1. Let $C \subset V_5$ be an irreducible $\mathfrak{A}_5$-invariant curve of degree 16 that is not contained in the surface $\mathscr{S}$. Then the following assertions hold:

(i) one has $C \cap \mathscr{S} = \Sigma'_{12} \cup \Sigma'_{20}$;

(ii) the curve C is disjoint from the $\mathfrak{A}_5$-orbits Σ_{12} and Σ_{20} that are contained in $\mathscr{S}$;

(iii) one has $C \cap \mathscr{C}' = \Sigma'_{12}$;

(iv) the curves C and $\mathscr{C}'$ intersect transversally at the points of Σ'_{12}.

Proof. Since $C \not\subset \mathscr{S}$, the intersection $C \cap \mathscr{S}$ consists of $C \cdot \mathscr{S} = 32$ points (counted with multiplicities). By Lemma 7.2.10 this means that $C \cap \mathscr{S}$ is actually a union of two $\mathfrak{A}_5$-orbits of lengths 12 and 20, and the surface $\mathscr{S}$ is smooth at the points of these orbits. By Lemma 7.2.2(i) the latter means that these orbits are Σ'_{12} and Σ'_{20}, respectively. This proves assertions (i) and (ii).

Recall from §11.3 that there is a unique surface $S_{\mathscr{C}'}$ in the pencil $\mathcal{Q}_2$ that contains the curve $\mathscr{C}'$, and the surface $S_{\mathscr{C}'}$ is smooth. We claim that the curve C is not contained in $S_{\mathscr{C}'}$. Indeed, suppose that $C \subset S_{\mathscr{C}'}$. Put $k = C \cdot \mathscr{C}'$. Then $k \geqslant 12$, since $\Sigma'_{12} \subset C \cap \mathscr{C}'$. Moreover, if $k \neq 12$, then $k \geqslant 24$ by Lemma 5.1.4. Furthermore, we have $C^2 \geqslant -2$ by the adjunction formula. On the other hand, we have

$$10k^2 - 192k + (56C^2 - 512) = 0$$

by Lemma 8.3.2. If $k \geqslant 24$, then

$$0 = 10k^2 - 192k + (56C^2 - 512) \geqslant$$
$$\geqslant 5760 - 4608 + (56C^2 - 512) =$$
$$= 640 + 56C^2 \geqslant 640 - 112 > 0,$$

which is absurd. If $k = 12$, then

$$0 = 10k^2 - 192k + (56C^2 - 512) =$$
$$= 1440 - 2304 + (56C^2 - 512) = 56C^2 - 1376,$$

which implies that $C^2 = \frac{172}{7}$. The latter is impossible, since C^2 is an integer. This shows that $C \not\subset S_{\mathscr{C}'}$.

We claim that $C \cap \mathscr{C}' = \Sigma'_{12}$ and the curves C and $\mathscr{C}'$ intersect transversally at the points of Σ'_{12}.

By Corollary 8.1.6 there exists a surface $S \in \mathcal{Q}_2$ such that $C \subset S$. Since C is not contained in $S_{\mathscr{C}'}$, we see that $S \neq S_{\mathscr{C}'}$. Hence $\mathscr{C}' \not\subset S$ by Corollary 8.1.9. We have

$$12 = \mathscr{C}' \cdot S \geqslant \sum_{P \in \mathscr{C}' \cap C} \mathrm{mult}_P(\mathscr{C}' \cdot S) =$$
$$= \sum_{P \in \Sigma'_{12}} \mathrm{mult}_P(\mathscr{C}' \cdot S) + \sum_{P \in (\mathscr{C}' \cap C) \setminus \Sigma'_{12}} \mathrm{mult}_P(\mathscr{C}' \cdot S) \geqslant$$
$$\geqslant \sum_{P \in \Sigma'_{12}} \mathrm{mult}_P(\mathscr{C}' \cdot S) \geqslant |\Sigma'_{12}| = 12,$$

CREMONA GROUPS AND THE ICOSAHEDRON   315

which implies that $C \cap \mathscr{C}' = \Sigma'_{12}$ and the curve $\mathscr{C}'$ intersects the surface S transversally at the points of Σ'_{12}. In particular, the curves C and $\mathscr{C}'$ intersect transversally at the points of Σ'_{12}. This proves assertions (iii) and (iv). □

Now we are ready to establish the following result whose proof is similar to the proof of Lemma 12.5.5. Recall from §12.5 that the $\mathfrak{A}_5$-invariant surface $\mathscr{F} \in |4H|$ is defined as an image via $\psi \circ \varepsilon$ of the exceptional divisor $\mathcal{F}_\mathcal{U}$ of the birational morphism $\psi \colon \mathcal{U} \to V_5$.

Lemma 13.1.2. *The surface $\mathscr{F}$ contains a unique $\mathfrak{A}_5$-invariant irreducible curve C'_{16} of degree 16. Moreover, C'_{16} is smooth and its genus is 5. Furthermore, the curve C'_{16} is hyperelliptic.*

Proof. Since the action of $\mathfrak{A}_5$ on $\mathcal{F}_\mathcal{U} \cong \mathbb{P}^1 \times \mathbb{P}^1$ is twisted diagonal by Lemma 11.4.8, there exists a unique effective divisor $\bar{Z}$ of bi-degree $(2,6)$ on $\mathcal{F}_\mathcal{U}$ that is $\mathfrak{A}_5$-invariant, and $\bar{Z}$ is a smooth irreducible hyperelliptic curve of genus 5 (see Lemma 6.4.11(i),(ii)).

We claim that $\bar{Z} \cdot \bar{\mathscr{F}} = 36$. Indeed, the divisor

$$\bar{\mathscr{F}}|_{\mathcal{F}_\mathcal{U}} \sim \left(\psi^*(4H) - 3\mathcal{F}_\mathcal{U} \right)|_{\mathcal{F}_\mathcal{U}}$$

is a divisor of bi-degree $(3,9)$ on $\mathcal{F}_\mathcal{U} \cong \mathbb{P}^1 \times \mathbb{P}^1$. Since $\bar{Z}$ is a divisor of bi-degree $(2,6)$ on $\mathcal{F}_\mathcal{U}$, this gives

$$\bar{Z} \cdot \bar{\mathscr{F}} = \bar{Z} \cdot \bar{\mathscr{F}}|_{\mathcal{F}_\mathcal{U}} = 2 \cdot 9 + 6 \cdot 3 = 36.$$

Moreover, $\bar{Z}$ is not contained in $\bar{\mathscr{F}}$, since otherwise $\bar{\mathscr{F}}|_{\mathcal{F}_\mathcal{U}} - \bar{Z}$ would be an effective $\mathfrak{A}_5$-invariant divisor on $\mathcal{F}_\mathcal{U}$ of bi-degree $(1,3)$, which is impossible by Lemma 6.4.11(v).

By Corollary 11.4.17 and Lemma 11.5.6(i),(iv), the involution ε acts biregularly on $\mathcal{U}$ outside of the curve $\bar{\mathcal{L}}_6$. Put $\bar{C}'_{16} = \varepsilon(\bar{Z})$ and put $C'_{16} = \psi(\bar{C}'_{16})$. Then $\bar{C}'_{16}$ and C'_{16} are irreducible $\mathfrak{A}_5$-invariant curves. Since $\varepsilon(\mathcal{F}_\mathcal{U}) = \bar{\mathscr{F}}$, we have $\bar{C}'_{16} \subset \bar{\mathscr{F}}$ and $C'_{16} \subset \mathscr{F}$. Note that $\bar{C}'_{16}$ is not contained in $\mathcal{F}_\mathcal{U}$, because $\bar{Z}$ is not contained in $\bar{\mathscr{F}}$. In particular, the induced morphism $\bar{C}'_{16} \to C'_{16}$ is birational. Hence, C'_{16} is birational to $\bar{Z}$. In the remaining part of the proof, we will show that the curve C'_{16} enjoys all desired properties.

We start with some preliminary computations that will be used to determine the degree of C'_{16}. Denote by $\tilde{Z}$ and $\tilde{C}'_{16}$ the proper transforms of the curves $\bar{Z}$ and $\bar{C}'_{16}$ on the threefold $\tilde{\mathcal{U}}$, respectively. Then $\tilde{Z} = \tilde{\varepsilon}(\tilde{C}'_{16})$.

Denote by $\tilde{\mathscr{F}}$ and $\tilde{\mathcal{F}}_\mathcal{U}$ the proper transforms of the surfaces $\bar{\mathscr{F}}$ and $\mathcal{F}_\mathcal{U}$ on the threefold $\tilde{\mathcal{U}}$, respectively. Then $\tilde{\mathscr{F}} = \tilde{\varepsilon}(\tilde{\mathcal{F}}_\mathcal{U})$ by construction. Thus, we have
$$\tilde{Z} \cdot \tilde{\mathscr{F}} = \tilde{C}'_{16} \cdot \tilde{\mathcal{F}}_\mathcal{U}, \qquad (13.1.3)$$
since $\tilde{\varepsilon}$ is biregular.

Denote by $L_1, \ldots, L_6$ the lines of $\mathcal{L}_6$ (recall that they are pairwise disjoint by Lemma 7.4.7). Denote by $\bar{L}_1, \ldots, \bar{L}_6$ their proper transforms on $\mathcal{U}$. Note that the curve $\mathscr{C}'$ intersects transversally the curve $\mathcal{L}_6$ at the points of Σ'_{12} by Lemma 11.3.4. Therefore, the proper transform $\bar{\mathcal{L}}_6 = \sum_{i=1}^{6} \bar{L}_i$ of the curve $\mathcal{L}_6$ intersects the surface $\mathcal{F}_\mathcal{U}$ transversally at 12 points, and $\bar{L}_i \cdot \mathcal{F}_\mathcal{U} = 2$ for any i.

We claim that the 12 points of the intersection $\bar{\mathcal{L}}_6 \cap \mathcal{F}_\mathcal{U}$ are contained in $\bar{Z}$. Indeed, let S_{Cl} be the unique $\mathfrak{A}_5$-invariant surface in $|3H|$ (see §12.2). By Lemma 12.2.5(i), the surface S_{Cl} contains both $\mathscr{C}'$ and $\mathcal{L}_6$. Denote by $\bar{S}_{Cl}$ the proper transform of the surface S_{Cl} on the threefold $\mathcal{U}$. Then $\bar{\mathcal{L}}_6 \subset \bar{S}_{Cl}$.

By Lemma 12.2.13(ii), the restriction $\bar{S}_{Cl}|_{\mathcal{F}_\mathcal{U}}$ is a smooth rational curve and, thus, contains a unique $\mathfrak{A}_5$-orbit of length 12 by Lemma 5.5.9(v), which must be the set $\bar{\mathcal{L}}_6 \cap \mathcal{F}_\mathcal{U}$, because $\bar{\mathcal{L}}_6 \subset \bar{S}_{Cl}$ and $\mathscr{C}' \cap \mathcal{L}_6 = \Sigma'_{12}$. On the other hand, we have
$$\bar{S}_{Cl}|_{\mathcal{F}_\mathcal{U}} \cdot \bar{Z} = 32,$$
because $\bar{S}_{Cl}|_{\mathcal{F}_\mathcal{U}}$ is a divisor of bi-degree $(1,13)$ on $\mathcal{F}_\mathcal{U}$ by Lemma 12.2.13(i), and $\bar{Z}$ is a divisor of bi-degree $(2,6)$ on $\mathcal{F}_\mathcal{U}$. By Lemma 5.1.4, this implies that the intersection
$$\bar{S}_{Cl}|_{\mathcal{F}_\mathcal{U}} \cap \bar{Z}$$
contains a unique $\mathfrak{A}_5$-orbit of length 12 in the curve $\bar{S}_{Cl}|_{\mathcal{F}_\mathcal{U}}$. Thus, this $\mathfrak{A}_5$-orbit must coincide with the intersection $\bar{\mathcal{L}}_6 \cap \mathcal{F}_\mathcal{U}$. Hence, the curve $\bar{\mathcal{L}}_6$ intersects the curve $\bar{Z}$ transversally at these 12 points, because $\bar{\mathcal{L}}_6$ intersects $\mathcal{F}_\mathcal{U}$ transversally.

Denote by $\tilde{E}_1, \ldots, \tilde{E}_6$ the r-exceptional divisors that are mapped to the curves $\bar{L}_1, \ldots, \bar{L}_6$, respectively. Then the set $\{\tilde{E}_1, \ldots, \tilde{E}_6\}$ is $\tilde{\varepsilon}$-invariant by Remark 12.5.1. Moreover, we have $\tilde{Z} \cdot \tilde{E}_i = 2$ because $\bar{\mathcal{L}}_6$ intersects the curve $\bar{Z}$ transversally at 12 points. Thus $\tilde{C}'_{16} \cdot \tilde{E}_i = 2$ as well. Since the multiplicity of the surface $\mathscr{F}$ at a general point of every line of $\mathcal{L}_6$ is 2 by

CREMONA GROUPS AND THE ICOSAHEDRON 317

Lemma 12.5.3, it follows from (13.1.3) that (cf. (12.5.6))

$$36 = \bar{Z} \cdot \bar{\mathcal{F}} = \tilde{Z} \cdot r^*(\bar{\mathcal{F}}) =$$
$$= \tilde{Z} \cdot \left(\tilde{\mathcal{F}} + 2\sum_{i=1}^{6} \tilde{E}_i\right) = \tilde{C}'_{16} \cdot \left(\tilde{\mathcal{F}}_{\mathcal{U}} + 2\sum_{i=1}^{6} \tilde{E}_i\right) =$$
$$= \tilde{C}'_{16} \cdot \tilde{\mathcal{F}}_{\mathcal{U}} + 24 = \tilde{C}'_{16} \cdot r^*(\mathcal{F}_{\mathcal{U}}) + 24 = \bar{C}'_{16} \cdot \mathcal{F}_{\mathcal{U}} + 24.$$

Thus, we see that $\bar{C}'_{16} \cdot \mathcal{F}_{\mathcal{U}} = 12$.

Let us show that C'_{16} is a curve of degree 16. We have (cf. (12.5.7))

$$20 = -K_{\mathcal{U}} \cdot \bar{Z} = (\zeta')^*(-K_{X'_{14}}) \cdot \bar{Z} =$$
$$= -K_{X'_{14}} \cdot \zeta'(\bar{Z}) = -K_{X'_{14}} \cdot \zeta'(\bar{C}'_{16}) =$$
$$= (\zeta')^*(-K_{X'_{14}}) \cdot \bar{C}'_{16} = -K_{\mathcal{U}} \cdot \bar{C}'_{16},$$

because $\iota \circ \tau \in \mathrm{Bir}(V_5)$ acts biregularly on X'_{14} by Corollary 11.4.17. Since $-K_{\mathcal{U}} \cdot \bar{C}'_{16} = 20$ and $\mathcal{F}_{\mathcal{U}} \cdot \bar{C}'_{16} = 12$, we have

$$20 = -K_{\mathcal{U}} \cdot \bar{C}'_{16} = \left(\psi^*(2H) - \mathcal{F}_{\mathcal{U}}\right) \cdot \bar{C}'_{16} =$$
$$= \psi^*(2H) \cdot \bar{C}'_{16} - \mathcal{F}_{\mathcal{U}} \cdot \bar{C}'_{16} = \psi^*(2H) \cdot \bar{C}'_{16} - 12 = 2H \cdot C'_{16} - 12,$$

which implies that $H \cdot C'_{16} = 16$, i.e., the curve C'_{16} is a curve of degree 16.

Now let us show that C'_{16} is smooth. By Corollary 8.1.6 there is a surface $S \in \mathcal{Q}_2$ such that $C'_{16} \subset S$. One has $S \neq \mathscr{S}$, since otherwise C'_{16} would be a rational curve by Lemma 8.1.16(i),(ii). Moreover, it follows from Lemma 8.3.3 that S is smooth. In particular, the surface S, and thus also the curve C'_{16}, does not pass through the $\mathfrak{A}_5$-orbits Σ_5, Σ_{10}, and Σ'_{10} by Lemma 8.2.1(iv).

Put $H_S = H|_S$. By the Hodge index theorem one has

$$\det\begin{pmatrix} H_S^2 & C'_{16} \cdot H_S \\ C'_{16} \cdot H_S & (C'_{16})^2 \end{pmatrix} = \det\begin{pmatrix} 10 & 16 \\ 16 & (C'_{16})^2 \end{pmatrix} \leqslant 0,$$

which implies that $(C'_{16})^2 \leqslant 25$. Recall that

$$(C'_{16})^2 = 2p_a(C'_{16}) - 2$$

by the adjunction formula, and thus $p_a(C) \leqslant 13$. Since C'_{16} is birational to $\bar{Z}$, one has

$$5 = g(\bar{Z}) \leqslant p_a(C'_{16}) - |\mathrm{Sing}(C'_{16})| \leqslant 13 - |\mathrm{Sing}(C'_{16})|$$

by Lemma 4.4.6. This shows that $|\mathrm{Sing}(C'_{16})| \leqslant 8$ so that C'_{16} is smooth by Theorem 7.3.5. In particular, the curves C'_{16} and $\bar{Z}$ are isomorphic, because they are both smooth and are birational by construction. Thus, C'_{16} is a hyperelliptic curve of genus 5, because $\bar{Z}$ has these properties.

To complete the proof, we must show that C'_{16} is the unique $\mathfrak{A}_5$-invariant irreducible curve of degree 16 contained in $\mathscr{F}$. Suppose that $\mathscr{F}$ contains some $\mathfrak{A}_5$-invariant irreducible curve of degree 16. Let us denote this curve by C and show that it coincides with C'_{16}.

By Lemma 13.1.1(iii), one has $C \cap \mathscr{C}' = \Sigma'_{12}$. Furthermore, the curves C and $\mathscr{C}'$ intersect transversally at the points of Σ'_{12} by Lemma 13.1.1(iv). By construction, we have $\varepsilon(\bar{\mathscr{F}}) = \mathcal{F}_\mathcal{U}$. Denote by $\bar{C}$ the proper transform of the curve C on the threefold $\mathcal{U}$. Then

$$\bar{C} \cdot \mathcal{F}_\mathcal{U} = |C \cap \mathscr{C}'| = 12,$$

because $C \cap \mathscr{C}' = \Sigma'_{12}$ and the curves C and $\mathscr{C}'$ intersect transversally at the points of Σ'_{12}. This gives, in particular, that

$$-K_\mathcal{U} \cdot \bar{C} = \left(\psi^*(2H) - \mathcal{F}_\mathcal{U}\right) \cdot \bar{C} = 2H \cdot C - |C \cap \mathscr{C}'| = 32 - 12 = 20.$$

By Corollary 11.4.17 and Lemma 11.5.6(i),(iv), the involution ε acts biregularly on $\mathcal{U}$ outside of the curve $\bar{\mathcal{L}}_6$. Thus, we can put $\bar{Z}' = \varepsilon(\bar{C})$. Then $\bar{Z}'$ is an irreducible $\mathfrak{A}_5$-invariant curve. Moreover, we have $\bar{Z}' \subset \mathcal{F}_\mathcal{U}$, since $\bar{C} \subset \bar{\mathscr{F}}$. Furthermore, we have

$$-K_\mathcal{U} \cdot \bar{Z}' = (\zeta')^*(-K_{X'_{14}}) \cdot \bar{Z}' = -K_{X'_{14}} \cdot \zeta'(\bar{Z}') =$$
$$= -K_{X'_{14}} \cdot \zeta'(\bar{C}) = (\zeta')^*(-K_{X'_{14}}) \cdot \bar{C} = -K_\mathcal{U} \cdot \bar{C} = 20,$$

because $\iota \circ \tau$ acts biregularly on X'_{14} by Corollary 11.4.17.

Recall that $\mathcal{F}_\mathcal{U} \cong \mathbb{P}^1 \times \mathbb{P}^1$. Let l be a fiber of the natural projection

$$\mathcal{F}_\mathcal{U} \to \mathscr{C}' \cong \mathbb{P}^1,$$

and let s be a section of this projection such that $s^2 = 0$. Then

$$\mathcal{F}_\mathcal{U}|_{\mathcal{F}_\mathcal{U}} \sim -s + 5l$$

and $\psi^*(H)|_{\mathcal{F}_\mathcal{U}} = 6l$. This implies that

$$-K_\mathcal{U}|_{\mathcal{F}_\mathcal{U}} \sim s + 7l.$$

We have $\bar{Z}' \sim as + bl$ for some non-negative integers a and b. Thus

$$20 = -K_{\mathcal{U}} \cdot \bar{Z}' = -K_{\mathcal{U}}|_{\mathcal{F}_{\mathcal{U}}} \cdot \bar{Z}' = (s + 7l) \cdot (as + bl) = b + 7a,$$

which implies that

$$(a, b) \in \{(0, 20), (1, 13), (2, 6)\}.$$

Since $\bar{Z}'$ is irreducible, we have either $(a, b) = (1, 13)$ or $(a, b) = (2, 6)$. In the former case, Lemmas 12.2.13(i) and 12.5.5(ii) imply that C is a curve of degree 22, which contradicts our assumption. Thus, we see that

$$(a, b) = (2, 6).$$

Hence $\bar{Z}' = \bar{Z}$, because $\bar{Z}$ is a unique $\mathfrak{A}_5$-invariant curve of bi-degree $(2, 6)$ on $\mathcal{F}_{\mathcal{U}}$ by Lemma 6.4.11(i). Therefore, one has $C = C'_{16}$ by construction of C'_{16}. This completes the proof of Lemma 13.1.2. □

By Lemma 5.4.3, the curve C'_{16} is the unique smooth irreducible curve of genus 5 that admits a faithful action of the group $\mathfrak{A}_5$.

Remark 13.1.4. One has

$$C'_{16} \cap \mathscr{S} = \Sigma'_{12} \cup \Sigma'_{20}$$

by Lemma 13.1.1(i). In particular, the curve C'_{16} is disjoint from the $\mathfrak{A}_5$-orbits Σ_{12} and Σ_{20}. Also, this implies that $C'_{16} \cap \mathcal{L}_{20} = \Sigma'_{20}$ by Lemma 7.2.12(ii),(iii). Moreover, we see that the curve C'_{16} is not contained in the surface S_{Cl}, since $\Sigma'_{20} \not\subset S_{Cl}$ by Corollary 12.2.7. Thus the intersection $C'_{16} \cap S_{Cl}$ consists of

$$C'_{16} \cdot S_{Cl} = 48$$

points (counted with multiplicities), and Theorem 7.3.5 implies that

$$C'_{16} \cap S_{Cl} = \Sigma'_{12}.$$

Finally, we note that the curve C'_{16} does not contain any $\mathfrak{A}_5$-orbits of length less than 20 except for Σ'_{12} by Lemma 5.1.5.

Lemma 13.1.5. *Let C be an $\mathfrak{A}_5$-invariant irreducible curve in V_5 of degree 16. Then either $C = C_{16}$ or $C = C'_{16}$.*

Proof. If $C \subset \mathscr{S}$, then $C = C_{16}$ by Lemma 8.1.16(i). If $C \subset \mathscr{F}$, then we have $C = C'_{16}$ by Lemma 13.1.2. Suppose that $C \not\subset \mathscr{S}$ and $C \not\subset \mathscr{F}$.

Since $C \not\subset \mathscr{F}$, the intersection $\mathscr{F} \cap C$ is an $\mathfrak{A}_5$-invariant set that consists of $\mathscr{F} \cdot C = 64$ points (counted with multiplicities). By Theorem 7.3.5, the intersection $\mathscr{F} \cap C$ contains the $\mathfrak{A}_5$-orbit of length 12 which is Σ'_{12} by Lemma 13.1.1(i). Thus the intersection $(\mathscr{F} \cap C) \setminus \Sigma'_{12}$ consists of

$$\mathscr{F} \cdot C - k|\Sigma'_{12}| = 64 - 12k$$

points (counted with multiplicities) for some positive integer $k \geqslant 3$, since $\Sigma'_{12} \subset \mathscr{C}'$ (see Lemma 11.3.1) and $\text{mult}_{\mathscr{C}'}(\mathscr{F}) = 3$ by Remark 12.5.2. The latter is impossible by Theorem 7.3.5. □

Lemma 13.1.6. *The curves C'_{16} and $\mathcal{L}_{15}$ are disjoint.*

Proof. By Corollary 8.1.7, there exists a surface S in $\mathcal{Q}_2$ that contains C'_{16}. Moreover, $S \neq \mathscr{S}$ by definition of the curve C'_{16}. Thus S is smooth by Lemma 8.3.3. Note that $\mathcal{L}_{15} \not\subset S$ by Lemma 10.2.6. Moreover, neither Σ_5 nor Σ_{15} is contained in S by Theorem 8.2.1(iv). Thus, it follows from Lemma 9.1.10(iv) that the intersection $\mathcal{L}_{15} \cap S$ is a single $\mathfrak{A}_5$-orbit of length

$$\mathcal{L}_{15} \cdot S = 30.$$

Hence, the intersection $\mathcal{L}_{15} \cap S$ is not contained in C'_{16} by Lemma 5.1.5, because C'_{16} is a smooth curve of genus 5. This shows that the curves C'_{16} and $\mathcal{L}_{15}$ are disjoint. □

13.2 Six twisted cubics

In this section we construct a unique $\mathfrak{A}_5$-invariant curve in V_5 that is a union of six twisted cubics and describe its properties.

Lemma 13.2.1. *There exists a unique $\mathfrak{A}_5$-invariant curve $\mathcal{T}_6$ on V_5 that is a union of 6 smooth rational normal cubic curves. The curve $\mathcal{T}_6$ is disjoint from the curve $\mathcal{L}_6$, while the intersection of $\mathcal{T}_6$ with the curve $\mathcal{L}_{15}$ is non-empty.*

Proof. Choose a line L_1 of $\mathcal{L}_6$. Let us use notation of §7.7. Let s and l be smooth rational curves in F_1 such that $s^2 = -1$, $l^2 = 0$, and $s \cdot l = 1$. Then there exists a curve $\check{C}_1 \in |s + l|$ that is invariant with respect to the stabilizer $G_1 \cong \mathfrak{D}_{10}$ of the line L_1. Indeed, the action of G_1 on F_1 is lifted from its action on $\mathbb{P}^2$ via the birational morphism $\alpha \colon F_1 \to \mathbb{P}^2$ that contracts

s to a point. Since the point $\alpha(s)$ is fixed by G_1, there exists a G_1-invariant line in $\mathbb{P}^2$ that does not pass through $\alpha(s)$. The proper transform of this line on F_1 is the required G_1-invariant curve $\check{C}_1 \in |s+l|$. Note that $\check{C}_1$ is a smooth rational curve.

We claim that $u(\check{C}_1)$ is a smooth rational cubic curve. If this is the case, then its orbit must be an $\mathfrak{A}_5$-invariant curve $\mathcal{T}_6$ on V_5 that is a union of 6 smooth rational normal cubic curves. Indeed, if $u(\check{C}_1)$ is stabilized by a subgroup of $\mathfrak{A}_5$ that is strictly larger than G_1, then it must be $\mathfrak{A}_5$-invariant by Lemma 5.1.1. The latter is impossible by Lemma 8.1.14 if $u(\check{C}_1)$ is a cubic curve.

Let us compute the class of the divisor $u^*(H)|_{F_1}$. One has

$$u^*(H)|_{F_1} \sim as + bl$$

for some non-negative integers a and b. Since l is contracted by v, we see that $u(l)$ is a line in V_5. Then

$$1 = u^*(H)|_{F_1} \cdot l = (as+bl) \cdot l = a,$$

which implies that $a = 1$. On the other hand, $F_1 \sim u^*(H) - 2E$ by Lemma 7.7.3. This gives

$$5 = H^3 = u^*(H) \cdot u^*(H) \cdot F_1 =$$
$$= u^*(H)|_{F_1} \cdot u^*(H)|_{F_1} = (s+bl)(s+bl) = 2b - 1,$$

which implies that $b = 3$. Thus, we have

$$u^*(H)|_{F_1} \cdot \check{C}_1 = (s+3l) \cdot (s+l) = 3,$$

which implies that either $u(\check{C}_1)$ is a cubic curve, or $\check{C}_1 \subset E_1$ and $u(\check{C}_1) = L_1$. The latter case is clearly impossible, because $\check{C}_1 \not\subset E_1 \cap F_1$. Indeed, we have

$$s + 3l - E_1|_{F_1} \sim (u^*(H) - E_1)|_{F_1} \sim 3l,$$

because $w = v \circ u^{-1}$ is a linear projection, and $v(F_1)$ is a smooth rational cubic curve. This implies that $E_1|_{F_1} = s$. Hence, $\check{C}_1 \not\subset E_1 \cap F_1$.

We see that $u(\check{C}_1)$ is a cubic curve. Denote this curve by C_1. Note that C_1 is smooth, since otherwise its singular point must be G_1-invariant, which is impossible by Theorem 7.3.5. Hence, $\mathcal{T}_6$ is a union of 6 rational normal cubic curves.

Let us show that $\mathcal{T}_6 \cap \mathcal{L}_6 = \varnothing$. Note that $E_1 \cap \check{C}_1 = \varnothing$, because $E_1|_{F_1} = s$ and $\check{C}_1 \in |s+l|$. Since $L_1 = u(E_1)$ and $C_1 = u(\check{C}_1)$, we see that $C_1 \cap L_1 = \varnothing$

as well. Thus, to prove that $\mathcal{T}_6 \cap \mathcal{L}_6 = \varnothing$ we have to show that C_1 does not intersect other irreducible components of $\mathcal{L}_6$.

Let $L_2, \ldots, L_6$ be irreducible components of $\mathcal{L}_6$ that are different from L_1. Suppose that some of them, say L_2, intersects C_1. Then all of them do by Lemma 5.1.3(ii). Recall that a line and a twisted cubic curve cannot have more than 2 points in common, because a twisted cubic curve is an intersection of quadrics.

Suppose that some, and thus any, of the lines $L_2, \ldots, L_6$ is either tangent to C_1 or intersects it in two points. Then all these 5 lines are contained in the linear span $\Pi \cong \mathbb{P}^3$ of C_1 in $\mathbb{P}^6$. This means that the curve $\mathcal{L}_6$ is contained in a proper linear subspace of $\mathbb{P}^6$. The latter contradicts Corollary 7.5.4. This shows that each of the lines $L_2, \ldots, L_6$ intersects C_1 transversally at one point. Since the lines $L_1, \ldots, L_6$ are disjoint by Lemma 7.4.7, we see that $C_1 \cap \mathcal{L}_6$ consists of 5 different points.

Recall from §11.1 that there exists a unique surface $S_{\mathcal{L}_6}$ in the pencil $\mathcal{Q}_2$ that contains $\mathcal{L}_6$. Moreover, the surface $S_{\mathcal{L}_6}$ is smooth. Note that

$$S_{\mathcal{L}_6} \cdot C_1 = 6.$$

Thus, if $C_1 \not\subset S_{\mathcal{L}_6}$, then the intersection $C_1 \cap S_{\mathcal{L}_6}$ consists of 6 points (counted with multiplicities). Since $C_1 \cap \mathcal{L}_6$ consists of 5 different points, we see that C_1 contains a G_1-invariant point provided that $C_1 \not\subset S_{\mathcal{L}_6}$. The latter is impossible by Lemma 5.5.9(ii), because the action of the group $G_1 \cong \mathfrak{D}_{10}$ on C_1 is faithful by construction (this also follows from Lemma 7.4.5).

Thus, we see that $\mathcal{T}_6 \subset S_{\mathcal{L}_6}$. Note that $\mathcal{T}_6 \cdot \mathcal{L}_6 = 30$ on the surface $S_{\mathcal{L}_6}$, because each of the lines $L_2, \ldots, L_6$ intersects C_1 transversally at one point, and $L_1 \cap C_1 = \varnothing$. Moreover, $\mathcal{L}_6^2 = -12$, since the lines $L_1, \ldots, L_6$ are disjoint. Then

$$\mathcal{T}_6^2 = \frac{114}{13}$$

by Lemma 8.3.2, which is absurd. The obtained contradiction shows that the curves $\mathcal{L}_6$ and $\mathcal{T}_6$ are disjoint.

Now let us show that the intersection $\mathcal{T}_6 \cap \mathcal{L}_{15}$ is non-empty. By Lemma 9.1.14 there is a line R of $\mathcal{L}_{15}$ such that

$$R \cap L_1 \neq \varnothing.$$

One has $R \subset H_1$, because H_1 is singular along L_1. Let $\check{R}$ be a proper transform of the line R on the threefold $\check{V}_5$. Then $\check{R}$ is contained in the surface F_1, and $\check{R}$ is contracted by the morphism v. Hence $\check{R} \sim l$ and the intersection of $\check{R}$ with

$$\check{C}_1 \sim s + l$$

is non-empty. This gives $C_1 \cap R \neq \emptyset$ and $\mathcal{T}_6 \cap \mathcal{L}_{15} \neq \emptyset$.

Finally, let us prove the uniqueness of the curve $\mathcal{T}_6$. Namely, let Z be some $\mathfrak{A}_5$-invariant curve on V_5 that is a union of 6 rational normal cubic curves. We are going to show that $Z = \mathcal{T}_6$.

Since Z consists of 6 irreducible components, one of them is G_1-invariant by Lemma 5.1.1. Denote it by C'_1. Let us show that $C'_1 = C_1$ (this would imply that $Z = \mathcal{T}_6$). We have $C'_1 \subset H_1$. Indeed, if $C'_1 \not\subset H_1$, then the G_1-invariant subset $C'_1 \cap H_1$ consists of three points (counted with multiplicities), which is impossible by Lemma 5.5.9(ii), since $G_1 \cong \mathfrak{D}_{10}$ acts on $C_1 \cong \mathbb{P}^1$ faithfully by Lemma 7.4.5.

Let $\check{C}'_1$ be the proper transform of the curve C'_1 on the threefold $\check{V}_5$. Then $\check{C}'_1 \subset F_1$. Thus, there are non-negative integers c and d such that

$$\check{C}'_1 \sim cs + dl.$$

Hence

$$3 = H \cdot C'_1 = u^*(H)|_{F_1} \cdot \check{C}'_1 = (s + 3l) \cdot (cs + dl) = 2c + d,$$

which implies that $c = 1$ and $d = 1$, since $\check{C}'_1$ is irreducible. Thus, one has

$$\check{C}'_1 \in |s + l|.$$

If $\check{C}'_1 \neq \check{C}_1$, then $\check{C}'_1 \cap \check{C}_1$ consists of a single point, which implies that the point $u(\check{C}'_1 \cap \check{C}_1)$ must be G_1-invariant. The latter is impossible by Theorem 7.3.5. Therefore, $\check{C}'_1 = \check{C}_1$, so that $C'_1 = C_1$. This completes the proof of Lemma 13.2.1. $\square$

Keeping in mind Lemma 13.2.1, in the rest of the book we will denote by $\mathcal{T}_6$ the $\mathfrak{A}_5$-irreducible curve that is a union of 6 twisted cubics on V_5, and will sometimes refer to these twisted cubics as the twisted cubics of $\mathcal{T}_6$ for brevity.

Remark 13.2.2 (cf. Remark 7.8.5). Note that one can prove Lemma 13.2.1 using the fact that there is an $\mathfrak{A}_5$-equivariant identification of a certain compactification of the variety of twisted cubics contained in V_5 with the Grassmannian $\mathrm{Gr}(2, \mathscr{W}_5)$ (see [110, Proposition 2.46]). Still we prefer to take a more straightforward way.

Lemma 13.2.3. *One has*

$$\mathcal{T}_6 \cap \mathscr{S} = \mathcal{T}_6 \cap \mathcal{L}_{12} = \mathcal{T}_6 \cap \mathcal{C}_{16} = \mathcal{T}_6 \cap \mathcal{C}_{20} = \Sigma_{12}.$$

In particular, the $\mathfrak{A}_5$-orbit Σ'_{12} and the curve $\mathcal{L}_{20}$ are disjoint from the curve $\mathcal{T}_6$.

Proof. One has $\Sigma'_{12} \subset \mathcal{L}_6$ by Lemma 7.4.3 and $\mathcal{T}_6 \cap \mathcal{L}_6 = \varnothing$ by Lemma 13.2.1. Therefore, one has $\Sigma'_{12} \not\subset \mathcal{T}_6$.

Note that $\mathcal{T}_6 \not\subset \mathscr{S}$ by Lemma 7.2.13(iii). Thus the intersection $\mathcal{T}_6 \cap \mathscr{S}$ consists of $\mathcal{T}_6 \cdot \mathscr{S} = 36$ points (counted with multiplicities). By Lemma 7.2.10 it means that
$$\mathcal{T}_6 \cap \mathscr{S} \subset \Sigma_{12} \cup \Sigma'_{12}.$$
Since we already know that $\Sigma'_{12} \not\subset \mathcal{T}_6$, we conclude that $\mathcal{T}_6 \cap \mathscr{S} = \Sigma_{12}$.

Recall that one has $\Sigma_{12} \subset \mathscr{C}$ by construction, one has $\Sigma_{12} \subset \mathcal{L}_{12}$ by Lemma 7.2.12(iii), one has $\Sigma_{12} \subset C_{16}$ by Lemma 8.1.16(iii), one has $\Sigma_{12} \subset C_{20}$ by Theorem 8.1.8(iii), and one has $\Sigma_{12} \not\subset \mathcal{L}_{20}$ by Lemma 7.2.12(ii). This implies the remaining assertions of the lemma. □

Lemma 13.2.4. *The curve $\mathcal{T}_6$ does not contain any of the orbits Σ_5, Σ_{10}, Σ'_{10}, or Σ_{15}. Moreover, the twisted cubics of $\mathcal{T}_6$ are pairwise disjoint.*

Proof. By Corollary 8.1.7 there is a surface $S \in \mathcal{Q}_2$ such that $\mathcal{T}_6 \subset S$. By Lemma 7.2.13(iii), one has $S \neq \mathscr{S}$. Moreover, it follows from Lemma 8.3.3 that S is smooth. Therefore, the surface S (and hence also the curve $\mathcal{T}_6$) does not contain the orbits Σ_5, Σ_{10}, Σ'_{10}, and Σ_{15} (see Theorem 8.2.1(iv)).

Let $C_1, \ldots, C_6$ be the twisted cubics of $\mathcal{T}_6$. Let G_1 be the stabilizer in $\mathfrak{A}_5$ of the curve C_1. Then $G_1 \cong \mathfrak{D}_{10}$ by Lemma 5.1.1. Recall that the action of G_1 on C_1 is faithful by Lemma 7.4.5. This implies that there exists exactly one G_1-orbit in C_1 of length 2, and the length of any other G_1-orbit in C_1 equals either 5 or 10 (see Lemma 5.5.9(ii)).

Suppose that the curves $C_1, \ldots, C_6$ are not disjoint so that two of them, say, C_1 and C_2, have a common point P. Let $\Lambda \subset C_1$ be the G_1-orbit of P, and let $\Sigma \subset \mathcal{T}_6$ be the $\mathfrak{A}_5$-orbit of P. Let c be the number of curves C_i that pass through the point P; then $c \geqslant 2$, and c equals the number of curves C_i that pass through any given point of Σ. Note that $|\Lambda|$ is the number of points of Σ lying on C_1, and it equals the number of points of Σ lying on any of the curves C_i. One has $|\Sigma|c = 6|\Lambda|$.

We claim that $|\Lambda| \neq 2$. Indeed, suppose that $|\Lambda| = 2$. Then $|\Sigma| = 12$ by Theorem 7.3.5. Hence $c = 1$, which contradicts the inequality $c \geqslant 2$.

We claim that $|\Lambda| \neq 5$. Indeed, suppose that $|\Lambda| = 5$. Then $|\Sigma|$ divides 30, so that either $\Sigma \in \{\Sigma_5, \Sigma_{10}, \Sigma'_{10}, \Sigma_{15}\}$, or $|\Sigma| = 30$ (see Theorem 7.3.5). Since we already proved that the former case is impossible, we conclude that $|\Sigma| = 30$ and $c = 1$. This again contradicts the inequality $c \geqslant 2$.

CREMONA GROUPS AND THE ICOSAHEDRON

Therefore, one has $|\Lambda| \geqslant 10$. Hence

$$\mathcal{T}_6^2 = \left(\sum_{i=1}^{6} C_i\right)^2 = \sum_{i=1}^{6} C_i^2 + \sum_{i=1}^{6}\left(C_i \cdot \left(\sum_{j \neq i} C_j\right)\right) \geqslant -12 + 6|\Lambda| \geqslant 48.$$

On the other hand, by the Hodge index theorem one has

$$\det\begin{pmatrix} H_S^2 & \mathcal{T}_6 \cdot H_S \\ \mathcal{T}_6 \cdot H_S & \mathcal{T}_6^2 \end{pmatrix} = \det\begin{pmatrix} 10 & 18 \\ 18 & \mathcal{T}_6^2 \end{pmatrix} \leqslant 0,$$

which implies that $\mathcal{T}_6^2 \leqslant 32$. The obtained contradiction shows that the curves $C_1, \ldots, C_6$ are pairwise disjoint. $\square$

Corollary 13.2.5. The curve $\mathcal{T}_6$ does not contain $\mathfrak{A}_5$-orbits of length 20.

It appears that $\mathcal{T}_6$ is the only reducible $\mathfrak{A}_5$-irreducible curve in V_5 of degree at least 16 and at most 19.

Lemma 13.2.6. Let Z be an $\mathfrak{A}_5$-irreducible curve of degree $16 \leqslant \deg(Z) \leqslant 19$. Suppose that Z is reducible. Then $Z = \mathcal{T}_6$.

Proof. Suppose that Z has $r > 1$ irreducible components, each having degree d. Since $16 \leqslant rd \leqslant 19$ and the curve Z is $\mathfrak{A}_5$-irreducible, Corollary 5.1.2 implies that $r = 6$ and $d = 3$. In particular, each irreducible component C_i of Z is a twisted cubic, since V_5 is an intersection of quadrics. Now the assertion follows by Lemma 13.2.1. $\square$

Lemma 13.2.7. The intersection $\mathcal{T}_6 \cap \mathcal{L}_{15}$ is an $\mathfrak{A}_5$-orbit of length 30 not contained in the surface $\mathscr{S}$.

Proof. By Corollary 8.1.7, there exists a surface S in $\mathcal{Q}_2$ that contains $\mathcal{T}_6$. Moreover, one has $S \neq \mathscr{S}$ by Lemma 7.2.13(iii). Thus S is smooth by Lemma 8.3.3. Note that $\mathcal{L}_{15} \not\subset S$ by Lemma 10.2.6. Moreover, neither Σ_5 nor Σ_{15} is contained in S by Theorem 8.2.1(iv). Thus, it follows from Lemma 9.1.10(iv) that the intersection $\mathcal{L}_{15} \cap S$ is a single $\mathfrak{A}_5$-orbit Σ of length

$$\mathcal{L}_{15} \cdot S = 30.$$

Since the intersection $\mathcal{T}_6 \cap \mathcal{L}_{15}$ is non-empty by Lemma 13.2.1, we see that

$$\mathcal{T}_6 \cap \mathcal{L}_{15} = \Sigma.$$

Finally, Σ is not contained in the surface $\mathscr{S}$, because $\mathcal{T}_6 \cap \mathscr{S} = \Sigma_{12}$ by Lemma 13.2.3. $\square$

13.3 Irreducible curves of degree 18

In this section we prove some auxiliary results about irreducible curves of degree 18 in V_5.

Lemma 13.3.1. *Let $C \subset V_5$ be a smooth irreducible $\mathfrak{A}_5$-invariant curve of genus 15 and degree d. Then $d \neq 18$.*

Proof. Suppose that $d = 18$. Corollary 8.1.6 implies that there is a surface $S \in \mathcal{Q}_2$ that contains C, and by Lemma 7.2.13(iii) one has $S \neq \mathscr{S}$. By Lemma 8.3.3 this implies that S is non-singular. Put $H_S = H|_S$.

Let us consider the linear system $|C - H_S|$. One has $(C - H_S)^2 = 2$ and

$$(C - H_S) \cdot H_S = 8,$$

which implies that

$$h^2(\mathcal{O}_S(C - H_S)) = h^0(\mathcal{O}_S(H_S - C)) = 0$$

by Serre duality. Hence

$$h^0(\mathcal{O}_S(C - H_S)) = h^1(\mathcal{O}_S(C - H_S)) + 3$$

by the Riemann–Roch formula. On the other hand, the linear system $|C - H_S|$ does not have fixed components. Indeed, if Z is the fixed part of the linear system $|C - H_S|$, then

$$H_S \cdot Z \leqslant H_S \cdot (C - H_S) = 8.$$

By Proposition 10.2.5 and Lemmas 10.2.6 and 11.3.2 we know that if Z is non-trivial, then Z must be one of the curves $\mathscr{C}'$ or $\mathcal{L}_6$. Moreover, one has $C^2 = 28$ by the adjunction formula. Using the intersection numbers listed in Table 10.1, and applying Lemma 8.3.2 we see that none of these cases is possible. The latter implies also that the linear system $|C - H_S|$ is free from base points by Theorem 7.3.5, because $(C - H_S)^2 = 2$.

We claim that $C - H_S$ is ample. Indeed, suppose that $C - H_S$ is not ample. Since $(C - H_S)^2 = 2$, it follows from Lemma 6.7.6 that S contains an irreducible curve F such that

$$20 + r(H_S \cdot F) = 2H_S^2 + r(H_S \cdot F) = \Big((C - H_S) \cdot H_S\Big)^2 = 64$$

for some $r \in \{6, 10, 15\}$. This gives

$$(H_S \cdot F)^2 = \frac{44}{r} \notin \mathbb{Z},$$

which is a contradiction.

Thus, we see that $C - H_S$ is ample, so that $h^1(\mathcal{O}_S(C - H_S)) = 0$ by Kodaira vanishing. Hence

$$h^0(\mathcal{O}_S(C - H_S)) = 3$$

and the linear system $|C - H_S|$ gives an $\mathfrak{A}_5$-equivariant morphism $\eta\colon S \to \mathbb{P}^2$. Since $C - H_S$ is ample, the morphism η is a finite morphism of degree 2 that is branched over a smooth sextic curve $B \subset \mathbb{P}^2$.

Let ς be the biregular involution of S that arises from the double cover η. Put $C' = \varsigma(C)$. Then $H_S \cdot C' = 18$ and $C^2 = (C')^2 = 28$. Put $k = C \cdot C'$. Applying Lemma 8.3.2, we conclude that

$$10k^2 - 648k + 10304 = 0,$$

which implies that either $k = 28$ or $k = \frac{184}{5}$. Since S is smooth, k must be an integer, so that $k = 28$. If $C \neq C'$, then the set $C' \cap C$ is an $\mathfrak{A}_5$-invariant set that consists of 28 points (counted with multiplicities). The latter is impossible by Lemma 5.1.4.

Thus, we see that $C = C'$, i.e., C is ς-invariant. Since $(C - H_S) \cdot C = 18$, we see that $\eta(C)$ is an irreducible plane curve of degree 9. Therefore,

$$28 = C^2 = 2\eta(C)^2 = 162,$$

which is absurd. The obtained contradiction completes the proof of Lemma 13.3.1. □

Lemma 13.3.2. Let $C \subset V_5$ be a smooth irreducible $\mathfrak{A}_5$-invariant curve of genus 10 and degree d. Then $d \neq 18$.

Proof. Suppose that $d = 18$. Corollary 8.1.6 implies that there is a surface $S \in \mathcal{Q}_2$ that contains C, and by Lemma 7.2.13(iii) one has $S \neq \mathscr{S}$. By Lemma 8.3.3 this implies that S is non-singular.

Put $H_S = H|_S$. One has $(3H_S - C)^2 = 0$ and $(3H_S - C) \cdot H_S = 12$, which implies that

$$h^2(\mathcal{O}_S(3H_S - C)) = h^0(\mathcal{O}_S(C - 3H_S)) = 0$$

by Serre duality. Hence

$$h^0(\mathcal{O}_S(3H_S - C)) = h^1(\mathcal{O}_S(3H_S - C)) + 2 \geqslant 2$$

by the Riemann–Roch formula. In particular, we see that the linear system $|3H_S - C|$ does not have fixed components.

Since $(3H_S - C)^2 = 0$ and the linear system $|3H_S - C|$ does not have fixed components, we see that the linear system $|3H_S - C|$ does not have base points and must be composed of a pencil, i.e., we have $3H_S - C = nE$ for some positive integer n and an effective divisor E such that $|E|$ is a pencil that does not have base points. We may assume that E is a generic curve in $|E|$. In this case, the curve E is a smooth elliptic curve. In particular, we have $H_S \cdot E \geqslant 3$. Since

$$12 = H_S \cdot (3H_S - C) = nH_S \cdot E \geqslant 3n,$$

we have $n \in \{1, 2, 3, 4\}$. Let us exclude all possibilities for n case by case.

Suppose that $n = 1$. Then

$$h^0\big(\mathcal{O}_S(3H_S - C)\big) = h^0\big(\mathcal{O}_S(2E)\big) = 2. \tag{13.3.3}$$

On the other hand, there is an exact sequence of $\mathfrak{A}_5$-representations

$$0 \to H^0\big(\mathcal{O}_{V_5}(H)\big) \to H^0\big(\mathcal{O}_{V_5}(3H)\big) \to H^0\big(\mathcal{O}_S(3H_S)\big) \to 0$$

since $h^1(\mathcal{O}_{V_5}(H)) = 0$ by Kodaira vanishing. In particular, it follows from Lemma 12.1.1 that there are no $\mathfrak{A}_5$-invariant two-dimensional subspaces in $H^0(\mathcal{O}_S(3H_S))$, which contradicts (13.3.3).

Suppose that either $n = 2$ or $n = 4$. Let us consider a divisor $H_S - \frac{n}{2}E$. We have $(H_S - \frac{n}{2}E)^2 = -2$ and $(H_S - \frac{n}{2}E) \cdot H_S = 4$, which implies that

$$h^2\big(\mathcal{O}_S(H_S - \frac{n}{2}E)\big) = h^0\big(\mathcal{O}_S(\frac{n}{2}E - H_S)\big) = 0$$

by Serre duality. Hence

$$h^0\big(\mathcal{O}_S(H_S - \frac{n}{2}E)\big) = h^1\big(\mathcal{O}_S(H_S - \frac{n}{2}E)\big) + 1$$

by the Riemann–Roch formula. In particular, we conclude that the linear system $|H_S - \frac{n}{2}E|$ is not empty. On the other hand, the linear system $|H_S - \frac{n}{2}E|$ does not have fixed components. Indeed, if Z is the fixed part of the linear system $|H_S - \frac{n}{2}E|$, then

$$H_S \cdot Z \leqslant H_S \cdot \left(H_S - \frac{n}{2}E\right) = 4.$$

The latter is impossible by Corollary 8.1.15, since Z is $\mathfrak{A}_5$-invariant. Thus, we see that $|H_S - \frac{n}{2}E|$ does not have fixed components. In particular, the divisor $H_S - \frac{n}{2}E$ is nef, which is absurd, since $(H_S - \frac{n}{2}E)^2 = -2$.

Thus, we see that $n = 3$. Then $3H_S - C \sim 3E$ and $H_S \cdot E = 4$. This implies that $(H_S - E)^2 = 2$ and $(H_S - E) \cdot H_S = 6$, so that

$$h^2\big(\mathcal{O}_S(H_S - E)\big) = h^0\big(\mathcal{O}_S(E - H_S)\big) = 0$$

by Serre duality. Hence

$$h^0\big(\mathcal{O}_S(H_S - E)\big) = h^1\big(\mathcal{O}_S(H_S - E)\big) + 3$$

by the Riemann–Roch formula. On the other hand, the linear system $|H_S - E|$ does not have fixed components. Indeed, if Z is the fixed part of the linear system $|H_S - E|$, then

$$H_S \cdot Z \leqslant H_S \cdot (H_S - E) = 6.$$

By Proposition 10.2.5 and Lemmas 10.2.6 and 11.3.2 we know that if Z is non-trivial, then Z must be one of the curves $\mathscr{C}'$ or $\mathcal{L}_6$. Moreover, one has $C^2 = 18$ by the adjunction formula. Using the intersection numbers listed in Table 10.1, and applying Lemma 8.3.2 we see that none of these cases is possible (cf. the proof of Lemma 13.3.1). The latter implies also that the linear system $|H_S - E|$ is free from base points by Theorem 7.3.5, because $(H_S - E)^2 = 2$.

We see that $|H_S - E|$ is free from base points. We claim that $H_S - E$ is ample. Indeed, suppose that $H_S - E$ is not ample. Since $(H_S - E)^2 = 2$, it follows from Lemma 6.7.6 that S contains an irreducible curve F such that

$$20 + r(H_S \cdot F) = 2H_S^2 + r(H_S \cdot F) = ((H_S - E) \cdot H_S)^2 = 36$$

for some $r \in \{6, 10, 15\}$. This gives

$$(H_S \cdot F)^2 = \frac{16}{r} \notin \mathbb{Z},$$

which is a contradiction.

Thus, we see that $H_S - E$ is ample, so that $h^1(\mathcal{O}_S(H_S - E)) = 0$ by Kodaira vanishing. Hence

$$h^0(\mathcal{O}_S(H_S - E)) = 3$$

and the linear system $|H_S-E|$ gives an $\mathfrak{A}_5$-equivariant morphism $\eta\colon S\to\mathbb{P}^2$. Since $H_S - E$ is ample, the morphism η is a finite morphism of degree 2 that is branched over a smooth sextic curve $B\subset\mathbb{P}^2$.

Let ς be the biregular involution of S that arises from the double cover η. Put $C'=\varsigma(C)$. Then $H_S\cdot C'=18$ and $C^2=(C')^2=18$. Put $k=C\cdot C'$. Applying Lemma 8.3.2, we conclude that

$$10k^2 - 648k + 8424 = 0,$$

which implies that either $k=18$ or $k=\frac{234}{5}$. Since k is an integer, we see that $k=18$. If $C\neq C'$, then the set $C'\cap C$ is an $\mathfrak{A}_5$-invariant set that consists of 18 points (counted with multiplicities). The latter is impossible by Lemma 5.1.4.

Thus, we see that $C = C'$, i.e., C is ς-invariant. Recall that E is a smooth elliptic curve, since we assume that E is a generic curve in $|E|$. Since

$$(H_S - E)\cdot E = 4,$$

we conclude that either $\eta(E)$ is a quartic curve and the induced morphism $E\to\eta(E)$ is an isomorphism, or $\eta(E)$ is a conic and the induced morphism $E\to\eta(E)$ is a double cover. The former case is impossible since a smooth plane quartic curve has genus 3. The latter case is impossible since otherwise one would have

$$0 = E^2 = 2\eta(E)^2 = 8.$$

The obtained contradiction completes the proof of Lemma 13.3.2. □

Lemma 13.3.4. Let $C\subset V_5$ be an irreducible $\mathfrak{A}_5$-invariant curve of degree 18. Then one of the following possibilities occurs:

(i) C is a smooth rational curve;

(ii) C is a smooth curve of genus 4;

(iii) C is a rational curve of arithmetic genus 12 such that $\mathrm{Sing}(C)=\Sigma'_{12}$, and every singular point of the curve C is an ordinary cusp.

Proof. Corollary 8.1.6 implies that there is a surface $S\in\mathcal{Q}_2$ such that $C\subset S$, and by Lemma 7.2.13(iii) one has $S\neq\mathscr{S}$. By Lemma 8.3.3 this implies that S is non-singular.

Put $H_S = H|_S$. Then $H_S \cdot (2H_S - C) = 2$ and one can easily deduce that
$$h^0(\mathcal{O}_S(2H_S - C)) = 0 \tag{13.3.5}$$
Indeed, the linear system $|2H_S - C|$ does not have fixed components, because there are no $\mathfrak{A}_5$-invariant curves of degree 1 and 2 on V_5 (see Corollary 8.1.15). On the other hand, $|2H_S - C|$ cannot be a non-empty linear system that is free from fixed curves, because the $K3$ surface S is not uniruled.

Using (13.3.5), we derive from the Riemann–Roch formula that
$$(2H_S - C)^2 \leqslant -4,$$
which implies that $C^2 \leqslant 28$. Recall that $C^2 = 2p_a(C) - 2$ by the adjunction formula, so that $p_a(C) \leqslant 15$.

Let $\bar{C}$ be the normalization of the curve C. Denote by g the genus of $\bar{C}$. Then
$$g \leqslant p_a(C) - |\mathrm{Sing}(C)| \leqslant 15 - |\mathrm{Sing}(C)|$$
by Lemma 4.4.6. By Theorem 7.3.5, the latter implies that either C is smooth, or $|\mathrm{Sing}(C)| \in \{5, 10, 12, 15\}$. Keeping in mind that S is smooth, we see that either C is smooth, or $|\mathrm{Sing}(C)| = 12$ by Theorems 7.3.5 and 8.2.1(iv).

Suppose that C is singular. Then $g \leqslant 15 - 12 = 3$, which implies that $g = 0$ by Lemma 5.1.5. By Lemma 4.4.6 one has $p_a(C) = 12$, and every singular point of the curve C is either an ordinary cusp, or a node. Keeping in mind that $\bar{C}$ contains a unique $\mathfrak{A}_5$-orbit of length 12 (also by Lemma 5.5.9(v)), we see that the latter case is impossible. Moreover, we have either $\mathrm{Sing}(C) = \Sigma_{12}$ or $\mathrm{Sing}(C) = \Sigma'_{12}$ by Theorem 7.3.5. If $\mathrm{Sing}(C) = \Sigma_{12}$, then
$$36 = 2\deg(C) = \mathscr{S} \cdot C \geqslant |\Sigma_{12}| \cdot \mathrm{mult}_{\Sigma_{12}}(\mathscr{S}) \cdot \mathrm{mult}_{\Sigma_{12}}(C) = 12 \cdot 2 \cdot 2 = 48,$$
which is a contradiction.

Thus, to complete the proof, we may assume that C is smooth and $g \notin \{0, 4\}$. By Lemma 5.1.5, this means
$$g \in \{5, 6, 9, 10, 11, 13, 15\}.$$

Moreover, Corollary 7.3.7 implies that $g \neq 13$.

The intersection $C \cap \mathscr{S}$ is an $\mathfrak{A}_5$-invariant set that consists of 36 points (counted with multiplicities). By Lemma 5.1.4, the set $C \cap \mathscr{S}$ must contain an $\mathfrak{A}_5$-orbit of length 12. Applying Lemma 5.1.5 again, we see that $g \neq 6$ and $g \neq 11$.

Let us show that the linear system $|3H_S - C|$ is either empty or does not have fixed components. Indeed, if Z is the fixed part of the linear system $|3H_S - C|$, then

$$H_S \cdot Z \leqslant H_S \cdot (3H_S - C) = 12.$$

By Proposition 10.2.5 and Lemmas 10.2.6 and 11.3.2 we know that if Z is non-trivial, then Z must be one of the curves $\mathscr{C}'$, $\mathcal{L}_6$, $\mathcal{L}_{10}$, $\mathcal{G}_5$, $\mathcal{G}_5'$, or $\mathcal{G}_6$ (cf. the proofs of Lemmas 13.3.1 and 13.3.2). Moreover, one has

$$C^2 = 2g - 2 \in \{8, 16, 18, 28\}$$

by the adjunction formula. Checking these possibilities case by case using the intersection numbers listed in Table 10.1 and applying Lemma 8.3.2, we see that none of these cases is possible (cf. the proof of Lemma 10.2.7).

Note that

$$(3H_S - C)^2 = 2(g - 10),$$

which implies that $|3H_S - C|$ is empty in the case when $g = 5$ or $g = 9$.

Suppose that $g = 5$ or $g = 9$. Recall from §12.2 there exists an $\mathfrak{A}_5$-invariant irreducible surface $S_{Cl} \in |3H|$. One has $C \not\subset S_{Cl}$, because otherwise $S_{Cl}|_S - C$ would be an effective divisor in $|3H_S - C|$. On the other hand, the set $S_{Cl} \cap C$ is an $\mathfrak{A}_5$-invariant set that consists of 54 points (counted with multiplicities). By Lemma 5.1.4, the set $S_{Cl} \cap C$ contains an $\mathfrak{A}_5$-orbit of length 30. The latter is impossible by Lemma 5.1.5.

Thus, we see that either $g = 15$ or $g = 10$. None of these cases is possible by Lemmas 13.3.1 and 13.3.2. The obtained contradiction completes the proof of Lemma 13.3.4. □

13.4 A singular curve of degree 18

In this section, we construct an $\mathfrak{A}_5$-invariant irreducible singular curve of degree 18 and describe its properties. We use notation of §11.4 and §13.1. Let $\bar{\mathscr{C}}$ be the proper transform of the curve $\mathscr{C}$ on the threefold $\mathcal{U}$. Note that

$$\bar{\mathscr{C}} \cap \bar{\mathcal{L}}_6 = \varnothing$$

by Lemma 7.5.6. Thus, the birational involution ε is biregular in a neighborhood of the curve $\bar{\mathscr{C}}$ by Lemma 11.5.6(iv). Put $\bar{C}_{18}^0 = \varepsilon(\bar{\mathscr{C}})$.

Lemma 13.4.1. One has $\bar{C}_{18}^0 \not\subset \mathcal{F}_\mathcal{U}$ and $\bar{C}_{18}^0 \cdot \psi^*(H) = 18$.

Proof. Note that $\bar{\mathscr{C}} \cdot \psi^*(H) = 6$. Furthermore, we know that $\bar{\mathscr{C}} \cdot \mathcal{F}_\mathcal{U} = 0$, because
$$\mathscr{C} \cap \mathscr{C}' = \varnothing$$
by Lemma 7.5.6. By Lemma 11.5.4, we have
$$\bar{C}_{18}^0 \cdot \left(3\psi^*(H) - 2\mathcal{F}_\mathcal{U}\right) = \bar{C}_{18}^0 \cdot \varepsilon^*(\psi^*(H)) = \bar{\mathscr{C}} \cdot \psi^*(H) = \mathscr{C} \cdot H = 6,$$
and
$$\bar{C}_{18}^0 \cdot \left(4\psi^*(H) - 3\mathcal{F}_\mathcal{U}\right) = \bar{C}_{18}^0 \cdot \varepsilon^*(\mathcal{F}_\mathcal{U}) = \bar{C} \cdot \mathcal{F}_\mathcal{U} = 0.$$
This gives $\bar{C}_{18}^0 \cdot \psi^*(H) = 18$ and $\bar{C}_{18}^0 \cdot \mathcal{F}_\mathcal{U} = 24$.

Suppose that $\bar{C}_{18}^0 \subset \mathcal{F}_\mathcal{U}$. Let l be a fiber of the natural projection
$$\mathcal{F}_\mathcal{U} \to \mathscr{C}' \cong \mathbb{P}^1,$$
and let s be a section of this projection such that $s \cdot l = 1$. Then
$$\bar{C}_{18}^0 \sim as + bl$$
for some non-negative integers a and b. We have
$$\mathcal{F}_\mathcal{U}|_{\mathcal{F}_\mathcal{U}} \sim -s + 5l$$
and $\psi^*(H)|_{\mathcal{F}_\mathcal{U}} \sim 6l$. Thus
$$18 = \bar{C}_{18}^0 \cdot \psi^*(H)|_{\mathcal{F}_\mathcal{U}} = 6l \cdot (as + bl),$$
which gives $a = 3$. Similarly, we have
$$24 = \bar{C}_{18}^0 \cdot \mathcal{F}_\mathcal{U}|_{\mathcal{F}_\mathcal{U}} = (-s + 5l) \cdot (as + bl) = -b + 5a = 15 - b,$$
which gives $b = -9$. The latter is absurd. □

In particular, we see that $\psi|_{\bar{C}_{18}^0}$ is a morphism that is birational on its image. Put $C_{18}^0 = \psi(\bar{C}_{18}^0)$.

Lemma 13.4.2. *The curve C_{18}^0 is a rational curve of degree 18. One has*
$$\mathrm{Sing}(C_{18}^0) = \Sigma_{12}',$$
and every singular point of the curve C_{18}^0 is an ordinary cusp.

Proof. The rationality of the curve C_{18}^0 follows from its construction. Moreover, the degree of C_{18}^0 is

$$C_{18}^0 \cdot H = \bar{C}_{18}^0 \cdot \psi^*(H) = 18.$$

The curve C_{18}^0 is an image of the curve $\mathscr{C}$ under the birational map $\iota \circ \tau$, which is well defined in a general point of every irreducible component of the curve $\mathcal{L}_{12}$. Moreover, all irreducible components of the curve $\mathcal{L}_{12}$ are contracted by the map $\iota \circ \tau$ by Lemma 11.5.8. The curve $\mathcal{L}_{12}$ is a disjoint union of 12 lines that are tangent to $\mathscr{C}$ at the points of Σ_{12} by Lemma 7.2.12(ii),(iv). Moreover, the map $\iota \circ \tau$ is well defined at every point of Σ_{12}, and the set $\iota \circ \tau(\Sigma_{12})$ consists of 12 points; in fact one can check that $\iota \circ \tau(\Sigma_{12}) = \Sigma'_{12}$. Therefore, the curve C_{18}^0 has 12 singular points. Now the assertion follows by Lemma 13.3.4. $\square$

Corollary 13.4.3. *The following assertions hold:*

(i) *the curve C_{18}^0 does not contain any $\mathfrak{A}_5$-orbits of length different from 12, 20, 30, and 60;*

(ii) *one has*

$$C_{18}^0 \cap \mathscr{S} = C_{18}^0 \cap \mathcal{L}_{12} = C_{18}^0 \cap C_{20} = \Sigma'_{12};$$

(iii) *the curve C_{18}^0 does not contain any of the $\mathfrak{A}_5$-orbits Σ_{12}, Σ_{20}, or Σ'_{20};*

(iv) *one has $C_{18}^0 \cap \mathcal{L}_{20} = \varnothing$.*

Proof. Lemma 13.4.2 implies that the normalization morphism $\mathbb{P}^1 \to C_{18}^0$ is bijective, so that assertion (i) follows from Lemma 5.1.4.

Since the curve C_{18}^0 is not contained in the surface $\mathscr{S}$ by Lemma 7.2.13(iii), we conclude that the intersection $C_{18}^0 \cap \mathscr{S}$ consists of

$$C_{18}^0 \cdot \mathscr{S} = 36$$

points (counted with multiplicities). Therefore, it follows from Lemma 7.2.10 that

$$C_{18}^0 \cap \mathscr{S} \subset \Sigma_{12} \cup \Sigma'_{12}.$$

On the other hand, we know from Lemma 13.4.2 that

$$\Sigma'_{12} \subset C_{18}^0 \cap \mathscr{S},$$

and that
$$\operatorname{mult}_{\Sigma'_{12}}(C^0_{18}) = 2.$$

Suppose that $\Sigma_{12} \subset C^0_{18}$. Then

$$36 = C^0_{18} \cdot \mathscr{S} \geqslant |\Sigma_{12}| \cdot \operatorname{mult}_{\Sigma_{12}}(\mathscr{S}) + |\Sigma'_{12}| \cdot \operatorname{mult}_{\Sigma'_{12}}(C^0_{18}) =$$
$$= 12 \cdot 2 + 12 \cdot 2 = 48$$

by Lemma 7.2.2(iii). The obtained contradiction shows that the $\mathfrak{A}_5$-orbit Σ_{12} is not contained in the curve C^0_{18}, and thus

$$C^0_{18} \cap \mathscr{S} = \Sigma'_{12}.$$

Recall that $\Sigma'_{12} \subset \mathcal{L}_{12}$ by Lemma 7.2.12(iii) and $\Sigma'_{12} \subset C_{20}$ by Theorem 8.1.8(iii), while $\Sigma'_{12} \not\subset \mathcal{L}_{20}$ by Lemma 7.2.12(ii). This gives assertions (ii) and (iv).

Finally, assertion (iii) is implied by assertion (ii) since the $\mathfrak{A}_5$-orbits Σ_{12}, Σ_{20}, and Σ'_{20} are contained in the surface $\mathscr{S}$. $\square$

Lemma 13.4.4 (cf. the proof of Lemma 11.5.9). Let C be an $\mathfrak{A}_5$-invariant irreducible curve in V_5 of degree 18. Suppose that C is singular. Then $C = C^0_{18}$.

Proof. By Lemma 13.3.4, we have $\operatorname{Sing}(C) = \Sigma'_{12}$, and every singular point of the curve C is an ordinary cusp. Denote by $\bar{C}$ the proper transform of the curve C on the threefold $\mathcal{U}$. Then the birational map

$$\varepsilon \colon \mathcal{U} \dashrightarrow \mathcal{U}$$

is biregular in a general point of the curve $\bar{C}$ by Corollary 11.4.17 and Lemma 11.5.6(i),(iv).

In the notation of §12.5, there exists a commutative diagram

$$\begin{array}{ccc} \tilde{\mathcal{U}} & \xrightarrow{\tilde{\varepsilon}} & \tilde{\mathcal{U}} \\ {\scriptstyle r}\downarrow & & \downarrow{\scriptstyle r} \\ \mathcal{U} & \dashrightarrow{\varepsilon} & \mathcal{U} \end{array}$$

where r is the blow-up of the proper transforms of the six lines of $\mathcal{L}_6$, and $\tilde{\varepsilon}$ is a biregular involution. Denote by $\tilde{C}$ the proper transform of the curve $\bar{C}$ on the threefold $\tilde{\mathcal{U}}$. Put $\tilde{C}' = \tilde{\varepsilon}(\tilde{C})$ and $\bar{C}' = r(\tilde{C}')$.

Let $\bar{H}$ be a sufficiently general surface in $|\psi^*(H)|$. Denote by $\tilde{H}$ its proper transform on $\tilde{\mathcal{U}}$ via r. Put $\check{M} = \tilde{\varepsilon}(\tilde{H})$ and $\bar{M} = r(\check{M})$. Then $\bar{M} \sim 3\bar{H} - 2\mathcal{F}_\mathcal{U}$ by Lemma 11.5.4. Thus, we have

$$\check{M} \sim r^*(3\bar{H} - 2\mathcal{F}_\mathcal{U}) - m\sum_{i=1}^{6} \tilde{E}_i,$$

where $\tilde{E}_1, \ldots, \tilde{E}_6$ are r-exceptional divisors, and m is some non-negative integer. Therefore, we have

$$\bar{H} \cdot \bar{C}' = \tilde{H} \cdot \tilde{C}' = \check{M} \cdot \tilde{C} =$$

$$= r^*(3\bar{H} - 2\mathcal{F}_\mathcal{U}) \cdot \tilde{C} - m\sum_{i=1}^{6} \tilde{E}_i \cdot \tilde{C} \leqslant r^*(3\bar{H} - 2\mathcal{F}_\mathcal{U}) \cdot \tilde{C} =$$

$$= 3\bar{H} \cdot C - 2\mathcal{F}_\mathcal{U} \cdot \bar{C} = 54 - 2 \cdot |\Sigma'_{12}| \cdot \mathrm{mult}_{\Sigma'_{12}}(C) = 54 - 48 = 6,$$

which implies that $\bar{H} \cdot \bar{C}' \leqslant 6$. Thus, if $\bar{C}' \not\subset \mathcal{F}_\mathcal{U}$, then $\psi(\bar{C}')$ is an irreducible $\mathfrak{A}_5$-invariant curve of degree

$$H \cdot \psi(\bar{C}') = \bar{H} \cdot \bar{C}' \leqslant 6$$

different from $\mathscr{C}'$. Hence, one has $\psi(\bar{C}') = \mathscr{C}$ by Lemma 11.3.2, which implies that $C = C_{18}^0$ by the construction of the curve C_{18}^0.

To complete the proof, we must show that $\bar{C}' \not\subset \mathcal{F}_\mathcal{U}$. There are two ways of doing this. The first way is to use arguments we used in the end of the proof of Lemma 13.1.5. Let us show another way.

Suppose that $\bar{C}' \subset \mathcal{F}_\mathcal{U}$. Then the surface $\mathscr{F} = \psi \circ \varepsilon(\mathcal{F}_\mathcal{U})$ contains the curve C. Recall from §12.5 that $\mathscr{F}$ is an $\mathfrak{A}_5$-invariant surface in $|4H|$, that $\mathscr{F}$ is singular along the curve $\mathscr{C}'$ and that $\mathrm{mult}_{\mathscr{C}'}(\mathscr{F}) = 3$.

By Corollary 8.1.6 there exists a surface $S \in \mathcal{Q}_2$ such that $C \subset S$. By Lemma 7.2.13(iii), one has $S \neq \mathscr{S}$. Thus, the surface S is smooth by Lemma 8.3.3.

We have $\mathscr{F}|_S = nC + \Omega$ for some positive integer n and some $\mathfrak{A}_5$-invariant effective divisor Ω on the surface S such that $C \not\subset \mathrm{Supp}(\Omega)$. Keeping in mind that

$$40 = nH \cdot C + H \cdot \Omega = 18n + H \cdot \Omega,$$

we see that $n = 1$ and $H \cdot \Omega = 22$ by Corollary 8.1.15. Moreover, we have

$$3 = \mathrm{mult}_{\mathscr{C}'}(\mathscr{F}) \leqslant \mathrm{mult}_{\Sigma'_{12}}(\mathscr{F}|_S) =$$

$$= \mathrm{mult}_{\Sigma'_{12}}(C) + \mathrm{mult}_{\Sigma'_{12}}(\Omega) = 2 + \mathrm{mult}_{\Sigma'_{12}}(\Omega),$$

which implies that $\Sigma'_{12} \subset \mathrm{Supp}(\Omega)$. Since C is a rational curve of arithmetic genus 12 by Lemma 13.3.4, we have $C^2 = 22$ on the surface S. Thus, we have
$$72 = 4H \cdot C = (C + \Omega) \cdot C = C^2 + C \cdot \Omega = 22 + C \cdot \Omega,$$
which implies that $C \cdot \Omega = 50$. Keeping in mind that C is singular at every point of the $\mathfrak{A}_5$-orbit Σ'_{12}, we see that
$$\Omega \cdot C = k\Sigma'_{12} + \Omega_C$$
for some integer $k \geqslant 2$ and some $\mathfrak{A}_5$-invariant effective zero-cycle Ω_C whose support does not contain Σ'_{12}. Thus the degree of Ω_C is $50 - 12k$, which is impossible by Theorem 7.3.5. $\square$

13.5 Bring's curve

We know from Lemma 12.3.6 that there is a smooth irreducible $\mathfrak{A}_5$-invariant curve $\mathcal{B}_{18}$ of genus 4 and degree 18, such that the union of $\mathcal{B}_{18}$ with the curve $\mathcal{G}_6$ is cut out on the surface S_{Cl} by some surface of the pencil $\mathcal{Q}_2$. The curve $\mathcal{B}_{18}$ is isomorphic to the Bring's curve (see Remark 5.4.2). The purpose of this section is to study the properties of the curve $\mathcal{B}_{18}$.

We are going to show that the curve $\mathcal{B}_{18}$ is the unique smooth irreducible $\mathfrak{A}_5$-invariant curve of degree 18 and genus 4 in V_5 (cf. Remark 13.5.4 below). To prove this we will need a preliminary computation that will be also used in the proof of Lemma 13.6.3.

Lemma 13.5.1. Let S be a surface from the pencil $\mathcal{Q}_2$. Suppose that S contains a smooth irreducible $\mathfrak{A}_5$-invariant curve C of degree 18 and genus 4. Then S does not contain any $\mathfrak{A}_5$-irreducible curve of degree at most 15 with the only possible exception the curve $\mathcal{G}_6$.

Proof. Suppose that S contains an $\mathfrak{A}_5$-irreducible curve Z of degree at most 15. By Proposition 10.2.5 and Lemma 10.2.6 we know that Z is one of the curves listed in Table 10.1. Going through all possible cases listed in Table 10.1, applying Lemma 8.3.2 and keeping in mind that $C^2 = 6$, we obtain the only a priori possible case $Z = \mathcal{G}_6$. $\square$

Now we are ready to prove the uniqueness of the $\mathfrak{A}_5$-invariant curve of degree 18 and genus 4 in V_5.

Lemma 13.5.2. Suppose that $C \subset V_5$ is a smooth irreducible $\mathfrak{A}_5$-invariant curve of degree 18 and genus 4. Then $C = \mathcal{B}_{18}$.

Proof. By Lemma 5.1.5 the curve C contains two $\mathfrak{A}_5$-orbits of length 12, which means that both $\mathfrak{A}_5$-orbits Σ_{12} and Σ'_{12} are contained in C by Theorem 7.3.5.

Suppose that C is contained in the surface S_{Cl}. Then

$$S_{Cl}|_S = C + \Omega,$$

where Ω is a one-cycle supported on a curve of degree at most

$$H \cdot S_{Cl} \cdot S - H \cdot C = 30 - 18 = 12.$$

By Lemma 13.5.1 one has $\Omega = \mathrm{Supp}(\Omega) = \mathcal{G}_6$, and by Lemma 12.3.6 one has $C = \mathcal{B}_{18}$.

Therefore, we may assume that $C \not\subset S_{Cl}$. We are going to obtain a contradiction in this case. Recall that the surface S_{Cl} contains the $\mathfrak{A}_5$-orbits Σ_{12} and Σ'_{12} by Lemma 12.2.5(i). Moreover, it is singular at the points of Σ_{12} by Lemma 12.2.14. Thus, one has

$$S_{Cl}|_C = m'\Sigma'_{12} + m\Sigma_{12} + \Xi,$$

where $m' \geqslant 1$, $m \geqslant 2$ and Ξ is an $\mathfrak{A}_5$-invariant zero-cycle of degree

$$\deg(\Xi) = S_{Cl} \cdot C - m'|\Sigma'_{12}| - m|\Sigma_{12}| = 54 - 12m' - 12m \in \{6, 18\}.$$

A contradiction with Theorem 7.3.5 completes the proof. $\square$

Remark 13.5.3. By Lemma 5.1.5 the curve $\mathcal{B}_{18}$ contains two $\mathfrak{A}_5$-orbits of length 12 and no other $\mathfrak{A}_5$-orbits of length less than 30. In particular, this implies that both $\mathfrak{A}_5$-orbits Σ_{12} and Σ'_{12} are contained in $\mathcal{B}_{18}$ by Theorem 7.3.5 (cf. the proof of Lemma 13.5.2). Since $\mathcal{B}_{18} \not\subset \mathscr{S}$ by Lemma 7.2.13(iii), the set $\mathcal{B}_{18} \cap \mathscr{S}$ consists of

$$\mathcal{B}_{18} \cdot \mathscr{S} = 36$$

points (counted with multiplicities). By Lemma 7.2.10 this means that

$$\mathcal{B}_{18} \cap \mathscr{S} = \Sigma_{12} \cup \Sigma'_{12}.$$

In particular, we see that the curve $\mathcal{B}_{18}$ is disjoint from the $\mathfrak{A}_5$-orbits Σ_{20} and Σ'_{20}, and

$$\mathcal{B}_{18} \cap C_{20} = \Sigma_{12} \cup \Sigma'_{12}$$

by Theorem 8.1.8(iii). Also, we have $\mathcal{B}_{18} \cap C_{16} = \Sigma_{12}$ by Lemma 8.1.16(iii), and $\mathcal{B}_{18} \cap \mathcal{L}_{20} = \varnothing$ by Lemma 7.2.12(ii).

Remark 13.5.4. There is another way to construct the curve $\mathcal{B}_{18}$. By Lemma 12.5.5(ii),(iii), the surface $\mathscr{F}$ contains an $\mathfrak{A}_5$-invariant rational curve C_{22} of degree 22 such that C_{22} has ordinary cusps at Σ'_{12} and is smooth elsewhere. In particular, one has $p_a(C_{22}) = 12$ by Lemma 4.4.6. By Corollary 8.1.6, this curve C_{22} is contained in some surface $S \in \mathcal{Q}_2$. One has $S \neq \mathscr{F}$ by Lemma 7.2.15, so that S is smooth by Lemma 8.3.3. We compute $C_{22}^2 = 22$ on S by the adjunction formula. Write

$$\mathscr{F}|_S = C_{22} + Z,$$

where Z is an effective $\mathfrak{A}_5$-invariant divisor on S, so that $H \cdot Z = 18$. We have

$$88 = C_{22} \cdot \mathscr{F}|_S = C_{22}^2 + Z \cdot C_{22} = 22 + Z \cdot C_{22},$$

which implies that $Z \cdot C_{22} = 66$. Since

$$72 = Z \cdot \mathscr{F}|_S = Z \cdot C_{22} + Z^2 = 66 + Z^2,$$

we see that $Z^2 = 6$.

We claim that Z is $\mathfrak{A}_5$-irreducible and reduced. Indeed, suppose that it is not. By Lemma 10.2.6, we know that S does not contain the curves $\mathscr{C}$ and $\mathcal{L}_{12}$. Moreover, by Lemma 10.2.7, we know that S contains at most one $\mathfrak{A}_5$-irreducible curve of degree 6. Thus, it follows from Proposition 10.2.5 and Lemma 11.3.2 that either $Z = 3\mathcal{L}_6$ or $Z = 3\mathscr{C}'$. In both cases, we have $Z^2 < 0$, which is a contradiction.

Thus, we see that Z is $\mathfrak{A}_5$-irreducible and reduced. This implies that Z is irreducible, because otherwise we must have $Z = \mathcal{T}_6$ by Lemma 13.2.6, so that $Z^2 = -12$ by Lemma 13.2.4. Since $Z^2 = 6$, we see that $p_a(Z) = 4$ by the adjunction formula. Hence, the curve Z is smooth by Corollary 7.3.6. Therefore, Z is a smooth irreducible $\mathfrak{A}_5$-invariant curve of genus 4. By Lemma 13.5.2, we have $Z = \mathcal{B}_{18}$.

13.6 Classification

We summarize our knowledge on the low-degree $\mathfrak{A}_5$-invariant curves in V_5 as follows.

Theorem 13.6.1 (cf. Proposition 10.2.5). *Let Z be an $\mathfrak{A}_5$-irreducible curve on V_5. Let r be the number of its irreducible components, and let d be the degree of any of these components. Suppose that $\deg(Z) \leqslant 19$. Then one of the following possibilities occurs:*

(i) $\deg(Z) = 6$, $r = 1$, and $d = 6$, so that Z is one of the rational sextic normal curves $\mathscr{C}$ or $\mathscr{C}'$;

(ii) $\deg(Z) = 6$, $r = 6$, and $d = 1$, so that Z is a union $\mathcal{L}_6$ of 6 lines;

(iii) $\deg(Z) = 10$, $r = 10$, and $d = 1$, so that Z is a union $\mathcal{L}_{10}$ of 10 lines;

(iv) $\deg(Z) = 10$, $r = 5$, and $d = 2$, so that Z is a union $\mathcal{G}_5$ or $\mathcal{G}_5'$ of 5 conics;

(v) $\deg(Z) = 12$, $r = 12$, and $d = 1$, so that Z is a union $\mathcal{L}_{12}$ of 12 lines;

(vi) $\deg(Z) = 12$, $r = 6$, and $d = 2$, so that Z is a union $\mathcal{G}_6$ of 6 conics;

(vii) $\deg(Z) = 15$, $r = 15$, and $d = 1$, so that Z is a union $\mathcal{L}_{15}$ of 15 lines;

(viii) $\deg(Z) = 16$, and $Z = C_{16}$ is a smooth rational curve;

(ix) $\deg(Z) = 16$, and $Z = C_{16}'$ is a smooth irreducible curve of genus 5 (cf. Lemma 5.4.3);

(x) $\deg(Z) = 18$, $r = 6$, and $d = 3$, so that Z is a union $\mathcal{T}_6$ of 6 twisted cubics;

(xi) $\deg(Z) = 18$, $Z = C_{18}^0$ is an irreducible rational curve of arithmetic genus 12 such that $\mathrm{Sing}(Z) = \Sigma_{12}'$, and every singular point of the curve Z is an ordinary cusp;

(xii) $\deg(Z) = 18$, and $Z = \mathcal{B}_{18}$ is a smooth irreducible curve of genus 4 (cf. Lemma 5.4.1);

(xiii) $\deg(Z) = 18$, and Z is a smooth rational curve (cf. Remark 13.6.2).

Moreover, the curve $\mathcal{L}_{15}$ is not contained in any surface $S \in \mathcal{Q}_2$. Each of the curves $\mathscr{C}$, $\mathcal{L}_{12}$, and C_{16} is contained in the surface $\mathscr{S}$, and is not contained in any other surface $S \in \mathcal{Q}_2$. Each of the other curves from the above list is contained in a unique surface $S \in \mathcal{Q}_2$, and the surface S is a smooth $K3$ surface.

Proof. If $\deg(Z) \leqslant 15$, then the classification given in the theorem follows from Proposition 10.2.5 and Lemma 11.3.2. If $16 \leqslant \deg(Z) \leqslant 19$ and Z is reducible, then $Z = \mathcal{T}_6$ by Lemma 13.2.6. Thus, to obtain the classification given in the theorem we may assume that $16 \leqslant \deg(Z) \leqslant 19$ and Z is irreducible. Then one has $\deg(Z) \in \{16, 18\}$ by Lemma 8.1.14. If $\deg(Z) = 16$, the result follows by Lemma 13.1.5. Finally, if $\deg(Z) = 18$, the result follows by Lemmas 13.3.4, 13.4.4, and 13.5.2.

We know from Lemma 10.2.6 that the curve $\mathcal{L}_{15}$ is not contained in any surface $S \in \mathcal{Q}_2$, and each of the curves $\mathscr{C}$ and $\mathcal{L}_{12}$ is contained in $\mathscr{S}$ but not in any other surface of $\mathcal{Q}_2$. The curve C_{16} is contained in $\mathscr{S}$ by construction (see Lemma 8.1.16) and is not contained in any other surface of the pencil $\mathcal{Q}_2$ by Corollary 8.1.9.

Suppose that Z is one of the curves listed in the assertion of the theorem that is different from $\mathcal{L}_{15}$, $\mathscr{C}$, $\mathcal{L}_{12}$, and C_{16}. Corollary 8.1.7. implies that there exists a surface S in $\mathcal{Q}_2$ that contains Z. The uniqueness of the surface S follows from Corollary 8.1.9. Moreover, one has $S \ne \mathscr{S}$ by Lemmas 7.2.13 and 8.1.16. If Z is different from $\mathcal{L}_{10}$, $\mathcal{G}_5$, and $\mathcal{G}_5'$, then Lemma 8.3.3 implies that S is non-singular.

Suppose that Z is one of the curves $\mathcal{L}_{10}$, $\mathcal{G}_5$, or $\mathcal{G}_5'$, and suppose that S is singular. Then S is a $K3$ surface with at most ordinary double points by Theorem 8.2.1(v), and one has $\operatorname{rk}\operatorname{Pic}(S)^{\mathfrak{A}_5} = 1$ by Lemma 6.7.3(i),(ii). Put $H_S = H|_S$. Then $H_S^2 = 10$, and $H_S \cdot Z = 10$. Recall that Z is a disjoint union of smooth rational curves by Lemmas 9.1.2 and 9.2.3. Thus $Z^2 < 0$, so that Z is not numerically proportional to H_S, which is a contradiction. □

Remark 13.6.2. We do not know if case (xiii) of Theorem 13.6.1 is actually possible. We also do not know if a curve satisfying the conditions of case (xiii) of Theorem 13.6.1 is unique (assuming that it exists) or not. Still we know that there is at most a finite number of curves of this type (see Theorem 8.3.1). We can also say something about $\mathfrak{A}_5$-orbits on a curve C described by case (xiii) of Theorem 13.6.1. Namely, intersecting C with the surface $\mathscr{S}$ and keeping in mind Lemma 5.5.9(v) it is easy to see that C contains one (and only one) of the $\mathfrak{A}_5$-orbits Σ_{12} and Σ_{12}', does not contain any other $\mathfrak{A}_5$-orbit of length less than 20, and is disjoint from the $\mathfrak{A}_5$-orbits Σ_{20}, Σ_{20}', Σ_{30}, and Σ_{30}' contained in $\mathscr{S}$. We will see later in Lemma 13.6.4 that C contains the $\mathfrak{A}_5$-orbit Σ_{12}, and thus C does not contain Σ_{12}'.

Now we are going to study $\mathfrak{A}_5$-invariant surfaces in the anticanonical linear system $|-K_{V_5}| = |2H|$ passing through $\mathfrak{A}_5$-invariant curves of low degree (cf. Lemma 10.2.7).

Lemma 13.6.3. *Let $Z \subset V_5$ be an $\mathfrak{A}_5$-irreducible curve of degree at most 19, and suppose that Z is different from $\mathscr{C}$, $\mathcal{L}_{12}$, $\mathcal{L}_{15}$, and C_{16}. Then there exists a unique surface $S \in \mathcal{Q}_2$ that contains Z. Moreover, the surface S is smooth. Furthermore, the surface S does not contain any other $\mathfrak{A}_5$-*

irreducible curve of degree at most 19, with the only exception the curves $\mathcal{G}_6$ and $\mathcal{B}_{18}$ that are contained in one surface of the pencil $\mathcal{Q}_2$.

Proof. By Theorem 13.6.1 there exists a unique surface $S \in \mathcal{Q}_2$ that contains the curve Z, and S is a smooth $K3$ surface. By Lemma 13.2.4, the curve Z is a union of 6 smooth rational curves provided that Z is as in case (x) of Theorem 13.6.1. Recall that

$$Z^2 = 2p_a(Z) - 2$$

by the adjunction formula. Using this together with the classification from Theorem 13.6.1 and the data listed in Table 10.1, we obtain the list of possible values of degrees and self-intersection numbers for the curve $Z \subset S$ given in Table 13.1.

Table 13.1: Intersection numbers for low degree curves

	$\mathscr{C}'$	$\mathcal{L}_6$	$\mathcal{L}_{10}$	$\mathcal{G}_5/\mathcal{G}_5'$	$\mathcal{G}_6$
$\deg(Z)$	6	6	10	10	12
Z^2	-2	-12	-20	-10	-12

	C'_{16}	$\mathcal{T}_6$	C^0_{18}	$\mathcal{B}_{18}$	$C_{18}(0)$
$\deg(Z)$	16	18	18	18	18
Z^2	8	-12	22	6	-2

The column of Table 13.1 labelled by $C_{18}(0)$ refers to the family of smooth rational $\mathfrak{A}_5$-invariant curves of degree 18 (cf. Remark 13.6.2).

Suppose that S contains another $\mathfrak{A}_5$-irreducible curve Z' of degree at most 19. Note that Z' is different from $\mathscr{C}$, $\mathcal{L}_{12}$, $\mathcal{L}_{15}$, and C_{16} by Theorem 13.6.1. Thus Z' is also one of the curves listed in Table 13.1, and the degree of Z' and the self-intersection of Z' are as in Table 13.1. Note also that it may happen that Z and Z' are both described by the last column of Table 13.1, so that they are both smooth rational curves of degree 18 (cf. Remark 13.6.2).

By Lemma 10.2.7 we may assume that the degree of at least one of the curves Z and Z' is greater than 15. Going through all remaining cases and applying Lemma 8.3.2, we conclude (like in the proofs of Lemmas 10.2.7 and 13.5.1) that the only possible case up to permutation is $Z = \mathcal{B}_{18}$ and $Z' = \mathcal{G}_6$ (so that $Z \cdot Z' = 48$). This case actually takes place by Lemma 12.3.6. □

As an easy application of Lemma 13.6.3 we can obtain some additional information about $\mathfrak{A}_5$-orbits on $\mathfrak{A}_5$-invariant curves of low degree.

Lemma 13.6.4. Let Z be an $\mathfrak{A}_5$-invariant smooth rational curve of degree 18 in V_5. Then Z is disjoint from the $\mathfrak{A}_5$-orbit Σ'_{12} and contains the $\mathfrak{A}_5$-orbit Σ_{12}.

Proof. The action of $\mathfrak{A}_5$ on Z is faithful by Lemma 7.3.8. By Lemma 5.5.9(v), the curve Z contains a unique $\mathfrak{A}_5$-orbit of length 12. Thus, it follows from Theorem 7.3.5 that it is enough to check that Σ'_{12} is disjoint from Z. Suppose that this is not the case. Let us show that this leads to a contradiction.

By Theorem 13.6.1, the curve Z is contained in a smooth surface $S \in \mathcal{Q}_2$. By Lemma 12.5.4, there exists an effective $\mathfrak{A}_5$-invariant divisor

$$D \in |4H|_S|$$

whose support contains the curve Z. Thus, we have $D = Z + R$ for some effective divisor R. Note that

$$H \cdot R = H \cdot D - H \cdot Z = 22.$$

This implies that R is an $\mathfrak{A}_5$-irreducible curve, because S does not contain any $\mathfrak{A}_5$-irreducible curve of degree less than 20 except Z by Lemma 13.6.3. Hence, R is irreducible by Corollary 5.1.2. In particular, the group $\mathfrak{A}_5$ acts faithfully on R by Lemma 7.3.8.

Since Z is a smooth rational curve, we have $Z^2 = -2$. Thus,

$$72 = 2C_{20} \cdot Z = (Z + R) \cdot Z = Z^2 + R \cdot Z = -2 + R \cdot Z,$$

which gives $Z \cdot R = 74$. We compute

$$88 = 2C_{20} \cdot R = (Z + R) \cdot R = Z \cdot R + R^2 = 74 + R^2,$$

which implies that $R^2 = 14$. By the adjunction formula, one has $p_a(R) = 8$. Hence, R is smooth by Lemma 4.4.6, because S does not contain $\mathfrak{A}_5$-orbits of length less than 12 by Lemma 6.7.1(ii). Therefore, R is an irreducible smooth curve of genus 8 with a faithful action of $\mathfrak{A}_5$. The latter is impossible by Lemma 5.1.5. □

Corollary 13.6.5. Suppose that C is an $\mathfrak{A}_5$-invariant smooth rational curve of degree 18 in V_5. Then $C \cap \mathscr{S} = \Sigma_{12}$.

Proof. The curve C is not contained in $\mathscr{S}$ by Theorem 13.6.1, so that the intersection of C and $\mathscr{S}$ consists of $C \cdot \mathscr{S} = 36$ points (counted with multiplicities). Now the assertion follows from Lemma 13.6.4. $\square$

Another application of Lemma 13.6.3 is the following information about intersections of $\mathfrak{A}_5$-invariant low degree curves (note that much more detailed information about similar intersections will be presented in §16.4).

Corollary 13.6.6. One has
$$C_{18}^0 \cap \mathscr{C}' = C_{18}^0 \cap \mathcal{L}_6 = \mathcal{B}_{18} \cap \mathcal{L}_6 = \Sigma'_{12}.$$

Proof. We know from Lemma 13.6.3 that there exists a surface in $\mathcal{Q}_2$ that contains the curve $\mathscr{C}'$ (cf. §11.3), and there exists a surface in $\mathcal{Q}_2$ that contains the curve C_{18}^0 but not the curve $\mathscr{C}'$. By Lemma 11.3.4 the curve $\mathscr{C}'$ contains the $\mathfrak{A}_5$-orbit Σ'_{12} and is disjoint from the $\mathfrak{A}_5$-orbits Σ_{12} and Σ'_{20}. By Lemma 13.4.2 the curve C_{18}^0 contains the $\mathfrak{A}_5$-orbit Σ'_{12}. Now the description of the intersection $C_{18}^0 \cap \mathscr{C}'$ follows from Corollary 8.1.12.

We know from Lemma 13.6.3 that there exists a surface in $\mathcal{Q}_2$ that contains the curve $\mathcal{L}_6$ (cf. §11.1), and there exists a surface in $\mathcal{Q}_2$ that contains the curve C_{18}^0 but not the curve $\mathcal{L}_6$. By Lemma 7.4.3 the curve $\mathcal{L}_6$ contains the $\mathfrak{A}_5$-orbit Σ'_{12} and is disjoint from the $\mathfrak{A}_5$-orbits Σ_{12} and Σ'_{20}. Now the description of the intersection $C_{18}^0 \cap \mathcal{L}_6$ follows from Corollary 8.1.12.

Similarly, we know from Lemma 13.6.3 that there exists a surface in $\mathcal{Q}_2$ that contains the curve $\mathcal{B}_{18}$ but not the curve $\mathcal{L}_6$ (cf. §12.3). By Remark 13.5.3 the curve $\mathcal{B}_{18}$ contains the $\mathfrak{A}_5$-orbit Σ'_{12}. Therefore, the description of the intersection $\mathcal{B}_{18} \cap \mathcal{L}_6$ follows from Corollary 8.1.12. $\square$

Corollary 13.6.7. One has
$$\mathcal{L}_{15} \cap C_{18}^0 = \varnothing.$$

Proof. Suppose that the intersection $C_{18}^0 \cap \mathcal{L}_{15}$ contains some point P. Note that P is not contained in $\mathscr{C}' \cup \mathcal{L}_6$, since
$$C_{18}^0 \cap \mathscr{C}' = C_{18}^0 \cap \mathcal{L}_6 = \Sigma'_{12}$$
by Corollary 13.6.6, while $\Sigma'_{12} \not\subset \mathcal{L}_{15}$ by Lemma 9.1.10(iv). Therefore, Lemma 11.5.6(iv) implies that the birational map $\iota \circ \tau$ is biregular in a neighborhood of P. Since one has
$$\iota \circ \tau(\mathcal{L}_{15}) = \mathcal{L}_{15}$$

by Lemma 11.5.9 (note that $\iota \circ \tau$ is well defined in a general point of $\mathcal{L}_{15}$, so that $\iota \circ \tau(\mathcal{L}_{15})$ does make sense), we conclude that $\mathscr{C} \cap \mathcal{L}_{15} \neq \varnothing$. This gives a contradiction with Corollary 9.1.11. □

Corollary 13.6.8. One has
$$\mathcal{L}_{15} \cap \mathcal{G}_6 = \mathcal{L}_{15} \cap \mathcal{B}_{18} = \varnothing.$$

Proof. Recall from Lemma 12.3.6 that the curves $\mathcal{G}_6$ and $\mathcal{B}_{18}$ are contained in the surface S_{Cl}. By Lemma 12.2.12 we have
$$\mathcal{L}_{15} \cap S_{Cl} = \Sigma_{15} \cup \Sigma,$$
where Σ is an $\mathfrak{A}_5$-orbit of length 30 contained in the curve $\mathcal{L}_6$. We know that the $\mathfrak{A}_5$-orbit Σ_{15} is disjoint from the curve $\mathcal{G}_6$ by Corollary 8.3.5 and is disjoint from the curve $\mathcal{B}_{18}$ by Lemma 5.1.4. Similarly, we know that the $\mathfrak{A}_5$-orbit Σ is disjoint from the curve $\mathcal{G}_6$ by Lemma 9.2.10 and is disjoint from the curve $\mathcal{B}_{18}$ by Corollary 13.6.6. □

Chapter 14

Orbits of small length

In this chapter we describe $\mathfrak{A}_5$-orbits of lengths 20 and 30 on V_5. Together with Theorem 7.3.5, this gives a complete classification of $\mathfrak{A}_5$-orbits on V_5. We show that all $\mathfrak{A}_5$-orbits of length 20 except for Σ'_{20} are contained in a certain $\mathfrak{A}_5$-irreducible curve $\mathcal{G}_{10}$ that is a union of 10 conics, and all $\mathfrak{A}_5$-orbits of length 30 are contained either in the curve $\mathcal{L}_{15}$ or in a certain $\mathfrak{A}_5$-irreducible curve $\mathcal{T}_{15}$ that is a union of 15 twisted cubics. We also study basic properties of the curves $\mathcal{G}_{10}$ and $\mathcal{T}_{15}$. It should be pointed out that many of these results can be proved by a direct computation (see Lemmas 5.5.13 and 5.5.14). Nevertheless, we prefer to use a slightly longer and more straightforward approach, due to our general policy to avoid using explicit equations of V_5 and related computations.

14.1 Orbits of length 20

Now we will apply the results on the structure of $\mathfrak{A}_5$-invariant anticanonical surfaces obtained in §8.2 to classify $\mathfrak{A}_5$-orbits of length 20 on V_5 (cf. Theorem 7.3.5).

Lemma 14.1.1. *Let $C \subset V_5$ be an $\mathfrak{A}_5$-invariant curve such that a stabilizer in $\mathfrak{A}_5$ of a general point in C is non-trivial. Suppose that C intersects the curve $C_{20} = \mathrm{Bs}(\mathcal{Q}_2)$ in a point P, and that both C_{20} and C are smooth at P. Then the curves C and C_{20} are transversal at P.*

Proof. Let $G \subset \mathfrak{A}_5$ be a stabilizer of a general point of C. Then P is fixed by G. Note that the action of G on the curve C_{20} is faithful, since otherwise $\mathscr{S}$ would contain an infinite number of $\mathfrak{A}_5$-orbits of length less than 60, which is impossible by Lemma 7.2.10. Now the assertion is implied by Lemma 4.4.2. □

Denote by $\Xi(20)$ the closure of the union of $\mathfrak{A}_5$-orbits of length 20 in V_5. Then a stabilizer of a general point (i.e., a point in a dense open subset) of $\Xi(20)$ is isomorphic to $\boldsymbol{\mu}_3$.

Lemma 14.1.2. *The following assertions hold:*

(i) *the length of the $\mathfrak{A}_5$-orbit of any point $P \in \Xi(20)$ divides 20;*

(ii) *the length of the $\mathfrak{A}_5$-orbit of any point of $\Xi(20)$ equals 20, with a possible exception of a finite number of points.*

Proof. Let $R \subset \Xi(20)$ be an irreducible component of $\Xi(20)$ such that $P \in R$, and P' be a general point of R. Then the length of the $\mathfrak{A}_5$-orbit of P' equals 20 by construction. Let $G_P \subset \mathfrak{A}_5$ and $G_{P'} \subset \mathfrak{A}_5$ be the stabilizers of the points P and P', respectively. Then $G_{P'} \subset G_P$, so that the length $|\mathfrak{A}_5|/|G_P|$ of the $\mathfrak{A}_5$-orbit of P divides the length $|\mathfrak{A}_5|/|G_{P'}|$ of the $\mathfrak{A}_5$-orbit of P'. This proves assertion (i).

Assertion (ii) follows from the fact that there is only a finite number of points in V_5 with $\mathfrak{A}_5$-orbits of length less than 20 by Theorem 7.3.5. □

Recall that the curve C_{20} contains a unique $\mathfrak{A}_5$-orbit Σ'_{20} of length 20 (see Theorem 8.1.8(vii)). The $\mathfrak{A}_5$-orbit Σ'_{20} is contained in any surface $S \in \mathcal{Q}_2$.

Lemma 14.1.3. *Let S be a surface in the pencil $\mathcal{Q}_2$. Then*

(o) *the surface S contains two $\mathfrak{A}_5$-orbits of length 20, if $S = \mathscr{S}$;*

(i) *the surface S contains one $\mathfrak{A}_5$-orbit of length 20, if $S = S_5$;*

(ii) *the surface S contains two $\mathfrak{A}_5$-orbits of length 20, if $S = S_{10}$ or $S = S'_{10}$;*

(iii) *the surface S contains three $\mathfrak{A}_5$-orbits of length 20, if S is different from $\mathscr{S}$, S_5, S_{10}, and S'_{10}.*

Proof. Assertion (o) follows from Lemma 7.2.10.

Let $S \in \mathcal{Q}_2$ be a singular surface such that $S \neq \mathscr{S}$. Then the singularities of S are ordinary double points by Theorem 8.2.1(v). Let $\theta \colon \tilde{S} \to S$ be the minimal resolution of singularities. Then $\tilde{S}$ contains exactly three $\mathfrak{A}_5$-orbits of length 20 by Lemma 6.7.1(iv).

Let P be a singular point of S, and let $E \subset \tilde{S}$ be the exceptional divisor contracted by θ to the point P. Then $E \cong \mathbb{P}^1$ since P is an ordinary double point of the surface S. Let $G_P \subset \mathfrak{A}_5$ be the stabilizer of the point P. Note that the action of G_P on E is faithful, since otherwise $\tilde{S}$ would contain infinitely many $\mathfrak{A}_5$-orbits of length less than 60, which is not the case by Lemma 6.7.1.

Suppose that $S = S_5$. Then $\mathrm{Sing}(S) = \Sigma_5$ by Theorem 8.1.8(iii),(iv), so that $G_P \cong \mathfrak{A}_4$ by Lemma 5.1.1. Thus, by Lemma 5.5.9(iii) there are two G_P-orbits of length 4 contained in E, that give rise to two $\mathfrak{A}_5$-orbits of length $4|\Sigma_5| = 20$ contained in $\tilde{S}$. Both of these $\mathfrak{A}_5$-orbits are mapped to Σ_5 by the morphism θ. Therefore, the surface S_5 contains a unique $\mathfrak{A}_5$-orbit of length 20, which gives assertion (i).

To prove assertion (ii), we may suppose that $S = S_{10}$, since the case $S = S'_{10}$ is treated similarly. Then $\mathrm{Sing}(S) = \Sigma_{10}$ by Theorem 8.1.8(iii),(iv), so that $G_P \cong \mathfrak{S}_3$ by Lemma 5.1.1. Thus, by Lemma 5.5.9(ii) there is a unique G_P-orbit of length 2 contained in E, that gives rise to an $\mathfrak{A}_5$-orbit of length $2|\Sigma_{10}| = 20$ contained in $\tilde{S}$. This $\mathfrak{A}_5$-orbit is mapped to Σ_{10} by the morphism θ. Therefore, the surface S_{10} contains two $\mathfrak{A}_5$-orbits of length 20, which gives assertion (ii).

If $S = S_{15}$, then none of the $\mathfrak{A}_5$-orbits of length 20 contained in $\tilde{S}$ is mapped to $\mathrm{Sing}(S) = \Sigma_{15}$. Thus in this case S and $\tilde{S}$ contain the same number of $\mathfrak{A}_5$-orbits of length 20, and this number equals three by Lemma 6.7.1(iv). Therefore, to prove assertion (iii) one may assume that $S \neq S_{15}$. By Theorem 8.2.1(ii) this means that S is smooth. Now assertion (iii) is again implied by Lemma 6.7.1(iv). □

Corollary 14.1.4. Let $P \in \Xi(20)$ be a point contained in the surface S_5. Then either $P \in \Sigma_5$, or $P \in \Sigma'_{20}$, where Σ'_{20} is the unique $\mathfrak{A}_5$-orbit of length 20 contained in the curve C_{20}.

Proof. The length l of the $\mathfrak{A}_5$-orbit of P is either 20, or 10, or 5, since l divides 20 by Lemma 14.1.2(i), and $l \notin \{1, 2, 4\}$ by Theorem 7.3.5. Neither of the $\mathfrak{A}_5$-orbits of length 10 in V_5 is contained in S_5 by Theorems 7.3.5 and 8.2.1(i). The unique $\mathfrak{A}_5$-orbit of length 5 in V_5 is Σ_5 by Theorem 7.3.5. Thus we may assume that the length of the $\mathfrak{A}_5$-orbit of P is 20. Now the assertion is implied by Lemma 14.1.3(i), since the $\mathfrak{A}_5$-orbit $\Sigma'_{20} \subset S_5$ is contained in the curve C_{20} (see Theorem 8.1.8(vi)). □

Remark 14.1.5. One has $\Xi(20) \cap C_{20} = \Sigma'_{20}$. Indeed, the length of the $\mathfrak{A}_5$-orbit of any point $P \in \Xi(20)$ divides 20 by Lemma 14.1.2(i), and the unique $\mathfrak{A}_5$-orbit of length dividing 20 contained in C_{20} is Σ'_{20} by Theorem 8.1.8(vii).

Lemma 14.1.6. Let $C \subset \Xi(20)$ be an irreducible curve. Let $G_C \subset V_5$ be the stabilizer of C. Then either $G_C \cong \boldsymbol{\mu}_3$, or $G_C \cong \mathfrak{S}_3$.

Proof. Let P be a general point of C, and let $G_P \subset \mathfrak{A}_5$ be the stabilizer of P. Then G_P is a normal subgroup of G_C. By Lemma 14.1.2(ii) the length of the $\mathfrak{A}_5$-orbit of P is 20, so that $G_P \cong \boldsymbol{\mu}_3$. Thus by Lemma 5.1.1 the stabilizer G_C is one of the groups $\boldsymbol{\mu}_3$ or $\mathfrak{S}_3$. □

Lemma 14.1.7. *Let $C_1, \ldots, C_r \subset \Xi(20)$, $r \geqslant 1$, be irreducible curves. Suppose that for each $1 \leqslant i \leqslant r-1$ the intersection $C_i \cap C_{i+1}$ contains a point Q_i such that $Q_i \notin \Sigma_5$. Then the linear span $\Pi \subset \mathbb{P}^6$ of $C_1 \cup \ldots \cup C_r$ is at most two-dimensional.*

Proof. Recall that we denote by $\mathscr{W}_7$ the seven-dimensional vector space such that $\mathbb{P}^6$ is identified with its projectivization, see §7.1. Let P_i be a general point of the curve C_i, and let $G_{P_i} \subset \mathfrak{A}_5$ be the stabilizer of P_i. Then $G_{P_i} \cong \boldsymbol{\mu}_3$ by Lemma 14.1.2(ii). Let $\Pi_i \subset \mathbb{P}^6$ be the linear span of the curve C_i. The group G_{P_i} acts trivially on the curve C_i and thus also on its span Π_i. This means that the corresponding vector subspace U_i of $\mathscr{W}_7$ is a sum of several isomorphic (one-dimensional) irreducible G_{P_i}-representations.

Denote by l_{Q_i} the length of the $\mathfrak{A}_5$-orbit of the point Q_i. Since l_{Q_i} divides 20 by Lemma 14.1.2(i), we see that either $l_{Q_i} = 10$ or $l_{Q_i} = 20$ by Theorem 7.3.5.

Let $G_{Q_i} \subset \mathfrak{A}_5$ be the stabilizer of the point Q_i. Note that $G_{P_i} \subset G_{Q_i}$, and $G_{P_{i+1}} \subset G_{Q_i}$. By Lemma 5.1.1 one has $G_{Q_i} \cong \mathfrak{S}_3$ if $l_{Q_i} = 10$, and $G_{Q_i} \cong \boldsymbol{\mu}_3$ if $l_{Q_i} = 20$. In both cases we see that the subgroups $G_{P_i} \subset G_{Q_i}$ and $G_{P_{i+1}} \subset G_{Q_i}$ must coincide. In particular, the vector subspaces $U_i, U_{i+1} \subset \mathscr{W}_7$ both split as sums of isomorphic irreducible representations of the group G_{P_i}. Since the intersection $U_i \cap U_{i+1}$ is non-trivial, we conclude that the irreducible representations of G_{P_i} that are contained in U_i and U_{i+1} are isomorphic, so that U_i and U_{i+1} split into sums of several copies of one and the same irreducible representations of G_{P_i}.

Applying the above argument for all $1 \leqslant i \leqslant r-1$, we see that the vector subspace $U \subset \mathscr{W}_7$ that corresponds to the linear span $\Pi \subset \mathbb{P}^6$ of $C_1 \cup \ldots \cup C_r$ is a sum of several isomorphic (one-dimensional) irreducible representations of the group $G_{P_1} \cong \boldsymbol{\mu}_3$. On the other hand, Lemma 7.1.6 and Corollaries 5.2.2(i) and 5.2.3(i) imply that $\dim(U) \leqslant 3$, which gives $\dim(\Pi) \leqslant 2$. □

Corollary 14.1.8. *Let C and C' be two curves contained in $\Xi(20)$. Suppose that at least one of these curves is not a line. Then any point of the intersection $C \cap C'$ is contained in the $\mathfrak{A}_5$-orbit Σ_5.*

Proof. Suppose that there is a point $P \in C \cap C'$ such that $P \notin \Sigma_5$. Then it follows from Lemma 14.1.7 that the linear span of $C \cup C'$ in $\mathbb{P}^6$ is at most two-dimensional. Since V_5 is an intersection of quadrics, this means that

$$\deg(C) + \deg(C') \leqslant 2,$$

which is not the case by assumption. □

Lemma 14.1.9. *There are no lines contained in* $\Xi(20)$.

Proof. Suppose that $L \subset \Xi(20)$ is a line. Then L is not a line of $\mathcal{L}_{15}$, since the stabilizer of a general point on a line of $\mathcal{L}_{15}$ does not contain a subgroup isomorphic to $\boldsymbol{\mu}_3$ by Lemma 7.4.6. In particular, one has $L \cap \Sigma_5 = \varnothing$ by Theorem 7.1.9(i) and Corollary 7.4.4.

Let $G_L \subset \mathfrak{A}_5$ be the stabilizer of the line L. Then G_L is not isomorphic to $\mathfrak{S}_3$, since otherwise L is a line of $\mathcal{L}_{10}$ by Proposition 7.4.1, which is impossible by Lemma 7.4.6. Thus $G_L \cong \boldsymbol{\mu}_3$ by Lemma 14.1.6.

Note that $L \not\subset S_5$ by Corollary 8.2.2. Since $L \cap \Sigma_5 = \varnothing$, one has

$$L \cap S_5 \subset \Sigma'_{20} \subset C_{20}$$

by Corollary 14.1.4. We claim that L has a unique common point with the $\mathfrak{A}_5$-orbit Σ'_{20}. Indeed, suppose that $L \cap \Sigma'_{20}$ consists of at least two points. Since

$$L \cap \Sigma'_{20} = L \cap S_5,$$

this means that $L \cap \Sigma'_{20}$ actually consists of two points. Let $L_1 = L, \ldots, L_{20}$ be the $\mathfrak{A}_5$-orbit of the line L. Then each of the lines L_i, $1 \leqslant i \leqslant 20$, has two common points with Σ'_{20}. Let c be the number of lines L_i, $1 \leqslant i \leqslant 20$, passing through some (and thus any) point of Σ'_{20}. Then one has

$$20c = 40,$$

so that $c = 2$. Let P_1 and P_2 be the points of Σ'_{20} contained in L. Since $c = 2$, there is a line in the $\mathfrak{A}_5$-orbit of L, say, L_2, that is different from L and that passes through the point P_2. Similarly, there is a point of Σ'_{20}, say, P_3, that is different from P_2 and that lies on L_2. Finally, there is a line in the $\mathfrak{A}_5$-orbit of L, say, L_3, that is different from L_2 and that passes through P_3. The three lines L_1, L_2, and L_3 are not coplanar because V_5 is an intersection of quadrics. Hence their linear span in $\mathbb{P}^6$ is at least three-dimensional. The latter contradicts Lemma 14.1.7.

Therefore, the line L has a unique common point with the $\mathfrak{A}_5$-orbit Σ'_{20}, and thus also with the curve C_{20}. Let us denote this point by P. The line L is tangent to the surface S_5 at P, since

$$S_5 \cdot L = 2 > 1 = |S_5 \cap L|.$$

Recall that C_{20} is smooth at the point P by Theorem 8.1.8(iv), and also the surface $\mathscr{S}$ is smooth at P (cf. Lemma 7.2.2(i)). Since L is not tangent to C_{20} at P by Lemma 14.1.1, we conclude that L is not tangent at P to any surface $S \in \mathcal{Q}_2$ different from S_5. In particular, L is not tangent to $\mathscr{S}$ at the point P, and thus the intersection $L \cap \mathscr{S}$ contains a point $P' \neq P$. Note that $P' \notin \Sigma'_{20}$, because we already know that L intersects Σ'_{20} only by the point P. The length of the $\mathfrak{A}_5$-orbit of the point P' divides 20 by Lemma 14.1.2(i). By Lemma 7.2.10 the latter means that $P' \in \mathscr{C}$. Since $\mathscr{S}$ is singular along the curve $\mathscr{C}$ by Lemma 7.2.2(i), we have

$$2 = \mathscr{S} \cdot L \geqslant \mathrm{mult}_P(\mathscr{S}) + \mathrm{mult}_{P'}(\mathscr{S}) = 1 + 2 = 3,$$

which is a contradiction. $\square$

Lemma 14.1.10. Let $C \subset \Xi(20)$ be an irreducible curve. Then

(i) the curve C is a conic;

(ii) the stabilizer of C in $\mathfrak{A}_5$ is isomorphic to $\mathfrak{S}_3$.

Proof. By Lemma 14.1.7 the curve C is contained in a plane in $\mathbb{P}^6$. This means that C is either a line or a conic since V_5 is an intersection of quadrics. Moreover, C cannot be a line by Lemma 14.1.9, so that C is a conic. This proves assertion (i).

Let $G_C \subset \mathfrak{A}_5$ be the stabilizer of the conic C. Then either $G_C \cong \boldsymbol{\mu}_3$ or $G_C \cong \mathfrak{S}_3$ by Lemma 14.1.6.

Assume that $G_C \cong \boldsymbol{\mu}_3$. Let P be some point of C that is contained in an $\mathfrak{A}_5$-orbit $\Sigma(P)$ of length 20. We claim that P is the unique common point of C and $\Sigma(P)$. Indeed, the $\mathfrak{A}_5$-orbit of C consists of 20 conics, because $G_C \cong \boldsymbol{\mu}_3$. The point P is not contained in any of these conics except for C by Corollary 14.1.8, so that any point of the $\mathfrak{A}_5$-orbit $\Sigma(P)$ except P is not contained in the conic C.

Suppose that $C \cap C_{20} = \varnothing$. Then a general surface $S \in \mathcal{Q}_2$ intersects C in four different points, that give rise to four different $\mathfrak{A}_5$-orbits of length 20 contained in S. The latter is impossible by Lemma 14.1.3(iii).

Therefore, the intersection $C \cap C_{20}$ contains some point P'. Since $C \cap C_{20} \subset \Sigma'_{20}$ by Remark 14.1.5, and P' must be a unique common point of C and Σ'_{20}, we conclude that $C \cap C_{20} = P'$. By Lemma 14.1.1 the conic C is not tangent to the curve C_{20} at P'. Thus C is not tangent to a general surface $S \in \mathcal{Q}_2$ at the point P', so that C intersects S in four different points (note that $C \not\subset S$ by Corollary 8.2.2). This is again impossible by Lemma 14.1.3(iii). The obtained contradiction shows that $G_C \cong \mathfrak{S}_3$ and gives assertion (ii). $\square$

CREMONA GROUPS AND THE ICOSAHEDRON 353

Lemma 14.1.11. Let $C \subset \Xi(20)$ be a conic. Then any $\mathfrak{A}_5$-orbit of length 20 that intersects C has exactly two common points with C.

Proof. Let $G_C \subset \mathfrak{A}_5$ be the stabilizer of the conic C. Then $G_C \cong \mathfrak{S}_3$ by Lemma 14.1.10(ii). Let P' be any point of the conic C, and $\Sigma(P')$ be the $\mathfrak{A}_5$-orbit of P'. Assume that $|\Sigma(P')| = 20$, so that the stabilizer $G_{P'} \subset \mathfrak{A}_5$ of the point P' is isomorphic to $\boldsymbol{\mu}_3$.

Let P be a general point of the conic C. Then the stabilizer $G_P \subset \mathfrak{A}_5$ of the point P is isomorphic to $\boldsymbol{\mu}_3$ by Lemma 14.1.2(ii). Since P is a general point of C, one has $G_P \subset G_{P'}$ and $G_P \subset G_C$. Thus $G_P \cong G_{P'}$, so that $G_{P'}$ is a subgroup of G_C. Therefore, the G_C-orbit of P' consists of

$$\frac{|G_C|}{|G_{P'}|} = \frac{|\mathfrak{S}_3|}{|\boldsymbol{\mu}_3|} = 2$$

points, that are contained in $\Sigma(P') \cap C$.

We claim that the intersection $\Sigma(P') \cap C$ does not contain any other points except for the two points that form the G_C-orbit of P'. Indeed, suppose that it does. Let $C_1 = C, C_2, \ldots, C_{10}$ be the $\mathfrak{A}_5$-orbit of the conic C. Then any of the conics C_i, $1 \leqslant i \leqslant 10$, contains more than two points of the $\mathfrak{A}_5$-orbit $\Sigma(P')$. This immediately implies that there is a point of $\Sigma(P')$ that lies on at least two conics among $C_1, \ldots, C_{10}$. Now a contradiction with Corollary 14.1.8 completes the proof. □

Lemma 14.1.12. Let $C \subset \Xi(20)$ be a conic. Then C is disjoint from the curve C_{20}.

Proof. Suppose that $C \cap C_{20} \neq \varnothing$. By Remark 14.1.5 one has $C \cap C_{20} \subset \Sigma'_{20}$, and by Lemma 14.1.11 we conclude that the intersection $C \cap C_{20}$ consists of two points, say, P_1 and P_2. By Lemma 14.1.1 the conic C is not tangent to the curve C_{20} at the points P_1 and P_2. Thus C is not tangent to a general surface $S \in \mathcal{Q}_2$ at the points P_1 and P_2, so that C intersects S in four different points (note that $C \not\subset S$ by Corollary 8.2.2). Hence it follows from Lemma 14.1.11 that there are two $\mathfrak{A}_5$-orbits of length 20 contained in S that intersect the curve C.

By Lemma 14.1.3(iii) we conclude that there exists a unique $\mathfrak{A}_5$-orbit of length 20 in S that is disjoint from C, and thus is also disjoint from C_{20}. This means that $\Xi(20)$ contains a curve C' such that the $\mathfrak{A}_5$-orbit of C' does not contain C. If C' intersects C_{20}, then by Remark 14.1.5 we know that $C' \cap C_{20} \subset \Sigma'_{20}$. Replacing C' by some other conic from its $\mathfrak{A}_5$-orbit if necessary, we may assume that C' contains at least one of the points P_1

or P_2. This means that C' intersects C at a point of the $\mathfrak{A}_5$-orbit Σ'_{20}. The latter is impossible by Corollary 14.1.8. Hence C' is disjoint from C_{20}.

By Lemma 14.1.10(i) the curve C' is a conic. Therefore, C' intersects a general surface $S \in \mathcal{Q}_2$ in four different points, that give rise to two different $\mathfrak{A}_5$-orbits of length 20 contained in S (see Lemma 14.1.11) that are disjoint from the conic C. This means that S contains at least four $\mathfrak{A}_5$-orbits of length 20. Now the contradiction with Lemma 14.1.3(iii) completes the proof. □

Lemma 14.1.13. Let $C \subset \Xi(20)$ be a conic. Then the intersection $C \cap \mathscr{S}$ consists of two points contained in the $\mathfrak{A}_5$-orbit $\Sigma_{20} \subset \mathscr{C}$ of length 20.

Proof. The conic C is disjoint from the curve C_{20} by Lemma 14.1.12. Since the only $\mathfrak{A}_5$-orbit of length 20 contained in $\mathscr{S} \setminus C_{20}$ is the $\mathfrak{A}_5$-orbit Σ_{20} of length 20 that lies on the curve $\mathscr{C} \subset \mathscr{S}$ by Lemma 7.2.10 and Theorem 8.1.8(vii), we see that the points of $C \cap \mathscr{S}$ are contained in $\mathscr{C}$. Since $\mathscr{S}$ is singular at the points of $\mathscr{C}$ by Lemma 7.2.2(i), we see that $|\mathscr{S} \cap C| \leqslant 2$. On the other hand, the $\mathfrak{A}_5$-orbit Σ_{20} has two common points with C by Lemma 14.1.11. Therefore, we see that $|\mathscr{S} \cap C| = 2$. □

Lemma 14.1.14. Let $C \subset \Xi(20)$ be a conic. Then the $\mathfrak{A}_5$-orbit of C consists of 10 conics $C_1 = C, C_2, \ldots, C_{10}$. One has

(i) for any $1 \leqslant i \leqslant 10$, the intersection $C_i \cap S_5$ consists of two points of the $\mathfrak{A}_5$-orbit Σ_5;

(ii) each point of the $\mathfrak{A}_5$-orbit Σ_5 is contained in exactly four conics among $C_1, \ldots, C_{10}$;

(iii) for any point $P \in \Sigma_5$, the tangent vectors to each three of the four conics among $C_1, \ldots, C_{10}$ passing through P are linearly independent.

Proof. Since the stabilizer of the conic C in $\mathfrak{A}_5$ is isomorphic to $\mathfrak{S}_3$ by Lemma 14.1.10(ii), the $\mathfrak{A}_5$-orbit of C indeed consists of 10 conics. By Lemma 14.1.12, for any $1 \leqslant i \leqslant 10$ the conic C_i is disjoint from the curve C_{20}. Thus Corollary 14.1.4 implies that for any $1 \leqslant i \leqslant 10$ one has $C_i \cap S_5 \subset \Sigma_5$. Let c be the number of the conics C_i, $1 \leqslant i \leqslant 10$, passing through some (and thus any) point of the $\mathfrak{A}_5$-orbit Σ_5, and p be the number of points of Σ_5 contained in some (and thus any) conic C_i, $1 \leqslant i \leqslant 10$. Then $p \leqslant 2$, since the surface S_5 does not contain any of the conics C_i by Corollary 8.2.2, and S_5 is singular at the points of Σ_5 by Theorem 8.2.1(iv). Moreover, one has $c \geqslant 3$ by Lemma 7.5.11. Since

$$10p = 5c,$$

CREMONA GROUPS AND THE ICOSAHEDRON   355

we conclude that $p = 2$ and $c = 4$, which proves assertions (i) and (ii). Assertion (iii) is implied by Lemma 7.5.9. □

Lemma 14.1.15. Let $C \subset \Xi(20)$ be a conic. Then

(i) the intersection $C \cap S_{10}$ (resp., the intersection $C \cap S'_{10}$) consists of one point of the $\mathfrak{A}_5$-orbit Σ_{10} (resp., of the $\mathfrak{A}_5$-orbit Σ'_{10}) and two smooth points of the surface S_{10} (resp., the surface S'_{10});

(ii) the conic C intersects a surface $S \in \mathcal{Q}_2$ in four distinct smooth points of S, provided that S is different from S_5, S_{10}, S'_{10}, and $\mathscr{S}$.

Proof. Consider the linear system $\mathcal{Q} = \mathcal{Q}_2|_C$. Since the base locus C_{20} of the pencil $\mathcal{Q}_2$ is disjoint from C by Lemma 14.1.12, the linear system $\mathcal{Q}$ is a pencil of divisors of degree 4 that has no base points, so that $\mathcal{Q}$ defines a four-to-one cover $t \colon C \to \mathbb{P}^1$. We are going to determine the ramification points of the morphism t.

Since the surface $\mathscr{S}$ is singular at the points of $\mathscr{C}$ by Lemma 7.2.2(i), it follows from Lemma 14.1.13 that $\mathscr{S}|_C$ is a divisor that is a sum of two points with multiplicity two. Similarly, since S_5 is singular at the points of Σ_5 by Theorem 8.2.1(iv), it follows from Lemma 14.1.14(i) that the restriction of S_5 to C is a divisor that is a sum of two points with multiplicity two.

Therefore, we already know four ramification points of the four-to-one cover t: two points coming from the intersection of C with S_5, and two points coming from the intersection of C with $\mathscr{S}$. Thus, the Riemann–Hurwitz formula gives

$$-2 = 2g(C) - 2 =$$
$$= 4\big(2g(\mathbb{P}^1) - 2\big) + \sum_{P \in \mathscr{S} \cap C} (e_P - 1) + \sum_{P \in S_5 \cap C} (e_P - 1) + \sum_{P \in C \setminus (\mathscr{S} \cup S_5)} (e_P - 1) =$$
$$= -4 + \sum_{P \in C \setminus (\mathscr{S} \cup S_5)} (e_P - 1),$$

where e_P denotes the ramification index of the morphism t at the point P. This implies that there are at most two ramification points of the morphism t not contained in $\mathscr{S} \cup S_5$, and if there are exactly two of them, then the ramification index at these points equals 2.

Suppose that C is disjoint from the $\mathfrak{A}_5$-orbit Σ_{10}. Then Lemmas 14.1.3(ii) and 14.1.11 imply that either C intersects the surface S_{10} at one point P so that the local intersection number $(C \cdot S_{10})_P = 4$, or C intersects the surface S_{10} at two points P_1 and P_2 so that the local intersection

numbers
$$(C \cdot S_{10})_{P_1} = (C \cdot S_{10})_{P_2} = 2.$$
In the former case P would be a ramification point of the morphism t with ramification index 3, so that this case is impossible. In the latter case each of the points P_1 and P_2 would be ramification points of t, so that t has no other ramification points except for P_1, P_2, the points of $C \cap \mathscr{S}$, and the points of $C \cap S_5$. In particular, the curve C intersects the surface $S'_{10} \in \mathcal{Q}_2$ transversally in four smooth points of S'_{10}, giving rise to two $\mathfrak{A}_5$-orbits of length 20 contained in $S'_{10} \setminus C_{20}$ (see Lemma 14.1.11). The latter is impossible by Lemma 14.1.3(ii).

Therefore, we have $C \cap \Sigma_{10} \neq \varnothing$. We claim that $|C \cap \Sigma_{10}| = 1$. Indeed, suppose that $|C \cap \Sigma_{10}| \geqslant 2$. Then the morphism t does not have ramification points outside of the union $\mathscr{C} \cup \Sigma_5 \cup \Sigma_{10}$. Therefore, C intersects the surface S'_{10} in four smooth points, giving rise to two $\mathfrak{A}_5$-orbits of length 20 contained in $S'_{10} \setminus C_{20}$ (see Lemma 14.1.11). The latter is again impossible by Lemma 14.1.3(ii).

We see that $|C \cap \Sigma_{10}| = 1$. By a similar argument one shows that $|C \cap \Sigma'_{10}| = 1$. In particular, the morphism t has no ramification points outside of the union
$$\mathscr{C} \cup \Sigma_5 \cup \Sigma_{10} \cup \Sigma'_{10},$$
so that all intersections of C with surfaces of $\mathcal{Q}_2$ outside this locus are transversal. In particular, C intersects the surface S_{10} (resp., the surface S'_{10}) transversally in two smooth points except for one point of Σ_{10} (resp., one point of Σ'_{10}), which gives assertion (i). Moreover, if $S \in \mathcal{Q}_2$ is a surface different from S_5, S_{10}, S'_{10}, and $\mathscr{S}$, then C intersects S transversally in four smooth points, which gives assertion (ii). $\square$

Now we are ready to give a complete description of the union of $\mathfrak{A}_5$-orbits of length 20 in V_5.

Theorem 14.1.16 (cf. Lemma 5.5.13). *The following assertions hold:*

(i) *the set $\Xi(20)$ is a union $\mathcal{G}_{10}$ of 10 conics $C_1, \ldots, C_{10}$ and the $\mathfrak{A}_5$-orbit $\Sigma'_{20} \subset C_{20}$ disjoint from these conics;*

(ii) *for any $1 \leqslant i \leqslant 10$, the intersection $C_i \cap \mathscr{S}$ consists of two points contained in the $\mathfrak{A}_5$-orbit $\Sigma_{20} \subset \mathscr{C}$ of length 20;*

(iii) *for any $1 \leqslant i \leqslant 10$, the intersection $C_i \cap S_5$ consists of two points of the $\mathfrak{A}_5$-orbit Σ_5;*

(iv) each point $P \in \Sigma_5$ is contained in exactly four conics among $C_1, \ldots, C_{10}$, and the tangent vectors to each three of the four of these conics passing through P are linearly independent;

(v) the conics $C_1, \ldots, C_{10}$ are disjoint outside of Σ_5;

(vi) for any $1 \leqslant i \leqslant 10$, the intersection $C_i \cap S_{10}$ (resp., the intersection $C_i \cap S'_{10}$) consists of one point of the $\mathfrak{A}_5$-orbit Σ_{10} (resp., of the $\mathfrak{A}_5$-orbit Σ'_{10}) and two smooth points of the surface S_{10} (resp., the surface S'_{10});

(vii) each point of the $\mathfrak{A}_5$-orbit Σ_{10} (resp., Σ'_{10}) is contained in a unique conic among $C_1, \ldots, C_{10}$;

(viii) for any $1 \leqslant i \leqslant 10$, the conic C_i intersects the surface $S \in \mathcal{Q}_2$ in four distinct smooth points of S provided that S is different from S_5, S_{10}, S'_{10}, and $\mathscr{S}$;

(ix) any point $P \in \Xi(20)$ such that $P \notin \Sigma_5 \cup \Sigma_{10} \cup \Sigma'_{10}$ is contained in an $\mathfrak{A}_5$-orbit of length 20.

Proof. Note that a general surface $S \in \mathcal{Q}_2$ contains three $\mathfrak{A}_5$-orbits of length 20 by Lemma 14.1.3(iii). In particular, $\Xi(20)$ contains some irreducible curve C, since the curve $C_{20} = \mathrm{Bs}(\mathcal{Q}_2)$ contains only one $\mathfrak{A}_5$-orbit of length 20 by Theorem 8.1.8(vii). By Lemma 14.1.10(i) the curve C is a conic.

Note that the stabilizer of C in $\mathfrak{A}_5$ is isomorphic to $\mathfrak{S}_3$ by Lemma 14.1.10(ii). Let $C_1 = C, C_2, \ldots, C_{10}$ be the $\mathfrak{A}_5$-orbit of the conic C. Lemma 14.1.13 implies assertion (ii). Assertion (iii) follows from Lemma 14.1.14(i), and assertion (iv) follows from Lemma 14.1.14(ii),(iii). Assertion (vi) is implied by Lemma 14.1.15(i), and assertion (viii) is implied by Lemma 14.1.15(ii). Applying Corollary 14.1.8, we obtain assertion (v). Combining it with assertion (vi), we get assertion (vii).

Using assertions (ii), (vi), and (viii), we see that the union Z of the conics C_i, $1 \leqslant i \leqslant 10$, covers one $\mathfrak{A}_5$-orbit of length 20 contained in the surface $\mathscr{S}$, one $\mathfrak{A}_5$-orbit of length 20 contained in the surface S_{10}, one $\mathfrak{A}_5$-orbit of length 20 contained in the surface S'_{10}, and two $\mathfrak{A}_5$-orbits of length 20 contained in a surface S for any $S \in \mathcal{Q}_2$ different from $\mathscr{S}$, S_5, S_{10}, and S'_{10}. Since Z is disjoint from C_{20} by Lemma 14.1.12, and C_{20} contains a unique $\mathfrak{A}_5$-orbit Σ'_{20} of length 20, we deduce from Lemma 14.1.3 that $\Xi(20)$ is a disjoint union of the curve Z and the $\mathfrak{A}_5$-orbit Σ'_{20}. This implies assertions (i) and (ix), and completes the proof. □

Keeping in mind Theorem 14.1.16, in the rest of the book we will denote the union of the ten conics contained in Ξ_{20} by $\mathcal{G}_{10}$, and will sometimes refer to these conics as the conics of $\mathcal{G}_{10}$ for brevity.

Remark 14.1.17. We have $\mathcal{G}_{10} \cap \mathscr{S} = \Sigma_{20}$ by Theorem 14.1.16(ii), and we have $\mathcal{G}_{10} \cap S_5 = \Sigma_5$ by Theorem 14.1.16(iii). Since the $\mathfrak{A}_5$-orbit Σ_{20} is not contained in C_{20} by Theorem 8.1.8(vii), we see that the intersection of the curve $\mathcal{G}_{10}$ with any surface $S \in \mathcal{Q}_2$ different from $\mathscr{S}$ does not contain Σ_{20}. Thus Theorem 14.1.16(vi) implies that the intersection $\mathcal{G}_{10} \cap S_{10}$ splits into a union of Σ_{10} and some $\mathfrak{A}_5$-orbit of length 20 different from Σ_{20} and Σ'_{20}, while the intersection $\mathcal{G}_{10} \cap S'_{10}$ splits into a union of Σ'_{10} and some $\mathfrak{A}_5$-orbit of length 20 different from Σ_{20} and Σ'_{20}. Also, Theorem 14.1.16(viii) implies that the intersection $\mathcal{G}_{10} \cap S_{15}$ splits into a union of two $\mathfrak{A}_5$-orbits of length 20 different from Σ_{20} and Σ'_{20}.

A useful consequence of Theorem 14.1.16 is the following:

Corollary 14.1.18. *Let $Z \subset V_5$ be an $\mathfrak{A}_5$-irreducible curve of degree $d < 30$. Suppose that $Z \neq \mathcal{L}_{15}$ and $Z \neq \mathcal{G}_{10}$. Then there exists a surface in the pencil $\mathcal{Q}_2$ that contains Z.*

Proof. By Corollary 8.1.7 we may assume that $20 \leqslant d \leqslant 29$. Let r be the number of irreducible components of Z. Then

$$r \in \{1, 5, 6, 10, 12, 20\}$$

by Corollary 5.1.2. If $r = 1$, then the assertion is implied by Corollary 8.1.6.

Let C be one of the irreducible components of Z, and $G_C \subset \mathfrak{A}_5$ be the stabilizer of C. If $r \in \{5, 6, 12\}$, then G_C acts on C faithfully by Lemma 7.4.5 and the assertion is implied by Lemma 8.1.5.

Thus, we may assume that either $r = 10$ or $r = 20$. In particular, one has $d = 20$. The stabilizer G_C is isomorphic either to $\mathfrak{S}_3$ or to $\boldsymbol{\mu}_3$ by Lemma 5.1.1. Since the stabilizer $G_P \subset \mathfrak{A}_5$ of a general point $P \in C$ is a normal subgroup of G_C, we see that either G_P is trivial, or $G_P \cong \boldsymbol{\mu}_3$. In the former case the assertion is again implied by Lemma 8.1.5. In the latter case a general point of the curve Z is contained in an $\mathfrak{A}_5$-orbit of length 20, so that $Z = \mathcal{G}_{10}$ by Theorem 14.1.16(i). □

By Theorem 8.2.1(v) and Lemmas 6.7.1 and 7.2.10, neither $\mathcal{L}_{15}$ nor $\mathcal{G}_{10}$ is contained in any surface of the pencil $\mathcal{Q}_2$. In fact, we can say more. We already know from Lemma 7.5.13 that $\mathcal{L}_{15}$ is not contained in any surface in the linear system $|2H|$. The same is true for $\mathcal{G}_{10}$ by:

Lemma 14.1.19. The curve $\mathcal{G}_{10}$ is not contained in any surface from the linear system $|2H|$.

Proof. Suppose that there is a surface $S \in |2H|$ passing through $\mathcal{G}_{10}$. Then $S \notin \mathcal{Q}_2$ by Corollary 8.2.2. By Lemma 8.1.1 this means that S is not $\mathfrak{A}_5$-invariant, so that there is a surface $S' \in |2H|$ different from S such that $\mathcal{G}_{10} \subset S \cap S'$. Since the linear span of the curve $\mathcal{G}_{10}$ is the whole $\mathbb{P}^6$ by Corollary 7.5.4, the surface S is irreducible. Thus $S \cap S'$ is a curve, and since
$$\mathcal{G}_{10} \cdot H = \deg(\mathcal{G}_{10}) = 20 = S \cdot S' \cdot H,$$
we see that $\mathcal{G}_{10} = S \cap S'$. Therefore, the surfaces S and S' generate an $\mathfrak{A}_5$-invariant pencil in $|2H|$. But the unique $\mathfrak{A}_5$-invariant pencil in $|2H|$ is $\mathcal{Q}_2$ by Lemma 8.1.1, which again gives a contradiction with Corollary 8.2.2. □

Remark 14.1.20. In §13.6, we classified all $\mathfrak{A}_5$-irreducible curves in V_5 of degree less than 20. Using Corollary 14.1.18, it seems possible to classify all $\mathfrak{A}_5$-irreducible curves of degree less than 30 in V_5. Recall that there is only a finite number of such curves by Theorem 8.3.1 (unlike $\mathfrak{A}_5$-irreducible curves of degree 30, that are infinitely numerous by Remark 7.4.2 and Corollary 12.4.7).

For example, let C be an $\mathfrak{A}_5$-irreducible curve of degree 20 in V_5 that is different from $\mathcal{G}_{10}$. By Corollary 8.1.10 the only $\mathfrak{A}_5$-irreducible curves of degree 20 that are contained in the surface $\mathscr{S}$ are $\mathcal{C}_{20}$ and $\mathcal{L}_{20}$. Therefore, one may assume that $C \not\subset \mathscr{S}$. Then the intersection $C \cap \mathscr{S}$ is an $\mathfrak{A}_5$-invariant set that consists of $C \cdot \mathscr{S} = 40$ points (counted with multiplicities), so that $C \cap \mathscr{S}$ is contained in the union $\Sigma_{20} \cup \Sigma'_{20}$ by Lemma 7.2.10.

Since all $\mathfrak{A}_5$-orbits of length 12 in V_5 are contained in $\mathscr{S}$ by Theorem 7.3.5, we see that C does not contain any of them. On the other hand, there is a surface $S \in \mathcal{Q}_2$ that contains C by Corollary 14.1.18. By Theorem 8.2.1(v), the surface S is a $K3$ surface with at most ordinary double points. Put $H_S = H|_S$. Recall that $\operatorname{rk Pic}(S)^{\mathfrak{A}_5} \leqslant 2$ by Lemma 6.7.3(i). If $\operatorname{rk Pic}(S)^{\mathfrak{A}_5} = 1$, then C is a Cartier divisor on S linearly equivalent to $2H_S$ by Remark 8.2.4, and thus $C = \mathcal{C}_{20}$ by Lemma 8.2.6(ii). Hence, we may assume that $\operatorname{rk Pic}(S)^{\mathfrak{A}_5} = 2$. Then S is smooth by Lemma 6.7.3(ii), so that $C^2 = 2p_a(C) - 2$ by the adjunction formula. As before, we may assume that C is not numerically proportional to H_S, since otherwise one would have $C \sim 2H_S$ by Lemma 6.7.3(i), and thus $C = \mathcal{C}_{20}$ by Lemma 8.2.6(ii). Hence we have
$$H_S^2 \cdot C^2 < (H_S \cdot C)^2$$

by the Hodge index theorem, which means that $C^2 \leqslant 38$ and, thus, $p_a(C) \leqslant 20$. Therefore, if C is an irreducible curve, then it must be smooth and

$$p_a(C) \in \{6, 11, 16\}$$

by Lemmas 5.1.5 and 4.4.6 and Theorem 7.3.5, because we already know that C does not contain $\mathfrak{A}_5$-orbits of length 12. We do not know whether any of these three cases are possible or not, but we expect that finding this out is an accessible task. Thus, let us assume that the curve C is reducible. Denote by r the number of its irreducible components. Then

$$r \in \{5, 10, 20\}$$

by Corollary 5.1.2. If $r = 20$, then C is a union of 20 lines, and thus $C = \mathcal{L}_{20}$ by Proposition 7.4.1. If $r = 10$, then C is a union of 10 conics, and one can show that the only $\mathfrak{A}_5$-irreducible union of 10 conics in V_5 is the curve $\mathcal{G}_{10}$ (cf. Remark 7.8.5). If $r = 5$, then C is a union of 5 curves of degree 4. We do not know whether this case is possible or not, but as above we expect that it is not difficult to find this out.

14.2 Ten conics

In this section we describe the intersections of the curve $\mathcal{G}_{10}$ with some $\mathfrak{A}_5$-invariant curves. We also describe which $\mathfrak{A}_5$-orbits of small length are contained in $\mathcal{G}_{10}$ and describe the $\mathcal{G}_{10} \cap S_{Cl}$.

Lemma 14.2.1. One has

$$\mathcal{G}_{10} \cap \mathscr{C} = \mathcal{G}_{10} \cap \mathcal{L}_{20} = \Sigma_{20}$$

and

$$\mathcal{G}_{10} \cap \mathcal{L}_{12} = \mathcal{G}_{10} \cap C_{16} = \mathcal{G}_{10} \cap C_{20} = \varnothing.$$

Proof. Recall that $\mathcal{G}_{10} \cap \mathscr{S} = \Sigma_{20}$ by Remark 14.1.17. The $\mathfrak{A}_5$-orbit Σ_{20} is contained in the curve $\mathscr{C}$ by construction, and is contained in the curve $\mathcal{L}_{20}$ by Lemma 7.2.12(iii). On the other hand, Σ_{20} is disjoint from the curve $\mathcal{L}_{12}$ by Lemma 7.2.12(ii), disjoint from the curve C_{16} by Lemma 8.1.16(iv), and disjoint from the curve C_{20} by Theorem 8.1.8(vii). □

Lemma 14.2.2. The intersection of the curves $\mathscr{C}'$ and $\mathcal{G}_{10}$ is a single $\mathfrak{A}_5$-orbit of length 20 different from Σ_{20} and Σ'_{20}.

Proof. By Lemma 5.1.5 we know that the curve $\mathscr{C}' \cong \mathbb{P}^1$ contains a unique $\mathfrak{A}_5$-orbit Σ of length 20. By Lemma 11.3.4 the $\mathfrak{A}_5$-orbit Σ is different from Σ_{20} and Σ'_{20}. By Lemma 5.1.4 the curve $\mathscr{C}'$ is disjoint from the $\mathfrak{A}_5$-orbits Σ_5, Σ_{10}, and Σ'_{10}. Now the assertion follows from Theorem 14.1.16(i),(ix). □

Lemma 14.2.3. *The curve $\mathcal{G}_{10}$ is disjoint from each of the curves $\mathcal{L}_6$, $\mathcal{G}_6$, and $\mathcal{T}_6$.*

Proof. Let C be one of the curves $\mathcal{L}_6$, $\mathcal{G}_6$, or $\mathcal{T}_6$. Then C does not contain the $\mathfrak{A}_5$-orbits Σ_5, Σ_{10}, and Σ'_{10}, and does not contain $\mathfrak{A}_5$-orbits of length 20. This follows from Corollary 7.4.9 if $C = \mathcal{L}_6$, from Remark 9.2.9 if $C = \mathcal{G}_6$, and from Lemma 13.2.4 and Corollary 13.2.5 if $C = \mathcal{T}_6$. Thus, the curves C and $\mathcal{G}_{10}$ are disjoint by Theorem 14.1.16(ix). □

Lemma 14.2.4. *The curves $\mathcal{L}_{10}$ and $\mathcal{G}_{10}$ are disjoint.*

Proof. A stabilizer $G_L \subset \mathfrak{A}_5$ of a line L of $\mathcal{L}_{10}$ is isomorphic to $\mathfrak{S}_3$ by Lemma 5.1.1, and G_L acts on L faithfully by Lemma 7.4.6. Therefore, there is a unique G_L-orbit of length 2 on L by Lemma 5.5.9(ii), so that there is a unique $\mathfrak{A}_5$-orbit Σ of length 20 contained in $\mathcal{L}_{10}$ by Lemma 9.1.2. By Remark 9.1.6 one has $\Sigma = \Sigma'_{20}$. Also, the $\mathfrak{A}_5$-orbits Σ_5, Σ_{10}, and Σ'_{10} are disjoint from $\mathcal{L}_{10}$ by Lemma 9.1.1 and Corollary 9.1.5. Thus the assertion follows from Theorem 14.1.16(i),(ix). □

Lemma 14.2.5. *The intersection of the curve $\mathcal{G}_5$ (resp., $\mathcal{G}'_5$) with the curve $\mathcal{G}_{10}$ is a single $\mathfrak{A}_5$-orbit of length 20 different from Σ_{20} and Σ'_{20}.*

Proof. It is enough to check the assertion for the curve $\mathcal{G}_5$. A stabilizer $G_C \subset \mathfrak{A}_5$ of a conic C of $\mathcal{G}_5$ is isomorphic to $\mathfrak{A}_4$ by Lemma 5.1.1, and G_C acts on C faithfully by Lemma 7.4.5. Therefore, there are two G_C-orbits of length 4 on C by Lemma 5.5.9(iii), so that there are two $\mathfrak{A}_5$-orbits of length 20 contained in $\mathcal{G}_5$ by Lemma 9.2.3. By Remark 9.2.5 one of these $\mathfrak{A}_5$-orbits is Σ'_{20}, and the other is some $\mathfrak{A}_5$-orbit different from Σ_{20} and Σ'_{20}. Note also that the $\mathfrak{A}_5$-orbits Σ_5, Σ_{10}, and Σ'_{10} are disjoint from $\mathcal{G}_5$ by Lemma 9.2.1 and Corollary 9.2.4. Therefore, the assertion follows from Theorem 14.1.16(i),(ix). □

Lemma 14.2.6. *One has $\mathcal{L}_{15} \cap \mathcal{G}_{10} = \Sigma_5$.*

Proof. By Theorem 14.1.16(ix) we know that any $\mathfrak{A}_5$-orbit contained in $\mathcal{G}_{10}$ either has length 20, or coincides with one of the $\mathfrak{A}_5$-orbits Σ_5, Σ_{10}, or Σ'_{10}. Now the assertion follows from Lemma 9.1.10(iv). □

The following refinement of Lemma 14.2.6 will be used in the proof of Lemma 17.5.10.

Lemma 14.2.7. *Let C be a conic of $\mathcal{G}_{10}$, let $P \in \Sigma_5$ be a point contained in C, and let L be a line of $\mathcal{L}_{15}$ passing through the point P. Then C and L are transversal at the point P.*

Proof. Recall that one has $\mathcal{G}_{10} \cap \mathcal{L}_{15} = \Sigma_5$ by Lemma 14.2.6. Recall from Theorem 14.1.16(iv) that P is contained in exactly 4 conics, say, $C = C_1$, C_2, C_3, and C_4, of $\mathcal{G}_{10}$. In particular, Lemma 7.5.9 implies that these conics are pairwise transversal at P. Moreover, the stabilizer $G_P \cong \mathfrak{A}_4$ of the point P acts transitively on the four one-dimensional Zariski tangent spaces of the conics C_1, C_2, C_3, and C_4 at the point P; note that these are naturally embedded into the Zariski tangent space $T_P(V_5)$. Thus, none of these tangent spaces can coincide with any of the three Zariski tangent spaces to the three lines, say, $L = L_1$, L_2, and L_3, of $\mathcal{L}_{15}$ passing through P (see Corollary 7.4.4). This shows that the conics C_i, $1 \leqslant i \leqslant 4$, are pairwise transversal to the lines L_j, $1 \leqslant j \leqslant 3$, and in particular C is transversal to L. □

Lemma 14.2.8. *The intersection of the curves C'_{16} and $\mathcal{G}_{10}$ is a single $\mathfrak{A}_5$-orbit of length 20 different from Σ_{20} and Σ'_{20}.*

Proof. The curve C'_{16} is disjoint from the $\mathfrak{A}_5$-orbits Σ_5, Σ_{10}, and Σ'_{10} by Lemma 5.1.4. On the other hand, C'_{16} contains two $\mathfrak{A}_5$-orbits of length 20 by Lemma 5.1.5. By Remark 13.1.4 one of these orbits is Σ'_{20}, and the other one is some $\mathfrak{A}_5$-orbit Σ different from Σ_{20}. Now the assertion follows from Theorem 14.1.16(i),(ix). □

Lemma 14.2.9. *The intersection of the curves C^0_{18} and $\mathcal{G}_{10}$ is a single $\mathfrak{A}_5$-orbit of length 20 different from Σ_{20} and Σ'_{20}.*

Proof. By Lemma 5.1.5 we know that the curve C^0_{18} contains a unique $\mathfrak{A}_5$-orbit Σ of length 20. By Corollary 13.4.3(iii) the $\mathfrak{A}_5$-orbit Σ is different from Σ_{20} and Σ'_{20}. Also, by Corollary 13.4.3(i) the curve C^0_{18} is disjoint from the $\mathfrak{A}_5$-orbits Σ_5, Σ_{10}, and Σ'_{10}. Therefore, the assertion follows from Theorem 14.1.16(i),(ix). □

Lemma 14.2.10. *The curves $\mathcal{B}_{18}$ and $\mathcal{G}_{10}$ are disjoint.*

Proof. Applying Remark 13.5.3 together with Theorem 14.1.16(ix) we obtain the required assertion. □

Remark 14.2.11. Assume that C is an $\mathfrak{A}_5$-invariant smooth rational curve of degree 18 in V_5 (cf. Remark 13.6.2). Then C contains a unique $\mathfrak{A}_5$-orbit Σ of length 20, and does not contain $\mathfrak{A}_5$-orbits of length 5 and 10, see Lemma 5.5.9(v). By Corollary 13.6.5 the $\mathfrak{A}_5$-orbit Σ is disjoint from the surface $\mathscr{S}$. Therefore, it follows from Theorem 14.1.16(i),(ix) that $C \cap \mathcal{G}_{10} = \Sigma$.

Lemma 14.2.12. One has

$$S_{Cl} \cap \mathcal{G}_{10} = \Sigma_{10} \cup \Sigma'_{10} \cup \Sigma_{20} \cup \Sigma,$$

where Σ is an $\mathfrak{A}_5$-orbit of length 20 different from Σ_{20} and Σ'_{20}.

Proof. One has $\Sigma_5 \subset \mathcal{G}_{10}$ by Theorem 14.1.16(iv), while $\Sigma_5 \not\subset S_{Cl}$ by Lemma 12.2.3. Thus the curve $\mathcal{G}_{10}$ is not contained in the surface S_{Cl}. Hence, the intersection $\Omega = S_{Cl} \cap \mathcal{G}_{10}$ consists of

$$S_{Cl} \cdot \mathcal{G}_{10} = 60$$

points (counted with multiplicities). Note that the $\mathfrak{A}_5$-orbit Σ_{20} is contained in Ω by Theorem 14.1.16(ii) and Lemma 12.2.5(v), while the $\mathfrak{A}_5$-orbits Σ_{10} and Σ'_{10} are contained in Ω by Theorem 14.1.16(vi) and Lemma 12.2.4. Also, the curve $\mathscr{C}'$ is contained in the surface S_{Cl} by Lemma 12.2.5(i), and $\mathscr{C}' \cong \mathbb{P}^1$ contains an $\mathfrak{A}_5$-orbit Σ of length 20 by Lemma 5.5.9(v). Thus, we have $\Sigma \subset \Omega$. It remains to notice that Σ is not contained in the surface $\mathscr{S}$ by Lemma 11.3.4. □

14.3 Orbits of length 30

By Lemma 9.1.10(iv), the $\mathfrak{A}_5$-orbit of every point in $\mathcal{L}_{15} \setminus (\Sigma_5 \cup \Sigma_{15})$ consists of 30 points. However, there are $\mathfrak{A}_5$-orbits of length 30 that are not contained in $\mathcal{L}_{15}$. In fact, Lemmas 6.7.1 and 7.2.10 together with Theorem 8.2.1 easily imply that $\mathfrak{A}_5$-orbits of length 30 that are not contained in $\mathcal{L}_{15}$ must form an $\mathfrak{A}_5$-invariant curve of degree 45 up to a finite number of sporadic $\mathfrak{A}_5$-orbits of length 30. In this section, we will construct this curve and prove that it contains all $\mathfrak{A}_5$-orbits of length 30 that are not contained in $\mathcal{L}_{15}$.

Let us use the notation and assumptions of §7.6. In particular, we denote by G_1 a subgroup in $\mathfrak{A}_5$ that is isomorphic to $\mathfrak{A}_4$, and by H_1 the unique hyperplane section of V_5 stabilized by G_1. The surface H_1 contains four disjoint lines L_1, L_2, L_3, and L_4 that are exceptional curves of the G_1-equivariant birational contraction $H_1 \to \mathbb{P}^2$.

The surface H_1 contains a G_1-irreducible curve that is a union of three twisted cubic curves (that are proper transforms on H_1 of the three lines in $\mathbb{P}^2$). We denote the latter twisted cubics by T_1, T_2, and T_3, and we denote by $\mathcal{T}_{15}$ the $\mathfrak{A}_5$-orbit of the cubics T_1, T_2, and T_3.

Lemma 14.3.1. The set

$$\bigl(T_1 \cap T_2\bigr) \cup \bigl(T_1 \cap T_3\bigr) \cup \bigl(T_2 \cap T_3\bigr) \cup \Bigl(\mathcal{L}_{15} \cap \bigl(L_1 \cup L_2 \cup L_3 \cup L_4\bigr)\Bigr)$$

consists of 15 points, and coincides with the intersection $\mathcal{L}_{15} \cap H_1$. The intersection $\Sigma_{15} \cap H_1$ consists of the points $T_1 \cap T_2$, $T_1 \cap T_3$, and $T_2 \cap T_3$, which form the unique G_1-orbit of length 3 in H_1.

Proof. By Lemma 6.2.5(ii), the points $T_1 \cap T_2$, $T_1 \cap T_3$, and $T_2 \cap T_3$ form the unique G_1-orbit of length 3 on H_1. In particular, the stabilizer in G_1 of each of these three points is $\boldsymbol{\mu}_2 \times \boldsymbol{\mu}_2$. Thus, Theorem 7.3.5 implies that they are contained in Σ_{15}, since $\Sigma_5 \cap H_1 = \varnothing$ by Lemma 7.6.1.

Recall that $\Sigma_{15} \subset \mathcal{L}_{15}$ by Lemma 9.1.10(ii), and the lines of $\mathcal{L}_{15}$ are disjoint away from Σ_5 by Lemma 9.1.10(i). Thus, the intersection $\mathcal{L}_{15} \cap H_1$ consists of 15 points. On the other hand, each intersection $\mathcal{L}_{15} \cap L_1$, $\mathcal{L}_{15} \cap L_2$, $\mathcal{L}_{15} \cap L_3$, and $\mathcal{L}_{15} \cap L_4$ consists of 3 points by Lemmas 7.6.6 and 9.1.15. Since L_1, L_2, L_3, and L_4 are lines of $\mathcal{L}_{10}$ by Lemma 7.6.6, and $\Sigma_{15} \not\subset \mathcal{L}_{10}$ by Lemma 9.1.2, we see that the intersection $\mathcal{L}_{15} \cap H_1$ consists of 12 points of

$$\mathcal{L}_{15} \cap \bigl(L_1 \cup L_2 \cup L_3 \cup L_4\bigr)$$

and three more points. These three points are $T_1 \cap T_2$, $T_1 \cap T_3$, and $T_2 \cap T_3$, because we already proved that they are contained in Σ_{15}, and $\Sigma_{15} \subset \mathcal{L}_{15}$. Since $\Sigma_{15} \not\subset \mathcal{L}_{10}$, we see that the intersection $\Sigma_{15} \cap H_1$ consists of the points $T_1 \cap T_2$, $T_1 \cap T_3$, and $T_2 \cap T_3$. $\square$

Now we are ready to prove

Theorem 14.3.2. *The following assertions hold:*

(o) *the curve $\mathcal{T}_{15}$ is a union of 15 smooth rational cubic curves;*

(i) *the $\mathfrak{A}_5$-orbits Σ_5, Σ_{12}, and Σ'_{12} are not contained in the curve $\mathcal{T}_{15}$;*

(ii) *each irreducible component of $\mathcal{T}_{15}$ contains two points of the $\mathfrak{A}_5$-orbit Σ_{10} (resp., Σ'_{10}), and each point of Σ_{10} (resp., Σ'_{10}) is contained in three irreducible components of $\mathcal{T}_{15}$;*

(iii) each irreducible component of $\mathcal{T}_{15}$ contains two points of the $\mathfrak{A}_5$-orbit Σ_{15}, and each point of Σ_{15} is contained in two irreducible components of $\mathcal{T}_{15}$;

(iv) one has $\Sigma_{15} = \mathcal{T}_{15} \cap \mathcal{L}_{15}$;

(v) the $\mathfrak{A}_5$-orbit of every point in $\mathcal{T}_{15}$ that is not contained in Σ_{10}, Σ'_{10}, and Σ_{15} has length 30;

(vi) the irreducible components of the curve $\mathcal{T}_{15}$ are disjoint away from $\Sigma_{15} \cup \Sigma_{10} \cup \Sigma'_{10}$;

(vii) the curve $\mathcal{T}_{15}$ is contained in S_{Cl};

(viii) one has $\Sigma'_{30} \not\subset \mathcal{T}_{15}$.

Proof. Recall that V_5 does not contain $\mathfrak{A}_5$-irreducible curves that are unions of 3 or 5 cubics by Lemmas 8.1.14 and 10.2.4, respectively. This implies assertion (o).

The surface H_1 is disjoint from the $\mathfrak{A}_5$-orbits Σ_5, Σ_{12}, and Σ'_{12} by Lemma 7.6.1. In particular, one has

$$\left(T_1 \cup T_2 \cup T_3\right) \cap \left(\Sigma_5 \cup \Sigma_{12} \cup \Sigma'_{12}\right) = \varnothing.$$

which implies that $\Sigma_5 \not\subset \mathcal{T}_{15}$, $\Sigma_{12} \not\subset \mathcal{T}_{15}$, and $\Sigma'_{12} \not\subset \mathcal{T}_{15}$, which is assertion (i).

Let us show that $\Sigma_{10} \subset \mathcal{T}_{15}$. By Corollary 7.6.12, the $\mathfrak{A}_5$-orbit Σ_{10} is contained in the union

$$H_1 \cup H_2 \cup H_3 \cup H_4 \cup H_5.$$

Thus, the intersection $H_1 \cap \Sigma_{10}$ is not empty. Moreover, since Σ_{10} is contained neither in $\mathcal{G}_5$ nor in $\mathcal{G}'_5$ by Corollary 9.2.4, the intersection $H_1 \cap \Sigma_{10}$ does not contain G_1-orbits of length 4 by Remark 7.6.9.

The intersection $H_1 \cap \Sigma_{10}$ does not contain the points $T_1 \cap T_2$, $T_1 \cap T_3$, and $T_2 \cap T_3$, since they are contained in Σ_{15} by Lemma 14.3.1. Thus, it follows from Lemma 6.2.5(ii), that $H_1 \cap \Sigma_{10}$ does not contain G_1-orbits of length 3. Hence, $H_1 \cap \Sigma_{10}$ consists of a single G_1-orbit of length 6 by Lemma 6.2.5(i),(ii). By Lemma 6.2.5(iii), all G_1-orbits of length 6 on H_1 are contained in $T_1 \cup T_2 \cup T_2$, so that each curve among T_1, T_2, and T_3 contains exactly two points in Σ_{10}. In particular, $\mathcal{T}_{15}$ contains the $\mathfrak{A}_5$-orbit Σ_{10}, each irreducible component of $\mathcal{T}_{15}$ contains exactly two points in Σ_{10}, and each point of Σ_{10} is contained in exactly three irreducible components of $\mathcal{T}_{15}$. Similarly, $\mathcal{T}_{15}$ contains the $\mathfrak{A}_5$-orbit Σ'_{10}, each irreducible component of $\mathcal{T}_{15}$ contains exactly two points in Σ'_{10}, and each point of Σ'_{10} is contained in exactly three irreducible components of $\mathcal{T}_{15}$. This is assertion (ii).

By Lemma 9.1.15, the intersection $\mathcal{L}_{15} \cap \mathcal{L}_{10}$ is an $\mathfrak{A}_5$-orbit of length 30, and each of the intersections $\mathcal{L}_{15} \cap L_1$, $\mathcal{L}_{15} \cap L_2$, $\mathcal{L}_{15} \cap L_3$, and $\mathcal{L}_{15} \cap L_4$ consists of 3 points. On the other hand,

$$\mathcal{L}_{15} \cap H_1 = (T_1 \cap T_2) \cup (T_1 \cap T_3) \cup (T_2 \cap T_3) \cup \Big(\mathcal{L}_{15} \cap (L_1 \cup L_2 \cup L_3 \cup L_4)\Big)$$

by Lemma 14.3.1. Since the curves T_1, T_2, and T_3 are disjoint from L_1, L_2, L_3, and L_4, we see that

$$(T_1 \cup T_2 \cup T_3) \cap \mathcal{L}_{15} \subset (T_1 \cap T_2) \cup (T_1 \cap T_3) \cup (T_2 \cap T_3) \subset \Sigma_{15}.$$

This gives assertion (iv), because $\Sigma_{15} \subset \mathcal{T}_{15}$ by Lemma 14.3.1. Moreover, this also shows that each curve among T_1, T_2, and T_3 (and thus also any other irreducible component of $\mathcal{T}_{15}$) contains exactly two points in Σ_{15}, because $\Sigma_{15} \subset \mathcal{L}_{15}$. Hence, each point in Σ_{15} is contained in exactly two irreducible components of $\mathcal{T}_{15}$. This is assertion (iii).

By Lemma 6.2.5(iii), all G_1-orbits on H_1 of length 6 are contained in $T_1 \cup T_2 \cup T_2$, and an $\mathfrak{A}_5$-orbit of every point in $T_1 \cup T_2 \cup T_3$ that is not of length 3 has length 6. Thus, the stabilizer in G_1 of every point in $T_1 \cup T_2 \cup T_3$ that is not contained in the $\mathfrak{A}_5$-orbits Σ_{10}, Σ'_{10}, and Σ_{15} is $\boldsymbol{\mu}_2$. Since $\Sigma_5 \not\subset \mathcal{T}_{15}$, we see that the stabilizer in $\mathfrak{A}_5$ of every point in $T_1 \cup T_2 \cup T_3$ that is not contained in the $\mathfrak{A}_5$-orbits Σ_{10}, Σ'_{10}, and Σ_{15} is also $\boldsymbol{\mu}_2$, which is assertion (v). This also implies that the irreducible components of the curve $\mathcal{T}_{15}$ are disjoint away from $\Sigma_{10} \cup \Sigma'_{10} \cup \Sigma_{15}$, because otherwise $\mathcal{T}_{15}$ would contain an $\mathfrak{A}_5$-orbit of length less than 30 that is different from Σ_{10}, Σ'_{10}, and Σ_{15}. But we already proved that the latter does not happen. This gives assertion (vi).

The restriction $S_{Cl}|_{H_1}$ is a G_1-invariant divisor in $|-3K_{H_1}|$. Thus, its support contains the curves T_1, T_2, and T_3 by Lemma 6.2.6. Hence, $\mathcal{T}_{15}$ is contained in S_{Cl}, which is assertion (vii). Since $\Sigma'_{30} \not\subset S_{Cl}$ by Corollary 12.2.7, we have $\Sigma'_{30} \not\subset \mathcal{T}_{15}$. This gives assertion (viii) and completes the proof of the Theorem 14.3.2. □

We keep the notation $\mathcal{T}_{15}$ for the curve introduced above in the rest of the book. We do not pursue a goal to unify this notation with the notation of §6.2. We will abuse terminology a little bit and refer to the irreducible components of the curve $\mathcal{T}_{15}$ as the twisted cubics of $\mathcal{T}_{15}$.

Remark 14.3.3. We constructed the curve $\mathcal{T}_6$ by applying the construction in §7.7 to a line of $\mathcal{L}_6$ (see the proof of Lemma 13.2.1). We can construct the curve $\mathcal{T}_{15}$ in a similar way as well. Namely, we just need to apply the construction from §7.7 to a line of $\mathcal{L}_{15}$. However, this way does not shed much light on the geometry of the curve $\mathcal{T}_{15}$.

Remark 14.3.4. Let T be an irreducible component of $\mathcal{T}_{15}$, and let G_1 be its stabilizer in $\mathfrak{A}_5$. Then $G_1 \cong \boldsymbol{\mu}_2 \times \boldsymbol{\mu}_2$ by Lemma 5.1.1, and T is the unique irreducible component of $\mathcal{T}_{15}$ that is G_1-invariant. Similarly, there exists a unique line L of $\mathcal{L}_{15}$ that is G_1-invariant. Although

$$\mathcal{T}_{15} \cap \mathcal{L}_{15} = \Sigma_{15}$$

by Theorem 14.3.2, the curves T and L do not intersect. This follows from the proof of Lemma 14.3.1. By Theorem 7.1.8, the same result was obtained in Lemma 5.5.14 by a direct computation.

Lemma 14.3.5 (cf. Lemma 14.1.3). *Let S be a surface in the pencil $\mathcal{Q}_2$. Then*

(o) *the surface S contains two $\mathfrak{A}_5$-orbits of length 30, if $S = \mathscr{S}$;*

(i) *the surface S contains three $\mathfrak{A}_5$-orbits of length 30, if $S = S_5$;*

(ii) *the surface S contains two $\mathfrak{A}_5$-orbits of length 30, if $S = S_{10}$ or $S = S'_{10}$;*

(iii) *the surface S contains one $\mathfrak{A}_5$-orbit of length 30, if $S = S_{15}$;*

(iv) *the surface S contains four $\mathfrak{A}_5$-orbits of length 30, if S is different from $\mathscr{S}$, S_5, S_{10}, S'_{10}, and S_{15}.*

Proof. Assertion (o) is contained in Lemma 7.2.10.

Let $S \in \mathcal{Q}_2$ be a singular surface such that $S \neq \mathscr{S}$. Then the singularities of S are ordinary double points by Theorem 8.2.1(v). Let $\theta \colon \tilde{S} \to S$ be the minimal resolution of singularities. Then $\tilde{S}$ contains exactly four $\mathfrak{A}_5$-orbits of length 30 by Lemma 6.7.1(v).

Let P be a singular point of S, and let $E \subset \tilde{S}$ be the exceptional divisor contracted by θ to the point P. Then $E \cong \mathbb{P}^1$ since P is an ordinary double point of the surface S. Let $G_P \subset \mathfrak{A}_5$ be the stabilizer of the point P. Note that the action of G_P on E is faithful, since otherwise $\tilde{S}$ would contain infinitely many $\mathfrak{A}_5$-orbits of length less than 60, which is not the case by Lemma 6.7.1.

Suppose that $S = S_5$. Then $\mathrm{Sing}(S) = \Sigma_5$ by Theorem 8.1.8(iii),(iv), so that $G_P \cong \mathfrak{A}_4$ by Lemma 5.1.1. Thus, by Lemma 5.5.9(iii) there is one G_P-orbits of length 6 contained in E, that gives rise to two $\mathfrak{A}_5$-orbits of length $6|\Sigma_5| = 30$ contained in $\tilde{S}$. This orbit is mapped to Σ_5 by the morphism θ. Therefore, the surface S_5 contains three $\mathfrak{A}_5$-orbits of length 30, which gives assertion (i).

To prove assertion (ii), we may suppose that $S = S_{10}$. Then $\mathrm{Sing}(S) = \Sigma_{10}$ by Theorem 8.1.8(iii),(iv), so that $G_P \cong \mathfrak{S}_3$ by Lemma 5.1.1. Thus, by Lemma 5.5.9(ii) there are two G_P-orbits of length 3 contained in E, that give rise to two $\mathfrak{A}_5$-orbit of length $3|\Sigma_{10}| = 30$ contained in $\tilde{S}$. These orbits are mapped to Σ_{10} by the morphism θ. Therefore, the surface S_{10} contains two $\mathfrak{A}_5$-orbits of length 30, which gives assertion (ii).

Suppose that $S = S_{15}$. Then $\mathrm{Sing}(S) = \Sigma_{15}$ by Theorem 8.1.8(iii),(iv), so that $G_P \cong \boldsymbol{\mu}_2 \times \boldsymbol{\mu}_2$ by Lemma 5.1.1. Thus, by Lemma 5.5.9(ii) there are three G_P-orbits of length 2 contained in E, that give rise to three $\mathfrak{A}_5$-orbit of length $2|\Sigma_{15}| = 30$ contained in $\tilde{S}$. All these orbits are mapped to Σ_{15} by the morphism θ. Therefore, the surface S_{15} contains a single $\mathfrak{A}_5$-orbit of length 30, which gives assertion (iii).

Now suppose that S is different from $\mathscr{S}$, S_5, S_{10}, S'_{10}, and S_{15}. By Theorem 8.2.1(ii) this means that S is smooth. Therefore, assertion (iv) is implied by Lemma 6.7.1(v). $\square$

Now we need the following two lemmas.

Lemma 14.3.6. Let C be an irreducible component of the curve $\mathcal{T}_{15}$. Then C intersects $\mathscr{S}$ in two points of the $\mathfrak{A}_5$-orbit Σ_{30}, and the local intersection multiplicity in each of these points is 3. In particluar, one has $\mathscr{S} \cap \mathcal{L}_{15} = \Sigma'_{30}$.

Proof. Recall that the curve C is not contained in the surface $\mathscr{S}$ by Lemma 7.2.10 and Theorem 14.3.2(v). Let P be a point in the intersection $C \cap \mathscr{S}$, and let Σ be its $\mathfrak{A}_5$-orbit. By Theorem 14.3.2(v), the length r of the $\mathfrak{A}_5$-orbit Σ equals either 10, or 15, or 30. By Lemma 7.2.10, this means that $r = 30$ and Σ is either Σ_{30} or Σ'_{30}. Since $\Sigma'_{30} \not\subset \mathcal{T}_{15}$ by Theorem 14.3.2(viii), we see that $\Sigma = \Sigma_{30}$.

Let c be a number of curves of $\mathcal{T}_{15}$ passing through some (and thus any) point of the $\mathfrak{A}_5$-orbit Σ_{30}, and p be the number of points of Σ_{30} that are contained in some (and thus any) curve of $\mathcal{T}_{15}$. One has

$$15p = 30c,$$

so that $p = 2c$. Also, we have $p \leqslant 3$ because the surface $\mathscr{S}$ is singular at the points of Σ_{30} by Lemma 7.2.2(i). This gives $p = 2$. We showed that C contains exactly two points of Σ_{30}. The local intersection index in each of them equals
$$\frac{1}{2} C \cdot \mathscr{S} = 3.$$

□

Lemma 14.3.7. Let C be an irreducible component of the curve $\mathcal{T}_{15}$. The following assertions hold:

(i) if S is a surface of the pencil $\mathcal{Q}_2$ different from $\mathscr{S}$, S_{10}, S'_{10}, and S_{15}, then C intersects S in six smooth points, giving rise to three $\mathfrak{A}_5$-orbits of length 30;

(ii) the curve C intersects the surface S_{10} (resp., S'_{10}) in two points of the $\mathfrak{A}_5$-orbit Σ_{10} (resp., Σ'_{10}), and in two smooth points contained in an $\mathfrak{A}_5$-orbit of length 30;

(iii) the curve C intersects the surface S_{15} in two points of the $\mathfrak{A}_5$-orbit Σ_{15}, and in two smooth points contained in an $\mathfrak{A}_5$-orbit of length 30.

Proof. Consider the linear system $\mathcal{Q} = \mathcal{Q}_2|_C$. Note that the base locus C_{20} of the pencil $\mathcal{Q}_2$ is disjoint from C by Theorems 8.1.8(vii) and 14.3.2(v). Thus the linear system $\mathcal{Q}$ is a pencil of divisors of degree 6 that has no base points, so that $\mathcal{Q}$ defines a six-to-one cover $t \colon C \to \mathbb{P}^1$. We are going to determine the ramification points of the morphism t.

By Theorem 14.3.2(ii),(iii) and Lemma 14.3.6, we already know eight ramification points of the six-to-one cover t: two points of the $\mathfrak{A}_5$-orbit Σ_{30} coming from the intersection of C with the surface $\mathscr{S}$, where the ramification index equals 3, two points of the $\mathfrak{A}_5$-orbit Σ_{10} coming from the intersection of C with the surface S_{10}, two points of the $\mathfrak{A}_5$-orbit Σ'_{10} coming from the intersection of C with the surface S'_{10}, and two points of the $\mathfrak{A}_5$-orbit Σ_{15} coming from the intersection of C with the surface S_{15}. Therefore, the

Riemann–Hurwitz formula gives

$$-2 = 2g(C) - 2 =$$
$$= 6\big(2g(\mathbb{P}^1) - 2\big) + \sum_{P \in \mathscr{S} \cap C} (e_P - 1) + \sum_{P \in S_{10} \cap C} (e_P - 1) + \sum_{P \in S'_{10} \cap C} (e_P - 1) +$$
$$+ \sum_{P \in S_{15} \cap C} (e_P - 1) + \sum_{P \in C \setminus (\mathscr{S} \cup S_{10} \cup S'_{10} \cup S_{15})} (e_P - 1) \geqslant$$
$$\geqslant -2 + \sum_{P \in C \setminus (\mathscr{S} \cup S_{10} \cup S'_{10} \cup S_{15})} (e_P - 1),$$

where e_P denotes the ramification index of the morphism t at the point P. This implies that there are no other ramification points of t except for those listed above. Now it is straightforward to derive the assertion of the lemma. □

Now we are ready to prove:

Theorem 14.3.8 (cf. Lemma 5.5.14). *Every $\mathfrak{A}_5$-orbit in V_5 of length 30 is contained in $\mathcal{L}_{15} \cup \mathcal{T}_{15}$. Vice versa, the $\mathfrak{A}_5$-orbit of every point in $\mathcal{L}_{15} \cup \mathcal{T}_{15}$ that is not contained in Σ_5, Σ_{10}, Σ'_{10}, and Σ_{15} has length 30.*

Proof. By Lemma 9.1.10(iv) and Theorem 14.3.2(v), the $\mathfrak{A}_5$-orbit of every point in $\mathcal{L}_{15} \cup \mathcal{T}_{15}$ that is not contained in Σ_5, Σ_{10}, Σ'_{10}, and Σ_{15} has length 30.

By Corollary 9.1.11, the intersection of the curve $\mathcal{L}_{15}$ with the surface $\mathscr{S}$ is a single $\mathfrak{A}_5$-orbit of length 30. By Remark 9.1.12, the intersections of $\mathcal{L}_{15}$ with the surfaces S_5 and S_{15} do not contain any $\mathfrak{A}_5$-orbits of length 30, and the intersection of $\mathcal{L}_{15}$ with each of the other surfaces of the pencil $\mathcal{Q}_2$ is a single $\mathfrak{A}_5$-orbit of length 30. On the other hand, by Lemma 14.3.6 the intersection of the curve $\mathcal{T}_{15}$ with the surface $\mathscr{S}$ is a single $\mathfrak{A}_5$-orbit of length 30. By Lemma 14.3.7, the intersection of $\mathcal{T}_{15}$ with each of the surfaces S_{10}, S'_{10}, and S_{15} contains a single $\mathfrak{A}_5$-orbit of length 30, and the intersection of $\mathcal{T}_{15}$ with any other surface of the pencil $\mathcal{Q}_2$ is a union of three $\mathfrak{A}_5$-orbits of length 30. Note also that the curves $\mathcal{T}_{15}$ and $\mathcal{L}_{15}$ do not have any common $\mathfrak{A}_5$-orbits of length 30 by Theorem 14.3.2(iv). Comparing the above numbers with those given in Lemma 14.3.5, we see that all $\mathfrak{A}_5$-orbits of length 30 in V_5 are covered by the union of the curves $\mathcal{T}_{15}$ and $\mathcal{L}_{15}$. □

14.4 Fifteen twisted cubics

In this section we collect information about the intersections of the curve $\mathcal{T}_{15}$ with $\mathfrak{A}_5$-invariant curves of small degree in V_5.

CREMONA GROUPS AND THE ICOSAHEDRON 371

Lemma 14.4.1. The curve $\mathcal{T}_{15}$ is disjoint from the curves $\mathcal{L}_{12}$, C_{16}, C'_{16}, C_{20}, and $\mathcal{L}_{20}$.

Proof. By Theorem 14.3.8, the length of the $\mathfrak{A}_5$-orbit of any point of the curve $\mathcal{T}_{15}$ is either 10, or 15, or 30. None of these orbits is contained in the curves $\mathcal{L}_{12}$ and $\mathcal{L}_{20}$ by Lemma 7.2.12(ii). Also, none of them is contained in the curve C_{20} by Theorem 8.1.8(vii), and none of them is contained in the curve C'_{16} by Lemma 5.1.5.

By Lemmas 5.5.9(v) and 8.1.16(iv) the $\mathfrak{A}_5$-orbit Σ'_{30} is the unique $\mathfrak{A}_5$-orbit of length 30 on the curve C_{16}, and C_{16} does not contain $\mathfrak{A}_5$-orbits of lengths 10 and 15. Thus, it follows from Theorem 14.3.2(viii) that the curves $\mathcal{T}_{15}$ and C_{16} are also disjoint. $\square$

Lemma 14.4.2. One has

$$\mathcal{T}_{15} \cap \mathscr{C} = \Sigma_{30}.$$

Proof. By Lemma 5.1.4 the curve $\mathscr{C}$ does not contain $\mathfrak{A}_5$-orbits of length 10 and 15. By Theorem 14.3.8 this means that the intersection of the curves $\mathcal{T}_{15}$ and $\mathscr{C}$ consists of all $\mathfrak{A}_5$-orbits of length 30 that are contained in $\mathscr{C}$ but not contained in $\mathcal{L}_{15}$. The only $\mathfrak{A}_5$-orbit of length 30 contained in $\mathscr{C}$ is Σ_{30}, and Σ_{30} is not contained in $\mathcal{L}_{15}$ by Corollary 9.1.11. $\square$

Lemma 14.4.3. The curve $\mathcal{T}_{15}$ intersects each of the curves $\mathcal{L}_6$ and $\mathcal{T}_6$ by a single $\mathfrak{A}_5$-orbit of length 30 disjoint from the surface $\mathscr{S}$. The intersection of the curves $\mathcal{T}_{15}$ and $\mathcal{G}_6$ is a union of two $\mathfrak{A}_5$-orbits of length 30 disjoint from the surface $\mathscr{S}$.

Proof. Let Z be one of the curves $\mathcal{L}_6$, $\mathcal{G}_6$, or $\mathcal{T}_6$, and C be an irreducible component of Z. Let G_C be the stabilizer of C in $\mathfrak{A}_5$. Then $G_C \cong \mathfrak{D}_{10}$ by Lemma 5.1.1, and G_C acts on C faithfully by Lemma 7.4.5. By Lemma 5.5.9(ii) this means that there are two G_C-orbits of length 5 on C. By Lemmas 7.4.7, 9.2.8, and 13.2.4 we conclude that there are two $\mathfrak{A}_5$-orbits of length 30 contained in Z, say, Σ and Σ'. Note that both Σ and Σ' are disjoint from the surface $\mathscr{S}$ by Lemmas 7.4.3 and 13.2.3 and Corollary 9.2.11.

By Corollary 8.3.4 we know that the curve Z does not contain $\mathfrak{A}_5$-orbits Σ_{10}, Σ'_{10}, and Σ_{15}. By Theorem 14.3.8 this means that the intersection of the curves $\mathcal{T}_{15}$ and Z consists of all $\mathfrak{A}_5$-orbits of length 30 that are contained in Z but not contained in $\mathcal{L}_{15}$. We know that the curves $\mathcal{G}_6$ and $\mathcal{L}_{15}$ are disjoint by Corollary 13.6.8, so that the intersection of $\mathcal{G}_6$ and $\mathcal{T}_{15}$ consists of both $\mathfrak{A}_5$-orbits of length 30 contained in $\mathcal{G}_6$. On the other hand, if Z is one of the curves $\mathcal{L}_6$ or $\mathcal{T}_6$, then we know from Lemmas 9.1.14 and 13.2.7 that Z intersects $\mathcal{L}_{15}$ by one $\mathfrak{A}_5$-orbit of length 30, so that the intersection of Z and $\mathcal{T}_{15}$ consists of a single $\mathfrak{A}_5$-orbit of length 30. □

Lemma 14.4.4. The curve $\mathcal{T}_{15}$ intersects each of the curves $\mathscr{C}'$, $\mathcal{L}_{10}$, $\mathcal{G}_5$, $\mathcal{G}'_5$, C^0_{18}, and $\mathcal{B}_{18}$ by a single $\mathfrak{A}_5$-orbit of length 30 disjoint from the surface $\mathscr{S}$.

Proof. Let Z be one of the curves $\mathscr{C}'$, $\mathcal{L}_{10}$, $\mathcal{G}_5$, $\mathcal{G}'_5$, C^0_{18}, or $\mathcal{B}_{18}$. Using Lemmas 5.1.4 and 9.2.1 together with Corollaries 9.1.3, 9.1.5, 9.2.4, and 13.4.3(i), we conclude that the curve Z does not contain $\mathfrak{A}_5$-orbits Σ_{10}, Σ'_{10}, and Σ_{15}. By Theorem 14.3.8 this means that the intersection of the curves $\mathcal{T}_{15}$ and Z consists of all $\mathfrak{A}_5$-orbits of length 30 that are contained in Z but not contained in $\mathcal{L}_{15}$.

Suppose that Z is one of the curves $\mathscr{C}'$, C^0_{18}, or $\mathcal{B}_{18}$. Then Z contains a unique $\mathfrak{A}_5$-orbit of length 30. This follows from Lemma 5.5.9(v) if $Z = \mathscr{C}'$, from Lemmas 5.5.9(v) and 13.4.2 if $Z = C^0_{18}$, and from Lemma 5.1.5 if $Z = \mathcal{B}_{18}$. Note that the latter $\mathfrak{A}_5$-orbit is disjoint from the surface $\mathscr{S}$ by Lemma 11.3.4, Corollary 13.4.3, and Remark 13.5.3. Also, we have $Z \cap \mathcal{L}_{15} = \varnothing$ by Lemma 11.3.7 and Corollaries 13.6.8 and 13.6.7, which proves the assertion for these cases.

Suppose that Z is one of the curves $\mathcal{G}_5$ or $\mathcal{G}'_5$. Let C be an irreducible component of Z, and G_C be the stabilizer of C in $\mathfrak{A}_5$. Then $G_C \cong \mathfrak{A}_4$ by Lemma 5.1.1, and G_C acts on C faithfully by Lemma 7.4.5. By Lemma 5.5.9(iii) this means that there is one G_C-orbit of length 6 on C. By Lemma 9.2.3 we conclude that there is one $\mathfrak{A}_5$-orbit of length 30 contained in Z. Note that this $\mathfrak{A}_5$-orbit is disjoint from the surface $\mathscr{S}$ by Remark 9.2.5. We have $Z \cap \mathcal{L}_{15} = \varnothing$ by Lemma 9.2.7, which proves the assertion for these cases.

Finally, consider the case $Z = \mathcal{L}_{10}$. Let L be an irreducible component of $\mathcal{L}_{10}$, and G_L be the stabilizer of L in $\mathfrak{A}_5$. Then $G_L \cong \mathfrak{S}_3$ by Lemma 5.1.1, and G_L acts on L faithfully by Lemma 7.4.6. Therefore, by Lemma 5.5.9(ii) there are two G_L-orbits of length 3 on L. By Lemma 9.1.2 this means that there are two $\mathfrak{A}_5$-orbits of length 30 contained in $\mathcal{L}_{10}$. Note that these $\mathfrak{A}_5$-orbits are disjoint from the surface $\mathscr{S}$ by Remark 9.1.6. By Lemma 9.1.15 the intersection of the curves $\mathcal{L}_{15}$ and $\mathcal{L}_{10}$ consists of a single $\mathfrak{A}_5$-orbit of length 30, which proves the assertion for $\mathcal{L}_{10}$ and completes the proof of the lemma. □

Lemma 14.4.5. One has

$$\mathcal{T}_{15} \cap \mathcal{G}_{10} = \Sigma_{10} \cup \Sigma'_{10}.$$

Proof. This follows from Theorems 14.1.16(vii),(ix), 14.3.2(ii), and 14.3.8. □

Chapter 15

Further properties of the invariant cubic

This chapter studies the geometry of the $\mathfrak{A}_5$-invariant cubic hypersurface S_{Cl} in more detail than was previously done in Chapter 12. In particular, we list all $\mathfrak{A}_5$-invariant curves of degree less than 20 contained in this surface, describe its singularities, prove that it is a surface of general type, find all lines contained in S_{Cl}, and bound its Picard rank.

15.1 Intersections with low degree curves

As an easy application of Lemma 13.6.3 we can describe all low degree $\mathfrak{A}_5$-invariant curves contained in the surface S_{Cl}.

Corollary 15.1.1. *The only $\mathfrak{A}_5$-irreducible curves of degree less than 20 that are contained in S_{Cl} are the curves $\mathscr{C}$, $\mathscr{C}'$, $\mathcal{L}_6$, $\mathcal{L}_{12}$, $\mathcal{G}_6$, and $\mathcal{B}_{18}$.*

Proof. By Lemma 12.2.5(i),(ii), the curves $\mathscr{C}$, $\mathscr{C}'$, $\mathcal{L}_6$, and $\mathcal{L}_{12}$ are contained in S_{Cl}. Moreover, the curve $\mathcal{G}_6$ is contained in S_{Cl} by Lemma 12.3.4. By the construction of the curve $\mathcal{B}_{18}$ in Lemma 12.3.6, it is also contained in S_{Cl}.

By Corollary 12.2.7, the surface S_{Cl} does not contain the curves $\mathcal{L}_{10}$, $\mathcal{G}_5$, $\mathcal{G}'_5$, and C_{16}. By Lemma 12.2.3, the curve $\mathcal{L}_{15}$ is not contained in S_{Cl}. As we know from Remark 13.1.4, the curve C'_{16} is also not contained in S_{Cl}.

Let C be an $\mathfrak{A}_5$-irreducible curve of degree less than 20 contained in the surface S_{Cl} that is different from $\mathscr{C}$, $\mathscr{C}'$, $\mathcal{L}_6$, $\mathcal{L}_{12}$, and $\mathcal{G}_6$. It follows from Theorem 13.6.1 that $\deg(C) = 18$. We know from Lemma 13.6.3 that C is contained in some surface $S \in \mathcal{Q}_2$. Thus the restriction $S_{Cl}|_S$ contains an

$\mathfrak{A}_5$-irreducible curve of degree at most
$$H \cdot S_{Cl}|_S - H \cdot C = 30 - 18 = 12.$$

Applying Lemma 13.6.3 once again we see that $C = \mathcal{B}_{18}$. □

Note that one can use Corollary 12.4.12 to simplify the proof of Corollary 15.1.1.

Corollary 15.1.2. One has
$$S_{Cl} \cap \mathcal{T}_6 = \Sigma_{12} \cup \Sigma,$$
where Σ is an $\mathfrak{A}_5$-orbit of length 30 different from Σ_{30} and Σ'_{30}.

Proof. By Corollary 15.1.1 we know that the curve $\mathcal{T}_6$ is not contained in the surface S_{Cl}. Thus the intersection $S_{Cl} \cap \mathcal{T}_6$ consists of $S_{Cl} \cdot \mathcal{T}_6 = 54$ points (counted with multiplicities). Now Theorem 7.3.5 together with Lemmas 13.2.3 and 13.2.4 implies that
$$S_{Cl} \cap \mathcal{T}_6 = \Sigma_{12} \cup \Sigma,$$
where Σ is some $\mathfrak{A}_5$-orbit of length 30. It remains to notice that Σ is not contained in $\mathscr{S}$, because $\mathcal{T}_6 \cap \mathscr{S} = \Sigma_{12}$ by Lemma 13.2.3. □

Corollary 15.1.3. Let Z be an $\mathfrak{A}_5$-invariant rational curve in V_5 of degree 18. Then
$$S_{Cl} \cap Z = \Sigma \cup \Sigma',$$
where Σ is an $\mathfrak{A}_5$-orbit of length 12, and Σ' is an $\mathfrak{A}_5$-orbit of length 30 different from Σ_{30} and Σ'_{30}. Moreover, if $Z = C_{18}^0$, then $\Sigma = \Sigma'_{12}$. Furthermore, if Z is smooth, then $\Sigma = \Sigma_{12}$.

Proof. The proof is similar to that of Corollary 15.1.2. Recall from Theorem 13.6.1 that either Z is smooth or $Z = C_{18}^0$. In both cases, the curve Z is not contained in the surface S_{Cl} by Corollary 15.1.1. Thus the intersection $S_{Cl} \cap C_{18}^0$ consists of
$$S_{Cl} \cdot C_{18}^0 = 54$$
points (counted with multiplicities). Now Lemma 5.1.4 and Corollary 13.4.3(i),(iii) imply that
$$S_{Cl} \cap Z = \Sigma \cup \Sigma',$$
where Σ is an $\mathfrak{A}_5$-orbit of length 12, and Σ' is an $\mathfrak{A}_5$-orbit of length 30. By Corollaries 13.4.3(ii) and 13.6.5, we conclude that Σ' is different from Σ_{30} and Σ'_{30}. Moreover, if $Z = C_{18}^0$, then $\Sigma = \Sigma'_{12}$ by Corollary 13.4.3(ii). Finally, if Z is smooth, then $\Sigma = \Sigma_{12}$ by Lemma 13.6.4. □

Corollary 15.1.4. Assume that C is an $\mathfrak{A}_5$-invariant smooth rational curve of degree 18 in V_5 (cf. Remark 13.6.2). Then C is disjoint from the curve $\mathcal{L}_{15}$, and $C \cap \mathcal{T}_{15} = \Sigma$, where Σ is an $\mathfrak{A}_5$-orbit of length 30 different from Σ_{30} and Σ'_{30}.

Proof. The curve C contains a unique $\mathfrak{A}_5$-orbit Σ of length 30, and does not contain $\mathfrak{A}_5$-orbits of length 10 and 15, see Lemma 5.5.9(v). Corollary 13.6.5 implies that Σ is different from Σ_{30} and Σ'_{30}.

Using Corollary 15.1.3 we conclude that Σ is contained in the surface S_{Cl}. By Corollary 13.6.5 we know that the $\mathfrak{A}_5$-orbits Σ'_{12} and Σ'_{20} are not contained in C, while the $\mathfrak{A}_5$-orbit Σ_{12} is not contained in $\mathcal{L}_6$ by Lemma 7.4.3. Applying Lemma 13.6.3 and Corollary 8.1.12, we see that the curves C and $\mathcal{L}_6$ are disjoint. Therefore, Lemma 12.2.12 implies that C is disjoint from the curve $\mathcal{L}_{15}$. Now it follows from Theorem 14.3.8 that $C \cap \mathcal{T}_{15} = \Sigma$. $\square$

15.2 Singularities of the invariant cubic

In this section, we describe the singularities of the surface S_{Cl} (see Theorem 15.2.11). Namely, we will prove that S_{Cl} is smooth away from Σ_{12}, and every point of Σ_{12} is an ordinary double point of the surface S_{Cl}. As an application, we deduce that S_{Cl} is a surface of general type (and in particular it is not rational).

The proof of Theorem 15.2.11 consists of five steps. First we prove that S_{Cl} has isolated singularities and its singular locus is a union of Σ_{12} and possibly some $\mathfrak{A}_5$-orbits of length 30 (see Lemmas 15.2.1 and 15.2.2). Then we show that each point in Σ_{12} is an isolated ordinary double point of S_{Cl} (see Lemma 15.2.3). After this, we prove that S_{Cl} has at most du Val singularities (see Lemma 15.2.4). Then we compute basic holomorphic and topological invariants of the minimal resolution of singularities of the surface S_{Cl} (see Theorem 15.2.6 and Corollary 15.2.8). Finally, we use basic properties of the pencil $\mathcal{Q}_2|_{S_{Cl}}$ to prove that the surface S_{Cl} is smooth away from the $\mathfrak{A}_5$-orbit Σ_{12}.

Let us start with

Lemma 15.2.1. Let P be a singular point of S_{Cl}. Denote by Σ its $\mathfrak{A}_5$-orbit. Then either $\Sigma = \Sigma_{12}$, or $|\Sigma| = 30$ and

$$\Sigma \not\subset \mathscr{C} \cup \mathscr{C}' \cup \mathcal{L}_6 \cup \mathcal{L}_{12} \cup \mathcal{G}_6 \cup \mathcal{B}_{18}.$$

Proof. Suppose that $\Sigma \ne \Sigma_{12}$. Note that $\Sigma \ne \Sigma_5$ by Lemma 12.2.3. Also, one has $\Sigma \ne \Sigma'_{12}$ by Lemma 12.2.5(iv), and Σ is none of the $\mathfrak{A}_5$-orbits Σ_{10}, Σ'_{10}, and Σ_{15} by Corollary 12.4.11. Hence, we have $|\Sigma| \geqslant 20$ by Theorem 7.3.5.

Let S be a surface in $\mathcal{Q}_2$ that contains Σ. We know that Σ is not contained in the curve $\mathscr{C}$ by Lemma 12.2.14, and Σ is not contained in the curve $\mathcal{L}_{12}$ by Lemma 12.2.5(iv). Thus $S \ne \mathscr{S}$ by Lemma 12.2.5(v). Hence, it follows from Lemmas 12.3.6, 12.4.1, and 12.4.2 together with Corollary 12.4.12 that $|\Sigma| = 30$ and S contains none of the curves $\mathscr{C}'$, $\mathcal{L}_6$, $\mathcal{L}_{12}$, $\mathcal{G}_6$, and $\mathcal{B}_{18}$. In particular, Σ is not contained in any of the latter curves. □

Lemma 15.2.2. *The surface S_{Cl} has isolated singularities.*

Proof. Suppose that the singular locus of the surface S_{Cl} contains an irreducible curve C. Denote by Z its $\mathfrak{A}_5$-orbit. Then Z is also contained in the singular locus of the surface S_{Cl}.

Let M be a general hyperplane section of V_5. Put $H_M = H|_M$. Then $S_{Cl}|_M$ is an irreducible curve contained in the linear system $|3H_M|$. Furthermore, the arithmetic genus of the curve $S_{Cl}|_M$ is

$$p_a\big(S_{Cl}|_M\big) = \frac{3H_M \cdot 2H_M + 2}{2} = 16$$

by the adjunction formula. On the other hand, the curve $S_{Cl}|_M$ has at least $\deg(Z)$ singular points, because $S_{Cl}|_M$ is singular at every point of the intersection $Z \cap M$. Thus, one has $\deg(Z) \leqslant 16$ by Lemma 4.4.6.

Since $\deg(Z) \leqslant 16$, the curve Z is one of the curves $\mathscr{C}$, $\mathscr{C}'$, $\mathcal{L}_6$, $\mathcal{L}_{12}$, or $\mathcal{G}_6$ by Corollary 15.1.1. The latter is impossible, since S_{Cl} is smooth at a general point of each of these curves by Lemma 15.2.1. □

Recall that the surface S_{Cl} is indeed singular: we already know that its singularities include the points of the $\mathfrak{A}_5$-orbit Σ_{12} by Lemma 12.2.14.

Lemma 15.2.3. *Each point in Σ_{12} is an isolated ordinary double point of S_{Cl}.*

Proof. Recall that V_5 contains the $\mathfrak{A}_5$-invariant curve $\mathcal{T}_6$ that is a union of six rational cubic curves (see §13.2). By Corollary 15.1.2, one has

$$S_{Cl} \cap \mathcal{T}_6 = \Sigma_{12} \cup \Omega_{30},$$

where Ω_{30} is an $\mathfrak{A}_5$-orbit of length 30 disjoint from $\mathscr{S}$. By Lemma 13.2.3 one has

$$\mathscr{S} \cap \mathcal{T}_6 = \mathscr{C} \cap \mathcal{T}_6 = \Sigma_{12}.$$

Let us use notation of §7.2. Denote by $\hat{S}_{Cl}$ and $\hat{T}_6$ the proper transforms of the surface S_{Cl} and the curve T_6 on the threefold $\mathcal{Y}$, respectively. By Lemma 12.2.14 the surface $\hat{S}_{Cl}$ is smooth at the points of the intersection $\hat{S}_{Cl} \cap E_{\mathcal{Y}}$, except possibly the points of the unique $\mathfrak{A}_5$-orbit $\hat{\Sigma}_{12}$ of length 12 contained in the curve $\hat{\mathscr{C}}$. Also, we have $\hat{T}_6 \cdot E_{\mathcal{Y}} = 12m$ for some positive integer m. Thus it follows from Lemma 7.2.3(i) that

$$0 \leqslant \hat{T}_6 \cdot \hat{\mathscr{S}} = \hat{T}_6 \cdot \left(\nu^*(2H) - 2E_{\mathcal{Y}}\right) = 36 - 2E_{\mathcal{Y}} \cdot \hat{T}_6 = 36 - 24m,$$

which implies that $m = 1$ and

$$\hat{T}_6 \cdot \hat{\mathscr{S}} = E_{\mathcal{Y}} \cdot \hat{T}_6 = 12.$$

Since $\hat{\Sigma}_{12}$ is the unique $\mathfrak{A}_5$-orbit of length 12 in $\hat{\mathscr{S}} \cap E_{\mathcal{Y}} = \hat{\mathscr{C}}$, we obtain

$$\hat{T}_6 \cap E_{\mathcal{Y}} = \hat{T}_6 \cap \hat{\mathscr{S}} = \hat{\Sigma}_{12}.$$

In particular,

$$\hat{\Sigma}_{12} \subset \hat{S}_{Cl} \cap \hat{T}_6.$$

Denote by $\hat{\Omega}_{30}$ the $\mathfrak{A}_5$-orbit in $\mathcal{Y}$ that is mapped to Ω_{30} by ν. Then

$$\hat{\Omega}_{30} \subset \hat{S}_{Cl} \cap \hat{T}_6$$

by construction. Since S_{Cl} is smooth at every point of $\mathscr{C} \setminus \Sigma_{12}$ (see Lemma 12.2.14), we have

$$\hat{S}_{Cl} \sim \nu^*(3H) - E_{\mathcal{Y}}.$$

Therefore

$$42 = 54 - E_{\mathcal{Y}} \cdot \hat{T}_6 = \hat{T}_6 \cdot \left(\nu^*(3H) - E_{\mathcal{Y}}\right) =$$
$$= \hat{T}_6 \cdot \hat{S}_{Cl} \geqslant \sum_{P \in \hat{\Omega}_{30}} \mathrm{mult}_P(\hat{S}_{Cl}) + \sum_{P \in \hat{\Sigma}_{12}} \mathrm{mult}_P(\hat{S}_{Cl}) \geqslant$$
$$\geqslant |\hat{\Omega}_{30}| + |\hat{\Sigma}_{12}| = 42,$$

which implies that $\hat{S}_{Cl}$ is smooth at the points of $\hat{\Sigma}_{12}$. Hence, the surface $\hat{S}_{Cl}$ is smooth at every point that is mapped to a point of Σ_{12} by ν.

The birational morphism

$$\nu|_{\hat{S}_{Cl}} \colon \hat{S}_{Cl} \to S_{Cl}$$

gives a minimal resolution of singularities in a neighborhood of the $\mathfrak{A}_5$-orbit $\Sigma_{12} \subset \mathrm{Sing}(S_{Cl})$. By the adjunction formula one has

$$K_{\hat{S}_{Cl}} \sim (K_\mathcal{Y} + \hat{S}_{Cl})|_{\hat{S}_{Cl}} \sim (-\nu^*(2H) + E_\mathcal{Y} + \nu^*(3H) - E_\mathcal{Y})|_{\hat{S}_{Cl}} \sim \nu^*(H)|_{\hat{S}_{Cl}},$$

which implies that the birational morphism $\nu|_{\hat{S}_{Cl}}$ is crepant. Since the morphism ν contracts a single smooth rational curve into each point of the curve $\mathscr{C}$, and in particular into each point of the $\mathfrak{A}_5$-orbit Σ_{12}, we conclude that the surface S_{Cl} has isolated ordinary double points at the points of Σ_{12}. $\square$

Now we are ready to prove:

Lemma 15.2.4. *The surface S_{Cl} has at most du Val singularities.*

Proof. Suppose that the singularities of S_{Cl} are not du Val. Let μ be the largest (positive rational) number such that the pair $(V_5, \mu S_{Cl})$ is log canonical. Note that $\mu \leqslant 1$. Let P be a minimal center in $\mathrm{LCS}(V_5, \mu S_{Cl})$ (see Definition 2.4.7). Then $P \neq S_{Cl}$ by Theorem 2.4.8, because the singularities of S_{Cl} are worse than canonical. Hence, P is either a curve or a point. Moreover, P cannot be a curve by Lemma 2.4.3, because S_{Cl} has isolated singularities by Lemma 15.2.2. Thus, we see that P is a point.

Denote by Σ the $\mathfrak{A}_5$-orbit of the point P. Then $\Sigma \neq \Sigma_{12}$ by Lemma 15.2.3, so that $|\Sigma| = 30$ by Lemma 15.2.1.

Choose any $\epsilon \in \mathbb{Q}$ such that $\frac{1}{3} > \epsilon > 0$. By Lemma 2.4.10, we can find an $\mathfrak{A}_5$-invariant effective $\mathbb{Q}$-divisor D such that

$$D \sim_\mathbb{Q} (1 + \epsilon) S_{Cl},$$

the log pair (V_5, D) is log canonical, and every center in $\mathrm{LCS}(V_5, D)$ is a point of Σ. Let $\mathcal{I}_\Sigma \subset \mathcal{O}_{V_5}$ be the ideal sheaf of the $\mathfrak{A}_5$-orbit Σ. Then $\mathcal{I}_\Sigma$ is the multiplier ideal sheaf of the log pair (V_5, D) by Remark 2.3.4. Since

$$2H \sim_\mathbb{Q} \left(K_{V_5} + D\right) + (1 - 3\epsilon)H,$$

it follows from Theorem 2.3.5 that

$$h^1\left(\mathcal{I}_\Sigma \otimes \mathcal{O}_{V_5}(2H)\right) = 0.$$

Thus, there is an exact sequence of $\mathfrak{A}_5$-representations

$$0 \to H^0\left(\mathcal{I}_\Sigma \otimes \mathcal{O}_{V_5}(2H)\right) \to H^0\left(\mathcal{O}_{V_5}(2H)\right) \to H^0\left(\mathcal{O}_\Sigma \otimes \mathcal{O}_{V_5}(2H)\right) \to 0.$$

Note that $h^0(\mathcal{O}_{V_5}(2H)) = 23$ by (7.1.3). This implies
$$30 = |\Sigma| = h^0\!\left(\mathcal{O}_\Sigma\right) = h^0\!\left(\mathcal{O}_\Sigma \otimes \mathcal{O}_{V_5}(2H)\right) \leqslant h^0\!\left(\mathcal{O}_{V_5}(2H)\right) = 23.$$
The obtained contradiction completes the proof. $\square$

Let $h\colon \hat{S}_{Cl} \to S_{Cl}$ be the minimal resolution of singularities of the surface S_{Cl}. Note that we do not claim now that h is just a blow-up of the points of Σ_{12}; this is actually the case, but we will be able to prove it only in Theorem 15.2.11 below. On the other hand, since the singularities of S_{Cl} are du Val by Lemma 15.2.4, we conclude that
$$K_{\hat{S}_{Cl}} \sim h^*(K_{S_{Cl}}) \sim h^*\!\left(H|_{S_{Cl}}\right).$$
In particular, we obtain:

Corollary 15.2.5. *The surface S_{Cl} is a surface of general type. In particular, it is not rational.*

We are able to compute basic invariants of the surface $\hat{S}_{Cl}$.

Theorem 15.2.6. *One has $K^2_{\hat{S}_{Cl}} = 15$ and*
$$h^2\!\left(\mathcal{O}_{\hat{S}_{Cl}}\right) = h^0\!\left(\mathcal{O}_{\hat{S}_{Cl}}(K_{\hat{S}_{Cl}})\right) = 7,$$
while $h^1(\mathcal{O}_{\hat{S}_{Cl}}) = 0$.

Proof. Recall that $K_{S_{Cl}} = H|_{S_{Cl}}$ by the adjunction formula. Thus, one has
$$K^2_{\hat{S}_{Cl}} = K^2_{S_{Cl}} = 3H^3 = 15.$$
Consider an exact sequence of sheaves
$$0 \to \mathcal{O}_{V_5}((n-3)H) \to \mathcal{O}_{V_5}(nH) \to \mathcal{O}_{S_{Cl}}(nH|_{S_{Cl}}) \to 0$$
for every integer n. It gives an exact sequence of cohomology groups
$$0 \to H^0\!\left(\mathcal{O}_{V_5}((n-3)H)\right) \to H^0\!\left(\mathcal{O}_{V_5}(nH)\right) \to$$
$$\to H^0\!\left(\mathcal{O}_{S_{Cl}}(nH|_{S_{Cl}})\right) \to H^1\!\left(\mathcal{O}_{V_5}((n-3)H)\right) \to$$
$$\to H^1\!\left(\mathcal{O}_{V_5}(nH)\right) \to H^1\!\left(\mathcal{O}_{S_{Cl}}(nH|_{S_{Cl}})\right) \to$$
$$\to H^2\!\left(\mathcal{O}_{V_5}((n-3)H)\right). \quad (15.2.7)$$

For $n = 0$, it follows from the exact sequence (15.2.7) and Kodaira vanishing that $h^1(\mathcal{O}_{S_{Cl}}) = 0$. Thus, $h^1(\mathcal{O}_{\hat{S}_{Cl}}) = 0$ as well, since the singularities of the surface S_{Cl} are rational.

For $n = 1$, it follows from (15.2.7) and Kodaira vanishing that

$$h^0(\mathcal{O}_{S_{Cl}}(K_{S_{Cl}})) = 7.$$

Hence, we have

$$h^0(\mathcal{O}_{\hat{S}_{Cl}}(K_{\hat{S}_{Cl}})) = h^0(\mathcal{O}_{S_{Cl}}(K_{S_{Cl}})) = 7,$$

because S_{Cl} has rational singularities and $K_{\hat{S}_{Cl}} \sim h^*(K_{S_{Cl}})$. By Serre duality, we conclude that $h^2(\mathcal{O}_{\hat{S}_{Cl}}) = 7$. □

Corollary 15.2.8. One has $\chi_{top}(\hat{S}_{Cl}) = 81$.

Proof. It follows from Theorem 15.2.6 and Noether's formula that

$$15 + \chi_{top}(\hat{S}_{Cl}) = K^2_{\hat{S}_{Cl}} + \chi_{top}(\hat{S}_{Cl}) = 12\chi(\mathcal{O}_{\hat{S}_{Cl}}) = 96,$$

which implies the assertion. □

Corollary 15.2.9. One has $\operatorname{rk}\operatorname{Pic}(\hat{S}_{Cl}) \leqslant 65$.

Proof. By Theorem 15.2.6 one has

$$h^1\left(\hat{S}_{Cl}, \mathbb{C}\right) = h^3\left(\hat{S}_{Cl}, \mathbb{C}\right) = 0.$$

Thus, Corollary 15.2.8 and Theorem 15.2.6 imply that

$$h^1\left(\Omega^1_{\hat{S}_{Cl}}\right) = h^2\left(\hat{S}_{Cl}, \mathbb{C}\right) - h^0\left(\mathcal{O}_{\hat{S}_{Cl}}(K_{\hat{S}_{Cl}})\right) - h^2\left(\mathcal{O}_{\hat{S}_{Cl}}\right) =$$
$$= \chi_{top}(\hat{S}_{Cl}) - 2 - 2h^0\left(\mathcal{O}_{\hat{S}_{Cl}}(K_{\hat{S}_{Cl}})\right) = 79 - 2h^0\left(\mathcal{O}_{\hat{S}_{Cl}}(K_{\hat{S}_{Cl}})\right) = 65,$$

which implies that $\operatorname{rk}\operatorname{Pic}(\hat{S}_{Cl}) \leqslant 65$. □

Corollary 15.2.10. All singular points of the surface S_{Cl} are ordinary double points. Moreover, the singular locus of S_{Cl} is either Σ_{12} or a union of Σ_{12} and an $\mathfrak{A}_5$-orbit of length 30 that is not contained in the curves $\mathscr{C}$, $\mathscr{C}'$, $\mathcal{L}_6$, $\mathcal{L}_{12}$, $\mathcal{G}_6$, and $\mathcal{B}_{18}$.

Proof. Apply Lemmas 15.2.1 and 15.2.4 together with Corollary 15.2.9. □

Now we are ready to give a complete description of singularities of the surface S_{Cl}.

Theorem 15.2.11. *The surface S_{Cl} is smooth away from the $\mathfrak{A}_5$-orbit Σ_{12}. Each point in Σ_{12} is an isolated ordinary double point of S_{Cl}.*

Proof. By Lemma 15.2.3, it is enough to prove that S_{Cl} is smooth away from the $\mathfrak{A}_5$-orbit Σ_{12}. To do this, we consider a pencil $\mathcal{P} = \mathcal{Q}_2|_{S_{Cl}}$. Its base locus is $\Sigma_{12} \cup \Sigma'_{12}$ by Corollary 12.2.11. Thus, the pencil $\mathcal{P}$ gives a rational map
$$\phi_{\mathcal{P}} \colon S_{Cl} \dashrightarrow \mathbb{P}^1$$
that is undefined only in the points of $\Sigma_{12} \cup \Sigma'_{12}$. Let us show how to resolve the indeterminacy of the map $\phi_{\mathcal{P}}$.

Recall that $K_{S_{Cl}} \sim H|_{S_{Cl}}$. Denote by $\hat{\mathcal{P}}$ the proper transform of the pencil $\mathcal{P}$ on the surface $\hat{S}_{Cl}$, and denote by $\hat{\mathscr{C}}$, $\hat{\mathcal{L}}_{12}$, $\hat{\mathcal{G}}_6$, and $\hat{\mathcal{B}}_{18}$ the proper transforms on the surface $\hat{S}_{Cl}$ of the curves $\mathscr{C}$, $\mathcal{L}_{12}$, $\mathcal{G}_6$, and $\mathcal{B}_{18}$, respectively. Denote by $\mathcal{E}_{12}$ the $\mathfrak{A}_5$-irreducible h-exceptional curve that is mapped to Σ_{12} by h. Recall from Lemma 12.2.5(v) that $\mathcal{L}_{12} + 3\mathscr{C} \sim 2K_{S_{Cl}}$, and recall from Lemma 12.3.6 that $\mathcal{B}_{18} + \mathcal{G}_6 \sim 2K_{S_{Cl}}$. Note that the multiplicity of the curve $\mathcal{B}_{18} + \mathcal{G}_6$ at the points of Σ_{12} equals 2, and the multiplicity of the divisor $3\mathscr{C} + \mathcal{L}_{12}$ at the points of Σ_{12} equals 4. Since each point in Σ_{12} is an isolated ordinary double point of S_{Cl}, we compute that
$$3\hat{\mathscr{C}} + \hat{\mathcal{L}}_{12} + \mathcal{E}_{12} \sim \hat{\mathcal{B}}_{18} + \hat{\mathcal{G}}_6 \sim h^*(2K_{S_{Cl}}) - \mathcal{E}_{12}.$$
Thus, the pencil $\hat{\mathcal{P}}$ is generated by the divisors $3\hat{\mathscr{C}} + \hat{\mathcal{L}}_{12} + \mathcal{E}_{12}$ and $\hat{\mathcal{B}}_{18} + \hat{\mathcal{G}}_6$, because $\mathcal{P}$ is generated by the divisors $\mathcal{B}_{18} + \mathcal{G}_6$ and $3\mathscr{C} + \mathcal{L}_{12}$.

Denote by $\hat{\Sigma}'_{12}$ the $\mathfrak{A}_5$-orbit of length 12 on the surface $\hat{S}_{Cl}$ that is mapped to Σ'_{12} by h. Put $\hat{\Sigma}_{12} = \mathcal{E}_{12} \cap \hat{\mathcal{B}}_{18}$. We claim that $\hat{\Sigma}_{12} \cup \hat{\Sigma}'_{12}$ is the base locus of the pencil $\hat{\mathcal{P}}$. Indeed, $\hat{\Sigma}'_{12}$ is contained in the base locus of the pencil $\hat{\mathcal{P}}$ by construction, and $\hat{\Sigma}_{12}$ is contained in the base locus of the pencil $\hat{\mathcal{P}}$, because $\hat{\mathcal{P}}$ is generated by the divisors $\hat{\mathcal{B}}_{18} + \hat{\mathcal{G}}_6$ and $3\hat{\mathscr{C}} + \hat{\mathcal{L}}_{12} + \mathcal{E}_{12}$ whose supports contain $\hat{\Sigma}_{12}$. Moreover, since $\Sigma_{12} \cup \Sigma'_{12}$ is the base locus of the pencil $\mathcal{P}$, we see that the base locus of the pencil $\hat{\mathcal{P}}$ consists of $\hat{\Sigma}'_{12}$ and some finite subset of $\mathcal{E}_{12}$. It follows from the adjunction formula that
$$6 = 2p_a(\mathcal{B}_{18}) - 2 = \left(K_{\hat{S}_{Cl}} + \hat{\mathcal{B}}_{18}\right) \cdot \hat{\mathcal{B}}_{18} =$$
$$= \left(h^*(H) + \hat{\mathcal{B}}_{18}\right) \cdot \hat{\mathcal{B}}_{18} = H \cdot \mathcal{B}_{18} + \hat{\mathcal{B}}_{18}^2 = 18 + \hat{\mathcal{B}}_{18}^2.$$

Hence, we compute

$$\hat{\mathcal{B}}_{18} \cdot \hat{\mathcal{G}}_6 = \hat{\mathcal{B}}_{18} \cdot \left(h^*(2K_{S_{Cl}}) - \mathcal{E}_{12} - \hat{\mathcal{B}}_{18} \right) =$$
$$= 36 - \hat{\mathcal{B}}_{18} \cdot \mathcal{E}_{12} - \hat{\mathcal{B}}_{18}^2 = 24 - \hat{\mathcal{B}}_{18}^2 = 12.$$

Therefore, we have
$$\hat{\mathcal{B}}_{18} \cap \hat{\mathcal{G}}_6 = \hat{\Sigma}_{12},$$
because $\mathcal{B}_{18} \cap \mathcal{G}_6 = \Sigma_{12}$ by Lemma 12.3.6, and the curves $\mathcal{B}_{18}$ and $\mathcal{G}_6$ are tangent at the points of Σ_{12}. This shows that
$$\hat{\mathcal{G}}_6 \cap \mathcal{E}_{12} = \hat{\Sigma}_{12}.$$

Thus, $\hat{\Sigma}_{12}$ is the unique $\mathfrak{A}_5$-orbit contained both in $\mathcal{E}_{12}$ and in the curve $\hat{\mathcal{B}}_{18} + \hat{\mathcal{G}}_6$. Therefore, $\hat{\Sigma}_{12}$ is exactly the intersection of $\mathcal{E}_{12}$ with the base locus of the pencil $\hat{\mathcal{P}}$, which implies that $\hat{\Sigma}_{12} \cup \hat{\Sigma}'_{12}$ is the base locus of the pencil $\hat{\mathcal{P}}$.

Recall from Lemma 12.3.6 that there is a smooth surface $S_{\mathcal{G}_6}$ in $\mathcal{Q}_2$ that contains both curves $\mathcal{G}_6$ and $\mathcal{B}_{18}$. By Corollary 8.1.13, the curve $\mathscr{C}$ intersects this surface transversally at the points of Σ_{12}. Thus, $\mathscr{C}$ also intersects the curves $\mathcal{G}_6$ and $\mathcal{B}_{18}$ transversally at these points (cf. Corollary 9.2.11 and Remark 13.5.3). On the other hand, $\mathscr{C}$ is tangent to the curve $\mathcal{L}_{12}$ at the points of Σ_{12} by Lemma 7.2.12(iv),(v), so that $\mathcal{L}_{12}$ is also transversal to the curves $\mathcal{G}_6$ and $\mathcal{B}_{18}$ at the points of Σ_{12}. This implies that
$$\hat{\Sigma}_{12} \not\subset \hat{\mathscr{C}} \cup \hat{\mathcal{L}}_{12}.$$

Let $g \colon \check{S}_{Cl} \to \hat{S}_{Cl}$ be the blow-up of the points of $\hat{\Sigma}_{12} \cup \hat{\Sigma}'_{12}$. Denote by $\mathcal{H}_{12}$ the $\mathfrak{A}_5$-irreducible g-exceptional curve that is mapped to $\hat{\Sigma}_{12}$, denote by $\mathcal{H}'_{12}$ the $\mathfrak{A}_5$-irreducible g-exceptional curve that is mapped to $\hat{\Sigma}'_{12}$, denote by $\check{\mathcal{P}}$ the proper transform of the pencil $\hat{\mathcal{P}}$ on the surface $\check{S}_{Cl}$, and denote by $\check{\mathscr{C}}, \check{\mathcal{L}}_{12}, \check{\mathcal{G}}_6, \check{\mathcal{B}}_{18}$ and $\check{\mathcal{E}}_{12}$ the proper transforms on the surface $\check{S}_{Cl}$ of the curves $\hat{\mathscr{C}}, \hat{\mathcal{L}}_{12}, \hat{\mathcal{G}}_6, \hat{\mathcal{B}}_{18}$, and $\mathcal{E}_{12}$, respectively. Then

$$3\check{\mathscr{C}} + \check{\mathcal{L}}_{12} + \check{\mathcal{E}}_{12} \sim \check{\mathcal{B}}_{18} + \check{\mathcal{G}}_6 + \mathcal{H}_{12} \sim$$
$$\sim (h \circ g)^*(2K_{S_{Cl}}) - g^*(\mathcal{E}_{12}) - \mathcal{H}_{12} - \mathcal{H}'_{12}.$$

This shows that the pencil $\check{\mathcal{P}}$ is generated by the divisors $3\check{\mathscr{C}} + \check{\mathcal{L}}_{12} + \check{\mathcal{E}}_{12}$ and $\check{\mathcal{B}}_{18} + \check{\mathcal{G}}_6 + \mathcal{H}_{12}$.

Put $\check{\Sigma}_{12} = \check{\mathcal{E}}_{12} \cap \mathcal{H}_{12}$. Then $\check{\Sigma}_{12}$ is an $\mathfrak{A}_5$-orbit of length 12 that is contained in the base locus of the pencil $\check{\mathcal{P}}$. In fact, the $\mathfrak{A}_5$-orbit $\check{\Sigma}_{12}$ is the base locus of the pencil $\check{\mathcal{P}}$, because

$$\left((h \circ g)^*(2K_{\check{S}_{Cl}}) - g^*(\check{\mathcal{E}}_{12}) - \mathcal{H}_{12} - \mathcal{H}'_{12}\right)^2 = 12.$$

This equality also implies that the base locus of the pencil $\check{\mathcal{P}}$ can be resolved by blowing up the $\mathfrak{A}_5$-orbit $\check{\Sigma}_{12}$. Thus, there exists a commutative diagram

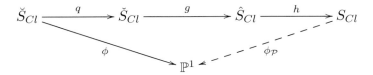

where $q\colon \check{\check{S}}_{Cl} \to \check{S}_{Cl}$ is the blow-up of the points of $\check{\Sigma}_{12}$, and ϕ is a surjective morphism. This commutative diagram resolves the indeterminacy of the rational map $\phi_{\mathcal{P}}$.

Let us compute the topological Euler characteristic $\chi_{top}(\check{S}_{Cl})$ of the surface $\check{S}_{Cl}$. Since $\chi_{top}(\hat{S}_{Cl}) = 81$ by Corollary 15.2.8, we have $\chi_{top}(\check{S}_{Cl}) = 96$ and

$$\chi_{top}(\check{\check{S}}_{Cl}) = 117.$$

On the other hand, we can compute $\chi_{top}(\check{\check{S}}_{Cl})$ using the morphish ϕ and the topological Euler characteristics of its singular fibers. Indeed, let R be a general fiber of ϕ, and let t be the number of singular fibers of the morphism ϕ. Then $\chi_{top}(R) = -66$ by Lemma 12.4.4, so that $\chi_{top}(\check{\check{S}}_{Cl})$ is a sum of

$$\left(\chi_{top}(\mathbb{P}^1) - t\right) \cdot \chi_{top}(R) = -66(2-t)$$

and the sum of the topological Euler characteristics of the singular fibers of ϕ. Luckily, the results obtained in §12.4 allow us to describe these singular fibers explicitly. This is what we are going to do now.

Let us use notation of §12.4. In particular, recall from Lemma 12.4.8 that by Z_{10}, Z'_{10}, and Z_{15} we denote the irreducible curves that are intersections of the surface S_{Cl} with the surfaces S_{10}, S'_{10}, and S_{15}, respectively. Denote by $\check{\mathscr{C}}$, $\check{\mathscr{C}}'$, $\check{\mathcal{L}}_6$, $\check{\mathcal{L}}_{12}$, $\check{\mathcal{G}}_6$, $\check{\mathcal{B}}_{18}$, $\check{Z}_{\mathscr{C}'}$, $\check{Z}_{\mathcal{L}_6}$, $\check{Z}_{10}$, $\check{Z}'_{10}$, $\check{Z}_{15}$, $\check{\mathcal{E}}_{12}$, and $\check{\mathcal{H}}_{12}$ the proper transforms on the surface $\check{S}_{Cl}$ of the curves $\mathscr{C}$, $\mathscr{C}'$, $\mathcal{L}_6$, $\mathcal{L}_{12}$, $\mathcal{G}_6$, $\mathcal{B}_{18}$, $Z_{\mathscr{C}'}$, $Z_{\mathcal{L}_6}$, Z_{10}, Z'_{10}, Z_{15}, $\mathcal{E}_{12}$, and $\mathcal{H}_{12}$, respectively. Then it follows from Lemmas 12.2.5(v), 12.3.6, 12.4.1, 12.4.2, 12.4.4, and 12.4.8 and Corollary 15.2.10

that the effective divisors

$$3\check{\mathscr{C}} + \check{\mathcal{L}}_{12} + \check{\mathcal{E}}_{12}, \ \check{\mathcal{B}}_{18} + \check{\mathcal{G}}_6 + \check{\mathcal{H}}_{12},$$
$$\check{\mathscr{C}}' + \check{Z}_{\mathscr{C}'}, \ \check{\mathcal{L}}_6 + \check{Z}_{\mathcal{L}_6}, \ \check{Z}_{10}, \ \check{Z}'_{10}, \ \check{Z}_{15} \quad (15.2.12)$$

are scheme fibers of the morphism ϕ. If all other fibers of ϕ are smooth, then, in particular, the surface S_{Cl} is smooth away from $\Sigma_{12} \cup \Sigma'_{12}$. Since S_{Cl} is smooth at the points of Σ'_{12} by Lemma 12.2.5(iv), we see that to complete the proof it is enough to show that all singular fibers of ϕ are listed in (15.2.12).

Before doing so, let us compute the topological Euler characteristics of the seven singular fibers of ϕ listed in (15.2.12). We have

$$\chi_{top}\left(\check{\mathscr{C}} \cup \check{\mathcal{L}}_{12} \cup \check{\mathcal{E}}_{12}\right) = 26$$

by construction. Moreover, it follows from Lemma 12.3.6 that

$$\chi_{top}\left(\check{\mathcal{B}}_{18} \cup \check{\mathcal{G}}_6 \cup \check{\mathcal{H}}_{12}\right) = 6,$$

it follows from Lemma 12.4.1 that

$$\chi_{top}\left(\check{\mathscr{C}}' \cup \check{Z}_{\mathscr{C}'}\right) = -46,$$

and it follows from Lemma 12.4.2 that

$$\chi_{top}\left(\check{\mathcal{L}}_6 \cup \check{Z}_{\mathcal{L}_6}\right) = -36.$$

Furthermore, it follows from Lemma 12.4.8 that

$$\chi_{top}(\check{Z}_{10}) = \chi_{top}(\check{Z}'_{10}) = 56$$

and $\chi_{top}(\check{Z}_{15}) = -51$. Observe that

$$(\chi_{top}(\mathbb{P}^1) - 7)\chi_{top}(R) + \chi_{top}\left(\check{\mathscr{C}} \cup \check{\mathcal{L}}_{12} \cup \check{\mathcal{E}}_{12}\right) +$$
$$+ \chi_{top}\left(\check{\mathcal{B}}_{18} \cup \check{\mathcal{G}}_6 \cup \check{\mathcal{H}}_{12}\right) + \chi_{top}\left(\check{\mathscr{C}}' \cup \check{Z}_{\mathscr{C}'}\right) + \chi_{top}\left(\check{\mathcal{L}}_6 \cup \check{Z}_{\mathcal{L}_6}\right) +$$
$$+ \chi_{top}(\check{Z}_{10}) + \chi_{top}(\check{Z}'_{10}) + \chi_{top}(\check{Z}_{15}) = 117 = \chi_{top}(\check{S}_{Cl}). \quad (15.2.13)$$

Our plan is to show that if there were singular fibers of ϕ not listed in (15.2.12), then they would contribute additional positive summands to the left-hand side of (15.2.13), which would break the equality and thus lead to a contradiction.

Suppose that ϕ has singular fibers that are not listed in (15.2.12). Denote them by $R_1, \ldots, R_s$, where $s = t - 7$, and denote their images on S_{Cl} by $T_1, \ldots, T_s$, respectively. Then it follows from Lemma 12.4.4 that each T_i is an irreducible curve that has ordinary cusps at the points of Σ_{12}, the curve T_i is singular at the points of an $\mathfrak{A}_5$-orbit of length 30, every point in this orbit is either a node or an ordinary cusp of the curve T_i, and the normalization of the curve T_i has genus 4. Thus, if T_i does not contain singular points of the surface S_{Cl} different from Σ_{12}, then R_i is an irreducible curve that has 30 singular points, and one of the following cases occurs:

$$\chi_{top}(R_i) = \begin{cases} -6 & \text{if } R_i \text{ has 30 ordinary cusps,} \\ -36 & \text{if } R_i \text{ has 30 nodes.} \end{cases}$$

Suppose that S_{Cl} is singular away from Σ_{12}. Then its singular locus consists of Σ_{12} and an $\mathfrak{A}_5$-orbit Σ of length 30 by Corollary 15.2.10. The $\mathfrak{A}_5$-orbit Σ is contained in one of the curves $T_1, \ldots, T_s$. Thus, we may assume that $\Sigma \subset T_1$. Then

$$R_1 = \check{T}_1 + \check{\mathcal{F}},$$

where $\check{T}_1$ is the proper transform of the curve T_1 on $\check{S}_{Cl}$, and $\check{\mathcal{F}}$ is the proper transform of the $\mathfrak{A}_5$-irreducible h-exceptional curve that is mapped to Σ by the morphism h. Note that $\check{\mathcal{F}}$ is a disjoint union of 30 smooth rational curves, because S_{Cl} has ordinary double points in the points of Σ by Corollary 15.2.10. This gives

$$\chi_{top}(R_1) = \begin{cases} 24 & \text{if } T_1 \text{ has ordinary cusps at the points of } \Sigma, \\ -6 & \text{if } T_1 \text{ has nodes at the points of } \Sigma. \end{cases}$$

We see that $66 + \chi_{top}(R_i) > 0$ in every possible case. Therefore, we have

$$117 = \chi_{top}(\check{S}_{Cl}) = \left(\chi_{top}(\mathbb{P}^1) - 7 - s\right) \cdot \chi_{top}(R) + \chi_{top}\left(\check{\mathscr{C}} \cup \check{\mathcal{L}}_{12} \cup \check{\mathcal{E}}_{12}\right) +$$
$$+ \chi_{top}\left(\check{\mathcal{B}}_{18} \cup \check{\mathcal{G}}_6 \cup \check{\mathcal{H}}_{12}\right) + \chi_{top}\left(\check{\mathscr{C}}' \cup \check{Z}_{\mathscr{C}'}\right) + \chi_{top}\left(\check{\mathcal{L}}_6 \cup \check{Z}_{\mathcal{L}_6}\right) +$$
$$+ \chi_{top}(\check{Z}_{10}) + \chi_{top}(\check{Z}'_{10}) + \chi_{top}(\check{Z}_{15}) + \sum_{i=1}^{s} \chi_{top}(R_i) =$$
$$= 117 + \sum_{i=1}^{s} \left(66 + \chi_{top}(R_i)\right) > 117.$$

The obtained contradiction shows that ϕ does not have singular fibers except those that are listed in (15.2.12), and thus the surface S_{Cl} has no singular points outside of Σ_{12}. This completes the proof of Theorem 15.2.11. □

Remark 15.2.14. Let S be a surface in $\mathcal{Q}_2$ that is different from $\mathscr{S}$, $S_{\mathcal{G}_6}$, $S_{\mathscr{C}'}$, $S_{\mathcal{L}_6}$, S_{10}, S'_{10}, and S_{15}. Put $Z = S|_{S_{Cl}}$. By Lemma 12.4.4(i),(iii) the divisor Z is an irreducible curve that has ordinary cusps at every point of Σ_{12}. The proof of Theorem 15.2.11 implies that Z is smooth away from the $\mathfrak{A}_5$-orbit Σ_{12} (cf. Remark 12.4.6). Thus, the normalization of the curve Z has genus 34 by Lemma 12.4.4(iv).

15.3 Projection to Clebsch cubic surface

The main tool we are going to use in this section is the $\mathfrak{A}_5$-equivariant surjective morphism

$$\theta_{\Pi_2}\colon V_5 \to \mathbb{P}^3 \cong \mathbb{P}(W_4)$$

that has been constructed in §7.5. Recall that the $\mathfrak{A}_5$-invariant plane $\Pi_2 \subset \mathbb{P}^6$ is disjoint from V_5 by Lemma 7.5.2, so that the linear projection from Π_2 induces an $\mathfrak{A}_5$-equivariant surjective morphism θ_{Π_2}.

Put $S_3 = \theta_{\Pi_2}(S_{Cl})$. By Remark 12.2.2 the surface S_3 is the Clebsch cubic surface (see §6.3), and the induced morphism

$$\theta_{\Pi_2}|_{S_{Cl}}\colon S_{Cl} \to S_3$$

is a finite morphism of degree 5. We will use the latter morphism together with the geometry of the lines on Clebsch cubic surface to study the geometry of curves on S_{Cl}.

Recall from Corollary 15.1.1 that the only $\mathfrak{A}_5$-irreducible curves of degree less than 20 contained in S_{Cl} are the curves $\mathscr{C}$, $\mathscr{C}'$, $\mathcal{L}_6$, $\mathcal{L}_{12}$, $\mathcal{G}_6$, and $\mathcal{B}_{18}$.

Similarly, the Clebsch cubic surface S_3 contains just five $\mathfrak{A}_5$-irreducible curves of degree less than 12. These five curves have been explicitly described in Theorem 6.3.18 as follows: two curves that are disjoint unions of six lines, two smooth rational curves of degree 6, and a smooth curve of genus 4 known as the Bring's curve (see Remark 5.4.2). It is quite natural to expect that the curves $\mathscr{C}$, $\mathscr{C}'$, $\mathcal{L}_6$, $\mathcal{L}_{12}$, $\mathcal{G}_6$, and $\mathcal{B}_{18}$ are mapped to some of these curves by θ_{Π_2}. This is indeed true by the following two lemmas.

Lemma 15.3.1. *The curves $\mathcal{L}_6$, $\mathcal{L}_{12}$, and $\mathcal{G}_6$ are mapped by the morphism θ_{Π_2} to one $\mathfrak{A}_5$-irreducible curve in S_3 that is a disjoint union of six lines. Moreover, one has $\theta_{\Pi_2}(\Sigma_{12}) = \theta_{\Pi_2}(\Sigma'_{12})$, and $\theta_{\Pi_2}(\Sigma_{12})$ is the unique $\mathfrak{A}_5$-orbit of length 12 contained in $\theta_{\Pi_2}(\mathcal{L}_6)$. Furthermore, the curve $\mathcal{L}_6 \cup \mathcal{L}_{12} \cup \mathcal{G}_6$ is the scheme theoretic preimage of $\theta_{\Pi_2}(\mathcal{L}_6)$ with respect to θ_{Π_2}.*

Proof. All irreducible components of $\mathcal{L}_6$ (resp., $\mathcal{L}_{12}$) are mapped by θ_{Π_2} to lines in S_3. Thus, the curves $\mathcal{L}_6$ and $\mathcal{L}_{12}$ are mapped by θ_{Π_2} to $\mathfrak{A}_5$-irreducible curves that are disjoint unions of six lines by Lemma 6.3.3(i).

The irreducible components of the curve $\mathcal{G}_6$ are mapped by θ_{Π_2} either to lines or to conics in S_3. Thus, it follows from Lemmas 6.3.3(i) and 6.3.14 that the curve $\mathcal{G}_6$ is mapped by θ_{Π_2} to an $\mathfrak{A}_5$-irreducible curve that is a disjoint union of six lines as well.

We have $\mathcal{L}_6 \cap \mathcal{L}_{12} = \Sigma'_{12}$ by Lemma 9.1.9 and $\mathcal{L}_{12} \cap \mathcal{G}_6 = \Sigma_{12}$ by Corollary 9.2.11. Hence, it follows from Lemma 6.3.12(i),(iii) that

$$\theta_{\Pi_2}(\Sigma'_{12}) = \theta_{\Pi_2}(\Sigma_{12})$$

is the unique $\mathfrak{A}_5$-orbit of length 12 contained in $\theta_{\Pi_2}(\mathcal{L}_{12})$. On the other hand, it follows from Lemma 6.3.12(iii),(vi),(vii) that for every $\mathfrak{A}_5$-orbit of length 12 on the surface S_3, there exists a unique $\mathfrak{A}_5$-irreducible curve in S_3 that is a union of 6 lines and that contains this orbit. Thus, one has

$$\theta_{\Pi_2}(\mathcal{L}_6) = \theta_{\Pi_2}(\mathcal{L}_{12}) = \theta_{\Pi_2}(\mathcal{G}_6).$$

Since the morphism

$$\theta_{\Pi_2}|_{S_{Cl}} \colon S_{Cl} \to S_3$$

is a finite morphism of degree 5 and $\mathcal{L}_6 \cup \mathcal{L}_{12} \cup \mathcal{G}_6$ is a curve of degree 30, we see that $\mathcal{L}_6 \cup \mathcal{L}_{12} \cup \mathcal{G}_6$ is actually a scheme-theoretic preimage of $\theta_{\Pi_2}(\mathcal{L}_6)$ with respect to θ_{Π_2}. □

By Theorem 6.3.18, the Clebsch cubic surface S_3 contains exactly two $\mathfrak{A}_5$-invariant smooth rational curves of degree 6. One of them intersects $\theta_{\Pi_2}(\mathcal{L}_6)$ by the unique $\mathfrak{A}_5$-orbit of length 12 contained in $\theta_{\Pi_2}(\mathcal{L}_6)$, and another one is disjoint from the curve $\theta_{\Pi_2}(\mathcal{L}_6)$.

Lemma 15.3.2. *The curves $\mathscr{C}$, $\mathscr{C}'$, and $\mathcal{B}_{18}$ are mapped by θ_{Π_2} to the irreducible $\mathfrak{A}_5$-invariant smooth rational sextic curve in S_3 that has a non-empty intersection with the curve $\theta_{\Pi_2}(\mathcal{L}_6)$. Moreover, the curve $\mathscr{C} \cup \mathscr{C}' \cup \mathcal{B}_{18}$ is a scheme-theoretic preimage of $\theta_{\Pi_2}(\mathscr{C})$ with respect to θ_{Π_2}.*

Proof. Since θ_{Π_2} is a linear projection, both curves $\theta_{\Pi_2}(\mathscr{C})$ and $\theta_{\Pi_2}(\mathscr{C}')$ are rational $\mathfrak{A}_5$-invariant curves of degree less or equal to 6. Hence, they are smooth and their degrees are exactly 6 by Theorem 6.3.18.

It follows from Theorem 6.3.18 and Lemma 6.3.17 that either

$$\theta_{\Pi_2}(\mathscr{C}) = \theta_{\Pi_2}(\mathscr{C}'),$$

or the intersection $\theta_{\Pi_2}(\mathscr{C}) \cap \theta_{\Pi_2}(\mathscr{C}')$ is the $\mathfrak{A}_5$-orbit of length 20. On the other hand,
$$\theta_{\Pi_2}(\Sigma_{12}) = \theta_{\Pi_2}(\Sigma'_{12})$$
is an $\mathfrak{A}_5$-orbit of length 12 by Lemma 15.3.1. This shows that
$$\theta_{\Pi_2}(\mathscr{C}) = \theta_{\Pi_2}(\mathscr{C}'),$$
since $\Sigma_{12} \subset \mathscr{C}$ by construction, and $\Sigma'_{12} \subset \mathscr{C}'$ by Lemma 11.3.1. Note that
$$\theta_{\Pi_2}(\mathscr{C}) \cap \theta_{\Pi_2}(\mathcal{L}_6) \neq \varnothing,$$
because $\theta_{\Pi_2}(\mathscr{C}) = \theta_{\Pi_2}(\mathscr{C}')$ and $\mathscr{C}' \cap \mathcal{L}_6 \neq \varnothing$.

Let $Z_1, \ldots, Z_r$ be $\mathfrak{A}_5$-irreducible curves in V_5 different from $\mathscr{C}$ and $\mathscr{C}'$ that are mapped by θ_{Π_2} to the curve $\theta_{\Pi_2}(\mathscr{C})$. Then it follows from Lemmas 15.3.1 and 6.3.17 that
$$3\left(a\mathscr{C} + b\mathscr{C}' + \sum_{i=1}^{r} m_i Z_i\right) \sim 10 K_{S_{Cl}} - 2\left(\mathcal{L}_6 + \mathcal{L}_{12} + \mathcal{G}_6\right) \qquad (15.3.3)$$
for some positive integers $a, b, m_1, \ldots, m_r$. In particular, we have
$$6(a+b) + \sum_{i=1}^{r} m_i \deg(Z_i) = a\deg(\mathscr{C}) + b\deg(\mathscr{C}') + \sum_{i=1}^{r} m_i \deg(Z_i) = 30,$$
which implies that the degree of each curve Z_i is at most
$$30 - 6(a+b) \leqslant 18.$$
On the other hand, $\mathscr{C}$, $\mathscr{C}'$, $\mathcal{L}_6$, $\mathcal{L}_{12}$, $\mathcal{G}_6$, and $\mathcal{B}_{18}$ are the only $\mathfrak{A}_5$-irreducible curves of degree less than 20 contained in S_{Cl} by Corollary 15.1.1, and the curve
$$\theta_{\Pi_2}(\mathcal{L}_6) = \theta_{\Pi_2}(\mathcal{L}_{12}) = \theta_{\Pi_2}(\mathcal{G}_6)$$
is different from $\theta_{\Pi_2}(\mathcal{L}_6)$ by Lemma 15.3.1. Therefore, one has either $r = 1$ and $Z_1 = \mathcal{B}_{18}$, or $r = 0$ and $\mathscr{C}$ and $\mathscr{C}'$ are the only curves that are mapped to $\theta_{\Pi_2}(\mathscr{C})$ by θ_{Π_2}. In the former case, the curve $\mathscr{C} \cup \mathscr{C}' \cup \mathcal{B}_{18}$ is a scheme-theoretic preimage of $\theta_{\Pi_2}(\mathscr{C})$ with respect to θ_{Π_2}, because the morphism
$$\theta_{\Pi_2}|_{S_{Cl}} \colon S_{Cl} \to S_3$$
is a finite morphism of degree 5, and $\mathscr{C} \cup \mathscr{C}' \cup \mathcal{B}_{18}$ is a curve of degree 30. Let us show that the latter case is impossible.

CREMONA GROUPS AND THE ICOSAHEDRON 391

Suppose that $\mathscr{C}$ and $\mathscr{C}'$ are the only curves that are mapped to $\theta_{\Pi_2}(\mathscr{C})$ by θ_{Π_2}. Then (15.3.3) implies that

$$3\left(a\mathscr{C} + b\mathscr{C}'\right) \sim 10K_{S_{Cl}} - 2\left(\mathcal{L}_6 + \mathcal{L}_{12} + \mathcal{G}_6\right),$$

where a and b are positive integers such that $a + b = 5$. We will show that this rational equivalence is impossible.

Recall that S_{Cl} is smooth away from Σ_{12} by Theorem 15.2.11. In particular, the curves $\mathcal{L}_6$ and $\mathscr{C}'$ are contained in the smooth locus of S_{Cl} by Lemmas 7.4.3 and 11.3.4. Thus, it follows from the adjunction formula that $\mathcal{L}_6^2 = -18$ on the surface S_{Cl}, since $\mathcal{L}_6$ is a disjoint union of six lines by Lemma 7.4.7. Similarly, we have $\mathscr{C}' \cdot \mathscr{C}' = -8$ on the surface S_{Cl}. Moreover, we know from Lemma 11.3.4 that $\Sigma'_{12} = \mathcal{L}_6 \cap \mathscr{C}'$, and the curves $\mathcal{L}_6$ and $\mathscr{C}'$ intersect transversally in the points of Σ'_{12}. Thus, we have $\mathcal{L}_6 \cdot \mathscr{C}' = 12$ on the surface S_{Cl}. Similarly, Lemmas 9.1.9 and 11.3.4 imply that

$$\mathcal{L}_6 \cdot \mathcal{L}_{12} = \mathscr{C}' \cdot \mathcal{L}_{12} = 12.$$

Also one has $\mathcal{L}_6 \cdot \mathcal{G}_6 = 0$ by Lemma 9.2.10 and $\mathscr{C}' \cdot \mathcal{G}_6 = 0$ by Lemma 11.3.6. Therefore, we compute

$$36b = 3b\mathcal{L}_6 \cdot \mathscr{C}' = 3a\mathcal{L}_6 \cdot \mathscr{C} + 3b\mathcal{L}_6 \cdot \mathscr{C}' =$$
$$= 3\mathcal{L}_6 \cdot \left(a\mathscr{C} + b\mathscr{C}'\right) = \left(10K_{S_{Cl}} - 2\left(\mathcal{L}_6 + \mathcal{L}_{12} + \mathcal{G}_6\right)\right) \cdot \mathcal{L}_6 =$$
$$= 10K_{S_{Cl}} \cdot \mathcal{L}_6 - 2\left(\mathcal{L}_6 + \mathcal{L}_{12} + \mathcal{G}_6\right) \cdot \mathcal{L}_6 =$$
$$= 60 - 2\mathcal{L}_6^2 - 2\mathcal{L}_6 \cdot \mathcal{L}_{12} - 2\mathcal{L}_6 \cdot \mathcal{G}_6 = 60 + 36 - 24 = 72,$$

which implies that $b = 2$. Thus

$$-48 = -24b = 3b\mathscr{C}' \cdot \mathscr{C}' = 3a\mathscr{C}' \cdot \mathscr{C} + 3b\mathscr{C}' \cdot \mathscr{C}' =$$
$$= 3\mathscr{C}' \cdot \left(a\mathscr{C} + b\mathscr{C}'\right) = \left(10K_{S_{Cl}} - 2\left(\mathcal{L}_6 + \mathcal{L}_{12} + \mathcal{G}_6\right)\right) \cdot \mathscr{C}' =$$
$$= 10K_{S_{Cl}} \cdot \mathscr{C}' - 2\left(\mathcal{L}_6 + \mathcal{L}_{12} + \mathcal{G}_6\right) \cdot \mathscr{C}' =$$
$$= 60 - 2\mathscr{C}' \cdot \mathcal{L}_6 - 2\mathscr{C}' \cdot \mathcal{L}_{12} - 2\mathscr{C}' \cdot \mathcal{G}_6 = 60 - 24 - 24 = 12,$$

which is absurd. $\square$

Recall from Theorem 14.3.2(o),(vii) that the surface S_{Cl} contains the $\mathfrak{A}_5$-irreducible curve $\mathcal{T}_{15}$ that is a union of 15 smooth rational cubic curves, and that contains all $\mathfrak{A}_5$-orbits of length 30 in V_5 that are not contained in the curve $\mathcal{L}_{15}$ by Theorem 14.3.8.

Lemma 15.3.4. *The surface S_{Cl} contains an $\mathfrak{A}_5$-irreducible curve $\mathcal{G}_{15}$ that is a union of 15 conics, such that the curves $\mathcal{G}_{15}$ and $\mathcal{T}_{15}$ are mapped by θ_{Π_2} to the $\mathfrak{A}_5$-invariant curve in $\mathbb{S}_3$ that is a union of 15 lines. Moreover, the following assertions hold:*

(i) *the curves $\mathcal{G}_{15}$ and $\mathcal{T}_{15}$ are the only $\mathfrak{A}_5$-irreducible curves in V_5 that are mapped to $\theta_{\Pi_2}(\mathcal{T}_{15})$ by θ_{Π_2};*

(ii) *one has $\mathcal{G}_{15} + \mathcal{T}_{15} \sim 5K_{S_{Cl}} \sim 5H|_{S_{Cl}}$;*

(iii) *the $\mathfrak{A}_5$-orbit Σ_{15} is not contained in the curve $\mathcal{G}_{15}$;*

(iv) *the $\mathfrak{A}_5$-orbits Σ_{12} and Σ'_{12} are not contained in $\mathcal{G}_{15} \cup \mathcal{T}_{15}$;*

(v) *the stabilizer in $\mathfrak{A}_5$ of a general point of $\mathcal{G}_{15}$ is trivial.*

Proof. Let us use the notation of §7.6 and the proof of Theorem 14.3.2. By Corollary 15.1.1, the intersection $S_{Cl} \cap H_1$ does not contain the curves C_1 and C'_1. Therefore, it follows from Lemma 6.2.6 that

$$S_{Cl}|_{H_1} = T_1 + T_2 + T_3 + Z$$

for some divisor Z in the pencil generated by $3C_1$ and $3C'_1$. Moreover, it follows from Lemma 6.2.2(v),(vi) that either Z is a disjoint union of three conics, or

$$Z = L_{12} + L_{13} + L_{14} + L_{23} + L_{24} + L_{34}.$$

Recall from §7.6 that we denote by $\mathcal{L}_{30}$ the $\mathfrak{A}_5$-orbit of the line L_{12} on V_5. We know from Lemma 7.6.7 that $\mathcal{L}_{30}$ is disjoint from the curves $\mathscr{C}$ and $\mathcal{L}_{12}$. Therefore, $\mathcal{L}_{30}$ is not contained in the surface S_{Cl}, because

$$S_{Cl} \cap \mathscr{S} = \mathscr{C} \cup \mathcal{L}_{12}$$

by Lemma 12.2.5(v), while $\mathcal{L}_{30} \cap \mathscr{S} \neq \varnothing$. Thus, the curve Z is not contained in $\mathcal{L}_{30}$, so that Z is a disjoint union of three conics. Denote by $\mathcal{G}_{15}$ the $\mathfrak{A}_5$-orbit of the curve Z. Then $\mathcal{G}_{15}$ is a union of 15 conics, since $\mathcal{G}_{15}$ cannot be a union of 5 conics by Corollary 15.1.1.

By construction, the curve $\mathcal{T}_{15}$ is the $\mathfrak{A}_5$-orbit of any of the curves T_1, T_2, and T_3 (see the proof of Theorem 14.3.2); thus, we have

$$\mathcal{G}_{15} + \mathcal{T}_{15} \sim 5H|_{S_{Cl}} \sim 5K_{S_{Cl}}.$$

This gives assertion (ii).

By Lemma 14.3.1, the intersection $\Sigma_{15} \cap H_1$ is the unique orbit of length 3 in H_1 of the stabilizer $G_1 \cong \mathfrak{A}_4$ of H_1. Note that

$$L_{12} + L_{13} + L_{14} + L_{23} + L_{24} + L_{34}$$

is the only curve in the pencil generated by $3C_1$ and $3C_1'$ that contains the intersection $\Sigma_{15} \cap H_1$ by Remark 6.2.7. In particular, Z does not contain the G_1-orbit $\Sigma_{15} \cap H_1$, which implies that $\mathcal{G}_{15}$ does not contain the $\mathfrak{A}_5$-orbit Σ_{15}. This gives assertion (iii). Similarly, neither Σ_{12} nor Σ_{12}' is contained in $\mathcal{G}_{15}$, because the surface H_1 is disjoint from the $\mathfrak{A}_5$-orbits Σ_{12} and Σ_{12}' by Lemma 7.6.1. This gives assertion (iv).

Recall from §7.6 that H_1 is the unique G_1-invariant hyperplane section of V_5, and thus it is mapped by θ_{Π_2} to the unique G_1-invariant hyperplane section of the cubic surface $S_3 \subset \mathbb{P}^3$. Hence, it follows from Lemma 6.3.15 that the curve

$$S_{Cl} \cap H_1 = Z + T_1 + T_2 + T_3$$

is mapped to a hyperplane section of the cubic surface $S_3 \subset \mathbb{P}^3$ that is a union of three lines whose $\mathfrak{A}_5$-orbit is a union of 15 lines. Thus, the curves $\mathcal{G}_{15}$ and $\mathcal{T}_{15}$ are mapped by θ_{Π_2} to the $\mathfrak{A}_5$-irreducible curve in S_3 that is a union of 15 lines.

The curve $\mathcal{G}_{15} + \mathcal{T}_{15}$ has degree 75, and $\theta_{\Pi_2}(\mathcal{T}_{15})$ is a curve of degree 15. Since

$$\theta_{\Pi_2}|_{S_{Cl}} \colon S_{Cl} \to S_3$$

is a finite morphism of degree 5, we see that the curves $\mathcal{G}_{15}$ and $\mathcal{T}_{15}$ are all $\mathfrak{A}_5$-irreducible curves in V_5 that are mapped to $\theta_{\Pi_2}(\mathcal{T}_{15})$ by θ_{Π_2}. This gives assertion (i).

Finally, it follows from Theorems 7.3.5, 14.1.16, and 14.3.8 that the stabilizer in $\mathfrak{A}_5$ of a general point in $\mathcal{G}_{15}$ is trivial. This gives assertion (v) and completes the proof of Lemma 15.3.4. □

By Lemma 6.3.3(i), twenty-seven lines on the Clebsch cubic surface $S_3 \subset \mathbb{P}^3$ split into three $\mathfrak{A}_5$-orbits that consist of 6, 6, and 15 lines. By Lemma 15.3.1, the curves $\mathcal{L}_6$, $\mathcal{L}_{12}$, and $\mathcal{G}_6$ are mapped by θ_{Π_2} to an $\mathfrak{A}_5$-irreducible curve that is a disjoint union of six lines, and their union is the whole preimage of this curve. Lemma 15.3.4 describes the curves in V_5 that are mapped by θ_{Π_2} to the $\mathfrak{A}_5$-invariant curve in S_3 that is a union of 15 lines. Let us describe the curves in V_5 that are mapped by θ_{Π_2} to the remaining six lines in S_3.

Lemma 15.3.5. Let $\mathcal{R}_6$ be an $\mathfrak{A}_5$-invariant curve in V_5 whose irreducible components are mapped by θ_{Π_2} to the $\mathfrak{A}_5$-irreducible curve in S_3 that is a union of six disjoint lines and that is different from $\theta_{\Pi_2}(\mathcal{L}_6)$. Then $\mathcal{R}_6$ is an $\mathfrak{A}_5$-irreducible curve $\mathcal{R}_6$ of degree 30, and $\mathcal{R}_6$ is a scheme-theoretic preimage of $\theta_{\Pi_2}(\mathcal{R}_6)$ with respect to θ_{Π_2}. Moreover, $\mathcal{R}_6$ is a disjoint union of six smooth quintic elliptic curves. Furthermore, the curve $\mathcal{R}_6$ is disjoint from the $\mathfrak{A}_5$-orbits Σ_{12} and Σ'_{12}.

Proof. By Lemma 15.3.1, the curve $\mathcal{L}_6 \cup \mathcal{L}_{12} \cup \mathcal{G}_6$ is the preimage via θ_{Π_2} of an $\mathfrak{A}_5$-irreducible curve in the Clebsch cubic surface S_3 that is a disjoint union of six lines. Let Z be a scheme-theoretic preimage via θ_{Π_2} of the other $\mathfrak{A}_5$-irreducible curve in S_3 that is a disjoint union of six lines, see Lemma 6.3.3(i),(ii). Then Z is a one-cycle of degree 30, because the morphism

$$\theta_{\Pi_2}|_{S_{Cl}} : S_{Cl} \to S_3$$

is a finite morphism of degree 5. If Z is not reduced or not $\mathfrak{A}_5$-irreducible, then it contains an $\mathfrak{A}_5$-irreducible curve of degree at most 15 that is different from $\mathscr{C}$, $\mathscr{C}'$, $\mathcal{L}_6$, $\mathcal{L}_{12}$, and $\mathcal{G}_6$. This is impossible by Corollary 15.1.1. Hence Z is an $\mathfrak{A}_5$-irreducible curve.

Put $\mathcal{R}_6 = Z$. Recall from Lemma 15.3.2 that the curves $\mathscr{C}$, $\mathscr{C}'$, and $\mathcal{B}_{18}$ are mapped by θ_{Π_2} to the irreducible $\mathfrak{A}_5$-invariant smooth rational sextic curve in S_3 that has a non-empty intersection with the curve $\theta_{\Pi_2}(\mathcal{L}_6)$, and hence has an empty intersection with $\theta_{\Pi_2}(\mathcal{R}_6)$ by Lemma 6.3.17. Thus, the curve $\mathcal{R}_6$ is disjoint from the curves $\mathscr{C}$, $\mathscr{C}'$, and $\mathcal{B}_{18}$. In particular, $\mathcal{R}_6$ is disjoint from the $\mathfrak{A}_5$-orbits Σ_{12} and Σ'_{12}.

Recall from Lemma 12.2.5(v) that $\mathcal{L}_{12} + 3\mathscr{C} \sim 2K_{S_{Cl}}$. This gives

$$\mathcal{L}_{12} \cdot \mathcal{R}_6 = \left(\mathcal{L}_{12} + 3\mathscr{C}\right) \cdot \mathcal{R}_6 = 2K_{S_{Cl}} \cdot \mathcal{R}_6 = 60,$$

because $\mathcal{R}_6$ is disjoint from the curve $\mathscr{C}$. In particular, the intersection $\mathcal{L}_{12} \cap \mathcal{R}_6$ is not empty. Similarly, we see that the intersection $\mathcal{L}_6 \cap \mathcal{R}_6$ is not empty. Indeed, we have

$$3\mathscr{C}' + 2\mathcal{L}_6 + \mathcal{L}_{12} + \mathcal{B}_{18} \sim 4K_{S_{Cl}}.$$

by Lemma 15.4.1. Thus, we have

$$2\mathcal{L}_6 \cdot \mathcal{R}_6 + 60 = 2\mathcal{L}_6 \cdot \mathcal{R}_6 + \mathcal{L}_{12} \cdot \mathcal{R}_6 =$$
$$= \left(3\mathscr{C}' + 2\mathcal{L}_6 + \mathcal{L}_{12} + \mathcal{B}_{18}\right) \cdot \mathcal{R}_6 = 4K_{S_{Cl}} \cdot \mathcal{R}_6 = 120,$$

because $\mathcal{R}_6$ is disjoint from the curves $\mathscr{C}'$ and $\mathcal{B}_{18}$. So, we have $\mathcal{L}_6 \cdot \mathcal{R}_6 = 30$. In particular, the intersection $\mathcal{L}_6 \cap \mathcal{R}_6$ is not empty. Note that the intersection

$$\theta_{\Pi_2}(\mathcal{L}_6) \cap \theta_{\Pi_2}(\mathcal{R}_6)$$

is an $\mathfrak{A}_5$-orbit of length 30 by Lemmas 15.3.1 and 6.3.12(vii). Hence it follows from $\mathcal{L}_6 \cdot \mathcal{R}_6 = 30$ that the intersection $\mathcal{L}_6 \cap \mathcal{R}_6$ an $\mathfrak{A}_5$-orbit of length 30.

By Theorem 14.3.8, one has either $\mathcal{L}_6 \cap \mathcal{R}_6 \subset \mathcal{L}_{15}$ or $\mathcal{L}_6 \cap \mathcal{R}_6 \subset \mathcal{T}_{15}$. The latter case is impossible by Lemmas 15.3.1, 15.3.4(i), and 6.3.12(vi),(vii). Thus, we see that

$$\mathcal{L}_6 \cap \mathcal{R}_6 \subset \mathcal{L}_{15}.$$

Therefore, the curve $\mathcal{R}_6$ has non-empty intersection with the curves $\mathcal{L}_6$, $\mathcal{L}_{12}$, and $\mathcal{L}_{15}$.

Denote by n the number of irreducible components of $\mathcal{R}_6$, and denote by d the degree of irreducible components of $\mathcal{R}_6$. Then n is a multiple of 6, since $\theta_{\Pi_2}(\mathcal{R}_6)$ is an $\mathfrak{A}_5$-irreducible curve that is a disjoint union of 6 lines. On the other hand, we have $dn = 30$, which implies that either $d = 1$ and $n = 30$, or $d = 5$ and $n = 6$. Since $\mathcal{R}_6$ has non-empty intersections with the curves $\mathcal{L}_6$, $\mathcal{L}_{12}$, and $\mathcal{L}_{15}$, the former case is impossible by Theorem 7.1.8 and Lemma 6.1.9. Thus, $d = 6$ and $n = 5$, so that $\mathcal{R}_6$ is a disjoint union of six irreducible quintic curves.

Let O be a sufficiently general point in $\mathcal{R}_6$, and let S be a surface in the pencil $\mathcal{Q}_2$ that contains O. Denote by Σ_O the $\mathfrak{A}_5$-orbit of the point O. Then $\Sigma_O \subset S$, and Σ_O is disjoint from the curve C_{20}. Also, we have $|\Sigma_O| = 60$ by Lemma 7.4.5. On the other hand, it follows from the description of intersections of S_{Cl} with the surfaces of the pencil $\mathcal{Q}_2$ given in Lemmas 12.2.5(v), 12.3.6, 12.4.1, 12.4.2, 12.4.4(i), and 12.4.8(i) that the curve $\mathcal{R}_6$ is not contained in the surface S. Hence, we have

$$60 = \mathcal{R}_6 \cdot S \geqslant \sum_{P \in \mathcal{R}_6 \cap S} \operatorname{mult}_P(\mathcal{R}_6) \operatorname{mult}_P(S) \geqslant$$
$$\geqslant \sum_{P \in \mathcal{R}_6 \cap C_{20}} \operatorname{mult}_P(\mathcal{R}_6) \operatorname{mult}_P(S) + \sum_{P \in \Sigma_O} \operatorname{mult}_P(\mathcal{R}_6) \operatorname{mult}_P(S) \geqslant$$
$$\geqslant |\mathcal{R}_6 \cap C_{20}| + |\Sigma_O| = |\mathcal{R}_6 \cap C_{20}| + 60,$$

which implies that $\mathcal{R}_6 \cap C_{20} = \varnothing$. In particular, we see that Σ_{12} and Σ'_{12} are not contained in $\mathcal{R}_6$ by Theorem 8.1.8(iii). Now Theorem 7.3.5 implies that every $\mathfrak{A}_5$-orbit contained in $\mathcal{R}_6$ consists of 30 or 60 points.

Denote by $R_1, \ldots, R_6$ the irreducible components of the curve $\mathcal{R}_6$. Denote by G_1 the stabilizer of R_1 in $\mathfrak{A}_5$. Then $G_1 \cong \mathfrak{D}_{10}$ by Lemma 5.1.1. To complete the proof, we must show that R_1 is a smooth elliptic curve.

Since Σ_{12} is not contained in $\mathcal{R}_6$, the curve R_1 is contained in the smooth locus of the surface S_{Cl} by Theorem 15.2.11. Hence, the self-intersection R_1^2 on the surface S_{Cl} is an integer. Moreover, one has $R_1^2 \leqslant 1$, because

$$15 R_1^2 \leqslant 25$$

by the Hodge index theorem. Now using the adjunction formula, we get

$$2 p_a(R_1) - 2 = R_1^2 + R_1 \cdot H \big|_{S_{Cl}} = R_1^2 + 5 \leqslant 6,$$

which implies that $p_a(R_1) \leqslant 4$. In particular, R_1 cannot have more than 4 singular points by Lemma 4.4.6. On the other hand, the curve $\mathcal{R}_6$ does not contain $\mathfrak{A}_5$-orbits of length less than 30, so that its irreducible component R_1 does not contain G_1-orbits of length less than 5. Therefore, R_1 is smooth.

Let g be the genus of the curve R_1. Then $g \leqslant 4$. On the other hand, the group G_1 acts faithfully on R_1 by Lemma 7.4.5. Since every $\mathfrak{A}_5$-orbit contained in $\mathcal{R}_6$ has length 30 or 60, we see that G_1-orbits in R_1 consist of either 5 or 10 points. Thus, $g = 1$ by Lemma 5.8.6. □

Corollary 15.3.6. *The surface S_{Cl} does not contain lines that are different from irreducible components of $\mathcal{L}_6$ and $\mathcal{L}_{12}$.*

Proof. Since lines in V_5 are mapped to lines in $\mathbb{P}^3$ by θ_{Π_2}, the assertion follows from Lemmas 15.3.1, 15.3.4(i), and 15.3.5. □

We will use the notation $\mathcal{G}_{15}$ for the curve described in Lemma 15.3.4 and the notation $\mathcal{R}_6$ for the curve described in Lemma 15.3.5 until the end of this chapter.

Remark 15.3.7. By Theorem 14.3.2(ii), the $\mathfrak{A}_5$-orbits Σ_{10} and Σ'_{10} are contained in the curve $\mathcal{T}_{15}$. One can use the projection θ_{Π_2} to give an alternative proof of this. Indeed, both Σ_{10} and Σ'_{10} are contained in S_{Cl} by Lemma 12.2.4. On the other hand, by Lemma 6.3.12(i),(ii) the surface S_3 contains a unique $\mathfrak{A}_5$-orbit of length 10, and does not contain $\mathfrak{A}_5$-orbits of smaller length. Hence,

$$\theta_{\Pi_2}(\Sigma_{10}) = \theta_{\Pi_2}(\Sigma'_{10})$$

is the unique $\mathfrak{A}_5$-orbit of length 10 in $\mathbb{S}_3$. Moreover, Lemma 6.3.12(v) also implies that $\theta_{\Pi_2}(\Sigma_{10})$ is contained in the $\mathfrak{A}_5$-irreducible curve in $\mathbb{P}^2$ that is a union of 15 lines. Thus, both Σ_{10} and Σ'_{10} are contained in $\mathcal{T}_{15} \cup \mathcal{G}_{15}$ by Lemma 15.3.4(i). On the other hand, one can easily show that Σ_{10} and Σ'_{10} are not contained in $\mathcal{G}_{15}$, so that they are contained in $\mathcal{T}_{15}$.

15.4 Picard group

We know from Theorem 15.2.6 that the group $\mathrm{Pic}^0(S_{Cl})$ is trivial. In this section we bound the Picard rank of the surface S_{Cl}.

Let us use notation of §12.5, §15.2, and §15.3.

Lemma 15.4.1. One has $\mathscr{F}|_{S_{Cl}} = 3\mathscr{C}' + 2\mathcal{L}_6 + \mathcal{L}_{12} + \mathcal{B}_{18}$.

Proof. The curves $\mathscr{C}'$, $\mathcal{L}_6$, $\mathcal{L}_{12}$, and $\mathcal{B}_{18}$ are contained in the surface $\mathscr{F}$ by Remarks 12.5.2 and 13.5.4 and Lemma 12.5.3. Also they are contained in the surface S_{Cl} by Corollary 15.1.1. Moreover, we have $\mathrm{mult}_{\mathscr{C}'}(\mathscr{F}) = 3$ and $\mathrm{mult}_{\mathcal{L}_6}(\mathscr{F}) = 2$ by Remark 12.5.2 and Lemma 12.5.3, respectively. Thus, the assertion follows from the fact that $\mathscr{F}|_{S_{Cl}}$ is an effective one-cycle of degree 60. $\square$

Denote by $h\colon \hat{S}_{Cl} \to S_{Cl}$ the minimal resolution of singularities of the surface S_{Cl}, i.e., the blow-up of S_{Cl} at the points of the $\mathfrak{A}_5$-orbit Σ_{12} (see Theorem 15.2.11). Denote by $\mathcal{E}_{12}$ the exceptional divisor of h. We compute $K_{\hat{S}_{Cl}} \cdot \mathcal{E}_{12} = 0$ and $\mathcal{E}_{12}^2 = -24$. Denote by $\hat{\mathscr{C}}$, $\hat{\mathscr{C}}'$, $\hat{\mathcal{L}}_6$, $\hat{\mathcal{L}}_{12}$, $\hat{\mathcal{G}}_6$, $\hat{\mathcal{B}}_{18}$, $\hat{\mathcal{G}}_{15}$, $\hat{\mathcal{T}}_{15}$, and $\hat{\mathcal{R}}_6$ the proper transforms on the surface $\hat{S}_{Cl}$ via the birational morphism h of the curves $\mathscr{C}$, $\mathscr{C}'$, $\mathcal{L}_6$, $\mathcal{L}_{12}$, $\mathcal{G}_6$, $\mathcal{B}_{18}$, $\mathcal{G}_{15}$, $\mathcal{T}_{15}$, and $\mathcal{R}_6$, respectively.

Lemma 15.4.2. *The following rational equivalences hold:*

(i) $\hat{\mathcal{L}}_{12} + 3\hat{\mathscr{C}} \sim 2K_{\hat{S}_{Cl}} - 2\mathcal{E}_{12}$,

(ii) $\hat{\mathcal{B}}_{18} + \hat{\mathcal{G}}_6 \sim 2K_{\hat{S}_{Cl}} - \mathcal{E}_{12}$,

(iii) $\hat{\mathcal{T}}_{15} + \hat{\mathcal{G}}_{15} \sim 5K_{\hat{S}_{Cl}}$,

(iv) $3\hat{\mathscr{C}}' + 2\hat{\mathcal{L}}_6 + \hat{\mathcal{L}}_{12} + \hat{\mathcal{B}}_{18} \sim 4K_{\hat{S}_{Cl}} - \mathcal{E}_{12}$,

(v) $\hat{\mathcal{R}}_6 + \hat{\mathcal{L}}_6 + \hat{\mathcal{L}}_{12} + \hat{\mathcal{G}}_6 \sim 4K_{\hat{S}_{Cl}} - \mathcal{E}_{12}$.

Proof. By Lemma 15.3.4(ii), we have

$$\mathcal{T}_{15} + \mathcal{G}_{15} \sim 5K_{S_{Cl}}.$$

By Lemma 15.3.4(iv), the $\mathfrak{A}_5$-orbit Σ_{12} is not contained in $\mathcal{G}_{15} \cup \mathcal{T}_{15}$. This gives assertion (iii), because $K_{S_{Cl}} \sim H|_{S_{Cl}}$.

By Lemma 12.2.5(v), we have $\mathcal{L}_{12} + 3\mathscr{C} \sim 2K_{S_{Cl}}$. By Lemma 12.3.6, we have $\mathcal{B}_{18} + \mathcal{G}_6 \sim 2K_{S_{Cl}}$. By Lemma 15.4.1, we have

$$3\mathscr{C}' + 2\mathcal{L}_6 + \mathcal{L}_{12} + \mathcal{B}_{18} \sim 4K_{S_{Cl}}.$$

Moreover, it follows from Corollary 6.3.7 and Lemmas 15.3.1 and 15.3.5 that

$$\mathcal{R}_6 + \mathcal{L}_6 + \mathcal{L}_{12} + \mathcal{G}_6 \sim 4K_{S_{Cl}}.$$

On the other hand, each point in Σ_{12} is an isolated ordinary double point of the surface S_{Cl} by Theorem 15.2.11. Furthermore, the supports of the divisors

$$\mathcal{L}_{12} + 3\mathscr{C}, \quad \mathcal{B}_{18} + \mathcal{G}_6, \quad 3\mathscr{C}' + 2\mathcal{L}_6 + \mathcal{L}_{12} + \mathcal{B}_{18}$$

and $\mathcal{R}_6 + \mathcal{L}_6 + \mathcal{L}_{12} + \mathcal{G}_6$ contain Σ_{12}. Thus, we have

$$\hat{\mathcal{L}}_{12} + 3\hat{\mathscr{C}} \sim 2K_{\hat{S}_{Cl}} - m_1 \mathcal{E}_{12},$$
$$\hat{\mathcal{B}}_{18} + \hat{\mathcal{G}}_6 \sim 2K_{\hat{S}_{Cl}} - m_2 \mathcal{E}_{12},$$
$$3\hat{\mathscr{C}}' + 2\hat{\mathcal{L}}_6 + \hat{\mathcal{L}}_{12} + \hat{\mathcal{B}}_{18} \sim 4K_{\hat{S}_{Cl}} - m_3 \mathcal{E}_{12},$$
$$\hat{\mathcal{R}}_6 + \hat{\mathcal{L}}_6 + \hat{\mathcal{L}}_{12} + \hat{\mathcal{G}}_6 \sim 4K_{\hat{S}_{Cl}} - m_4 \mathcal{E}_{12}$$

for some positive integers m_1, m_2, m_3, and m_4, because $K_{\hat{S}_{Cl}} \sim h^*(K_{S_{Cl}})$.

Note that the curves $\mathscr{C}$, $\mathcal{L}_{12}$, $\mathcal{G}_6$, and $\mathcal{B}_{18}$ are smooth; moreover, the curve $\mathscr{C}$ contains the $\mathfrak{A}_5$-orbit Σ_{12} by construction, the curve $\mathcal{L}_{12}$ contains Σ_{12} by Lemma 7.2.12(v), the curve $\mathcal{G}_6$ contains Σ_{12} by Corollary 9.2.11, and the curve $\mathcal{B}_{18}$ contains Σ_{12} by Lemma 12.3.6. On the other hand, the curve $\mathcal{L}_6$ is disjoint from Σ_{12} by Lemma 7.4.3, the curve $\mathscr{C}'$ is disjoint from Σ_{12} by Lemma 11.3.4, and the curve $\mathcal{R}_6$ is disjoint from Σ_{12} by Lemma 15.3.5. Therefore, we compute

$$48 = \hat{\mathcal{L}}_{12} \cdot \mathcal{E}_{12} + 3 = \hat{\mathcal{L}}_{12} \cdot \mathcal{E}_{12} + 3\hat{\mathscr{C}} \cdot \mathcal{E}_{12} =$$
$$= \left(\hat{\mathcal{L}}_{12} + 3\hat{\mathscr{C}}\right) \cdot \mathcal{E}_{12} = \left(2K_{\hat{S}_{Cl}} - m_1 \mathcal{E}_{12}\right) \cdot \mathcal{E}_{12} =$$
$$= 2K_{\hat{S}_{Cl}} \cdot \mathcal{E}_{12} - m_1 \mathcal{E}_{12}^2 = -m_1 \mathcal{E}_{12}^2 = 24m_1.$$

Thus, we have $m_1 = 2$, which proves assertion (i). Similarly, we have

$$24 = \hat{\mathcal{B}}_{18} \cdot \mathcal{E}_{12} + \hat{\mathcal{G}}_6 \cdot \mathcal{E}_{12} =$$
$$= \left(\hat{\mathcal{B}}_{18} + \hat{\mathcal{G}}_6\right) \cdot \mathcal{E}_{12} = \left(2K_{\hat{S}_{Cl}} - m_2\mathcal{E}_{12}\right) \cdot \mathcal{E}_{12} =$$
$$= 2K_{\hat{S}_{Cl}} \cdot \mathcal{E}_{12} - m_2\mathcal{E}_{12}^2 = -m_2\mathcal{E}_{12}^2 = 24m_2,$$

which implies that $m_2 = 1$. This proves assertion (ii). Similarly, $m_3 = 1$, since

$$24 = \hat{\mathcal{L}}_{12} \cdot \mathcal{E}_{12} + \hat{\mathcal{B}}_{18} \cdot \mathcal{E}_{12} = 3\hat{\mathscr{C}}' \cdot \mathcal{E}_{12} + 2\hat{\mathcal{L}}_6 \cdot \mathcal{E}_{12} + \hat{\mathcal{L}}_{12} \cdot \mathcal{E}_{12} + \hat{\mathcal{B}}_{18} \cdot \mathcal{E}_{12} =$$
$$= \left(3\hat{\mathscr{C}}' + 2\hat{\mathcal{L}}_6 + \hat{\mathcal{L}}_{12} + \hat{\mathcal{B}}_{18}\right) \cdot \mathcal{E}_{12} = \left(4K_{\hat{S}_{Cl}} - m_3\mathcal{E}_{12}\right) \cdot \mathcal{E}_{12} =$$
$$= 4K_{\hat{S}_{Cl}} \cdot \mathcal{E}_{12} - m_3\mathcal{E}_{12}^2 = -m_3\mathcal{E}_{12}^2 = 24m_3.$$

This proves assertion (iv). Finally, we have

$$24 = \hat{\mathcal{G}}_6 \cdot \mathcal{E}_{12} + \hat{\mathcal{L}}_{12} \cdot \mathcal{E}_{12} = \hat{\mathcal{R}}_6 \cdot \mathcal{E}_{12} + \hat{\mathcal{L}}_6 \cdot \mathcal{E}_{12} + \hat{\mathcal{L}}_{12} \cdot \mathcal{E}_{12} + \hat{\mathcal{G}}_6 \cdot \mathcal{E}_{12} =$$
$$= \left(\hat{\mathcal{R}}_6 + \hat{\mathcal{L}}_6 + \hat{\mathcal{L}}_{12} + \hat{\mathcal{G}}_6\right) \cdot \mathcal{E}_{12} = \left(4K_{\hat{S}_{Cl}} - m_4\mathcal{E}_{12}\right) \cdot \mathcal{E}_{12} =$$
$$= 4K_{\hat{S}_{Cl}} \cdot \mathcal{E}_{12} - m_4\mathcal{E}_{12}^2 = -m_4\mathcal{E}_{12}^2 = 24m_4,$$

which implies $m_4 = 1$. This proves assertion (v). $\square$

Using Lemma 15.4.2 and the adjunction formula one can find the intersection form of the irreducible components of the curves $\hat{\mathscr{C}}$, $\hat{\mathscr{C}}'$, $\hat{\mathcal{L}}_6$, $\hat{\mathcal{L}}_{12}$, $\hat{\mathcal{G}}_6$, $\hat{\mathcal{B}}_{18}$, $\hat{\mathcal{G}}_{15}$, $\hat{\mathcal{T}}_{15}$, $\hat{\mathcal{R}}_6$, and $\mathcal{E}_{12}$ on the surface $\hat{S}_{Cl}$. Consider the linear system

$$\mathcal{P}_{\hat{S}_{Cl}} = \left|4K_{\hat{S}_{Cl}} - \hat{\mathcal{L}}_{12} - \hat{\mathcal{B}}_{18} - \mathcal{E}_{12}\right|.$$

Lemma 15.4.3. *The linear system $\mathcal{P}_{\hat{S}_{Cl}}$ is a pencil that is free from base points. This pencil contains divisors $3\hat{\mathscr{C}}' + 2\hat{\mathcal{L}}_6$ and $3\hat{\mathscr{C}} + \hat{\mathcal{G}}_6 + 2\mathcal{E}_{12}$, and every divisor in this pencil is $\mathfrak{A}_5$-invariant. A general divisor in $\mathcal{P}_{\hat{S}_{Cl}}$ is a smooth irreducible curve of genus 16.*

Proof. It follows from Lemma 15.4.2(i),(ii),(iv) that

$$3\hat{\mathscr{C}}' + 2\hat{\mathcal{L}}_6 \sim 4K_{\hat{S}_{Cl}} - \hat{\mathcal{L}}_{12} - \hat{\mathcal{B}}_{18} - \mathcal{E}_{12} \sim 3\hat{\mathscr{C}} + \hat{\mathcal{G}}_6 + 2\mathcal{E}_{12},$$

which implies that the dimension of the linear system $\mathcal{P}_{\hat{S}_{Cl}}$ is at least one.

By Lemma 11.3.4, we have $\Sigma_{12} \not\subset \mathscr{C}'$. Moreover, the curves $\mathscr{C}$ and $\mathcal{L}_6$ are disjoint by Lemma 7.4.3, and the curves $\mathcal{G}_6$ and $\mathcal{L}_6$ are disjoint by Lemma 9.2.10. Therefore, the curves $\hat{\mathscr{C}}'$ and $\hat{\mathcal{L}}_6$ are disjoint from the curves $\hat{\mathscr{C}}$, $\hat{\mathcal{G}}_6$, and $\mathcal{E}_{12}$. This means that the supports of the divisors $3\hat{\mathscr{C}}' + 2\hat{\mathcal{L}}_6$ and $3\hat{\mathscr{C}} + \hat{\mathcal{G}}_6 + 2\mathcal{E}_{12}$ are disjoint, and thus the linear system $\mathcal{P}_{\hat{S}_{Cl}}$ is composed of a pencil. In fact, it is a pencil, since the divisor $3\hat{\mathscr{C}} + \hat{\mathcal{G}}_6 + 2\mathcal{E}_{12}$ has connected support and is reduced at a general point of $\hat{\mathcal{G}}_6$.

Since both divisors $3\hat{\mathscr{C}}' + 2\hat{\mathcal{L}}_6$ and $3\hat{\mathscr{C}} + \hat{\mathcal{G}}_6 + 2\mathcal{E}_{12}$ are $\mathfrak{A}_5$-invariant, the action of the group $\mathfrak{A}_5$ on the pencil $\mathcal{P}_{\hat{S}_{Cl}}$ is trivial by Lemma 5.5.9(v). Thus, each divisor in $\mathcal{P}_{\hat{S}_{Cl}}$ is $\mathfrak{A}_5$-invariant.

Let $\hat{Z}$ be a sufficiently general curve in the pencil $\mathcal{P}_{\hat{S}_{Cl}}$. Then $\hat{Z}$ is a smooth irreducible curve by Bertini's theorem. Let g be the genus of $\hat{Z}$. We compute

$$2g - 2 = \left(K_{\hat{S}_{Cl}} + \hat{Z}\right) \cdot \hat{Z} = K_{\hat{S}_{Cl}} \cdot \hat{Z} =$$
$$= h^*(H|_{S_{Cl}}) \cdot (3\hat{\mathscr{C}}' + 2\hat{\mathcal{L}}_6) = H \cdot (3\mathscr{C}' + 2\mathcal{L}_6) = 30.$$

This gives $g = 16$. $\square$

Lemma 15.4.4. The pencil $\mathcal{P}_{\hat{S}_{Cl}}$ contains the curve $\hat{\mathcal{G}}_{15}$.

Proof. By Lemma 14.4.4, the curve $\mathcal{T}_{15}$ intersects the curve $\mathscr{C}'$ by a single $\mathfrak{A}_5$-orbit of length 30. In particular, this gives $\hat{\mathscr{C}}' \cdot \hat{\mathcal{T}}_{15} \geqslant 30$. Thus, it follows from Lemma 15.4.2(iii) that

$$30 + \hat{\mathscr{C}}' \cdot \hat{\mathcal{G}}_{15} \leqslant \hat{\mathscr{C}}' \cdot \hat{\mathcal{T}}_{15} + \hat{\mathscr{C}}' \cdot \hat{\mathcal{G}}_{15} =$$
$$= \hat{\mathscr{C}}' \cdot \left(\hat{\mathcal{T}}_{15} + \hat{\mathcal{G}}_{15}\right) = 5K_{\hat{S}_{Cl}} \cdot \hat{\mathscr{C}}' = 5H \cdot \mathscr{C}' = 30,$$

which implies that $\hat{\mathscr{C}}' \cdot \hat{\mathcal{G}}_{15} = 0$. Similarly, $\mathcal{T}_{15}$ intersects the curve $\mathcal{L}_6$ by a single $\mathfrak{A}_5$-orbit of length 30 disjoint by Lemma 14.4.3. This gives

$$30 + \hat{\mathcal{L}}_6 \cdot \hat{\mathcal{G}}_{15} \leqslant \hat{\mathcal{L}}_6 \cdot \hat{\mathcal{T}}_{15} + \hat{\mathcal{L}}_6 \cdot \hat{\mathcal{G}}_{15} =$$
$$= \hat{\mathcal{L}}_6 \cdot \left(\hat{\mathcal{T}}_{15} + \hat{\mathcal{G}}_{15}\right) = 5K_{\hat{S}_{Cl}} \cdot \hat{\mathcal{L}}_6 = 5H \cdot \mathcal{L}_6 = 30,$$

by Lemma 15.4.2(iii). Therefore, we also have $\hat{\mathcal{L}}_6 \cdot \hat{\mathcal{G}}_{15} = 0$. Thus, we have

$$\hat{\mathcal{G}}_{15} \cdot \left(3\hat{\mathscr{C}}' + 2\hat{\mathcal{L}}_6\right) = 3\hat{\mathcal{G}}_{15} \cdot \hat{\mathscr{C}}' + 2\hat{\mathcal{G}}_{15} \cdot \hat{\mathcal{L}}_6 = 0.$$

By Lemma 15.4.3, the curve $\hat{\mathcal{G}}_{15}$ is contained in the support of some divisor of the pencil $\mathcal{P}_{\hat{S}_{Cl}}$. On the other hand, one has $\mathcal{E}_{12} \cdot \hat{\mathcal{G}}_{15} = 0$ by Lemma 15.3.4(iv), and

$$h^*(H|_{S_{Cl}}) \cdot \hat{Z} = h^*(H|_{S_{Cl}}) \cdot \hat{\mathcal{G}}_{15} = 30.$$

This means that the degrees of $\hat{Z}$ and $\hat{\mathcal{G}}_{15}$ with respect to some ample divisor on $\hat{S}_{Cl}$ are equal to each other. Therefore, $\hat{\mathcal{G}}_{15}$ is an element of the pencil $\mathcal{P}_{\hat{S}_{Cl}}$. □

Remark 15.4.5. It follows from Theorem 14.1.16(i) that a general curve $\hat{Z}$ of the pencil $\mathcal{P}_{\hat{S}_{Cl}}$ does not contain $\mathfrak{A}_5$-orbits of length 20, since the curve $\mathcal{G}_{10}$ is not contained in the surface S_{Cl} by Corollary 15.1.1, and $\mathcal{P}_{\hat{S}_{Cl}}$ has no fixed points by Lemma 15.4.3. Moreover, the intersection $\hat{Z} \cap \hat{\mathcal{T}}_{15}$ consists of

$$\hat{Z} \cdot \hat{\mathcal{T}}_{15} = \hat{Z} \cdot \left(5K_{\hat{S}_{Cl}} - \hat{\mathcal{G}}_{15}\right) = 150$$

points by Lemma 15.4.2(iii), which split into five $\mathfrak{A}_5$-orbits of length 30. All other $\mathfrak{A}_5$-orbits contained in Z are of length 60 (cf. Lemma 5.1.5).

Now we are ready to prove the main result of this section.

Theorem 15.4.6. *One has*

$$44 \leqslant \operatorname{rk} \operatorname{Pic}(\hat{S}_{Cl}) \leqslant 65.$$

Proof. By Corollary 15.2.9, we have $\operatorname{rk} \operatorname{Pic}(\hat{S}_{Cl}) \leqslant 65$.

The pencil $\mathcal{P}_{\hat{S}_{Cl}}$ defines a surjective morphism

$$\hat{\phi} \colon \hat{S}_{Cl} \to \mathbb{P}^1.$$

By Lemmas 15.4.3 and 15.4.4, three reducible fibers of $\hat{\phi}$ are the curves

$$\hat{\mathscr{C}}' + \hat{\mathcal{L}}_6, \ \hat{\mathscr{C}} + \hat{\mathcal{G}}_6 + \mathcal{E}_{12}$$

and $\hat{\mathcal{G}}_{15}$. Moreover, the pencil $\mathcal{P}_{\hat{S}_{Cl}}$ contains one more reducible curve that was not discussed yet. Let us describe it.

Let $\hat{\Sigma}_{15}$ be the proper transform of the $\mathfrak{A}_5$-orbit Σ_{15} on $\hat{S}_{Cl}$ via h. Note that Σ_{15} is not contained in any of the curves $\mathscr{C}$, $\mathscr{C}'$, $\mathcal{L}_6$, $\mathcal{L}_{12}$, $\mathcal{G}_6$, $\mathcal{B}_{18}$, and $\mathcal{G}_{15}$ by Lemmas 5.1.4 and 15.3.4(iii) and Corollary 8.3.5. Hence $\hat{\Sigma}_{15}$ is not contained in the curves $\hat{\mathscr{C}}$, $\hat{\mathscr{C}}'$, $\hat{\mathcal{L}}_6$, $\hat{\mathcal{L}}_{12}$, $\hat{\mathcal{G}}_6$, $\hat{\mathcal{G}}_{15}$, $\hat{\mathcal{B}}_{18}$, and $\mathcal{E}_{12}$. Let $\hat{N}_{15}$ be a divisor in $\mathcal{P}_{\hat{S}_{Cl}}$ that contains $\hat{\Sigma}_{15}$. Then $\hat{N}_{15}$ is an $\mathfrak{A}_5$-irreducible curve by Corollary 15.1.1, and $\hat{N}_{15}$ is different from $\hat{\mathcal{G}}_{15}$.

We claim that the curve $\hat{N}_{15}$ is reducible. Indeed, suppose that it is irreducible. Then $\hat{N}_{15}$ is singular at the points of $\hat{\Sigma}_{15}$ by Lemma 5.1.4. The arithmetic genus of $\hat{N}_{15}$, as well as of any other curve in $\mathcal{P}_{\hat{S}_{Cl}}$, equals 16 by Lemma 15.4.3. This gives a contradiction with Lemmas 4.4.6 and 5.1.5.

Denote by r the number of irreducible components of the curve $\hat{N}_{15}$. Then $r \geqslant 5$ by Corollary 5.1.2. Thus, the irreducible components of the curves $\hat{\mathcal{L}}_6$, $\hat{\mathcal{G}}_6$, $\mathcal{E}_{12}$, $\hat{\mathcal{G}}_{15}$, and $\hat{N}_{15}$ generate a sublattice Λ in the Neron–Severi lattice of the surface $\hat{S}_{Cl}$ such that the intersection form on Λ has signature $(0, 1, 37 + r)$. Keeping in mind that there is also some ample divisor on S_{Cl}, we get
$$\operatorname{rk} \operatorname{Pic}\left(\hat{S}_{Cl}\right) \geqslant 39 + r \geqslant 44.$$
$\square$

As Oscar Zariski remarked in [126, p. 110], "the evaluation of the Picard number for a given surface presents in general grave difficulties." Nevertheless, we believe that one can approach the problem in our case using the information on low degree $\mathfrak{A}_5$-invariant curves contained in S_{Cl}, and in particular the bound $\operatorname{rk} \operatorname{Pic}(\hat{S}_{Cl}) \geqslant 44$ given by Theorem 15.4.6 can be improved by taking into account the irreducible components of the curves $\mathcal{R}_6$, $\hat{\mathcal{L}}_{12}$, and $\hat{\mathcal{T}}_{15}$. It would be interesting to know whether $\operatorname{rk} \operatorname{Pic}(\hat{S}_{Cl}) = 65$ or not (cf. [116, Theorem 2]), because surfaces having maximal Picard rank are fun (see [7]). We expect that $\operatorname{rk} \operatorname{Pic}(\hat{S}_{Cl}) = 65$. Unfortunately, we were unable to prove this. It will also be interesting to find out whether the Picard group of the surface S_{Cl} has torsion or not.

Remark 15.4.7. One can use the properties of the pencil $\mathcal{P}_{\hat{S}_{Cl}}$ to give an alternative proof of Theorem 15.2.11. Indeed, we know from Corollary 15.2.10 that all singular points of the surface S_{Cl} are ordinary double points, and either $\operatorname{Sing}(S_{Cl}) = \Sigma_{12}$, or $\operatorname{Sing}(S_{Cl}) = \Sigma_{12} \cup \Sigma$ for some $\mathfrak{A}_5$-orbit of length 30 that is not contained in the curves $\mathscr{C}$, $\mathscr{C}'$, $\mathcal{L}_6$, $\mathcal{L}_{12}$, $\mathcal{G}_6$, and $\mathcal{B}_{18}$. In particular, replacing the morphism $h \colon \hat{S}_{Cl} \to S_{Cl}$ (about which we do not know a priori that it is just the blow-up of the points of Σ_{12}) by the minimal resolution of singularities of the surface S_{Cl}, we can construct the linear system $\mathcal{P}_{\hat{S}_{Cl}}$ exactly as we did above. The proof of Lemma 15.4.3 implies that $\mathcal{P}_{\hat{S}_{Cl}}$ is a base point free pencil that is generated by the divisors $3\hat{\mathscr{C}}' + 2\hat{\mathcal{L}}_6$ and $3\hat{\mathscr{C}} + \hat{\mathcal{G}}_6 + 2\mathcal{E}_{12}$. Moreover, the proof of Lemma 15.3.4 implies that S_{Cl} is smooth at every point of the curve $\mathcal{G}_{15}$, so that the proof of Lemma 15.4.4 gives $\hat{\mathcal{G}}_{15} \in \mathcal{P}_{\hat{S}_{Cl}}$. Now arguing as in the proof of

Theorem 15.4.6, we see that

$$\operatorname{rk}\operatorname{Pic}(\hat{S}_{Cl}) \geqslant 39 + |\operatorname{Sing}(S_{Cl}) \setminus \Sigma_{12}|.$$

The latter implies that S_{Cl} is smooth away from Σ_{12}, since $\operatorname{rk}\operatorname{Pic}(\hat{S}_{Cl}) \leqslant 65$ by Corollary 15.2.9.

Chapter 16

Summary of orbits, curves, and surfaces

In this chapter we summarize information about the incidence relations between some $\mathfrak{A}_5$-orbits of small length (those described in Theorem 7.3.5, and also the $\mathfrak{A}_5$-orbits Σ_{20}, Σ'_{20}, Σ_{30}, and Σ'_{30}), some $\mathfrak{A}_5$-irreducible curves of low degree (those described in Theorem 13.6.1, and also the curves C_{20}, $\mathcal{L}_{20}$, $\mathcal{G}_{10}$, and $\mathcal{T}_{15}$) and some $\mathfrak{A}_5$-invariant surfaces of low degree contained in V_5 (the singular surfaces $\mathscr{S}$, S_5, S_{10}, S'_{10}, and S_{15} of the pencil $\mathcal{Q}_2$, and the surface S_{Cl}). As in Table 13.1, by $C_{18}(0)$ we denote smooth rational $\mathfrak{A}_5$-invariant curves of degree 18 (whose existence is not known to us; see Remark 13.6.2).

16.1 Orbits vs. curves

In Table 16.1 we collect what we know about the incidence relation between $\mathfrak{A}_5$-orbits of small length and $\mathfrak{A}_5$-irreducible curves of low degree in V_5. A "+" sign in the table means that a curve labelling the corresponding row contains an $\mathfrak{A}_5$-orbit labelling the corresponding column. A "−" sign means that the corresponding curve does not contain the corresponding $\mathfrak{A}_5$-orbit. The information about the curve $\mathscr{C}$ is contained in Lemma 7.2.10. The information about the curve $\mathscr{C}'$ follows from Lemma 11.3.4 together with Lemmas 7.2.10 and 5.1.4. The information about the curve $\mathcal{L}_6$ follows from Lemma 7.4.3 and Corollary 7.4.9. The information about the curve $\mathcal{L}_{10}$ follows from Lemma 9.1.1, Corollaries 9.1.3 and 9.1.5, and Remark 9.1.6. The information about the curves $\mathcal{G}_5$ and $\mathcal{G}'_5$ follows from Lemmas 9.2.1 and 9.2.2, Corollary 9.2.4, and Remark 9.2.5. The information about the curve $\mathcal{L}_{12}$ follows from Lemma 7.2.12(iii) and Remark 8.3.6. The information about the curve $\mathcal{G}_6$ follows from Corollaries 8.3.5 and 9.2.11. The information about the curve $\mathcal{L}_{15}$ follows from Corollaries 7.4.4 and 9.1.11 and Lemma 9.1.10(ii),(iv). The information about the curve C_{16} follows from Lemmas 7.2.10 and 8.1.16(iii),(iv). The information about the curve C'_{16} is contained in Remark 13.1.4. The information about the curve $\mathcal{T}_6$ follows from Lemmas 13.2.3 and 13.2.4. The information about the curve C_{18}^0 is contained in Corollary 13.4.3(i),(ii). The information about the curve $\mathcal{B}_{18}$ is contained in Remark 13.5.3. The information about the curves of type $C_{18}(0)$ follows from Lemma 5.1.4 together with Corollaries 15.1.3 and 13.6.5. Note that we do not know if the curves of this type exist (see Remark 13.6.2), so that the corresponding row of the table describes the $\mathfrak{A}_5$-orbits on such curves *assuming* their existence. The information about the curve C_{20} follows from assertions (iii), (vi), and (vii) of Theorem 8.1.8. The information about the curve $\mathcal{L}_{20}$ follows from Lemma 7.2.12(ii),(iii). The information about the curve $\mathcal{G}_{10}$ is implied by assertions (i), (ii), (iii), (vi), and (ix) of Theorem 14.1.16. Finally, the information about the curve $\mathcal{T}_{15}$ is contained in assertions (i), (ii), (iii), (v), and (viii) of Theorem 14.3.2 together with Lemma 14.3.6 and Theorem 14.3.8.

16.2 Orbits vs. surfaces

In Table 16.2 we collect what we know about the incidence relation between $\mathfrak{A}_5$-orbits of small length and $\mathfrak{A}_5$-invariant surfaces of low degree in V_5. A "+" sign in the table means that a surface labelling the corresponding column contains an $\mathfrak{A}_5$-orbit labelling the corresponding row. A "−" sign means that the corresponding surface does not contain the corresponding $\mathfrak{A}_5$-orbit. The information about the surface $\mathscr{S}$ follows from Lemma 7.2.10. The information about the surfaces S_5, S_{10}, S'_{10}, and S_{15} follows from Theorem 8.2.1(i) and assertions (iii), (vi), and (vii) of Theorem 8.1.8. The information about the surface S_{Cl} follows from Lemmas 12.2.3, 12.2.4, and 12.2.5(i) together with Corollary 12.2.7.

16.3 Curves vs. surfaces

In Table 16.3 we collect what we know about the incidence relation between $\mathfrak{A}_5$-irreducible curves and $\mathfrak{A}_5$-invariant surfaces of low degree in V_5. A "+" sign in the table means that a surface S labelling the corresponding column

Table 16.1: Incidence relation for orbits and curves

	Σ_5	Σ_{10}	Σ'_{10}	Σ_{12}	Σ'_{12}	Σ_{15}	Σ_{20}	Σ'_{20}	Σ_{30}	Σ'_{30}
$\mathscr{C}$	−	−	−	+	−	−	+	−	+	−
$\mathscr{C}'$	−	−	−	−	+	−	−	−	−	−
$\mathcal{L}_6$	−	−	−	−	+	−	−	−	−	−
$\mathcal{L}_{10}$	−	−	−	−	−	−	−	+	−	−
$\mathcal{G}_5$	−	−	−	−	−	−	−	+	−	−
$\mathcal{G}'_5$	−	−	−	−	−	−	−	+	−	−
$\mathcal{L}_{12}$	−	−	−	+	+	−	−	−	−	−
$\mathcal{G}_6$	−	−	−	+	−	−	−	−	−	−
$\mathcal{L}_{15}$	+	−	−	−	−	+	−	−	−	+
C_{16}	−	−	−	+	−	−	−	+	−	−
C'_{16}	−	−	−	−	+	−	−	+	−	−
$\mathcal{T}_6$	−	−	−	+	−	−	−	−	−	−
C^0_{18}	−	−	−	−	+	−	−	−	−	−
$\mathcal{B}_{18}$	−	−	−	+	+	−	−	−	−	−
$C_{18}(0)$	−	−	−	+	−	−	−	−	−	−
C_{20}	−	−	−	+	+	−	−	+	−	−
$\mathcal{L}_{20}$	−	−	−	−	−	−	+	+	−	−
$\mathcal{G}_{10}$	+	+	+	−	−	−	+	−	−	−
$\mathcal{T}_{15}$	−	+	+	−	−	+	−	−	+	−

contains a curve C labelling the corresponding row. Otherwise we list the $\mathfrak{A}_5$-orbits whose union is the intersection $S \cap C$. To make the notation more compact we denote the orbits Σ_5, Σ_{10}, Σ'_{10}, Σ_{12}, Σ'_{12}, Σ_{15}, Σ_{20}, Σ'_{20}, Σ_{30}, and Σ'_{30} by the symbols 5, 10, 10', 12, 12', 15, 20, 20', 30, and 30', respectively. By [20] and [30] we denote any of the $\mathfrak{A}_5$-orbits of length 20 and 30, respectively, that are different from Σ_{20}, Σ'_{20}, Σ_{30}, and Σ'_{30}. We know that the curves $\mathscr{C}$, $\mathcal{L}_{12}$, C_{16}, and $\mathcal{L}_{20}$ are contained in the surface $\mathscr{S}$, and also that the curve C_{20} is contained in the surfaces $\mathscr{S}$, S_5, S_{10}, S'_{10}, and S_{15}, from the constructions of these curves. By Theorem 13.6.1 the curves $\mathscr{C}'$, $\mathcal{L}_6$, $\mathcal{L}_{10}$, $\mathcal{G}_5$, $\mathcal{G}'_5$, $\mathcal{G}_6$, C'_{16}, $\mathcal{T}_6$, C^0_{18}, $\mathcal{B}_{18}$, and any curve of type $C_{18}(0)$ are not contained in any of the surfaces $\mathscr{S}$, S_5, S_{10}, S'_{10}, and S_{15}, while the curves $\mathscr{C}$, $\mathcal{L}_{12}$, and C_{16} are not contained in any of the surfaces S_5, S_{10}, S'_{10}, and S_{15}. Also, the curve $\mathcal{L}_{20}$ is not contained in any of the

Table 16.2: Incidence relation for orbits and surfaces

	$\mathscr{S}$	S_5	S_{10}	S'_{10}	S_{15}	S_{Cl}
Σ_5	−	+	−	−	−	−
Σ_{10}	−	−	+	−	−	+
Σ'_{10}	−	−	−	+	−	+
Σ_{12}	+	+	+	+	+	+
Σ'_{12}	+	+	+	+	+	+
Σ_{15}	−	−	−	−	+	+
Σ_{20}	+	−	−	−	−	+
Σ'_{20}	+	+	+	+	+	−
Σ_{30}	+	−	−	−	−	+
Σ'_{30}	+	−	−	−	−	−

surfaces S_5, S_{10}, S'_{10}, and S_{15} since it is not contained in the base locus C_{20} of the pencil $\mathcal{Q}_2$. Therefore, the information about the splitting of the intersections of the curves $\mathscr{C}$, $\mathscr{C}'$, $\mathcal{L}_6$, $\mathcal{L}_{10}$, $\mathcal{G}_5$, $\mathcal{G}'_5$, $\mathcal{L}_{12}$, $\mathcal{G}_6$, C_{16}, C'_{16}, $\mathcal{T}_6$, C^0_{18}, $\mathcal{B}_{18}$, and $\mathcal{L}_{20}$ with the surfaces $\mathscr{S}$, S_5, S_{10}, S'_{10}, and S_{15} into $\mathfrak{A}_5$-orbits follows from Corollary 8.1.11 and the information collected in Table 16.1. Note that some of these intersections were already described earlier, see, e.g., Lemmas 7.4.3 and 11.3.4. The same works for curves of type $C_{18}(0)$, under an assumption that curves of this type do exist. The information about the intersections of the curve $\mathcal{L}_{15}$ with the surfaces $\mathscr{S}$, S_5, S_{10}, S'_{10}, and S_{15} is contained in Corollary 9.1.11 and Remark 9.1.12. The information about the intersections of the curve $\mathcal{G}_{10}$ with the surfaces $\mathscr{S}$, S_5, S_{10}, S'_{10}, and S_{15} is contained in Remark 14.1.17. The intersections of the curve $\mathcal{T}_{15}$ with the surfaces $\mathscr{S}$, S_5, S_{10}, S'_{10}, and S_{15} are described in Lemmas 14.3.6 and 14.3.7. We know that the surface S_{Cl} contains the curves $\mathscr{C}$, $\mathscr{C}'$, $\mathcal{L}_6$, $\mathcal{L}_{12}$, $\mathcal{B}_{18}$, and $\mathcal{T}_{15}$ by Corollary 15.1.1 and Theorem 14.3.2(vii). The intersections of the surface S_{Cl} with the curves $\mathcal{L}_{10}$, $\mathcal{G}_5$, $\mathcal{G}'_5$, C_{16}, $\mathcal{T}_6$, C^0_{18}, $\mathcal{L}_{20}$, C_{20}, $\mathcal{G}_{10}$, and any curve of type $C_{18}(0)$ are described in Corollaries 12.2.8, 12.2.9, 15.1.2, 15.1.3, 12.2.10, and 12.2.11 and Lemma 14.2.12. Finally, the intersections of S_{Cl} with the curves $\mathcal{L}_{15}$ and C'_{16} are described in Lemma 12.2.12 and Remark 13.1.4.

Table 16.3: Incidence relation for curves and surfaces

	$\mathscr{S}$	S_5	S_{10}	S'_{10}	S_{15}	S_{Cl}
$\mathscr{C}$	+	12	12	12	12	+
$\mathscr{C}'$	12'	12'	12'	12'	12'	+
$\mathcal{L}_6$	12'	12'	12'	12'	12'	+
$\mathcal{L}_{10}$	20'	20'	20'	20'	20'	[30]
$\mathcal{G}_5$	20'	20'	20'	20'	20'	[30]
$\mathcal{G}'_5$	20'	20'	20'	20'	20'	[30]
$\mathcal{L}_{12}$	+	12, 12'	12, 12'	12, 12'	12, 12'	+
$\mathcal{G}_6$	12	12	12	12	12	+
$\mathcal{L}_{15}$	30'	5	[30]	[30]	15	15, [30]
C_{16}	+	12, 20'	12, 20'	12, 20'	12, 20'	12
C'_{16}	12', 20'	12', 20'	12', 20'	12', 20'	12', 20'	12'
$\mathcal{T}_6$	12	12	12	12	12	12, [30]
C^0_{18}	12'	12'	12'	12'	12'	12', [30]
$\mathcal{B}_{18}$	12, 12'	12, 12'	12, 12'	12, 12'	12, 12'	+
$C_{18}(0)$	12	12	12	12	12	12, [30]
C_{20}	+	+	+	+	+	12, 12'
$\mathcal{L}_{20}$	+	20'	20'	20'	20'	20
$\mathcal{G}_{10}$	20	5	10, [20]	10', [20]	[20], [20]	10, 10', 20, [20]
$\mathcal{T}_{15}$	30	[30], [30], [30]	10, [30]	10', [30]	15, [30]	+

16.4 Curves vs. curves

In Table 16.4 we collect what we know about the intersections between $\mathfrak{A}_5$-irreducible curves of low degree in V_5. In each cell we describe the intersection of the curves labeling the corresponding row and column of the table. As in Table 16.3, we denote the orbits Σ_5, Σ_{10}, Σ'_{10}, Σ_{12}, Σ'_{12}, Σ_{15}, Σ_{20}, Σ'_{20}, Σ_{30}, and Σ'_{30} by the symbols 5, 10, 10′, 12, 12′, 15, 20, 20′, 30, and 30′, respectively. By [20] and [30] we denote any of the $\mathfrak{A}_5$-orbits of length 20 and 30, respectively, that are different from Σ_{20}, Σ'_{20}, Σ_{30}, and Σ'_{30}. As above, we know that the curves $\mathscr{C}$, $\mathcal{L}_{12}$, C_{16}, $\mathcal{L}_{20}$, and C_{20} are contained in the surface $\mathscr{S}$. Their pairwise intersections are described by Lemma 7.2.12(v),(vi), assertions (iii), (viii), and (ix) of Theorem 8.1.8 and assertions (v), (vi), (vii), and (viii) of Lemma 8.1.16. By Theorem 13.6.1 the curves $\mathscr{C}'$, $\mathcal{L}_6$, $\mathcal{L}_{10}$, $\mathcal{G}_5$, $\mathcal{G}'_5$, $\mathcal{G}_6$, C'_{16}, $\mathcal{T}_6$, C^0_{18}, $\mathcal{B}_{18}$, and any curve of type $C_{18}(0)$ are contained in the surfaces of the pencil $\mathcal{Q}_2$ different from $\mathscr{S}$, while the curves $\mathscr{C}$, $\mathcal{L}_{12}$, and C_{16} are not contained in any surface of $\mathcal{Q}_2$ except S. Also, suppose that C and Z are different curves such that each of them is one of the curves $\mathscr{C}'$, $\mathcal{L}_6$, $\mathcal{L}_{10}$, $\mathcal{G}_5$, $\mathcal{G}'_5$, $\mathcal{G}_6$, C'_{16}, $\mathcal{T}_6$, C^0_{18}, $\mathcal{B}_{18}$, or a curve of type $C_{18}(0)$; then we know from Lemma 13.6.3 that C and Z are not contained in the same surface of the pencil $\mathcal{Q}_2$, unless C and Z are the curves $\mathcal{G}_6$ and $\mathcal{B}_{18}$ (or another way around). The curve $\mathcal{L}_{20}$ is not contained in any of the surfaces of $\mathcal{Q}_2$ except $\mathscr{S}$ since it is not contained in the base locus C_{20} of the pencil $\mathcal{Q}_2$. Therefore, the remaining information about the splitting of the pairwise intersections of the curves $\mathscr{C}$, $\mathscr{C}'$, $\mathcal{L}_6$, $\mathcal{L}_{10}$, $\mathcal{G}_5$, $\mathcal{G}'_5$, $\mathcal{L}_{12}$, $\mathcal{G}_6$, C_{16}, C'_{16}, $\mathcal{T}_6$, C^0_{18}, $\mathcal{B}_{18}$, and $\mathcal{L}_{20}$ into $\mathfrak{A}_5$-orbits follows from Corollary 8.1.12 and the information collected in Table 16.1, with the only exception being the intersection $\mathcal{G}_6 \cap \mathcal{B}_{18}$ that is described by Lemma 12.3.6. Note that some of the above intersections were already explicitly described earlier, see, e.g., Corollaries 10.2.8 and 13.6.6. In the same way we obtain the information about the intersections of the curves of type $C_{18}(0)$, assuming their existence, with the curves $\mathscr{C}$, $\mathscr{C}'$, $\mathcal{L}_6$, $\mathcal{L}_{10}$, $\mathcal{G}_5$, $\mathcal{G}'_5$, $\mathcal{L}_{12}$, $\mathcal{G}_6$, C_{16}, C'_{16}, $\mathcal{T}_6$, C^0_{18}, $\mathcal{B}_{18}$, and $\mathcal{L}_{20}$. We emphasize that we were not able to prove that the curves of type $C_{18}(0)$ do exist, nor to find out if it there is a unique curve of this type assuming its existence (see Remark 13.6.2). The "?" sign in the cells of the table corresponds to the self-intersection of the curves of this type. The information about the intersection of the curve C_{20} with the curves $\mathscr{C}'$, $\mathcal{L}_6$, $\mathcal{L}_{10}$, $\mathcal{G}_5$, $\mathcal{G}'_5$, $\mathcal{G}_6$, C'_{16}, $\mathcal{T}_6$, C^0_{18}, $\mathcal{B}_{18}$, and the curves of type $C_{18}(0)$ follows from Corollary 8.1.11. The information about the intersections of the curve $\mathcal{L}_{15}$ with the curves $\mathscr{C}$, $\mathcal{L}_{12}$, C_{16}, $\mathcal{L}_{20}$, and C_{20} is contained in Corollary 9.1.11. The information about the intersections of the curve $\mathcal{L}_{15}$ with the curves $\mathscr{C}'$, $\mathcal{L}_6$, $\mathcal{L}_{10}$, $\mathcal{G}_5$, $\mathcal{G}'_5$, $\mathcal{G}_6$, C'_{16}, $\mathcal{T}_6$, C^0_{18}, $\mathcal{B}_{18}$, and $\mathcal{G}_{10}$ is contained in Lemmas 11.3.7, 9.1.14, 9.1.15, 9.2.7, 13.1.6, 13.2.7, and 14.2.6 together with Corollaries 13.6.8 and 13.6.7. The intersections of the curve $\mathcal{G}_{10}$ with the curves $\mathscr{C}$, $\mathcal{L}_{12}$, C_{16}, $\mathcal{L}_{20}$, and C_{20} are described by Lemma 14.2.1. The intersections of the curve $\mathcal{G}_{10}$ with the curves $\mathscr{C}'$, $\mathcal{L}_6$, $\mathcal{L}_{10}$, $\mathcal{G}_5$, $\mathcal{G}'_5$, $\mathcal{G}_6$, C'_{16}, $\mathcal{T}_6$, C^0_{18}, and $\mathcal{B}_{18}$ are described by Lemmas 14.2.2, 14.2.3, 14.2.4, 14.2.5, 14.2.8, 14.2.9, and 14.2.10. The intersection of the curves $\mathcal{T}_{15}$ and $\mathcal{L}_{15}$ is described in Theorem 14.3.2(iv). The intersections of the curve $\mathcal{T}_{15}$ with the curves $\mathscr{C}$, $\mathscr{C}'$, $\mathcal{L}_6$, $\mathcal{L}_{10}$, $\mathcal{G}_5$, $\mathcal{G}'_5$, $\mathcal{L}_{12}$, $\mathcal{G}_6$, C_{16}, C'_{16}, $\mathcal{T}_6$, C^0_{18}, $\mathcal{B}_{18}$, $\mathcal{L}_{20}$, C_{20}, and $\mathcal{G}_{10}$ are described by Lemmas 14.4.1, 14.4.2, 14.4.3, 14.4.4, and 14.4.5. Finally, the intersections of any curve of type $C_{18}(0)$, under an assumption that such curves exist, with the curves $\mathcal{L}_{15}$, $\mathcal{G}_{10}$, and $\mathcal{T}_{15}$ are described by Corollary 15.1.4 and Remark 14.2.11.

Table 16.4: Incidence relation for curves

	$\mathcal{C}$	$\mathcal{C}'$	$\mathcal{C}''$	$\mathcal{L}_6$	$\mathcal{L}_{10}$	$\mathcal{G}_5$	$\mathcal{G}_5'$	$\mathcal{L}_{12}$	$\mathcal{G}_6$	$\mathcal{L}_{15}$	C_{16}	C_{16}'	$\mathcal{T}_6$	C_{18}^0	$\mathcal{B}_{18}$	$C_{18}(0)$	C_{20}	$\mathcal{L}_{20}$	$\mathcal{G}_{10}$	$\mathcal{T}_{15}$
$\mathcal{C}$	$\mathcal{C}$	$\varnothing$	$\varnothing$	$\varnothing$	$\varnothing$	$\varnothing$	$\varnothing$	12	12	$\varnothing$	12	$\varnothing$	12	$\varnothing$	12	12	12	20	20	30
$\mathcal{C}'$	$\varnothing$	$\mathcal{C}'$	12'	$\varnothing$	$\varnothing$	$\varnothing$	$\varnothing$	12'	$\varnothing$	$\varnothing$	$\varnothing$	12'	$\varnothing$	12'	12'	$\varnothing$	12'	$\varnothing$	[20]	[30]
$\mathcal{C}''$	$\varnothing$	12'	$\mathcal{C}''$	$\varnothing$	$\varnothing$	$\varnothing$	$\varnothing$	12'	$\varnothing$	[30]	$\varnothing$	12'	$\varnothing$	12'	12'	$\varnothing$	12'	$\varnothing$	$\varnothing$	[30]
$\mathcal{L}_6$	$\varnothing$	$\varnothing$	$\varnothing$	$\mathcal{L}_6$	$\varnothing$	20'	20'	$\varnothing$	$\varnothing$	[30]	20'	20'	$\varnothing$	$\varnothing$	$\varnothing$	$\varnothing$	20'	20'	$\varnothing$	[30]
$\mathcal{L}_{10}$	$\varnothing$	$\varnothing$	$\varnothing$	$\varnothing$	$\mathcal{L}_{10}$	$\mathcal{G}_5$	20'	$\varnothing$	$\varnothing$	$\varnothing$	20'	20'	$\varnothing$	$\varnothing$	$\varnothing$	$\varnothing$	20'	20'	$\varnothing$	[30]
$\mathcal{G}_5$	$\varnothing$	$\varnothing$	$\varnothing$	20'	20'	$\mathcal{G}_5$	20'	$\varnothing$	$\varnothing$	$\varnothing$	20'	20'	$\varnothing$	$\varnothing$	$\varnothing$	$\varnothing$	20'	20'	[20]	[30]
$\mathcal{G}_5'$	$\varnothing$	$\varnothing$	$\varnothing$	20'	20'	20'	$\mathcal{G}_5'$	$\varnothing$	$\varnothing$	$\varnothing$	20'	20'	$\varnothing$	$\varnothing$	$\varnothing$	$\varnothing$	20'	20'	[20]	[30]
$\mathcal{L}_{12}$	12	12'	12'	$\varnothing$	$\varnothing$	$\varnothing$	$\varnothing$	$\mathcal{L}_{12}$	12	$\varnothing$	12	12'	12	12'	12,12'	12	12,12'	$\varnothing$	$\varnothing$	$\varnothing$
$\mathcal{G}_6$	12	$\varnothing$	$\varnothing$	$\varnothing$	$\varnothing$	$\varnothing$	$\varnothing$	12	$\mathcal{G}_6$	$\varnothing$	12	$\varnothing$	12	$\varnothing$	12	12	30'	$\varnothing$	5	[30],[30]
$\mathcal{L}_{15}$	$\varnothing$	$\varnothing$	$\varnothing$	[30]	$\varnothing$	$\varnothing$	$\varnothing$	$\varnothing$	$\varnothing$	$\mathcal{L}_{15}$	30'	$\varnothing$	[30]	$\varnothing$	$\varnothing$	$\varnothing$	12,20'	20'	$\varnothing$	15
C_{16}	12	$\varnothing$	$\varnothing$	20'	20'	20'	20'	12	12	30'	C_{16}	20'	12	$\varnothing$	12	12	12,20'	$\varnothing$	$\varnothing$	$\varnothing$
C_{16}'	$\varnothing$	12'	12'	20'	20'	20'	20'	12'	$\varnothing$	$\varnothing$	20'	C_{16}'	$\varnothing$	12'	12'	$\varnothing$	12'	20'	[20]	$\varnothing$
$\mathcal{T}_6$	12	$\varnothing$	$\varnothing$	$\varnothing$	$\varnothing$	$\varnothing$	$\varnothing$	12	12	[30]	12	$\varnothing$	$\mathcal{T}_6$	$\varnothing$	12	12	12	$\varnothing$	5	$\varnothing$
C_{18}^0	$\varnothing$	12'	12'	$\varnothing$	$\varnothing$	$\varnothing$	$\varnothing$	12'	$\varnothing$	$\varnothing$	$\varnothing$	12'	$\varnothing$	C_{18}^0	12'	$\varnothing$	12'	$\varnothing$	$\varnothing$	[30]
$\mathcal{B}_{18}$	12	12'	12'	$\varnothing$	$\varnothing$	$\varnothing$	$\varnothing$	12	12	$\varnothing$	12	12'	12	12'	$\mathcal{B}_{18}$	12	12,12'	$\varnothing$	[20]	[30]
$C_{18}(0)$	12	$\varnothing$	$\varnothing$	$\varnothing$	$\varnothing$	$\varnothing$	$\varnothing$	12	12	$\varnothing$	12	$\varnothing$	12	$\varnothing$	12	?	12	$\varnothing$	$\varnothing$	[30]
C_{20}	12	12'	12'	20'	20'	20'	20'	12,12'	12	30'	12,20'	12'	12	12'	12,12'	12	C_{20}	20'	$\varnothing$	$\varnothing$
$\mathcal{L}_{20}$	20	$\varnothing$	$\varnothing$	20'	20'	20'	20'	$\varnothing$	$\varnothing$	5	$\varnothing$	20'	$\varnothing$	$\varnothing$	$\varnothing$	$\varnothing$	20'	$\mathcal{L}_{20}$	20	$\varnothing$
$\mathcal{G}_{10}$	20	[20]	$\varnothing$	$\varnothing$	[20]	[20]	[20]	$\varnothing$	[30],[30]	15	$\varnothing$	$\varnothing$	[30]	[20]	[20]	[20]	$\varnothing$	20	$\mathcal{G}_{10}$	10,10'
$\mathcal{T}_{15}$	30	[30]	[30]	[30]	[30]	[30]	[30]	$\varnothing$	[30],[30]	15	$\varnothing$	$\varnothing$	15	$\varnothing$	[30]	[30]	[30]	$\varnothing$	10,10'	$\mathcal{T}_{15}$

Part V

Singularities of linear systems

Chapter 17

Base loci of invariant linear systems

In this chapter we study $\mathfrak{A}_5$-invariant linear systems on V_5 with special prescribed base loci. In particular, we completely describe the base locus of the unique $\mathfrak{A}_5$-invariant two-dimensional linear subsystem in $|2H|$, and establish other technical results that will be used in the proof of Theorem 1.4.1 in Chapter 18.

17.1 Orbits of length 10

In this section, we prove that the points of each of the $\mathfrak{A}_5$-orbits of length 10 in V_5 impose independent linear conditions on the elements of $H^0(\mathcal{O}_{V_5}(2H))$. We will use this result in §17.2. The main tool will be the following theorem of Jack Edmonds.

Theorem 17.1.1 ([41, Theorem 2]). *Let $d \geqslant 2$, and Σ is a finite subset $\mathbb{P}^n$. If any linear subspace $\mathbb{P}^k \subset \mathbb{P}^n$ contains at most $kd+1$ points of Σ, then the points of Σ impose independent linear conditions on the forms of degree d in $\mathbb{P}^n$.*

Remark 17.1.2. By Corollary 7.5.4, the linear span of the $\mathfrak{A}_5$-orbit Σ_{10} is the whole $\mathbb{P}^6$. In particular, each linear subspace $\mathbb{P}^k \subset \mathbb{P}^6$, $1 \leqslant k \leqslant 6$, contains at most $k+4$ points of Σ_{10}. The same applies to the $\mathfrak{A}_5$-orbit Σ'_{10}.

Using Theorem 17.1.1 we prove the following.

Lemma 17.1.3. *The points of each of the $\mathfrak{A}_5$-orbits Σ_{10} or Σ'_{10} impose independent linear conditions on the elements of $H^0(\mathcal{O}_{V_5}(2H))$.*

Proof. It is sufficient to prove that the points of the $\mathfrak{A}_5$-orbit Σ_{10} impose independent linear conditions on the elements of $H^0(\mathcal{O}_{\mathbb{P}^6}(2))$. By Theorem 17.1.1 it is enough to check that a line (resp., $\mathbb{P}^2$, $\mathbb{P}^3$, $\mathbb{P}^4$) cannot contain 3 (resp., 5, 7, 9) points of Σ_{10}.

Suppose that there is a line $L \subset \mathbb{P}^6$ that contains $p \geqslant 3$ points of Σ_{10}. Let $L_1 = L, L_2, \ldots, L_r$ be the $\mathfrak{A}_5$-orbit of L. Each of the lines L_i contains p points of Σ_{10}, and through each point of Σ_{10} there passes an equal number (say, c) of the lines L_i. One has $c \leqslant 3$ by Theorem 7.1.9(i). Since

$$rp = 10c,$$

we obtain the inequality $r \leqslant 10$, so that either $r = 6$ or $r = 10$ by Proposition 7.4.1. In the former case, the lines $L_1, \ldots, L_6$ are the lines of $\mathcal{L}_6$ by Proposition 7.4.1, which is impossible by Corollary 7.4.9. In the latter case, the lines $L_1, \ldots, L_{10}$ are the lines of $\mathcal{L}_{10}$ by Proposition 7.4.1, which is impossible by Corollary 9.1.5.

Suppose that there is a plane $\Pi \subset \mathbb{P}^6$ that contains $p \geqslant 5$ points of Σ_{10}. Let $\Pi' \neq \Pi$ be some other plane from the $\mathfrak{A}_5$-orbit of Π. Then Π' contains p points of Σ_{10} as well. If $\Pi \cap \Pi' = \varnothing$, then the union $\Pi \cup \Pi'$ spans a hyperplane in $\mathbb{P}^6$, and $\Sigma_{10} \subset \Pi \cup \Pi'$, which is impossible by Remark 17.1.2. If $\Pi \cap \Pi'$ is a point, then the union $\Pi \cup \Pi'$ spans a four-dimensional subspace of $\mathbb{P}^6$ and contains at least 9 points of Σ_{10}, which is impossible by Remark 17.1.2. If $\Pi \cap \Pi' = L$ is a line, then L contains at most two points of Σ_{10}, so that the union $\Pi \cup \Pi'$ spans a three-dimensional subspace of $\mathbb{P}^6$ and contains at least 8 points of Σ_{10}, which is again impossible by Remark 17.1.2.

Suppose that there is a three-dimensional subspace $\mathbb{P} \subset \mathbb{P}^6$ that contains $p \geqslant 7$ points of Σ_{10}. Let $\mathbb{P}' \neq \mathbb{P}$ be some other subspace from the $\mathfrak{A}_5$-orbit of $\mathbb{P}$. Then $\mathbb{P}'$ also contains p points of Σ_{10}. The intersection $\mathbb{P} \cap \mathbb{P}'$ must contain at least 4 points of Σ_{10}, so that $\mathbb{P} \cap \mathbb{P}' = \Pi$ is a plane. Since Π cannot contain more than 4 points of Σ_{10}, we see that the union $\mathbb{P} \cup \mathbb{P}'$ spans a four-dimensional subspace of $\mathbb{P}^6$ and contains the whole Σ_{10}, which is impossible by Remark 17.1.2.

Finally, suppose that some four-dimensional subspace of $\mathbb{P}^6$ contains at least 9 points of Σ_{10}. An immediate contradiction with Remark 17.1.2 completes the proof. □

17.2 Linear system $\mathcal{Q}_3$

In this section, we determine the base locus of the unique $\mathfrak{A}_5$-invariant two-dimensional linear system $\mathcal{Q}_3$ contained in $|2H|$.

Remark 17.2.1. Consider the rational $\mathfrak{A}_5$-invariant map

$$\xi\colon V_5 \dashrightarrow \mathbb{P}^2 \cong \mathbb{P}(W_3')$$

given by the linear system $\mathcal{Q}_3$. Since there are no $\mathfrak{A}_5$-orbits of length at most 5 in $\mathbb{P}^2$ by Lemma 5.3.1(i), we see that the map ξ must be undefined in the points of Σ_5. Therefore, one has $\Sigma_5 \subset \mathrm{Bs}(\mathcal{Q}_3)$.

Lemma 17.2.2. *The base locus of the linear system $\mathcal{Q}_3$ does not contain rational normal sextic curves.*

Proof. Suppose that C is a rational normal sextic curve that is contained in the base locus of the linear system $\mathcal{Q}_3$. Let $\mathcal{I}_C \subset \mathcal{O}_{V_5}$ be the corresponding ideal sheaf. Since a rational normal curve is projectively normal, there is an exact sequence of $\mathfrak{A}_5$-representations

$$0 \to H^0\big(\mathcal{I}_C \otimes \mathcal{O}_{V_5}(2H)\big) \longrightarrow H^0\big(\mathcal{O}_{V_5}(2H)\big) \xrightarrow{\alpha} H^0\big(\mathcal{O}_C \otimes \mathcal{O}_{V_5}(2H)\big) \to 0.$$

By Lemma 8.1.1, the vector space $H^0(\mathcal{O}_{V_5}(2H))$ contains a unique three-dimensional $\mathfrak{A}_5$-subrepresentation. Since C is contained in the base locus of the linear system $\mathcal{Q}_3$, this three-dimensional $\mathfrak{A}_5$-subrepresentation must be contained in the kernel of α. On the other hand,

$$\dim\big(\ker(\alpha)\big) = h^0(\mathcal{O}_{V_5}(2H)) - h^0(\mathcal{O}_C \otimes \mathcal{O}_{V_5}(2H)) = 23 - 13 = 10.$$

Therefore, all one-dimensional subrepresentations of $H^0(\mathcal{O}_{V_5}(2H))$ are also contained in the kernel of α by Corollary 8.1.2. This means that $C \subset \mathrm{Bs}(\mathcal{Q}_2)$, which is not the case by Theorem 8.1.8(i). $\square$

Lemma 17.2.3. *The base locus of the linear system $\mathcal{Q}_3$ does not contain the curve $\mathcal{L}_{12}$.*

Proof. Suppose that $\mathcal{L}_{12}$ is contained in the base locus of the linear system $\mathcal{Q}_3$. Let us use notation of §7.2. Let $\hat{\mathcal{L}}_{12} \subset \hat{\mathscr{S}}$ be the proper transform of the curve $\mathcal{L}_{12} \subset \mathscr{S}$ on the threefold $\mathcal{Y}$. Let M be a general surface in $\mathcal{Q}_3$. Denote by $\hat{M}$ its proper transform on the threefold $\mathcal{Y}$. Since the rational normal sextic curve $\mathscr{C}$ is not contained in the base locus of the linear system $\mathcal{Q}_3$ by Lemma 17.2.2, the curve $\mathscr{C}$ is not contained in M. Thus, we have $\hat{M} \sim \nu^*(2H)$. Hence, $\hat{M}|_{\hat{\mathscr{S}}}$ is an effective divisor of bi-degree $(2, 10)$ on $\hat{\mathscr{S}} \cong \mathbb{P}^1 \times \mathbb{P}^1$ by Lemma 7.2.3(ii). On the other hand, its support contains the curve $\hat{\mathcal{L}}_{12}$, which is a divisor of bi-degree $(0, 12)$ on $\hat{\mathscr{S}}$. This is a contradiction. $\square$

Lemma 17.2.4. The base locus of the linear system $\mathcal{Q}_3$ does not contain the curves $\mathcal{L}_{10}$, $\mathcal{G}_5$, $\mathcal{G}_5'$, and $\mathcal{G}_6$.

Proof. Let C be one of the curves $\mathcal{L}_{10}$, $\mathcal{G}_5$, $\mathcal{G}_5'$, and $\mathcal{G}_6$. Suppose that C is contained in the base locus of the linear system $\mathcal{Q}_3$. By Theorem 13.6.1 there is a smooth surface $S \in \mathcal{Q}_2$ such that S contains the curve C.

Let M be a general surface in $\mathcal{Q}_3$. Put $M_S = M|_S$. Then

$$h^0(\mathcal{O}_S(M_S - C)) \geq 3$$

because the linear system $\mathcal{Q}_3$ is two-dimensional and does not contain the surface S. In particular, the linear system $|M_S - C|$ is not empty.

Denote by k the number of irreducible components of C. Then $C^2 = -2k$ by Lemmas 9.1.2, 9.2.3 and 9.2.8, so that

$$\left(M_S - C\right)^2 = M_S^2 - 2M_S \cdot C + C^2 = 40 - 4\deg(C) - 2k < 0.$$

One has

$$M_S - C \sim Z + F$$

for some effective divisors F and Z on S such that the linear system $|Z|$ is free from base components, and F is the fixed part of the linear system $|M_S - C|$. Since $(M_S - C)^2 < 0$, we see that F is non-trivial. On the other hand, we have

$$\deg(F) = 20 - \deg(C) \leq 10.$$

Since S does not contain $\mathfrak{A}_5$-invariant curves of degree strictly less than 18 different from C by Lemma 13.6.3, we conclude that $\mathrm{Supp}(F) = C$. This means that for some $m \geq 2$ the linear system $|M_S - mC|$ is mobile. On the other hand, we compute

$$\left(M_S - mC\right)^2 = M_S^2 - 2mM_S \cdot C + m^2 C^2 = 40 - 4m\deg(C) - 2km^2 < 0.$$

The obtained contradiction shows that the curve C is not contained in the base locus of the linear system $\mathcal{Q}_3$. $\square$

Now we are ready to prove that $\mathrm{Bs}(\mathcal{Q}_3)$ is zero-dimensional.

Lemma 17.2.5. The base locus of the linear system $\mathcal{Q}_3$ is a finite collection of points.

Proof. Suppose that there is an $\mathfrak{A}_5$-irreducible curve C contained in $\mathrm{Bs}(\mathcal{Q}_3)$. Then $\deg(C) \leqslant 19$, so that in particular $\deg(C) \neq 17$ and $\deg(C) \neq 19$ by Lemma 8.1.14. By Remark 17.2.1, one has $\Sigma_5 \subset \mathrm{Bs}(\mathcal{Q}_3)$. Let us consider two cases $\Sigma_5 \subset C$ and $\Sigma_5 \not\subset C$ separately.

Suppose that $\Sigma_5 \subset C$. Then for any point $P \in \Sigma_5$ the multiplicity $\mathrm{mult}_P(C) \geqslant 3$ by Lemma 7.5.11. Consider the rational $\mathfrak{A}_5$-invariant map

$$\theta_{\Pi_3} \colon V_5 \dashrightarrow \mathbb{P}^2 \cong \mathbb{P}(W_3)$$

given by the projection from the linear span $\Pi_3 \cong \mathbb{P}^3$ of Σ_5 (see §7.5). Let $C' \subset \mathbb{P}^2$ be the image of the curve C under θ_{Π_3}. Then

$$0 \leqslant \deg(C') \leqslant \deg(C) - \sum_{P \in \Sigma_5} \mathrm{mult}_P(C) \leqslant \deg(C) - 15 \leqslant 19 - 15 = 4.$$

By Lemma 5.3.1(iv) this implies that either $\deg(C) = 19$ and θ_{Π_3} induces a double cover from C to the $\mathfrak{A}_5$-invariant conic C', or $\deg(C) = 17$ and θ_{Π_3} is a birational map from C to the $\mathfrak{A}_5$-invariant conic C', or $\deg(C) = 15$ and C' is an $\mathfrak{A}_5$-invariant collection of points $\Sigma \subset \mathbb{P}^2$. The former two cases are impossible since $\deg(C) \notin \{17, 19\}$. In the latter case $|\Sigma| = 15$ by Lemma 5.3.1(i), and $C = \mathcal{L}_{15}$ is the $\mathfrak{A}_5$-invariant union of 15 lines by Proposition 10.2.5, which contradicts Lemma 7.5.13.

Therefore, we see that $\Sigma_5 \not\subset C$. Choose general surfaces M and M' in the linear system $\mathcal{Q}_3$. Then

$$M \cdot M' = mC + \sum_{i=1}^{r} n_i C_i$$

for some positive integer m, some non-negative integer r, some irreducible (not necessarily $\mathfrak{A}_5$-invariant) curves $C_1, \ldots, C_r$, and some positive integers $n_1, \ldots, n_r$. Put $d_i = \deg(C_i)$ for every i.

Since Σ_5 is contained in the base locus of the linear system $\mathcal{Q}_3$ by Remark 17.2.1 and $\Sigma_5 \not\subset C$, we see that Σ_5 is contained in the union of the curves $C_1, \ldots, C_r$. In particular, the number r is actually positive. If $\sum_{i=1}^{r} d_i \leqslant 4$, then some curve C_i contains more than d_i points of Σ_5; this means that C_i is contained in the linear span $\Pi_3 \subset \mathbb{P}^6$ of Σ_5, which is impossible by Lemma 7.5.3. Thus $\sum_{i=1}^{r} d_i \geqslant 5$, so that

$$\deg(C) \leqslant m \deg(C) \leqslant 20 - \sum_{i=1}^{r} n_i d_i \leqslant 20 - \sum_{i=1}^{r} d_i \leqslant 15.$$

Note that $C \neq \mathcal{L}_{15}$ by Lemma 7.5.13. Moreover, $C \neq \mathcal{L}_6$, since the linear subsystem in $|2H|$ consisting of surfaces that pass through $\mathcal{L}_6$ does not contain $\mathcal{Q}_3$ by Lemma 11.1.10. Thus, it follows from Proposition 10.2.5 that either C is a rational normal sextic, or

$$C \in \{\mathcal{L}_{10}, \mathcal{L}_{12}, \mathcal{G}_5, \mathcal{G}_5', \mathcal{G}_6\}.$$

This contradicts Lemmas 17.2.2, 17.2.3, and 17.2.4. □

In particular, Lemma 17.2.5 implies that the linear system $\mathcal{Q}_3$ is not composed of a pencil and, hence, gives a dominant rational map

$$\xi \colon V_5 \dashrightarrow \mathbb{P}^2 \cong \mathbb{P}(W_3').$$

Lemma 17.2.6. *The base locus of the linear system $\mathcal{Q}_3$ contains the $\mathfrak{A}_5$-orbits Σ_{10} and Σ_{10}'.*

Proof. Let us show that the base locus of the linear system $\mathcal{Q}_3$ contains Σ_{10}. The restriction map

$$H^0(\mathcal{O}_{V_5}(2H)) \to H^0(\mathcal{O}_{\Sigma_{10}} \otimes \mathcal{O}_{V_5}(2H))$$

is surjective by Lemma 17.1.3. Denote its kernel by $\mathcal{K}$. Then $\mathcal{K}$ is a 13-dimensional $\mathfrak{A}_5$-invariant subspace of $H^0(\mathcal{O}_{V_5}(2H))$. Since there exists a surface in $\mathcal{Q}_2$ passing through Σ_{10} (which is the surface S_{10} in the notation of Theorem 8.2.1), the kernel $\mathcal{K}$ contains a one-dimensional $\mathfrak{A}_5$-subrepresentation I. By Corollary 8.1.2, the quotient $\mathcal{K}/I$ contains a three-dimensional $\mathfrak{A}_5$-invariant subspace, which corresponds to $\mathcal{Q}_3$. This shows that the base locus of the linear system $\mathcal{Q}_3$ contains Σ_{10}. Similarly, we see that the base locus of the linear system $\mathcal{Q}_3$ contains Σ_{10}'. □

Lemma 17.2.7. *The $\mathfrak{A}_5$-orbit Σ_5 is cut out by $\mathcal{Q}_3$ scheme-theoretically in some neighborhood of Σ_5.*

Proof. Let S, S', S'' be general surfaces in $\mathcal{Q}_3$. By Remark 17.2.1 the $\mathfrak{A}_5$-orbit Σ_5 is contained in the base locus of $\mathcal{Q}_3$. Let us first show that the surfaces S, S', and S'' are smooth at every point of Σ_5.

Suppose that some (and thus any) of the surfaces S, S', and S'' is singular at some point in Σ_5. Since S, S', and S'' are general, the multiplicity of S, S', and S'' in any point of Σ_5 equals $m = \mathrm{mult}_{\Sigma_5}(\mathcal{Q}_3)$. By Lemma 17.2.5

CREMONA GROUPS AND THE ICOSAHEDRON   421

the intersection of the surfaces S, S', and S'' consists of 40 points (counted with multiplicities). Thus

$$40 = S \cdot S' \cdot S'' \geqslant \sum_{P \in \Sigma_5} \mathrm{mult}_P(S) \cdot \mathrm{mult}_P(S') \cdot \mathrm{mult}_P(S'') \geqslant$$

$$\geqslant \sum_{P \in \Sigma_5} m^3 = 5m^3 \geqslant 40,$$

which implies that $m = 2$ and $S \cap S' \cap S'' = \Sigma_5$. The latter is impossible by Lemma 17.2.6.

Now we know that all surfaces S, S', and S'' are smooth at the points of Σ_5. Choose a point $P \in \Sigma_5$. Denote by T, T', and T'' the Zariski tangent planes at the point P of the surfaces S, S', and S'', respectively. We claim that the intersection

$$T \cap T' \cap T'' \subset T_P(V_5)$$

is trivial. Indeed, otherwise $T \cap T' \cap T''$ would be a non-trivial proper sub-representation of the stabilizer $G_P \subset \mathfrak{A}_5$ of the point P. This is impossible by Lemma 7.5.9. □

Proposition 17.2.8. *The base locus of the linear system $\mathcal{Q}_3$ is a union of the $\mathfrak{A}_5$-orbits Σ_5, Σ_{10}, Σ'_{10}, and Σ_{15}.*

Proof. By Lemma 17.2.5 it is enough to show that Σ_5, Σ_{10}, Σ'_{10}, and Σ_{15} are contained in $\mathrm{Bs}(\mathcal{Q}_3)$. By Remark 17.2.1 one has $\Sigma_5 \subset \mathrm{Bs}(\mathcal{Q}_3)$. By Lemma 17.2.6 one has $\Sigma_{10} \subset \mathrm{Bs}(\mathcal{Q}_3)$ and $\Sigma'_{10} \subset \mathrm{Bs}(\mathcal{Q}_3)$.

Suppose that $\Sigma_{15} \not\subset \mathrm{Bs}(\mathcal{Q}_3)$. By Lemma 17.2.5 the intersection of three general surfaces $S, S', S'' \in \mathcal{Q}_3$ consists of 40 points (counted with multiplicities). By Theorem 7.3.5 this implies that the multiplicity of the zero-cycle $S \cdot S' \cdot S''$ is larger than 1 at the points of Σ_5, which contradicts Lemma 17.2.7. □

Corollary 17.2.9. *The rational map*

$$\xi \colon V_5 \dashrightarrow \mathbb{P}^2 \cong \mathbb{P}(W'_3)$$

given by the linear system $\mathcal{Q}_3$ becomes regular after a blow-up at the points of Σ_5, Σ_{10}, Σ'_{10}, and Σ_{15}. Moreover, the linear system $\mathcal{Q}_3$ is not composed of a pencil, and the map ξ is dominant.

We conclude this section with an application of Proposition 17.2.8 that will be used later in the proof of Theorem 19.2.1.

Lemma 17.2.10. Let Σ be one of the $\mathfrak{A}_5$-orbits Σ_{10} or Σ'_{10}, and let S be a surface in $\mathcal{Q}_2$ that contains Σ (so that either $S = S_{10}$ or S'_{10} in the notation of Theorem 8.2.1). Let D be an effective $\mathbb{Q}$-divisor on V_5 such that $D \sim_{\mathbb{Q}} H$ and $S \not\subset \mathrm{Supp}(D)$, and let P be a point in Σ. Then $\mathrm{mult}_P(D) < 1$.

Proof. Put $m = \mathrm{mult}_P(D)$. Suppose that $m \geqslant 1$. Let $u\colon \check{V}_5 \to V_5$ be the blow-up of Σ. Denote by $N_1, \ldots, N_{10}$ the exceptional divisors of u. Denote by $\check{D}$ the proper transform of the divisor D on the threefold $\check{V}_5$. Then

$$\check{D} \sim_{\mathbb{Q}} u^*(H) - m \sum_{i=1}^{10} N_i,$$

Denote by $\check{S}$ the proper transform of the surface S on the threefold $\check{V}_5$. Then $\check{S}$ is smooth and

$$\check{S} \sim u^*(2H) - 2\sum_{i=1}^{10} N_i,$$

since S has ordinary double points in every point of Σ (see Theorem 8.2.1).

By Proposition 17.2.8, the base locus of the linear system $\mathcal{Q}_3$ is a union of the $\mathfrak{A}_5$-orbits Σ_5, Σ_{10}, Σ'_{10}, and Σ_{15}. Denote by $\check{\mathcal{Q}}_3$ the proper transform of the linear system $\mathcal{Q}_3$ on the threefold $\check{V}_5$. By Corollary 17.2.9, one has

$$\check{\mathcal{Q}}_3 \sim u^*(2H) - \sum_{i=1}^{10} N_i,$$

the base locus of the linear system $\check{\mathcal{Q}}_3$ does not contain curves, and the base points of the linear system $\check{\mathcal{Q}}_3$ are proper transforms of the $\mathfrak{A}_5$-orbits Σ_5, Σ', and Σ_{15} on the threefold $\check{V}_5$, where Σ' is the $\mathfrak{A}_5$-orbit among Σ_{10} and Σ'_{10} that is different from Σ. In particular, the linear system $\check{\mathcal{Q}}_3|_{\check{S}}$ is free from base points, since the $\mathfrak{A}_5$-orbits Σ_5, Σ', and Σ_{15} are not contained in the surface S by Theorem 8.2.1(i).

Put $D_{\check{S}} = \check{D}|_{\check{S}}$. Put $H_{\check{S}} = u^*(H)|_{\check{S}}$ and put $\check{C}_i = N_i|_{\check{S}}$. Then

$$D_{\check{S}} \sim_{\mathbb{Q}} 2H_{\check{S}} - \breve{m}\sum_{i=1}^{10} \check{C}_i,$$

for some positive rational number $\breve{m} \geqslant m$. Note that each $\check{C}_i$ is a smooth rational curve such that $\check{C}_i^2 = -2$. Moreover, since

$$\check{\mathcal{Q}}_3|_{\check{S}} \subset \left| 2H_{\check{S}} - \sum_{i=1}^{10} \check{C}_i \right|,$$

we conclude that the linear system $|2H_{\check{S}} - \sum_{i=1}^{10} \check{C}_i|$ is base point free. On the other hand, we have $H_{\check{S}} \cdot \check{C}_i = 0$, which gives

$$0 \leqslant \left(2H_{\check{S}} - \sum_{i=1}^{10} \check{C}_i\right) \cdot D_{\check{S}} = \left(2H_{\check{S}} - \sum_{i=1}^{10} \check{C}_i\right) \cdot \left(H_{\check{S}} - \check{m}\sum_{i=1}^{10} \check{C}_i\right) =$$
$$= 20 + \check{m}\sum_{i=1}^{10} \check{C}_i^2 = 20 - 20\check{m}.$$

This implies that $m = \check{m} = 1$ and the divisor $2H_{\check{S}} - \sum_{i=1}^{10} \check{C}_i$ is not ample, because $D_{\check{S}}$ is an effective non-zero $\mathbb{Q}$-divisor.

We know that the linear system $|2H_{\check{S}} - \sum_{i=1}^{10} \check{C}_i|$ is base point free, and

$$\left(2H_{\check{S}} - \sum_{i=1}^{10} \check{C}_i\right)^2 = 20 > 0,$$

but the divisor $2H_{\check{S}} - \sum_{i=1}^{10} \check{C}_i$ is not ample. Hence, it follows from Lemma 6.7.6 that there exists a (not necessarily $\mathfrak{A}_5$-invariant) smooth rational curve $\check{F}$ on the surface $\check{S}$ such that

$$20\left(20 + r(H_{\check{S}} \cdot \check{F})^2\right) = \left(2H_{\check{S}} - \sum_{i=1}^{10} \check{C}_i\right)^2 \cdot \left(2H_{\check{S}}^2 + r(H_{\check{S}} \cdot \check{F})^2\right) =$$
$$= 2\left(H_{\check{S}} \cdot \left(2H_{\check{S}} - \sum_{i=1}^{10} \check{C}_i\right)\right)^2 = 800$$

for some $r \in \{1, 5, 6, 10, 12, 15\}$, and the $\mathfrak{A}_5$-orbit of the curve $\check{F}$ consists of r irreducible components (the curve $\check{F}$ being one of them). Thus, we have

$$r(H_{\check{S}} \cdot \check{F})^2 = 20$$

for some $r \in \{1, 5, 6, 10, 12, 15\}$, which implies that $r = 5$ and $H_{\check{S}} \cdot \check{F} = 2$.

Therefore, the curve $u(\check{F})$ is a conic and its $\mathfrak{A}_5$-orbit consists of 5 conics. All of them must be contained in the surface S. The latter is impossible by Theorem 13.6.1, because the surface S is singular. $\square$

17.3 Isolation of orbits in $\mathscr{S}$

Let Σ be an $\mathfrak{A}_5$-orbit in V_5 such that $\Sigma \subset \mathscr{S} \setminus \mathscr{C}$ and $|\Sigma| \in \{20, 30, 60\}$. Put

$$n = \lfloor \frac{|\Sigma|}{5} \rfloor - 1,$$

so that
$$n = \begin{cases} 3 \text{ if } |\Sigma| = 20, \\ 5 \text{ if } |\Sigma| = 30, \\ 11 \text{ if } |\Sigma| = 60. \end{cases}$$

Let $\mathcal{M}$ be the linear system on V_5 consisting of all surfaces in $|nH|$ that contain Σ. Denote by $\mathcal{I}_\Sigma \subset \mathcal{O}_{V_5}$ the ideal sheaf of the subset Σ. By (7.1.3), we have

$$h^0\big(\mathcal{I}_\Sigma \otimes \mathcal{O}_{V_5}(nH)\big) \geqslant h^0\big(\mathcal{O}_{V_5}(nH)\big) - |\Sigma| =$$
$$= \frac{5n(n+1)(n+2)}{6} + n + 1 - |\Sigma| > 0,$$

which implies that $\mathcal{M}$ is not empty. Note that the linear system $\mathcal{M}$ has base points, since its base locus contains Σ by construction. The goal of this section is to prove:

Proposition 17.3.1. *The base locus of the linear system $\mathcal{M}$ does not contain curves.*

Let us prove Proposition 17.3.1. Suppose that the base locus of the linear system $\mathcal{M}$ contains curves. Then it must contain some $\mathfrak{A}_5$-irreducible curve C. Let us show that the latter is impossible.

Lemma 17.3.2. *One has $C \subset \mathscr{S}$.*

Proof. Since $n \geqslant 2$ and $|H|$ is free from base points, the base locus of the linear system $\mathcal{M}$ is contained in the surface $\mathscr{S}$. In particular, we have $C \subset \mathscr{S}$. $\square$

Since $C \subset \mathscr{S}$ and C is $\mathfrak{A}_5$-invariant, it follows from Lemma 7.2.13(i) that either $C = \mathscr{C}$ or $\deg(C) \geqslant 12$. The former case is impossible by:

Lemma 17.3.3. *One has $C \neq \mathscr{C}$.*

Proof. Suppose that $C = \mathscr{C}$. Let $\mathcal{I}_\mathscr{C} \subset \mathcal{O}_{V_5}$ be the ideal sheaf of $\mathscr{C}$, and let

$$\mathcal{I}_{\mathscr{C} \cup \Sigma} \subset \mathcal{O}_{V_5}$$

be the ideal sheaf of the subset $\mathscr{C} \cup \Sigma$. Then

$$h^0\big(\mathcal{I}_\Sigma \otimes \mathcal{O}_{V_5}(nH)\big) = h^0\big(\mathcal{I}_{\mathscr{C} \cup \Sigma} \otimes \mathcal{O}_{V_5}(nH)\big),$$

CREMONA GROUPS AND THE ICOSAHEDRON 425

since $\mathscr{C}$ is contained in the base locus of the linear system $\mathcal{M}$. On the other hand,
$$h^0\Big(\mathcal{I}_{\mathscr{C}\cup\Sigma}\otimes\mathcal{O}_{V_5}(nH)\Big) \leqslant h^0\Big(\mathcal{I}_{\mathscr{C}}\otimes\mathcal{O}_{V_5}(nH)\Big).$$
Since $\mathscr{C}$ is projectively normal, we have
$$h^0\Big(\mathcal{I}_{\mathscr{C}}\otimes\mathcal{O}_{V_5}(nH)\Big) = h^0\Big(\mathcal{O}_{V_5}(nH)\Big) - h^0\Big(\mathcal{O}_{\mathscr{C}}\otimes\mathcal{O}_{V_5}(nH)\Big) =$$
$$= h^0\Big(\mathcal{O}_{V_5}(nH)\Big) - 6n - 1,$$

Thus, we obtain
$$h^0\Big(\mathcal{O}_{V_5}(nH)\Big) - 6n - 1 = h^0\Big(\mathcal{I}_{\mathscr{C}}\otimes\mathcal{O}_{V_5}(nH)\Big) \geqslant$$
$$\geqslant h^0\Big(\mathcal{I}_{\mathscr{C}\cup\Sigma}\otimes\mathcal{O}_{V_5}(nH)\Big) = h^0\Big(\mathcal{I}_{\Sigma}\otimes\mathcal{O}_{V_5}(nH)\Big) \geqslant$$
$$\geqslant h^0\Big(\mathcal{O}_{V_5}(nH)\Big) - |\Sigma|,$$

which implies that $|\Sigma| \geqslant 6n + 1$. Thus, we proved that
$$|\Sigma| \geqslant 6n + 1 = \begin{cases} 19 \text{ if } |\Sigma| = 20, \\ 31 \text{ if } |\Sigma| = 30, \\ 67 \text{ if } |\Sigma| = 60, \end{cases}$$

which implies that $|\Sigma| = 20$ and $n = 3$. Then it follows from Lemmas 7.2.10 and 8.1.8(vi) that Σ is contained in the base curve C_{20} of the pencil $\mathcal{Q}_2$. In particular, the linear system $\mathcal{M}$ contains plenty of divisors of the form $S \cup H$, where S is a general surface in $\mathcal{Q}_2$ and H is a general surface in $|H|$. Since $\mathscr{C}$ is not contained in the base locus of the pencil $\mathcal{Q}_2$ by Theorem 8.1.8(i) and the linear system $|H|$ is base point free, we see that $\mathscr{C}$ is not contained in the base locus of $\mathcal{M}$. □

Let us use notation of §7.2. Denote by $\hat{\mathcal{M}}$, $\hat{C}$, and $\hat{\Sigma}$ the proper transforms on the threefold $\mathcal{Y}$ of the linear system $\mathcal{M}$, the curve C, and the $\mathfrak{A}_5$-orbit Σ, respectively. Then the surface $\hat{\mathscr{S}}$ contains the curve $\hat{C}$ and the $\mathfrak{A}_5$-orbit $\hat{\Sigma}$. Moreover, it follows from Lemma 17.3.3 that $\hat{\mathcal{M}} \sim \nu^*(nH)$. Note that $\hat{\mathcal{M}}$ is the linear system on $\mathcal{Y}$ consisting of *all* surfaces in $|\nu^*(nH)|$ that contain $\hat{\Sigma}$. In particular, every curve in the linear system $\hat{\mathcal{M}}|_{\hat{\mathscr{S}}}$ is a curve in $|\nu^*(nH)|_{\hat{\mathscr{S}}}|$ that contains $\hat{\Sigma}$.

Remark 17.3.4. We do not claim that the linear system $\hat{\mathcal{M}}|_{\hat{\mathscr{S}}}$ consists of all curves in $|\nu^*(nH)|_{\hat{\mathscr{S}}}|$ that contain $\hat{\Sigma}$. Indeed, note that

$$h^1\Big(\mathcal{O}_Y(\nu^*(nH))\Big) = 0$$

by Theorem 2.3.5, and

$$\nu^*(nH) - \hat{\mathscr{S}} \sim \nu^*(n-2H) + 2E_Y.$$

Thus the exact sequence of sheaves

$$0 \to \mathcal{O}_Y(\nu^*(nH) - \hat{\mathscr{S}}) \to \mathcal{O}_Y(\nu^*(nH)) \to \mathcal{O}_{\hat{\mathscr{S}}}(\nu^*(nH)|_{\hat{\mathscr{S}}}) \to 0$$

gives an exact sequence of cohomology groups

$$0 \longrightarrow H^0\Big(\mathcal{O}_Y(\nu^*(nH) - \hat{\mathscr{S}})\Big) \longrightarrow H^0\Big(\mathcal{O}_Y(\nu^*(nH))\Big) \xrightarrow{\alpha}$$
$$\xrightarrow{\alpha} H^0\Big(\mathcal{O}_{\hat{\mathscr{S}}}(\nu^*(nH)|_{\hat{\mathscr{S}}})\Big) \longrightarrow H^1\Big(\mathcal{O}_Y(\nu^*((n-2)H) + 2E_Y)\Big) \longrightarrow 0.$$

However, the restriction map α is not surjective, because

$$h^1\Big(\mathcal{O}_Y(\nu^*(n-2H) + 2E_Y)\Big) =$$
$$= h^0\Big(\mathcal{O}_Y(\nu^*(nH) - \hat{\mathscr{S}})\Big) - h^0\Big(\mathcal{O}_Y(\nu^*(nH))\Big) + h^0\Big(\mathcal{O}_{\hat{\mathscr{S}}}(\nu^*(nH)|_{\hat{\mathscr{S}}})\Big) =$$
$$= h^0\Big(\mathcal{O}_Y(\nu^*(n-2H) + 2E_Y)\Big) - h^0\Big(\mathcal{O}_Y(\nu^*(nH))\Big) + h^0\Big(\mathcal{O}_{\mathbb{P}^1 \times \mathbb{P}^1}(5n, n)\Big) =$$
$$= h^0\Big(\mathcal{O}_{V_5}((n-2)H)\Big) - h^0\Big(\mathcal{O}_{V_5}(nH)\Big) + 5n^2 + 6n + 1 =$$
$$= \frac{5n(n-1)(n-2)}{6} + n - 1 - \left(\frac{5n(n+1)(n+2)}{6} + n + 1\right)$$
$$+ 5n^2 + 6n + 1 = 6n - 1$$

by (7.1.3). Note that Theorem 2.3.5 is not applicable to get the vanishing

$$h^1\Big(\mathcal{O}_Y(\nu^*(n-2H) + 2E_Y)\Big) = 0,$$

because the big divisor

$$\nu^*((n-2)H) + 2E_Y - K_Y \sim \nu^*(nH) + E_Y$$

is not nef.

Contrary to Remark 17.3.4, we know that the divisor

$$\nu^*((n-2)H) + E_{\mathcal{Y}} - K_{\mathcal{Y}} \sim \nu^*(nH)$$

is nef and big. Thus, applying Theorem 2.3.5, we get

$$h^1\Big(\mathcal{O}_{\mathcal{Y}}(\nu^*((n-2)H) + E_{\mathcal{Y}})\Big) = 0.$$

Therefore, one has an exact sequence of cohomology groups

$$0 \longrightarrow H^0\Big(\mathcal{O}_{\mathcal{Y}}(\nu^*(nH) - E_{\mathcal{Y}} - \hat{\mathscr{S}})\Big) \longrightarrow$$
$$\longrightarrow H^0\Big(\mathcal{O}_{\mathcal{Y}}(\nu^*(nH) - E_{\mathcal{Y}})\Big) \longrightarrow$$
$$\longrightarrow H^0\Big(\mathcal{O}_{\hat{\mathscr{S}}}(\nu^*(nH)|_{\hat{\mathscr{S}}} - E_{\mathcal{Y}}|_{\hat{\mathscr{S}}})\Big) \longrightarrow 0, \quad (17.3.5)$$

because

$$\nu^*(nH) - E_{\mathcal{Y}} - \hat{\mathscr{S}} \sim \nu^*((n-2)H) + E_{\mathcal{Y}}.$$

To use (17.3.5), we must replace $\mathcal{M}$ by its linear subsystem that consists of all surfaces containing both Σ and $\mathscr{C}$. This subsystem is a proper linear subsystem by Lemma 17.3.3. Let us denote it by $\mathcal{M}'$, i.e., $\mathcal{M}'$ is a linear system on V_5 consisting of all surfaces in $|nH|$ that contain both $\mathscr{C}$ and Σ.

Lemma 17.3.6. The linear system $\mathcal{M}'$ is not empty. Moreover, its base locus is contained in $\mathscr{S}$.

Proof. The linear system $\mathcal{M}'$ is not empty, since it contains plenty of surfaces that look like $\mathscr{S} \cup R$, where R is any surface in $|(n-2)H|$. The very same reason implies that the base locus of the linear system $\mathcal{M}'$ is contained in $\mathscr{S}$. $\square$

Denote by $\hat{\mathcal{M}}'$ the proper transform of the linear system $\mathcal{M}'$ on the threefold $\mathcal{Y}$. Then the base locus of $\hat{\mathcal{M}}'$ is contained in $\hat{\mathscr{S}} \cup E_{\mathcal{Y}}$. Put

$$m' = \mathrm{mult}_{\mathscr{C}}(\mathcal{M}').$$

Then

$$\hat{\mathcal{M}}' \sim \nu^*(nH) - m'E_{\mathcal{Y}},$$

where $m' \geqslant 1$ by construction. Thus, $\hat{\mathcal{M}}' + (m'-1)E_{\mathcal{Y}}$ is a linear subsystem in $|\nu^*(nH) - E_{\mathcal{Y}}|$ that consists of all surfaces passing through $\hat{\Sigma}$.

Lemma 17.3.7. The linear system $\hat{\mathcal{M}}'$ does not have fixed components, so that $\mathcal{M}'$ does not have fixed components either.

Proof. By (17.3.5), the base locus of the linear system $|\nu^*(nH) - E_\mathcal{Y}|$ does not contain the surface $\hat{\mathscr{S}}$. Let $\mathcal{I}_{\hat{\Sigma}} \subset \mathcal{O}_{\hat{\mathscr{S}}}$ be the ideal sheaf of the subset

$$\hat{\Sigma} \subset \hat{\mathscr{S}} \cong \mathbb{P}^1 \times \mathbb{P}^1.$$

Then

$$h^0\Big(\mathcal{I}_{\hat{\Sigma}} \otimes \mathcal{O}_{\hat{\mathscr{S}}}\big(\nu^*(nH)|_{\hat{\mathscr{S}}} - E_\mathcal{Y}|_{\hat{\mathscr{S}}}\big)\Big) = h^0\Big(\mathcal{I}_{\hat{\Sigma}} \otimes \mathcal{O}_{\mathbb{P}^1 \times \mathbb{P}^1}(5n-2, n-2)\Big) \geqslant$$
$$\geqslant h^0\Big(\mathcal{O}_{\mathbb{P}^1 \times \mathbb{P}^1}(5n-2, n-2)\Big) - |\hat{\Sigma}| = 5n^2 - 6n + 1 - |\hat{\Sigma}| > 0.$$

Thus, it follows from (17.3.5) that there exists an effective divisor in the linear system $|\nu^*(nH) - E_\mathcal{Y}|$ whose support contains the subset $\hat{\Sigma} \subset \mathcal{Y}$ and does not contain the surface $\hat{\mathscr{S}}$. The proper transform of this divisor on V_5 belongs to the linear system $\mathcal{M}'$ by construction. Thus, the base locus of the linear system $\mathcal{M}'$ does not contain the surface $\mathscr{S}$, which implies that $\mathcal{M}'$ does not have base components by Lemma 17.3.6. □

On the other hand, the linear system $\mathcal{M}'$ does have base points, since Σ and $\mathscr{C}$ are both contained in its base locus by construction. Similarly, its base locus must contain the curve C, because $\mathcal{M}'$ is a subsystem of the linear system $\mathcal{M}$. Thus, the curve $\hat{C}$ is contained in the base locus of $\hat{\mathcal{M}}'$. By (17.3.5), every curve in the linear system $\big|(\nu^*(nH) - E_\mathcal{Y})|_{\hat{\mathscr{S}}}\big|$ is cut out by a divisor in the linear system $|\nu^*(nH) - E_\mathcal{Y}|$. In particular, every curve in

$$\big|(\nu^*(nH) - E_\mathcal{Y})|_{\hat{\mathscr{S}}}\big|$$

that passes through $\hat{\Sigma}$ is cut out by a surface in $|\nu^*(nH) - E_\mathcal{Y}|$ that passes through $\hat{\Sigma}$. Since $\hat{C} \subset \hat{\mathscr{S}}$, $\hat{C} \neq \hat{\mathscr{C}}$ and $\hat{C}$ is contained in the base locus of the linear system $\hat{\mathcal{M}}'$, the curve $\hat{C}$ must be contained in the base locus of the linear subsystem in $\big|(\nu^*(nH) - E_\mathcal{Y})|_{\hat{\mathscr{S}}}\big|$ that consists of curves passing through $\hat{\Sigma}$.

Recall that $\mathfrak{A}_5$ acts on $\hat{\mathscr{S}} \cong \mathbb{P}^1 \times \mathbb{P}^1$ diagonally by Corollary 7.2.5. Thus, $\hat{C}$ is a unique $\mathfrak{A}_5$-invariant curve of bi-degree $(1,1)$ on $\hat{\mathscr{S}}$ by Lemma 6.4.14. Also, $\hat{\mathscr{C}}$ is an $\mathfrak{A}_5$-invariant curve of bi-degree $(1,1)$ on $\hat{\mathscr{S}}$ by Lemma 7.2.3(iii). On the other hand, we have $C \neq \mathscr{C}$ by Lemma 17.3.3. The obtained contradiction completes the proof of Proposition 17.3.1.

17.4 Isolation of arbitrary orbits

Let Σ be an $\mathfrak{A}_5$-orbit in V_5. Then

$$|\Sigma| \in \{5, 10, 12, 15, 20, 30, 60\}$$

by Lemma 7.3.5. Put $n = \lfloor \frac{|\Sigma|}{5} \rfloor$ if $|\Sigma| \leqslant 12$, and put $n = \lfloor \frac{|\Sigma|}{5} \rfloor - 1$ in the remaining cases, so that

$$n = \begin{cases} 1 & \text{if } |\Sigma| = 5, \\ 2 & \text{if } |\Sigma| = 10, \\ 2 & \text{if } |\Sigma| = 12, \\ 2 & \text{if } |\Sigma| = 15, \\ 3 & \text{if } |\Sigma| = 20, \\ 5 & \text{if } |\Sigma| = 30, \\ 11 & \text{if } |\Sigma| = 60. \end{cases}$$

Let $\mathcal{M}$ be the linear system on V_5 consisting of all surfaces in $|nH|$ that contain Σ. Denote by $\mathcal{I}_\Sigma \subset \mathcal{O}_{V_5}$ the ideal sheaf of the subset Σ. By (7.1.3), we have

$$h^0\Big(\mathcal{I}_\Sigma \otimes \mathcal{O}_{V_5}(nH)\Big) \geqslant h^0\Big(\mathcal{O}_{V_5}(nH)\Big) - |\Sigma| = \\ = \frac{5n(n+1)(n+2)}{6} + n + 1 - |\Sigma| > 0,$$

which implies that $\mathcal{M}$ is not empty. Note that the linear system $\mathcal{M}$ has base points, since its base locus contains Σ by construction. The goal of this section is to prove:

Proposition 17.4.1. If the base locus of the linear system $\mathcal{M}$ contains an irreducible curve C, then $|\Sigma| = 20$, one has $\Sigma \subset C$, and either $C = \mathscr{C}$ or $C = \mathscr{C}'$.

Note that if $|\Sigma| = 20$ and $\Sigma \subset \mathscr{C}$, then $\mathscr{C}$ must be a base curve of the linear system $\mathcal{M}$. Similarly, if $|\Sigma| = 20$ and $\Sigma \subset \mathscr{C}'$, then $\mathscr{C}'$ must be a base curve of the linear system $\mathcal{M}$.

An immediate consequence of Proposition 17.4.1 is the following.

Corollary 17.4.2. Put $m = \lfloor \frac{|\Sigma|}{5} \rfloor$. Let $\mathcal{B}$ be the linear system on V_5 consisting of all surfaces in $|mH|$ that contain Σ. Then the base locus of the linear system $\mathcal{B}$ does not contain curves.

Proof. Suppose that the base locus of the linear system $\mathcal{B}$ contains an irreducible curve C. By Proposition 17.4.1, we have $m = 4$, $|\Sigma| = 20$, $\Sigma \subset C$, and C is an $\mathfrak{A}_5$-invariant smooth rational sextic curve in V_5. Since C is an intersection of quadrics, the restriction map

$$H^0\Big(\mathcal{O}_{V_5}(4H)\Big) \to H^0\Big(\mathcal{O}_C(4H|_C)\Big)$$

is surjective. On the other hand, the divisor $4H|_C$ is a divisor of degree 24 on $C \cong \mathbb{P}^1$. This implies that the base locus of the linear system $\mathcal{B}$ is Σ, because $24 \geqslant 20 = |\Sigma|$. □

We will prove Proposition 17.4.1 in several steps. We start with:

Lemma 17.4.3. Suppose that $|\Sigma| \in \{5, 10, 15\}$. Then the base locus of the linear system $\mathcal{M}$ does not contain curves.

Proof. By Theorem 7.3.5, we have

$$\Sigma \in \{\Sigma_5, \Sigma_{10}, \Sigma'_{10}, \Sigma_{15}\}.$$

If $\Sigma = \Sigma_5$, then $n = 1$ and the required assertion follows from Lemma 7.5.3. In the remaining cases, we have $n = 2$ and the required assertion follows from Proposition 17.2.8. □

So, to prove Proposition 17.4.1, we may assume that

$$|\Sigma| \in \{12, 20, 30, 60\}$$

by Theorem 7.3.5. In particular, $n \geqslant 2$. On the other hand, there exists a surface $S \in \mathcal{Q}_2$ that contains Σ. Note that S is irreducible by Remark 8.1.4.

Lemma 17.4.4. The linear system $\mathcal{M}$ does not have fixed components.

Proof. Since $n \geqslant 2$ and $|H|$ is free from base points, the base locus of $\mathcal{M}$ must be contained in S. Let $\mathcal{I}_S \subset \mathcal{O}_{V_5}$ be the ideal sheaf of the surface S. Then (7.1.3) gives

$$h^0\Big(\mathcal{I}_\Sigma \otimes \mathcal{O}_{V_5}(nH)\Big) \geqslant h^0\Big(\mathcal{O}_{V_5}(nH)\Big) - |\Sigma| =$$
$$= \frac{5n(n+1)(n+2)}{6} + n + 1 - |\Sigma| > \frac{5n(n-1)(n-2)}{6} + n - 1 =$$
$$= h^0\Big(\mathcal{O}_{V_5}((n-2)H)\Big) = h^0\Big(\mathcal{I}_S \otimes \mathcal{O}_{V_5}(nH)\Big),$$

which implies that S is not a fixed component of the linear system $\mathcal{M}$. This means that $\mathcal{M}$ does not have fixed components, since S is irreducible. □

Lemma 17.4.5. Suppose that either $\Sigma \subset \mathscr{C}$ or $\Sigma \subset \mathscr{C}'$. If $|\Sigma| \neq 20$, then the base locus of the linear system $\mathcal{M}$ contains no curves.

CREMONA GROUPS AND THE ICOSAHEDRON 431

Proof. If $\Sigma \subset \mathscr{C}$, then we put $C = \mathscr{C}$. If $\Sigma \subset \mathscr{C}'$, then we put $C = \mathscr{C}'$. Since C is an intersection of quadrics, $n \geqslant 2$ and $\Sigma \subset C$, we see that the base locus of the linear system $\mathcal{M}$ is contained in C.

Suppose that $|\Sigma| \neq 20$. Then $6n \geqslant |\Sigma|$. Since C is an intersection of quadrics, the restriction map

$$H^0\big(\mathcal{O}_{V_5}(nH)\big) \to H^0\big(\mathcal{O}_C(nH|_C)\big)$$

is surjective. On the other hand, the divisor $nH|_C$ is a divisor of degree $6n$ on $C \cong \mathbb{P}^1$. This implies that the base locus of the linear system $\mathcal{M}$ is Σ, because $6n \geqslant |\Sigma|$. $\square$

If $\Sigma = \Sigma_{12}$, then $\Sigma \subset \mathscr{C}$ by construction. If $\Sigma = \Sigma'_{12}$, then $\Sigma \subset \mathscr{C}'$ by Lemma 11.3.1. Thus, Lemma 17.4.5 implies:

Corollary 17.4.6. If either $\Sigma = \Sigma_{12}$ or $\Sigma = \Sigma'_{12}$, then the base locus of the linear system $\mathcal{M}$ contains no curves.

By Corollary 17.4.6, we may assume that $|\Sigma| \in \{20, 30, 60\}$. By Lemma 17.4.5, we may assume that $\Sigma \not\subset \mathscr{C}$ and $\Sigma \not\subset \mathscr{C}'$ in order to prove Proposition 17.4.1. In particular, we may assume that $S \neq \mathscr{S}$ by Proposition 17.3.1. By Theorem 8.2.1(v), the surface S is a $K3$ surface with at most ordinary double points. Moreover, one has

$$|\mathrm{Sing}(S)| \in \{0, 5, 10, 15\}$$

by Theorem 8.2.1(ii),(iii). In particular, the surface S is smooth at every point of Σ.

Put $H_S = H|_S$. Let $\mathcal{M}_S$ be the linear system on S consisting of all curves in $|nH_S|$ that contain Σ. Note that the restriction map

$$H^0(\mathcal{O}_{V_5}(nH)) \to H^0\big(\mathcal{O}_S(nH_S)\big)$$

is surjective by Lemma 8.2.5, which implies that $\mathcal{M}_S = \mathcal{M}|_S$. Thus, to prove Proposition 17.4.1, it is enough to show that the base locus of the linear system $\mathcal{M}_S$ contains no curves.

Suppose that the base locus of the linear system $\mathcal{M}_S$ contains curves. Denote by F the fixed part of the linear system $\mathcal{M}_S$, and denote by $\mathcal{R}_S$ its mobile part. Let R be a general curve in $\mathcal{R}_S$. Then R is nef, since $\mathcal{R}_S$ is a linear system without fixed curves. One has $nH_S - F \sim R$. Moreover, the divisor R is Cartier by Lemma 6.7.4. Put $d = F \cdot H_S$. Then

$$10n - d = nH_S \cdot H_S - F \cdot H_S = R \cdot H_S \geqslant 0,$$

which implies that
$$d \leqslant 10n \tag{17.4.7}$$
On the other hand, we have
$$10F^2 \leqslant d^2 \tag{17.4.8}$$
by the Hodge index theorem. Since
$$h^0\bigl(\mathcal{O}_S(nH_S)\bigr) = 5n^2 + 2$$
by the Riemann–Roch formula and Theorem 2.3.5, we have
$$h^0(\mathcal{O}_S(R)) \geqslant h^0(\mathcal{O}_S(nH_S)) - |\Sigma| = 5n^2 + 2 - |\Sigma|. \tag{17.4.9}$$

Lemma 17.4.10. *The divisor R is big.*

Proof. Since R is nef, it is enough to show that $R^2 > 0$. Note that $R^2 \geqslant 0$. Suppose that $R^2 = 0$. Then the linear system $|R|$ does not have base points. This implies that $R \sim kE$ for some positive integer k and some irreducible curve E such that $E^2 = 0$. Hence, $h^0(\mathcal{O}_S(E)) = 2$ by [109, Proposition 2.6]. On the other hand, we have $H_S \cdot E \geqslant 3$, because S is not uniruled. Thus, we have
$$10n \geqslant 10n - F \cdot H_S = nH_S \cdot H_S - F \cdot H_S = R \cdot H_S = kE \cdot H_S \geqslant 3k,$$
so that $k \leqslant \frac{10n}{3}$.

We have $h^0(\mathcal{O}_S(R)) = k + 1$ by [109, Proposition 2.6]. By (17.4.9), we have
$$k + 1 \geqslant 5n^2 + 2 - |\Sigma|.$$
Suppose that $|\Sigma| = 20$. Then $n = 3$ and
$$k + 1 \geqslant 5n^2 + 2 - |\Sigma| = 27,$$
which implies that $k \geqslant 26$. This inequality contradicts $k \leqslant \frac{10n}{3} = 10$.

Suppose that $|\Sigma| = 30$. Then $n = 5$ and
$$k + 1 \geqslant 5n^2 + 2 - |\Sigma| = 97,$$
which implies that $k \geqslant 96$. This inequality contradicts $k \leqslant \frac{10n}{3} = \frac{50}{3}$.

Suppose that $|\Sigma| = 60$. Then $n = 11$ and
$$k + 1 \geqslant 10n^2 + 2 - |\Sigma| = 1152,$$
which implies that $k \geqslant 1151$. This inequality contradicts $k \leqslant \frac{10n}{3} = \frac{110}{3}$. □

Lemma 17.4.11. One has $|\Sigma| = 20$ and $F = \mathscr{C}'$. Moreover, the surface S is smooth.

Proof. By Lemma 17.4.10, the divisor R is nef and big. Then

$$h^0(\mathcal{O}_S(R)) = 2 + \frac{R^2}{2} = 2 + \frac{10n^2 - 2nd + F^2}{2}$$

by the Riemann–Roch formula and Theorem 2.3.5. By (17.4.9), we have

$$2 + \frac{10n^2 - 2nd + F^2}{2} \geqslant 5n^2 + 2 - |\Sigma|,$$

which implies that

$$F^2 \geqslant 2nd - 2|\Sigma|. \qquad (17.4.12)$$

Thus, we have

$$d^2 - 20nd + 20|\Sigma| \geqslant 0$$

by (17.4.8). Moreover, we have $d \leqslant 10n$ by (17.4.7). The last two inequalities imply that $d \leqslant 7$, because $|\Sigma| \in \{20, 30, 60\}$ and $n = \lfloor \frac{|\Sigma|}{5} \rfloor - 1$.

Since F is $\mathfrak{A}_5$-invariant and $S \neq \mathscr{S}$, it follows from Theorem 13.6.1 that F is one of the curves $\mathscr{C}'$ or $\mathcal{L}_6$, and the surface S is smooth. This gives $F^2 = -2$ in the case when $F = \mathscr{C}'$, and $F^2 = -12$ in the case when $F = \mathcal{L}_6$ (see Lemma 7.4.7). Plugging this into (17.4.12), we see that $|\Sigma| = 20$ and $F = \mathscr{C}'$. $\square$

By Lemma 17.4.11, we see that $|\Sigma| = 20$, $n = 3$, and the fixed part of the linear system $\mathcal{M}_S$ is the curve $\mathscr{C}'$. In particular, the surface S is the smooth surface $S_{\mathscr{C}'}$ (see §11.3), and the divisor $R \sim 3H_S - \mathscr{C}'$ is very ample on S by Lemma 11.4.7.

Recall that we assumed earlier that $\Sigma \not\subset \mathscr{C}'$. Let $\mathcal{I}_\Sigma \subset \mathcal{O}_S$ be the ideal sheaf of the subset Σ, let $\mathcal{I}_{\mathscr{C}'} \subset \mathcal{O}_S$ be the ideal sheaf of the curve $\mathscr{C}'$, and let $\mathcal{I}_{\mathscr{C}' \cup \Sigma} \subset \mathcal{O}_S$ be the ideal sheaf of the subset $\mathscr{C}' \cup \Sigma$. Then

$$h^0\big(\mathcal{I}_\Sigma \otimes \mathcal{O}_S(3H_S)\big) = h^0\big(\mathcal{I}_{\mathscr{C}' \cup \Sigma} \otimes \mathcal{O}_S(3H_S)\big) = h^0\big(\mathcal{I}_\Sigma \otimes \mathcal{O}_S(3H_S - \mathscr{C}')\big),$$

since $\mathscr{C}'$ is contained in the base locus of the linear system $\mathcal{M}_S$ and $\Sigma \not\subset \mathscr{C}'$. Moreover, we have

$$h^0\big(\mathcal{I}_\Sigma \otimes \mathcal{O}_S(3H_S - \mathscr{C}')\big) \geqslant h^0\big(\mathcal{O}_S(3H_S - \mathscr{C}')\big) - 2,$$

since $3H_S - \mathscr{C}'$ is very ample on S. On the other hand, we have

$$h^0\big(\mathcal{I}_{\mathscr{C}'} \otimes \mathcal{O}_S(3H_S)\big) = h^0\big(\mathcal{O}_S(3H_S - \mathscr{C}')\big) = 2 + \frac{(3H_S - \mathscr{C}')^2}{2} = 28$$

by the Riemann–Roch formula and Kodaira vanishing. Note that

$$h^0(\mathcal{O}_S(3H_S)) = 47$$

by the Riemann–Roch formula and Kodaira vanishing. Thus, we obtain

$$26 = h^0\Big(\mathcal{I}_{\mathscr{C}'} \otimes \mathcal{O}_S(3H_S)\Big) - 2 \geqslant$$
$$\geqslant h^0\Big(\mathcal{I}_{\mathscr{C}' \cup \Sigma} \otimes \mathcal{O}_S(3H_S)\Big) = h^0\Big(\mathcal{I}_\Sigma \otimes \mathcal{O}_S(3H_S)\Big) \geqslant$$
$$\geqslant h^0\Big(\mathcal{O}_S(3H_S)\Big) - |\Sigma| = 47 - |\Sigma| = 27.$$

The obtained contradiction completes the proof of Proposition 17.4.1.

17.5 Isolation of the curve $\mathcal{L}_{15}$

Let $\mathcal{M}$ be the linear system on V_5 consisting of all surfaces in $|3H|$ that contain the curve $\mathcal{L}_{15}$. In this section we will obtain a *partial* description of the base locus of $\mathcal{M}$ (see Proposition 17.5.6).

Lemma 17.5.1. The dimension of the linear system $\mathcal{M}$ is at least 3. In particular, it is not empty. Moreover, $\mathcal{M}$ does not have fixed components.

Proof. Let $L_1, \ldots, L_{15}$ be the lines of $\mathcal{L}_{15}$, so that $\mathcal{L}_{15} = \sum_{i=1}^{15} L_i$. For each line L_i of $\mathcal{L}_{15}$, let us fix 3 sufficiently general points P_i, O_i, and Q_i in L_i. Put

$$\Xi = \Sigma_5 \sqcup \{P_1, O_1, Q_1, P_2, O_2, Q_2, \ldots, P_{15}, O_{15}, Q_{15}, \}.$$

Then $|\Xi| = 50$. On the other hand, $h^0(\mathcal{O}_{V_5}(3H)) = 54$ by (7.1.3). Thus the linear subsystem in $|3H|$ consisting of all surfaces passing through Ξ is at least three-dimensional. On the other hand, every surface $M \in |3H|$ that passes through Ξ must contain $\mathcal{L}_{15}$, since otherwise we would have

$$3 = M \cdot L_i \geqslant \mathrm{mult}_{\Sigma_5}(M) + \mathrm{mult}_{P_i}(M) + \mathrm{mult}_{O_i}(M) + \mathrm{mult}_{Q_i}(M) \geqslant 4.$$

Thus, the dimension of the linear system $\mathcal{M}$ is at least 3.

We claim that the linear system $\mathcal{M}$ does not have fixed components (cf. the proof of Lemma 8.1.3). Indeed, suppose that $\mathcal{M}$ has fixed components. Then a general surface of the linear system $\mathcal{M}$ is a union $F \cup M$ of the fixed part F of $\mathcal{M}$ and a surface M from the mobile part of $\mathcal{M}$. Each of the surfaces F and M is contained either in the linear system $|H|$ or in the linear system $|2H|$. Hence, neither F nor M contains the curve $\mathcal{L}_{15}$ by Lemma 7.5.13, which is a contradiction. □

Let M be a general surface in $\mathcal{M}$.

Lemma 17.5.2. One has $\operatorname{mult}_{\mathcal{L}_{15}}(M) = 1$.

Proof. Let M' be a general surface in $\mathcal{M}$. Then
$$M \cdot M' = \delta \mathcal{L}_{15} + \Upsilon,$$
where δ is an integer such that
$$\delta \geqslant \operatorname{mult}_{\mathcal{L}_{15}}(M) \cdot \operatorname{mult}_{\mathcal{L}_{15}}(M') = \operatorname{mult}^2_{\mathcal{L}_{15}}(M),$$
and Υ is an effective one-cycle whose support does not contain any irreducible component of $\mathcal{L}_{15}$. Then
$$45 = H \cdot M \cdot M' = \delta H \cdot \mathcal{L}_{15} + H \cdot \Upsilon \geqslant \delta H \cdot \mathcal{L}_{15} = 15\delta,$$
which implies that $\delta \leqslant 3$. Hence, we have $\operatorname{mult}_{\mathcal{L}_{15}}(M) = 1$. $\square$

Let $f \colon \bar{V}_5 \to V_5$ be the blow-up of V_5 at Σ_5, and let $B_1, \ldots, B_5$ be f-exceptional divisors that are contracted to the points of Σ_5 by f (see §7.5). By Lemma 7.5.8, the linear system $|f^*(H) - \sum_{i=1}^{5} B_i|$ is free from base points and gives an $\mathfrak{A}_5$-equivariant morphism
$$\bar{\theta}_{\Pi_3} \colon \bar{V}_5 \to \mathbb{P}^2$$
whose general fiber is an elliptic curve.

Denote by $\bar{M}$ the proper transform of M on $\bar{V}$, and denote the proper transform of $\mathcal{L}_{15}$ on $\bar{V}_5$ by $\bar{\mathcal{L}}_{15}$. Note that $\bar{\mathcal{L}}_{15}$ is an $\mathfrak{A}_5$-invariant curve that is a disjoint union of 15 smooth rational curves (cf. Corollary 7.4.4 and Lemma 9.1.10(i)).

Lemma 17.5.3. One has
$$\bar{M} \sim f^*(3H) - 2\sum_{i=1}^{5} B_i.$$

Proof. We have
$$\bar{M} \sim f^*(3H) - m\sum_{i=1}^{5} B_i$$
for some positive integer m. Note that the surface M must be singular at every point of Σ_5 (see Remark 7.5.12), which implies $m \geqslant 2$. Let N be a general fiber of the elliptic fibration $\bar{\theta}_{\Pi_3}$. Then
$$\bar{M} \cdot N = \left(f^*(3H) - m\sum_{i=1}^{5} B_i \right) \cdot N = 15 - 5m,$$

because $B_i \cdot N = 1$ by Lemma 7.5.8. Thus, we see that $m \leqslant 3$. Let us show that $m \neq 3$.

Suppose that $m = 3$. Then $\bar{M}$ is contracted by the morphism $\bar{\theta}_{\Pi_3}$ to a cubic curve $C_3 \subset \mathbb{P}^2$. By Corollary 7.4.4, the curve $\bar{\mathcal{L}}_{15}$ is contracted by the morphism $\bar{\theta}_{\Pi_3}$ to the unique $\mathfrak{A}_5$-orbit on $\mathbb{P}^2$ that consists of 15 points (see Lemma 5.3.1(i)). By construction, one has $\bar{\mathcal{L}}_{15} \subset \bar{M}$. Thus, C_3 contains the unique $\mathfrak{A}_5$-orbit on $\mathbb{P}^2$ that consists of 15 points, which is impossible by Lemma 5.3.1(v). The obtained contradiction shows that $m = 2$. □

Let $g \colon \tilde{V}_5 \to \bar{V}_5$ be the blow-up of $\bar{V}_5$ at $\bar{\mathcal{L}}_{15}$. Then the action of $\mathfrak{A}_5$ lifts to $\tilde{V}_5$. Let $F_1, \ldots, F_{15}$ be g-exceptional divisors that are contracted to the irreducible components of $\bar{\mathcal{L}}_{15}$ by g. Put $h = f \circ g$. Denote by $\tilde{M}$ the proper transform of M on $\tilde{V}$. Using Lemmas 17.5.2 and 17.5.3, we get

Corollary 17.5.4. One has

$$\tilde{M} \sim 3h^*(H) - 2\sum_{i=1}^{5} g^*(B_i) - \sum_{i=1}^{15} F_i.$$

Let $\tilde{\mathcal{L}}_6$ be the proper transform on $\tilde{V}_5$ of the curve $\mathcal{L}_6$. Then

$$\tilde{M} \cdot \tilde{\mathcal{L}}_6 = -12$$

by Lemma 9.1.14 and Corollary 17.5.4. In particular, the divisor $\tilde{M}$ is not nef, and the curve $\mathcal{L}_6$ is contained in the base locus of $\mathcal{M}$. Note that the curve $\mathcal{L}_{15}$ is also contained in the base locus of the linear system $\mathcal{M}$ by construction.

Lemma 17.5.5. Let $\tilde{C}$ be an irreducible curve in $\tilde{V}_5$ such that $\tilde{C}$ is contained in some h-exceptional surface. Then $\tilde{M} \cdot \tilde{C} \geqslant 0$.

Proof. Let $\tilde{B}_i$ be the proper transform on $\tilde{V}_5$ of the surface B_i, and let

$$v_i \colon \tilde{B}_i \to B_i$$

be the restriction of the morphism $g \colon \tilde{V}_5 \to \bar{V}_5$ to $\tilde{B}_i$. Then v_i is a blow-up of three non-collinear points in $B_i \cong \mathbb{P}^2$. Moreover, the divisor $\tilde{M}|_{\tilde{B}_i}$ is linearly equivalent to the proper transform via v_i of a general conic in $B_i \cong \mathbb{P}^2$ that passes thought the points of the blow-up. The latter implies that $\tilde{M}|_{\tilde{B}_i}$ is nef, so that $\tilde{M} \cdot \tilde{C} \geqslant 0$ if $\tilde{C} \subset \tilde{B}_i$.

Therefore, to prove that $\tilde{M} \cdot \tilde{C} \geqslant 0$, we may assume that $\tilde{C}$ is contained in one of the exceptional divisors of g, say, divisor F_1. Since $\mathcal{L}_{15} \not\subset \mathscr{S}$ by

CREMONA GROUPS AND THE ICOSAHEDRON 437

Lemma 7.2.13(ii), we conclude that $F_1 \cong \mathbb{P}^1 \times \mathbb{P}^1$ by Theorem 7.7.1 and Remark 4.2.7. Since
$$-K_{\tilde{V}_5} \cdot g(F_1) = 0,$$
we see that $F_1|_{F_1}$ is a divisor on $F_1 \cong \mathbb{P}^1 \times \mathbb{P}^1$ of bi-degree $(-1, -1)$ by the adjunction formula. Therefore, the divisor $\tilde{M}|_{F_1}$ is a divisor on F_1 of bi-degree $(2, 1)$, which implies that $\tilde{M} \cdot \tilde{C} \geqslant 0$. □

In particular, if the base locus of the linear system $\mathcal{M}$ does not contain curves besides the irreducible components of $\mathcal{L}_{15}$ and $\mathcal{L}_6$, then Lemma 17.5.5 implies that $\tilde{M} \cdot \tilde{C} \geqslant 0$ for every irreducible curve $\tilde{C}$ in $\tilde{V}_5$ that is not contained in $\mathrm{Supp}(\tilde{\mathcal{L}}_6)$. We believe that former condition is indeed true. Unfortunately, we do not know how to prove this. So, instead of this we prove the following assertion, avoiding the complete description of the base curves of $\mathcal{M}$.

Proposition 17.5.6. For every irreducible curve $\tilde{C} \subset \tilde{V}_5$ such that $\tilde{C} \not\subset \tilde{\mathcal{L}}_6$, one has $\tilde{M} \cdot \tilde{C} \geqslant 0$.

Let us prove Proposition 17.5.6. Suppose that it is not true. Then there exists an $\mathfrak{A}_5$-irreducible curve $\tilde{Z} \subset \tilde{V}_5$ such that $\tilde{M} \cdot \tilde{Z} < 0$ and $\tilde{Z} \neq \tilde{\mathcal{L}}_6$.

By Lemma 17.5.5, the curve $\tilde{Z}$ is not contracted by h. Thus, $h(\tilde{Z})$ is an $\mathfrak{A}_5$-irreducible curve. Put $Z = h(\tilde{Z})$. Moreover, $Z \neq \mathcal{L}_{15}$ by Lemma 17.5.5. Furthermore, we have $Z \neq \mathcal{L}_6$, since $\tilde{Z} \neq \tilde{\mathcal{L}}_6$ by assumption. Every irreducible component of the curve Z lies in the base locus of the linear system $\mathcal{M}$, because every irreducible component of the curve $\tilde{Z}$ is contained in the base locus of the linear system $\tilde{\mathcal{M}}$.

Lemma 17.5.7. One has $6 \leqslant \deg(Z) \leqslant 24$ and $\deg(Z) \neq 15$.

Proof. Let M' be a general surface in $\mathcal{M}$. Then
$$M \cdot M' = Z + \mathcal{L}_{15} + \mathcal{L}_6 + \Omega,$$
where Ω is an effective one-cycle on V_5. Note that we allow Ω to be a zero one-cycle, since $\mathcal{M}$ could be a priori composed of a pencil. One has
$$45 = H \cdot M \cdot M' = H \cdot Z + H \cdot \mathcal{L}_{15} + H \cdot \mathcal{L}_6 + H \cdot \Omega =$$
$$= \deg(Z) + 21 + H \cdot \Omega \geqslant \deg(Z) + 21,$$
which implies that $\deg(Z) \leqslant 24$. By Lemma 8.1.14, one has $\deg(Z) \geqslant 6$. Moreover, we have $\deg(Z) \neq 15$, since otherwise $Z = \mathcal{L}_{15}$ by Theorem 13.6.1. □

Lemma 17.5.8. Suppose that $\Sigma_5 \not\subset Z$. Then $\deg(Z) \leqslant 19$.

Proof. Suppose that $\deg(Z) \geqslant 20$. Let M' be a general surface in $\mathcal{M}$. Then

$$M \cdot M' = Z + \mathcal{L}_{15} + \mathcal{L}_6 + \Omega,$$

where Ω is an effective one-cycle on V_5. Note that $\Sigma_5 \subset \operatorname{Supp}(\Omega)$. Indeed, both M and M' are singular at every point of Σ_5 by Lemma 17.5.3. On the other hand, $\mathcal{L}_6$ does not contain Σ_5 by Corollary 7.4.9, and Z does not contain Σ_5 by assumption. This shows that the multiplicity of the one-cycle $M \cdot M'$ at every point of Σ_5 must be at least 4. The latter implies that $\Sigma_5 \subset \operatorname{Supp}(\Omega)$, since $\operatorname{mult}_{\Sigma_5}(\mathcal{L}_{15}) = 3$ by Corollary 7.4.4.

Let H_{Σ_5} be a general hyperplane section of V_5 that passes though Σ_5. Then the intersection $\operatorname{Supp}(\Omega) \cap H_{\Sigma_5}$ consists of finitely many points by Lemma 7.5.3. Thus, we have

$$45 = H_{\Sigma_5} \cdot M \cdot M' = H_{\Sigma_5} \cdot Z + H_{\Sigma_5} \cdot \mathcal{L}_{15} + H_{\Sigma_5} \cdot \mathcal{L}_6 + H_{\Sigma_5} \cdot \Omega \geqslant$$
$$\geqslant 41 + H_{\Sigma_5} \cdot \Omega \geqslant 41 + \sum_{P \in \Sigma_5} \operatorname{mult}_P(\Omega) \geqslant 46,$$

which is absurd. □

Let r be the number of irreducible components of Z, and let $C_1, \ldots, C_r$ be its irreducible components. Then $\deg(Z) = r\deg(C_i)$.

Lemma 17.5.9. One has $r \in \{1, 5, 6, 10, 12\}$.

Proof. It follows from Corollary 5.1.2 that

$$r \in \{1, 5, 6, 10, 12, 15, 20, 30, 60\}.$$

Thus, one has $r \in \{1, 5, 6, 10, 12, 20\}$ by Lemma 17.5.7. We also claim that $r \neq 20$. Indeed, suppose that $r = 20$. Then Z is a union of 20 lines, since $\deg(Z) \leqslant 24$ by Lemma 17.5.7. On the other hand, there are exactly three lines in V_5 passing through each point in Σ_5 by Theorem 7.1.9(i), and these lines are the lines of $\mathcal{L}_{15}$ (see Corollary 7.4.4). Therefore, if $r = 20$, then $\Sigma_5 \not\subset Z$, which contradicts Lemma 17.5.8. □

Lemma 17.5.10. There exists a unique surface in the pencil $\mathcal{Q}_2$ that contains Z.

Proof. If $\deg(Z) \leqslant 19$, then the assertion follows from Theorem 13.6.1, since $\deg(Z) \neq 15$ by Lemma 17.5.7. Therefore, we may assume that $\deg(Z)$ is at least 20. Then $\Sigma_5 \subset Z$ by Lemma 17.5.8. Since Σ_5 is not contained in the base locus C_{20} of the pencil $\mathcal{Q}_2$ by Theorem 8.1.8(vii), we have $Z \neq C_{20}$, which implies the uniqueness part of the assertion.

Let us prove the existence part of the assertion. Recall that $\deg(Z) \leqslant 24$ by Lemma 17.5.7. Thus, by Corollary 14.1.18 it is enough to prove that the curve Z does not coincide with the curve $\mathcal{G}_{10}$.

Suppose that $Z = \mathcal{G}_{10}$. Let C be an irreducible component of Z. Let $\tilde{C}$ be the proper transform of the curve C on $\tilde{V}_5$. Recall that $\tilde{M} \cdot \tilde{Z} < 0$, so that $\tilde{M} \cdot \tilde{C} < 0$. On the other hand, we have

$$\tilde{M} \sim 3h^*(H) - 2\sum_{i=1}^{5} g^*(B_i) - \sum_{i=1}^{15} F_i,$$

by Corollary 17.5.4. Recall that C is a conic by Theorem 14.1.16(i), so that

$$h^*(H) \cdot \tilde{C} = H \cdot C = 2.$$

By Theorem 14.1.16(iii) one has

$$\left(\sum_{i=1}^{5} g^*(B_i)\right) \cdot \tilde{C} = 2.$$

By Lemma 14.2.7 one has

$$\left(\sum_{i=1}^{15} F_i\right) \cdot \tilde{C} = 0.$$

Therefore, we compute

$$0 > \tilde{M} \cdot \tilde{C} = 3h^*(H) \cdot \tilde{C} - 2\sum_{i=1}^{5} g^*(B_i) \cdot \tilde{C} - \sum_{i=1}^{15} F_i \cdot \tilde{C} = 2,$$

which is absurd. $\square$

Keeping in mind Lemma 17.5.10, denote by S the unique surface in $\mathcal{Q}_2$ that contains Z. By Lemma 7.5.13, one has $\mathcal{L}_{15} \not\subset S$. Note that

$$\mathcal{L}_{15} \cap S \neq \Sigma_5 \cup \Sigma_{15},$$

because S cannot contain both $\mathfrak{A}_5$-orbits Σ_5 and Σ_{15} by Theorem 8.2.1(i). Since $S \cdot \mathcal{L}_{15} = 30$, it follows from Lemma 9.1.10(iv) that either $\mathcal{L}_{15} \cap S = \Sigma_5$, or $\mathcal{L}_{15} \cap S = \Sigma_{15}$, or $\mathcal{L}_{15} \cap S$ is an $\mathfrak{A}_5$-orbit of length 30.

Since $\tilde{Z} \cdot \tilde{M} < 0$ by assumption, we see that the intersection $Z \cap \mathcal{L}_{15}$ is not empty. Since $Z \subset S$ and $\mathcal{L}_{15} \cap S$ consists of one $\mathfrak{A}_5$-orbit, we see that

$$Z \cap \mathcal{L}_{15} = S \cap \mathcal{L}_{15}.$$

In particular, we conclude that either $Z \cap \mathcal{L}_{15} = \Sigma_5$, or $Z \cap \mathcal{L}_{15} = \Sigma_{15}$, or $Z \cap \mathcal{L}_{15}$ is an $\mathfrak{A}_5$-orbit of length 30.

Lemma 17.5.11. One has $\Sigma_5 \not\subset Z$.

Proof. Suppose that $\Sigma_5 \subset Z$. By Theorem 8.2.1(iv), the surface S has an ordinary double point at every point of Σ_5 (so that we have $S = S_5$ in the notation of Theorem 8.2.1). Let $\bar{S}$ be the proper transform of the surface S on the threefold $\bar{V}_5$. Then $\bar{\mathcal{L}}_{15} \cdot \bar{S} = 0$, which implies that $\bar{\mathcal{L}}_{15}$ is disjoint from $\bar{S}$. In particular, the curve $\tilde{Z}$ is disjoint from the union of the divisors $F_1, \ldots, F_{15}$. Recall that the divisor $f^*(H) - \sum_{i=1}^{5} B_i$ is nef on $\bar{V}$, because the linear system $|f^*(H) - \sum_{i=1}^{5} B_i|$ is free from base points by Lemma 7.5.8. Hence, one has

$$\tilde{M} \cdot \tilde{Z} = \left(3h^*(H) - 2\sum_{i=1}^{5} g^*(B_i)\right) \cdot \tilde{Z} =$$

$$= h^*(H) \cdot \tilde{Z} + 2g^*\left(f^*(H) - \sum_{i=1}^{5} B_i\right) \cdot \tilde{Z} \geqslant 0$$

by Corollary 17.5.4. Thus, we have $\tilde{M} \cdot \tilde{Z} \geqslant 0$, which contradicts our initial assumption that $\tilde{M} \cdot \tilde{Z} < 0$. □

In particular, Lemma 17.5.11 implies that $S \neq S_5$. Moreover, we know that $\deg(Z) \leqslant 19$ by Lemma 17.5.8. By Theorem 13.6.1, the curve Z must be one of the curves $\mathscr{C}$, $\mathscr{C}'$, $\mathcal{L}_{10}$, $\mathcal{G}_5$, $\mathcal{G}'_5$, $\mathcal{L}_{12}$, $\mathcal{G}_6$, C_{16}, C'_{16}, $\mathcal{T}_6$, C^0_{18}, $\mathcal{B}_{18}$, or a smooth rational curve of degree 18.

Lemma 17.5.12. One has $Z \neq \mathscr{C}'$.

Proof. The curve $\mathscr{C}'$ is disjoint from $\mathcal{L}_{15}$ by Lemma 11.3.7. On the other hand, we already proved that Z does intersect $\mathcal{L}_{15}$. Thus, one has $Z \neq \mathscr{C}'$. □

Lemma 17.5.13. The curve Z is none of the curves $\mathscr{C}$, $\mathcal{L}_{12}$, or C_{16}.

Proof. Suppose that Z is one of the curves $\mathscr{C}$, $\mathcal{L}_{12}$, or C_{16}. Then $S = \mathscr{S}$ by Theorem 13.6.1. In particular, $\Sigma_{15} \not\subset Z$ by Lemma 7.2.10, which implies that the intersection $Z \cap \mathcal{L}_{15}$ consists of 30 points. Thus, we see that $Z \neq \mathcal{L}_{12}$ by Remark 8.3.6. Since $S \cdot \mathcal{L}_{15} = 30$, the intersection $Z \cap \mathcal{L}_{15}$ consists of 30 smooth points of the surface S. Now Lemma 7.2.2(i) implies that $Z \neq \mathscr{C}$, and Z intersects each line of $\mathcal{L}_{15}$ transversally. Thus, we see that $Z = C_{16}$. By Corollary 17.5.4, we have

$$\tilde{M} \cdot \tilde{Z} = \left(3h^*(H) - 2\sum_{i=1}^{5} g^*(B_i) - \sum_{i=1}^{15} F_i\right) \cdot \tilde{Z} =$$

$$= 3H \cdot Z - \sum_{i=1}^{15} F_i \cdot \tilde{Z} = 48 - 30 = 18 > 0,$$

which contradicts our initial assumption that $\tilde{M} \cdot \tilde{Z} < 0$. □

Lemma 17.5.13 and Theorem 13.6.1 imply that the surface S must be smooth. In particular, we see that $S \neq S_{15}$ in the notation of Theorem 8.2.1, and $Z \cap \mathcal{L}_{15}$ is an $\mathfrak{A}_5$-orbit of length 30.

Corollary 17.5.14. *The curve $\mathcal{L}_{15}$ intersects the surface S transversally.*

Lemma 17.5.15. *The curve Z is singular at every point of $Z \cap \mathcal{L}_{15}$.*

Proof. Suppose that Z is smooth at some (and thus every) point of $Z \cap \mathcal{L}_{15}$. Since S intersects $\mathcal{L}_{15}$ transversally by Corollary 17.5.14, the curve Z intersects $\mathcal{L}_{15}$ transversally at every point of $Z \cap \mathcal{L}_{15}$ as well. Thus one has

$$\tilde{M} \cdot \tilde{Z} = 3\deg(Z) - 30 \geqslant 0,$$

because $\deg(Z) \geqslant 10$ by Lemmas 17.5.12 and 17.5.13. The latter contradicts our initial assumption that $\tilde{M} \cdot \tilde{Z} < 0$. □

Finally, recall that the curves $\mathcal{L}_{10}$, $\mathcal{G}_5$, $\mathcal{G}'_5$, $\mathcal{G}_6$, C'_{16}, $\mathcal{T}_6$, and $\mathcal{B}_{18}$ are smooth, and the curve C^0_{18} has exactly 12 singular points (see Theorem 13.6.1). Thus, the curve Z cannot be singular at every point of $Z \cap \mathcal{L}_{15}$, since we already proved that $Z \cap \mathcal{L}_{15}$ is an $\mathfrak{A}_5$-orbit of length 30. The obtained contradiction completes the proof of Proposition 17.5.6.

Chapter 18

Proof of the main result

In this chapter we prove Theorem 1.4.1, which is the main result of this book. In §18.1, we show that Theorem 1.4.1 follows from a very technical Theorem 18.1.1. In §18.2, we prove three auxiliary results that are used in the proof of Theorem 18.1.1. In §18.3 and §18.4, we prove Theorem 18.1.1. In §18.5, we give an alternative proof of one technical assertion of §18.3 that is based on the technique of multiplier ideal sheaves. The advantage of this alternative proof is that it does not use the results of Chapter 17 as much as the original proof does. In §18.6, we give an alternative proof of the main result of §18.4 that uses a "multiplication by two" trick (see Lemma 2.4.5) and the technique of multiplier ideal sheaves.

18.1 Singularities of linear systems

The purpose of this chapter is to prove:

Theorem 18.1.1. Let $\mathcal{D}$ be an $\mathfrak{A}_5$-invariant mobile linear system on V_5. Choose $\lambda \in \mathbb{Q}$ such that
$$K_{V_5} + \lambda \mathcal{D} \sim_{\mathbb{Q}} 0.$$
Suppose that $\mathrm{mult}_{\mathcal{L}_6}(\mathcal{D}) \leqslant \frac{1}{\lambda}$ and $\mathrm{mult}_{\mathscr{C}'}(\mathcal{D}) \leqslant \frac{1}{\lambda}$. Then the log pair $(V_5, \lambda \mathcal{D})$ has canonical singularities.

Using Theorem 18.1.1 we can easily give:

Proof of Theorem 1.4.1. Let $\mathcal{M}$ be any mobile linear system on V_5. By Lemma 11.5.5, there is a birational selfmap $f \in \langle \tau, \iota \rangle$ such that
$$\mathrm{mult}_{\mathcal{L}_6}\left(f^{-1}(\mathcal{M})\right) \leqslant \frac{1}{\mu_f}$$

and
$$\mathrm{mult}_{\mathscr{C}'}\left(f^{-1}(\mathcal{M})\right) \leqslant \frac{1}{\mu_f},$$
where μ_f is a positive rational number such that
$$K_{V_5} + \mu_f f^{-1}(\mathcal{M}) \sim_{\mathbb{Q}} 0.$$
The log pair
$$\left(V_5, \mu_f f^{-1}(\mathcal{M})\right)$$
has canonical singularities by Theorem 18.1.1. By Theorem 3.3.1, the threefold V_5 is $\mathfrak{A}_5$-birationally rigid, and $\mathrm{Bir}^{\mathfrak{A}_5}(V_5)$ is generated by its subgroup $\mathrm{Aut}^{\mathfrak{A}_5}(V_5)$ and the elements ι and τ. Thus,
$$\mathrm{Bir}^{\mathfrak{A}_5}(V_5) \cong \mathfrak{S}_5 \times \boldsymbol{\mu}_2$$
by Corollary 11.4.4. $\square$

In the remaining part of this chapter we prove Theorem 18.1.1.

18.2 Restricting divisors to invariant quadrics

In this section we will prove three auxiliary results that will be used later in the proof of Theorem 18.1.1. We start with the following lemma that was implicitly proved in the proof of [22, Lemma 4.3].

Lemma 18.2.1. Let $\mathcal{R}$ be an $\mathfrak{A}_5$-invariant linear system on V_5 such that $\mathscr{S}$ is not its fixed component, and let λ be a positive rational number such that
$$K_{V_5} + \lambda \mathcal{R} \sim_{\mathbb{Q}} 0.$$
Let R be a general surface in $\mathcal{R}$. Then
$$\mathrm{mult}_{\mathscr{C}}(R) \leqslant \frac{1}{\lambda}.$$

Proof. Suppose that $\mathrm{mult}_{\mathscr{C}}(R) > \frac{1}{\lambda}$. Let us use notation of §7.2. Denote by $\hat{R}$ the proper transform of the surface R on $\mathcal{Y}$. Then
$$\lambda \hat{R} \sim_{\mathbb{Q}} \nu^*(2H) - \lambda \mathrm{mult}_{\mathscr{C}}(R) E_{\mathcal{Y}}.$$

By Lemma 7.2.3(ii), the latter implies that $\lambda\hat{R}|_{\hat{\mathscr{S}}}$ is a $\mathbb{Q}$-divisor of bi-degree
$$\Big(2 - 2\lambda\mathrm{mult}_{\mathscr{C}}(R), 10 - 2\lambda\mathrm{mult}_{\mathscr{C}}(R)\Big)$$
on $\hat{\mathscr{S}} \cong \mathbb{P}^1 \times \mathbb{P}^1$. This is impossible, because
$$2 - 2\lambda\mathrm{mult}_{\mathscr{C}}(R) < 0,$$
while $\lambda\hat{R}|_{\hat{\mathscr{S}}}$ is obviously an effective $\mathbb{Q}$-divisor. $\square$

Lemma 18.2.2. Let C be an irreducible curve in $\mathscr{S}$ such that $C \neq \mathscr{C}$, let $\mathcal{R}$ be an $\mathfrak{A}_5$-invariant linear system on V_5 such that $\mathscr{S}$ is not its fixed component, and let λ be a positive rational number such that
$$K_{V_5} + \lambda\mathcal{R} \sim_{\mathbb{Q}} 0.$$
Let R be a general surface in $\mathcal{R}$. Then
$$\mathrm{mult}_C\Big(R \cdot \mathscr{S}\Big) \leqslant \frac{1}{\lambda}.$$
Moreover, if $C \neq C_{20}$, then
$$\mathrm{mult}_C\Big(R \cdot \mathscr{S}\Big) < \frac{1}{\lambda}.$$

Proof. Let us use notation of §7.2. Denote by $\hat{R}$ the proper transform of the surface R on $\mathcal{Y}$. Then
$$\lambda\hat{R} \sim_{\mathbb{Q}} \nu^*(2H) - \lambda\mathrm{mult}_{\mathscr{C}}(\mathcal{R})E_{\mathcal{Y}},$$
where $\lambda\mathrm{mult}_{\mathscr{C}}(\mathcal{R}) \leqslant 1$ by Lemma 18.2.1. Thus $\lambda\hat{R}|_{\hat{\mathscr{S}}}$ is an effective $\mathbb{Q}$-divisor of bi-degree
$$\Big(2 - 2\lambda\mathrm{mult}_{\mathscr{C}}(\mathcal{R}), 10 - 2\lambda\mathrm{mult}_{\mathscr{C}}(\mathcal{R})\Big)$$
on $\hat{\mathscr{S}} \cong \mathbb{P}^1 \times \mathbb{P}^1$ by Lemma 7.2.3(ii).

Let Z be the $\mathfrak{A}_5$-orbit of the curve C. Denote by $\hat{Z}$ the proper transform of the curve Z on $\mathcal{Y}$. Then $\hat{Z}$ is an $\mathfrak{A}_5$-irreducible curve that is contained in $\hat{\mathscr{S}} \cong \mathbb{P}^1 \times \mathbb{P}^1$. Denote by (a,b) the bi-degree of the curve $\hat{Z}$. Note that one has $Z \neq \mathscr{C}$ by assumption. Thus, we have
$$a + b \geqslant 12 \qquad (18.2.3)$$

by Corollary 7.2.5 and Lemma 6.4.4(i).

Put
$$\lambda \hat{R}|_{\hat{\mathscr{S}}} = \epsilon \hat{Z} + \Omega,$$
where ϵ is a non-negative rational number, and Ω is an effective $\mathbb{Q}$-divisor on $\hat{\mathscr{S}}$ whose support does not contain the curve $\hat{Z}$. Note that $\epsilon = \mathrm{mult}_C(R \cdot \mathscr{S})$. Moreover, Ω is a $\mathbb{Q}$-divisor of bi-degree
$$\Big(2 - 2\lambda\mathrm{mult}_{\mathscr{C}}(\mathcal{R}) - \epsilon a, 10 - 2\lambda\mathrm{mult}_{\mathscr{C}}(\mathcal{R}) - \epsilon b\Big)$$
on $\hat{\mathscr{S}}$. Thus, we have
$$\begin{cases} 2 \geqslant 2 - 2\lambda\mathrm{mult}_{\mathscr{C}}(\mathcal{R}) \geqslant \epsilon a, \\ 10 \geqslant 10 - 2\lambda\mathrm{mult}_{\mathscr{C}}(\mathcal{R}) \geqslant \epsilon b. \end{cases} \qquad (18.2.4)$$

Now the inequalities (18.2.3) and (18.2.4) imply that $\epsilon \leqslant 1$.

Suppose that $\epsilon = 1$. Then $a = 2$ and $b = 10$ by (18.2.3) and (18.2.4). Hence, the curve $\hat{Z}$ is irreducible by Corollary 7.2.5 and Lemma 6.4.8. Also, we have $\deg(Z) = 20$ by Lemma 7.2.3(ii). Therefore, one has $Z = C = C_{20}$ by Corollary 8.1.10. □

The following result is an analog of Lemma 18.2.2 that can be applied to surfaces in $\mathcal{Q}_2$ different from $\mathscr{S}$.

Lemma 18.2.5. Let C be an irreducible curve in V_5 such that $C \neq \mathscr{C}'$ and C is not a line of $\mathcal{L}_6$. Suppose that there exists a surface $S \in \mathcal{Q}_2$ that contains C and is different from $\mathscr{S}$. Let $\mathcal{R}$ be an $\mathfrak{A}_5$-invariant linear system on V_5 such that S is not its fixed component, and let λ be a positive rational number such that
$$K_{V_5} + \lambda \mathcal{R} \sim_{\mathbb{Q}} 0.$$
Let R be a general surface in $\mathcal{R}$. Then
$$\mathrm{mult}_C\Big(R \cdot S\Big) \leqslant \frac{1}{\lambda}.$$
Moreover, if $C \neq C_{20}$, then
$$\mathrm{mult}_C\Big(R \cdot S\Big) < \frac{1}{\lambda}.$$

Proof. Recall that S is a $K3$ surface with at most ordinary double points by Theorem 8.2.1(v). Put $R_S = R|_S$ and $H_S = H|_S$. Let Z be the $\mathfrak{A}_5$-orbit of the curve C. Put
$$\lambda R_S = \epsilon Z + \Omega$$
for some positive rational number ϵ and some $\mathfrak{A}_5$-invariant effective $\mathbb{Q}$-divisor Ω on S such that $\mathrm{Supp}(\Omega)$ does not contain Z. Note that $\epsilon = \mathrm{mult}_C(R \cdot S)$. We are going to prove that $\epsilon \leqslant 1$, and that $\epsilon < 1$ unless $Z = C = C_{20}$.

Suppose that $\epsilon \geqslant 1$. Then $\deg(Z) \leqslant 20$. Indeed, this follows from
$$20 = \lambda R_S \cdot H_S = \epsilon H_S \cdot Z + H_S \cdot \Omega \geqslant \epsilon H_S \cdot Z \geqslant H_S \cdot Z = \deg(Z).$$

To start with, suppose that Z is a curve of degree 20. Then
$$20 = 2H_S \cdot H_S = \lambda R_S \cdot H_S = \epsilon Z \cdot H_S + \Omega \cdot H_S =$$
$$= \epsilon \deg(Z) + \Omega \cdot H_S = 20\epsilon + \Omega \cdot H_S \geqslant 20\epsilon \geqslant 20,$$
which implies that Ω is a zero divisor and $\epsilon = 1$. Thus, we have
$$\lambda R_S = Z \sim_\mathbb{Q} 2H_S \sim_\mathbb{Q} C_{20}.$$

Recall that
$$|\mathrm{Sing}(S)| \in \{0, 5, 10, 15\}$$
by Theorem 8.2.1(ii),(iii), and that $H_S^2 = 10$. Hence, Z is a Cartier divisor on S by Lemma 6.7.4, because Z is $\mathfrak{A}_5$-invariant by construction. Since $\mathrm{Pic}(S)^{\mathfrak{A}_5}$ has no torsion by Lemma 6.7.3(i), we see that $Z \sim C_{20}$. By Lemma 8.2.6(ii), we have $Z = C_{20}$. Since C_{20} is irreducible, we also have $Z = C$.

Thus, to complete the proof, we may assume that $\deg(Z) \leqslant 19$. Note that $Z \neq \mathcal{L}_{12}$, $Z \neq C_{16}$, and $Z \neq \mathscr{C}$, because these curves are contained in $\mathscr{S}$ and are not contained in any other surface in $\mathcal{Q}_2$ by Theorem 13.6.1. Since $C \neq \mathscr{C}'$ and C is not a line of $\mathcal{L}_6$ by assumption, it follows from Theorem 13.6.1 that Z must be one of the curves $\mathcal{L}_{10}$, $\mathcal{G}_5$, $\mathcal{G}_5'$, $\mathcal{G}_6$, C_{16}', $\mathcal{T}_6$, C_{18}^0, $\mathcal{B}_{18}$, or a smooth rational curve of degree 18.

By Lemma 13.6.3, the surface S is smooth, and S does not contain any other $\mathfrak{A}_5$-irreducible curve of degree at most 19 unless $Z = \mathcal{G}_6$ or $Z = \mathcal{B}_{18}$. In the latter case, the surface S contains both curves $\mathcal{G}_6$ and $\mathcal{B}_{18}$.

Let us first exclude the possibilities $Z = \mathcal{L}_{10}$, $Z = \mathcal{G}_5$, and $Z = \mathcal{G}_5'$. Suppose that Z is one of these curves. Consider the linear system $|3H_S - 2Z|$. Since the number $(3H_S - 2Z)^2$ is negative (it equals -110 if $Z = \mathcal{L}_{10}$, and -70 if $Z = \mathcal{G}_5$ or $Z = \mathcal{G}_5'$), the linear system $|3H_S - 2Z|$ is either empty or has a fixed component.

Now let us consider the linear system $|3H_S - Z|$. Note that the number $(3H_S - Z)^2$ is positive (it equals 10 if $Z = \mathcal{L}_{10}$, and 20 if $Z = \mathcal{G}_5$ or $Z = \mathcal{G}_5'$). Thus, it follows from the Riemann–Roch formula and Serre duality that

$$h^0\Big(\mathcal{O}_S(3H_S - Z)\Big) \geqslant h^0\Big(\mathcal{O}_S(3H_S - Z)\Big) - h^1\Big(\mathcal{O}_S(3H_S - Z)\Big) =$$
$$= \chi\Big(\mathcal{O}_S(3H_S - Z)\Big) = 2 + \frac{(3H_S - Z)^2}{2} > 2.$$

So, the mobile part of the linear system $|3H_S - Z|$ is not empty. Let us show that its fixed part is empty. Since

$$(3H_S - Z) \cdot H_S = 20,$$

the fixed part (if it exists) of the linear system $|3H_S - Z|$ must be an effective one-cycle on V_5 of degree at most 19. On the other hand, the surface S does not contain any $\mathfrak{A}_5$-invariant curves of degree at most 19 that are different from Z by Lemma 13.6.3. So, the fixed part (if it exists) of the linear system $|3H_S - Z|$ must coincide with the curve Z. The latter implies that the linear system $|3H_S - 2Z|$ is mobile, which is not the case as we have just shown. Therefore, the linear system $|3H_S - Z|$ does not have a fixed part.

Thus $|3H_S - Z|$ is mobile. Hence, the divisor $3H_S - Z$ is nef. Therefore, we have

$$0 \leqslant (3H_S - Z) \cdot \Omega = (3H_S - Z) \cdot (2H_S - \epsilon Z) =$$
$$= 60 - (3\epsilon + 2)\deg(Z) + \epsilon Z^2 = \begin{cases} 40 - 50\epsilon & \text{if } Z = \mathcal{L}_{10}, \\ 40 - 40\epsilon & \text{if } Z = \mathcal{G}_5 \text{ or } Z = \mathcal{G}_5', \end{cases}$$

which implies that $\epsilon = 1$, and either $Z = \mathcal{G}_5$ or $Z = \mathcal{G}_5'$. Without loss of generality, we may assume that $Z = \mathcal{G}_5$.

Recall that $2H_S - Z \sim_\mathbb{Q} \Omega$, so that

$$H_S \cdot \Omega = H_S \cdot (2H_S - Z) = 10,$$

which implies that Ω is a non-zero effective $\mathbb{Q}$-divisor on the surface S. Thus, the divisor $3H_S - Z$ is not ample, since

$$(3H_S - Z) \cdot \Omega = (3H_S - Z) \cdot (2H_S - Z) = 0.$$

Hence, it follows from Lemma 6.7.6 that S contains a smooth rational curve T such that $(3H_S - Z) \cdot T = 0$ and

$$(3H_S - Z)^2 \left(2H_S^2 + r(H_S \cdot T)^2\right) = 2\Big(H_S \cdot (3H_S - Z)\Big)^2,$$

where $r \in \{1, 5, 6, 10, 12, 15\}$ is the number of curves in the $\mathfrak{A}_5$-orbit of the curve T. This gives

$$r(H_S \cdot T)^2 = 20,$$

which implies that $r = 5$ and $H_S \cdot T = 2$. This means that T is a conic, and the $\mathfrak{A}_5$-orbit of the curve T consists of 5 conics.

Note that T is not an irreducible component of $Z = \mathcal{G}_5$, since

$$(3H_S - Z) \cdot Z = 40 > 0 = (3H_S - Z) \cdot T.$$

Hence, the $\mathfrak{A}_5$-orbit of the curve T must be the curve $\mathcal{G}_5'$ by Lemma 7.8.1. On the other hand, the curve $\mathcal{G}_5'$ cannot be contained in the surface S by Lemma 13.6.3. The obtained contradiction shows that $Z \neq \mathcal{L}_{10}$, $Z \neq \mathcal{G}_5$ and $Z \neq \mathcal{G}_5'$.

Therefore, Z must be one of the curves $\mathcal{G}_6$, C_{16}', $\mathcal{T}_6$, C_{18}^0, $\mathcal{B}_{18}$, or a smooth rational curve of degree 18. If $4H_S - Z$ is nef, then

$$0 \leqslant (4H_S - Z) \cdot \Omega = (4H_S - Z) \cdot (2H_S - \epsilon Z) =$$
$$= 80 - (4\epsilon + 2)\deg(Z) + \epsilon Z^2, \quad (18.2.6)$$

which immediately leads to a contradiction in all cases. Indeed, if $Z = \mathcal{G}_6$, then (18.2.6) becomes

$$0 \leqslant 56 - 60\epsilon.$$

If $Z = C_{16}'$, then (18.2.6) becomes

$$0 \leqslant 48 - 56\epsilon.$$

If $Z = \mathcal{T}_6$, then (18.2.6) becomes

$$0 \leqslant 32 - 72\epsilon.$$

If $Z = C_{18}^0$, then (18.2.6) becomes

$$0 \leqslant 44 - 50\epsilon.$$

If $Z = \mathcal{B}_{18}$, then (18.2.6) becomes

$$0 \leqslant 44 - 66\epsilon.$$

If Z is a smooth rational curve of degree 18, then (18.2.6) becomes

$$0 \leqslant 44 - 74\epsilon.$$

All these inequalities are contradictory since $\epsilon \geqslant 1$.

Thus, we see that $4H_S - Z$ is not nef. In particular, this implies that neither $Z = \mathcal{G}_6$ nor $Z = \mathcal{B}_{18}$. Indeed, if $Z = \mathcal{G}_6$, then S contains $\mathcal{B}_{18}$ and

$$3H_S - Z \sim \mathcal{B}_{18}$$

by Lemma 12.3.6, which implies that $3H_S - Z$ is nef, because $\mathcal{B}_{18}^2 = 6 > 0$ by the adjunction formula. If $Z = \mathcal{B}_{18}$, then S contains $\mathcal{G}_6$ and

$$3H_S - Z \sim \mathcal{G}_6$$

by Lemma 12.3.6, which implies that $4H_S - Z$ is nef, since

$$(H_S + \mathcal{G}_6) \cdot \mathcal{G}_6 = 12 + \mathcal{G}_6^2 = 0$$

by the adjunction formula and Lemma 9.2.8.

We see that Z is one of the curves C'_{16}, $\mathcal{T}_6$, C^0_{18}, or a smooth rational curve of degree 18. In all of these cases, we know the self-intersection number Z^2 (see Table 13.1). This gives

$$(4H_S - Z)^2 = 160 - 8\deg(Z) + Z^2 > 0.$$

Thus, it follows from the Riemann–Roch formula and Serre duality that

$$h^0\big(\mathcal{O}_S(4H_S - Z)\big) \geqslant h^0\big(\mathcal{O}_S(4H_S - Z)\big) - h^1\big(\mathcal{O}_S(4H_S - Z)\big) =$$
$$= \chi\big(\mathcal{O}_S(4H_S - Z)\big) = 2 + \frac{(4H_S - Z)^2}{2} > 2.$$

In particular, the mobile part of the linear system $|4H_S - Z|$ is non-trivial. Moreover, since the divisor $4H_S - Z$ is not nef, the fixed part of the linear system $|4H_S - Z|$ is also non-trivial. Hence

$$4H_S - Z \sim F + M,$$

for some effective divisors F and M on S such that the linear system $|M|$ is free from base components, and F is the fixed part of the linear system $|4H_S - Z|$. Note that F is $\mathfrak{A}_5$-invariant. Moreover, the surface S does not contain any $\mathfrak{A}_5$-irreducible curve of degree at most 19 that is different from Z by Lemma 13.6.3. Since S is not covered by rational curves, we have $\deg(M) \geqslant 3$. Therefore,

$$\deg(F) = 40 - \deg(Z) - \deg(M) \leqslant 37 - \deg(Z) = 19.$$

Thus, we have $\mathrm{Supp}(F) = Z$, because S does not contain any $\mathfrak{A}_5$-invariant curves of degree at most 19 that are different from Z by Lemma 13.6.3. Since $\deg(F) < 2\deg(Z)$, this actually gives $F = Z$. Therefore, the linear system $|4H_S - 2Z|$ is mobile. In particular, one has

$$(4H_S - 2Z)^2 \geqslant 0.$$

On the other hand, using Table 13.1, we see that

$$(4H_S - 2Z)^2 = 160 - 16\deg(Z) + 4Z^2 < 0.$$

The obtained contradiction completes the proof of Lemma 18.2.5. □

18.3 Exclusion of points and curves different from $\mathcal{L}_{15}$

Let $\mathcal{D}$ be an $\mathfrak{A}_5$-invariant mobile linear system on V_5, and λ be a positive rational number such that $K_{V_5} + \lambda \mathcal{D} \sim_{\mathbb{Q}} 0$. To prove Theorem 18.1.1 we must show that centers of non-canonical singularities of the log pair $(V_5, \lambda \mathcal{D})$ do not exist provided that $\mathrm{mult}_{\mathcal{L}_6}(\mathcal{D}) \leqslant \frac{1}{\lambda}$ and $\mathrm{mult}_{\mathscr{C}'}(\mathcal{D}) \leqslant \frac{1}{\lambda}$. Let us start with:

Theorem 18.3.1. *No point in V_5 is a center of non-canonical singularities of the log pair $(V_5, \lambda \mathcal{D})$.*

Proof. Suppose that there exists a point $P \in V_5$ that is a center of non-canonical singularities of the log pair $(V_5, \lambda \mathcal{D})$. Then

$$\mathrm{mult}_P\left(D_1 \cdot D_2\right) > \frac{4}{\lambda^2} \tag{18.3.2}$$

by Theorem 2.5.2. Denote by Σ the $\mathfrak{A}_5$-orbit of the point P. Then

$$|\Sigma| \in \left\{5, 10, 12, 15, 20, 30, 60\right\}$$

by Theorem 7.3.5. Put $n = \lfloor \frac{|\Sigma|}{5} \rfloor$. Let $\mathcal{M}$ be the linear system on V_5 consisting of all surfaces in $|nH|$ that contain Σ. Then the base locus of the linear system $\mathcal{M}$ does not contain curves by Corollary 17.4.2.

Let D_1 and D_2 be general members of the linear system $\mathcal{D}$. Let M be a general surface in $\mathcal{M}$. By (18.3.2), we obtain

$$\frac{4|\Sigma|}{\lambda^2} \geqslant \frac{20n}{\lambda^2} = D_1 \cdot D_2 \cdot M \geqslant \operatorname{mult}_P\Big(D_1 \cdot D_2\Big)|\Sigma| > \frac{4}{\lambda^2}|\Sigma|,$$

which is a contradiction. □

Thus, to prove Theorem 18.1.1, it is enough to show that no curve in V_5 can be a center of non-canonical singularities of the log pair $(V_5, \lambda\mathcal{D})$ provided that $\operatorname{mult}_{\mathcal{L}_6}(\mathcal{D}) \leqslant \frac{1}{\lambda}$ and $\operatorname{mult}_{\mathscr{C}'}(\mathcal{D}) \leqslant \frac{1}{\lambda}$.

Lemma 18.3.3. Let Z be an $\mathfrak{A}_5$-irreducible curve on V_5. Suppose that every irreducible component of Z is a center of non-canonical singularities of the log pair $(V_5, \lambda\mathcal{D})$. Then $\deg(Z) \leqslant 19$.

Proof. Recall that an irreducible curve $C \subset V_5$ is a center of non-canonical singularities of the log pair $(V_5, \lambda\mathcal{D})$ if and only if $\operatorname{mult}_C(\mathcal{D}) > \frac{1}{\lambda}$ by Lemma 2.4.4(ii). Thus, we have $\operatorname{mult}_Z(\mathcal{D}) > \frac{1}{\lambda}$.

Let D_1 and D_2 be general surfaces in $\mathcal{D}$. Then

$$D_1 \cdot D_2 = \operatorname{mult}_Z\Big(D_1 \cdot D_2\Big) Z + \Omega,$$

where Ω is an effective one-cycle whose support does not contain irreducible components of the curve Z. One has

$$\operatorname{mult}_Z\Big(D_1 \cdot D_2\Big) \geqslant \operatorname{mult}_Z(D_1) \cdot \operatorname{mult}_Z(D_2) > \frac{1}{\lambda^2},$$

since $\operatorname{mult}_Z(\mathcal{D}) > \frac{1}{\lambda}$. Thus

$$\frac{20}{\lambda^2} = D_1 \cdot D_2 \cdot H = \Big(\operatorname{mult}_Z\Big(D_1 \cdot D_2\Big) Z + \Omega\Big) \cdot H \geqslant$$
$$\geqslant \deg(Z) \operatorname{mult}_Z\Big(D_1 \cdot D_2\Big) > \frac{\deg(Z)}{\lambda^2},$$

which implies that $\deg(Z) \leqslant 19$. □

Now we are ready to prove:

Proposition 18.3.4. Let C be an irreducible curve in V_5 that is a center of non-canonical singularities of the log pair $(V_5, \lambda\mathcal{D})$. Then either C is a curve $\mathscr{C}'$, or a line of $\mathcal{L}_6$, or a line of $\mathcal{L}_{15}$.

Proof. By Lemma 18.3.3 the curve C is an irreducible component of some $\mathfrak{A}_5$-irreducible curve of degree at most 19. Moreover, one has $C \neq \mathscr{C}$ by Lemmas 2.4.4(ii) and 18.2.1.

Suppose that C is not a line of $\mathcal{L}_{15}$. Then C is contained in some surface $S \in \mathcal{Q}_2$ by Corollary 8.1.7 (or by Theorem 13.6.1). Let D be a general surface in $\mathcal{D}$. If $S = \mathscr{S}$, then Lemma 18.2.2 gives

$$\mathrm{mult}_C(D) \leqslant \mathrm{mult}_C\big(D \cdot \mathscr{S}\big) \leqslant \frac{1}{\lambda},$$

because $C \neq \mathscr{C}$. Similarly, if $S \neq \mathscr{S}$, then

$$\mathrm{mult}_C(D) \leqslant \mathrm{mult}_C\big(D \cdot S\big) \leqslant \frac{1}{\lambda}$$

by Lemma 18.2.5. Therefore, in both cases, we obtain a contradiction with Lemma 2.4.4(ii), because C is a center of non-canonical singularities of the log pair $(V_5, \lambda \mathcal{D})$. $\square$

Remark 18.3.5. Using Lemma 11.5.1, one can construct plenty of examples of linear systems $\mathcal{D}$ such that every line of $\mathcal{L}_6$ is a center of non-canonical singularities of the log pair $(V_5, \lambda \mathcal{D})$. Furthermore, using Lemma 11.5.4, one can construct plenty of examples of linear systems $\mathcal{D}$ such that $\mathscr{C}'$ is a center of non-canonical singularities of the log pair $(V_5, \lambda \mathcal{D})$. Thus, Theorem 18.1.1 fails to hold without the assumption that the curve $\mathscr{C}'$ and the lines of $\mathcal{L}_6$ are not centers of non-canonical singularities of the log pair $(V_5, \lambda \mathcal{D})$.

By Lemma 2.4.4(ii), Theorem 18.3.1, and Proposition 18.3.4, to prove Theorem 18.1.1 it is enough to show that no line of $\mathcal{L}_{15}$ is a center of non-canonical singularities of the log pair $(V_5, \lambda \mathcal{D})$ *assuming*, if necessary, that $\mathrm{mult}_{\mathcal{L}_6}(\mathcal{D}) \leqslant \frac{1}{\lambda}$ and $\mathrm{mult}_{\mathscr{C}'}(\mathcal{D}) \leqslant \frac{1}{\lambda}$.

18.4   Exclusion of the curve $\mathcal{L}_{15}$

Let $\mathcal{D}$ be an $\mathfrak{A}_5$-invariant mobile linear system on V_5, and λ be a positive rational number such that $K_{V_5} + \lambda \mathcal{D} \sim_{\mathbb{Q}} 0$.

Proposition 18.4.1. *One has* $\mathrm{mult}_{\mathcal{L}_{15}}(\mathcal{D}) < \frac{1}{\lambda}$.

Proof. Let $f \colon \bar{V}_5 \to V_5$ be the blow-up of V_5 at Σ_5, and let $B_1, \ldots, B_5$ be f-exceptional divisors that are contracted to the points of Σ_5 by f. By Lemma 7.5.8 the linear system $|f^*(H) - \sum_{i=1}^5 B_i|$ is free from base points and gives an $\mathfrak{A}_5$-equivariant morphism $\bar{\theta}_{\Pi_3} \colon \bar{V}_5 \to \mathbb{P}^2$.

Let $\bar{\mathcal{L}}_{15}$ be the proper transform on $\bar{V}_5$ of the curve $\mathcal{L}_{15}$. Then $\bar{\mathcal{L}}_{15}$ is smooth $\mathfrak{A}_5$-invariant curve that is a disjoint union of 15 smooth rational curves by Corollary 7.4.4 and Lemma 9.1.10(i). Let $g\colon \tilde{V}_5 \to \bar{V}_5$ be the blow-up of $\bar{V}_5$ at $\bar{\mathcal{L}}_{15}$, and let $F_1, \ldots, F_{15}$ be g-exceptional divisors that are contracted to the irreducible components of $\bar{\mathcal{L}}_{15}$ by g. Put $h = f \circ g$ and

$$\tilde{M} = 3h^*(H) - 2\sum_{i=1}^{5} g^*(B_i) - \sum_{i=1}^{15} F_i.$$

Let $\tilde{\mathcal{L}}_6$ be the proper transform of the curve $\mathcal{L}_6$ on $\tilde{V}_5$. Then

$$\tilde{M} \cdot \tilde{\mathcal{L}}_6 = -12$$

by Lemma 9.1.14. In particular, the divisor $\tilde{M}$ is not nef. However, the irreducible components of the curve $\tilde{\mathcal{L}}_6$ are the only irreducible curves in $\tilde{V}_5$ that have negative intersection with $\tilde{M}$ by Proposition 17.5.6.

Put $n = \lambda \mathrm{mult}_{\mathcal{L}_{15}}(\mathcal{D})$. We must show that $n < 1$. Suppose that $n \geq 1$. Denote by $\tilde{\mathcal{D}}$ the proper transform of the linear system $\mathcal{D}$ on the threefold $\tilde{V}$. Put $m = \lambda \mathrm{mult}_{\Sigma_5}(\mathcal{D})$, put $\tilde{H} = h^*(H)$, put $\tilde{E} = \sum_{i=1}^{5} g^*(B_i)$, and put $\tilde{F} = \sum_{i=1}^{15} F_i$. Then

$$\lambda \tilde{\mathcal{D}} \sim_{\mathbb{Q}} 2\tilde{H} - m\tilde{E} - n\tilde{F},$$

and the linear system $\tilde{\mathcal{D}}$ is free from base components. Let $\tilde{D}_1$ and $\tilde{D}_2$ be two sufficiently general surfaces in $\tilde{\mathcal{D}}$. One has

$$\lambda^2 \tilde{D}_1 \cdot \tilde{D}_2 = \delta \tilde{\mathcal{L}}_6 + \tilde{\Omega},$$

where δ is a non-negative rational number, and $\tilde{\Omega}$ is an effective one-cycle (with coefficients in $\mathbb{Q}$) on $\tilde{V}_5$ whose support does not contain any irreducible component of the curve $\tilde{\mathcal{L}}_6$. We have $\tilde{M} \cdot \tilde{\Omega} \geq 0$ by Proposition 17.5.6.

Let us express $\tilde{M} \cdot \tilde{\Omega}$ in terms of m, n, and δ. Since all irreducible components of the curve $\bar{\mathcal{L}}_{15}$ are disjoint and have trivial intersection with $-K_{\bar{V}_5}$, we obtain $\tilde{F}^3 = 30$. Moreover, we have

$$\tilde{H} \cdot \tilde{E}^2 = 0, \ \tilde{H} \cdot \tilde{E} \cdot \tilde{F} = 0, \ \tilde{H}^2 \cdot \tilde{E} = 0, \ \tilde{H}^2 \cdot \tilde{F} = 0, \ \tilde{E}^2 \cdot \tilde{F} = 0.$$

Furthermore, we have

$$\tilde{E}^3 = 5, \ \tilde{H} \cdot \tilde{F}^2 = -15, \ \tilde{E} \cdot \tilde{F}^2 = -15.$$

This gives

$$0 \leqslant \tilde{M} \cdot \tilde{\Omega} = \lambda^2 \tilde{M} \cdot \tilde{D}_1 \cdot \tilde{D}_2 - \delta \tilde{M} \cdot \tilde{L}_6 = \lambda^2 \tilde{M} \cdot \tilde{D}_1 \cdot \tilde{D}_2 + 12\delta =$$
$$= \left(3\tilde{H} - 2\tilde{E} - \tilde{F}\right) \cdot \left(2\tilde{H} - m\tilde{E} - n\tilde{F}\right)^2 + 12\delta =$$
$$= 60 - 10m^2 - 45n^2 - 60n + 30mn + 12\delta. \quad (18.4.2)$$

We claim that
$$6\delta \leqslant 20 - 5m^2. \quad (18.4.3)$$

Indeed, since the divisor
$$\tilde{H} - \tilde{E} = g^*\left(f^*(H) - \sum_{i=1}^{5} B_i\right)$$

is nef, we have

$$\lambda^2 \left(\tilde{H} - \tilde{E}\right) \cdot \tilde{D}_1 \cdot \tilde{D}_2 =$$
$$= \delta \left(\tilde{H} - \tilde{E}\right) \cdot \tilde{\mathcal{L}}_6 + \left(\tilde{H} - \tilde{E}\right) \cdot \tilde{\Omega} \geqslant \delta \left(\tilde{H} - \tilde{E}\right) \cdot \tilde{L}_6 = 6\delta.$$

On the other hand, we have
$$\lambda^2 \left(\tilde{H} - \tilde{E}\right) \cdot \tilde{D}_1 \cdot \tilde{D}_2 = \left(\tilde{H} - \tilde{E}\right) \cdot \left(2\tilde{H} - m\tilde{E} - n\tilde{F}\right)^2 = 20 - 5m^2,$$

which gives (18.4.3).

Combining (18.4.2) and (18.4.3), we get
$$100 - 20m^2 - 45n^2 - 60n + 30mn \geqslant$$
$$\geqslant 60 - 10m^2 - 45n^2 - 60n + 30mn + 12\delta \geqslant 0,$$

which, after division by 5, gives
$$20 - 4m^2 - 9n^2 - 12n + 6mn \geqslant 0. \quad (18.4.4)$$

We have $m \geqslant \frac{3n}{2}$. Indeed, let $\bar{\mathcal{D}}$ be the proper transform of the linear system $\mathcal{D}$ on the threefold $\bar{V}$, and let $\bar{C}$ be a general conic in $B_1 \cong \mathbb{P}^2$ that passes through the three points of intersection $B_1 \cap \bar{L}_{15}$ (that are non-collinear by Lemma 7.5.9). Then $\bar{C}$ is not contained in the surface $\bar{D}$. Thus, we have
$$2m = \lambda \bar{D} \cdot \bar{C} \geqslant 3n,$$

which implies that $m \geqslant \frac{3n}{2}$.

Combining (18.4.4) with inequalities $n \geqslant 1$ and $m \geqslant 3n/2$, we easily obtain a contradiction. Indeed, if

$$m < 3n - \frac{5}{4},$$

then

$$20 - 4m^2 - 9n^2 - 12n + 6mn = 20 - (3n-m)^2 - 3m^2 - 12n <$$
$$< 20 - \frac{25}{16} - 3m^2 - 12n \leqslant 20 - \frac{25}{16} - \frac{27}{4} - 12 < 0.$$

On the other hand, if

$$m \geqslant 3n - \frac{5}{4} > \frac{7}{4},$$

then

$$20 - 4m^2 - 9n^2 - 12n + 6mn = 20 - (3n-m)^2 - 3m^2 - 12n \leqslant$$
$$\leqslant 20 - 3m^2 - 12n \leqslant 20 - \frac{147}{16} - 12 < 0.$$

This completes the proof of Proposition 18.4.1. □

Now using Lemma 2.4.4(ii) we obtain:

Corollary 18.4.5. *The lines of $\mathcal{L}_{15}$ are not centers of non-canonical singularities of the log pair $(V_5, \lambda \mathcal{D})$.*

Combining Theorem 18.3.1, Proposition 18.3.4, and Corollary 18.4.5, we obtain the proof of Theorem 18.1.1.

18.5 Alternative approach to exclusion of points

In the previous sections, we proved Theorem 18.1.1. A crucial role in this proof is played by Theorem 18.3.1. The proof of Theorem 18.3.1 heavily relies on Corollary 17.4.2. The goal of this section is to prove Theorem 18.3.1 and at the same time avoid using Corollary 17.4.2. Note that in spite of this, we will use Proposition 17.2.8 and Lemma 17.4.5 that constitute an easy part of the proof of Corollary 17.4.2. The essential replacement is a new approach to exclusion of *long* $\mathfrak{A}_5$-orbits (see Lemmas 18.5.6 and 18.5.7 below).

CREMONA GROUPS AND THE ICOSAHEDRON 457

Let $\mathcal{D}$ be an $\mathfrak{A}_5$-invariant mobile linear system on V_5, and let λ be a positive rational number such that

$$K_{V_5} + \lambda\mathcal{D} \sim_{\mathbb{Q}} 0.$$

Lemma 18.5.1. Let P be a point in V_5. Suppose that

$$P \in \Sigma_5 \cup \Sigma_{10} \cup \Sigma'_{10} \cup \Sigma_{12} \cup \Sigma'_{12} \cup \Sigma_{15}.$$

Then P is not a center of non-canonical singularities of the log pair $(V_5, \lambda\mathcal{D})$.

Proof. Suppose that P is a center of non-canonical singularities of the log pair $(V_5, \lambda\mathcal{D})$. Let D and D' be general surfaces in $\mathcal{D}$. Then

$$\operatorname{mult}_P(D \cdot D') > \frac{4}{\lambda^2} \qquad (18.5.2)$$

by Theorem 2.5.2.

By Lemma 7.5.3, there exists a surface $H_{\Sigma_5} \in |H|$ that contains Σ_5 and contains no curves in the intersection $D \cap D'$. Thus, if $P \in \Sigma_5$, then (18.5.2) implies that

$$\frac{20}{\lambda^2} = D \cdot D' \cdot H_{\Sigma_5} \geqslant \operatorname{mult}_P(D \cdot D')|\Sigma_5| > \frac{4}{\lambda^2}|\Sigma_5| = \frac{20}{\lambda^2}.$$

This shows that $P \notin \Sigma_5$. Thus, the point P is contained in one of the $\mathfrak{A}_5$-orbits Σ_{10}, Σ'_{10}, Σ_{12}, Σ'_{12}, or Σ_{15}.

Denote by Σ the $\mathfrak{A}_5$-orbit of the point P. By Proposition 17.2.8 and Lemma 17.4.5, there exists a surface $Q_\Sigma \in |2H|$ that contains Σ and contains no curves of the intersection $D \cap D'$. If Σ is one of the $\mathfrak{A}_5$-orbits Σ_{10}, Σ'_{10}, or Σ_{15}, this follows from Proposition 17.2.8. If Σ is one of the $\mathfrak{A}_5$-orbits Σ_{12} or Σ'_{12}, this follows from Lemma 17.4.5, because $\Sigma_{12} \subset \mathscr{C}$ by construction, and $\Sigma'_{12} \subset \mathscr{C}'$ by Lemma 11.3.1. Therefore

$$\frac{40}{\lambda^2} = D \cdot D' \cdot Q_\Sigma \geqslant \operatorname{mult}_P(D \cdot D')|\Sigma| > \frac{4}{\lambda^2}|\Sigma| \geqslant \frac{40}{\lambda^2}$$

by Theorem 2.5.2. The obtained contradiction completes the proof of the lemma. $\square$

Lemma 18.5.3. Let P be a point in V_5 such that $P \in \mathscr{C} \cup \mathscr{C}' \cup \mathcal{L}_6$. Then P is not a center of canonical singularities of the log pair $(V_5, \lambda\mathcal{D})$.

Proof. Suppose that P is a center of non-canonical singularities of the log pair $(V_5, \lambda \mathcal{D})$. Denote by C a curve among $\mathscr{C}$, $\mathscr{C}'$, and $\mathcal{L}_6$ that contains P. Let D and D' be general surfaces in $\mathcal{D}$. Write

$$D \cdot D' = \epsilon C + \Omega,$$

where ϵ is a non-negative integer, and Ω is an effective one-cycle whose support Ω does not contain C. Then

$$\mathrm{mult}_P(\Omega) = \mathrm{mult}_P(D \cdot D') - \epsilon \mathrm{mult}_P(C) \geqslant \frac{4}{\lambda^2} - \epsilon \qquad (18.5.4)$$

by Theorem 2.5.6. On the other hand,

$$\frac{20}{\lambda^2} = H \cdot D \cdot D' = \epsilon C \cdot H + \Omega \cdot H \geqslant \epsilon C \cdot H = 6\epsilon,$$

so that

$$\epsilon \leqslant \frac{10}{3\lambda^2}. \qquad (18.5.5)$$

Let Σ be the $\mathfrak{A}_5$-orbit of P. Then $|\Sigma| \geqslant 12$. Indeed, this is implied by Lemma 5.1.4 if $C = \mathscr{C}$ or $C = \mathscr{C}'$, and by Corollary 7.4.9 if $C = \mathcal{L}_6$.

Let $\mathcal{Q}$ be the linear subsystem in $|2H|$ that consists of all surfaces containing C. Then the base locus of $\mathcal{Q}$ coincides with the curve C. This is implied by Corollary 11.1.6 if $C = \mathcal{L}_6$, and by the fact that a sextic rational normal curve is an intersection of quadrics if $C = \mathscr{C}$ or $C = \mathscr{C}'$. Let Q be a general member of $\mathcal{Q}$. Then Q does not contain irreducible components of $\mathrm{Supp}(\Omega)$. Thus (18.5.4) and (18.5.5) give

$$\frac{40}{\lambda^2} = Q \cdot D \cdot D' = \epsilon C \cdot Q + \Omega \cdot Q = 12\epsilon + \Omega \cdot Q \geqslant$$

$$\geqslant 12\epsilon + \sum_{O \in \Sigma} O\big(\Omega \cdot Q\big) \geqslant 12\epsilon + \sum_{O \in \Sigma} O(\Omega) \geqslant$$

$$\geqslant 12\epsilon + |\Sigma|\left(\frac{4}{\lambda^2} - \epsilon\right) \geqslant 12\epsilon + 12\left(\frac{4}{\lambda^2} - \epsilon\right) = \frac{48}{\lambda^2},$$

which is absurd. $\square$

Lemma 18.5.6. *Let P be a point in V_5. Suppose that $P \in \mathscr{S}$. Then P is not a center of non-canonical singularities of the log pair $(V_5, \lambda \mathcal{D})$.*

Proof. Suppose that P is a center of non-canonical singularities of the log pair $(V_5, \lambda \mathcal{D})$. Then $\mathscr{S}$ is smooth at P, because $P \notin \mathscr{C}$ by Lemma 18.5.3 and $\mathscr{C}$ is the singular locus of $\mathscr{S}$ by Lemma 7.2.2(i).

Let us use notation of §7.2. Let D be a general surface in $\mathcal{D}$. Denote by $\hat{D}$ its proper transform on $\mathcal{Y}$. Then

$$\lambda \hat{D} \sim \nu^*(2H) - \lambda \mathrm{mult}_{\mathscr{C}}(D) E_{\mathcal{Y}}.$$

By Lemma 7.2.3(ii),(iii), the effective $\mathbb{Q}$-divisor $\lambda \hat{D}|_{\hat{\mathscr{S}}}$ has bi-degree

$$\Big(2 - 2\lambda \mathrm{mult}_{\mathscr{C}}(D),\, 10 - 2\lambda \mathrm{mult}_{\mathscr{C}}(D)\Big)$$

on $\hat{\mathscr{S}} \cong \mathbb{P}^1 \times \mathbb{P}^1$.

Let $\hat{P}$ be a point in $\hat{\mathscr{S}}$ such that $\nu(\hat{P}) = P$. Then ν is an isomorphism in a neighborhood of the point $\hat{P}$, so that the log pair $(\mathcal{Y}, \lambda \hat{D})$ is not canonical at $\hat{P}$. Hence, the log pair

$$\Big(\mathcal{Y}, \lambda \hat{D} + \hat{\mathscr{S}}\Big)$$

is not log canonical at $\hat{P}$. By Theorem 2.3.7, the log pair

$$\Big(\hat{\mathscr{S}}, \lambda \hat{D}|_{\hat{\mathscr{S}}}\Big)$$

is not log canonical at $\hat{P}$ either. On the other hand, the log pair $(\hat{\mathscr{S}}, \lambda \hat{D}|_{\hat{\mathscr{S}}})$ is log canonical outside of the curve $\hat{\mathscr{C}}$ and some finite set that contains $\hat{P}$. This follows from Lemma 18.2.2, because $P \notin \mathscr{C}$. Write

$$\lambda \hat{D}|_{\hat{\mathscr{S}}} = \epsilon \hat{\mathscr{C}} + \Omega,$$

where ϵ is a non-negative rational number, and Ω is an effective $\mathbb{Q}$-divisor on $\hat{\mathscr{S}}$ whose support does not contain the curve $\hat{\mathscr{C}}$. Then the log pair $(\hat{\mathscr{S}}, \Omega)$ is not log canonical at $\hat{P}$, and is log canonical outside of finitely many points in $\hat{\mathscr{S}}$.

Let δ be the largest rational number such that the log pair $(\hat{\mathscr{S}}, \delta\Omega)$ has log canonical singularities at $\hat{P}$, i.e., δ is the log canonical threshold of the log pair $(\hat{\mathscr{S}}, \delta\Omega)$ at $\hat{P}$ (see [73, Definition 8.1]). Then $\delta < 1$, because $(\hat{\mathscr{S}}, \Omega)$ is not log canonical at $\hat{P}$. Let $\mathcal{I}$ be the multiplier ideal sheaf (see §2.3) of the log pair $(\hat{\mathscr{S}}, \delta\Omega)$, and let $\mathcal{Z}$ be the subscheme of $\hat{\mathscr{S}}$ defined by $\mathcal{I}$. Then $\mathrm{Supp}(\mathcal{Z})$ is a finite set that contains $\hat{P}$.

Let R be a Cartier divisor of bi-degree $(0, 8)$ on the surface $\hat{\mathscr{S}} \cong \mathbb{P}^1 \times \mathbb{P}^1$. Put $B = R - K_{\hat{\mathscr{S}}} - \delta\Omega$. Then B is a $\mathbb{Q}$-divisor of bi-degree

$$\Big(2\delta\lambda\mathrm{mult}_{\mathscr{C}}(D) + \delta\epsilon + 2 - 2\delta,\, 2\delta\lambda\mathrm{mult}_{\mathscr{C}}(D) + \delta\epsilon + 10 - 10\delta\Big)$$

on the surface $\hat{\mathscr{S}}$ (cf. §7.2). This implies that the $\mathbb{Q}$-divisor B is ample. Hence
$$h^1\Big(\mathcal{O}_{\hat{\mathscr{S}}}(R) \otimes \mathcal{I}\Big) = 0$$
by Theorem 2.3.5. This gives
$$h^0\Big(\mathcal{O}_{\mathcal{Z}}\Big) = h^0\Big(\mathcal{O}_{\hat{\mathscr{S}}}(R)\Big) - h^0\Big(\mathcal{O}_{\hat{\mathscr{S}}}(R) \otimes \mathcal{I}\Big) \leqslant h^0\Big(\mathcal{O}_{\hat{\mathscr{S}}}(R)\Big) = 9,$$
which implies that $\mathrm{Supp}(\mathcal{Z})$ consists of at most 9 points. The latter is impossible by Lemma 7.2.10. $\square$

Lemma 18.5.7. Let P be a point in V_5. Suppose that $P \notin \mathscr{S}$. Then P is not a center of non-canonical singularities of the log pair $(V_5, \lambda\mathcal{D})$.

Proof. Suppose that P is a center of non-canonical singularities of the log pair $(V_5, \lambda\mathcal{D})$. Let S be a surface in $\mathcal{Q}_2$ that contains P. Then S is a $K3$ surface that has at most ordinary double points by Theorem 8.2.1(v). Moreover, the $\mathfrak{A}_5$-orbit of the point P consists of at least 20 points by Lemma 18.5.1 and Theorem 7.3.5. Thus, the surface S is smooth at P by Theorem 8.2.1(ii),(iii). Furthermore, we have $P \notin \mathscr{C}' \cup \mathcal{L}_6$ by Lemma 18.5.3.

Let D be a general surface in $\mathcal{D}$. Then the log pair
$$\Big(V_5, \lambda D + S\Big)$$
is not log canonical at P. Put $D_S = D|_S$ and $H_S = H|_S$. Then $(S, \lambda D_S)$ is not log canonical at P by Theorem 2.3.7. On the other hand, the log pair $(S, \lambda D_S)$ is log canonical in a punctured neighborhood of the point P by Lemma 18.2.5, because $P \notin \mathscr{C}' \cup \mathcal{L}_6$.

Let δ be the largest rational number such that the log pair $(S, \delta\lambda D_S)$ has log canonical singularities at $\hat{P}$, let $\mathcal{I}_S$ be the multiplier ideal sheaf of the log pair $(S, \delta\lambda D_S)$, let $\mathcal{Z}_S$ be a subscheme in S defined by $\mathcal{I}_S$, and let Z be the $\mathfrak{A}_5$-orbit of the point P. Since the log pair $(S, \lambda D_S)$ is log canonical in a punctured neighborhood of P, the subscheme $\mathcal{Z}_S$ is a disjoint union of subschemes $\mathcal{Z}'_S$ and $\mathcal{Z}''_S$ such that $\mathrm{Supp}(\mathcal{Z}'_S) = Z$ and $Z \cap \mathrm{Supp}(\mathcal{Z}''_S) = \varnothing$. It follows from Theorem 2.3.5 that
$$h^1\Big(\mathcal{I}_S \otimes \mathcal{O}_S(2H_S)\Big) = 0.$$
Hence there is an exact sequence of $\mathfrak{A}_5$-representations
$$0 \to H^0\Big(\mathcal{I}_S \otimes \mathcal{O}_S(2H_S)\Big) \to H^0\Big(\mathcal{O}_S(2H_S)\Big) \to$$
$$\to H^0\Big(\mathcal{O}_{\mathcal{Z}'_S}\Big) \oplus H^0\Big(\mathcal{O}_{\mathcal{Z}''_S}(2H_S)\Big) \to 0. \quad (18.5.8)$$

Since $h^0(\mathcal{O}_S(2H_S)) = 22$ by the Riemann–Roch formula and Kodaira vanishing, it follows from (18.5.8) that $h^0(\mathcal{O}_{\mathcal{Z}'_S}) \leqslant 22$. Keeping in mind that $|Z| \geqslant 20$ and $\mathrm{Supp}(\mathcal{Z}'_S) = Z$, we see that $|Z| = 20$ by Theorem 7.3.5. Since $\mathrm{Supp}(\mathcal{Z}'_S) = Z$ and

$$22 \geqslant h^0(\mathcal{O}_{\mathcal{Z}'_S}) \geqslant |Z| = 20,$$

we see that $h^0(\mathcal{O}_{\mathcal{Z}'_S}) = 20$, because $h^0(\mathcal{O}_{\mathcal{Z}'_S})$ is divisible by $|Z| = 20$. Thus, the exact sequence (18.5.8) implies that there is a surjective morphism of $\mathfrak{A}_5$-representations

$$H^0\big(\mathcal{O}_S(2H_S)\big) \to H^0\big(\mathcal{O}_{\mathcal{Z}'_S}\big),$$

so that $H^0(\mathcal{O}_S(2H_S))$ has a two-dimensional $\mathfrak{A}_5$-subrepresentation. The latter is impossible by Lemma 8.2.6(iii). □

Theorem 18.3.1 follows from Lemmas 18.5.6 and 18.5.7.

18.6 Alternative approach to the exclusion of $\mathcal{L}_{15}$

Let $\mathcal{D}$ be an $\mathfrak{A}_5$-invariant mobile linear system on V_5, and let λ be a positive rational number such that

$$K_{V_5} + \lambda \mathcal{D} \sim_\mathbb{Q} 0.$$

Suppose that $\mathrm{mult}_{\mathcal{L}_6}(\mathcal{D}) \leqslant \frac{1}{\lambda}$ and $\mathrm{mult}_{\mathscr{C}'}(\mathcal{D}) \leqslant \frac{1}{\lambda}$. Then neither the curve $\mathscr{C}'$ nor a line of $\mathcal{L}_6$ is a center of non-canonical singularities of $(V_5, \lambda\mathcal{D})$ by Lemma 2.4.4(ii). Note that $\mathscr{C}$ is not a center of non-canonical singularities of $(V_5, \lambda\mathcal{D})$ by Lemmas 2.4.4(ii) and 18.2.1. Moreover, it follows from Theorem 18.3.1 that no point in V_5 is a center of non-canonical singularities of $(V_5, \lambda\mathcal{D})$. Furthermore, it follows from Proposition 18.3.4 that no irreducible curve in V_5 is a center of non-canonical singularities of $(V_5, \lambda\mathcal{D})$ except possibly for a line of $\mathcal{L}_{15}$.

By Corollary 18.4.5, the lines of $\mathcal{L}_{15}$ are not centers of non-canonical singularities of the log pair $(V_5, \lambda\mathcal{D})$. The latter implies Theorem 18.1.1 (which implies Theorem 1.4.1 as explained in §18.1). Now we present an alternative proof of the fact that the lines of $\mathcal{L}_{15}$ are not centers of non-canonical singularities of the log pair $(V_5, \lambda\mathcal{D})$. The main benefit of this approach is that it allows us to avoid using very involved Proposition 18.4.1, which is replaced by a much easier Lemma 18.6.6.

Suppose that the log pair $(V_5, \lambda\mathcal{D})$ is not canonical, so that the lines of $\mathcal{L}_{15}$ are centers of non-canonical singularities of the log pair $(V_5, \lambda\mathcal{D})$. It follows from Lemma 2.4.5 that there exists a positive rational number $\mu < 2\lambda$ such that the log pair $(V_5, \mu\mathcal{D})$ is not log canonical, and every center of non-canonical singularities of $(V_5, \lambda\mathcal{D})$ is a center of non log canonical singularities of $(V_5, \mu\mathcal{D})$. Let $\mathcal{I}$ be the multiplier ideal sheaf of the log pair $(V_5, \mu\mathcal{D})$. Then

$$h^1\left(\mathcal{I} \otimes \mathcal{O}_{V_5}(2H)\right) = 0 \tag{18.6.1}$$

by Theorem 2.3.5.

Remark 18.6.2. The ideal $\mathcal{I}$ is not trivial, since $(V_5, \mu\mathcal{D})$ is not log canonical. Thus, it defines a (non-empty) subscheme $\mathcal{Z}$ of V_5. It may be non-reduced, not equidimensional and not irreducible. However, it follows from the movability of $\mathcal{D}$ that $\mathcal{Z}$ has no two-dimensional components.

Let $\mathcal{Z}^0$ and $\mathcal{Z}^1$ be subschemes of $\mathcal{Z}$ such that $\mathcal{Z}^0$ is a zero-dimensional scheme, the support of $\mathcal{Z}_1$ is a curve, and

$$\mathcal{Z} = \mathcal{Z}^0 \sqcup \mathcal{Z}^1,$$

so that in particular the supports of the subschemes $\mathcal{Z}^0$ and $\mathcal{Z}^1$ are disjoint. Namely, $\mathrm{Supp}(\mathcal{Z}^1)$ consists of all one-dimensional components of $\mathrm{Supp}(\mathcal{Z})$, and $\mathrm{Supp}(\mathcal{Z}^0)$ consists of all (isolated) points in $\mathrm{Supp}(\mathcal{Z})$ that are not contained in $\mathcal{Z}^1$.

Lemma 18.6.3. *The support of $\mathcal{Z}^0$ consists of at most 23 points, and the points of the support of $\mathcal{Z}^0$ impose independent linear conditions on sections of $H^0(\mathcal{O}_{V_5}(2H))$.*

Proof. Note that

$$H^0\left(\mathcal{O}_{\mathcal{Z}} \otimes \mathcal{O}_{V_5}(2H)\right) = H^0\left(\mathcal{O}_{\mathcal{Z}^0}\right) \oplus H^0\left(\mathcal{O}_{\mathcal{Z}^1} \otimes \mathcal{O}_{V_5}(2H)\right)$$

because $\mathrm{Supp}(\mathcal{Z}^0)$ and $\mathrm{Supp}(\mathcal{Z}^1)$ are disjoint. By (18.6.1), there is an exact sequence

$$0 \to H^0\left(\mathcal{I} \otimes \mathcal{O}_{V_5}(2H)\right) \to H^0\left(\mathcal{O}_{V_5}(2H)\right) \to$$
$$\to H^0\left(\mathcal{O}_{\mathcal{Z}^0}\right) \oplus H^0\left(\mathcal{O}_{\mathcal{Z}^1} \otimes \mathcal{O}_{V_5}(2H)\right) \to 0,$$

and the assertion follows. $\square$

Lemma 18.6.4. *The support of $\mathcal{Z}^0$ does not contain $\mathfrak{A}_5$-orbits of length 20.*

Proof. Suppose that the support of $\mathcal{Z}^0$ does contain an $\mathfrak{A}_5$-orbit of length 20. By Lemma 18.6.3 and Theorem 7.3.5 this implies that $\mathrm{Supp}(\mathcal{Z}^0)$ is a single $\mathfrak{A}_5$-orbit of length 20. Moreover, by Lemma 18.6.3 the points of $\mathrm{Supp}(\mathcal{Z}^0)$ impose independent linear conditions on sections of $H^0(\mathcal{O}_{V_5}(2H))$. Since

$$h^0\Big(\mathcal{O}_{V_5}(2H)\Big) = 23$$

by (7.1.3), there exists a three-dimensional $\mathfrak{A}_5$-invariant subspace

$$U \subset H^0\Big(\mathcal{O}_{V_5}(2H)\Big)$$

that consists of sections vanishing at the points of $\mathrm{Supp}(\mathcal{Z}^0)$. By Lemma 8.1.1 one has $U \cong W_3'$. Thus, the linear subsystem in $|2H|$ that consists of all surfaces passing through $\mathrm{Supp}(\mathcal{Z}^0)$ is the linear system $\mathcal{Q}_3$. On the other hand, by Proposition 17.2.8 the base locus of $\mathcal{Q}_3$ does not contain $\mathfrak{A}_5$-orbits of length 20, which is a contradiction. □

Let $Z_1, \ldots, Z_s$ be $\mathfrak{A}_5$-irreducible curves on V_5 such that

$$\mathrm{Supp}(\mathcal{Z}^1) = Z_1 \cup \ldots \cup Z_s.$$

Put $d_i = \deg(Z_i)$.

Lemma 18.6.5 (cf. Lemma 18.3.3). *One has $\sum_{i=1}^{s} d_i \leqslant 19$.*

Proof. Let D and D' be general surfaces in $\mathcal{D}$. Then

$$D \cdot D' = \sum_{i=1}^{s} m_i Z_i + \Omega,$$

where Ω is an effective one-cycle whose support does not contain components of the curves Z_i. One has

$$m_i > \frac{4}{\mu^2} > \frac{1}{\lambda^2}$$

by Theorem 2.5.1. Thus

$$\frac{20}{\lambda^2} = D \cdot D' \cdot H = \Big(\sum_{i=1}^{s} m_i Z_i + \Omega\Big) \cdot H \geqslant \sum_{i=1}^{s} m_i d_i > \frac{1}{\lambda^2} \sum_{i=1}^{s} d_i,$$

and the assertion follows. □

Now we are ready to prove:

Lemma 18.6.6. The curve $\mathcal{L}_{15}$ is not contained in the support of $\mathcal{Z}^1$.

Proof. The curve $\mathcal{L}_{15}$ is a disjoint union $\bigsqcup_{i=1}^{5} \Delta_i$ of five triples of non-coplanar lines, each triple having one point in common, which is a point of the $\mathfrak{A}_5$-orbit Σ_5 (see Corollary 7.4.4 and Lemma 9.1.10(i)). Thus one computes that
$$h^0\left(\mathcal{O}_{\Delta_i} \otimes \mathcal{O}_{V_5}(2H)\right) = 7.$$
Therefore, we have
$$h^0\left(\mathcal{O}_{\mathcal{L}_{15}} \otimes \mathcal{O}_{V_5}(2H)\right) = 35.$$

Suppose that $\mathcal{L}_{15}$ is contained in the support of $\mathcal{Z}^1$. Then the support of $\mathcal{Z}^1$ consists of the curve $\mathcal{L}_{15}$ by Lemma 18.6.5 and Corollary 8.1.15.

Let μ' be the greatest positive rational number such that the log pair $(V_5, \mu'\mathcal{D})$ is log canonical at a generic point of some (and thus any) line of $\mathcal{L}_{15}$, and let $\mathcal{I}'$ be the multiplier ideal sheaf of the log pair $(V_5, \mu'\mathcal{D})$. Then $\mu' \leqslant \mu$ by construction (see Remark 2.3.4). Moreover, one has $\mathcal{I} \subset \mathcal{I}'$ by Definition 2.3.2, and one has
$$h^1\left(\mathcal{I}' \otimes \mathcal{O}_{V_5}(2H)\right) = 0 \tag{18.6.7}$$
by Theorem 2.3.5.

Let $\mathcal{Z}'$ be the subscheme of V_5 that is defined by $\mathcal{I}'$. Then $\mathrm{Supp}(\mathcal{Z}')$ contains $\mathcal{L}_{15}$ by construction. Moreover, the subscheme $\mathcal{Z}'$ is reduced at a generic point of some (and thus any) line of $\mathcal{L}_{15}$. Furthermore, the support of $\mathcal{Z}'$ is a disjoint union of the curve $\mathcal{L}_{15}$ and some (possibly empty) finite set, because the support of $\mathcal{Z}^1$ consists of the curve $\mathcal{L}_{15}$, and $\mathcal{I} \subset \mathcal{I}'$. Thus, applying (18.6.7) together with (7.1.3) and Corollary 4.1.5, we get
$$35 = h^0\left(\mathcal{O}_{\mathcal{L}_{15}} \otimes \mathcal{O}_{V_5}(2H)\right) \leqslant h^0\left(\mathcal{O}_{\mathcal{Z}'} \otimes \mathcal{O}_{V_5}(2H)\right) \leqslant h^0\left(\mathcal{O}_{V_5}(2H)\right) = 23$$
as above. The obtained contradiction completes the proof of the lemma. □

Thus, no line of $\mathcal{L}_{15}$ is a center of non-canonical singularities of $(V_5, \lambda\mathcal{D})$.

Chapter 19

Halphen pencils and elliptic fibrations

In this chapter we describe all $\mathfrak{A}_5$-invariant Halphen pencils on V_5 (see Definition 3.4.1). The same method gives a complete classification of $\mathfrak{A}_5$-rational maps from V_5 whose general fiber is an elliptic curve. Furthermore, we describe all $\mathfrak{A}_5$-birational maps from V_5 to Fano threefolds with canonical singularities. The proofs of these results are similar to the proof of Theorem 1.4.1 and are obtained by strengthening Theorem 18.1.1.

19.1 Statement of results

Recall from §8.1 that the anticanonical linear system $|-K_{V_5}|$ contains a unique $\mathfrak{A}_5$-invariant pencil $\mathcal{Q}_2$. Moreover, a general surface in $\mathcal{Q}_2$ is a smooth $K3$ surface by Theorem 8.2.1(ii).

Theorem 19.1.1. Let $\mathcal{P}$ be an $\mathfrak{A}_5$-invariant pencil on V_5 whose general member is an irreducible surface that is birational to a smooth surface of Kodaira dimension zero. Then $\mathcal{P} = \alpha^{-1}\mathcal{Q}_2$ for some birational self-map $\alpha \in \mathrm{Bir}^{\mathfrak{A}_5}(V_5)$.

Recall from §7.5 that there is an $\mathfrak{A}_5$-equivariant projection

$$\theta_{\Pi_3} \colon V_5 \dashrightarrow \mathbb{P}^2 \cong \mathbb{P}(W_3)$$

from the linear subspace $\Pi_3 \subset \mathbb{P}^6$. Let $f \colon \bar{V}_5 \to V_5$ be the blow-up of the

$\mathfrak{A}_5$-orbit Σ_5. By Lemma 7.5.8, there exists a commutative diagram

$$\begin{array}{ccc} & \bar{V}_5 & \\ {\scriptstyle f} \swarrow & & \searrow {\scriptstyle \bar{\theta}_{\Pi_3}} \\ V_5 & \dashrightarrow{\theta_{\Pi_3}} & \mathbb{P}^2 \end{array}$$
(19.1.2)

where $\bar{\theta}_{\Pi_3}$ is a morphism whose general fiber is an elliptic curve.

Theorem 19.1.3. *Let $h\colon V_5 \dashrightarrow S$ be an $\mathfrak{A}_5$-rational map whose general fiber is birational to a smooth elliptic curve. Then there exists an $\mathfrak{A}_5$-birational map $t\colon S \dashrightarrow \mathbb{P}^2$ that fits into the commutative diagram*

$$\begin{array}{ccc} V_5 & \dashrightarrow{\alpha} & V_5 \\ {\scriptstyle h} \downarrow & & \downarrow {\scriptstyle \theta_{\Pi_3}} \\ S & \dashrightarrow{t} & \mathbb{P}^2 \end{array}$$

for some birational selfmap $\alpha \in \mathrm{Bir}^{\mathfrak{A}_5}(V_5)$.

Let $\nu\colon \mathcal{Y} \to V_5$ be the blow-up of the curve $\mathscr{C}$ (see §7.2), let $\pi\colon \mathcal{W} \to V_5$ be the blow-up of the curve $\mathcal{L}_6$, let $\psi\colon \mathcal{U} \to V_5$ be the blow-up of the curve $\mathscr{C}'$, and let $\bar{\psi}\colon \bar{\mathcal{U}} \to \mathcal{W}$ be the blow-up of the proper transform of the curve $\mathscr{C}'$ on the threefold $\mathcal{W}$ (see Chapter 11). By Proposition 7.2.4 and Lemmas 11.4.10 and 11.2.1, the linear systems $|-K_\mathcal{Y}|$, $|-K_\mathcal{U}|$ and $|-K_\mathcal{W}|$ are free from base points and give birational morphisms

$$\zeta\colon \mathcal{Y} \to X_{14},$$
$$\zeta'\colon \mathcal{U} \to X'_{14},$$
$$\varphi\colon \mathcal{W} \to X_4,$$

respectively, where X_{14}, X'_{14}, and X_4 are Fano threefolds with canonical Gorenstein singularities. Moreover, it follows from (11.4.12) and Remark 11.4.13 that the linear system $|-K_{\bar{\mathcal{U}}}|$ is also free from base points and gives a morphism $\bar{\chi}\colon \bar{\mathcal{U}} \to \mathbb{P}^3$ whose Stein factorization is

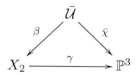

where γ is a double cover branched along a singular sextic surface (defined in the appropriate coordinates by the equation (11.2.17)), and β is a birational morphism given by the linear system $|-2K_{\bar{\mathcal{U}}}|$.

Theorem 19.1.4. Let $g\colon V_5 \dashrightarrow V$ be an $\mathfrak{A}_5$-birational map such that V is a Fano variety with canonical singularities. Then V is isomorphic to one of the threefolds V_5, X_{14}, X'_{14}, X_4, or X_2.

Theorems 19.1.1, 19.1.3, and 19.1.4 constitute the main result of this chapter. Their proofs rely on the following technical result.

Theorem 19.1.5. Let $\mathcal{D}$ be an $\mathfrak{A}_5$-invariant mobile linear system on V_5, and let λ be a positive rational number such that

$$K_{V_5} + \lambda \mathcal{D} \sim_{\mathbb{Q}} 0.$$

Denote by $\mathcal{D}_{\mathcal{Y}}$, $\mathcal{D}_{\mathcal{U}}$, $\mathcal{D}_{\mathcal{W}}$, $\mathcal{D}_{\bar{\mathcal{U}}}$, and $\mathcal{D}_{\bar{V}_5}$ the proper transforms of the linear system $\mathcal{D}$ on the threefolds $\mathcal{Y}$, $\mathcal{U}$, $\mathcal{W}$, $\bar{\mathcal{U}}$, and $\bar{V}_5$, respectively. Suppose that the singularities of the log pair $(V_5, \lambda \mathcal{D})$ are canonical but not terminal. Then exactly one of the following possibilities holds:

(i) $\mathbb{CS}(V_5, \lambda \mathcal{D})$ consists of points in Σ_5, one has $\mathrm{mult}_{\Sigma_5}(\mathcal{D}) = \frac{2}{\lambda}$, the log pair $(\bar{V}_5, \lambda \mathcal{D}_{\bar{V}_5})$ has terminal singularities, and $\mathcal{D}_{\bar{V}_5}$ is a pull back of a mobile linear system on $\mathbb{P}^2$ via the elliptic fibration $\bar{\theta}_{\Pi_3}\colon \bar{V}_5 \to \mathbb{P}^2$;

(ii) $\mathbb{CS}(V_5, \lambda \mathcal{D}) = \{\mathscr{C}\}$, one has $\mathrm{mult}_{\mathscr{C}}(\mathcal{D}) = \frac{1}{\lambda}$, and the log pair $(\mathcal{Y}, \lambda \mathcal{D}_{\mathcal{Y}})$ has terminal singularities;

(iii) $\mathbb{CS}(V_5, \lambda \mathcal{D}) = \{\mathscr{C}'\}$, one has $\mathrm{mult}_{\mathscr{C}'}(\mathcal{D}) = \frac{1}{\lambda}$, and the log pair $(\mathcal{U}, \lambda \mathcal{D}_{\mathcal{U}})$ has terminal singularities;

(iv) the set $\mathbb{CS}(V_5, \lambda \mathcal{D})$ consists of lines of $\mathcal{L}_6$, one has $\mathrm{mult}_{\mathcal{L}_6}(\mathcal{D}) = \frac{1}{\lambda}$, and the log pair $(\mathcal{W}, \lambda \mathcal{D}_{\mathcal{W}})$ has terminal singularities;

(v) the set $\mathbb{CS}(V_5, \lambda \mathcal{D})$ consists of the curve $\mathscr{C}'$ and the lines of $\mathcal{L}_6$, one has
$$\mathrm{mult}_{\mathcal{L}_6}(\mathcal{D}) = \mathrm{mult}_{\mathscr{C}'}(\mathcal{D}) = \frac{1}{\lambda},$$
and the log pair $(\bar{\mathcal{U}}, \lambda \mathcal{D}_{\bar{\mathcal{U}}})$ has terminal singularities;

(vi) $\mathbb{CS}(V_5, \lambda \mathcal{D}) = \{C_{20}\}$, one has $\mathrm{mult}_{C_{20}}(\mathcal{D}) = \frac{1}{\lambda}$, and $\mathcal{D}$ is composed of the pencil $\mathcal{Q}_2$.

Let us show how to derive Theorems 19.1.1, 19.1.3, and 19.1.4 from Theorem 19.1.5. Let $\mathcal{B}$ be an $\mathfrak{A}_5$-invariant mobile linear system on V_5. Then it follows from Lemma 11.5.5 and Theorem 18.1.1 that there exists a birational selfmap $\alpha \in \mathrm{Bir}^{\mathfrak{A}_5}(V_5)$ such that the singularities of the log

pair $(V_5, \lambda\alpha^{-1}\mathcal{B})$ are canonical, where λ is a positive rational number defined by
$$K_{V_5} + \lambda\alpha^{-1}\mathcal{B} \sim_{\mathbb{Q}} 0.$$
Put $\mathcal{D} = \alpha^{-1}\mathcal{B}$. Denote by $\mathcal{D}_{\mathcal{Y}}, \mathcal{D}_{\mathcal{U}}, \mathcal{D}_{\mathcal{W}}, \mathcal{D}_{\bar{\mathcal{U}}}$, and $\mathcal{D}_{\bar{V}_5}$ the proper transforms of the linear system $\mathcal{D}$ on the threefolds $\mathcal{Y}, \mathcal{U}, \mathcal{W}, \bar{\mathcal{U}}$, and $\bar{V}_5$, respectively. In the proofs of Theorems 19.1.1, 19.1.3, and 19.1.4, we will choose the linear system $\mathcal{B}$ appropriately in each case and use this notation. Recall that the *log canonical divisor* corresponding to the log pair $(X, \mathcal{D}_X)$, where $\mathcal{D}_X$ is a linear system on a variety X, is a divisor $K_X + D$ for $D \in \mathcal{D}_X$.

Proof of Theorem 19.1.1. Let $\mathcal{P}$ be an $\mathfrak{A}_5$-invariant pencil on V_5 whose general member is an irreducible surface that is birational to a smooth surface of Kodaira dimension zero. Put $\mathcal{B} = \mathcal{P}$. Then the singularities of the log pair $(V_5, \lambda\mathcal{D})$ are not terminal by Theorem 3.4.2.

If $\mathbb{CS}(V_5, \lambda\mathcal{D})$ contains the curve C_{20}, then $\mathcal{D}$ is composed of the pencil $\mathcal{Q}_2$ by Theorem 19.1.5, which implies that $\mathcal{D} = \mathcal{Q}_2$, because a general member of the pencil $\mathcal{D}$ is an irreducible surface. Thus, we may assume that C_{20} is not contained in $\mathbb{CS}(V_5, \lambda\mathcal{D})$. It follows from Theorem 19.1.5 that the singularities of one of the log pairs $(\bar{V}_5, \lambda\mathcal{D}_{\bar{V}_5})$, $(\mathcal{Y}, \lambda\mathcal{D}_{\mathcal{Y}})$, $(\mathcal{U}, \lambda\mathcal{D}_{\mathcal{U}})$, $(\mathcal{W}, \lambda\mathcal{D}_{\mathcal{W}})$, or $(\bar{\mathcal{U}}, \lambda\mathcal{D}_{\bar{\mathcal{U}}})$ are terminal, and the corresponding log canonical divisor is $\mathbb{Q}$-rationally equivalent to 0. This is impossible by Theorem 3.4.2. $\square$

Proof of Theorem 19.1.3. Suppose that there exists an $\mathfrak{A}_5$-rational map
$$h: V_5 \dashrightarrow S$$
whose general fiber is birational to a smooth elliptic curve. Then there exists a commutative diagram

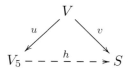

such that V is a smooth threefold, and u is an $\mathfrak{A}_5$-birational map.

Let B be a very ample divisor on S whose class in $\mathrm{Pic}(S)$ is $\mathfrak{A}_5$-invariant. Let $\mathcal{B}$ be the proper transform on V_5 of the linear system $|v^*(B)|$. By Theorem 3.2.6(i), the singularities of the log pair $(V_5, \lambda\mathcal{D})$ are not terminal.

If $\mathbb{CS}(V_5, \lambda\mathcal{D})$ contains a point of Σ_5, then Theorem 19.1.5 implies that the linear system $\mathcal{D}_{\bar{V}_5}$ is a pull back of a mobile linear system on $\mathbb{P}^2$ via the elliptic fibration
$$\bar{\theta}_{\Pi_3}: \bar{V}_5 \to \mathbb{P}^2,$$

which is exactly what we want to prove. Thus, we may assume that the set $\mathbb{CS}(V_5, \lambda\mathcal{D})$ does not contain points of Σ_5. Since $\mathcal{D}$ is not composed of a pencil by construction, it follows from Theorem 19.1.5 that the singularities of one of the log pairs $(\mathcal{Y}, \lambda\mathcal{D}_\mathcal{Y})$, $(\mathcal{U}, \lambda\mathcal{D}_\mathcal{U})$, $(\mathcal{W}, \lambda\mathcal{D}_\mathcal{W})$, or $(\bar{\mathcal{U}}, \lambda\mathcal{D}_{\bar{\mathcal{U}}})$ are terminal, and the corresponding log canonical divisor is $\mathbb{Q}$-rationally equivalent to 0. This contradicts Theorem 3.2.6(i). □

Proof of Theorem 19.1.4. Suppose that there exists an $\mathfrak{A}_5$-birational map

$$g\colon V_5 \dashrightarrow V$$

such that V is a Fano variety with canonical singularities. Let n be a positive integer such that the divisor $-nK_V$ is very ample, and let $\mathcal{B}$ be the proper transform of the linear system $|-nK_V|$ on the threefold V_5 via g. If the log pair $(V_5, \lambda\mathcal{D})$ has terminal singularities, then it follows from Theorem 3.2.1(i) that $V \cong V_5$ and the composition $g \circ \alpha$ is biregular. Thus, we may assume that the singularities of the log pair $(V_5, \lambda\mathcal{D})$ are not terminal.

Note that $\mathcal{D}$ is not composed of a pencil by construction. Similarly, it is not a pull back of a mobile linear system on $\mathbb{P}^2$ via the elliptic fibration $\bar{\theta}_{\Pi_3}\colon \bar{V}_5 \to \mathbb{P}^2$. Thus, by Theorem 19.1.5, one of the log pairs $(\mathcal{Y}, \lambda\mathcal{D}_\mathcal{Y})$, $(\mathcal{U}, \lambda\mathcal{D}_\mathcal{U})$, $(\mathcal{W}, \lambda\mathcal{D}_\mathcal{W})$, or $(\bar{\mathcal{U}}, \lambda\mathcal{D}_{\bar{\mathcal{U}}})$ has terminal singularities, and the corresponding log canonical divisor is $\mathbb{Q}$-rationally equivalent to 0. Applying Theorem 3.2.1(i), we see that one of the following birational maps is biregular:

$$\zeta \circ \nu^{-1} \circ \alpha^{-1} \circ g^{-1} \colon V \dashrightarrow X_{14},$$
$$\zeta' \circ \psi^{-1} \circ \alpha^{-1} \circ g^{-1} \colon V \dashrightarrow X'_{14},$$
$$\varphi \circ \pi^{-1} \circ \alpha^{-1} \circ g^{-1} \colon V \dashrightarrow X_4,$$
$$\beta \circ \bar{\psi}^{-1} \circ \pi^{-1} \circ \alpha^{-1} \circ g^{-1} \colon V \dashrightarrow X_2.$$

In particular, V is isomorphic to one of the threefolds V_5, X_{14}, X'_{14}, X_4, or X_2. □

The rest of this chapter is devoted to the proof of Theorem 19.1.5. Throughout this proof we will use the notation introduced above.

19.2 Exclusion of points

Let $\mathcal{D}$ be an $\mathfrak{A}_5$-invariant mobile linear system on V_5, let λ be a positive rational number such that $K_{V_5} + \lambda\mathcal{D} \sim_\mathbb{Q} 0$. Denote by $\mathcal{D}_{\bar{V}_5}$ the proper transform of the linear system $\mathcal{D}$ on the threefold $\bar{V}_5$. Suppose that the singularities of the log pair $(V_5, \lambda\mathcal{D})$ are canonical. Let P be a point in V_5, and let Σ be its $\mathfrak{A}_5$-orbit. Suppose that $P \in \mathbb{CS}(V_5, \lambda\mathcal{D})$. The goal of this section is to prove the following:

Theorem 19.2.1. The $\mathfrak{A}_5$-orbit Σ coincides with Σ_5, one has $\mathrm{mult}_\Sigma(\mathcal{D}) = \frac{2}{\lambda}$, and $\mathcal{D}_{\bar{V}_5}$ is a pull back of a mobile linear system on $\mathbb{P}^2$ via the elliptic fibration $\bar{\theta}_{\Pi_3} \colon \bar{V}_5 \to \mathbb{P}^2$.

Let D and D' be general members of the linear system $\mathcal{D}$. Then

$$\mathrm{mult}_\Sigma\left(D \cdot D'\right) \geqslant \frac{4}{\lambda^2} \qquad (19.2.2)$$

by Theorem 2.5.6. This easily implies:

Lemma 19.2.3. Suppose that $\Sigma = \Sigma_5$. Then $\mathrm{mult}_\Sigma(\mathcal{D}) = \frac{2}{\lambda}$ and $\mathcal{D}_{\bar{V}_5}$ is a pull back of a mobile linear system on $\mathbb{P}^2$ via the elliptic fibration $\bar{\theta}_{\Pi_3} \colon \bar{V}_5 \to \mathbb{P}^2$.

Proof. Let M be a general hyperplane section of V_5 that contains Σ_5. Then M does not contain curves in the intersection $D \cap D'$ by Lemma 7.5.3. Thus, we have

$$\frac{20}{\lambda^2} = M \cdot D \cdot D' \geqslant \mathrm{mult}_{\Sigma_5}\left(D \cdot D'\right)|\Sigma_5| \geqslant \frac{4}{\lambda^2}|\Sigma_5| = \frac{20}{\lambda^2}$$

by (19.2.2). This implies, in particular, that

$$\mathrm{mult}_{\Sigma_5}\left(D \cdot D'\right) = \frac{4}{\lambda^2}.$$

Thus, one has $\mathrm{mult}_{\Sigma_5}(D) = \frac{2}{\lambda}$ by Theorem 2.5.6. This gives

$$\mathcal{D}_{\bar{V}_5} \sim_{\mathbb{Q}} f^*\left(\frac{2}{\lambda}H\right) - \frac{2}{\lambda}\sum_{i=1}^{5} B_i,$$

where $B_1, \ldots, B_5$ are exceptional divisors of the blow-up $f \colon \bar{V}_5 \to V_5$ of the $\mathfrak{A}_5$-orbit Σ_5.

Let $\bar{F}$ be a general fiber of the elliptic fibration $\bar{\theta}_{\Pi_3}$ (see (19.1.2)). Then

$$\mathcal{D}_{\bar{V}_5} \cdot \bar{F} = \left(f^*\left(\frac{2}{\lambda}H\right) - \frac{2}{\lambda}\sum_{i=1}^{5} B_i\right) \cdot \bar{F} = 0,$$

which implies that $\mathcal{D}_{\bar{V}_5}$ is a pull back of a mobile linear system on $\mathbb{P}^2$ via $\bar{\theta}_{\Pi_3}$. $\square$

By Lemma 19.2.3, to prove Theorem 19.2.1 it is enough to show that the $\mathfrak{A}_5$-orbit Σ coincides with Σ_5. Suppose that this is not the case. Then

$$|\Sigma| \in \{10, 12, 15, 20, 30, 60\}$$

by Theorem 7.3.5. Moreover, one has $|\Sigma| \neq 12$ by Theorem 7.3.5 and Lemma 18.5.3, because $\Sigma_{12} \subset \mathscr{C}$ by construction, and $\Sigma'_{12} \subset \mathscr{C}'$ by Lemma 7.5.6.

Put $n = \frac{|\Sigma|}{5}$ if $|\Sigma| = 10$, and put $n = \frac{|\Sigma|}{5} - 1$ in the remaining cases, so that

$$n = \begin{cases} 2 \text{ if } |\Sigma| = 10, \\ 2 \text{ if } |\Sigma| = 15, \\ 3 \text{ if } |\Sigma| = 20, \\ 5 \text{ if } |\Sigma| = 30, \\ 11 \text{ if } |\Sigma| = 60. \end{cases}$$

Let $\mathcal{M}$ be the linear system on V_5 consisting of all surfaces in $|nH|$ that contain Σ. By Lemma 18.5.3, we have $\Sigma \not\subset \mathscr{C} \cup \mathscr{C}'$. Thus, Proposition 17.4.1 implies that the base locus of the linear system $\mathcal{M}$ does not contain curves.

Let M be a general surface in $\mathcal{M}$. Then

$$\frac{20n}{\lambda^2} = D \cdot D' \cdot M \geqslant \sum_{O \in \Sigma} \mathrm{mult}_O \left(D \cdot D' \right) =$$

$$= |\Sigma| \mathrm{mult}_P \left(D \cdot D' \right) \geqslant \frac{4}{\lambda^2} |\Sigma| \quad (19.2.4)$$

by (19.2.2). This implies that $|\Sigma| = 10$, $n = 2$ and

$$\mathrm{mult}_\Sigma \left(D \cdot D' \right) = \frac{4}{\lambda^2}.$$

Thus, either $\Sigma = \Sigma_{10}$ or $\Sigma = \Sigma'_{10}$ by Theorem 7.3.5, and $\mathrm{mult}_\Sigma(D) = \frac{2}{\lambda}$ by Theorem 2.5.6. Now we can apply Lemma 17.2.10 to obtain a contradiction, which completes the proof of Theorem 19.2.1.

19.3 Exclusion of curves

Let $\mathcal{D}$ be an $\mathfrak{A}_5$-invariant mobile linear system on V_5, let λ be a positive rational number such that $K_{V_5} + \lambda \mathcal{D} \sim_\mathbb{Q} 0$. Suppose that the singularities of the log pair $(V_5, \lambda \mathcal{D})$ are canonical. Let C be an irreducible curve in V_5. Suppose that C is a center in $\mathbb{CS}(V_5, \lambda \mathcal{D})$. The main result of this section is:

Theorem 19.3.1. Either C is a line of $\mathcal{L}_6$, or C is one of the curves $\mathscr{C}$, $\mathscr{C}'$, or C_{20}. Moreover, one has

$$\mathrm{mult}_C(\mathcal{D}) = \frac{1}{\lambda}.$$

Furthermore, if $C = C_{20}$, then $\mathbb{CS}(V_5, \lambda\mathcal{D}) = \{C_{20}\}$ and $\mathcal{D}$ is composed of the pencil $\mathcal{Q}_2$.

Let Z be the $\mathfrak{A}_5$-orbit of the curve C. Let r be the number of irreducible components of the curve Z. Denote by $C_1 = C, C_2, \ldots, C_r$ the irreducible components of Z. Then each C_i is a center in $\mathbb{CS}(V_5, \lambda\mathcal{D})$. Thus, we have

$$\mathrm{mult}_{C_i}(\mathcal{D}) = \frac{1}{\lambda}$$

by Lemma 2.4.4(ii),(iii), because we assumed that the log pair $(V_5, \lambda\mathcal{D})$ has canonical singularities.

Let D and D' be general surfaces in the linear system $\mathcal{D}$. Then

$$D \cdot D' = \sum_{i=1}^{r} \mathrm{mult}_{C_i}\!\left(D \cdot D'\right) C_i + \Omega,$$

where Ω is an effective one-cycle whose support does not contain the curves $C_1, \ldots, C_r$. One has

$$\mathrm{mult}_{C_i}\!\left(D \cdot D'\right) \geqslant \mathrm{mult}_{C_i}(D)\mathrm{mult}_{C_i}(D') = \frac{1}{\lambda^2},$$

since $\mathrm{mult}_{C_i}(\mathcal{D}) = \frac{1}{\lambda}$ for every i. Thus, we have

$$\frac{20}{\lambda^2} = D \cdot D' \cdot H = \left(\sum_{i=1}^{r} \mathrm{mult}_{C_i}\!\left(D \cdot D'\right) C_i + \Omega\right) \cdot H \geqslant$$

$$\geqslant \sum_{i=1}^{r} \mathrm{mult}_{C_i}\!\left(D \cdot D'\right)\!\left(H \cdot C_i\right) \geqslant \frac{r(H \cdot C)}{\lambda^2} = \frac{\deg(Z)}{\lambda^2}, \quad (19.3.2)$$

which implies that $\deg(Z) \leqslant 20$.

Suppose that $Z \neq \mathscr{C}$, $Z \neq \mathscr{C}'$, and $Z \neq \mathcal{L}_6$. Then $\deg(Z) \geqslant 10$ by Theorem 13.6.1. To prove Theorem 19.3.1, we must show that $Z = C = C_{20}$, that $\mathbb{CS}(V_5, \lambda\mathcal{D}) = \{C_{20}\}$, and that $\mathcal{D}$ is composed of the pencil $\mathcal{Q}_2$.

It follows from Proposition 18.4.1 and Lemma 2.4.4(iii) that $Z \neq \mathcal{L}_{15}$. Similarly, we can prove:

Lemma 19.3.3. One has $Z \neq \mathcal{G}_{10}$.

Proof. Suppose that $Z = \mathcal{G}_{10}$. Then $r = 10$ and $C_1, \ldots, C_{10}$ are conics (see Theorem 14.1.16(i)). Moreover, it follows from (19.3.2) that Ω is a zero one-cycle and

$$\text{mult}_{C_i}\left(D \cdot D'\right) = \text{mult}_{C_i}(D)\text{mult}_{C_i}(D') = \frac{1}{\lambda^2}. \tag{19.3.4}$$

In particular, the base locus of the linear system $\mathcal{D}$ is contained in $\mathcal{G}_{10}$.

By Theorem 14.1.16(iv), each point $P \in \Sigma_5$ is contained in exactly four conics among $C_1, \ldots, C_{10}$. Let $f \colon \bar{V}_5 \to V_5$ be the blow-up of V_5 at Σ_5, and let $B_1, \ldots, B_5$ be f-exceptional divisors. Denote by $\bar{C}_i$ the proper transform of the conic C_i on the threefold $\bar{V}_5$. By Theorem 14.1.16(iv) the intersection

$$B_i \cap \left(\bar{C}_1 \cup \bar{C}_2 \cup \ldots \cup \bar{C}_{10}\right) \tag{19.3.5}$$

consists of four points in $B_i \cong \mathbb{P}^2$ such that no three of these points lie on one line. Moreover, the curves $\bar{C}_1, \ldots, \bar{C}_{10}$ are disjoint by Theorem 14.1.16(v).

Let $D_{\bar{V}_5}$ be the proper transform of the divisor D on $\bar{V}_5$. Then

$$D_{\bar{V}_5} \sim_{\mathbb{Q}} f^*\left(\frac{2}{\lambda}H\right) - \text{mult}_{\Sigma_5}(D) \sum_{i=1}^{5} B_i.$$

Therefore, there exists an irreducible conic $\bar{C} \subset B_i \cong \mathbb{P}^2$ that contains all four points in the intersection (19.3.5) and is not contained in the support of the divisor $D_{\bar{V}_5}|_{B_i}$. Thus

$$2\text{mult}_{\Sigma_5}(D) = \lambda D_{\bar{V}_5}|_{B_i} \cdot \bar{C} \geqslant 4\text{mult}_{\mathcal{G}_{10}}(D),$$

which implies that

$$\text{mult}_{\Sigma_5}(D) \geqslant 2\text{mult}_{\mathcal{G}_{10}}(D) = \frac{2}{\lambda}.$$

In particular, the points of Σ_5 are also centers in $\mathbb{CS}(V_5, \lambda\mathcal{D})$.

Denote by $\mathcal{D}_{\bar{V}_5}$ the proper transform of the linear system $\mathcal{D}$ on the threefold $\bar{V}_5$. Then it follows from Theorem 19.2.1 that $\text{mult}_{\Sigma_5}(\mathcal{D}) = \frac{2}{\lambda}$ and $\mathcal{D}_{\bar{V}_5}$ is a pull back of a mobile linear system on $\mathbb{P}^2$ via the elliptic fibration $\bar{\theta}_{\Pi_3} \colon \bar{V}_5 \to \mathbb{P}^2$. In particular, the linear system $\mathcal{D}_{\bar{V}_5}$ does not have base curves that are contracted by f, since its exceptional surfaces $B_1, \ldots, B_5$ are sections of the elliptic fibration $\bar{\theta}_{\Pi_3}$ (see Lemma 7.5.8). Thus, the curves

$\bar{C}_1, \ldots, \bar{C}_{10}$ are the only curves contained in the base locus of the linear system $\mathcal{D}_{\bar{V}_5}$.

Let $g\colon \acute{V}_5 \to \bar{V}_5$ be the blow-up of the curves $\bar{C}_1, \ldots, \bar{C}_{10}$. Denote by $F_1, \ldots, F_{10}$ the g-exceptional surfaces. Put

$$T = (f \circ g)^*(2H) - 2\sum_{i=1}^{5} g^*(B_i) - \sum_{i=1}^{10} F_i,$$

and denote by $\mathcal{D}_{\acute{V}_5}$ the proper transform of the linear system $\mathcal{D}$ on the threefold $\acute{V}_5$. Then

$$\lambda \mathcal{D}_{\acute{V}_5} \sim_{\mathbb{Q}} T,$$

which implies that T has non-negative intersection with any curve in $\acute{V}_5$ that is not contained in the base locus of the linear system $\mathcal{D}_{\acute{V}_5}$.

We claim that the divisor T is nef. Indeed, suppose that T is not nef. Then there is an irreducible curve $\acute{C} \subset \acute{V}_5$ such that $T \cdot \acute{C} < 0$. Since $\acute{C}$ is contained in the base locus of the linear system $\mathcal{D}_{\acute{V}_5}$, we have $\acute{C} \subset F_k$ for some $1 \leqslant k \leqslant 10$, because $\bar{C}_1, \ldots, \bar{C}_{10}$ are the only curves contained in the base locus of the linear system $\mathcal{D}_{\bar{V}_5}$. Moreover, it follows from (19.3.4) that the curve $\acute{C}$ must be contracted by g to a point in $\bar{C}_k$. Hence

$$T \cdot \acute{C} = \left((f \circ g)^*(2H) - 2\sum_{i=1}^{5} g^*(B_i) - \sum_{i=1}^{10} F_i\right) \cdot \acute{C} = -F_k \cdot \acute{C} = 1,$$

which contradicts the inequality $T \cdot \acute{C} < 0$. This shows that T is nef.

Denote by $D_{\acute{V}_5}$ and $D'_{\acute{V}_5}$ the proper transforms of the surfaces D and D' on the threefold $\acute{V}_5$, respectively. Then

$$T \cdot D_{\acute{V}_5} \cdot D'_{\acute{V}_5} = \frac{1}{\lambda^2}\left((f \circ g)^*(2H) - 2\sum_{i=1}^{5} g^*(B_i) - \sum_{i=1}^{10} F_i\right)^3 =$$

$$= -\sum_{i=1}^{10} F_i^3 = -20,$$

which is impossible, since T is nef. The obtained contradiction completes the proof of Lemma 19.3.3. □

Since $Z \neq \mathcal{L}_{15}$ and $Z \neq \mathcal{G}_{10}$, it follows from Corollary 14.1.18 that there exists a surface $S \in \mathcal{Q}_2$ that contains Z, because $\deg(Z) \leqslant 20$. We may assume that $S \not\subset D$, since D is a general surface in $\mathcal{D}$. Thus

$$\mathrm{mult}_C\left(D \cdot S\right) \leqslant \frac{1}{\lambda}$$

and one has $Z = C = C_{20}$ by Lemmas 18.2.2 and 18.2.5, because we assumed that $Z \ne \mathscr{C}$, $Z \ne \mathscr{C}'$ and $Z \ne \mathcal{L}_6$. Hence, we have

$$\mathrm{mult}_C(D \cdot S) = \frac{1}{\lambda},$$

because $\mathrm{mult}_C(D) \geqslant \frac{1}{\lambda}$ by Lemma 2.4.4(iii).

Since C_{20} is the base curve of the pencil $\mathcal{Q}_2$, we may assume that S is a general surface in $\mathcal{Q}_2$. In particular, the surface S is smooth by Lemma 8.2.1(ii). Put $D_S = D|_S$ and $H_S = H|_S$. Then

$$\lambda D_S = Z + \Upsilon$$

for some effective $\mathbb{Q}$-divisor Υ whose support does not contain the curve Z. Since

$$20 = \lambda D_S \cdot H_S = Z \cdot H_S + \Omega \cdot H_S \geqslant Z \cdot H_S = 20,$$

we see that $\Upsilon = 0$. In particular, we have

$$\mathrm{Supp}\left(D \cdot Q\right) = C_{20},$$

which implies that $\mathcal{D}$ is composed of the pencil $\mathcal{Q}_2$ by Lemma 4.3.2. We conclude that $\mathbb{CS}(V_5, \lambda\mathcal{D}) = \{C_{20}\}$ by Theorem 19.2.1, because $\Sigma_5 \not\subset C_{20}$ by Theorem 8.1.8(vii).

19.4 Description of non-terminal pairs

Let $\mathcal{D}$ be an $\mathfrak{A}_5$-invariant mobile linear system on V_5. Let λ be a positive rational number such that

$$K_{V_5} + \lambda\mathcal{D} \sim_\mathbb{Q} 0.$$

Suppose that the singularities of the log pair $(V_5, \lambda\mathcal{D})$ are canonical. Then it follows from Theorems 19.2.1 and 19.3.1 that the set $\mathbb{CS}(V_5, \lambda\mathcal{D})$ does not contain centers of canonical singularities of the log pair $(V_5, \lambda\mathcal{D})$ possibly excepting the following ones: points of Σ_5, lines of $\mathcal{L}_6$, the curve $\mathscr{C}$, the curve $\mathscr{C}'$, or the curve C_{20}. Moreover, if $C = C_{20}$, then

$$\mathbb{CS}(V_5, \lambda\mathcal{D}) = \{C_{20}\}$$

and $\mathcal{D}$ is composed of the pencil $\mathcal{Q}_2$. Now we describe the set $\mathbb{CS}(V_5, \lambda\mathcal{D})$ in a similar way in the remaining cases. Denote by $\mathcal{D}_\mathcal{Y}$ and $\mathcal{D}_{\bar{V}_5}$ the proper transforms of the linear system $\mathcal{D}$ on the threefolds $\mathcal{Y}$ and $\bar{V}_5$, respectively.

Lemma 19.4.1. Suppose that the set $\mathbb{CS}(V_5, \lambda\mathcal{D})$ contains the points of Σ_5. Then it does not contain anything else. Moreover, the log pair $(\bar{V}_5, \lambda\mathcal{D}_{\bar{V}_5})$ has terminal singularities.

Proof. By Theorem 19.2.1, the linear system $\mathcal{D}_{\bar{V}_5}$ is a pull back of a mobile linear system on $\mathbb{P}^2$ via the elliptic fibration $\bar{\theta}_{\Pi_3} \colon \bar{V}_5 \to \mathbb{P}^2$. Since $\mathcal{D}_{\bar{V}_5}$ does not contain fixed components, we see that every irreducible base curve of the linear system $\mathcal{D}_{\bar{V}_5}$ must be contracted by $\bar{\theta}_{\Pi_3}$ to a point in $\mathbb{P}^2$. The latter implies that the curves $\mathcal{L}_6$, $\mathscr{C}$, and $\mathscr{C}'$ are not contained in the base locus of the linear system $\mathcal{D}$, since they do not contain the $\mathfrak{A}_5$-orbit Σ_5 by Corollary 7.4.9 and Lemma 5.1.4. In particular, we see that the set $\mathbb{CS}(V_5, \lambda\mathcal{D})$ consists of the points of the $\mathfrak{A}_5$-orbit Σ_5, and no curve contained in the exceptional locus $B_1 \cup B_2 \cup B_3 \cup B_4 \cup B_5$ of the blow-up $f \colon \bar{V}_5 \to V_5$ of Σ_5 can be a center in $\mathbb{CS}(\bar{V}_5, \lambda\mathcal{D}_{\bar{V}_5})$. Therefore, if the set $\mathbb{CS}(\bar{V}_5, \lambda\mathcal{D}_{\bar{V}_5})$ is not empty, then $\mathbb{CS}(\bar{V}_5, \lambda\mathcal{D}_{\bar{V}_5})$ contains only points in $B_1 \cup B_2 \cup B_3 \cup B_4 \cup B_5$. Let us show that the latter is impossible.

Suppose that there exists a point P contained in some divisor

$$B_k, \quad 1 \leqslant k \leqslant 5,$$

such that $P \in \mathbb{CS}(\bar{V}_5, \lambda\mathcal{D}_{\bar{V}_5})$. Let $D_{\bar{V}_5}$ and $D'_{\bar{V}_5}$ be general surfaces in the linear system $\mathcal{D}_{\bar{V}_5}$. Then

$$\mathrm{mult}_P\left(D_{\bar{V}_5} \cdot D'_{\bar{V}_5}\right) \geqslant \frac{4}{\lambda^2}$$

by (19.2.4). Let $G_k \subset \mathfrak{A}_5$ be the stabilizer of the point $f(B_k)$, and let Σ be the G_k-orbit of the point P in $\bar{V}_5$. Then $|\Sigma| \geqslant 3$ by Lemma 7.5.9. Thus, we have

$$\frac{4}{\lambda^2} = B_k \cdot D_{\bar{V}_5} \cdot D'_{\bar{V}_5} \geqslant \sum_{O \in \Sigma} \mathrm{mult}_O\left(D_{\bar{V}_5} \cdot D'_{\bar{V}_5}\right) \geqslant \sum_{O \in \Sigma} \frac{4}{\lambda^2} = \frac{4|\Sigma|}{\lambda^2} \geqslant \frac{12}{\lambda^2},$$

which is absurd. This shows that the log pair $(\bar{V}_5, \lambda\mathcal{D}_{\bar{V}_5})$ has terminal singularities. $\square$

Lemma 19.4.2. Suppose that the set $\mathbb{CS}(V_5, \lambda\mathcal{D})$ contains the curve $\mathscr{C}$. Then it does not contain anything else. Moreover, the log pair $(\mathcal{Y}, \lambda\mathcal{D}_\mathcal{Y})$ has terminal singularities.

Proof. By Lemmas 19.2.1 and 19.4.1, the set $\mathbb{CS}(V_5, \lambda\mathcal{D})$ does not contain points in V_5. Similarly, it follows from Theorem 19.3.1 that the set $\mathbb{CS}(V_5, \lambda\mathcal{D})$ does not contain curves that are different from the curve $\mathscr{C}'$

and from the lines of $\mathcal{L}_6$. We must show that the set $\mathbb{CS}(V_5, \lambda \mathcal{D})$ does not contain the curve $\mathscr{C}'$ and the lines of $\mathcal{L}_6$ either.

Let us use notation of §7.2. Let D be a general surface in $\mathcal{D}$. Denote by $D_\mathcal{Y}$ its proper transform on $\mathcal{Y}$. Then $\mathrm{mult}_\mathscr{C}(D) = \frac{1}{\lambda}$ by Lemma 2.4.4(ii),(iii), since we assume that $(V_5, \lambda \mathcal{D})$ has canonical singularities. Thus, we have

$$\lambda D_\mathcal{Y} \sim_\mathbb{Q} \nu^*(2H) - E_\mathcal{Y},$$

which implies that $\lambda D_\mathcal{Y}|_{\hat{\mathscr{S}}}$ is an effective $\mathbb{Q}$-divisor on $\hat{\mathscr{S}} \cong \mathbb{P}^1 \times \mathbb{P}^1$ of bi-degree $(0, 8)$. In particular, the support of the divisor $D_\mathcal{Y}$ does not contain the curve $\hat{\mathscr{C}}$, since $\hat{\mathscr{C}}$ is a curve of bi-degree $(1, 1)$ on the surface $\hat{\mathscr{S}}$ (see Lemma 7.2.3(iii)).

Let $\hat{\Sigma}'_{12}$ be the $\mathfrak{A}_5$-orbit in $\hat{\mathscr{S}}$ that is mapped to Σ'_{12} by ν. Denote by $\hat{\mathscr{C}}'$ and $\hat{\mathcal{L}}_6$ the proper transforms of the curves $\mathscr{C}'$ and $\mathcal{L}_6$ on the threefold $\mathcal{Y}$, respectively. Then

$$\hat{\mathscr{C}}' \cap \hat{\mathscr{S}} = \hat{\mathcal{L}}_6 \cap \hat{\mathscr{S}} = \hat{\Sigma}'_{12}$$

by Lemmas 7.4.3 and 11.3.4. Since $\lambda D_\mathcal{Y}|_{\hat{\mathscr{S}}}$ is an effective $\mathbb{Q}$-divisor of bi-degree $(0, 8)$ on $\hat{\mathscr{S}}$, and $\hat{\Sigma}'_{12}$ is not contained in any curve of bi-degree $(0, 1)$ on $\hat{\mathscr{S}}$, we have

$$\mathrm{mult}_{\hat{\Sigma}'_{12}}\left(\lambda D_\mathcal{Y}|_{\hat{\mathscr{S}}}\right) \leqslant 1.$$

We also have

$$\mathrm{mult}_{\hat{\Sigma}'_{12}}\left(\lambda D_\mathcal{Y}|_{\hat{\mathscr{S}}}\right) \geqslant \mathrm{mult}_{\hat{\Sigma}'_{12}}(\lambda \mathcal{D}_\mathcal{Y}) \geqslant$$
$$\geqslant \max\Big(\mathrm{mult}_{\hat{\mathscr{C}}'}(\lambda \mathcal{D}_\mathcal{Y}), \mathrm{mult}_{\hat{\mathcal{L}}_6}(\lambda \mathcal{D}_\mathcal{Y})\Big) =$$
$$= \max\Big(\mathrm{mult}_{\mathscr{C}'}(\lambda \mathcal{D}), \mathrm{mult}_{\mathcal{L}_6}(\lambda \mathcal{D})\Big),$$

which implies that neither $\mathscr{C}'$ nor $\mathcal{L}_6$ is contained in $\mathbb{CS}(V_5, \lambda \mathcal{D})$. Thus, we see that $\mathbb{CS}(V_5, \lambda \mathcal{D}) = \{\mathscr{C}\}$.

Let us show that the log pair $(\mathcal{Y}, \lambda \mathcal{D}_\mathcal{Y})$ has terminal singularities. Suppose that this is not the case. Since

$$K_\mathcal{Y} + \frac{2}{\lambda} \mathcal{D}_\mathcal{Y} \sim_\mathbb{Q} \nu^*\left(K_{V_5} + \frac{2}{\lambda}\mathcal{D}\right) \sim_\mathbb{Q} 0,$$

we see that every center in $\mathbb{CS}(\mathcal{Y}, \lambda \mathcal{D}_\mathcal{Y})$ is mapped by ν to a center in $\mathbb{CS}(V_5, \lambda \mathcal{D})$. Since we already proved that $\mathbb{CS}(V_5, \lambda \mathcal{D})$ does not contain centers different from the curve $\mathscr{C}$, we conclude that $\mathbb{CS}(\mathcal{Y}, \lambda \mathcal{D}_\mathcal{Y})$ contains some curve $C \subset E_\mathcal{Y}$ such that $\nu(C) = \mathscr{C}$.

Let Z be the $\mathfrak{A}_5$-orbit of the curve C. Then $Z \subset E_\mathcal{Y}$ and

$$\frac{1}{\lambda} = \mathrm{mult}_{\mathscr{C}}(D) \geqslant \mathrm{mult}_Z(D_\mathcal{Y}) \geqslant \frac{1}{\lambda}$$

by Lemma 2.4.4(ii),(iii). This implies that $\mathrm{mult}_Z(D_\mathcal{Y}) = \frac{1}{\lambda}$.

Recall that $E_\mathcal{Y} \cong \mathbb{P}^1 \times \mathbb{P}^1$ by Lemma 7.2.8. Let l be a fiber of the natural projection $E_\mathcal{Y} \to \mathscr{C}$, and let s be the section of this projection such that $s^2 = 0$. Then

$$Z \sim as + bl$$

for some integers a and b such that $a \geqslant 1$ and $b \geqslant 0$. On the other hand, we have $\hat{\mathscr{C}} \sim s + l$ and $\mathrm{mult}_Z(D_\mathcal{Y}) = \frac{1}{\lambda}$. Thus, we obtain

$$\lambda D_\mathcal{Y}|_{E_\mathcal{Y}} = \delta Z + \Omega \sim_\mathbb{Q} \left(\nu^*(2H) - E_\mathcal{Y}\right)|_{E_\mathcal{Y}} \sim s + 7l$$

for some positive rational number $\delta \geqslant 1$ and some effective $\mathbb{Q}$-divisor Ω on $E_\mathcal{Y}$ whose support does not contain the irreducible components of the curve Z. This implies that $\delta = a = 1$ and $b \leqslant 7$. Hence $Z = \hat{\mathscr{C}}$ by Corollary 7.2.9 and Lemma 6.4.4(i). The latter is impossible, since we already proved that the support of the divisor $D_\mathcal{Y}$ does not contain the curve $\hat{\mathscr{C}}$. □

19.5 Completing the proof

In this section we prove Theorem 19.1.5. Let us use its assumptions and notation. It follows from Theorems 19.2.1 and 19.3.1 that the set $\mathbb{CS}(V_5, \lambda \mathcal{D})$ does not contain centers except possibly the points of Σ_5, the lines of $\mathcal{L}_6$, the curve $\mathscr{C}$, the curve $\mathscr{C}'$, or the curve C_{20}.

By Theorem 19.3.1, we may assume that the curve C_{20} is not contained in $\mathbb{CS}(V_5, \lambda \mathcal{D})$. By Lemmas 19.4.1 and 19.4.2, we may assume that the set $\mathbb{CS}(V_5, \lambda \mathcal{D})$ does not contain points in Σ_5 and does not contain the curve $\mathscr{C}$. Thus, we have one of the following possibilities:

- the set $\mathbb{CS}(V_5, \lambda \mathcal{D})$ consists of the curve $\mathscr{C}'$;
- the set $\mathbb{CS}(V_5, \lambda \mathcal{D})$ consists of the lines of $\mathcal{L}_6$;
- the set $\mathbb{CS}(V_5, \lambda \mathcal{D})$ consists of the curve $\mathscr{C}'$ and the lines of $\mathcal{L}_6$.

Now Theorem 19.1.5 is implied by the following three lemmas whose proof occupies the remaining part of the current section.

Recall that $\pi\colon \mathcal{W} \to V_5$ denotes the blow-up of the curve $\mathcal{L}_6$ (see §11.1).

CREMONA GROUPS AND THE ICOSAHEDRON 479

Lemma 19.5.1. Suppose that the set $\mathbb{CS}(V_5, \lambda\mathcal{D})$ contains the lines of $\mathcal{L}_6$. Then either the log pair $(\mathcal{W}, \lambda\mathcal{D}_\mathcal{W})$ has terminal singularities, or the set $\mathbb{CS}(\mathcal{W}, \lambda\mathcal{D}_\mathcal{W})$ consists of the proper transform of the curve $\mathscr{C}'$ on the threefold $\mathcal{W}$. In the former case, the set $\mathbb{CS}(V_5, \lambda\mathcal{D})$ consists of irreducible components of the curve $\mathcal{L}_6$. In the latter case, the set $\mathbb{CS}(V_5, \lambda\mathcal{D})$ consists of the curve $\mathscr{C}'$ and the lines of $\mathcal{L}_6$.

Proof. Let $S_{\mathcal{L}_6}$ be the surface in $\mathcal{Q}_2$ that contains $\mathcal{L}_6$ (see §11.1). Denote by $\check{S}_{\mathcal{L}_6}$ the proper transform of $S_{\mathcal{L}_6}$ on the threefold $\mathcal{W}$. Then

$$\lambda\mathcal{D}_\mathcal{W} \sim_\mathbb{Q} \check{S}_{\mathcal{L}_6} \sim -K_\mathcal{W} \sim \pi^*(2H) - \sum_{i=1}^6 E_i,$$

where $E_1, \ldots, E_6$ are π-exceptional surfaces. By Corollary 11.1.4, the surface $\check{S}_{\mathcal{L}_6}$ is a smooth $K3$ surface and the divisor $-K_\mathcal{W}|_{\check{S}_{\mathcal{L}_6}}$ is very ample.

Denote by $L_1, \ldots, L_6$ the lines of $\mathcal{L}_6$. We may assume that $\pi(E_i) = L_i$ for every $i \in \{1, \ldots, 6\}$. Recall that $E_i \cong \mathbb{P}^1 \times \mathbb{P}^1$ (see §11.1).

Let $l_i \subset E_i$ be a fiber of the natural projection $E_i \to L_i$, and s_i be a section of this projection such that $s_i^2 = 0$. Then

$$-E_i|_{E_i} \sim s_i$$

for every $i \in \{1, \ldots, 6\}$. Put

$$\check{L}_i = \check{S}_{\mathcal{L}_6} \cap E_i$$

for every $i \in \{1, \ldots, 6\}$ (note that $\check{S}_{\mathcal{L}_6}$ intersects each surface E_i transversally along $\check{L}_i$). Then $\check{L}_i \sim s_i + 2l_i$.

If the singularities of the log pair $(\mathcal{W}, \lambda\mathcal{D}_\mathcal{W})$ are terminal, then we are done. If the set $\mathbb{CS}(\mathcal{W}, \lambda\mathcal{D}_\mathcal{W})$ consists of the proper transform of the curve $\mathscr{C}'$ on the threefold $\mathcal{W}$, then we are also done. Suppose that there exists a center in $\mathbb{CS}(\mathcal{W}, \lambda\mathcal{D}_\mathcal{W})$ that is different from $\mathscr{C}'$. Since

$$K_\mathcal{W} + \frac{2}{\lambda}\mathcal{D}_\mathcal{W} \sim_\mathbb{Q} \pi^*\left(K_{V_5} + \frac{2}{\lambda}\mathcal{D}\right) \sim_\mathbb{Q} 0$$

by Theorem 19.3.1, we see that every center in $\mathbb{CS}(\mathcal{W}, \lambda\mathcal{D}_\mathcal{W})$ is mapped by π to a center in $\mathbb{CS}(V_5, \lambda\mathcal{D})$. Therefore, for some $1 \leqslant k \leqslant 6$ the set $\mathbb{CS}(\mathcal{W}, \lambda\mathcal{D}_\mathcal{W})$ contains a curve C such that $\pi(C) = L_k$, because $\mathbb{CS}(V_5, \lambda\mathcal{D})$ does not contain centers that are different from the curve $\mathscr{C}'$ and the lines of $\mathcal{L}_6$. In particular, one has $C \subset E_k$.

Let D be a general surface in the linear system $\mathcal{D}$, and let $D_{\mathcal{W}}$ be its proper transform on the threefold $\mathcal{W}$. Then

$$\frac{1}{\lambda} = \mathrm{mult}_{L_k}(D) \geqslant \mathrm{mult}_C(D_{\mathcal{W}}) \geqslant \frac{1}{\lambda}$$

by Lemma 2.4.4(iii) and Theorem 19.3.1, which implies that

$$\mathrm{mult}_C(D_{\mathcal{W}}) = \frac{1}{\lambda}.$$

Let $G_k \subset \mathfrak{A}_5$ be a stabilizer of the line L_k; then $G_k \cong \mathfrak{D}_{10}$ by Lemma 5.1.1. Let Z be the G_k-orbit of the curve C. Then $Z \subset E_k$. We have

$$\lambda D_{\mathcal{W}}|_{E_k} = \epsilon Z + \Omega \sim_{\mathbb{Q}} \check{S}_{\mathcal{L}_6}|_{E_k} = \check{L}_k \sim s_k + 2l_k$$

for some positive rational number $\epsilon \geqslant 1$ and some effective $\mathbb{Q}$-divisor Ω on E_k whose support does not contain the irreducible components of the curve Z. Since $\pi(C) = L_k$ and $\epsilon \geqslant 1$, we have

$$Z = C = \check{L}_k$$

by Lemmas 11.1.9 and 6.4.13(i),(ii).

We proved that $\mathbb{CS}(\mathcal{W}, \lambda \mathcal{D}_{\mathcal{W}})$ contains all curves $\check{L}_1, \ldots, \check{L}_6$ (in fact, we proved that the set $\mathbb{CS}(\mathcal{W}, \lambda \mathcal{D}_{\mathcal{W}})$ consists exactly of these curves). Thus

$$\lambda D_{\mathcal{W}}|_{\check{S}_{\mathcal{L}_6}} = \delta \sum_{i=1}^{6} \check{L}_i + \Upsilon,$$

for some non-negative rational number δ and some effective $\mathbb{Q}$-divisor Υ on the surface $\check{S}_{\mathcal{L}_6}$ whose support does not contain the curves $\check{L}_1, \ldots, \check{L}_6$. One has $\delta \geqslant 1$ by Lemma 2.4.4(iii). We compute

$$-K_{\mathcal{W}}|_{\check{S}_{\mathcal{L}_6}} \cdot \Upsilon = -K_{\mathcal{W}}|_{\check{S}_{\mathcal{L}_6}} \cdot \left(\lambda D_{\mathcal{W}}|_{\check{S}_{\mathcal{L}_6}} - \delta \sum_{i=1}^{6} \check{L}_i \right) =$$

$$= -K_{\mathcal{W}}|_{\check{S}_{\mathcal{L}_6}} \cdot \left(-K_{\mathcal{W}}|_{\check{S}_{\mathcal{L}_6}} - \delta \sum_{i=1}^{6} \check{L}_i \right) = -K_{\mathcal{W}}^3 - \delta \sum_{i=1}^{6} (-K_{\mathcal{W}}) \cdot \check{L}_i =$$

$$= 4 - \delta \sum_{i=1}^{6} (-K_{\mathcal{W}}) \cdot \check{L}_i = 4 - 24\delta \leqslant -20,$$

which is impossible, since the divisor $-K_{\mathcal{W}}|_{\check{S}_{\mathcal{L}_6}}$ is very ample. The obtained contradiction completes the proof of Lemma 19.5.1. $\square$

CREMONA GROUPS AND THE ICOSAHEDRON 481

Recall that $\psi\colon \mathcal{U} \to V_5$ denotes the blow-up of the curve $\mathscr{C}'$ (see §11.3).

Lemma 19.5.2. Suppose that the set $\mathbb{CS}(V_5, \lambda\mathcal{D})$ contains the curve $\mathscr{C}'$. Then either the log pair $(\mathcal{U}, \lambda\mathcal{D}_\mathcal{U})$ has terminal singularities, or the set $\mathbb{CS}(\mathcal{U}, \lambda\mathcal{D}_\mathcal{U})$ consists of the proper transforms of the lines of $\mathcal{L}_6$ on the threefold $\mathcal{U}$. In the former case the set $\mathbb{CS}(V_5, \lambda\mathcal{D})$ consists of the curve $\mathscr{C}'$. In the latter case the set $\mathbb{CS}(V_5, \lambda\mathcal{D})$ consists of the curve $\mathscr{C}'$ and the lines of $\mathcal{L}_6$.

Proof. Suppose that the singularities of the log pair $(\mathcal{U}, \lambda\mathcal{D}_\mathcal{U})$ are worse than terminal. It is enough to show that the set $\mathbb{CS}(\mathcal{U}, \lambda\mathcal{D}_\mathcal{U})$ consists of the proper transforms of the lines of $\mathcal{L}_6$ on the threefold $\mathcal{U}$. Suppose that this is not the case. Since

$$K_\mathcal{U} + \frac{2}{\lambda}\mathcal{D}_\mathcal{U} \sim_\mathbb{Q} \psi^*\left(K_{V_5} + \frac{2}{\lambda}\mathcal{D}\right) \sim_\mathbb{Q} 0,$$

we see that every center in $\mathbb{CS}(\mathcal{U}, \lambda\mathcal{D}_\mathcal{U})$ is mapped by ψ to a center in $\mathbb{CS}(V_5, \lambda\mathcal{D})$. Since $\mathbb{CS}(V_5, \lambda\mathcal{D})$ does not contain centers that are different from the curve $\mathscr{C}'$ and the lines of $\mathcal{L}_6$, we conclude that $\mathbb{CS}(\mathcal{U}, \lambda\mathcal{D}_\mathcal{U})$ contains a curve C such that $\psi(C) = \mathscr{C}'$. Denote by $\mathcal{F}_\mathcal{U}$ the ψ-exceptional surface (see §11.3). Then $C \subset \mathcal{F}_\mathcal{U}$.

Recall that there exists a surface $S_{\mathscr{C}'} \in \mathcal{Q}_2$ that contains $\mathscr{C}'$, and the surface $S_{\mathscr{C}'}$ is smooth (see §11.3). Denote by $\bar{S}_{\mathscr{C}'}$ the proper transform of the surface $S_{\mathscr{C}'}$ on the threefold $\mathcal{U}$. Put

$$\bar{\mathscr{C}}' = \bar{S}_{\mathscr{C}'} \cap \mathcal{F}_\mathcal{U}.$$

Then $\bar{\mathscr{C}}'$ is a (smooth) rational $\mathfrak{A}_5$-invariant curve. Note that $\mathcal{F}_\mathcal{U}$ is smooth, and the surfaces $\bar{S}_{\mathscr{C}'}$ and $\mathcal{F}_\mathcal{U}$ intersect transversally along the curve $\bar{\mathscr{C}}'$.

Let D be a general surface in the linear system $\mathcal{D}$, and let $D_\mathcal{U}$ be its proper transform on the threefold $\mathcal{U}$. Then

$$\frac{1}{\lambda} = \operatorname{mult}_{\mathscr{C}'}(D) \geqslant \operatorname{mult}_C(D_\mathcal{U}) \geqslant \frac{1}{\lambda}$$

by Lemma 2.4.4(iii) and Theorem 19.3.1. Thus, one has

$$\operatorname{mult}_C(D_\mathcal{U}) = \frac{1}{\lambda}.$$

We have $\mathcal{F}_\mathcal{U} \cong \mathbb{P}^1 \times \mathbb{P}^1$ by Lemma 11.3.3. Furthermore, the action of $\mathfrak{A}_5$ on $\mathcal{F}_\mathcal{U}$ is twisted diagonal by Lemma 11.4.8. Let l be a fiber of the natural projection $\mathcal{F}_{\mathcal{U}} \to \bar{\mathscr{C}}'$, let s be a section of this projection such that $s^2 = 0$, and let Z be the $\mathfrak{A}_5$-orbit of the curve C. Then $Z \subset \mathcal{F}_{\mathcal{U}}$ and

$$\lambda \mathcal{D}_{\mathcal{U}}|_{\mathcal{F}_{\mathcal{U}}} = \epsilon Z + \Omega \sim_{\mathbb{Q}} \bar{S}_{\mathscr{C}'}|_{\mathcal{F}_{\mathcal{U}}} = \bar{\mathscr{C}}' \sim s + 7l$$

for some positive rational number $\epsilon \geqslant 1$ and some effective $\mathbb{Q}$-divisor Ω on $\mathcal{F}_{\mathcal{U}}$ whose support does not contain the irreducible components of the curve Z. This implies that $\epsilon = 1$ and $Z \sim s + bl$ for some non-negative integer $b \leqslant 7$. Now we conclude that $Z = C = \bar{\mathscr{C}}'$ by Lemmas 5.6.4(i) and 6.4.11(o).

We see that $\mathbb{CS}(\mathcal{U}, \lambda \mathcal{D}_{\mathcal{U}})$ contains the curve $\bar{\mathscr{C}}'$ (in fact, we proved that it consists exactly of the curve $\bar{\mathscr{C}}'$). Thus

$$\lambda \mathcal{D}_{\mathcal{U}}|_{\bar{S}_{\mathscr{C}'}} = \delta \bar{\mathscr{C}}' + \Upsilon,$$

for some positive rational number $\delta \geqslant 1$ and some effective $\mathbb{Q}$-divisor Υ on the surface $\bar{S}_{\mathscr{C}'}$ whose support does not contain the curve $\bar{\mathscr{C}}'$. We compute

$$-K_{\mathcal{U}}|_{\bar{S}_{\mathscr{C}'}} \cdot \Upsilon = -K_{\mathcal{U}}|_{\bar{S}_{\mathscr{C}'}} \cdot \left(\lambda \mathcal{D}_{\mathcal{U}}|_{\bar{S}_{\mathscr{C}'}} - \delta \bar{\mathscr{C}}'\right) =$$
$$= -K_{\mathcal{U}}|_{\bar{S}_{\mathscr{C}'}} \cdot \left(-K_{\mathcal{U}}|_{\bar{S}_{\mathscr{C}'}} - \delta \bar{\mathscr{C}}'\right) = -K_{\mathcal{U}}^3 + \delta K_{\mathcal{U}} \cdot \bar{\mathscr{C}}' = 14 - 14\delta \leqslant 0.$$

This implies that $\delta = 1$ and $\Upsilon = 0$, because the divisor $-K_{\mathcal{U}}|_{\bar{S}_{\mathscr{C}'}}$ is very ample by Lemma 11.4.7. Hence

$$\bar{\mathscr{C}}' = \lambda \mathcal{D}_{\mathcal{U}}|_{\bar{S}_{\mathscr{C}'}} \sim_{\mathbb{Q}} \left(\psi^*(2H) - \mathcal{F}_{\mathcal{U}}\right)|_{\bar{S}_{\mathscr{C}'}} \sim_{\mathbb{Q}} \psi^*(2H)|_{\bar{S}_{\mathscr{C}'}} - \bar{\mathscr{C}}',$$

since $\mathcal{F}_{\mathcal{U}}|_{\bar{S}_{\mathscr{C}'}} = \bar{\mathscr{C}}'$. Therefore, we have

$$2\bar{\mathscr{C}}' \sim_{\mathbb{Q}} \psi^*(2H)|_{\bar{S}_{\mathscr{C}'}}.$$

The latter is absurd, since $(2\bar{\mathscr{C}}')^2 = -8$, while

$$\left(\psi^*(2H)|_{\bar{S}_{\mathscr{C}'}}\right)^2 = (2H)^2 \cdot S_{\mathscr{C}'} = 40.$$

The obtained contradiction completes the proof of Lemma 19.5.2. □

Lemma 19.5.3. Suppose that the set $\mathbb{CS}(V_5, \lambda \mathcal{D})$ consists of the curve $\mathscr{C}'$ and the lines of $\mathcal{L}_6$. Then the log pair $(\bar{\mathcal{U}}, \lambda \mathcal{D}_{\bar{\mathcal{U}}})$ has terminal singularities.

Proof. Recall from Lemma 11.3.4 that the curve $\mathscr{C}'$ intersects transversally the curve $\mathcal{L}_6$ at the points of Σ'_{12}. By blowing up V_5 at $\mathcal{L}_6$, we obtain the threefold $\mathcal{W}$. Similarly, the threefold $\mathcal{U}$ is the blow-up of the threefold V_5 at the curve $\mathscr{C}'$. Moreover, the threefold $\bar{\mathcal{U}}$ is the blow-up of the proper transform of the curve $\mathscr{C}'$ on the threefold $\mathcal{W}$ (see §11.4).

Denote by $\bar{\mathcal{L}}_6$ the proper transform of the curve $\mathcal{L}_6$ on the threefold $\mathcal{U}$, and let $\bar{\pi} \colon \bar{\mathcal{W}} \to \mathcal{U}$ be the blow-up of the curve $\bar{\mathcal{L}}_6$. The threefolds $\bar{\mathcal{U}}$ and $\bar{\mathcal{W}}$ differ by Atiyah flops (see §4.2). Indeed, denote by $\bar{Z}_1, \ldots, \bar{Z}_{12}$ the fibers of the natural projection of the ψ-exceptional divisor $\mathcal{F}_\mathcal{U}$ to the curve $\mathscr{C}'$ that are mapped to the points of the $\mathfrak{A}_5$-orbit $\Sigma'_{12} \subset \mathscr{C}'$. Denote by $\tilde{Z}_1, \ldots, \tilde{Z}_{12}$ the proper transforms of the curves $\bar{Z}_1, \ldots, \bar{Z}_{12}$ on the threefold $\bar{\mathcal{W}}$. Then there exists a commutative diagram

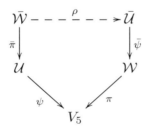

where ρ is an Atiyah flop in the curves $\tilde{Z}_1, \ldots, \tilde{Z}_{12}$.

Suppose that the singularities of the log pair $(\bar{\mathcal{U}}, \lambda \mathcal{D}_{\bar{\mathcal{U}}})$ are worse than terminal, i.e., the set $\mathbb{CS}(\bar{\mathcal{U}}, \lambda \mathcal{D}_\mathcal{U})$ is not empty. Since $\mathbb{CS}(V_5, \lambda \mathcal{D})$ consists of the curve $\mathscr{C}'$ and the lines of $\mathcal{L}_6$, we have

$$\mathrm{mult}_{\mathscr{C}'}(\mathcal{D}) = \mathrm{mult}_{\mathcal{L}_6}(\mathcal{D}) = \frac{1}{\lambda}$$

by Lemma 2.4.4(iii) and Theorem 19.3.1. This shows that

$$K_{\bar{\mathcal{U}}} + \frac{2}{\lambda}\mathcal{D}_{\bar{\mathcal{U}}} \sim_\mathbb{Q} \bar{\psi}^*\left(K_\mathcal{W} + \frac{2}{\lambda}\mathcal{D}_\mathcal{W}\right) \sim_\mathbb{Q} (\pi \circ \bar{\psi})^*\left(K_{V_5} + \frac{2}{\lambda}\mathcal{D}\right) \sim_\mathbb{Q} 0.$$

Thus, we conclude that every center in $\mathbb{CS}(\bar{\mathcal{U}}, \lambda \mathcal{D}_\mathcal{U})$ is mapped by $\bar{\psi}$ to a center in $\mathbb{CS}(\mathcal{W}, \lambda \mathcal{D}_\mathcal{W})$. In particular, the set $\mathbb{CS}(\mathcal{W}, \lambda \mathcal{D}_\mathcal{W})$ is not empty. By Lemma 19.5.1, the set $\mathbb{CS}(\mathcal{W}, \lambda \mathcal{D}_\mathcal{W})$ consists of the proper transform $\check{\mathscr{C}}'$ of the curve $\mathscr{C}'$ on the threefold $\mathcal{W}$. Thus, the set $\mathbb{CS}(\bar{\mathcal{U}}, \lambda \mathcal{D}_\mathcal{U})$ contains a curve C such that $\bar{\psi}(C) = \check{\mathscr{C}}'$.

The birational map ρ is an isomorphism in a general point of the curve C. Let $C_{\bar{\mathcal{W}}}$ be the proper transform of the curve C on the threefold $\bar{\mathcal{W}}$. Denote by $\mathcal{D}_{\bar{\mathcal{W}}}$ the proper transform of the linear system $\mathcal{D}$ on the threefold $\bar{\mathcal{W}}$. Then

$$K_{\bar{\mathcal{W}}} + \frac{2}{\lambda}\mathcal{D}_{\bar{\mathcal{W}}} \sim_{\mathbb{Q}} \bar{\pi}^*\left(K_{\mathcal{U}} + \frac{2}{\lambda}\mathcal{D}_{\mathcal{U}}\right) \sim_{\mathbb{Q}} (\psi \circ \bar{\pi})^*\left(K_{V_5} + \frac{2}{\lambda}\mathcal{D}\right) \sim_{\mathbb{Q}} 0,$$

and $C_{\bar{\mathcal{W}}} \in \mathbb{CS}(\bar{\mathcal{W}}, \lambda\mathcal{D}_{\bar{\mathcal{W}}})$. Moreover, $\bar{\pi}(C_{\bar{\mathcal{W}}})$ is a curve contained in the ψ-exceptional divisor $\mathcal{F}_{\mathcal{U}}$. Thus, we have $\psi \circ \bar{\pi}(C_{\bar{\mathcal{W}}}) = \mathscr{C}'$, and $\bar{\pi}(C_{\bar{\mathcal{W}}})$ is contained in $\mathbb{CS}(\mathcal{U}, \lambda\mathcal{D}_{\mathcal{U}})$, which is impossible by Lemma 19.5.2. The obtained contradiction completes the proof of Lemma 19.5.3. □

Therefore, we have completed the proof of Theorem 19.1.5. As was explained in §19.1, it implies Theorems 19.1.1, 19.1.3, and 19.1.4.

Bibliography

[1] M. Artin, D. Mumford, Some elementary examples of unirational varieties which are not rational, *Proc. London Math. Soc.* **25** (1972), 75–95.

[2] M. Atiyah, On analytic surfaces with double points, *Proceedings of the Royal Society. London. Series A. Mathematical, Physical and Engineering Sciences* **247** (1958), 237–244.

[3] E. Arbarello, M. Cornalba, P. Griffiths, J. Harris, *Geometry of Algebraic Curves. Vol. I.*, Grundlehren der Mathematischen Wissenschaften, **267**. Springer-Verlag, New York, 1985.

[4] S. Bannai, H. Tokunaga, A note on embeddings of $\mathfrak{S}_4$ and $\mathfrak{A}_5$ into the two-dimensional Cremona group and versal Galois covers, *Publ. Res. Inst. Math. Sci.* **43** (2007), 1111–1123.

[5] W. Barth, Two projective surfaces with many nodes, admitting the symmetries of the icosahedron, *J. Alg. Geom.* **5** (1996), 173–186.

[6] W. Barth, C. Peters, A. Van de Ven, *Compact Complex Surfaces*, Ergebnisse der Mathematik und ihrer Grenzgebiete 4 (1984), Springer–Verlag, Berlin.

[7] A. Beauville, Some surfaces with maximal Picard number, *Journal de l'Ecole Polytechnique* **1** (2014), 101–116.

[8] J. Blanc, A. Beauville, On Cremona transformations of prime order, *C.R. Acad. Sci. Paris Ser. I* **339** (2004), 257–259.

[9] J. Blanc, S. Lamy, Weak Fano threefolds obtained by blowing-up a space curve and construction of Sarkisov links, *Proceedings of the London Mathematical Society* **104** (2012), 1047–1075.

[10] H. Blichfeldt, *Finite Collineation Groups*, University of Chicago Press, Chicago, 1917.

[11] F. Bogomolov, Yu. Prokhorov, On stable conjugacy of finite subgroups of the plane Cremona group I, *Cent. Eur. J. Math.* **11** (2013), no. 12, 2099–2105.

[12] H. Braden, T. Northover, Bring's curve: its period matrix and the vector of Riemann constants, *SIGMA Symmetry Integrability Geom. Methods Appl.* **8** (2012), Paper 065.

[13] L. Brenton, On singular complex surfaces with negative canonical bundle, with applications to singular compactifications of $\mathbb{C}^2$ and to 3-dimensional rational singularities, *Math. Ann.* **248** (1980), 117–124.

[14] Th. Breuer, *Characters and Automorphism Groups of Compact Riemann Surfaces*, LMS Lecture Note Series. **280**. Cambridge: Cambridge University Press. xii, 199p.

[15] I. Cheltsov, Del Pezzo surfaces with nonrational singularities, *Math. Notes* **62** (1997), 377–389.

[16] I. Cheltsov, Birationally rigid Fano varieties, *Russ. Math. Surv.* **60** (2005), 875–965.

[17] I. Cheltsov, Non-rational nodal quartic threefolds, *Pacific Journal of Math.* **226** (2006), 65–82.

[18] I. Cheltsov, Two local inequalities, *Izv. Math.* **78** (2014), 375–426.

[19] I. Cheltsov, J. Park, Halphen pencils on weighted Fano threefold hypersurfaces, *Cent. Eur. J. Math.* **7** (2009), 1–45.

[20] I. Cheltsov, J. Park, Sextic double solids, Cohomological and Geometric Approaches to Rationality Problems, *Progr. Math.* **282** (2010), 75–132.

[21] I. Cheltsov, C. Shramov, Log canonical thresholds of smooth Fano threefolds, *Russ. Math. Surv.* **63** (2008), 859–958.

[22] I. Cheltsov, C. Shramov, Extremal metrics on del Pezzo threefolds, *Proc. Steklov Inst. Math.* **264** (2009), no. 1, 30–44.

[23] I. Cheltsov, C. Shramov, Five embeddings of one simple group, *Transactions of the AMS* **366** (2014), 1289–1331.

[24] I. Cheltsov, C. Shramov, Three embeddings of the Klein simple group into the Cremona group of rank three, *Transformation Groups* **17** (2012), 303–35.

[25] A. Clebsch, Ueber die Anwendung der quadratischen Substitution auf die Gleichungen 5ten Grades und die geometrische Theorie des ebenen Fünfseits, *Math. Ann.* **4** (1871), 284–345.

[26] H. Clemens, J. Kollár, S. Mori, Higher Dimensional Complex Geometry, *Asterisque* **166** (1988), 144 pp. (1989).

[27] J. Conway, R. Curtis, S. Norton, R. Parker, R. Wilson, *Atlas of finite groups*, Clarendon Press, Oxford, 1985.

[28] A. Corti, Factorizing birational maps of threefolds after Sarkisov, *J. Alg. Geom.* **4** (1995), 223–254.

[29] A. Corti, Singularities of linear systems and threefold birational geometry, *LMS Lecture Note Series* **281** (2000), 259–312.

[30] A. Corti, J. Kollár, K. Smith *Rational and Nearly Rational Varieties*, Cambridge University Press, 2003.

[31] J. Cutrone, N. Marshburn, Towards the classification of weak Fano threefolds with $\rho = 2$, *Cent. Eur. J. Math.* **11** (2013), no. 9, 1552–1576.

[32] I. Dolgachev, Rational surfaces with a pencil of elliptic curves, *Math. USSR Izv.* **30** (1966), 1073–1100.

[33] I. Dolgachev, Special algebraic $K3$ surfaces I, *Math. USSR Izv.* **37** (1973), 833–846.

[34] I. Dolgachev, *Classical Algebraic Geometry: A Modern View*, Cambridge University Press (2012)

[35] I. Dolgachev, V. Iskovskikh, Finite subgroups of the plane Cremona group, Birkhauser Boston, *Progr. Math.* **269** (2009), 443–548.

[36] R. Dye, A plane sextic curve of genus 4 with A_5 for collineation group, *Journal of the LMS* **52** (1995), 97–110.

[37] W. Edge, Bring's curve, *Journal of the LMS* **18** (1978), 539–545.

[38] W. Edge, Tritangent planes of Bring's curve, *Journal of the LMS* **23** (1981), 215–222.

[39] W. Edge, A pencil of four-nodal plane sextics, *Mathematical Proceedings of the Cambridge Philosophical Society* **89** (1981), 413–421.

[40] D. Eisenbud, J. Harris, On varieties of minimal degree (a centennial account). Algebraic geometry, Bowdoin, 1985 (Brunswick, Maine, 1985), *Proc. Sympos. Pure Math.* **46**, Part 1, Amer. Math. Soc., Providence, RI (1987), 3–13.

[41] D. Eisenbud, J.-H. Koh, *Remarks on Points in a Projective Space*, MSRI Publications **15**, Springer, New York, 157–172.

[42] N. Elkies, On some points-and-lines problems and configurations, *Period. Math. Hungar.* **53** (2006), 133–148.

[43] S. Endraß, On the divisor class group of double solids, *Manuscripta Math.* **99** (1999), 341–358.

[44] H. Finkelnberg, Small resolutions of the Segre cubic, *Indagationes Math.* **90** (1987), 261–277.

[45] H. Flenner, M. Zaidenberg, Locally nilpotent derivations on affine surfaces with a $\mathbb{C}^*$-action, *Osaka J. Math.* **42** (2005), 931–974.

[46] W. Fulton, J. Harris, *Representation Theory. A First Course.* Graduate Texts in Mathematics **129**, New York: Springer-Verlag (1991).

[47] M. Furushima, N. Nakayama, The family of lines on the Fano threefold V_5, *Nagoya Math. J.* **116** (1989), 111–122.

[48] The GAP Group, GAP — Groups, Algorithms, and Programming, Version 4.7.4; 2014. http://www.gap-system.org

[49] P. Griffiths, J. Harris, *Principles of Algebraic Geometry*, 2nd edition. Wiley Classics Library, New York (1994).

[50] M. Green, On the analytic solution of the equation of fifth degree, *Compositio Math.* **37** (1978), 233–241.

[51] G. Halphen, Sur les courbes planes du sixieme degre a neuf points doubles, *Bulletin de la Société Mathématique de France* **10** (1882), 162–172.

[52] J. Harris, *Algebraic Geometry. A First Course*, Graduate Texts in Mathematics **133**. Springer-Verlag, New York, 1992.

[53] R. Hartshorne, *Algebraic Geometry*, Graduate Texts in Mathematics **52** (1977) Springer-Verlag, New York-Heidelberg.

BIBLIOGRAPHY

[54] K. Hashimoto, Period map of a certain $K3$ family with an S_5-action, *J. Reine Angew. Math.* **652** (2011), 1–65.

[55] K. Hashimoto, Finite symplectic actions on the $K3$ lattice, *Nagoya Math. J.* **206** (2012), 99–153.

[56] F. Hidaka, L. Watanabe, Normal Gorenstein surfaces with ample anticanonical divisor, *Tokyo J. Math.* **4** (1981), 319–330.

[57] N. Hitchin, Spherical harmonics and the icosahedron, Groups and symmetries, *CRM Proc. Lecture Notes* **47**, Amer. Math. Soc., Providence, RI (2009), 215–231.

[58] N. Hitchin, Vector bundles and the icosahedron, Vector bundles and complex geometry, *Contemp. Math.* **522**, Amer. Math. Soc., Providence, RI (2010), 71–87.

[59] B. Hunt, *The Geometry of Some Special Arithmetic Quotients*, Lecture Notes in Math. **1637** (Springer–Verlag, New York, 1996).

[60] A. Iliev, The Fano surface of the Gushel' threefold, *Compositio Math.* **94** (1994) no. 1, 81–107.

[61] N. Inoue, F. Kato, On the geometry of Wiman's sextic, *Journal of Mathematics of Kyoto University* **45** (2005), 743–757.

[62] V. Iskovskih, Fano threefolds I, *Math. USSR Izv.* **41** (1977), no. 3, 485–527.

[63] V. Iskovskikh, Birational automorphisms of three-dimensional algebraic varieties, *J. Soviet Math.* **13** (1980), 815–868.

[64] V. Iskovskikh, Birational rigidity of Fano hypersurfaces in the framework of Mori theory, *Russ. Math. Surv.* **56** (2001), 207–291.

[65] V. Iskovskikh, Ju. I. Manin, Three-dimensional quartics and counterexamples to the Lüroth problem, *Math. USSR Sb.* **86** (1971), 141–166.

[66] V. Iskovskikh, Yu. Prokhorov, *Fano Varieties*, Encyclopaedia of Mathematical Sciences **47** (1999) Springer, Berlin.

[67] P. Jahnke, T. Peternell, I. Radloff, Threefolds with big and nef anticanonical bundles I, *Math. Ann.* **333** (2005), 569–631.

[68] Y. Kawamata, On Fujita's freeness conjecture for 3-folds and 4-folds, *Math. Ann.* **308** (1997), 491–505.

[69] Y. Kawamata, Subadjunction of log canonical divisors II, *American J. Math.* **120** (1998), 893–899.

[70] S. Katz, Small resolutions of Gorenstein threefold singularities, Algebraic geometry: Sundance 1988, *Contemp. Math.* **116**, Amer. Math. Soc., Providence, RI, (1991), 61–70.

[71] F. Klein, *Lectures on the Icosahedron and the Solution of Equations of the Fifth Degree*, Dover Phoenix Editions, New York: Dover Publications, 2003.

[72] J. Kollár et al., Flips and abundance for algebraic threefolds, *Astérisque* **211** (1992).

[73] J. Kollár, Singularities of pairs, *Proceedings of Symposia in Pure Math.* **62** (1997), 221–287.

[74] J. Kollár, S. Mori, *Birational Geometry of Algebraic Varieties*, Cambridge University Press (1998).

[75] V. Kulikov, Degenerations of $K3$ surfaces and Enriques surfaces, *Math. USSR Izv.* **41** (1977), no. 5, 957–989.

[76] A. Kuribayashi, H. Kimura, Automorphism groups of compact Riemann surfaces of genus five, *J. Algebra* **134** (1990), no. 1, 80–103.

[77] R. Lazarsfeld, *Positivity in Algebraic Geometry* **II**, Springer-Verlag, Berlin, 2004.

[78] M. van Leeuwen, A. Cohen, B. Lisser, *LiE, A Package for Lie Group Computations*, Computer Algebra Nederland, Amsterdam, 1992.

[79] Yu. Manin, *Cubic Forms*, 2nd Edition, Algebra, Geometry, Arithmetic, North Holland, 1986.

[80] J. McKelvey, The groups of birational transformations of algebraic curves of genus 5, *Amer. J. Math.* **34** (1912), no. 2, 115–146.

[81] M. Mella, Birational geometry of quartic 3-folds. II. The importance of being $\mathbb{Q}$-factorial, *Math. Ann.* **330** (2004), 107–126.

BIBLIOGRAPHY

[82] F. Melliez, Duality of $(1,5)$-polarized abelian surfaces, *Math. Nachr.* **253** (2003), 55–80.

[83] G. Miller, H. Blichfeldt, L. Dickson, *Theory and Applications of Finite Groups*, Dover, New York, 1916.

[84] S. Mori, S. Mukai, On Fano 3-folds with $B_2 \geqslant 2$, Advanced Studies in Pure Mathematics, *Algebraic Varieties and Analytic Varieties* **1** (1983), 101–129.

[85] S. Mukai, Curves, $K3$ surfaces and Fano 3-folds of genus $\leqslant 10$, *Algebraic Geometry and Commutative Algebra*, in Honor of Masayoshi Nagata, Vol. I, 357–377 (1988).

[86] S. Mukai, H. Umemura, Minimal rational threefolds, *Lecture Notes in Math.* **1016** (1983), 490–518.

[87] J. Müller, S. Sarkar, *On finite groups of symmetries of surfaces*, preprint, arXiv:1308.3180.

[88] V. Nikulin, Kählerian $K3$ surfaces and Niemeier lattices. I, *Izv. Math.*, 77:5 (2013), 954–997.

[89] K. Oguiso, Local families of $K3$ surfaces and applications, *J. Alg. Geom.* **12** (2003), 405–433.

[90] Y. Ohshima, Lines on del Pezzo surfaces with Gorenstein singularities, *Kumamoto J. Math.* **18** (2005), 93–98.

[91] D. Orlov, Exceptional set of vector bundles on the variety V_5, *Moscow Univ. Math. Bull.* **46**:5 (1991), 48–50.

[92] K. Petterson, *On Nodal Determinantal Quartic Hypersurfaces in $\mathbb{P}^4$*, Thesis, University of Oslo (1998)

[93] H. Pinkham, Simple elliptic singularities, del Pezzo surfaces and Cremona transformations, *Proceedings of Symposia in Pure Math.* **30** (1977), 69–71.

[94] V. Popov, Jordan groups and automorphism groups of algebraic varieties, Automorphisms in Birational and Affine Geometry, *Springer Proceedings in Mathematics and Statistics*, **79** (2014), 185–213.

[95] Yu. Prokhorov, Automorphism groups of Fano 3-folds, *Russ. Math. Surv.* **45** (1990), 222–223

[96] Yu. Prokhorov, Simple finite subgroups of the Cremona group of rank 3, *J. Alg. Geom.* **21** (2012), 563–600.

[97] Yu. Prokhorov, Fields of invariants of finite linear groups, Cohomological and geometric approaches to rationality problems, 245–273, *Progr. Math.* **282**, Birkhäuser Boston, Inc., Boston, MA, 2010.

[98] Yu. Prokhorov, On birational involutions of $\mathbb{P}^3$, *Izv. Math.* **77** (2013), 627–648.

[99] Yu. Prokhorov, G-Fano threefolds I, *Adv. Geom.* **13** (2013), 389–418.

[100] Yu. Prokhorov, *On stable conjugacy of finite subgroups of the plane Cremona group II*, preprint, arXiv:1308.5698 (2013).

[101] Z. Ran, Normal bundles of rational curves in projective spaces, *Asian J. Math.* **11** (2007), no. 4, 567–608.

[102] J. Rauschning, P. Slodowy, An aspect of icosahedral symmetry, *Canad. Math. Bull.* **45** (2002), no. 4, 686–696.

[103] Z. Reichstein, B. Youssin, Essential dimensions of algebraic groups and a resolution theorem for G-varieties, with an appendix by J. Kollár and E. Szabo, *Canad. J. Math.* **52** (2000), no. 5, 1018–1056.

[104] M. Reid, Hyperelliptic linear systems on a $K3$ surface, *J. London Math. Soc. (2)* **13** (1976), no. 3, 427–437.

[105] M. Reid, Minimal models of canonical 33-folds, Algebraic Varieties and Analytic Varieties (Tokyo, 1981), *Advanced Studies in Pure Math.* **1**, North-Holland, 131–180.

[106] M. Reid, Nonnormal del Pezzo surfaces, *Publ. Res. Inst. Math. Sci.* **30** (1994), 695–727.

[107] M. Reid, *Chapters on algebraic surfaces*, Complex algebraic geometry (Park City, UT, 1993), IAS/Park City Math. Ser. **3**, Amer. Math. Soc., Providence, RI (1997), 3–159.

[108] G. Riera, R. Rodriguez, The period matrix of Bring's curve, *Pacific J. Math.* **154** (1992), 179–200.

[109] B. Saint-Donat, Projective models of $K-3$ surfaces, *American J. Math.* **96** (1974), 602–639.

[110] G. Sanna, *Rational curves and instantons on the Fano threefold Y_5*, arXiv:1411.7994 (2014).

[111] N. Shepherd-Barron, Invariant theory for S_5 and the rationality of M_6, *Compositio Math.* **70** (1989), no. 1, 13–25.

[112] V. Shokurov, Prym varieties: Theory and applications, *Math. USSR Izv.* **23** (1984), 93–147.

[113] K. Shramov, $\mathbb{Q}$-factorial quartic threefolds, *Sb. Math.* **198** (2007), 1165–1174.

[114] J. Sekiguchi, The birational action of S_5 on $\mathbb{P}^2(\mathbb{C})$ and the icosahedron, *J. Math. Soc. Japan* **44** (1992), no. 4, 567–589.

[115] T. Springer, *Invariant Theory*, Lecture Notes in Math. **585** (1977), Springer-Verlag, Berlin-New York.

[116] M. Stoll, D. Testa, *The surface parametrizing cuboids*, preprint, arXiv:1009.0388 (2010).

[117] M. Szurek, J. Wiśniewski, Fano bundles of rank 2 on surfaces, *Compositio Math.* **76** (1990), 295–305.

[118] K. Takeuchi, Some birational maps of Fano 3-folds, *Compositio Math.* **71** (1989), 265–283.

[119] H. Tokunaga, Two-dimensional versal G-covers and Cremona embeddings of finite groups, *Kyushu Journal of Math.* **60** (2006), 439–456.

[120] C. Voisin, Sur la jacobienne intermediaire du double solide d'indice deux, *Duke Math. J.* **57** (1988), 629–646.

[121] C. Voisin, Unirational threefolds with no universal codimension 2 cycle, to appear in *Invent. Math.*

[122] A. Wiman, Ueber die algebraischen Curven von den Geschlechtern $p = 4$, 5, und 6, welche eindeutige Transformationen in sich besitzen, *Bihang till K. Svenska Vet.-Akad. Handlingar* **21** (1895), no. 3, 1–41.

[123] A. Wiman, Zur Theorie der endlichen Gruppen von birationalen Transformationen in der Ebene, *Math. Ann.* **48** (1897), 195–240.

[124] G. Xiao, Galois covers between $K3$ surfaces, *Annales de l'Institut Fourier* **46** (1996), 73–88.

[125] Q. Ye, On Gorenstein log del Pezzo surfaces, *Japan. J. Math.* (N.S.) **28** (2002), no. 1, 87–136.

[126] O. Zariski, *Algebraic Surfaces*, 2nd edition, Springer, 1971.

Index

2.$\mathfrak{A}_5$, see Binary icosahedral group
2.$\mathfrak{D}_{10}$, see Binary dihedral group

$\mathfrak{A}_4$-orbits
   in $\mathbb{P}^2$, 103, 106
   in the quintic del Pezzo surface, 108, 110, 182

$\mathfrak{A}_4$-orbits of lines
   in $\mathbb{P}^2$, 103
   in the quintic del Pezzo surface, 106

$\mathfrak{A}_5$
   action on curves, 69
   permutation representations, 68
   representation theory, 71
   subgroups, 67

$\mathfrak{A}_5$-birational superrigidity, see G-birational superrigidity

$\mathfrak{A}_5$-invariant cubic in the quintic del Pezzo threefold, see Surface S_{Cl}

$\mathfrak{A}_5$-invariant low degree curves
   in $\mathbb{P}^2$, see Conic $\mathfrak{C}$, curves $\mathcal{L}_6$, $\mathcal{L}_{10}$, $\mathcal{L}_{12}$, $\mathcal{L}_{15}$, $\mathcal{L}_{20}$
   in the Clebsch cubic surface, see Curves $\mathcal{L}_6$, $\mathcal{L}'_6$, $\mathcal{L}_{15}$, $\mathcal{B}_6$
   in the quintic del Pezzo surface, see Curves $\mathcal{L}_{10}$, $\mathcal{G}_5$, $\mathcal{G}'_5$, $\mathcal{T}_{15}$, $\mathcal{R}$, $\mathcal{R}'$
   in the quintic del Pezzo threefold, 166, 170, 203, 302, see also Curves $\mathscr{C}$, $\mathscr{C}'$, $\mathcal{L}_6$, $\mathcal{L}_{10}$, $\mathcal{L}_{12}$, $\mathcal{L}_{15}$, $\mathcal{L}_{20}$, $\mathcal{L}_{30}$, $\mathcal{G}_5$, $\mathcal{G}'_5$, $\mathcal{G}_{10}$, $\mathcal{G}_{15}$, $\mathcal{T}_6$, $\mathcal{T}_{15}$, $\mathcal{R}_6$, $\mathcal{B}_{18}$, C_{16}, C'_{16}, C^0_{18}, C_{20}, C_{22}, curves of type $C_{18}(0)$

$\mathfrak{A}_5$-orbit Σ_5 in the quintic del Pezzo threefold, 168, 169, 171, 175, 176, 178, 179, 185, 198, 206, 207, 210, 212, 213, 217, 250, 259, 278, 283, 285, 317, 320, 324, 325, 349, 350, 354–356, 361–364, 368, 370, 406, 407, 410, 417, 420, 421, 430, 435, 438, 440, 453, 457, 464, 466, 467, 470, 471, 473, 475, 476, 478

$\mathfrak{A}_5$-orbit Σ_6
   in $\mathbb{P}^2$, 73, 74, 95, 96, 98, 99, 101, 102, 117, 118, 142, 143, 170
   in the quintic del Pezzo surface, 111, 115, 116

$\mathfrak{A}_5$-orbit Σ'_6 in the quintic del Pezzo surface, 111, 115, 116

$\mathfrak{A}_5$-orbit Σ_{10}
   in $\mathbb{P}^2$, 73, 74, 95, 96, 99, 101, 102, 170
   in the Clebsch cubic surface, 119–121
   in the quintic del Pezzo surface, 111, 113, 115, 116, 180
   in the quintic del Pezzo threefold, 168, 169, 182, 183, 198, 206, 208, 209, 215, 217, 283, 285, 302, 306, 317, 324, 349, 357, 358, 361–364, 369, 370, 372, 373, 396, 406, 407, 410, 415, 420–422, 430, 457, 471

$\mathfrak{A}_5$-orbit Σ'_{10}
   in the quintic del Pezzo surface, 111, 113, 115, 116, 180
   in the quintic del Pezzo threefold, 168, 169, 182, 183, 198, 206, 208, 209, 215, 217, 283, 285, 302, 306, 317, 324, 349, 357, 358, 361–364, 369, 370, 372, 373, 396, 406, 407, 410, 415,

420–422, 430, 457, 471
$\mathfrak{A}_5$-orbit Σ_{12}
   in $\mathbb{P}^2$, 73, 74, 95, 102, 170
   in the Clebsch cubic surface, 119–122, 389, 390
   in the quintic del Pezzo threefold, 164, 168–170, 175, 179, 191, 194–196, 209, 214, 218, 245, 258, 283, 285–288, 291, 293, 295, 299, 300, 302, 306, 319, 323, 325, 331, 334, 338, 341, 343, 364, 376–378, 381–383, 388, 392, 394, 397, 402, 406, 407, 410, 431, 457, 471

$\mathfrak{A}_5$-orbit Σ'_{12}
   in the Clebsch cubic surface, 119–122, 389, 390
   in the quintic del Pezzo threefold, 164, 168–170, 175, 179, 185, 191, 194–196, 209, 210, 214, 218, 245, 250, 255, 258, 274, 283, 286, 296, 305, 307–309, 313, 316, 318–320, 323, 330, 333–341, 343, 344, 364, 376, 388, 392, 394, 406, 407, 410, 431, 457, 471, 477, 483

$\mathfrak{A}_5$-orbit Σ_{15}
   in $\mathbb{P}^2$, 73, 74, 95, 96, 99, 100, 102, 170, 212
   in the Clebsch cubic surface, 119–121
   in the quintic del Pezzo surface, 111, 113, 115, 116
   in the quintic del Pezzo threefold, 168, 169, 198, 206, 209, 210, 212, 215, 217, 283, 285, 286, 303, 306, 320, 324, 325, 345, 349, 363–365, 367–370, 372, 392, 406, 407, 410, 421, 430, 440, 457

$\mathfrak{A}_5$-orbit Σ_{20}
   in $\mathbb{P}^2$, 73, 74, 95, 170
   in the Clebsch cubic surface, 119, 121, 122
   in the quintic del Pezzo surface, 111–113, 115
   in the quintic del Pezzo threefold, 164, 168, 170, 194, 258, 283, 286, 313, 334, 338, 341, 354, 356, 358–363, 406, 407, 410

$\mathfrak{A}_5$-orbit Σ'_{20} in the quintic del Pezzo threefold, 164, 168, 170, 191, 196, 216, 236, 258, 285, 313, 319, 334, 338, 341, 347–349, 352, 353, 356, 358–363, 406, 407, 410

$\mathfrak{A}_5$-orbit Σ_{30}
   in the Clebsch cubic surface, 120, 121
   in the quintic del Pezzo threefold, 164, 213, 286, 341, 368, 371, 376, 377, 406, 407, 410

$\mathfrak{A}_5$-orbit Σ'_{30} in the quintic del Pezzo threefold, 164, 196, 212, 213, 285, 286, 341, 365, 368, 371, 376, 377, 406, 407, 410

$\mathfrak{A}_5$-orbits
   in $\mathbb{P}^2$, see $\mathfrak{A}_5$-orbits Σ_6, Σ_{10}, Σ_{12}, Σ_{15}, Σ_{20}
   in the Clebsch cubic surface, see $\mathfrak{A}_5$-orbits Σ_{10}, Σ_{12}, Σ'_{12}, Σ_{15}, Σ_{20}, Σ_{30}
   in the quintic del Pezzo surface, see $\mathfrak{A}_5$-orbits Σ_6, Σ'_6, Σ_{10}, Σ'_{10}, Σ_{15}, Σ_{20}
   in the quintic del Pezzo threefold, see $\mathfrak{A}_5$-orbits Σ_5, Σ_{10}, Σ'_{10}, Σ_{12}, Σ'_{12}, Σ_{15}, Σ_{20}, Σ'_{20}, Σ_{30}, Σ'_{30}

$\mathfrak{A}_5$-orbits of conics
   in the quintic del Pezzo surface, see Curves $\mathcal{G}_5$, $\mathcal{G}'_5$
   in the quintic del Pezzo threefold, see Curves $\mathcal{G}_5$, $\mathcal{G}'_5$, $\mathcal{G}_6$, $\mathcal{G}_{10}$, $\mathcal{G}_{15}$

$\mathfrak{A}_5$-orbits of lines
   in $\mathbb{P}^2$, see Curves $\mathcal{L}_6$, $\mathcal{L}_{10}$, $\mathcal{L}_{12}$, $\mathcal{L}_{15}$, $\mathcal{L}_{20}$
   in the Clebsch cubic surface, see Curves $\mathcal{L}_6$, $\mathcal{L}'_6$, $\mathcal{L}_{15}$
   in the quintic del Pezzo surface, see Curve $\mathcal{L}_{10}$
   in the quintic del Pezzo threefold, see Curves $\mathcal{L}_6$, $\mathcal{L}_{10}$, $\mathcal{L}_{12}$, $\mathcal{L}_{15}$, $\mathcal{L}_{20}$, $\mathcal{L}_{30}$

$\mathfrak{A}_5$-orbits of twisted cubics
   in the quintic del Pezzo surface, see Curve $\mathcal{T}_{15}$
   in the quintic del Pezzo threefold,

INDEX 497

see Curves $\mathcal{T}_6$, $\mathcal{T}_{15}$
$\mathfrak{A}_5$-birational rigidity, see G-birational rigidity
$\mathfrak{A}_6$
    action on $\mathbb{P}^2$, 49
    action on $\mathbb{P}^3$, 47
    action on the Segre cubic, 48
    subgroup of $\mathrm{Cr}_2(\mathbb{C})$, 49
    subgroup of $\mathrm{Cr}_3(\mathbb{C})$, xii
$\mathfrak{A}_7$, xii
Anticanonical linear system, see Linear system $|2H|$ on the quintic del Pezzo threefold
Atiyah flop, 7, 8, 54, 269, 273, 307, 483
Automorphism group of the quintic del Pezzo threefold, 158

$\mathcal{B}_6$, see Curve $\mathcal{B}_6$
$\mathcal{B}_{18}$, see Curve $\mathcal{B}_{18}$
Bannai, 140
Barth sextic, 9
Binary dihedral group, 82, 91, 92, 94, 290
Binary icosahedral group, 5, 82, 87, 124, 139
Birational automorphism group, xx, 1, 11, 260, 271, 444
Birational rigidity, see G-birational rigidity
Birational superrigidity, see G-birational superrigidity
Bogomolov, 3
Braden, 78
Bring's curve, 74, 78, see also Curves $\mathcal{B}_6$, $\mathcal{B}_{18}$

$\mathscr{C}$, see Curve $\mathscr{C}$
$\mathscr{C}'$, see Curve $\mathscr{C}'$
C_{16}, see Curve C_{16}
C'_{16}, see Curve C'_{16}
C^0_{18}, see Curve C^0_{18}
$C_{18}(0)$, see Curve of type $C_{18}(0)$
C_{20}, see Curve C_{20}
C_{22}, see Curve C_{22}
Canonical singularities, 23, 24, 26, 27, 30, 31, 33, 40, 41, 45–48, 143–148, 179, 199, 221, 264, 265, 380, 443, 444, 451–453, 456–458, 460, 461, 465, 467, 470–472, 475, see also Du Val singularity
Castelnuovo bound, 232
Center of canonical singularities, 30, 31, 38, 457, 471, 472, 476, 479, 481
Center of log canonical singularities, 30–32, 199, 380
Center of non-canonical singularities, 30, 31, 33, 451–453, 457, 458, 460–462
Center of non-log canonical singularities, 30, 31, 33
Centralizer, 1, 82
Clebsch diagonal cubic surface, 8, 117–120, 122, 142, 388, 389, 393
Compound du Val singularity, 254, 266
Conic $\mathfrak{C}$, 73, 74, 99, 102, 157, 158, 213
Conics, see $\mathfrak{A}_5$-orbits of conics
Conjugacy in Cremona groups, 1, 6, 11
Corti's inequality, 33, 34, 38, 463
Cremona groups
    conjugacy in, 1, 6, 11, 140
    finite subgroups, 1, 49
    icosahedral subgroups of $\mathrm{Cr}_2(\mathbb{C})$, 119, 140
    icosahedral subgroups of $\mathrm{Cr}_3(\mathbb{C})$, 5, 8, 11
$\mathbb{CS}(X, B_X)$, see Set of centers of canonical singularities
Curve $\mathcal{B}_6$, 121, 122, 388
Curve $\mathcal{B}_{18}$, 291, 293, 337–340, 342, 344, 345, 362, 372, 375, 377, 382, 388, 389, 397, 402, 406, 408, 411, 440, 441
Curve $\mathscr{C}$
    in the Clebsch cubic surface, 8, 122, 123, 388, 389, 394
    in the quintic del Pezzo threefold, 156–160, 162, 164, 165, 169, 175, 176, 181, 191, 195, 196, 200, 209, 212, 215, 216, 235, 255, 256, 258, 269, 282, 283, 286, 287, 293, 313, 332, 334, 336, 339, 340, 354, 356, 360, 371, 375, 377, 382, 388, 389, 397, 399, 402, 406–408, 410, 411, 423, 424, 427–431, 440, 444, 457, 461, 466, 467, 471, 472, 476, 478

Curve $\mathscr{C}'$
   in the Clebsch cubic surface, 122, 123, 388, 389
   in the quintic del Pezzo threefold, 175, 176, 235, 255–259, 261, 265, 269, 272, 275, 277, 283, 287, 293, 306, 307, 310, 311, 313, 316, 320, 336, 340, 344, 360, 372, 375, 377, 382, 388, 389, 397, 399, 402, 406–408, 410, 411, 429–431, 433, 440, 443, 452, 453, 461, 466, 467, 471, 472, 478, 479, 481, 482

Curve C_{16} in the quintic del Pezzo threefold, 196, 198, 212, 285, 313, 319, 323, 338, 340, 341, 360, 371, 406, 408, 411, 440

Curve C_{16}' in the quintic del Pezzo threefold, 313, 315, 319, 320, 340, 362, 371, 440, 441

Curve C_{18}^0 in the quintic del Pezzo threefold, 333–335, 340, 344, 362, 372, 376, 406, 408, 411, 440, 441

Curve C_{20} in the quintic del Pezzo threefold, 191, 192, 194, 196, 202–205, 212, 258, 283, 285, 286, 307, 308, 311, 323, 334, 338, 347–349, 353, 356, 358–360, 371, 406, 408, 411, 445, 446, 467, 472, 475, 478

Curve C_{22} in the quintic del Pezzo threefold, 309, 339

Curve $\mathcal{G}_5$
   in the quintic del Pezzo surface, 110, 111, 180
   in the quintic del Pezzo threefold, 181, 182, 185, 188, 213–216, 230, 234, 236, 285, 340, 361, 372, 406, 408, 411, 418, 440, 441

Curve $\mathcal{G}_5'$
   in the quintic del Pezzo surface, 110, 111, 180
   in the quintic del Pezzo threefold, 181, 182, 185, 188, 213–216, 230, 234, 236, 285, 340, 361, 372, 406, 408, 411, 418, 440, 441

Curve $\mathcal{G}_6$, 188, 206, 217–219, 232, 234, 259, 289–291, 293, 337, 340, 342, 345, 361, 371, 373, 375, 377, 382, 388, 394, 397, 402, 406, 408, 411, 418, 440, 441

Curve $\mathcal{G}_{10}$ in the quintic del Pezzo threefold, 347, 356, 358–363, 401, 406, 408, 411, 473, 474

Curve $\mathcal{G}_{15}$ in the quintic del Pezzo threefold, 392, 396, 397, 400, 401, 403

Curve $\mathcal{L}_6$
   $\mathfrak{A}_6$-invariant curve in $\mathbb{P}^3$, 47
   in $\mathbb{P}^2$, 73, 74, 95, 96, 99, 102
   in the Clebsch cubic surface, 117–122, 286, 388, 394
   in the quintic del Pezzo threefold, 170–175, 184, 185, 204, 206, 209, 210, 213, 216, 218, 234, 239–244, 250, 255, 256, 258, 260, 269, 272, 273, 283, 286, 293, 295, 307, 316, 320, 335, 339, 340, 344, 361, 367, 371, 375, 377, 382, 388, 389, 391, 394, 396, 397, 402, 406, 408, 411, 416, 437, 438, 443, 446, 451–453, 457, 461, 466, 467, 472, 475, 478, 479, 481, 482

Curve $\mathcal{L}_6'$
   $\mathfrak{A}_6$-invariant curve in $\mathbb{P}^3$, 47
   in the Clebsch cubic surface, 8, 117–122, 286, 388, 394

Curve $\mathcal{L}_{10}$
   in $\mathbb{P}^2$, 73, 74, 95, 96, 117
   in the quintic del Pezzo surface, 110, 111, 115–117, 180
   in the quintic del Pezzo threefold, 170–172, 180, 207, 209, 213, 216, 219, 230, 234, 236, 285, 340, 361, 372, 406, 408, 411, 416, 418, 440, 441

Curve $\mathcal{L}_{12}$
   in $\mathbb{P}^2$, 73, 74, 95, 99
   in the quintic del Pezzo threefold, 164, 165, 170–172, 192, 195, 196, 206, 210, 218, 232, 234, 235, 245, 250, 258, 261, 275, 283, 293, 307, 323, 334, 339, 340, 342, 360, 371, 375, 377,

INDEX

382, 388, 391, 392, 394, 396, 397, 402, 406, 408, 411, 417, 440

Curve $\mathcal{L}_{15}$
  in $\mathbb{P}^2$, 73, 74, 95, 96, 99, 102
  in the Clebsch cubic surface, 117, 118, 120, 121, 392
  in the quintic del Pezzo threefold, 170–172, 178, 191, 210, 212, 213, 216, 233–235, 250, 253, 254, 259, 276, 283, 286, 320, 325, 340, 342, 344, 345, 347, 358, 361–365, 367, 368, 370, 377, 392, 406, 408, 411, 434–441, 451–453, 456, 461, 464, 474

Curve $\mathcal{L}_{20}$
  in $\mathbb{P}^2$, 73, 74, 95
  in the quintic del Pezzo threefold, 164, 165, 170–172, 192, 194, 196, 286, 319, 334, 338, 359, 360, 371, 406, 408, 411

Curve $\mathcal{L}_{30}$ in the quintic del Pezzo threefold, 181, 392

Curve $\mathcal{R}_6$ in the quintic del Pezzo threefold, 394

Curve $\mathcal{T}_6$ in the quintic del Pezzo threefold, 320, 323–325, 339, 340, 361, 367, 371, 376, 379, 406, 408, 411, 440, 441

Curve $\mathcal{T}_{15}$
  in the quintic del Pezzo surface, 115, 116
  in the quintic del Pezzo threefold, 347, 364–373, 377, 392, 396, 397, 401, 406, 408, 411

Curves of type $C_{18}(0)$ in the quintic del Pezzo threefold, 340–343, 363, 377, 406, 408, 411, 449, 450

Curves $\mathcal{R}$ and $\mathcal{R}'$ in the quintic del Pezzo surface, 111, 116

$\mathfrak{D}_{10}$, see Dihedral group

Del Pezzo surface, see Clebsch cubic surface, quintic del Pezzo surface, $\mathbb{P}^1 \times \mathbb{P}^1$, $\mathbb{P}^2$

Diagonal action of $\mathfrak{A}_5$ on $\mathbb{P}^1 \times \mathbb{P}^1$, 123, 126, 138, 140, 161, 163

Dihedral group, xix, 82, 91, 133, 167, 290, see also Curves $\mathcal{L}_6$, $\mathcal{G}_6$, $\mathcal{T}_6$

Dolgachev, 50

Double quadric, 10

Du Val singularity, 24, 144–148, 179, 199, 221, 380

Edge pencil on the surface S_5, 111, 115, 116

Elliptic fibration, 40, 45, 50, 143, 176, 185, 259, 435, 453, 465–467, 469, 470

Exclusion of canonical centers, 470, 472, 476, 479, 481, 482

Exclusion of non-canonical centers, 451–453, 457, 458, 460, 461

$\mathscr{F}$, see Surface $\mathscr{F}$

Fano threefold, see Double quadric, Mukai–Umemura threefold, quartic double solid, quartic threefold X_4, quintic del Pezzo threefold, $\mathbb{P}^3$, Segre cubic, sextic double solid, three-dimensional quadric, threefolds X_{14}, X'_{14} and X_2

Fano variety, see Del Pezzo surface, Fano threefold, G-Fano variety

$\mathcal{F}_\mathcal{U}$, see Surface $\mathcal{F}_\mathcal{U}$

$\mathcal{G}_5$, see Curve $\mathcal{G}_5$
$\mathcal{G}'_5$, see Curve $\mathcal{G}'_5$
$\mathcal{G}_6$, see Curve $\mathcal{G}_6$
$\mathcal{G}_{10}$, see Curve $\mathcal{G}_{10}$
$\mathcal{G}_{15}$, see Curve $\mathcal{G}_{15}$

G-Fano variety, 3, 39, 46
G-Minimal Model Program, 3, 4, 140
G-Mori fiber space, 3, 6, 39
G-birational map, xix, xx
G-birational rigidity, 3, 11, 39, 40, 46, 140, 261, 272, 443
G-birational superrigidity, 40, 47, 140
G-equivariant
  rational map, xix
  vector bundle, xix
G-irreducible subvariety, xviii
G-morphism, xix
G-rational map, xix
G-variety, xix

Gorenstein singularity, 161, 254, 264, 466

Gr(2, 5)

action of PSL$_2$($\mathbb{C}$), 85
construction of the quintic del Pezzo threefold, 11, 153
twisted cubics on the quintic del Pezzo threefold, 323
Grassmannian, see Gr(2, 5)

Halphen pencil, 49, 50, 60, 143, 189–191, 198, 201, 465, 467, 468
Hirzebruch surface, xviii, 136
Hyperelliptic curve, 78, 129, 132, 191, 315
Hyperplane sections of the quintic del Pezzo threefold, 156, 178, 179, 185

$\mathcal{I}(X, B_X)$, see Multiplier ideal sheaf
Icosahedral group, see $\mathfrak{A}_5$- entries
Inoue, 111
Irreducible divisor, xvii
Isolation of centers, 134, 424, 429, 434

Jahnke, 161

$K3$ surface, 10, 143, 198, see also Surfaces S_5, S_{10}, S'_{10}, S_{15}, $S_{\mathcal{L}_6}$, $S_{\mathscr{C}'}$, $S_{\mathcal{G}_6}$
Kato, 111
Kawamata log terminal singularities, 23–26, 28–32, 34, 199, 380
Kawamata's subadjunction theorem, 32, 199, 380
Kawamata–Shokurov contraction theorem, 40
Kawamata–Shokurov trick, 33, 199, 380
Klein, 78
Klein simple group, see PSL$_2$($\mathbf{F}_7$)
Kollár, 1
Kuznetsovschina, 83–86, 154, 155, 172, 188, 323, 347, 367

$\mathcal{L}(X, B_X)$, see Log canonical singularities subscheme
$\mathcal{L}_6$, see Curve $\mathcal{L}_6$
$\mathcal{L}'_6$, see Curve $\mathcal{L}'_6$
$\mathcal{L}_{10}$, see Curve $\mathcal{L}_{10}$
$\mathcal{L}_{12}$, see Curve $\mathcal{L}_{12}$
$\mathcal{L}_{15}$, see Curve $\mathcal{L}_{15}$
$\mathcal{L}_{20}$, see Curve $\mathcal{L}_{20}$
$\mathcal{L}_{30}$, see Curve $\mathcal{L}_{30}$

$\mathbb{LCS}(X, B_X)$, see Set of centers of log canonical singularities
LCS(X, B_X), see Locus of log canonical singularities
Linear projection
of the quartic threefold X_4, 253
of the quintic del Pezzo surface, 116
of the quintic del Pezzo threefold, 174, 183, 320, 388
Linear system $\mathcal{Q}_2$, see Pencil $\mathcal{Q}_2$
Linear system $\mathcal{Q}_3$
application to isolation of points, 463
base locus, 416, 421
construction, 189
Linear system $|2H|$ on the quintic del Pezzo threefold
$\mathfrak{A}_5$-invariant subsystems, see Linear system $\mathcal{Q}_3$, pencil $\mathcal{Q}_2$
action of $\mathfrak{A}_5$, 189
action of PSL$_2$($\mathbb{C}$), 159
curve $\mathcal{G}_{10}$, 359
curve $\mathcal{L}_{15}$, 178
Linear system $|3H|$ on the quintic del Pezzo threefold
$\mathfrak{A}_5$-invariant surface, see Surface S_{Cl}
action of $\mathfrak{A}_5$, 281
action of PSL$_2$($\mathbb{C}$), 281
Linear system $|4H|$ on the quintic del Pezzo threefold
$\mathfrak{A}_5$-invariant surface, see Surface $\mathscr{F}$
restriction to anticanonical surfaces, 308
Lines, see $\mathfrak{A}_5$-orbits of lines
Locus of log canonical singularities, 29, 30
Log canonical singularities, 23, 29–31, 33, 380
Log canonical singularities subscheme, 29
Log terminal singularities, see Kawamata log terminal singularities

McKelvey, 79
Minimal center of log canonical singularities, 32, 199, 380
Minimal Model Program, see G-Minimal Model Program

INDEX

Mobile linear system, xvii, 26, 28, 31, 33, 34, 38, 41, 60, 272, 430, 443
Mori fiber space, see G-Mori fiber space
Mukai–Umemura threefold, 8
Multiplier ideal sheaf, 28, 29, 380, 459, 460, 462, 464

Nadel's vanishing theorem, 29
$\mathbb{NCS}(X, B_X)$, see Set of centers of non-canonical singularities
Neron–Severi lattice, 402
$\mathbb{NLCS}(X, B_X)$, see Set of centers of non-log canonical singularities
Node, xviii, 63, 99, 114, 295, 298, 300, 302, 304, 305, 331, 387
Noether's theorem, 78
Noether–Fano inequalities, 39, 41, 45, 50
Normalizer, 1, 82, 167

Octahedral group, 83, 156
Ohshima, 180
One-dimensional linear system, 59, see also Pencils
Orbit of a subvariety, xviii
Orbits, see $\mathfrak{A}_4$-orbits, $\mathfrak{A}_5$-orbits
Ordinary cusp, xviii, 63, 114, 159, 160, 191, 193, 293, 295, 298, 300, 302, 305, 309, 312, 330, 331, 333, 335, 339, 340, 387, 388

$\mathbb{P}^1$, action of finite groups, 82
$\mathbb{P}^1 \times \mathbb{P}^1$
　action of $\mathfrak{A}_5$, 123–131, 134, 135, 139, 140
　action of $\mathfrak{D}_{10}$, 133
　exceptional divisor, 162, 183, 242, 256
　normalization of the surface $\mathscr{S}$, 160
　rational surface, 5, 136
$\mathbb{P}^2$
　action of $\mathfrak{A}_4$, 102–104, 106, 108
　action of $\mathfrak{A}_5$, 4, 73, 95, 96, 98–102, 107, 118, 119, 140, 157
　action of $\mathrm{PSL}_2(\mathbb{C})$, 157
　finite morphism from S_5, 116
　lines on the quintic del Pezzo threefold, 157
　rational map from the quintic del Pezzo threefold, 176, 185, 212, 214, 234, 419, 465

$\mathbb{P}^3$
　action of $\mathfrak{A}_5$, 5, 6, 9, 10, 74, 75
　action of $\mathfrak{A}_6$, 47
　action of $\mathfrak{D}_{10}$, 92, 186
　action of $\mathrm{PSL}_2(\mathbf{F}_7)$, 48
　finite morphism from V_5, 174, 282, 283, 286, 388, 389, 394
$\mathbb{P}^4$
　action of $\mathfrak{A}_5$, 8–10, 76, 78, 188
　action of $\mathfrak{D}_{10}$, 92
　action of $\mathrm{PSL}_2(\mathbb{C})$, 92
　action of $\mathfrak{S}_5$, 260
　conics on the quintic del Pezzo threefold, 188
　rational map from the quintic del Pezzo threefold, 244
$\mathbb{P}^6$
　action of $\mathfrak{A}_5$, 156, 174
　action of $\mathfrak{A}_5$ on cubics, 281
　action of $\mathfrak{A}_5$ on quadrics, 189
　action of $\mathrm{PSL}_2(\mathbb{C})$, 156, 162
　action of $\mathrm{PSL}_2(\mathbb{C})$ on cubics, 281
　action of $\mathrm{PSL}_2(\mathbb{C})$ on quadrics, 159
　construction of the quintic del Pezzo threefold, 11
Pencil $\mathcal{Q}_2$
　$K3$ surfaces, 198
　base locus, see Curve C_{20}
　construction, 189
　low degree curves, 190, 191, 194, 203
　restriction to the surface S_{Cl}, 293
　singular surfaces, see Surfaces $\mathscr{S}$, S_5, S_{10}, S'_{10} and S_{15}
Pencils, see also Halphen pencil
　$\mathfrak{A}_4$-invariant curves of degree 15 in the quintic del Pezzo surface, 109
　$\mathfrak{A}_4$-invariant pencil of conics on the quintic del Pezzo surface, 106
　$\mathfrak{A}_5$-invariant anticanonical surfaces in the quintic del Pezzo threefold, see Pencil $\mathcal{Q}_2$
　$\mathfrak{A}_5$-invariant curves of degree 10 in the quintic del Pezzo surface, see Edge pencil on the surface S_5
　$\mathfrak{A}_5$-invariant curves of degree 12 in the Clebsch cubic surface, 122

$\mathfrak{A}_5$-invariant curves of degree 30 in the surface S_{Cl}, 399
$\mathfrak{A}_5$-invariant curves of degree 40 in an anticanonical surface, 308
$\mathfrak{A}_5$-invariant quadrics in $\mathbb{P}^4$, 8
$\mathfrak{A}_5$-invariant quartics in $\mathbb{P}^3$, 10
$\mathfrak{A}_5$-invariant sextic curves in $\mathbb{P}^2$, 73, 99
restriction of $\mathcal{Q}_2$ to the surface S_{Cl}, 293
Peternell, 161
PGL$_3(\mathbb{C})$, 95
PGL$_4(\mathbb{C})$, 5
Picard group, 144, 146, 201, 269, 397, 401, 402
Plücker embedding, 11, 153
Projection, see Linear projection
Projective plane, see $\mathbb{P}^2$
Projectivization, xviii
Prokhorov, 3
PSL$_2(\mathbb{C})$
    action on Gr$(2,5)$, 83
    action on Mukai–Umemura threefold, 8
    automorphism group of the quintic del Pezzo threefold, 158
    finite subgroups, 82
    invariant anticanonical surface, see Surface $\mathscr{S}$
    invariant curve, see Curve $\mathscr{C}$ in the quintic del Pezzo threefold
    representation theory, 79
PSL$_2(\mathbb{C})$-orbit in V_5, closure of unique two-dimensional, see Surface $\mathscr{S}$
PSL$_2(\mathbb{C})$-orbit in V_5, unique one-dimensional, see Curve $\mathscr{C}$
PSL$_2(\mathbf{F}_7)$
    action on $\mathbb{P}^2$, 49
    action on $\mathbb{P}^3$, 48
    subgroup of Cr$_2(\mathbb{C})$, 49
    subgroup of Cr$_3(\mathbb{C})$, xii
PSp$_4(\mathbf{F}_3)$, xii

$\mathcal{Q}_2$, see Pencil $\mathcal{Q}_2$
$\mathcal{Q}_3$, see Linear system $\mathcal{Q}_3$
Quartic double solid, 10
Quartic threefold X_4, 244–247, 249–251, 254, 255, 260, 261, 265, 266, 467

Quintic del Pezzo surface, 4, 7, 10, 105–111, 115, 116, 140–143, 179, 180
Quintic del Pezzo threefold
    $\mathfrak{A}_5$-orbits, 169, 356, see also $\mathfrak{A}_5$-orbits Σ_5, Σ_{10}, Σ'_{10}, Σ_{12}, Σ'_{12}, Σ_{15}, Σ_{20}, Σ'_{20}, Σ_{30}, Σ'_{30}
    $\mathfrak{A}_5$-orbits, 370
    action of $\mathfrak{A}_5$, 156, 158, 161, 163
    action of PSL$_2(\mathbb{C})$, 156–159
    anticanonical linear system, 156, 159, 178, 182, 189, 239, 342, 359
    automorphism group, 158
    birational rigidity, 261, 272, 443, 451–453, 456–458, 460
    conics, see Curves $\mathcal{G}_5$, $\mathcal{G}'_5$, $\mathcal{G}_{10}$, $\mathcal{G}_{15}$
    construction, 153–155
    elliptic fibration, 176, 185, 259, 435, 453, 466, 468
    Halphen pencil, 189–191, 198, 201, 465, 468
    linear system $|H|$, 155, 156, 174, 178, 182
    linear system $|3H|$, 156, 182, 281, 282, 289, 434
    lines, 157, 170, 183, see also Curves $\mathcal{L}_6$, $\mathcal{L}_{10}$, $\mathcal{L}_{12}$, $\mathcal{L}_{15}$, $\mathcal{L}_{20}$, $\mathcal{L}_{30}$
    low degree curves, see Curves $\mathscr{C}$, $\mathscr{C}'$, $\mathcal{L}_6$, $\mathcal{L}_{10}$, $\mathcal{L}_{12}$, $\mathcal{L}_{15}$, $\mathcal{L}_{20}$, $\mathcal{L}_{30}$, $\mathcal{G}_5$, $\mathcal{G}'_5$, $\mathcal{G}_{10}$, $\mathcal{G}_{15}$, $\mathcal{T}_6$, $\mathcal{T}_{15}$, $\mathcal{R}_6$, $\mathcal{B}_{18}$, C_{16}, C'_{16}, C^0_{18}, C_{20}, C_{22}, curves of type $C_{18}(0)$
    Main Result, 11, 443
    twisted cubics, see Curves $\mathcal{T}_6$, $\mathcal{T}_{15}$

$\mathcal{R}$, $\mathcal{R}'$, see Curves $\mathcal{R}$ and $\mathcal{R}'$
$\mathcal{R}_6$, see Curve $\mathcal{R}_6$
Radloff, 161

$\mathfrak{S}_4$, permutation group, see Octahedral group
$\mathfrak{S}_5$, permutation group, 11, 76, 87, 89, 90, 95, 105, 114, 115, 117–119, 123, 140, 142, 260, 261, 266, 286, 307, 444
$\mathfrak{S}_6$, permutation group, 48
S$_3$, see Clebsch diagonal cubic surface
S$_5$, see Quintic del Pezzo surface
$\mathscr{S}$, see Surface $\mathscr{S}$

INDEX

$\hat{\mathscr{S}}$, see $\mathbb{P}^1 \times \mathbb{P}^1$ as normalization of the surface $\mathscr{S}$
S_5, see Surface S_5
S_{10}, S'_{10}, see Surfaces S_{10} and S'_{10}
S_{15}, see Surface S_{15}
$S_{\mathscr{C}'}$, see Surface $S_{\mathscr{C}'}$
$S_{\mathcal{G}_6}$, see Surface $S_{\mathcal{G}_6}$
$S_{\mathcal{L}_6}$, see Surface $S_{\mathcal{L}_6}$
S_{Cl}, see Surface S_{Cl}
Sarkisov link, 239, 260, 269
Segre cubic, 7, 9, 48
Sekiguchi, 110
Set of centers of canonical singularities, 30
Set of centers of log canonical singularities, 30
Set of centers of non-canonical singularities, 30
Set of centers of non-log canonical singularities, 30
Sextic double solid, 9
Singular curve of degree 18, see Curve C_{18}^0
Singularities of pairs, see Terminal, canonical, Kawamata log terminal, log canonical singularities
Singularities of the surface S_{Cl}, 287, 306, 383, 403
Singularity of type cD$_4$, 254, 266
$\mathfrak{sl}_2(\mathbb{C})$, 83
SL$_2(\mathbb{C})$
  finite subgroups, 82
  representation theory, 79
SL$_2(\mathbf{F}_8)$, xii
SO$_3(\mathbb{R})$, 100
Subadjunction, see Kawamata's subadjunction theorem
Surface $\mathscr{F}$, 307–309, 315, 336, 339, 397
Surface $\mathcal{F}_\mathcal{U}$, 256, 263, 271, 275, 287, 307, 309, 315, 333
Surface $\mathscr{S}$
  $\mathfrak{A}_5$-orbits, 164, 348, 367
  PSL$_2(\mathbb{C})$-orbit, 156
  isolation of points, 134, 424, 431
  lines, 157, 183
  low degree curves, 164–166, 194, 196
  normalization, 159
  singularities, 159
Surface S_5, 198, 212, 300, 302, 348, 349, 354–358, 367, 406, 408, 440
Surface S_{15}, 198, 212, 300, 302, 306, 367, 369, 406, 408, 441
Surface $S_{\mathscr{C}'}$, 256, 257, 261, 293, 312, 314, 433, 481
Surface $S_{\mathcal{G}_6}$, 289, 291, 293, 384
Surface $S_{\mathcal{L}_6}$, 240–243, 258, 293, 295, 308, 322, 479
Surface S_{Cl}, see also Clebsch diagonal cubic surface
  Bring's curve, 288
  definition, 283
  intersections with $\mathfrak{A}_5$-invariant anticanonical surfaces, 293
  lines, 388, 396
  low degree curves, 283, 285, 288, 343, 375, 394, 397
  pencil $\mathcal{P}_{\hat{S}_{Cl}}$, 399, 400
  Picard group, 397, 401
  projection to Clebsch cubic surface, 388, 389
  singularities, 287, 306, 383, 403
  surface of general type, 381
Surfaces S_{10} and S'_{10}, 198, 300, 302, 306, 348, 355, 357, 358, 367, 369, 406, 408, 422
Surfaces with icosahedral symmetry, see Clebsch cubic surface, Hirzebruch surface, $K3$ surface, quintic del Pezzo surface, $\mathbb{P}^1 \times \mathbb{P}^1$, $\mathbb{P}^2$
Szabó, 1

$\mathcal{T}_6$, see Curve $\mathcal{T}_6$
$\mathcal{T}_{15}$, see Curve $\mathcal{T}_{15}$
Terminal singularities, 3, 23, 24, 30, 31, 38, 40, 41, 50, 246, 467, 469–471, 476, 479, 481, 482
Tetrahedral group, see $\mathfrak{A}_4$- entries
Three-dimensional projective space, see $\mathbb{P}^3$, Clebsh diagonal cubic surface
Three-dimensional quadric, 6, 8–10
Threefold $\mathcal{U}$, 256, 263, 265–267, 269, 271, 273, 275, 276, 287, 306, 307, 309, 315, 332, 466–468, 481, 482

Threefold $\mathcal{W}$, 241–245, 247, 254, 260, 261, 265, 269, 466–468, 478, 479
Threefold X_2, 265, 266, 466, 467
Threefold X_{14}, 161, 466, 467
Threefold X'_{14}, 264–269, 273, 275, 306, 317, 318, 466, 467
Threefold $\mathcal{Y}$, 159, 161, 163, 197, 284, 285, 287, 379, 417, 425–427, 444, 445, 459, 466–468, 475
Tjurina number, 254
Tokunaga, 140
Twisted cubic curves, see Curves $\mathcal{T}_6$, $\mathcal{T}_{15}$
Twisted diagonal action on $\mathbb{P}^1 \times \mathbb{P}^1$
 of $\mathfrak{A}_5$, 123, 126, 138, 263
 of $\mathfrak{D}_{10}$, 133, 243
Two-dimensional quadric, see $\mathbb{P}^1 \times \mathbb{P}^1$

$\mathcal{U}$, see Threefold $\mathcal{U}$
Untwisting, 119, 143, 272, 443

V_5, see Quintic del Pezzo threefold
V_{22}, see Mukai–Umemura threefold

$\mathcal{W}$, see Threefold $\mathcal{W}$
Wemyss, 254
Wiman's sextic curve, 115

X_2, see Threefold X_2
X_4, see Quartic threefold X_4
X_{14}, see Threefold X_{14}
X'_{14}, see Threefold X'_{14}
Xiao, 144

$\mathcal{Y}$, see Threefold $\mathcal{Y}$

Zariski, 402
Zariski tangent space, 2, 61, 69, 105, 141, 142, 173, 176–178, 258, 259, 276, 362, 421
Zariski's main theorem, 255